LIPID PEROXIDATION

Edited by **Angel Catala**

Lipid Peroxidation
http://dx.doi.org/10.5772/2929
Edited by Angel Catala

Contributors

Marisa Repetto, Jimena Semprine, Alberto Boveris, Vessela D. Kancheva and Olga T. Kasaikina, Hanaa Ali Hassan Mostafa Abd El-Aal, Paula M. González, Natacha E. Piloni, Susana Puntarulo, Mihaela Ilie, Denisa Margină, Jiri Sochor, Branislav Ruttkay-Nedecky, Petr Babula, Vojtech Adam, Jaromir Hubalek, Rene Kizek, Biljana Kaurinovic, Mira Popovic, Natalia Fagali, Angel Catalá, Neda Mimica-Dukić, Nataša Simin, Emilija Svirčev, Dejan Orčić, Ivana Beara, Marija Lesjak, Biljana Božin, Kamsiah Jaarin, Yusof Kamisah, Toshiyuki Toyosaki, Gema Nieto, Gaspar Ros, Juliann G. Kiang, Risaku Fukumoto, Nikolai V. Gorbunov, Maria Armida Rossi, Yaşar Gül Özkaya, María Cristina Carrillo, María de Luján Alvarez, Juan Pablo Parody, Ariel Darío Quiroga, María Paula Ceballos, Teresa Sousa, Joana Afonso, Félix Carvalho, António Albino-Teixeira, Juliana C. Fantinelli, Ignacio A. Pérez Núñez, Luisa F. González Arbeláez, Susana M. Mosca, Nicolas J. Pillon, Christophe O. SoulageAlba Naudí, Mariona Jové, Victòria Ayala, Omar Ramírez, Rosanna Cabré, Joan Prat, Manuel Portero-Otin, Isidre Ferrer, Reinald Pamplona, Bojana Kisic, Dijana Miric, Lepsa Zoric, Aleksandra Ilic, Ichiro Shimizu, Noriko Shimamoto, Katsumi Saiki, Mai Furujo, Keiko Osawa, Nina P. Kanunnikova, Natalya Z. Bashun, Andrey G. Moiseenok, Mary C. Vagula, Elisa M. Konieczko

Published by InTech

Janeza Trdine 9, 51000 Rijeka, Croatia

Publishing Process Manager Oliver Kurelic
Technical Editor InTech Prepress, Novi Sad
Cover InTech Design team

Additional hard copies can be obtained from orders@intechopen.com

Lipid Peroxidation, Edited by Angel Catala
p. cm.
ISBN 978-953-51-0716-3

For EU product safety concerns:
IN TECH d.o.o., Prolaz Marije Krucifikse Kozulić 3, 51000 Rijeka, Croatia,
info@intechopen.com or visit our website at intechopen.com.

Meet the editor

Dr Angel Catalá received a Ph.D. in 1965 at the UNLP, Argentina. From 1964 to 1974 he worked as Assistant in Biochemistry at the School of Medicine, UNLP. From 1974 to 1976 he was a Fellow of the NIH at the University of Connecticut, USA. From 1985 to 2004 he served as a Full Professor of Biochemistry at the UNLP. His laboratory has been interested for many years in the lipid peroxidation process. Professor Catalá has published over 100 papers in peer reviewed journals, several chapters in books, three edited books including this one. He received awards at 40th International Conference Biochemistry of Lipids 1999, Pan-American Nutrition Food Science 2006, Pharmacology Bernardo Houssay, 2006. Angel Catalá belongs to the Editorial Board of many prestigious journals.

Contents

Preface

Lipids containing polyunsaturated fatty acids are susceptible to free radical-initiated oxidation and can participate in chain reactions that increase damage to biomolecules. Lipid peroxidation, which leads to lipid hydroperoxide formation often, occurs in response to oxidative stress. Hydroperoxides are usually reduced to their corresponding alcohols by glutathione peroxidases. However, these enzymes are decreased in certain diseases resulting in a temporary increase of lipid hydroperoxides that favors their degradation into several compounds, including hydroxy-alkenals. The best known of these are: 4-hydroxy-2-nonenal (4-HNE) and 4-hydroxy-2-hexenal (4-HHE), which derive from lipid peroxidation of n-6 and n-3 fatty acids, respectively. Compared to free radicals, these aldehydes are relatively stable and can diffuse within or even escape from the cell and attack targets far from the site of the original event. These aldehydes exhibit great reactivity with biomolecules, such as proteins, DNA, and phospholipids, generating a variety of intra and intermolecular covalent adducts. At the membrane level, proteins and amino lipids can be covalently modified by lipid peroxidation products (hydoxy-alkenals). These aldehydes can also act as bioactive molecules in physiological and/or pathological conditions.

The purpose of this book is to concentrate on recent developments on lipid peroxidation. The articles collected in this book are contributions by invited researchers with a long-standing experience in different research areas. We hope that the material presented here is understandable to a broad audience, not only scientists but also people with general background in many different biological sciences. This volume offers you up-to-date, expert reviews of the fast-moving field of Lipid Peroxidation. The book is divided in four mayor sections: Lipid peroxidation: chemical mechanisms, antioxidants, biological implications; Evaluation of lipid peroxidation processes; Lipid peroxidation in vegetables, oils, plants and meats and Lipid peroxidation in health and disease.

Angel Catala
Facultad de Ciencias Exactas,
Universidad Nacional de La Plata,
Argentina

Section 1

Lipid Peroxidation: Chemical Mechanisms, Antioxidants, Biological Implications

Lipid Peroxidation: Chemical Mechanism, Biological Implications and Analytical Determination

Marisa Repetto, Jimena Semprine and Alberto Boveris

Additional information is available at the end of the chapter

http://dx.doi.org/10.5772/45943

1. Introduction

Currently, lipid peroxidation is considered as the main molecular mechanisms involved in the oxidative damage to cell structures and in the toxicity process that lead to cell death. First, lipid peroxidation was studied for food scientists as a mechanism for the damage to alimentary oils and fats, nevertheless other researchers considered that lipid peroxidation was the consequence of toxic metabolites (e.g. CCl_4) that produced highly reactive species, disruption of the intracellular membranes and cellular damage (Dianzani & Barrera, 2008).

Lipid peroxidation is a complex process known to occur in both plants and animals. It involves the formation and propagation of lipid radicals, the uptake of oxygen, a rearrangement of the double bonds in unsaturated lipids and the eventual destruction of membrane lipids, with the production of a variety of breakdown products, including alcohols, ketones, alkanes, aldehydes and ethers (Dianzani & Barrera, 2008).

In pathological situations the reactive oxygen and nitrogen species are generated at higher than normal rates, and as a consequence, lipid peroxidation occurs with α-tocopherol deficiency. In addition to containing high concentrations of polyunsaturated fatty acids and transition metals, biological membranes of cells and organelles are constantly being subjected to various types of damage (Chance et al., 1979; Halliwell & Gutteridge, 1984). The mechanism of biological damage and the toxicity of these reactive species on biological systems are currently explained by the sequential stages of reversible oxidative stress and irreversible oxidative damage. Oxidative stress is understood as an imbalance situation with increased oxidants or decreased antioxidants (Sies, 1991a; Boveris et al., 2008). The concept implies the recognition of the physiological production of oxidants (oxidizing free-radicals and related species) and the existence of operative antioxidant defenses. The imbalance concept recognizes the physiological effectiveness of the antioxidant defenses in maintaining both oxidative stress and cellular damage at a minimum level in physiological conditions (Boveris et al., 2008).

Lipid peroxidation is a chain reaction initiated by the hydrogen abstraction or addition of an oxygen radical, resulting in the oxidative damage of polyunsaturated fatty acids (PUFA). Since polyunsaturated fatty acids are more sensitive than saturated ones, it is obvious that the activated methylene (RH) bridge represents a critical target site. The presence of a double bond adjacent to a methylene group makes the methylene C-H bond weaker and therefore the hydrogen in more susceptible to abstraction. This leaves an unpaired electron on the carbon, forming a carbon-centered radical, which is stabilized by a molecular rearrangement of the double bonds to form a conjugated diene which then combines with oxygen to form a peroxyl radical. The peroxyl radical is itself capable of abstracting a hydrogen atom from another polyunsaturated fatty acid and so of starting a chain reaction (Halliwell & Gutteridge, 1984) (Fig. 1).

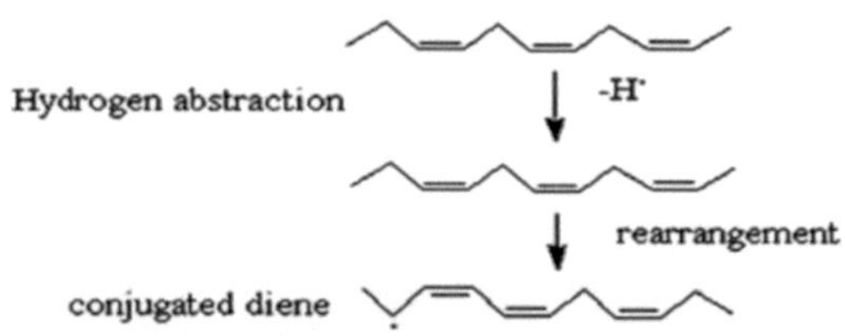

Figure 1. Initiation step of lipid peroxidation process.

Molecular oxygen rapidly adds to the carbon-centered radicals (R·) formed in this process, yielding lipid peroxyl radicals (ROO·). Decomposition of lipid peroxides is catalyzed by transition metal complexes yielding alcoxyl (RO·) or hydroxyl (HO·) radicals. These participate in chain reaction initiation that in turn abstract hydrogen and perpetuate the chain reaction of lipid peroxidation. The formation of peroxyl radicals leads to the production of organic hydroperoxides, which, in turn, can subtract hydrogen from another PUFA. This reaction is termed propagation, implying that one initiating hit can result in the conversion of numerous PUFA to lipid hydroperoxides. In sequence of their appearance, alkyl, peroxyl and alkoxyl radicals are involved. The resulting fatty acid radical is stabilized by rearrangement into a conjugated diene that retains the more stable products including hydroperoxides, alcohols, aldehydes and alkanes. Lipid hydroperoxide (ROOH) is the first, comparatively stable, product of the lipid peroxidation reaction (Halliwell & Gutteridge, 1984) (Fig. 2).

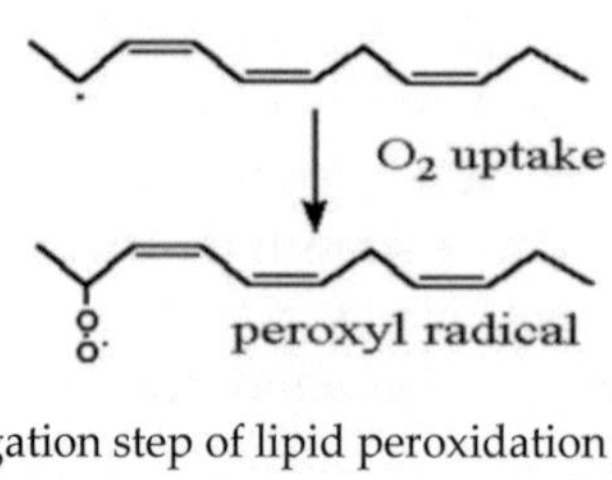

Figure 2. Initial phase of the propagation step of lipid peroxidation process indicating the oxygen uptake.

Reduced iron complexes (Fe^{2+}) react with lipid peroxides (ROOH) to give alkoxy radicals, whereas oxidized iron complexes (Fe^{3+}) react more slowly to produce peroxyl radicals. Both radicals can take part in the propagation of the chain reaction. The end products of these complex metal ion-catalyzed breakdowns of lipid hydroperoxides include the cytotoxic aldehydes and hydrocarbon gases such as ethane.

The free radical chain reaction propagates until two free radicals conjugate each other to terminate the chain. The reaction can also terminate in the presence of a chain-breaking anti-oxidant such as vitamin E (α-tocopherol) (Halliwell & Gutteridge, 1984).

In conditions in which lipid peroxidation is continuously initiated it gives non-radical products destroying two radicals at a time. In the presence of transition metal ions, ROOH can give rise to the generation of radicals capable of re-initiating lipid peroxidation by redox-cycling of these metal ions (Halliwell & Gutteridge, 1984).

Lipid peroxidation causes a decrease in membrane fluidity and in the barrier functions of the membranes. The many products of lipid peroxidation such as hydroperoxides or their aldehyde derivatives inhibit protein synthesis, blood macrophage actions and alter chemotactic signals and enzyme activity (Fridovich & Porter, 1981).

2. Biological implications of lipid peroxidation

The biological production of reactive oxygen species primarily superoxide anion ($O_2^{\cdot-}$) and hydrogen peroxide (H_2O_2) is capable of damaging molecules of biochemical classes including nucleic acids and aminoacids. Exposure of reactive oxygen to proteins produces denaturation, loss of function, cross-linking, aggregation and fragmentation of connective tissues as collagen (Chance et al., 1979). However, the most damaging effect is the induction of lipid peroxidation. The cell membrane which is composed of poly-unsaturated fatty acids is a primary target for reactive oxygen attack leading to cell membrane damage.

The lipid peroxidation of polyunsaturated fatty acids may be enzymatic and non-enzymatic. Enzymatic lipid peroxidation is catalyzed by the lipoxygenases family, a family of lipid peroxidation enzymes that oxygenates free and esterified PUFA generating as a consequence, peroxy radicals. Non enzymatic lipid peroxidation and formation of lipid-peroxides are initiated by the presence of molecular oxygen and is facilitated by Fe^{2+} ions (Repetto et al., 2010a).

Oxidative breakdown of biological phospholipids occurs in most cellular membranes including mitochondria, microsomes, peroxisomes and plasma membrane. The toxicity of lipid peroxidation products in mammals generally involves neurotoxicity, hepatotoxicity and nephrotoxicity (Boveris et al., 2008). The principal mechanism involves detoxification process in liver. Toxicity from lipid peroxidation affect the liver lipid metabolism where cytochrome P-450s is an efficient catalyst in the oxidative transformation of lipid derived aldehydes to carboxylic acids adding a new facet to the biological activity of lipid oxidation metabolites. Cytochrome P-450-mediated metabolism operates in parallel with other metabolic transformations of aldehydes; hence, the P450s could serve as reserve or compensatory mechanisms when other high capacity pathways of aldehyde elimination are compromised due to disease or toxicity. Finally, 4-hydroxynonenal (HNE), unsaturated aldehydes, such as acrolein, trans-2-hexenal, and crotonaldehyde, are also food constituents or environmental pollutants, P-450s may be significant in favoring lipid peroxidation that has significant downstream effects and possibly play a major role in cell signaling pathways. Oxidized lipids appear to have a signaling function in pathological situations, are pro-inflammatory agonists and contribute to neuronal death under conditions in which membrane lipid peroxidation occurs. For example, mitochondrial lipid cardiolipin makes up to 18% of the total phospholipids and 90% of the fatty acyl chains are unsaturated. Oxidation of cardiolipin may be one of the critical factors initiating apoptosis by liberating cytochrome c from the mitochondrial inner membrane and facilitating permeabilization of the outer membrane. The release of cytochrome c activates a proteolytic cascade that culminates in apoptotic cell death (Navarro & Boveris, 2009).

Previous results indicate that lipid peroxidation has a role in the pathogenesis of several pathologies as neurodegenerative (Dominguez et al., 2008; Famulari et al., 1996; Fiszman et al., 2003), inflammatory (Farooqui & Farooqui, 2011), infectious (Repetto et al., 1996), gastric (Repetto et al., 2003) and nutritional diseases (Repetto et al., 2010b).

Oxidative damage in liver is associated with hepatic lipid metabolism, and may be affecting the absorption and transport mechanisms of α-tocopherol in this organ. In the liver, the morphological damage is previous to the lipid peroxidation and the consumption of endogenous antioxidants. In kidney and heart, indeed, lipid peroxidation and oxidative damage preceded necrosis (Repetto et al., 2010b).

Lipid peroxidation is a chain reaction process characterized by repetitive hydrogen abstraction by HO· and RO·, and addition of O_2 to alkyl radicals (R·) resulting in the generation of ROO· and in the oxidative destruction of polyunsaturated fatty acids, in which the methylene group (=RH-) is the main target (Halliwell & Gutteridge, 1984).

The association between increased phospholipid oxidation, free-radical mediated reactions and pathological states was early recognized (Cadenas, 1989; Verstraeten et al., 1997; Liu et al., 2003). The contribution by Sies of the concept of oxidative stress followed (Sies, 1991a,1991b) with the implication that increased free-radical mediated reactions, basically by HO· and RO·, would produce phospholipid, protein, lipid, DNA, RNA or carbohydrate oxidation, whatever is close (Halliwell & Gutteridge, 1984). The increased oxidation of the cell biochemical constituents is associated with ultra structural changes in mitochondrial morphology with mitochondrial swelling and increased matrix volume (Boveris et al., 2008). In human liver, the morphological changes can affect the organ structure and function as it is the case for the bile canaliculi that are damaged in liver transplanted patients; a fact that is interpreted as consequence of the oxidative damage that is associated to ischemia-reperfusion (Cutrin et al., 1996). Interestingly, there are reports in rat liver experimental models, of increased peroxidation secondary to increased mitochondrial production of O_2^- and H_2O_2 (Fridovich, 1978; Navarro &Boveris, 2007; Navarro et al., 2009).

3. Chemical mechanisms for lipid peroxidation process

The spectrum of oxygen reactive species that are considered responsible for biological oxygen toxicity include the intermediates of the partial reduction of oxygen, superoxide radical ($O_2^{\cdot-}$), hydrogen peroxide (H_2O_2), and other reactive species as hydroxyl radicals (HO·), peroxyl radical (ROO·), nitric oxide (NO), peroxinitrite (ONOO-) and singlet oxygen ($^1O_2$).

The biological effects of excess levels of the spectrum of these species are quite similar, and that is the reason they are collectively called reactive oxygen species (ROS). The main free-radical mediated chain reactions in biological systems are summarized in Fig. 3. The Beckman-Radi-Freeman pathway and the Cadenas-Poderoso shunt have been added to the original consecutive reactions of the Fenton/Haber-Weiss pathway and lipid peroxidation process to incorporate NO and ONOO· to the biochemical free-radical mediated chain reaction (Moncada et al., 1991; Boveris et al., 2008) (Fig. 3).

In the last years the denominations "reactive oxygen species" (ROS) and "reactive nitrogen species" (RNS) had became very popular. The ROS denomination involves the three chemical species of the Fenton/Haber-Weiss pathway (O_2^-, H_2O_2 and HO·), the products of the partial reduction of oxygen. Similarly, the RNS denomination is loosely referring to the three chemical species of the Beckman-Radi-Freeman pathway (NO, ONOO·, and NO_2) (Moncada et al., 1991). The reference as a whole to either group, ROS and RNS, is usually made to explain or to refer to their biological activity, what reflects the fact that each group, ROS and RNS, are auto-propagated in biological systems from their promoters, O_2^- and NO. Nevertheless, the advantage and facility in referring to the biological effects implies the ignorance of the biochemistry of the process.

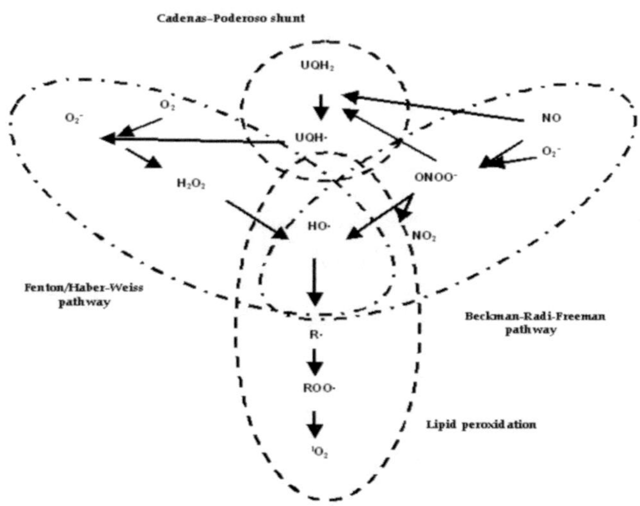

Figure 3. The free-radical mediated chain reaction in biochemistry. $O_2^{\cdot-}$, superoxide radical; H_2O_2, hydrogen peroxide, HO·, hydroxyl radical; NO, nitric oxide; ONOO·, peroxinitrite; ·NO2, nitrogen dioxide; UQH2, ubiquinol; UQH·, ubisemiquinone; R·, alkyl radical; ROO·, peroxyl radical; $^1O_2$, singlet oxygen.

The individual steps of the free-radical mediated chain reaction of biological systems (Fig. 3) are in majority non-enzymatic second order reactions with fast reaction rates, about 10^7 M^{-1} s^{-1}. The exceptions are the enzymatic dismutation of O_2^- (10^{10} M^{-1} s^{-1}, catalyzed by the antioxidant enzyme superoxide dismutase, SOD), the first order reaction of decomposition of $ONOO^-$, and the relatively lower rate (10^5 M^{-1} s^{-1}) of the homolysis of H_2O_2 catalyzed by Fe^{2+} (Boveris et al., 2008).

Concerning the molecular mechanisms that produces lipid peroxidation in biological systems previous, it is accepted that lipid peroxidation may be a consequence of a) intermediates of the partial reduction of oxygen (homolysis of H_2O_2 and $HO\cdot$ generation), b) direct autoxidation of lipids, c) intermediates of the nitric oxide metabolism, and d) modifications of lipid membrane surface structure (Fridovich & Porter, 1981; Boveris et al., 2008; Navarro & Boveris, 2009; Repetto et al., 2010a;).

The lipid peroxidation process is induced for the pro-oxidant effect of transition metals. A vast evidence supports the occurrence of reactions of metal ions with H_2O_2, and hydroperoxides in the cytosol and in biological membranes. The latter ones are the main target of oxidative damage. In other words, by one mechanism, transition metals produce lipid peroxidation by stimulation of the oxidative capacity of H_2O_2 by promoting free-radical mediated processes (Fridovich, 1978; Moncada et al., 1991; Verstraeten et al., 1997; Repetto et al., 2010a; Repetto & Boveris, 2012), and by another mechanism, they bind to negatively charged phospholipids which alters the physical properties of the bilayer and favors the initiation and propagation reactions of lipid peroxidation (Repetto et al., 2010a; Repetto & Boveris, 2012).

Lipid peroxidation is a chain reaction initiated by hydrogen abstraction or by addition of an oxygen radical, resulting in the oxidative damage of polyunsaturated fatty acids (PUFA). Since polyunsaturated fatty acids are more sensitive than saturated ones, it is obvious that the activated methylene (RH) bridge represents a critical target site. This initiation is usually performed by a radical of sufficient reactivity (Eq.1):

$$R_1H + R\cdot \rightarrow R_1\cdot + RH \qquad (1)$$

Molecular oxygen rapidly adds to the carbon-centred radical ($R\cdot$) formed in this process, yielding the lipid peroxyl radical (ROO) (Eq. 2):

$$R\cdot + O_2 \rightarrow ROO \qquad (2)$$

The formation of peroxyl radicals leads to the production of organic hydroperoxides, which, in turn, can abstract hydrogen from another PUFA, analogous to reaction 1:

$$R_1H + ROO\cdot \rightarrow R_1\cdot + ROOH \qquad (3)$$

This reaction is termed propagation, implying that one initiating hit results in the conversion of numerous PUFA to lipid hydroperoxides.

In the sequence of their appearance, alkyl, peroxyl, and alkoxyl radicals are generated in the free radical chain reaction.

The alkyl radical is stabilized by rearrangement into a conjugated diene that is a relatively stable product.

Lipid hydroperoxide (ROOH) is the first stable product of the lipid peroxidation reaction. Under conditions where lipid peroxidation is continuously initiated, radical anhilation or termination occurs with the destroying of two radicals at once:

$$ROO\cdot + ROO\cdot \rightarrow ROH + RO\cdot + {}^1O_2 \quad (4)$$

In the presence of transition metal ions, ROOH gives raise to the generation of radicals capable of (re-)initiating the lipid peroxidation by redox-cycling of the metal ions (Repetto et al., 2010a; Repetto & Boveris, 2012):

$$ROOH + Me^{n+} \rightarrow RO\cdot + Me^{(n-1)+} \quad (5)$$

$$ROOH + Me^{(n-1)+} \rightarrow ROO\cdot + Me^{n+} \quad (6)$$

3.1. Autoxidation of lipids: Non-enzymatic lipid peroxidation

Non-enzymatic lipid peroxidation is a free radical driven chain reaction in which one free radical induces the oxidation of lipids, mainly phospholipids containing polyunsaturated fatty acids. Autoxidation of lipids in biological systems is a direct process that occurs by homolysis of endogenous hydroperoxides by scission of ROOH and production of RO· and ROO·.

The polyunsaturated fatty acids such as linoleic and arachidonic acids, which are present as phosphoglyceride esters in lipid membranes, are particularly susceptible to autoxidation. Moreover, autoxidation in biological systems has been associated with such important pathological events as damage to cellular membranes in the process of aging and the action of certain toxic substance. The autoxidation of most organic substrates in homogeneous solution is a spontaneous free-radical chain process at oxygen partial pressures above 100 torr (Repetto et al., 2010a).

Lipid hydroperoxides, in presence or absence of catalytic metal ions, produce a large variety of products including short and long chain aldehydes and phospholipids and cholesterol ester aldehydes, which provide an equivalent hydrogen abstraction from an unsaturated fatty acid and formation of free radical. The secondary products can be used to assess the degree of lipid peroxidation in a system (Sies, 1991a) (Eq. 7 to 9).

Eq. 7 requires some comments. As written is thermodinamically non spontaneous since it involves the breaking of a C-H bond (435 kJ/mol). However, polyunsaturated fatty acids in solutions are readily autooxidized, likely catalized by transition metal ions. The R· radicals reaction with O_2 yielding ROO·.

$$RH \rightarrow R\cdot + H\cdot \quad (7)$$

$$R\cdot + O_2 \rightarrow ROO\cdot \quad (8)$$

$$ROO\cdot + RH \rightarrow ROOH + R\cdot \quad (9)$$

Transition metal ions Fe^{2+} and Cu^{+} stimulate lipid peroxidation by the reductive cleavage of endogenous lipid hydroperoxides (ROOH) of membrane phospholipids to the corresponding alkoxyl (RO·) and peroxyl (ROO·) radicals in a process that is known as ROOH-dependent lipid peroxidation (Eqs. 10 and 11):

$$Fe^{2+} + ROOH \rightarrow RO\cdot + OH^{-} + Fe^{3+} \quad (10)$$

$$Fe^{3+} + ROOH \rightarrow RO_2\cdot + H^{+} + Fe^{2+} \quad (11)$$

The mechanisms of these two reactions appear to involve the formation of Fe(II)-Fe(III) or Fe(II)-O_2-Fe(III) complexes with maximal rates of HO· radical formation at a ratio Fe(II)/Fe(III) of 1 (Repetto et al., 2010a; Repetto & Boveris, 2012).

Cu^{2+} and Cu^{+} are known for their capacity to decompose organic hydroperoxides (ROOH) to form RO· and ROO· (Eqs. 12 and 13) (Sies, 1991a; Repetto et al., 2010a; Repetto & Boveris, 2012).

$$Cu^{+} + ROOH \rightarrow RO\cdot + OH^{-} + Cu^{2+} \quad (12)$$

$$Cu^{2+} + ROOH \rightarrow RO_2\cdot + H^{+} + Cu^{+} \quad (13)$$

3.2. Lipid peroxidation generated for intermediates of the partial reduction of oxygen

The physiological generation of the products of the partial reduction of oxygen, O_2^{-} and H_2O_2, constitute the biological basis of the process of lipid peroxidation in mammalian aerobic cells. From a molecular point of view hydroxyl radical (HO·) generation, formed from H_2O_2 and Fe^{2+} by the Fenton reaction, has been considered for a long time as the likely rate-limiting step for physiological lipid peroxidation (Verstraeten et al., 1997; Repetto & Boveris, 2012). The Fenton reaction and Fenton-like reactions (Eq. 14) are frequently used to explain the toxic effects of redox-active metals (Eq. 5), where $M^{(n)+}$ is usually a transition metal ion:

$$Fe^{2+} + H_2O_2 \rightarrow [Fe(II)H_2O_2] \rightarrow Fe^{3+} + HO^{-} + HO \quad (14)$$

Trace (nM) levels of cellular and circulating active transition metal ions seem enough for the catalysis of a slow Fenton reaction *in vivo*, at the physiological levels of H_2O_2 (0.1-1.0 μM) (Chance et al., 1979).

Reactive oxygen species mainly include $O_2^{\cdot-}$ and H_2O_2, which are physiologically generated as by-products of mitochondrial electron transfer. The formation of $O_2^{\cdot-}$ is originated from the auto-oxidation of the ubisemiquinone of complexes I and III and the production of H_2O_2 occurs by intramitochondrial Mn-SOD catalysis (Navarro & Boveris, 2004; Navarro et al., 2007, 2010). When the electron transfer process is blocked at complexes I and III, electrons pass directly to O_2 producing $O_2^{\cdot-}$. The reactive oxygen and nitrogen species, although kept in low steady-state concentrations by antioxidant systems, are able to react and damage biomolecules (Fig. 3). Mitochondria are considered the main intracellular source of oxidizing reactive oxygen species (Navarro Boveris, 2004; Navarro et al., 2005, 2009, 2010).

At low level of H_2O_2, Fe^{2+} induces lipid peroxide decomposition, generating peroxyl and alkoxyl radicals and favoring lipid peroxidation. These results indicate that the onset of the Fe^{3+} stimulatory effect on Fe^{2+}-dependent lipid peroxidation is due to reactive oxygen species production via Fe^{2+} oxidation with endogenous ROOH (Repetto & Boveris, 2012).

The Cu^+ ion is considered an effective catalyst for the Fenton reaction (Eq. 15) [3].

$$Cu^+ + H_2O_2 \rightarrow [Cu(I)\text{-}H_2O_2] \rightarrow Cu^{2+} + HO^- + HO^\cdot \quad (15)$$

The process of lipid peroxidation has been recognized as a free radical-mediated and physiologically occurring with the supporting evidence of in situ organ chemiluminescence (Boveris et al., 1980). The main initiation reaction is understood to be mediated by $HO^\cdot$ or by a ferryl intermediate, both with the equivalent potential for hydrogen abstraction from an unsaturated fatty acid, with formation of an alkyl radical ($R^\cdot$) (Repetto & Boveris, 2012) (Eq. 16):

$$HO^\cdot + RH \rightarrow H_2O + R^\cdot \quad (16)$$

One effect of the reaction of hydroxyl radicals, their formation catalyzed by iron ions, with lipids is to make those lipids insoluble or fibrotic that can be considered causative of membrane disruption and oxidative damage associated in different pathologies.

3.3. Lipid peroxidation generated from intermediates of the nitric oxide metabolism

An area of interest that has currently increased over the past decades is the study of nitric oxide (NO) since the demonstration, in 1987, of its formation by the enzyme NO synthase in vascular endothelial cells. This NO radical accounts for the properties of the called endothelial derived relaxing factor, is the endogenous stimulator of the soluble guanylate cyclase and is a potent vasodilator *in vitro* (Moncada et al., 1991). Unsaturated fatty acids are susceptible to nitration reactions. The nitric oxide (NO)-derived species are diffusible across membranes, their concentration in the hydrophobic core of membranes and lipoproteins lead to react fast with fatty acids and lipid peroxyl radicals ($ROO^\cdot$) during the lipid oxidadation process generating oxidized and nitrated products of free lipids (arachidonic acid, arachidonate oleate, linoleate) and esterified (cholesteryl linoleate). Lipid nitration process includes *in vivo* different molecular mechanisms: a) NO autooxidation to nitrite, which has oxidant and nitrating properties, b) electrophilic addition of NO relates species to unsaturated fatty acids, c) radical reactions between $ROO^\cdot$ and NO, d) peroxynitrite ($ONOO^\cdot$) derived free radicals mediate oxidation, nitrosation and nitration reactions. These species are considered currently as mediators of adaptative inflammatory responses.

NO is an endogenous mediator of many physiological functions through stimulation of the guanylate cyclase enzyme including the regulation of vascular relaxing, post-traslational protein changes, gene expression and inflammatory cell function (Moncada et al., 1991). Free and esterified fatty acids as arachidonic and linoleic acids are important components of lipoproteins and membranes that may be oxidized for different compounds. The NO and NO-derived radicals react with fatty acids generating oxidized and nitrated species as nitroalkenes and consequently, nitroalcohols. At low oxygen concentrations the most important biological NO derivatives is $ONOO^{\cdot}$. The nitroalkylation process occurs *in vitro* and in vivo, is involved in redox processes and cell signaling through the reversible covalent bound and post-traslational modifications responsible for structure, function and subcellular distribution of proteins (Valdez et al., 2011) and regulating the pro-inflammatory effect of oxidant exposure (Nair et al., 2007).

A novel mechanism for hydroxyl radical production, which is not dependent on the presence of transition metals, has recently been proposed. This involves the production of peroxynitrite (Beckman et al., 1990, 1994; Rachmilewitz et al., 1993) which has proinflammatory effects *in vitro* (Moncada et al., 1991), from the reaction of NO with $O_2^{\cdot-}$ (Eqs. 17 to 20):

$$NO + O_2^{\cdot-} \rightarrow ONOO^- \quad (17)$$

$$ONOO^- + H^+ \rightarrow ONOOH \quad (18)$$

$$ONOOH \rightarrow HO^{\cdot} + NO_2 \quad (19)$$

$$2\,H^+ + O_2^{\cdot-} + O_2^{\cdot-} \rightarrow H_2O_2 + O_2^{\cdot-} \quad (20)$$

In pathological situations, macrophages and neutrophils, recruited to a site of injury, are activated to produce NO as part of the inflammatory response. Furthermore, SOD activity rapidly scavenges $O_2^{\cdot-}$ and also prolongs the vaso-relaxant effects of NO (Murphy & Sies, 1991; Hogg et al., 1992; Rachmilewitz et al., 1993).

3.4. Modifications of lipid membrane structure

The presence of cholesterol in cell surface membranes influences their susceptibility to peroxidation, probably both by intercepting some of the radicals present and by affecting the internal structure of the membrane by interaction of its large hydrophobic ring structure with fatty acid-side-chains. As lipid peroxidation precedes in any membrane, several of the products produced have a detergent-like activity, specially released fatty acids or phospholipids with one of their fatty-acid side-chains removed. This will contribute to increased membrane disruption and further peroxidation.

The onset of lipid peroxidation within biological membranes is associated with changes in their physicochemical properties and with alteration of biological function of lipids and proteins. Polyunsaturated fatty acids and their metabolites play physiological roles: energy provision, membrane structure, fluidity, flexibility and selective permeability of cellular membranes, and cell signaling and regulation of gene expression (Catala, 2006). The hydroxyl radical generated as a consequence of the Fenton reaction, oxidizes the cellular components of biological membranes (Fig. 4).

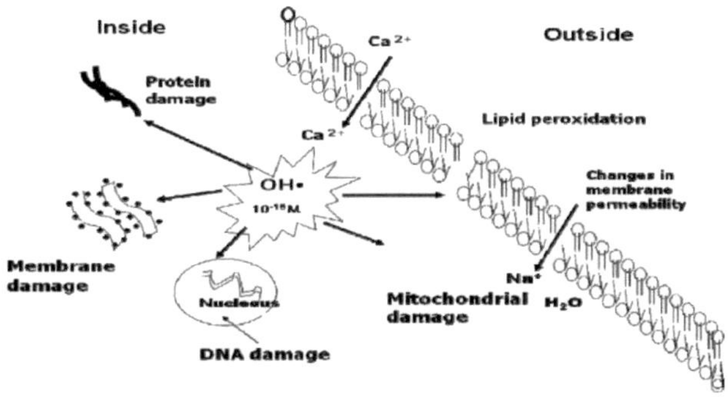

Figure 4. Lipid, DNA and protein oxidative damage from reactive hydroxyl radical.

The binding of positively charged species to a membrane (to the negatively-charged head-groups of phospholipids) can alter the susceptibility of the membrane to oxidative damage. This can be seen as either an enhancement or an inhibition of the rate of lipid peroxidation. Several metal ions such as Ca^{2+}, Co^{2+}, Cd^{2+}, Al^{3+}, Hg^{2+} and Pb^{2+} alter the rate of peroxidation in liposomes, erythrocytes and microsomal membranes, often stimulating the peroxidation induced by iron ions.

In the lipid peroxidation of the brain phosphatidylcholine-phosphatidylserine (PC-PS) liposomes (Repetto et al., 2010a) hydrogen abstraction occurred at the allilic carbons 9 and 10 of the oleic acid chain. Secondary initiation reactions are provided by hydrogen abstraction by RO· and ROO· (Eqs. 21 to 23) at the mentioned tertiary carbons:

$$RO\cdot + RH \rightarrow ROH + R \tag{21}$$

$$RO_2\cdot + RH \rightarrow ROOH + R\cdot \tag{22}$$

$$R\cdot + O_2 \rightarrow RO_2\cdot \tag{23}$$

The R· and ROO· radicals (Eqs.21-23) are central to the free radical-mediated process of lipid peroxidation. The addition reaction of R· with O_2 to yield ROO· (Eq. 23) yield a product that is able to abstract hydrogen atoms and to regenerate R· for a new cycle of the free-radical chain-reaction [38]. The whole process, by repetition of reaction 23, consumes O_2 and produces malondialdehyde (O=HC-CH_2-CH=O), 4-hydroxynonenal and other dialdehydes as secondary and end products of lipid peroxidation. The process produces TBARS at an approximate ratio of 0.12 TBARS/O_2 and normally utilized as measurement of the rate and extent of lipid peroxidation (Junqueira et al., 2004).

There are two consequences of lipid peroxidation: structural damage to membranes and generation of secondary products. Membrane damage derives from the production of broken fatty acyl chains, lipid-lipid or lipid-protein cross-links, and endocyclization reactions to produce isoprostanes and neuroprostanes (Catala, 2006). This effect is severe for biological systems, produce damage of membrane function, enzymatic inactivation and toxic effects on cellular division and function.

4. Role of transition metal on lipid peroxidation process

Studies in the past two decades have shown that redox active metals undergo redox cycling reactions and possess the ability to produce reactive radicals such as superoxide anion radical and nitric oxide in biological systems. Disruption of metal ion homeostasis leads to oxidative stress, a state with increased formation of reactive oxygen species that overwhelms antioxidant protection and subsequently induces DNA damage, lipid peroxidation, protein modification and other effects, all symptomatic for numerous diseases, involving cancer, cardiovascular disease, diabetes, atherosclerosis, neurological disorders, and chronic inflammation.

The mechanism of lipid peroxidation in biological systems caused by free radicals has been the focus of scientific interest for many years (Chance et al., 1976; Fridovich & Porter, 1981; Fraga et al., 1988; Gonzalez-Flecha et al., 1991a, 1991b; Famulari et al., 1996; Fiszman et al., 2003; Junqueira et al., 2004; Catala, 2006; Boveris et al., 2008; Dianzani & Barrera, 2008; Dominguez et al., 2008; Repetto, 2008; Repetto et al., 2010b). Currently, it is known that the $OH^{\cdot}$ radical, is formed mainly by the Haber-Weiss reaction, and it is responsible for the biological damage (Repetto et al., 2010a; Repetto et al., 2010b) (Eq.24):

$$O_2^{-} + H_2O_2 \rightarrow O_2 + HO\cdot + HO^{-} \quad (24)$$

However, this reaction would not proceed significantly *in vivo* because the rate constant for the reaction is lower than that of the dismutation reaction. Nevertheless, a modification of the Haber-Weiss reaction, the Fenton reaction and Fenton-like reactions, utilizes the redox cycling ability of iron to increase the rate of reaction, is more feasible *in vivo* (Chance et al., 1979; Boveris et al., 1980; Gonzalez Flecha et al., 1991b), and is frequently used to explain the toxic effects of redox-active metals where $M^{(n)+}$ is usually a transition metal ion.

As a transition metal that can exist in several valences and that can bind up to six ligands, iron is an important component of industrial catalysts in the chemical industry especially for redox reactions (Repetto et al., 2010a; Repetto & Boveris, 2012).

There are several reports on the role of transition metals in lipid peroxidation process associated with cellular toxicities, because once they enter our physiological systems, these metals play a role in oxidative adverse effects. Some transition metals including iron, chromium, lead, and cadmium generate lipid peroxidation *in vitro* e *in vivo*: fatty acids, cod liver oil, biological membranes, tissues and organs, suggesting that metals contribute to the oxidative effects of lipid peroxidation observed in various diseases (Repetto et al., 2010a; Repetto & Boveris, 2012).

The Fenton reaction occurs *in vivo* at a very low rate, and hence cannot account for any substantial production of $OH^{\cdot}$ radicals in biology. On the other hand, when catalysed by transition metal ions, $OH^{\cdot}$ radicals can be formed through reactions 25 and 26:

$$M^{(n)+} + O_2^{-} \rightarrow M^{(n-1)+} + O_2 \quad (25)$$

$$M^{(n-1)+} + H_2O_2 \rightarrow M^{(n)+} + HO\cdot + HO^{-} \quad (26)$$

The concentration of intracellular redox active transition metals is either low or negligible: free Fe^{2+} is 0.2-0.5 μM and the pool of free Cu^{2+} is about a single ion per cell. However, trace (nM) levels of cellular and circulating active transition metal ions seem enough for the catalysis of a slow Fenton reaction *in vivo* at the physiological levels of hydrogen peroxide (H_2O_2, 0.1-1.0 μM) (Repetto et al., 2010a; Repetto & Boveris, 2012).

It is well known that iron serves as a catalyst for the formation of the highly reactive hydroxyl radical via Fenton reaction. In addition to ferrous ion, many metal ions including Cu (I), Cr (II), and Co (II) were found to have the oxidative features of the Fenton reagent. Therefore, the mixtures of these metal compounds with H_2O_2 were named "Fenton like reagents". In actual in vivo systems, once organic peroxides (ROOH) are formed by the action of ROS, heat, and/or photo-irradiation, ROOH can be substituted for $HO^{\cdot}$, where ROOH reacts with metal ions to form alkoxyl radicals. Subsequently, a chain reaction of lipid peroxidation occurs.

The mechanisms for metal transition ions promoted lipid peroxidation are H_2O_2 decomposition and direct homolysis of endogenous hydroperoxides. The Fe^{2+}-H_2O_2-mediated lipid peroxidation takes place by a pseudo-second order process, and the Cu^{2+}-mediated process by a pseudo-first order reaction. Co^{2+} and Ni^{2+} alone, do not induce lipid peroxidation. Nevertheless, when they are combined with Fe^{2+}, Fe^{2+}-H_2O_2-mediated lipid peroxidation is stimulated in the presence of Ni^{2+} and is inhibited in the presence of Co^{2+} (Fig. 5) (Repetto et al., 2010a).

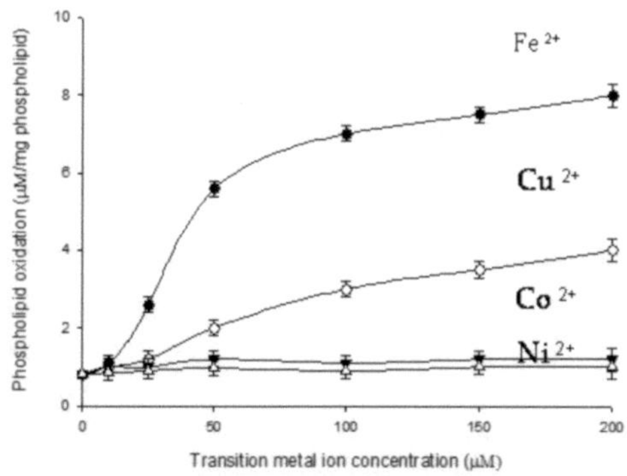

Figure 5. Phospholipid oxidation at different concentrations of transition metals.

There are many factors influencing lipid peroxidation products formation from lipids catalyzed by various metals. For example, the quantitative measurement of the reaction of Fe (II) and H_2O_2 has shown that a stoichiometric amount of hydroxyl radical is spin-trapped when ion concentration was less than 1 μM, suggesting that the strength of the Fenton system depended on the metal concentration. Since Fenton reported that a mixture of hydrogen peroxide and ferrous salts was an effective oxidant of a large variety of organic substrates in 1894 this reagent (the Fenton's reagent) has been used to investigate many subjects related to *in vitro* oxidation of organic substrates including lipids (Repetto et al., 2010a; Repetto & Boveris, 2012).

In the *in vitro* model of phosphatidylcholine/phosphatidyserine (60:40) liposomes and hydrogen peroxide (H_2O_2), Fe and Cu promote lipid peroxidation, interpreted as the consequence of the homolytic scission of H_2O_2 and of endogenous hydroperoxides (ROOH) and of the generation of hydroxyl ($HO^•$) and alcoxyl ($RO^•$) radicals (Cadenas, 1989) depending strictly on the participation of Fe and Cu as redox-reactive metals . However, Co^{2+} and Ni^{2+} alone, do not induce lipid peroxidation. Nevertheless, when they are combined with Fe^{2+}, Fe^{2+}-H_2O_2-mediated lipid peroxidation is stimulated in the presence of Ni^{2+} and inhibited in the presence of Co^{2+} (Repetto et al., 2010a; Repetto & Boveris, 2012).

Cr(III) occurs in nature and is an essential trace element utilized in the regulation of blood glucose levels. Cr(III) reacts with superoxide, subsequently Cr(II) yields hydroxyl radical via Fenton-like reaction with H_2O_2 to initiate lipid peroxidation.

Cadmium intoxication was shown to increase lipid peroxidation in rat liver, kidney and heart. However, the mechanisms of cadmium toxicity are not fully understood. Cadmium indirectly affects the generation of various radicals including superoxide and hydroxyl radical. The generation of hydrogen peroxide by cadmium ion may become a source of radicals in the Fenton system (Jomova & Valko, 2011).

5. Toxic effects of secondary products of lipid peroxidation

Many aldehydes are produced during the peroxidative decomposition of unsaturated fatty acids. Compared with free radicals, aldehydes are highly stable and diffuse out from the cell and attack targets far from the site of their production. About 32 aldehydes were identified as products of lipid peroxidation: a) saturated aldehydes (propanal, butanal, hexanal, octanal, being the decanal the most important); b) 2,3-trans-unsaturated-aldehydes (hexenal, octenal, nonenal, decenal and undecenal); c) a series of 4-hydroxylated,2,3-trans-unsaturated aldehydes: 4-hydroxyundecenal, being 4-hydroxinonenal (HNE) the most important quantitatively. Malonyldialdehyde (MDA) was considered for a long time as the most important lipid peroxidation metabolite. However, MDA is practically no toxic.

Recent studies have demonstrated that the most effective product of lipid peroxidation causing cellular damage is HNE. HNE produces different effects: acts as an intracellular signal able to modulate gene expression, cell proliferation, differentiation and apoptosis. The hydroxyl-group close to a carbonyl group present in HNE chemical structure is related to its high reactivity with different targets (thiol and amine groups). HNE is easily diffusible specie, but its biological effect depends on the molecule target and behavior as a signal to produce the damage.

Oxidative stress is a well known mechanism of cellular injury that occurs with increased lipoperoxidation of cell phospholipids and that has been implicated in various cell dysfunctions (Sies, 1991a,b; Catala, 2006). Aldehydes exhibit high reactivity with bio-molecules, such as proteins, DNA and phospholipids generating intra and intermolecular adducts.

The physiological concentrations of these products are low; however, higher concentrations correspond to pathological situations. Therefore, DNA damage caused by lipid peroxidation end products could provide promising markers for risk prediction and targets for preventive measures. DNA-reactive aldehydes can damage DNA either by reacting directly with DNA bases or by generating more reactive bifunctional intermediates, which form exocyclic DNA adducts. Of these, HNE and MDA, acrolein, and crotonaldehyde have been shown to modify DNA bases, yielding promutagenic lesions and to contribute to the mutagenic and carcinogenic effects associated with oxidative stress-induced lipid peroxidation and HNE and MDA implicated carcinogenesis.

The end-products of lipid peroxidation (HNE and MDA) cause protein damage by addition reactions with lysine amino groups, cysteine sulfhydryl groups, and histidine imidazole groups (Esterbauer et al., 1991; Esterbauer, 1996). Modifications of protein by aldehyde products of lipid peroxidation contribute to neurodegenerative disorders, activation of kinases (Uchida et al., 1999; Uchida, 2003) and inhibition of the nuclear transcription factor (Camandola et al., 2000).

6. Lipid peroxidation of subcellular fragments

6.1. Microsomes

Microsomes isolated from liver have been shown to catalyze an NADPH-dependent peroxidation of endogenous unsaturated fatty acids in the presence of ferric ions and metal chelators, such as ADP or pyrophosphates. Microsomal membranes are particularly susceptible to lipid peroxidation owing to the presence of high concentrations of polyunsaturated fatty acids (Poyer & McCay, 1971). The mechanism involved in the initiation of peroxidation in the NADPH-dependent microsomal system do not appear to involve neither superoxide nor hydrogen peroxide, since neither superoxide dismutase nor catalase cause inhibition of peroxidation. Nevertheless, reduced iron plays an important role in both the initiation and propagation of NADPH-dependent microsomal lipid peroxidation (Shires, 1975).

Microsomal membrane lipids, particularly the polyunsaturated fatty acids, undergo degradation during NADPH-dependent lipid peroxidation. The degradation of membrane lipids during lipid peroxidation has been observed to result in the production of singlet oxygen, which is detected as chemiluminescence (Boveris et al., 1980).

Nonenzymatic peroxidation of microsomal membranes also occurs and is probably mediated in part by endogenous hemoproteins and transition metals. High concentrations of transition metals (50 μM) promote auto-oxidation of phospholipids (Repetto et al., 2010a).

6.2. Mitochondria

It is currently accepted that mitochondrial complex I is particularly sensitive to inactivation by oxygen free radicals and reactive nitrogen species. This special characteristic is frequently referred as complex I syndrome, with the symptoms of reduced mitochondrial respiration with malate-glutamate and ADP and of reduced complex I activity. This complex I syndrome has been observed in aging (Navarro et al., 2005; Navarro & Boveris, 2004, 2008), in ischemia-reperfusion (Gonzalez-Flecha et al., 1993), in Parkinson's disease, and in other neurodegenerative diseases (Schapira et al., 1990a, 1990b; Sayre et al., 1999; Carreras et al., 2004; Schapira, 2008; Navarro et al., 2009), and in this study, with the addition of the increased rates of production of $O_2^{\cdot -}$ and H_2O_2 by complex I mediated reactions, reactions with the free radicals intermediates of the lipid peroxidation process (mainly ROO·), and amine-aldehyde adduction reactions. It is now understood that the three processes above mentioned alter the native non-covalent polypeptide interactions of complex I and promote synergistically protein damage and inactivation by shifting the noncovalent bonding to covalent cross linking (Navarro et al., 2005). Complex I oxidative protein damage has also been considered the result of protein modification by reaction with malonaldehyde and 4-HO-nonenal (Sayre et al., 1999). It was hypothesized that protein damage in the subunits of complexes I and IV follows to free radical-mediated cross-linking and inactivation. The subunits that are normally held together by noncovalent forces are shifted to covalent cross-linking after reaction with the hydroperoxyl radicals (ROO·) and the stable aldehydes produced during the lipid peroxidation process.

The hypothesis that cumulative free radical-mediated protein damage is the chemical basis of respiratory complexes I and IV inactivation (Berlett & Stadtman, 1997) offers the experimental approach of the chronic use of vitamin E, as an antioxidant for the lipid phase of the inner mitochondrial membrane and for the prevention of the mitochondrial /damage associated with aging. The adduction reactions of malonaldehyde and 4-HO-nonenal with protein evolve to stable advanced lipid peroxidation products (Sayre et al., 1999) and protein carbonyls (Nair et al., 2007; Navarro et al., 2008). The molecular mechanism involved in the inactivation of complex I is likely accounted for by ROO· and ONOO-. Upon aging, frontal cortex and hippocampal mitochondria show a decreased rate of respiration, especially marked with NAD-dependent substrates, and decreased enzymatic activities of complexes I and IV associated with an increase in the content of oxidation products (TBARS and protein carbonyls) (Navarro et al., 2008) (Fig. 6).

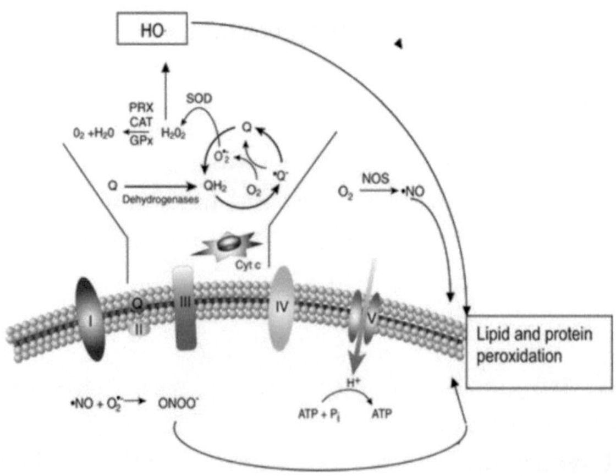

Figure 6. Lipid peroxidation and protein peroxidation by secondary products of lipid peroxidation in mitochondria.

7. Lipid peroxidation and human pathologies

The organism must confront and control the balance of both pro-oxidants and antioxidants continuously. The balance between these is tightly regulated and extremely important for maintaining vital cellular and biochemical functions. This balance often referred to as the redox potential, is specific for each organelle and biological site, and any interference of the balance in any direction might be deleterious for the cell and organism. Changing the balance towards an increase in the pro-oxidant over the capacity of the antioxidant is defined as oxidative stress and might lead to oxidative damage. Changing the balance towards an increase in the reducing power, or the antioxidant, might also cause damage and can be defined as reductive stress.

Oxidative stress and damage have been implicated in numerous disease processes, including inflammation, degenerative diseases, and tumor formation and involved in physiological phenomena, such as aging and embryonic development. The dual nature of these species with their beneficial and deleterious characteristics implies the complexities of their effects at a biological site.

Lipid peroxidation has been pointed out as a key chemical event in the oxidative stress associated with several inborn and acquired pathologies. Disruption of organelle and cell membranes together with calcium homeostasis alterations are the main supramolecular events linked to lipid peroxidation. However, it is not clear if lipid peroxidation process is a cause, triggering step of the clinical manifestations of the disease, or a consequence of toxic effects of lipid peroxidation products.

In pathological situations the reactive oxygen species are generated and as a consequence lipid peroxidation occurs with α-tocopherol deficiency. In addition to containing high concentrations of polyunsaturated fatty acids and transitional metals, red blood cells are constantly being subjected to various types of oxidative stress. Red blood cells however are protected by a variety of antioxidant systems which are capable of preventing most of the adverse effects under normal conditions. Among the antioxidant systems in the red cells, α-tocopherol possesses an important and unique role. α-tocopherol may protect the red cells from oxidative damage via a free radical scavenging mechanism and as a structural component of the cell membrane (Chitra & Shyamaladevi, 2011).

Levels of Met-Hb· are regarded as an index of intracellular damage to the red cell and it is increased when α-tocopherol is consumed and the rate of lipid peroxidation is increased. Scavenging of free radicals by α-tocopherol is the first and the most critical step in defending against oxidative damage to the red cells. When α-tocopherol is adequate, GSH and ascorbic acid may complement the antioxidant functions of α-tocopherol by providing reducing equivalents necessary for its recycling/regeneration.

On the other hand, when α-tocopherol is absent, GSH and ascorbic acid release transitional metals from the bound forms and/or maintain metal ions in a catalytic state. Free radical generation catalysed by transition metal ions in turn initiates oxidative damage to cell membranes. Membrane damage can lead to release of heme compounds from erythrocytes. The heme compounds released may further promote oxidative damage especially when reducing compounds are present (Boveris et al., 2008).

8. Lipid peroxidation and aging

Aging is a process directly related to systemic oxidative stress. Two components of the oxidative stress situation have been recognized in human aging: a decrease in availability of nutritional molecular antioxidants and an accumulation of products derived from the oxidation of biological structures. Oxidation of biomolecules is related to susceptibility to diseases, such as cancer and heart disease, as well as associated with the process of aging (Navarro et al., 2005; Navarro & Boveris, 2007, 2008).

The products derived from lipid peroxidation, measured in plasma by Junqueira et al., (2004) as fluorescent products, were higher in elderly than younger human subjects and even higher in disabled octogenarians and nonagenarians. This increase in lipid peroxidation products was directly correlated with age, and was associated with decreases in vitamin E and C.

9. Analytical determination of lipid peroxidation

Since the acceptation of the oxidative stress concept, scientists and physicians have been searching for a simple assay or a small group of determination that would result useful for the assessment of oxidative stress and lipid peroxidation in clinical situations. The determinations of marker metabolites are usually performed in blood, red blood cells or plasma. The markers for systemic oxidative stress are normally present in healthy humans and the assays for systemic oxidative stress are comparative, which makes necessary to have reference values from normal individuals.

At present, the plasma levels of oxidation products derived from free-radical mediated reactions and of antioxidants are used as indicators of systemic oxidative stress in humans and experimental animals. The more utilized determination of an oxidation product is MDA, which is determined with low specificity but with great efficiency by the simple and useful assay of TBARS with measurements made by spectrophotometry or spectrofluorometry. The normal plasma levels of TBARS are 2-3 μM (Junqueira et al., 2004).

Oxidative damage is characterized by increases in the levels of the oxidation products of macromolecules, such as thiobarbituric acid reactive substances (TBARS), and protein carbonyls. Many of these products can be found in biological fluids, as well as addition-derivatives of these reactive end-products. As a result of lipid peroxidation a great variety of aldehydes can be produced, including hexanal, malondialdehyde (MDA) and 4-hydroxynonenal (Catala, 2006).

Oxidation of an endogenous antioxidant reflects an oxidative stress that is evaluated by measuring the decrease in the total level of the antioxidant or the increase in the oxidative form. The only way not to be influenced by nutritional status is to measure the ratio between oxidized and reduced antioxidants present in blood. The published literature provides compelling evidence that a) MDA represents a side product of enzymatic PUFA-oxygenation and a secondary end product of no enzymatic (autoxidative) fatty peroxide formation and decomposition and b) sensitive analytical methods exist for the unambiguous isolation and direct quantification of MDA. Conceptually, these two facts indicate that MDA is an excellent index of lipid peroxidation. However, this conclusion is limited in practice by several important consideration: a) MDA yield as a result of lipid peroxidation varies with the nature of the PUFA peroxidised (specially its degree of instauration) and the peroxidation stimulus, b) only certain lipid oxidation products decompose yield MDA, c) MDA is only one of several end product of fatty peroxide formation and decomposition, d) the peroxidation environment influences both the formation of lipid-derived precursors and their decomposition to MDA, e) MDA itself is a reactive substance which can be oxidative and metabolically degraded, f) oxidative injury to no lipid biomolecules has the potential to generate MDA. With biological materials, it appears prudent to consider the TBARS test more than an empirical indicator of the potential occurrence of peroxidative lipid damage and not as a measure of lipid peroxidation (Repetto, 2008). The thiobarbituric acid test (TBARS) has been employed to a uniquely great degree over the last five decades to detect and quantify lipid peroxidation in a variety of chemical as well as biological material. Two underlying assumptions are implicit from the widespread use of the TBARS test to assess lipid peroxidation: a) an operative and quantitative relationship exists between lipid peroxidation and MDA, b) product formation during the TBARS test is diagnostic of the presence and amount of fatty peroxides.

Lipid peroxidation proceeds by a free-radical mediated chain reaction that includes initiation, propagation and termination reactions. The chain reaction is initiated by the abstraction of a hydrogen atom from a methylene group of an unsaturated fatty acid. Propagation is cycled through rounds of lipid peroxyl radical abstraction of the bis-methylene hydrogen atoms of a polyunsaturated fatty acyl chain to generate new radicals, after O_2 addition, resulting in the conversion of alkyl radical in hydroperoxyl radical. Termination involves the reaction of two hydroperoxyl radicals to form non-radical products. This reaction is particularly interesting since it is accompanied, although at low yield, by emission of light or chemiluminiscence. Some lipid peroxidation products are light-emitting species and their luminescence is used as an internal marker of oxidative stress (Chance et al., 1979; Boveris et al., 1980, Gonzalez-Flecha et al., 1991b; Sies, 1991a; Repetto, 2008). The measurement of light emission derived from $^1O_2$ and excited triplet carbonyl compounds, which are the most important chemiluminiscent species in the lipid peroxidation of biological systems, is directly related to the rate of lipid peroxidation and allows an indirect assay of the content of lipophilic antioxidants in the sample (Gonzalez-Flecha et al., 1991a). Lipophilic antioxidants react with lipid peroxyl radicals and lower antioxidant content is associated with higher chemiluminescence (Repetto, 2008).

The low-level chemiluminescence which accompanies the peroxidation of polyunsaturated fatty acids has been used as a tool in kinetic and mechanistic studies of biological samples to estimate the extent of the reactions and even to indicate tissue damage promoted by oxidants. Triplet carbonyls and singlet oxygen formed in the annihilation of intermediate peroxyl radicals (ROO·) have been identified as the chemiluminescence emitters.

Chemiluminescence is a very interesting way to evaluate an oxidative stress and lipid peroxidation in biological samples and living systems. The emission of light has been observed during stress in different experimental models. Chemiluminescence is very sensitive and thus can be applied to measure free radical production in human tissues.

Chemiluminescent systems may be classified in two classes based on the origin of the emitting molecule. In the first class, the emitter is a product of the chemical reaction (direct chemiluminescence). In the second class, there is energy transfer between an electronically excited product molecule and a second substance which then becomes the emitter (sensitized chemiluminescence) (Boveris et al., 1980; Gonzalez-Flecha et al., 1991b; Repetto, 2008).

The chemical mechanism responsible for spontaneous organ light emission is provided by the Russell's reaction in which two secondary or tertiary peroxyl radicals (ROO•) yield $^1O_2$ and excited carbonyl groups (=CO*) as products. In turn, two $^1O_2$, through dimol emission, lead to photoemission at 640 and 670 nm, whereas =CO* yields photons at the 460-470 nm band (Boveris et al., 1980). The main sources of the chemiluminescence detected in the direct and sensitized chemiluminescence is the dimol emission of $^1O_2$ (reaction 27) and the photon emission from excited carbonyl groups (reaction 28) (Boveris et al., 1980).

$$2\ ^1O_2 \rightarrow 2\ O_2 + h\upsilon\ (634\text{-}703\ \text{nm}) \qquad (27)$$

$$RO^* \rightarrow RO + h\upsilon\ (380\text{-}460\ \text{nm}) \qquad (28)$$

These reactions are accompanied by chemiluminescence whose intensity may serve as an indirect measure of peroxide free radical and α-tocopherol concentration in the sample.

Lipid peroxidation has been recognized as free radical-mediated and physiologically occurring (Navarro & Boveris, 2004, Navarro et al., 2010; Repetto & Boveris, 2012) with the supporting evidence of *in situ* organ chemiluminescence (Repetto, 2008). Spontaneous chemiluminescence of *in situ* organs directly reports the intracellular formation of singlet oxygen ($^1O_2$) (Boveris et al., 1980) and represents an issue of direct chemiluminescence. The generation of $^1O_2$ implies the collision of two peroxyl radicals (ROO·) with formation of excited species, $^1O_2$ itself and excited carbonyls, followed by photoemission. Light emission from *in situ* organs is a physiological phenomenon that provides a determination of the steady state concentration of singlet oxygen and indirectly of the rate of oxidative free radical reactions (Boveris et al., 1980). *In situ* liver chemiluminescence has been recognized as a reliable indicator of oxidative stress and damage in rat liver upon hydroperoxide infusion (Gonzalez-Flecha et al., 1991b), ischemia-reperfusion (Gonzalez-Flecha et al., 1993), and chronic and acute alcohol intoxication (Videla et al., 1983). The increases in photoemission observed were parallel to increased contents of indicators of lipid peroxidation (malonaldehyde and 4-HO-nonenal) but with a higher experimental/control ratio in organ chemiluminescence (Boveris et al., 1980).

Tert-butyl hydroperoxide initiated chemiluminescence is an example of sensitized chemiluminescence, and it has been used to enhance the chemiluminescence accompanying lipid peroxidation and the α-tocopherol content of tissues. This method has been successfully utilized to detect the existence of oxidative damage associated to experimental or pathological situations in tissue homogenates, subcellular fractions, and in human heart, liver and muscle biopsies (Gonzalez-Flecha et al., 1991b).

Tissue homogenates or blood samples are subjected to *in vitro* oxidative damage by supplementation with tert-butyl hydroperoxide. It reacts with hemoproteins and Fe^{2+} producing peroxyl and alcoxyl free radicals, which enter to the propagation phase of the lipid peroxidation radical chain reaction. The termination steps of the chain reaction generate compounds in an excited state: singlet oxygen and carbonyl groups. This assay is useful to evaluate the integral level of the non-enzymatic antioxidant defenses of a tissue (Gonzalez-Flecha et al., 1991a, 1993).

The increase of tert-butyl hydroperoxide-initiated chemiluminescence is indicative that α-tocopherol is the antioxidant consumed in erythrocytes and suggest that reactive oxygen species and lipid peroxidation catalyzed by reduced transition metals may be responsible for the onset of oxidative damage and the occurrence of systemic oxidative stress in patients suffering oxidative damage associated to neurological pathologies as Parkinson (Famulari et al., 1996, Dominguez et al., 2008), Alzheimer disease (Famulari et al., 1996; Repetto et al., 1999; Dominguez et al., 2008; Serra et al., 2009), and vascular dementia (Famulari et al., 1996, Dominguez et al., 2008; Serra et al., 2009); immunological diseases as HIV infection and AIDS (Repetto et al., 1996), hyperthyroidism and hypothyroidism (Abalovich et al., 2003). These methods were used to evaluate lipid peroxidation and oxidative damage in experimental models of oxidative stress in rats (Repetto et al., 2003, 2010; Ossani et al., 2007; Repetto & Ossani, 2008; Repetto & Boveris, 2010).

A common question of the researchers in the field is which the method of choice is. The answer is: none of them, and all of them. Each assay measures something different. Diene conjugation tells one about the early stages of peroxidation, as a direct measurement of lipid peroxides. In the absence of metal ions to decompose lipid peroxides there will be little formation of hydrocarbon gases, carbonyl compounds, or their fluorescent complexes, which does not necessarily mean therefore that nothing is happening. Even if peroxides do not decompose, the TBARS test can still detect them because of decomposition of peroxides. Changes in the mechanism of peroxide decomposition might alter the amount generated without any change in the overall rate of lipid peroxidation. Whatever method is chosen, one should think clearly what is being measured and how it relates to the overall lipid peroxidation process. Whatever possible, two or more different assay methods should be used.

10. Conclusion

Lipid peroxidation is a physiological process that takes place in all aerobic cells. Unsaturated fatty acids which are structural part of cell membranes are subjected to lipid peroxidation by a non enzymatic and free-radical mediated reaction chain. The molecular mechanisms of the lipid peroxidation process are known and it can be estimated that about 1 % of the total oxygen uptake of cells, organs and bodies in taken up by the reactions of lipid peroxidation. The initiation reactions are provided by the transition-metal catalyzed hemolytic scission of H_2O_2 and ROOH. In turn, H_2O_2 is mainly generated from the mitochondrial dismutation of superoxide radical ($O_2^{\cdot-}$). The products and by-products of lipid peroxidation are cytotoxic and lead in successive steps to oxidative stress, oxidative damage and apoptosis. In a long series of physiological and pathophysiological processes, including aging and neurodegenerative diseases, the rates of mitochondrial $O_2^{\cdot-}$ and H_2O_2 are increased with a parallel increase in the rate of the lipid peroxidation process. It is expected that supplementation with adequate antioxidants, as for instance, α-tocopherol, will keep sensitive cells and organs in healthy conditions and increase lifespan.

Author details

Marisa Repetto, Jimena Semprine and Alberto Boveris
University of Buenos Aires, School of Pharmacy and Biochemistry,
General and Inorganic Chemistry,
Institute of Biochemistry and Molecular Medicine (IBIMOL-UBA-CONICET), Argentina

Acknowledgement

We thank to Dr. Jorge Serra for helping in the revision of this version.

11. References

Abalovich, M.; Llesuy, S.; Gutierrez, S. & Repetto, M. (2003) Peripheral markers of oxidative stress in Graves´ disease. The effects of methimazole and 131 Iodine treatments. *Clinical Endocrinology*. Vol. 59, pp. 321-327, ISSN: 1365-2265

Beckman, J.; Beckman, T.; Chen, J.; Marshall, P. & Freeman, B. (1990) Apparent hydroxyl radical production from peroxynitrite. Implications for endothelial injury from nitric oxide oxide and superoxide. *Proceeding of the National Academy of Sciences of the United States*. Vol. 87, pp. 1620-1624, ISSN: 0027-8424

Beckman, J.; Chen, J.; Ischiropulos, H. & Crow, J. (1994) Oxidative chemistry of peroxynitrite. *Methods in Enzymology*. Vol. 233, pp. 229-240, ISSN: 0076-6879

Berlett, B.S. & Stadtman, E.R. (1997) Protein oxidation in aging, disease, and oxidative stress. *The Journal of Biological Chemistry*. Vol. 272, pp. 20313–20316, ISSN: 0021-9258

Boveris, A.; Cadenas, E.; Reiter, R.; Filipkowski, M.; Nakase, Y. & Chance, B. (1980) Organ chemiluminescence: noninvasive assay for oxidative radical reactions *Proceeding of the National Academy of Sciences of the United States.* Vol. 177, pp. 347-351, ISSN: 0027-8424

Boveris, A.; Fraga, C.; Varsavsky, A. & Koch, O. (1983) Increased chemiluminescence and superoxide production in the liver of chronically ethanol-treated rats. *Archives of Biochemistry and Biophysics.* Vol. 227, pp. 534-541, ISSN: 0003-9861

Boveris, A. & Navarro, A. (2008) Brain mitochondrial dysfunction in aging. *Life,* Vol. 60, No.5, pp. 308-314, ISSN: 1521-6543

Boveris, A.; Repetto, M.G.; Bustamante, J.; Boveris, A.D. & Valdez, L.B. (2008). The concept of oxidative stress in pathology. In: Álvarez, S.; Evelson, P. (ed.), *Free Radical Pathophysiology,* pp. 1-17, Transworld Research Network: Kerala, India, ISBN: 978-81-7895-311-3

Cadenas, E. (1989) Biochemistry of oxygen toxicity. *Annual Review of Biochemistry.* Vol. 58, pp. 79-110, ISSN:0066-4154

Camandola, S.; Poli, G. & Mattson, M. (2000) The lipid peroxidation product 4-hydroxy-2,3-nonenal inhibits constitutive and inducible activity of nuclear factor-bin neurons. *Molecular Brain Research.* Vol. 85, pp. 53–60, ISSN: 0021-9258

Carreras, M.C.; Franco, M.C.; Peralta, J.G. & Poderoso, J.J. (2004) Nitric oxide, complex I, and the modulation of mitochondrial reactive species in biology and disease. *Molecular Aspects of Medicine.* Vol. 25, pp. 125–139, ISSN: 0098-2997

Catala, A. (2006) An overview of lipid peroxidation with emphasis in outer segments of photoreceptors and the chemiluminescence assay. *The International Journal of Biochemistry and Cell Biology.* Vol. 38, pp. 1482-1495, ISSN: 1357-2725

Chance, B.; Sies, H. & Boveris, A. (1979) Hydroperoxide metabolism in mammalian organs. *Physiological Reviews.* Vol. 59, pp. 527-605, ISSN:@0031-9333

Cutrin, JC.; Cantino, D.; Biasi, F.; Chiarpotto, E.; Salizzoni, M.; Andorno, E.; Massano, G.; Lanfranco, G.; Rizetto, M.; Boveris, A. & Poli, G. (1996) Reperfusion damage to the bile canaliculi in transplanted human liver. *Hepatology.* Vol. 24, pp. 1053-1057, ISSN: 1527-3350

Dianzani, M. & Barrera, G. (2008) Pathology and physiology of lipid peroxidation and its carbonyl products. In: Álvarez, S.; Evelson, P. (ed.), *Free Radical Pathophysiology,* pp. 19-38, Transworld Research Network: Kerala, India, ISBN: 978-81-7895-311-3

Domínguez, R.O.; Marschoff, E.R.; Guareschi, E.M.; Repetto, M.G.; Famulari, A.L.; Pagano, M.A. & Serra, J.A. (2008). Insulin, glucose and glycated haemoglobin in Alzheimer's and vascular dementia with and without superimposed Type II diabetes mellitus condition. *Journal of Neural Transmission,* Vol. 115, pp. 77-84, ISSN: 0300-9564.

Esterbauer, H.; Schaur, J. & Zollner, H. (1991) Chemistry and biochemistry of 4-hydroxynonenal, malondialdehyde and related aldehydes. *Free Radical in Biology & Medicine.* Vol. 11, pp. 81–128, ISSN: 0891-5849

Esterbauer, H. (1996) Estimation of peroxidative damage. A critical review. *Pathologie Biologie.* Vol. 44, pp. 25–28, ISSN: 0031-3009

Famulari, A.; Marschoff, E.; Llesuy, S.; Kohan, S.; Serra, J.; Domínguez, R.; Repetto, M.G.; Reides, C. & Lustig, E.S. de (1996). Antioxidant enzymatic blood profiles associated with risk factors in Alzheimer's and vascular diseases. A predictive assay to differentiate demented subjects and controls. *Journal of the Neurological Sciences*, Vol. 141, pp. 69-78, ISSN: 0022-510X

Farooqui, T. & Farooqui, A. (2011) Lipid-mediated oxidative stress and inflammation in the pathogenesis of Parkinson´s disease. *Parkinson´s disease*. DOI: 10.4061/2011/247467

Fiszman, M.; D´Eigidio, M.; Ricart, K.; Repetto, M.G.; Llesuy, S.; Borodinsky, L.; Trigo, R.; Riedstra, S.; Costa, P.; Saizar, R.; Villa, A. & Sica, R. (2003). Evidences of oxidative stress in Familial Amyloidotic Polyneuropathy Type 1. *Archives of Neurology*, Vol. 60, pp. 593-597, ISSN 0003-9942

Fraga, C.; Leibovitz, B. & Tappel, A. (1988). Lipid peroxidation measured as thiobarbituric acid-reactive substances in tissue slices: characterization and comparison with homogenates and microsomes. *Free Radicals in Biology and Medicine*, Vol. 4, pp. 155-161, ISSN: 0891-5849

Fridovich, I. (1978) Superoxide radicals, superoxide dismutases and the aerobic lifestyle. *Photochemistry and Photobiology*. Vol. 28, pp. 733-741, ISSN: 1010-6030

Fridovich, S. & Porter, N. (1981) Oxidation of arachidonic acid in micelles by superoxide and hydrogen peroxide. *The Journal of Biological Chemistry*. Vol. 256, pp. 260-265, ISSN: 0021-9258

Gatto, E.; Carreras, M.C.; Pargament, G.; Reides, C.; Repetto, M.G.; Llesuy, S.; Fernández Pardal, M. & Poderoso, J. (1996). Neutrophil function nitric oxide and blood oxidative stress in Parkinson's disease. *Movement Disorders*, Vol. 11, pp. 261-267, ISSN: 0885-3185

Gatto, E.; Carreras, C.; Pargament, G.; Riobó, N.; Reides, C.; Repetto, M.; Fernández Pardal, N.; Llesuy, S. & Poderoso, J. (1997). Neutrophyl function nitric oxide and blood oxidative stress in Parkinson´s Disease. *Focus Parkinson´s Disease,* Vol. 9, pp. 12-14

Gonzalez Flecha, B., Repetto, M.; Evelson, P. & Boveris, A. (1991a) Inhibition of microsomal lipid peroxidation by α -tocopherol and α -tocopherol acetate. *Xenobiotica*. 21: 1013–1022, ISSN: 0049-8254

González Flecha, B.; Llesuy, S. & Boveris, A. (1991b). Hydroperoxide-initiated chemiluminescence: assay for oxidative stress in biopsies of heart, liver and muscle. Free Radicals in Biology and Medicine, Vol. 10, pp. 93-100, ISSN: 0891-5849

Gonzalez-Flecha, B.; Cutrin, J.C. & Boveris, A. (1993) Time course and mechanism of oxidative stress and tissue damage in rat liver subjected to in vivo ischemia-reperfusion. *Journal of Clinical Investigation*. Vol. 91, pp. 456–464, ISSN: 0021-9738

Halliwell, B. & Gutteridge, J.M.C. (1984). Oxygen toxicity, oxygen radicals, transition metals and disease. *Biochemical Journal*, Vol. 218, pp. 1-14, ISSN: 0264-6021

Hogg, N.; Darley-Usmar, V.; Wilson, M. & Moncada, S. (1992) Production of hydroxyl radicals from the simultaneous generation of superoxide and nitric oxide. *Biochemical Journal*. Vol. 281, pp. 419-424, ISSN: 0264-6021

Jomova, K. & Valko, M. (2011) Advances in metal-induced oxidative stress and human disease. *Toxicology.* Vol. 283, pp. 65-87, ISSN: 0300-483X.

Junqueira, V.; Barros, S.; Chan, S.; Rodríguez, L.; Giavarotti, L.; Abud, R. & Deucher, G. (2004) Aging and oxidative stress. *Molecular Aspects of Medicine.* Vol. 25, pp. 5–16, ISSN: 0098-2997

Liu, Q.; Raina, A.K.; Smith, M.A.; Sayre,. LM. & Perry, G. (2003) Hydroxynonenal, toxic carbonyls, and Alzheimer disease. *Molecular Aspects of Medicine.* Vol. 24, pp. 305–313, ISSN: 0098-2997

Moncada, S.; Palmer, R. & Higgs, E. (1991) Nitric oxide: Physiology, patophysiology and pharmacology. *Pharmaceutical Reviews.* Vol. 43, pp. 109-141, ISSN:◎1918-5561

Murphy, M. & Sies, H. (1991) Reversible conversion of nitroxyl anion to oxide by superoxide dismutase. *Proceeding of the National Academy of Sciences of the United States.* Vol. 88, pp. 10860-10864, ISSN: 0027-8424

Nair, U.; Barstsch, H. & Nair, J. (2007) Lipid peroxidation-induced DNA damage in cancer-prone inflammatory diseases: a review of published adduct types and levels in humans. *Free Radical in Biology & Medicine.* Vol. 43, pp. 1109-1120, ISSN: 0891-5849

Navarro, A. & Boveris, A. (2004). Rat brain and liver mitochondria develop oxidative stress and lose enzymatic activities on aging. *American Journal of Physiology - Regulatory, Integrative and Comparative Physiology,* Vol. 287, pp. 1244-1249, ISSN: 0363-6119

Navarro, A.; Gomez, C.; Sanchez-Pino, MJ.; Gonzalez, H.;, Bandez, MJ.; Boveris, AD.; Boveris, A. (2005) Vitamin E at high doses improves survival, neurological performance, and brain mitochondrial function in aging male mice. *American Journal of Physiology - Regulatory, Integrative and Comparative Physiology,* Vol. 289, pp. 1392–1399, ISSN: 0363-6119

Navarro, A.; Boveris, A. (2007) The mitochondrial energy transduction system and the aging process. *American Journal of Physiology - Regulatory, Integrative and Comparative Physiology,* Vol. 292, pp. 670-686, ISSN: 0363-6119

Navarro, A.; Lopez-Cepero, JM.; Bandez, MJ.; Sanchez-Pino, MJ.; Gomez, C.; Cadenas, E.; Boveris, A. (2008) Hippocampal mitochondrial dysfunction in rat aging. *American Journal of Physiology - Regulatory, Integrative and Comparative Physiology,* Vol. 294, pp. 501-509, ISSN: 0363-6119

Navarro, A. & Boveris, A. (2009). Brain mitochondrial dysfunction and oxidative damage in Parkinson's disease. *Journal of Bioenergetics and Biomembranes,* Vol. 41, pp. 517-521, ISSN: 0145-479X

Navarro, A.; Boveris, A.; Bández, M.J.; Sánchez-Pino, M.J.; Gómez, C.; Muntane, G. & Ferrer, I. (2009). Human brain cortex: mitochondrial oxidative damage and adaptive response in Parkinson's disease and in dementia with Lewy bodies. *Free Radicals in Biology and Medicine,* Vol. 46, pp. 1574-1580, ISSN: 0891-5849

Navarro, A.; Bández, M.; Gómez, C.; Sánchez-Pino, M.; Repetto, M.G. & Boveris, A. (2010). Effects of rotenone and pyridaben on complex I electron transfer and on mitochondrial nitric oxide synthase functional activity. *Journal of Bioenergetics and Biomembranes*, Vol. 42, pp. 405-412, ISSN: 0145-479X

Ossani, G.; Dalghi, M. & Repetto, M. (2007) Oxidative damage and lipid peroxidation in the kidney of choline-defficient rats. *Frontiers in Bioscience.* Vol. 12, pp. 1174-1183, ISSN:1093-9946

Poyer, J. & McCay, P. (1971) Reduced triphosphopyridine nucleotide oxidase-catalyzed alterations of membrane phospholipids. Dependence on Fe^{3+}. *The Journal of Biological Chemistry.* Vol. 246, pp. 263-269, ISSN: 0021-9258

Rachmilewitz, D.; Stamler, J.; Karmeli, F.; Mollins, M.; Singel, D.; Loscalo, J.; Xavier, R. & Podolsky, D. (1993) Peroxynitrite induced rat colitis-a new model of colonic inflammation. *Gastroenterology.* Vol. 105. pp. 1681-1688, ISSN: 0016-5085

Repetto, M.; Reides, C.; Gomez Carretero, M.; Costa, M.; Griemberg, G., & Llesuy S. (1996) Oxidative Stress in Erythrocytes of HIV infected patients. *Clinica Chimica Acta.* Vol. 255, pp. 107-117, ISSN: 0009-8981

Repetto, M.G.; Reides, C.; Evelson, P.; Kohan, S.; Lustig, E.S. de & Llesuy, S. (1999). Peripheral markers of oxidative stress in probable Alzheimer patients. *European Journal of Clinical Investigation,* Vol. 29, pp. 643-649, ISSN: 0014-2972

Repetto, M.; María, A.; Giordano, O.; Guzmán,J.; Guerreiro, E. & Llesuy, S. (2003) Protective effect of Artemisia douglasiana Besser extracts on ethanol induced oxidative stress in gastric mucosal injury. *Journal of Pharmacy and Pharmacology.* Vol. 55, pp. 551-557, ISSN: 0022-3573

Repetto, M.G. (2008). Clinical use of chemiluminescence assays for the determination of systemic oxidative stress. In: Popov, I.; Lewin, G. (ed.), *Handbook of chemiluminescent methods in oxidative stress assessment*. Transworld Research Network: Kerala, India; pp. 163-194, ISBN: 978-81-7895-334-2

Repetto, M.G. & Ossani, G. (2008) Sequential histopathological and oxidative damage in different organs in choline deficient rats. In: Álvarez, S.; Evelson P. (ed.), *Free Radical Pathophysiology*. Transworld Research Network: Kerala, India; pp. 433-450, ISBN: 978-81-7895-311-3

Repetto, M.G.; Ferrarotti, N.F. & Boveris, A. (2010a) The involvement of transition metal ions on iron- dependent lipid peroxidation. *Archives of Toxicology.* Vol. 84, pp. 255-262, ISSN: 0340-5761

Repetto, M.; Ossani, G.; Monserrat, A. & Boveris, A. (2010b) Oxidative damage: The biochemical mechanism of cellular injury and necrosis in choline deficiency. *Experimental and Molecular Pathology.* Vol. 88, pp. 143-149. ISSN: 0014-4800.

Repetto, M. & Boveris, A. (2010) Bioactivity of sesquiterpenes: novel compounds that protect from alcohol-induced gastric mucosal lesions and oxidative damage. *Mini Reviews in Medicinal Chemistry.* Vol. 10, pp. 615-623. ISSN: 1389-5575

Repetto, M.G. & Boveris A. (2012). Transition metals: bioinorganic and redox reactions in biological systems. In: *Transition metals: uses and characteristics.* Nova Science Publishers Inc (ed.): New York, USA. pp. 349-370., ISBN: 978-1-61761-110-0

Sayre, L.M.; Perry, G. & Smith, M.A. (1999) In situ methods for detection and localization of markers of oxidative stress: application in neurodegenerative disorders. *Methods in Enzymology.* Vol. 309, pp. 133–152, ISSN: 0076-6879

Sayre, L.M.; Sha, W.; Xu, G.; Kaur, K.; Nadkarni, D.; Subbanagounder, G. & Salomon, R.G. (1996) Immunochemical evidence supporting 2-pentylpyrrole formation on proteins exposed to 4-hydroxy-2-nonenal. *Chemical Research in Toxicology.* Vol. 9, pp. 1194–1201, ISSN: 0893-228X

Schapira, A.H. (2008) Mitochondria in the aetiology and pathogenesis of Parkinson's disease. *The Lancet Neurology.* Vol. 7, pp. 97–109, ISSN:1474-4422

Schapira, A.H.; Cooper, J.M.; Dexter, D.; Clark, J.B.; Jenner, P. & Marsden, C.D. (1990) Mitochondrial complex I deficiency in Parkinson's disease. *Journal of Neurochemistry.* Vol. 54, pp. 823–827, ISSN: 0022-3042

Schapira, AH.; Mann, V.M.; Cooper,. JM.; Dexter, D.; Daniel, S.E.; Jenner, P.; Clark, J.B. & Marsden, C.D. (1990) Anatomic and disease specificity of NADH CoQ1 reductase (complex I) deficiency in Parkinson's disease. *Journal of Neurochemistry.* Vol. 55, pp. 2142–2145, ISSN: 0022-3042

Serra, J.A.; Domínguez, R.O.; Marschoff, E.R.; Guareschi E.M.; Famulari, A.L. & Boveris, A. (2009) Systemic oxidative stress associated with the neurological diseases of aging. *Neurochemical Research,* Vol. 34, pp. 2122–2132, ISSN/ISBN: 03643190

Shires, T. (1975) Inhibition by lipoperoxidation of amino acid incorporation by rough microsomal membranes *in vitro* and its partial reversibility. *Archives of Biochemistry and Biobiophys.* Vol. 171, pp. 695-707. ISSN: 0003-9861

Sies, H. (1991a) Oxidative stress: from basic research to clinical application. *American Journal of Medicine.* Vol. 91, pp. 31-38, ISSN: 0002-9343

Sies, H. (1991b). Role of reactive oxygen species in biological processes. *Wiener Klinische Wochenschrift,* Vol. 69, pp. 965–968, ISSN: 1613-7671

Valdez, L.B.; Zaobornij, T.; Bombicino, S.; Iglesias, D.E.; Boveris, A.; Donato, M.; D'Annunzio, V.; Buchholz, B. & Gelpi, R.A. (2011) Complex I syndrome in myocardial stunning and the effect of adenosine. *Free Radical in Biology & Medicine.* Vol. 51, pp. 1203-1212, ISSN: 0891-5849

Verstraeten, S.; Nogueira, L., Schreier, S. & Oteiza, P. (1997) Effect of trivalent metal ions on phase separation and membrane lipid packing: role in lipid peroxidation. *Archives of Biochemistry and Biobiophys.* Vol. 338, pp. 121-127, ISSN: 0003-9861

Videla, L.; Fraga, C.; Koch, O. & Boveris, A. (1983) Chemiluminescence of the in situ rat liver alter acute ethanol intoxication-effect of (+)-cyanidanol-3. *Biochemical Pharmacology.* Vol. 32, pp. 2822-2825, ISSN: 0006-2952

Uchida, K.; Shiraishi, M.; Naito, Y.; Tori, Y.; Nakamura, Y. & Osawa, T. (1999) Activation of stress signaling pathways by the end product of lipid peroxidation, 4-hydroxy-2-nonenal is a potential inducer of intracellular peroxide production. *Journal of Biological Chemistry.* Vol. 274, pp. 2234–2242. ISSN: 0021-9258

Uchida, K. (2003) 4-Hydroxy-2-nonenal: A product and mediator of oxidative stress. *Progress in Lipid Research*. Vol. 42, pp. 318–343. ISSN: 0163-7827

Yin, H.; Xu, L. & Porter, N.A. (2011) Free radical lipid peroxidation: mechanisms and analysis. *Chemical Reviews*. 111: 5944-5972. ISSN: 0009-2665

Chapter 2

Lipid Oxidation in Homogeneous and Micro-Heterogeneous Media in Presence of Prooxidants, Antioxidants and Surfactants

Vessela D. Kancheva and Olga T. Kasaikina

Additional information is available at the end of the chapter

http://dx.doi.org/10.5772/46021

1. Introduction

The human body is constantly subjected to a significant oxidative stress as a result of the misbalance between antioxidative protective systems and the formation of strong oxidizing substances, including free radicals. The stress can damage DNA, proteins, lipids and carbohydrates and could cause negative effect to intracellular signal transmission. Antioxidants could be promising agents for management of oxidative stress-related diseases. Oxygen is essential for all living organisms, but at the same time it is a source of constant aggression for them. In its ground triplet state ($^3O_2$) oxygen has weak reactivity, but it can produce strongly aggressive and reactive particles such as singlet state oxygen ($^1O_2$), hydroperoxides (H_2O_2), superoxide anion ($O_2^{\cdot -}$), hydroxylic radical ($OH^{\cdot}$) and various peroxide ($LO_2^{\cdot}$) and alkoxy radicals ($LO^{\cdot}$). It is well known that the latter lead to an oxidative degradation of biological macromolecules, changing their properties and thus the cell structure and functionality. The free radicals formation in the hydrophobic parts of the biological membranes initiates radical disintegration of the hydrocarbon "tails" of the lipids. This process is known as lipid peroxidation (Figure 1) [1-3].

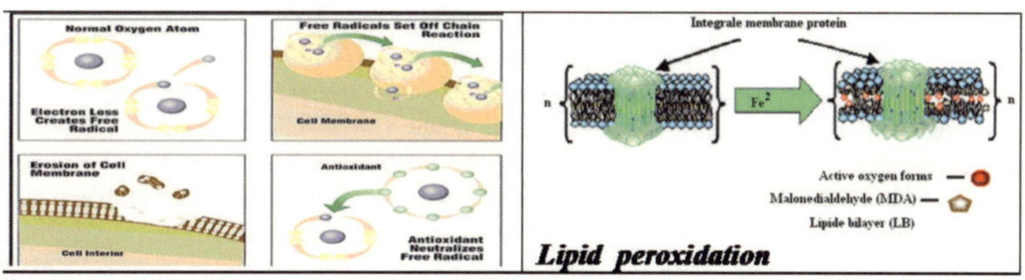

Figure 1. Erosion of cell membrane, antioxidant neutralizes free radicals and lipid peroxidation

In recent decades, many communications have been devoted to the significant role of physical factors, which control the structure of nutrition systems, in the chemistry of lipid oxidation in oil/water (O/W) emulsions [4,5]. In this case, the rate of lipid oxidation strongly depends on the physical properties of interfaces, because they affect the character of the interaction between water-soluble compounds of transition metals and hydroperoxides located inside and on the surface of emulsion droplets. For example, positively charged and high viscosity interfaces that hinder the contact between iron ions and hydroperoxides inhibit oxidation of fat emulsions [6,7]. Another example of the influence of physical factors on the oxidation is the antioxidant polar paradox [6-9], which is based on the fact that nonpolar antioxidants are efficient in O/W emulsions, because they are located in emulsion droplets together with oxidizable lipids. Polar antioxidants are more efficient in W/O emulsions because they are concentrated at the interfaces [10-12]. In the frame of this chapter, the main features of lipid oxidation in homogeneous and micro - heterogeneous oil media, formed by surfactants (W/O microemulsions) will be discussed.

2. Kinetic model of homogeneous lipid oxidation

The kinetic model is based (Table 1) on the reactions and corresponding rate constants known for the oxidation of methyl linoleate (MeLi), because linoleic esters are easily oxidizable components of many natural lipid systems and thus determines the oxidizability of the lipid substrates.

The kinetic scheme of liquid-phase (homogeneous) oxidation of lipids (LH) includes reactions 1-12.

In the presence of an initiator (I), i.e. initiated oxidation, the formation of radicals occurs with a constant initiation rate $R_{IN}=2k_1[I]$. Under autoxidation conditions ([I]=0), the rates of radical formation in the reactions of LH with O_2 (reaction 6) and decomposition of lipid hydroperoxides (LOOH) (reactions 7 and 8) increase as LOOH is accumulated. The chain termination occurs due to recombination or disproportionation of radicals (reactions 9, 11 and 12). The scheme of inhibited oxidation includes reactions 13-19, known for the phenolic antioxidants, AH.

The calculations was performed for three groups of AH, which differ in their activity in reaction 13 with peroxyl radicals LOO: group I (of the type of alpha-tocopherol) with k_{13}=1.5 $10^6 M^{-1}s^{-1}$; group II (AH of type of unhindered phenols, e.g. hydroquinone), with k_{13}=1.5 $10^5 M^{-1}s^{-1}$; and group III (AH of the type of sterically hindered phenols, e.g. butylated hydroxyl toluene, BHT with k_{13}=1.5 $10^4 M^{-1}s^{-1}$. The effect of AH regeneration in reaction 16 was considered for rapidly (k_{16}=4 $10^7 M^{-1}s^{-1}$) and slowly (k_{16}=3 $10^3 M^{-1}s^{-1}$) reacting phenoxyl radicals A·. Reaction 19 is the chain transfer by the inhibitor radical. It can be a hydrogen abstraction from the substrate molecule with regeneration of the inhibitor. Chain transfer can occur as the addition of A· to unsaturated bonds of polyene compounds, in this case, the inhibitor is not regenerated [13]. We examined both cases, rate constants k_{19} were varied within the 0-100 $M^{-1}s^{-1}$ range (taking into account the published data). In this series of calculations, we accepted that reaction 16 occurs as disproportionation.

No	Reaction	$k/M^{-1}s^{-1}$, MeLi	Refs.	$k/M^{-1}s^{-1}$, Limonene
1	$I \rightarrow I\cdot + I\cdot$	$5\ 10^{-6}$	13	$5\ 10^{-6}$
2	$I\cdot + O_2 \rightarrow IO_2\cdot$	$5\ 10^{6}$	13	$5\ 10^{6}$
3	$IO_2\cdot + LH \rightarrow LO_2\cdot + IOOH$	$1\ 10^{3}$	13	$1\ 10^{3}$
4	$L\cdot + O_2 \rightarrow LO_2\cdot$	$5\ 10^{6}$ $1\ 10^{8}$	13 15	$1.5\ 10^{7}$
5	$LO_2\cdot + LH \rightarrow LO_2\cdot + LOOH$	90 100	13 15,16	14
6	$LH + O_2 \rightarrow L\cdot + HO_2\cdot$	$5.8\ 10^{-11}$	13	$1.2\ 10^{-14}$
7	$LOOH + LH \rightarrow L\cdot + LO_2 + H_2O$	$2.3\ 10^{-7}$	13	$4\ 10^{-8}$
8	$2LOOH \rightarrow LO\cdot + LO_2 + H_2O$	$2.4\ 10^{-6}$	13	$1\ 10^{-6}$
9	$2\ LO_2\cdot \rightarrow Alc + Ket$	$1\ 10^{5}$ $4.4\ 10^{6}$ $1\ 10^{7}$	17 13,18 15	$3.5\ 10^{6}$
10	$LO\cdot + LH \rightarrow Alc + Ket$	$1\ 10^{7}$ $1\ 10^{5}$	19 13	$1.7\ 10^{7}$
11	$LO\cdot + LO_2\cdot \rightarrow Ket + LOOH$	$5\ 10^{6}$	13	
12	$LO_2\cdot + IO_2\cdot \rightarrow Alc + Ket$	$5\ 10^{6}$	13	$5\ 10^{6}$
13	$AH + LO_2\cdot \rightarrow A\cdot + LOOH$	$1.5\ 10^{4}$ -$1.5\ 10^{6}$ $2\ 10^{6}$	13,16 15	
14	$AH + LO\cdot \rightarrow A\cdot + Alc$	$1\ 10^{7}$	13	
15	$AH + IO_2\cdot \rightarrow A\cdot + IOOH$	$1.5\ 10^{5}$	13	
16	$2A\cdot \rightarrow P1\ (+AH)$	$3\ 10^{3}$ – $4\ 10^{7}$	13,16,18	
17	$A\cdot + LO_2\cdot \rightarrow P2$	$2.5\ 10^{6}$ $3\ 10^{8}$	20 13,16	
18	$A\cdot + LO\cdot \rightarrow P3 + AH + Ket$	$3\ 10^{8}$	13	
19	$A\cdot + LH \rightarrow L\cdot + (AH/P4)$	0-100 0.07	13 21	

Note: The rate constants (k_0 correspond to the oxidation of MeLi at 60°C; in reaction 1, 2 and 4, k are presented in s^{-1}. Initial concentrations: [LH]=2.9M, $[LOOH]_0=10^{-5}M$, $[I]=4\ 10^{-3}M$, $[AH]_0=10^{-4}M$, $[O_2]=10^{-3}M$=const; oxidation usually occurs at a constant oxygen pressure, therefore $[O_2]$ is included in the corresponding rate constants: $k_2=k_4=k_6=k_i[O_2]$.

Table 1. The Approximate Rate Constants of the Different Reactions Involved in the Autoxidation of Methyl Linoleate [13] and Limonene [14] in initiated oxidation, autoxidation and inhibited oxidation (at 60°C).

The main kinetic parameters	Initiated oxidation	Lipid autoxidation
Rate of initiation (R_{IN})	Constant and well-controlled $R_{IN}=2k_1[I]$	$R_{IN}=2k_6[LH][O_2]$ + $2k_7[LH][LOOH]+2k_8[LOOH]^2$
Rate of oxidation (R_0) and (R_A)	$R_0=k_p\ [LH](R_{IN}/2k_t)^{0.5}$	
Rate of non-inhibited oxidation (R_0)	$R_A=k_p[LH]R_{IN}/nk_A[AH]_0$	$R_0=k_p\ [LH](R_{IN}/k_t)^{0.5}$
	$k_p = k_5$; $k_t = k_9$	$R_A=k_p[LH]R_{IN}/nk_A[AH]_0$
Rate of inhibited oxidation (R_A)		
Oxidizability parameter	$a = k_p/(2k_t)^{0.5}$	$a = k_p/(2k_t)^{0.5}$
Inhibition degree (ID)	$ID = v_0/v_A$	$ID = R_0/R_A$
Induction period (IP)	$IP=n[AH]_0/R_{IN}$	$IP=n[AH]_0/R_{IN}$
Antioxidant efficiency	nk_A	$PF=IP_A/IP_0$ and $RAE=(IP_A-IP_0)/IP_0$

Table 2. The main kinetic parameters of initiated oxidation and lipid autoxidation

Under other equivalent conditions, the bimolecular decay of 2A· by disproportionation in which AH is regenerated gives a considerable advantage in retardation effects as compared with the situation where no regeneration occurs (recombination of 2A·). The presence of the second hydroxyl group in the aromatic ring results in higher k_{16}. In this case, an increase in the induction period related to AH regeneration is most pronounced.

Lipid oxidation is one of the important reactions in biology. Chemical reaction kinetics considers two aspects: the rate of reaction and effective factors – temperature concentration of reactants and products. This knowledge is an essential prerequisite for modeling the lipid oxidation, the shelf life of stored foods, durability of low density proteins, and so on.

3. Effect of pro-oxidants (ROH) leading to acceleration of lipid hydroperoxides (LOOH) decomposition

3.1. Kinetic modeling of lipid oxidation for different mechanism of LOOH decomposition

A kinetic analysis of non-inhibited lipid (LH) autoxidation for different mechanisms of hydroperoxides (LOOH) decay is proposed [22]. It is based on using of mathematical simulation methods of LH autoxidation kinetics. Kinetic schemes of LH autoxidation for some different ways of hydroperoxides decay - mono-molecular, pseudo-mono-molecular and/or bimolecular mechanism are presented. This analysis permits establishing the influence degree of different hydroperoxides decay mechanisms on the kinetic parameters, characterizing the substrate oxidizability. The proposed kinetic analysis has been applied to the methyl linoleate, MeLi) autoxidation at 60°C.

The kinetic model that describes the lipid hydroperoxides decomposition taking into account the possibility of monomolecular (LOOH), pseudo-monomolecular (LOOH + LH) and bimolecular (2 LOOH) mechanisms in both cases: in presence of an oxygen (O_2) and in its absence, i.e. in an inert atmosphere (N_2) is illustrated by Scheme 1. In these equations:

$$[LOOH] = \frac{1+K_1[LH]}{2K_2}\left[\sqrt{1+\frac{4K_2[T]}{(1+K_1[LH]}}-1\right]$$

$$[T] = \left(\sqrt{[To]} + \frac{1}{2}\frac{kp}{\sqrt{kt}}[LH]\sqrt{k_i\, t}\right)^2$$

$$C = \frac{4K_2}{(1+K_1[LH])^2}$$

$$d = \frac{e_o k_{3o} + e_1 k_{31} K_1[LH]}{1+K_1[LH]} = k_i$$

$$[O_2] = \frac{k_p}{\sqrt{k_t}}[\mathrm{LH}]\sqrt{k_i[T_o]}\; t + \frac{1}{4}(\frac{k_p}{\sqrt{k_t}}[LH])^2 k_i t$$

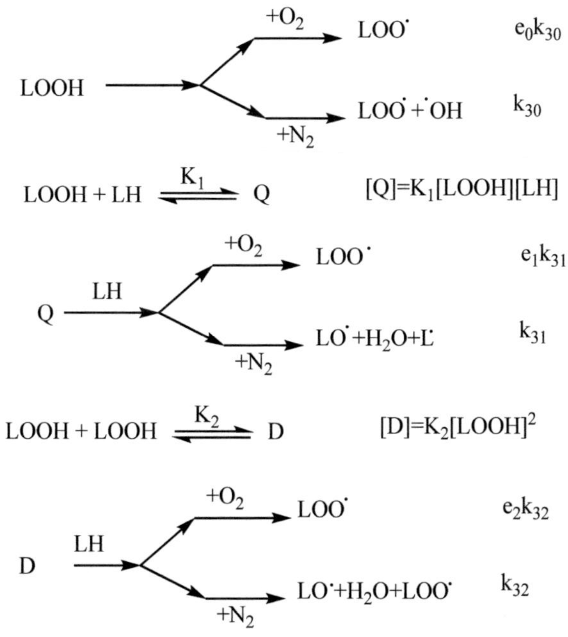

$$[T]=[LOOH]+[Q]+[D]=[LOOH](1+K_1[LH]+K_2[LOOH])$$

LH: is linoleic acid with its allylic hydrogen
LOO•: peroxide radical
LOOH: lipid hydroperoxides
K_1 and K_2: are the equilibrium constants for complexes Q and D, respectively
[T]: summary concentration of LOOH
k_{30}, k_{31} and k_{32}: are the corresponding rate constants
e_0, e_1, e_2: are the corresponding radicals yield

Scheme 1. Kinetic scheme of lipid hydroperoxide decomposition reactions

The kinetic scheme 2 is significantly simplified and readily solved assuming a quasi-steady-state for LOO•, rapid achievement of equilibrium and neglected of the loss of Q and LOOH since their decomposition rate constants are low. There are marked: **C** - is the ratio between the equilibrium constants of bi- and pseudo-mono-molecular mechanisms of LOOH decomposition, needed to be marked for the solution of the equation.

$$\text{Chain Generation} \xrightarrow{+O_2,\ +LH} LOO^{\bullet} + HO_2^{\bullet} \qquad R_{IN}$$

$$LOO^{\bullet} + LH \xrightarrow{+O_2} LOO^{\bullet} + LOOH \qquad k_p$$

$$LOOH \xrightarrow{+O_2,\ +LH} LOO^{\bullet} + \text{products} \qquad e_0k_{30}$$

$$LOOH + LH \rightleftharpoons Q \xrightarrow{+O_2,\ +LH} LOO^{\bullet} + \text{products} \qquad e_1k_{31}$$

$$LOOH + LOOH \rightleftharpoons D \xrightarrow{+O_2,\ +LH} LOO^{\bullet} + \text{products} \qquad e_2k_{32}$$

$$2\ LOO^{\bullet} \longrightarrow \text{products} \qquad k_t$$

R_{IN}: the rate of chain generation, k_p: rate constants of chain propagation, k_t: rate constants of chain

Scheme 2. Kinetic scheme of lipid autoxidation by Kancheva and Belyakov [22]

Figures 2-5 presents kinetics of different mechanisms of lipid hydroperixides decomposition. In Figure 6 it is shown, that k_i doesn't change with growing of MeLi concentration from 0.3 to 1.7 M, when the concentration of MeLi hydroperoxydes is smaller than 5 10^{-3} M. It is established, that MeLi hydroperoxides decay is in agreement with a first order reaction and pseudo-mono-molecular mechanism (a reaction between hydroperoxides and non-oxidized lipid substrate; LOOH + LH).

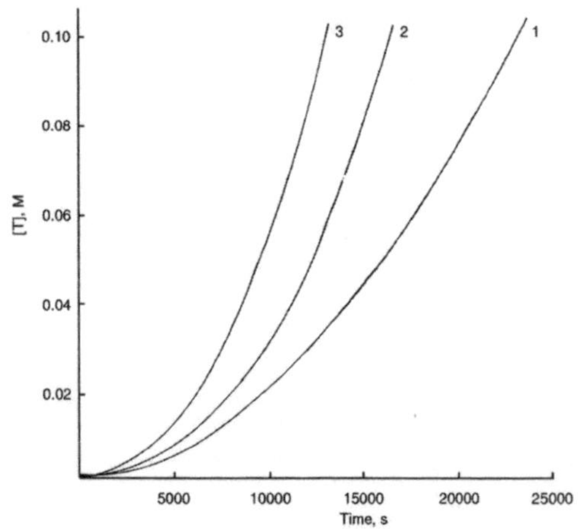

No	W_i, M/s ($\times 10^{-8}$)	W_i^o, M/s ($\times 10^{-11}$)	$(ek_3)_{eff}$ ($\times 10^{-7}$)	C
1	2.0	2.0	2.0	10^{-4}
2	6.3	2.0	4.7	10^{2}
3	8.6	2.2	7.6	10^{3}

Figure 2. Influence of dimer formation equilibrium constant (K_2) on the kinetics of MeLi autoxidation at 60°C, when e_2k_{32} has a great value (e_2k_{32}=2 10^{-6}), $[T_0]=10^{-4}$M, d=2 10^{-7} and $k_p/(k_t)^{1/2} = 6\ 10^{-2}$

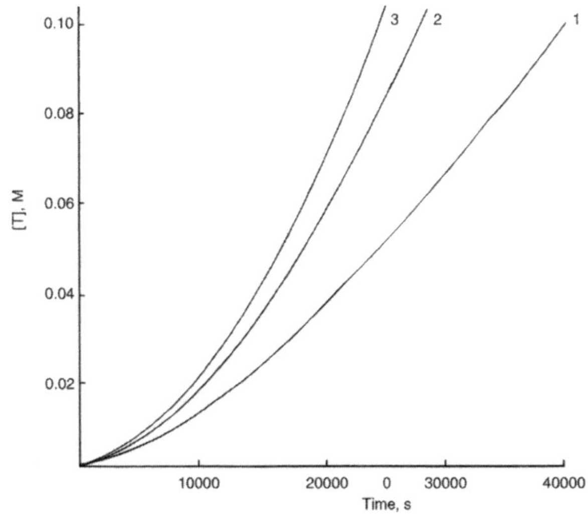

No	c	$(ek_3)_{eff}$ ($\times 10^{-7}$)
1	10^3	5.54
2	10^2	1.25
3	20	1.71

Figure 3. Influence of dimer formation equilibrium constant (K_2) on the kinetics of MeLi autoxidation at 60°C, when e_2k_{32} has a small value ($e_2k_{32}= 10^{-10}$), $[T_0]=10^{-4}$M, d=2 10^{-7} and $k_p/(k_t)^{1/2} = 6\ 10^{-2}$(i.e. very small value of e_2)

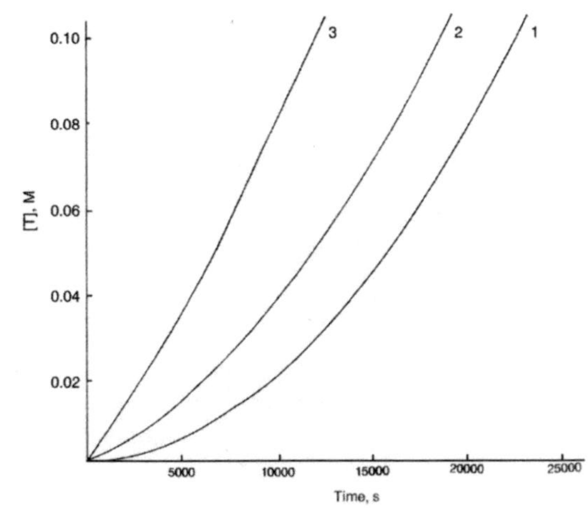

$$\frac{[T]-[T_0]}{t} = A \cdot t + B$$

$$A = \frac{1}{4}\left(\frac{k_p}{\sqrt{k_6}}[LH]\right)^2 \cdot d$$

$$B = \frac{k_2}{\sqrt{k_6}}[LH]\sqrt{W_i^O + d \cdot [T_0]}$$

No	W_i^o, M/s	A ($\times 10^{-10}$)	B
1	0	1.8	2.68 10^{-7}
2	10^{-9}	1.8	1.92 10^{-6}
3	10^{-8}	1.8	6.01 10^{-6}

Figure 4. Influence of the substrate (MeLi) diluting with an inert solvent (concentrations of 25, 50 and 100%) at 60°C (e_2k_{32}=2 10^{-6})

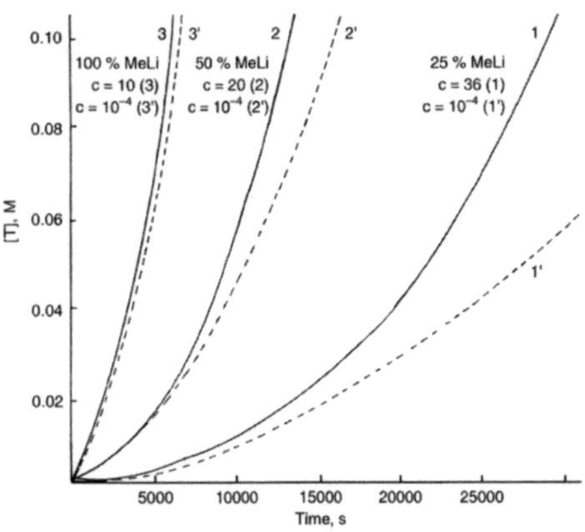

No	[LH] MeLi %	c	d $(ek_3)_{eff}$ ($\times 10^{-7}$)	$\frac{k_p}{\sqrt{k_t}}[LH]$	W_i [T] = 0.1 M ($\times 10^{-8}$)
1	25	36	1.3	4.0 10^{-2}	4.5
1'	25	10^{-4}	3.0	4.0 10^{-2}	1.3
2	50	20	3.1	8.0 10^{-2}	4.1
2'	50	10^{-4}	2.0	8.0 10^{-2}	2.0
3	100	10	3.2	1.6 10^{-1}	3.8
3'	100	10^{-4}	2.7	1.6 10^{-1}	2.7

Figure 5. Kinetic curves of inhibited oxidation and autoxidation of MeLi at 60°C, when there is no dimerization of lipid hydroperoxides (K_2=0)

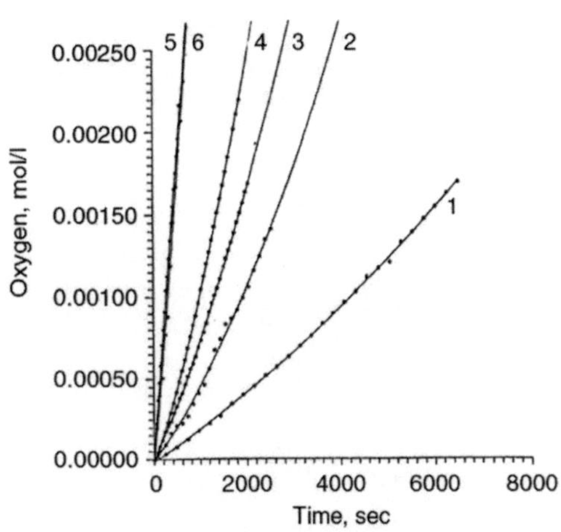

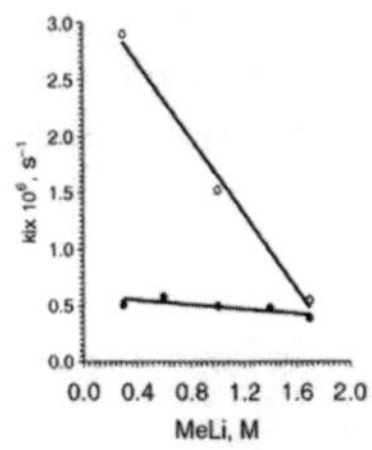

$$LOOH + LH \underset{}{\overset{K_1}{\rightleftharpoons}} Q \qquad LOOH + ROH \underset{}{\overset{K_S}{\rightleftharpoons}} S$$

No	[MeLi] M	[1 – OD] M
1	0.3	0
2	0.3	0.1
3	1.0	0
4	1.0	0.1
5	1.7	0
6	1.7	0.1

Figure 6. Effect of ROH (0.1M, 1-Octadecanol, 1-OD) on the kinetics of hydroperoxide accumulation of MeLi at 60°C, at different MeLi concentrations (0.3, 1.0 and 1.7M) and dependence of the effective initiation rate constants k_i on [LH] in the absence (1) and in presence of 0.1M 1-Octadecanol (2).

It is shown, that in presence of a lipid hydroxyl compound k_i^{ROH} is strongly growing with the decrease of MeLi (LH) concentration (Figure 6). This is explained with the competition of reactions (LOOH + LH and LOOH + ROH). Some different mechanisms, which are possible for reaction between LOOH and ROH, were discussed (Scheme 3).

$$LOOH + ROH \overset{K_S}{\rightleftharpoons} S$$

$$[S] = K_s[LOOH][ROH]$$

$$S \xrightarrow{+O_2} LOO^{\bullet} + products \qquad e_s k_s$$

$$[T]=[LOOH]+[Q]+[S] =[LOOH](1+K_1[[LH]+K_s[ROH])$$

$$[LOOH] = [T]/(1+K_1[LH] +K_s[ROH])$$

$$[Q] = K_1[LH]\,[T]/(1+K_1[LH]+K_s[ROH]$$

$$[S] =K_s[ROH]\,[T]\,/(1+K_1[LH]+K_s[ROH])$$

Ks: is the equilibrium constant for complex S, initial rate of decomposition of complex S ($e_s k_s$)
Lipid hydroxy compounds (LOH) from the oxidized lipid substrate (LH) is formed during the whole oxidation process:
LOH and H_2O with the rate e_0 k_{30} [LOOH] from the hydroperoxides of substrate LH
LOH and H_2O with the rate e_1 k_{31} [Q] from the Q|
LOH and H_2O with the rate e_3 k_{33} [S] from the S

Scheme 3. Kinetic scheme of lipid hydroperoxides (LOOH) decay in presence of ROH

$$A = \frac{1}{4}\left(\frac{k_p}{\sqrt{k_t}}[LH]\right)^2 k_i$$

$$k_i = \frac{4A}{(\frac{k_p}{\sqrt{k_t}}[LH])^2}$$

$$B = \frac{k_p}{\sqrt{k_t}}\sqrt{k_i[T_o]}$$

$$[T_o] = (\frac{B}{\frac{k_p}{\sqrt{k_t}}[LH]\sqrt{k_i}})^2$$

There are presented some different mechanisms of the interaction between lipid hydroperoxides (LOOH) and hydroxy compounds (ROH):

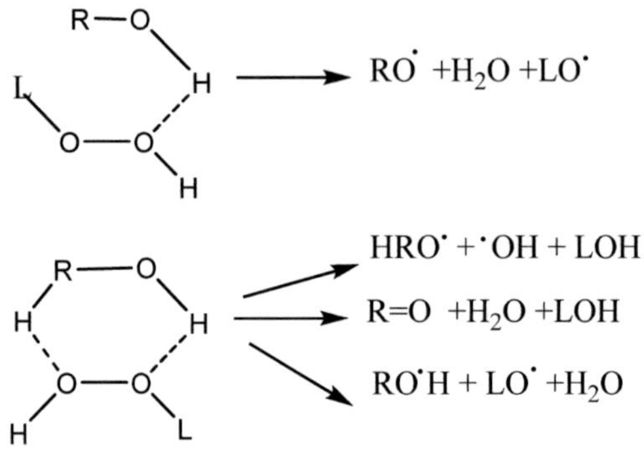

Scheme 4. Additional lipid hydroperoxides decomposition in presence of an antioxidant (AH)

$$\mathrm{LOOH + AH} \underset{}{\overset{K_P}{\rightleftharpoons}} \mathrm{P} \xrightarrow[\mathrm{LH,\,O_2}]{k_{iP}} \mathrm{LO_2^{\bullet}}$$

$$K_P = \frac{[\mathrm{P}]}{[\mathrm{LOOH}][\mathrm{AH}]}$$

K_P: the equilibrium constant for complex P,
k_{iP}-initiation rate constant of P decomposition
Total hydroperoxides concentration [T] in presence of ROH and AH
[T] = [LOOH]+[Q]+[S]+[P]= [LOOH](1+K1[LH]+KS[ROH]+KP[AH])

Scheme 5. Equilibrium constant of complex formation between an antioxidant (AH) and lipid hydroperoxides (ROH)

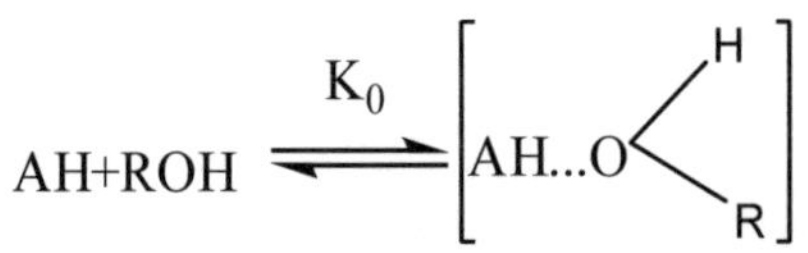

$$K_0 = \frac{[\mathrm{AH}]_0 - [\mathrm{AH}]}{[\mathrm{AH}][\mathrm{ROH}]}$$

$$[\mathrm{AH}] = \frac{[\mathrm{AH}]_0}{1 + K_0[\mathrm{ROH}]}$$

Scheme 6.

It has been proven [23,24] that fatty alcohols with different chain length, mono- and diacylglycerols increase the rate of LOOH decomposition into free radicals and thus accelerated lipid oxidation in absence of an antioxidant. In presence of phenolic antioxidants ROH make complexes basing on H bond formation and thus decrease the antioxidant efficiency of them [25,26]. DL-alpha –tocopherol and butylated hydroxyl toluene demonstrate the best antioxidant efficiency in presence of ROH [26]. Taking into account that ROH are formed during the proceeding, transportation and storage of lipids and lipid containing products as a result of hydrolysis, it is of importance to know how to improve their oxidative stability.

4. Antioxidants – Inhibitors of lipid oxidation

The introduction of antioxidants in the affected body normalizes not only the peroxide oxidation, but also the lipid content. Antioxidants used in oncology are effective in the first stages as mono-therapy with antioxidants at high concentrations and at the last stages mainly as additives in the complex tumor therapy - the antioxidant is in low concentrations. In this respect the medical treatment of most of diseases includes formulations based on a combination of traditional drugs with targeted functionality and different antioxidants [3,27].

The activity of antioxidants depends on complex factors including the nature of the antioxidants, the condition of oxidation, the properties of substrate, being oxidized and the stage of oxidation [2,3,27-33].

Capacity of antioxidants has at least two sides: the antioxidant potential, determined by its composition and properties of constituents and is the subject of food chemistry, and the biological effects, depending, among other things, on bioavailability of antioxidants, and is a medico-biological problem.

4.1. Classification of antioxidants [2,16,30-36]

Depending on their mechanism of action:

Antioxidants, inhibiting lipid oxidation by trapping lipid peroxide radicals– they are aromatic compounds with a weak O-H, N-H bonds (phenols, amines, aminophenols, diamines etc.

Antioxidants, inhibiting the oxidation process by trapping alkyl radicals – they are quinones, methylene quinones, which are effective in low oxygen concentration.

Hydroperoxide decomposers – these compounds react with hydroperoxides without formation of free radicals.

Metal chelators – oxidation process can be inhibited by addition of compounds, forming complexes with metal ions and thus made them inactive towards hydroperoxides. In this groups are hydroxyl acids, flavonoids etc.

Antioxidants with multistage action – systems containing such kind of compounds (alcohols and amines) inhibitors can be regenerated during the oxidation process.

Inhibitors with combined action – inhibitor molecule has two or more functional groups, each of them react in different reactions.

Depending on their nature:

Natural antioxidants – usually with low toxicity (with some exception), wide spectrum of biological and antioxidant activities.

Synthetic - they are with a high antioxidant activity. However, antioxidants for application in foods and additives or supplements they must pass additional criteria (no toxicity, safety, healthy, low cost, etc.)

Depending of their biological activity:

Bio-antioxidants – compounds with both biological and antioxidant activities. Last decades there is a growing interest to the nature-like bio-antioxidants. The most important known bio-antioxidants are flavonoids and phenolic acids.

Antioxidants without a biological activity - some even natural antioxidants can show a toxic activity and for that reason they must be tested.

Depending on the number of phenolic groups:

Monophenols – the known compounds are butylated hydroxyl toluene (BHT), Tocopherol (TOH), p-coumaric acid (p-CumA), ferulic acid (FA), sinapic acid (SA) etc.

Biphenols – the known compounds are caffeic acid (CA), hydroquinone (HQ), tertbutylated hydroquinone (TBHQ) etc.

Polyphenols –flavonoids: quercetin (Qu), rutin (Ru), luteolin (Lu), kampferol (Kf), isorhamnetin (Isorh) etc.

The inherent compositional and structural complexity of real foods and in vivo studied means that systematic studies of lipid oxidation must first be carried out in model systems. The following models were applied to explain the structure-activity relationship of different phenolic antioxidants: model 1, a DPPH assay used for the determination of the radical scavenging capacity ($AH+DPPH^{\bullet} \rightarrow A^{\cdot}+DPPH\text{-}H$); model 2, chemiluminescence (CL) of a model substrate RH (cumene and diphenylmethane) used for determination of the rate constant of a reaction with model peroxyl radicals ($AH+RO_2^{\cdot} \rightarrow A^{\cdot}+ROOH$); model 3, lipid autoxidation (LAO) used for the determination of the chain-breaking antioxidant efficiency and reactivity ($AH+LOO^{\cdot} \rightarrow A^{\cdot}+LOOH$; $A^{\cdot}+LH(+O_2) \rightarrow AH+LOO^{\cdot}$); and model 4, theoretical methods used for predicting the activity (predictable activity by statistical and/or quantum-chemical calculations).

4.2. Structure of the antioxidants

By combination of different experimental methods: DPPH test, lipid autoxidation kinetics, chemiluminescence kinetics and quantum chemical calculations it has been proven that the prooxidant activity of chalcones is due to the possible reaction of phenoxyl radicals formed with oxygen and formation of dioxiethanes, [37]Vasil'ev *et al.*, 2009:

New bis-coumarins are found to have anti-HIV activity [38]. Together with their antioxidant capacity (Fig. 8C) they are one of the most important bio-antioxidants nowadays.

The studied simple dihydroxy-coumarins are natural (Cum0) and nature-like synthetic compounds with a wide range of biological activities against cancer, inflammatory, cardio-vascular diseases, diabetes etc. Together with the strong antioxidant activity and synergistic effect with Tocopherol, they are very important for the practical application, [40]Kancheva *et al.*, 2010a.

	Abbr.	R3		R5	Activity	Methods
A. Benzoic acids 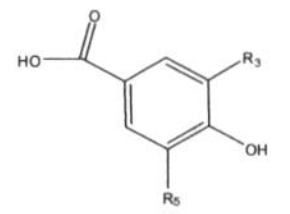	HBA	H		H	Weak	LAO,CL,Theor
	VanA	H		OCH_3	Weak	LAO,CL,Theor
	SyrA	OCH_3		OCH_3	Moderate	LAO,CL,Theor
	DHBA	OH		H	Strong	LAO,CL,Theor
	GA	OH		OH	Strong	LAO, DPPH,Theor
B.Cinnamic acids 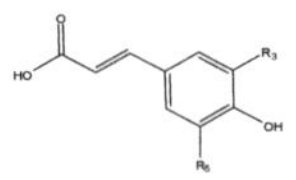	p-CA	H		H	Weak	LAO, CL,DPPH, Theor
	FA	OCH_3		H	Moderate	LAO, CL,DPPH, Theor
	SA	OCH_3		OCH3	Moderate	LAO, CL,DPPH, Theor
	CA	OH		OH	Strong	LAO, CL, DPPH, Theor.
	PHC	Prenyl		H	Strong	LAO
	DPHC	Prenyl		Prenyl	Moderate	LAO
C. N-cinnamic acids amides 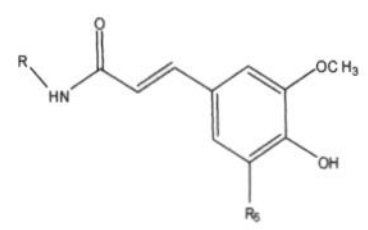	Abbr.	R		R5	Activity	Method
	N1	$CH(CH_2C_6H_5)$-$COOC(CH_3)_3$		H	Moderate	LAO
	N2	$CH(CH_2C_6H_4$-F-m)-$COOCH_3$		H	Moderate	LAO
	N3	$CH(CH_3)$-$COOC(CH_3)_3$		H	Moderate	LAO
	N4	$CH(CH_2C_6H_4$-OH-p)-$COOCH_3$		H	Moderate	LAO
	N5	$CH(CH_2C_6H_4$-OH-p)-$COOCH_3$		OCH_3	Strong	LAO
	N6	$CH(CH_2C_6H_4$-F-m)-$COOCH_3$		OCH_3	Strong	LAO
	N7	$CH(CH_2C_6H_5)$-$COOC(CH_3)_3$		OCH_3	Strong	LAO
D.Hydroxy-chalcones 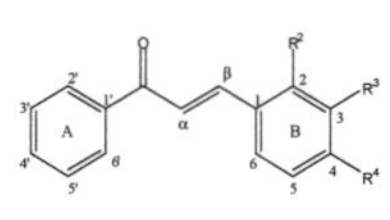	Abbr.	R2	R3	R4	Activity	Methods
	Ch1	H	OH	H	Weak	LAO, CL, DPPH, Theor
	Ch2	H	H	OH	Moderate	LAO, CL, DPPH,Theor
	Ch3	H	OH	OH	Strong	LAO, CL, DPPH Theor
	Ch4	H	OH	OCH_3	Weak	LAO, CL, DPPH,Theor
	Ch5	OH	H	H	Weak	LAO, CL, DPPH,Theor
	Ch6	OH	OCH_3	H	Weak	LAO, CL, DPPH,Theor

Table 3. The main structures and activities with methods for benzoic acids, cinnamic acids, N-cinnamic acids amides and chalcones

Chalcone Ch3(ΔH_f^0 = -58.3 kcal/mol) *Radical formed from Ch 3 (ΔH_f^0 = -29.0 kcal/mol)*

(a)

D1 *D2*

Dioxiethanes D1 and D2
$\Delta H_f^0(D_1^\bullet)$ = -19. 9 kcal/mol, $\Delta H_f^0(D_2^\bullet)$ = -27.1 kcal/mol

(b)

Figure 7. a) Optimized structures of Chalcone Ch3 (7a) and its aryl radical (7b); b) Optimized structures of Dioxiethanes D1 and D2 formed from Ch3 radical and oxygen.

E.Simple Coumarins	**Abbr.**	**R3**	**R4**	**R5**	**R6**	**R8**	**Activity**	**Method**
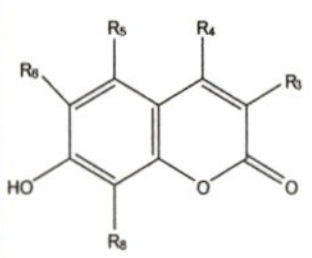	Cum0	H	H	H	OH	H	Strong	LAO
	Cum1	H	CH_3	H	OH	H	Strong	LAO
	Cum2	H	CH_3	H	H	OH	Strong	LAO
	Cum3	EtCOOMe	CH_3	H	H	OH	Strong	LAO
	Cum4	MeCOOEt	CH_3	H	H	OH	Strong	LAO
	Cum5	H	CH_3	OH	H	H	Weak	LAO
	Cum6	H	CH_3	H	H	H	Weak	LAO
	Cum7	H	OH	H	H	H	Weak	LAO
F.Bis-Coumarins	**Abbr.**	**R3**	**R4**	**R5**	**Activity**	**Method**	**Abbr.**	**R3**
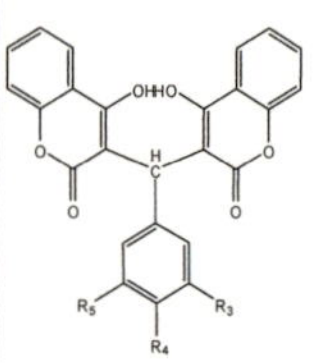	Bis-Cum1	OH	OH	H	Strong	LAO	Bis-Cum1	OH
	Bis-Cum2	OCH_3	OH	OCH_3	Moderate	LAO	Bis-Cum2	OCH_3
	Bis-Cum3	OCH_3	OH	NO_2	Weak	LAO	Bis-Cum3	OCH_3
	Bis-Cum4	OCH_3	OCH_3	H	Weak	LAO	Bis-Cum4	OCH_3
	Bis-Cum5	OCH_3	OCH_3	OCH_3	Weak	LAO	Bis-Cum5	OCH_3

Table 4. Simple and Bis-Coumarins

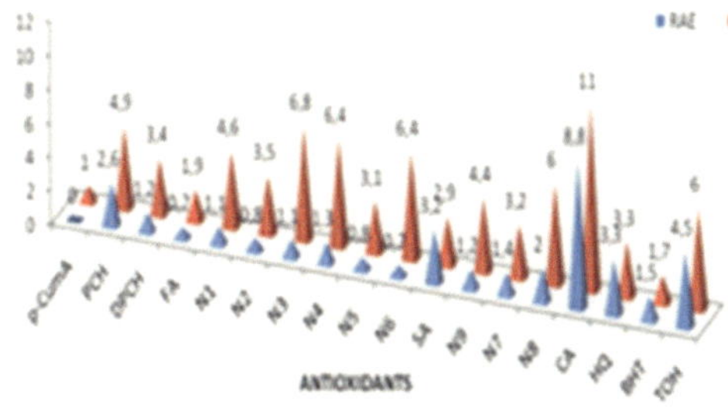

(a) [30]Kancheva, 2009

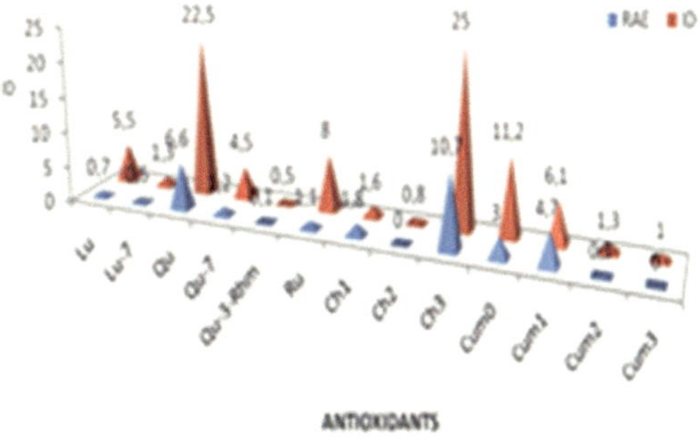

(b) [30]Kancheva, 2009

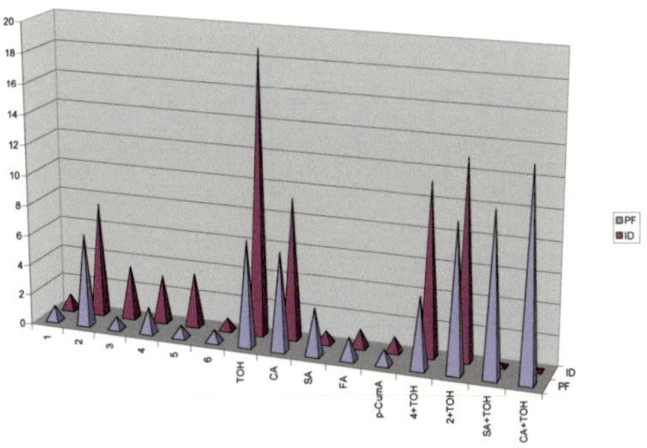

1=Cum7; 2-6 corresponding bis-coumarins

(c) [38]Kancheva et al, 2010b

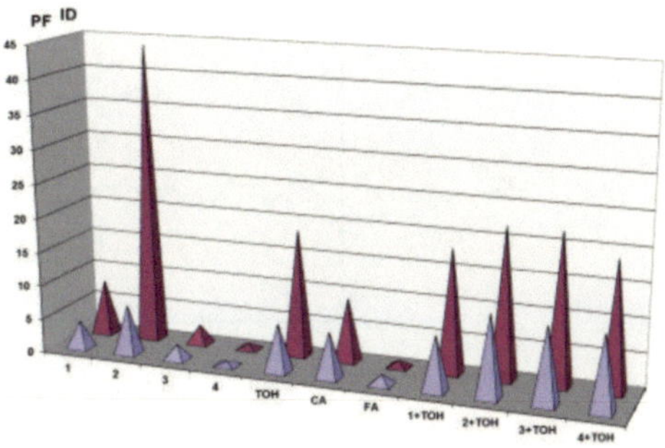

Xanthene(1,2) and neo-lignans (3,4)

(d) [39]Kancheva et al, 2012

Figure 8. The main kinetic parameters PF, RAE (antioxidant efficiency) and ID (inhibition degree) of lipid autoxidation in presence of different antioxidants (for abbreviation see corresponding tables)

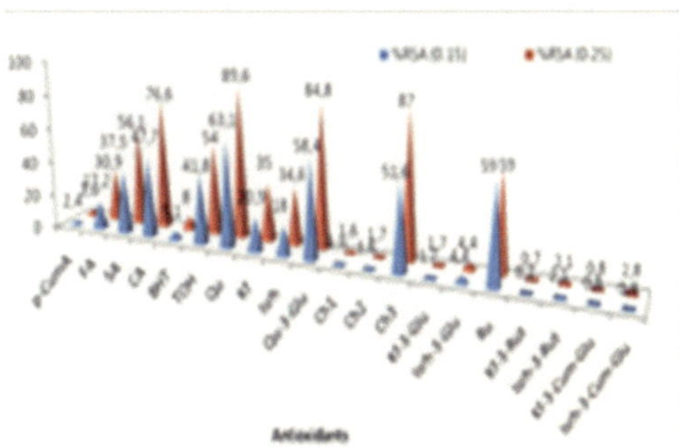

(a) [30] Kancheva, 2009
[DPPH]/[AH]=0.15 and 0.25,

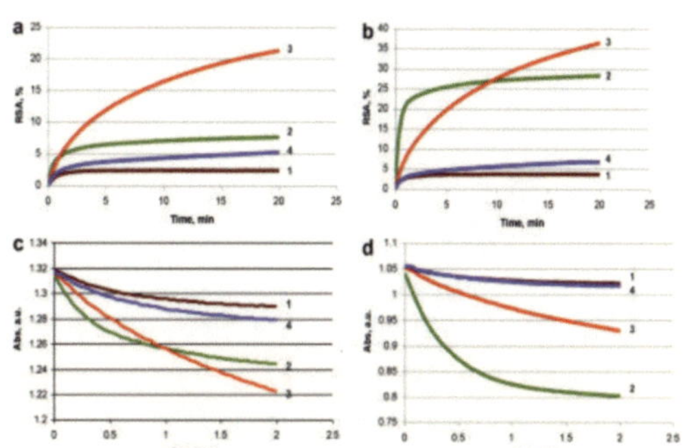

(b) [39]Kancheva et al, 2012
[DPPH]/[AH]=0.25 (a,c) and 0.40 (b,d)

Figure 9. Radical scavenging activity (%) of studied compounds at different [DPPH]/[AH] ratio

Nature-like neo- and xhanthene-lignans recently synthesized showed activity agains cardio-vascular, inflammatory and cancer diseases. Together with their excellent capacity to scavenge free radicals and to inhibit lipid autoxidation these bio-antioxidants are of great importance for the practie, as individuals and in binary mixtures with TOH [39].

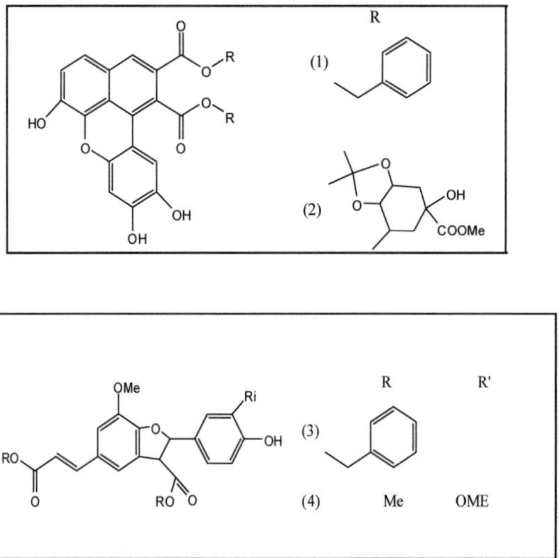

Figure 10. Structures of Xanthene (MF1, MF2) and neo-lignans (MF3, MF4)

The highest values of radical scavenging activity (%RSA_{max}) and largest rate constants for reaction with DPPH radical were obtained for xanthenes and neo-lignans (compounds 2 and 3, Fig. 7B). Comparison of %RSA_{max} with that of standard antioxidants DL-a-tocopherol (TOH), caffeic acid (CA) and butylated hydroxyl toluene (BHT) give the following new order of %RSA max: TOH(61.1%) > CA(58.6%) > 3(36.3%) > 2(28.1%) > 4(6.7%) > 1(3.6%) = BHT(3.6%). On the basis of a comparable kinetic analysis with standard antioxidants a new order of the antioxidant efficiency were obtained: **PF:** 2(7.2) ≥ TOH(7.0) > CA(6.7) > 1(3.1) > 3(2.2) > FA(1.5) > 4(0.6); and of the antioxidant reactivity: **ID:** 2(44.0) >> TOH(18.7) >> CA(9.3) >> 1(8.4) > 3(2.8) >FA(1.0) > 4(0.9) [36].

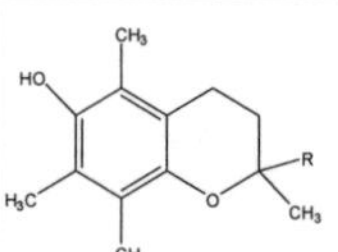 *R=CH_3 2,2,5,7,8-pentamethyl-chroman-3-ol (Chroman C1)* *R=Phytyl; alpha-tocoperol (TOH)*	*R_1=H; R_2=OH –tert-butylated-hydroquinone (TBHQ)* *R_1=t-But; R_2=CH_3 –tert-butyl-hydroxytoluene (BHT)* 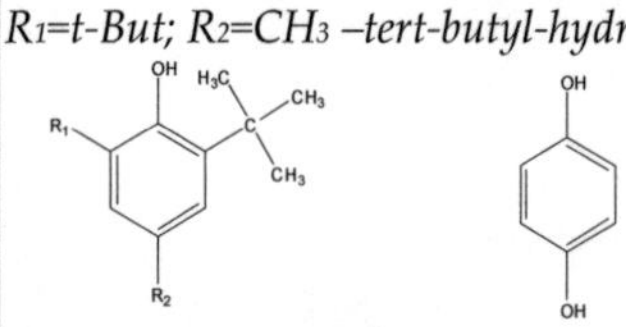 *Hydroquinone (HQ)*
ChrC1 and TOH -Strong activity, *LAO, CL, DPPH,Theor*	*TBHQ, HQ -strong activity,* *BHT –weak/moderate activity;* *LAO, CL, DPPH,Theor*

Table 5. Standard Antioxidants

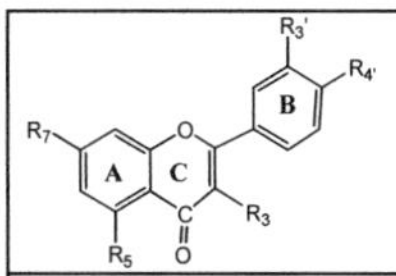	R3′	R4′	R3	R5	R7	%RSA exper	%QSAR theor	LAO Activity
Quercetin (Qu)	OH	OH	OH	OH	OH	62.2	88.40	Strong
Qu-3-O-Glu	OH	OH	O-Glu	OH	OH	63.9	88.40	Strong
Qu-3-O-Rhm	OH	OH	O-Rhm	OH	OH	59.0	88.40	Strong
Qu-3-O-Rut	OH	OH	O-Rut	OH	OH	62.2	88.40	Strong
Qu-7-O-Glu	OH	OH	OH	OH	O-Glu	nd	nd	Strong
Luteolin (Lu)	OH	OH	H	OH	OH	nd	nd	Strong
Lu-7-O-Glu	OH	OH	H	OH	O-Glu	nd	nd	Strong
Kampferol (Kf)	H	OH	OH	OH	OH	54.4	88.40	Strong
Kf-3-O-Glu	H	OH	O-Glu	OH	OH	1.7	12.75	Weak
Kf-3-O-Rut	H	OH	O-Rut	OH	OH	0.7	12.75	Weak
Kf-3-O-Cum-Glu	H	OH	O-Cum-Glu	OH	OH	0.8	12.75	Weak
Isorhamnetin (Isorh)	OCH3	OH	OH	OH	OH	19.2	88.40	Strong
Isorh-3-O-Glu	OCH3	OH	O-Glu	OH	OH	4.4	12.75	Weak
Isorh-3-O-Rut	OCH3	OH	O-Rut	OH	OH	2.1	12.75	Weak
Isorh-3-O-Cum-Glu	OCH3	OH	O-Cum-Glu	OH	OH	2.8	12.75	Weak

Glu: D-glucoside; Rut: rutinoside; Glu-Com: p-coumaroyl-glucosides;
%QSARtheor=3.954+75.950.$I_{3',4'\text{-di-OH or 3-OH}}$ + 8.499.$I_{5\text{-OH}}$ – by statistical analysis (QSAR) of Amic *et al* [43]
(I=1 for 3′,4′-di-OH and/or3-OH) and I=1 for 5-OH); %QSAR=3.95+8.5+75.95 (for Qu all derivatives, Kf, Isrh) – 88.40;
%QSAR=3.95+8.5 (for Kf 3Oderivatives and Isrh 3Oderivatives) – 12.75

Table 6. Substitution pattern of the series of flavonoids examined for their radical scavenging activity [41-43]

4.3. Synergism, additivism and/or antagonism of binary mixtures of phenolic antioxidants [30,31,35-37,41-45]

It is known that in the literature usually are published data about mixtures without or with synergism between the components. Separation of different effects of binary mixtures (synergism, additivism and/or antagonism) of different antioxidants was made for the first time by Denisov [32]. The latest gives possibility to make differences about different effects of binary mixtures, not only to be separated as mixtures without or with a synergism.

Synergism – is observed when the inhibiting effect of the binary mixtures (IP_{1+2}) is higher than the sum of the induction periods of the individual phenolic antioxidants ($IP_1 + IP_2$) i.e. $IP_{1+2} > IP_1 + IP_2$. The percent of the synergism is presented by the following formulae % $Synergism = 100[IP_{1+2} - (IP_1 + IP_2)]/ (IP_1 + IP_2)$.

Additivism - is observed when the inhibiting effect of the binary mixtures (IP_{1+2}) is equal to the sum of the induction periods of the phenolic antioxidants alone ($IP_1 + IP_2$) i.e. $IP_{1+2} = IP_1 + IP_2$.

Antagonism - is observed when the inhibiting effect of the binary mixtures (IP_{1+2}) is lower than the sum of the induction periods of the individual phenolic antioxidants ($IP_1 + IP_2$) i.e. $IP_{1+2} < IP_1 + IP_2$.

Binary mixtures (1:1)	*[AH] mM*	*IP_{1+2} h*	*IP_1 h*	*IP_2 h*	*Effects, %*	*Ref*
Qu (1) + Lu(2)	0.1	7.5±0.8	9.9±0.9	2.2±0.2	Antagonism	41
	0.5	12.3±0.9	24.5±0.6	6.3±0.4	Antagonism,	41
Qu (1) + Ru(2)	0.1	8.3±0.8	9.9±0.9	2.7±0.2	Antagonism	41
	0.5	21.5±0.6	24.5±0.6	2.8±0.2	Antagonism	41
Qu-7(1) + Lu-7(2)	0.5	2.0±0.2	1.5±0.2	1.8±0.2	Antagonism	41
	1.0	2.1±0.2	1.0±0.2	1.4±0.2	Additivism	41
Qu (1)+ α-TOH(2)	0.1	29.7±1.5	9.9±0.9	10.5±0.9	Synergism,46%	41
Ru (1) + α-TOH(2)	0.1	24.9±1.5	2.7±0.2	10.5±0.9	Synergism,87%	41
#Myr (1) +α-TOH(2)	0.1	10.5±0.9	4.7±0.3	3.2±0.2	Synergism,33%	3
	0.3	20.5±1.5	8.9±0.9	5.5±0.5	Synergism,42%	3
	0.6	31.1±1.5	16.3±0.9	7.4±0.5	Synergism,14%	3
CA (1) + α-TOH(2)	0.1	20.4±1.5	9.8±0.9	10.5±0.9	Additivism	3
SA(1) +α- TOH(2)	0.1	16.1±0.9	5.3±0.5	10.5±0.9	Additivism	3
BHT(1) +α- TOH(2)	0.1	21.5±1.5	7.5±0.5	10.5±0.9	Synergism,19%	3
TBHQ(1) + αTOH(2)	0.1	26.1±1.5	7.9±0.5	10.5±0.9	Synergism,42%	3
Cum_1(1) + αTOH(2)	0.1	11.8±0.9	1.5±0.2	10.5±0.9	Additivism	40
Cum_6 (1) + αTOH(2)	0.1	14.2±0.9	2.0±0.2	10.5±0.9	Synergism,14%	40
Cum_4 (1) + αTOH(2)	0.1	12.7±0.9	7.1±0.5	10.5±0.9	Antagonism	40
BisCum1(1) + αTOH(2)	0.1	12.6±0.9	7.9±0.5	10.5±0.9	Antagonism	38
BisCum3(1) + αTOH(2)	0.1	6.1±0.5	2.2±0.2	10.5±0.9	Antagonism	38
MF1(1)+α-TOH(2)	0.1	10.0±0.9	4.0±0.3	10.5±0.9	Antagonism	39
MF2(1)+α-TOH(2)	0.1	15.0±0.9	9.2±0.9	10.5±0.9	Antagonism	39
MF3(1)+α-TOH(2)	0.1	14.0±0.9	2.8±0.2	10.5±0.9	Synergism,5.3%	39
MF4(1)+α-TOH(2)	0.1	13.8±0.9	0.75±0.05	10.5±0.9	Synergism,22%	39
#SA(1) + α-TOH(2)	0.1	45.0±1.0	8.5±0.5	21.0±1.5	Synergism,52%	44
Lipid substrate oxidized TGSO, 80°C , only # TGL,100°C						

Table 7. Effects of equimolar (1:1) binary mixtures of studied antioxidants without and with alpha-tocopherol (α-TOH)

Synergism obtained for different binary mixtures are explained taking into account that the during the oxidation process the antioxidant molecules of both strong antioxidants may be regenerated, which leads to higher antioxidant efficiency of the mixture, than of the individual compounds. The regeneration of both antioxidant molecules of compounds with catecholic moiety, QH_2 (4-hydroxy-bis-coumarin, caffeic acid (CA) and MF1-MF3) and of tocopherol (TOH) is possible as a result of the following possible reactions:

a. reaction of an antioxidant radical ($QH^{\bullet}$ or $TO^{\bullet}$) with other antioxidant molecule:

- semiquinone radical ($QH^{\bullet}$) and tocopherol (TOH)

$QH^{\bullet} + TOH \rightarrow QH_2 + TO^{\bullet}$ (regeneration of QH_2 by H transfer)

- between tocopheryl radical ($TO^{\bullet}$) and QH_2

$TO^{\bullet} + QH_2 \rightarrow TOH + QH^{\bullet}$ (regeneration of TOH by H transfer)

b. homo-disproportionation reaction of two equal radicals:

$2QH^{\bullet} \rightarrow QH_2 + Q$ regeneration of QH_2 (Q is quinone)

$2TO^{\bullet} \rightarrow TOH + T{=}O$ regeneration of TOH (T=O is tocopheryl quinone)

c. cross-disproportionation reaction of different radicals:

$QH^{\bullet} + TO^{\bullet} \rightarrow QH_2 + T{=}O$ regeneration of QH_2

$TO^{\bullet} + QH^{\bullet} \rightarrow TOH + Q$ regeneration of TOH.

As a result the oxidation stability of lipid sample increases, because the both antioxidants with strong efficiency are regenerated during the oxidation process.

In case of binary mixture of TOH with monophenolic antioxidants (AH), predominantly TOH molecule will be regenerated during the following reactions:

- tocopheryl radical ($TO^{\bullet}$) with monophenolic antioxidant (AH)

$TO^{\bullet} + AH \rightarrow TOH + A^{\bullet}$

- homo-disproportionation of tocopheryl radicals ($TO^{\bullet}$)

$2TO^{\bullet} \rightarrow TOH + T{=}O$

- cross-disproportionation of phenoxyl radical ($A^{\bullet}$) and tocopheryl radical ($TO^{\bullet}$)

$A^{\bullet} + TO^{\bullet} \rightarrow TOH + A_{-H}$ or $A^{\bullet} + TO^{\bullet} \rightarrow T{=}O + AH$ (depending on AH structure)

These reactions demonstrate that during oxidation process initial molecules of individual antioxidants are regenerated by different mechanisms in the binary mixtures. Nevertheless both binary mixtures may be used as effective antioxidant compositions. It is proven that the positions of phenolic hydroxyl groups in 4-hydroxy-bis-coumarins are of significance for their antioxidant activity and mechanism of action. Comparable kinetic analysis showed that the antioxidant efficiency (PF) and reactivity (ID) depend significantly from the substitution of the phenolic ring.

5. Surface-active compounds – surfactants (S)

Surfactants (S) are amphiphilic substances which adsorb at interface and decrease an excess of free energy (surface tension, γ) of interface. Surfactants may act as detergents, wetting agents, emulsifiers, foaming agents, and dispersants. A surfactant molecule contains both a water insoluble hydrophobic component and a water soluble hydrophilic component (polar head).

Surfactant solutions are one of the simplest examples of self-assembling soft nano-systems, whose micro-aggregates (micelles) are of 1–500 nm in size [10,11]. Micelles are prevalent in naturally occurring and biological catalytic reactions Micelles are formed by those surfactants which possess rather long bulky hydrophobic part along with strong hydrophilic head. Such surfactants form direct micelles in water and other polar solvent, and reverse micelles in organic solution.

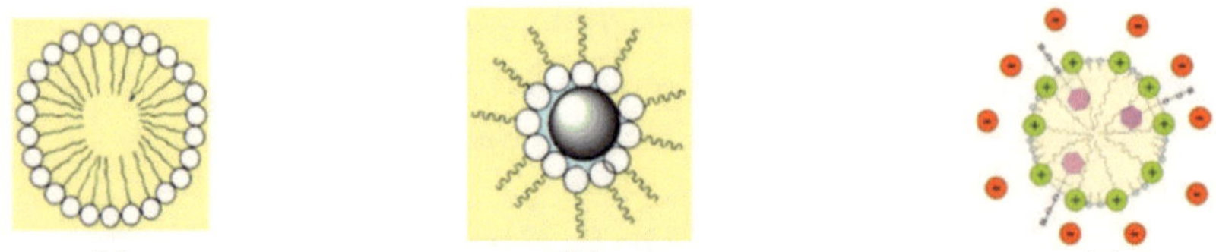

(a) Direct micelle in water | (b) Reverse micelle in oil | (c) Mixed micelle {nROOH…mS}

Figure 11. Micelles formed by surfactants in polar and nonpolar media

The phenomenon of micellar catalysis has been known for a rather long time and applied in many processes, but the significant influence of surfactants on the lipid and hydrocarbon oxidation has been found and studied only in recent decades [6-8, 46-48]. Specific features of micellar catalysis for oxidation processes have two causes: (1) Hydroperoxides (LOOH), which are formed as the primary oxidation products, are amphiphilic and surface active, in contrast to initial oils; (2) There is spontaneous allocation of amphiphilic compounds in every heterogeneous and colloid system resulting in the reduction in the total free energy of a system, including the interface boundaries (the rule of polarity equalization). In the presence of surfactants in oxidized oil hydroperoxides and surfactants form mixed micelles {nLOOH...mS}(Fig. 11c). Using the measurement of the interphase tension [49], nuclear magnetic resonance (NMR), and dynamic light scattering [50], it was shown the association of LOOH and a surfactant in combined micelles in which LOOH plays the role of a co-surfactant. The average self-diffusion coefficient for hydroperoxide decreases with growth of the surfactant (CTAC) concentration up to the equalization with the surfactant diffusion coefficient, when all LOOH is bound in mixed micelles {nLOOH···mS}. The mixed micelle effective size calculated by the Stocks–Einstein equation was ~2 nm [50]. The size determined by DLS for mixture cumene hydroperoxide and CTAB are about 20 nm. Hydroperoxide facilitates the colloid dilution of CTAB in organic medium.

5.1. Effect of surfactants on lipid oxidation.

The comparison of the effects of different surfactants on lipid and hydrocarbon oxidation reveals that cationic surfactants (CS) promote hydroperoxide destruction resulting in the formation of free radicals [46,47] and the oxidation as a whole (Fig.12a,b).

By means of NMR and GC–MS, it is shown that in the presence of CS (CTAC and CTAB)) cumene hydroperoxide decomposes into dimethyl phenylcarbinol, acetophenon, and dicumylperoxide which are known as resulting from the radical decomposition of hydroperoxide. In the presence of anionic SDS, cumene hydroperoxide decomposes without radical formation into phenol and acetone [48].

The kinetics of oxidation of sunflower (TGSO) and olive (TGOO) oil triacylglycerols (Fig.12a) and natural olefin (limonene) (Fig.12b) in the presence of surfactants show that cationic surfactants (CS) promote the oxidation, whereas the anionic sodium dodecylsulfate (SDS) has no influence in the case of limonene and SDS demonstrates a weak retardation in TGOO oxidation (Fig.12a.) The chain breaking inhibitor α-tocopherol completely suppress the limonene oxidation accelerated by CTAC and CTAB (Fig.12b). Before and after the induction periods, the oxidation rate is described by the well known equation for the liquid-phase chain oxidation of hydrocarbons and lipids (see above):

$$R_{02} = a\left[LH\right]\cdot R_{IN}^{\ \ 0.5} \quad (1)$$

where R_{IN} is the radical initiation rate. R_i can be calculated from the duration of the induction period (τ), caused with α-tocopherol (Fig.12) as follows: $R_{IN} = 2[InH]/\tau$.

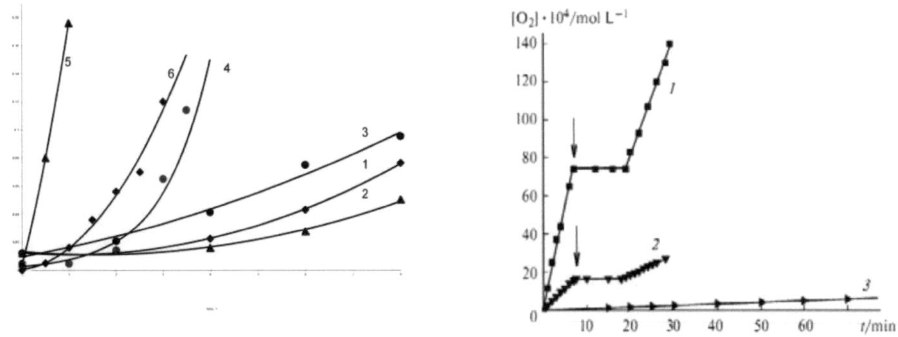

(a) (b)

Figure 12. (a). Effect of surfactant on the LOOH formation during autoxidation of TGOO at 100oC and TGSO at 80oC: 1, 4- without additives; 3, 5 - 0.1M CTAB; 2,4- 0.1 M SDS; 6 - 0.04 M 1-OD [48]; (b) Effect of surfactant on the oxygen absorption during 1M limonene oxidation at 60oC; 1 – 1mM CTAC; 2 – 0,8 mM CTAB; 3 – 1mM SDS; 3- without additives; [LOOH]=22mM; Arrows show the moments of introducing α-tocopherol: 1-0,8 mM; 2 – 0,2 mM [51].

The rate constants for the propagation (k_p) and termination (k_t) of the oxidation chain for limonene and its oxidizability parameter $a = k_p/(2k_t)^{0.5}$ are known at the temperatures 30-80°C [52]. So, the radical initiation rate can be calculated on the base of the measured value of oxygen uptake rate: $R_i = (R_{02}/a[LH])^2$. In the case of limonene both formulas give very close R_i values. These data show that the mechanism of the catalytic action of CS on the oxidation processes consists in the increase of the chain initiation rate, caused by acceleration of hydroperoxides decomposition into free radicals.

To estimate the mixed micelles {nLOOH… mS} as a free radical initiator quantitatively, cumene hydroperoxide and hydrogen peroxide, which are produced in industrial scale, and limonene hydroperoxide were taken, and natural polyphenol quercetin was used as a radical acceptor. The interaction of LOOH with quercetin (Qu) in the presence of surfactants (S) is described by reactions:

$nLOOH + mS ⊚ \{nLOOH … mS\} \rightarrow LO_2^{\cdot}$
$Qu + LO_2^{\cdot} \rightarrow$ products,

The initiation rate is equal to: R_i = -2d[Qu]/dt. Quercetin is characterized by an intense absorption band in a visible region of the electronic spectrum, it is soluble in water and organic solvents and readily reacts with free radicals; i.e. it is a convenient kinetic probe to study the processes of radical generation in various media. The generated peroxyl radicals come into the volume and can initiate chain oxidation, polymerization, or other radical processes. In the presence of oxygen the concentrations of hydroperoxide and other polar products increase during the accelerated oxidation, and this, in its turn, influences the structure and properties of micelles. The instability of micelles and also the varied composition of their polar cores do not allow applying the known and frequently used pseudophase approach to the analysis of the lipid oxidation kinetics in the presence of surfactants. The comparison of the activities of LOOH and cationic surfactants in the generation of radicals can be conducted on the basis of the specific rates of radical initiation $\varpi_i = R_i / ([LOOH]\cdot[S])$. Data in the Table 6 show that cationic surfactants catalyze free radical initiation from hydroperoxide both in water and in organic solutions that is both in direct and reverse mixed micelles.

It means that the CS–hydroperoxide system can be used both as a lipophilic and a hydrophilic initiator. Small amounts of LOOH and CS provide significant radical generation rates (10^{-8}–10^{-7} Ms^{-1}), which are inaccessible at low temperatures for the known azoinitiators. By their activity in the generation of radicals in organic media the surfactants can be arranged in the following order, which indicates the essential role of counter ions in the catalytic action of CS:

CTAC ≈ TDTAC > CTAB ≈ CPB > DCDMAB > CTAHS

Surfactant	Cumene hydroperoxide, 37°C		Hydrogen peroxide, 37°C		Limonene hydroperoxide, 60 °C
	ϖ_i, $(M\cdot s)^{-1}$ Organics	ϖ_i, $(M\cdot s)^{-1}$ Water	ϖ_i, $(M\cdot s)^{-1}$ Organics	ϖ_i, $(M\cdot s)^{-1}$ Water	ϖ_i, $(M\cdot s)^{-1}$ Organics
CTAC	$2{,}1\cdot10^{-3}$	$3{,}7\cdot10^{-4}$	$0{,}67\cdot10^{-3}$	$0{,}14\cdot10^{-3}$	$27\cdot10^{-3}$
CTAB	$1{,}9\cdot10^{-3}$	$1{,}9\cdot10^{-4}$	$0{,}2\cdot10^{-3}$	$0{,}14\cdot10^{-3}$	$3{,}6\cdot10^{-3}$
CTAHS	$0{,}17\cdot10^{-3}$	$1{,}1\cdot10^{-4}$	≈0	≈0	$2{,}5\cdot10^{-3}$
DCDMAB	$1{,}5\cdot10^{-3}$	$2{,}8\cdot10^{-4}$	$2{,}1\cdot10^{-3}$	$0{,}37\cdot10^{-3}$	$3{,}6\cdot10^{-3}$
CPB	$1{,}9\cdot10^{-3}$	$3{,}3\cdot10^{-4}$	$0{,}2\cdot10^{-3}$	$0{,}14\cdot10^{-3}$	$3{,}6\cdot10^{-3}$
TDTAC	$2{,}1\cdot10^{-3}$	$3{,}7\cdot10^{-4}$	$0{,}47\cdot10^{-3}$	$0{,}13\cdot10^{-3}$	-
SDS	≈0	≈0	≈0	≈0	≈0
Lecithin	≈0	≈0	≈0	≈0	≈0

Table 8. Specific rates of radical initiation in the system: cationic surfactant + hydroperoxide in chlorbenzene and in water solution [47]

Along with hydroperoxide, water, and other polar oxidation products, catalytic and inhibiting components can be concentrated in mixed micelles {nLOOH·mS}. The combination of cationic surfactants with transition metal compounds known as homogeneous catalysts of the hydrocarbon oxidation was found to demonstrate synergism, i.e., for the mixture of components the oxidation rate (R_{Σ}) exceeds the sum of the rates in the experiments with separately used components (R_{Me} and R_S): $\beta = R_{\Sigma}/(R_{Me} + R_S) > 1$. The ethylbenzene is oxidized selectively into acetophenon and water catalyzed with the combination of CTAB and cobalt acetylacetonate [53]. Under similar conditions limonene is oxidized with the primary formation of a carbonyl compound (carvon) [51].

Let us look on the mixed micelles of hydroperoxide and cationic surfactant once more. In mixed micelle, peroxide bond is localized in the interphase which has very strong intensity of electric field, about $5 \cdot 10^5$V/m. It affects peroxide bond, weakens it and facilitates decomposition into free radicals. Apparent activation energies of hydroperoxide decay decrease to 50-60 kJ/mol in mixed micelles from ~ 100 kJ/mol for thermal decay [51].

In the case of anionic surfactant the direction of electric field is different and decomposition into radicals is not facilitated. On the contrary, alkali metal alkyl sulfates [47,48] and alkyl phosphates [54,55] act as antioxidants to retard or completely suppress the oxidation process.

Nonionic surfactants form neutral micelles which have no electric field. May be, by that reason nonionic surfactants do not affect free radical formation in hydroperoxide decay, although they form mixed micelles {LOOH…S} with nonionic surfactant as well.

5.2. Phospholipid oxidation.

Phospholipids (PL) are natural surfactants, which are widely used in the production of food, drug, and cosmetics. PL are the basic lipid components of plasmatic cell membranes and membranes of subcellular organelles of animals, plants, and microorganisms. Phosphatidylcholines (1,2-diacyl-sn-glycero-3-phosphocholines, lecithins) are the most widely used; they are present in large amounts in myocardium, liver, kidneys, and egg yolk [56]. In lecithin molecules, anionic phosphate and cationic choline (tetraalkylammonium) groups are connected via a zwitterionic bond to form a neutral polar head. Hydrocarbon moiety represents residues of fatty acids, whose composition depends on the type of PLs (egg, soybean, fish, etc.). Lecithins have a zwitterionic structure in a wide pH range.

Unsaturated fatty acid residues of PL are readily oxidized with atmospheric oxygen as well as nonpolar unsaturated lipids. The primary products of PL oxidation are mainly isomeric hydroperoxides [9,12,57-59]. Lecithins are easily dissolved in organic solvents to yield compact reverse micelles [60,61]. In aqueous solutions, lecithin forms multilamellar liposomes or vesicles under the action of ultrasound dispersion [12,56,62]. Using the DLS method, it was found that, at egg lecithin concentrations 10–90 mg/mL, the size of micro-aggregates observed are equal to 5-6 nm in organic solvents and in water, liposomes are formed with a wide size distribution of 60–1000 nm.

PC oxidation in the presence of azoinitiators or transition metals occurs via free radical chain mechanism. The formation of micro-aggregates both in organic and water media results in a nonlinear dependence of the rate of oxygen absorption on substrate concentration (at constant initiation rate). The deviations from the linearity were observed at concentrations of egg lecithin above 5 mg/mL, corresponding to the formation of micro-aggregates. The rate increment caused by a further increase in the concentration markedly decreases [60,61]. It is possible that a partial shielding of active C–H bonds, which interact with peroxyl radicals from an initiator in solvent bulk results in a relative decrease in the oxidation rate.

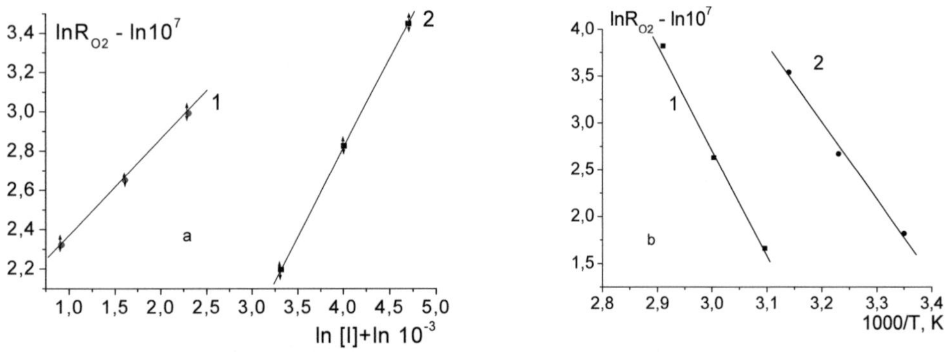

Figure 13. a) Dependences of lecithin (45mg/mL) oxidation rates on the initiator concentration in logarithmic coordinates: 1 – in n-decane solution, 60°C, I – azobisisobutyronitrile, AIBN; 2 – in water, 37°C, I – azodiiso-butyramidine-dihydrochloride (AAPH); b) Temperature dependence of the rate of lecithin (45mg/mL) oxidation plotted in Arrhenius coordinates: 1 – in chlorbenzene, [AIBN]= 5mM; 2 – in water, [AAPH]=55mM.

It was shown in [60,61] that the dependences of PC oxidation rates on initiation rates differ in organic and water solutions. In organic solvents, R_{O2} is proportional square root of R_i, whereas in water, $R_{O2} \sim R_i$. It can be seen from Fig.13a, where the dependences of egg lecithin oxidation rates on the corresponding initiator concentration (it means $R_{O2} – R_{IN}^n$, because $R_{IN} = k_i$[initiator]) are presented: in organics (chlorobenzene) n = 0,5 and n ≈ 1 in water media.

So, the rate of PL oxidation in an aqueous medium cannot be described by Eq. (1), common for lipid and hydrocarbon oxidation. Therefore, even in a narrow concentration range, it is unreasonable to compare the oxidizability of PL in water and an organic medium using parameter $a = k_p/(2k_t)^{0.5}$. Nevertheless, in many studies devoted to the oxidation of phospholipids in various media, Eq. (1) was applied to describe the rate of oxidation (absorption of oxygen [63-66] or accumulation of hydroperoxides [9,58,67]) and to determine the oxidizability of PLs or individual phosphatidylcholines [68]. The majority of these works was carried out at the physiological temperature (37°C). The measurements were performed in different ranges of the overall concentrations of PL and with different initiators and inhibitors used to determine the initiation rates; therefore, the conclusions were very different right up to the opposite ones.

According to [58,68] the oxidizability of PLs in aqueous dispersions is lower than that in organic solvents by an order of magnitude; it is higher in reverse micelles than in molecular alcohol solutions [58]. In [65,68], it was assumed that the micro-heterogeneity of PL solutions and the dispersity of colloidal solutions do not influence the oxidizability of unsaturated lipids in both aqueous and organic media.

A comparison of the experimentally measured rate of O_2 absorption during PL oxidation in aqueous solutions with the corresponding values obtained in an organic solvent at the same temperature, mass concentration of PL, and the initiation rate, which is governed by the contents of water- (AAPH) and oi-soluble (AIBN) azo-initiators, respectively, in the volumes of the solvents demonstrates the following [61]. At a temperature of 45°C, radical initiation rate of $22.5 \cdot 10^{-8}$ M/s, and PL concentration of 45 mg/mL, the rates of oxygen absorption in water and chlorobenzene are $3.5 \cdot 10^{-6}$ and $2.1 \cdot 10^{-6}$ M/s, respectively. A comparison suggests that, in the presence of a source of radicals, PL organized into multilamellar liposomes is ~1.5_fold faster oxidized in water than in the organic solution of reverse micelles. In order to explain this result, for micro-heterogeneous systems, one must introduce the concept of the effective (apparent) concentration of an oxidized substrate. In a system of multilamellar liposomes, the effective concentration of the oxidized substrate is higher than that in a system of reversed micelles occurring in an organic medium; therefore, a higher oxidation rate is observed at the same temperature and the rate of radical initiation. The rates of PL oxidation both in organics and in water, initiated by corresponding initiator, increase with temperature according to the Arrhenius equation (Fig.10b). The effective activation energy of AAPH-initiated PL oxidation in an aqueous solution (74 kJ/mol) is lower than the activation energy of AAPH decomposition (112 kJ/mol). Hence, a radical chain mechanism of PL oxidation in water is more complex than described above mechanism of model oil oxidation (Table 1). In a micro-heterogeneous medium, in addition to individual radicals and molecules, reagents that are included into microaggregates (liposomes) and characterized by reactivity different from that of molecular_dispersed particles in corresponding reactions are involved in the stages of chain initiation, propagation, and termination. Cross-disproportionation reactions of different radicals occur to result in the imitation of a linear chain termination.

α-Tocopherol is well known the most effective lipid antioxidant [1-3]. Lecithin liposome oxidation in the presence of α-tocopherol demonstrates that marked induction periods may be observed when α-tocopherol is added inside liposome during preparation. The incomplete suppression of O_2 absorption by α-tocopherol may be indicative of the PC oxidation inside of liposome without migration of peroxyl radicals into the bulk solvent.

It turns out that catecholamines dopamine, adrenaline and noradrenaline are much more strong and effective inhibitors for PC oxidation than α-tocopherol (Compare Fig.14 and Fig.15). Evidently, the positive charge of catecholamines at neutral pH facilitates their adsorption and protective action on the surface of negative charged liposomes [61].

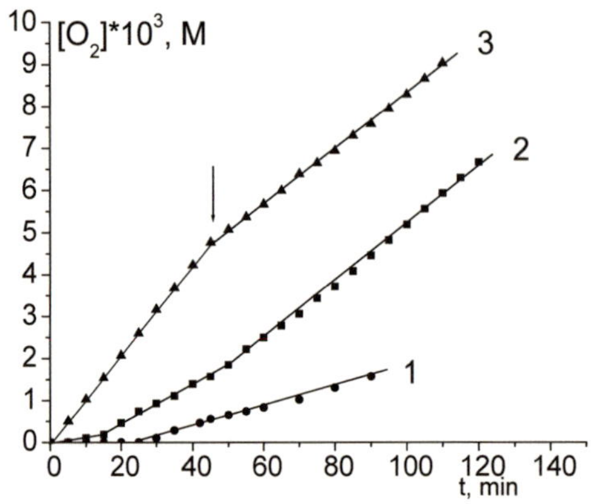

Figure 14. Kinetic curves for O_2 absorption during AAPH- initiated (55 mM) oxidation of PL (45 mg/ml) in water at 37°C in the presence of (1) 0.33 and (2) 0.05 mM α-tocopherol incorporated upon preparation of liposomes and (3)0.05 mM α-tocopherol introduced directly into solution

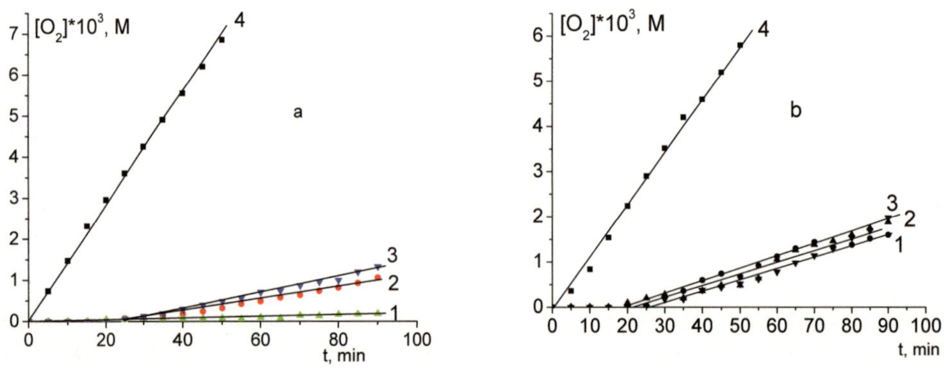

Figure 15. Effect of 0.1 mM (1) adrenalin, (2) dopamine, (3) noradrenalin on AAPH-initiated (55 mM) oxidation of PL (45 mg/ml) at 37°C in (a) water and (b) in phosphate buffer with pH 7.4; (4) no additives.

It must be noted that in the phosphate buffer, induction periods τ are nearly equal for all of the catecholamines (Fig.15b), while, in an aqueous solution, adrenalin provides a longer inhibition of the oxidation than dopamine and noradrenalin do (Fig.15a). The analysis of the ratios between the rates of radical initiation and the durations of the induction periods testified that, for all catecholamines in the buffer solution, the stoichiometry of inhibition, which is numerically equal to the number of radicals corresponding to one acceptor molecule, is $n = (R_{IN} \cdot \tau)/[CA]_0 = 2$, which is characteristic of catechols. In an aqueous and a physiological solution (0.9% NaCl), dopamine and noradrenalin exhibit n= 2, while for adrenaline, n= 4. Moreover, the adrenalin-containing mixture acquires a pink color in water.

Figure 16a illustrates variations in the optical absorption spectra of adrenalin solutions in water and a physiological solution (0.9% NaCl) during its free-radical oxidation initiated by AAPH. It can be seen that the oxidation results in the formation of a colored product with an absorption maximum at 480 nm. In the phosphate buffer of pH 7.4, adrenalin also undergoes transformations (Fig. 16b); however, they yield no colored product. The spectral characteristics of the pink product ($\varepsilon = 4.02 \times 10^3$ M^{-1} cm^{-1} at 480 nm) correspond to adrenochrome (3-hydroxy-1-methyl-2,3-dihydro-1H-indole-5,6-dion), which is formed via the abstraction of four hydrogen atoms from adrenaline [68].

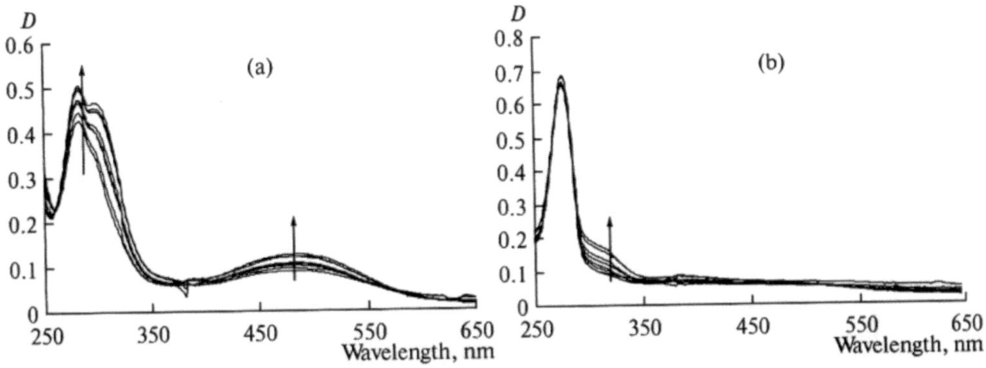

Figure 16. Variations in UV spectrum of 0.1 mM adrenalin solution in the process of its oxidation at 37°C (a) in water and 0.9% NaCl solution and (b) in phosphate buffer with pH 7.2.

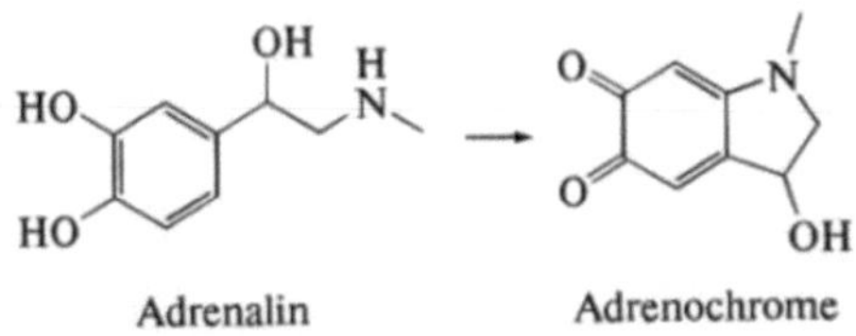

Figure 17. Structures of adrenalin and adrenochrome

It is interesting that, in an aqueous solution, under the conditions of AAPH-initiated free-radical oxidation, adrenalin is quantitatively transformed into adrenochrome. However, in the phosphate buffer solution adrenochrome is not formed.

6. Concluding remarks

Radical scavenging activity towards DPPH radical gives information only about the H-donating capacity of the studied compounds and some preliminary information for their possibility to be used as antioxidants. Antioxidant activity is capacity of the compound to shorten the oxidation chain length as a result of its reaction with peroxyl radicals. For that reason we mean as antioxidant activity the chain-breaking activity of the compounds. This comparable study showed a good correlation between experimental antioxidant activity of compounds under study and their predictable activity by using TLC DPPH radical test.

It has been demonstrated that phenolic compounds with catecholic moiety are the most powerful scavengers of free radicals and they may be used as effective chain-breaking antioxidants. The highest antiradical and antioxidant activity of phenolic antioxidants with catecholic moiety is explained by possible mechanism of homo-disproportionation of their semiquinone radicals formed.

Thus regeneration of the antioxidant molecule during the oxidation process is possible. It has been found for the first time that only substitution in the aromatic nucleus of the studied bis-coumarins and xanthenes-lignans is responsible for their antioxidant activity.

It must be noted that antioxidants' activity depends significantly not only on their structural characteristics, but also on the properties of the substrate being oxidized and the experimental conditions applied. Structural characteristics of the complex system: oxidizing substrate - antioxidant must be considered. On the basis of this comparable analysis, the most effective individual antioxidants and binary mixtures were proposed for highest and optimal lipid oxidation stability.

Author details

Vessela D. Kancheva
Lipid Chemistry Department, Institute of Organic Chemistry with Centre of Phytochemistry, Bulgarian Academy of Sciences, Sofia, Bulgaria

Olga T. Kasaikina
N.N. Semenov Institute of Chemical Physics, Russian Academy of Sciences, Moscow, Russia

Acknowledgement

The financial support of Project No BG051PO001/3.3-05-0001 "Science and Business" - Ministry of Education and Sciences in Bulgaria for publishing this chapter in the Lipid Peroxidation monograph open access is gratefully acknowledged.

7. References

[1] Kancheva V (2010a) Antioxidants. Structure - activity relationship. In: "Antioxidants - Prevention and Healthy Aging", Ed. by F. Ribarova, SIMELPRESS Publ., Sofia, Bulgaria, Chapter 1, pp 56-72.

[2] Kancheva V (2010b) Oxidative stress and lipid oxidation. In: "Antioxidants - Prevention and Healthy Aging", Ed. by F.Ribarova, SIMELPRESS Publ., Sofia, Bulgaria, Chapter 3 pp 233-238.

[3] Kancheva VD (2012) Phenolic antioxidants of natural origin – structure activity relationship and their beneficial effect on human health. In: "Phytochemicals and Human Health: Pharmacological and Molecular Aspects", Nova Science Publishers Inc., USA, Ed. A.A.Farooqui, Chapter I, pp 1-45.

[4] Huang SW, Frankel EN, Schwarz K, Aeschbach R, German JB (1996) Antioxidant activity of alpha-tocopherol and Trolox in different lipid substrates: Bulk oils and oil-in-water emulsions, *J.Agric. Food Chem*. 44: 444–452.

[5] Huang S W and Frankel EN (1997) Antioxidant activity of tea catechins in different lipid systems, *J. Agric. Food Chem.* 45:3033–30383.

[6] Mc'Clements DJ Food Emulsions: Principles, Practice and Techniques, Boca Raton, FL: Taylor and Francis/CRC Press, 2005.

[7] Chaiyasit W, Elias R J, Mc'Clements DJ, Decker EA (2007) Role of Physical Structures in Bulk Oils on Lipid Oxidation, *Critical Reviews in Food Science and Nutrition.* 47: 299-317.

[8] Chen B, McClements DJ, Decker EA (2011) Minor Components in Food Oils: A Critical Review of their Roles on Lipid Oxidation Chemistry in Bulk Oils and Emulsions , *Critical Reviews in Food Science and Nutrition.* 51: 901-916. http://dx.doi.org/10.1080/10408398.2011.606379

[9] Porter NA, Weber BA, Weenen H, Khan JA. (1980) Autoxidation of Polyunsaturated Lipids. Factors Controlling the Stereochemistry of Product Hydroperoxides, *J. Am. Chem. Soc.*102: 5597-5601.

[10] Dynamics of Surfactant Self-Assemblies. Micelles, Microemulsions, Vesicles and Lyotropic Phases, Zana, R., Ed., Boca Raton, FL: Taylor and Francis/CRC Press, 2005.

[11] Hamley I.W. (2000) Introduction to Soft Matter: Polymers, Colloids, Amphiphils and Liquid Crystals, Chichester: Wiley.

[12] Roschek BJ, Tallman KA, Rector CL, Gillmore JG, Pratt DA, Punta C, Porter NA. (2006) Peroxyl Radical Clocks, J. Org. Chem. 71: 3527–3532.

[13] Kasaikina OT, Kortenska VD, Yanishlieva NV (1999) Effect of chain transfer and recombination/disproportionation of inhibitor radicals on inhibited oxidation of lipids, *Russ. Chem. Bull.* 48: 1891-1996.

[14] Pisarenko LM, Krugovov DA, Kasaikina OT (2008) A kinetic model for limonene oxidation. *Russ. Chem. Bull.* 57:80-86.

[15] Naumov VV, Vasil'ev RF (2003) Antioxidant and prooxidant effect of tocopherol. *Kinet. Catal.* 44: 101-105.

[16] Roginsky VA (1988) Phenolic antioxidant-efficiency and reactivity, Naukam Moscow, Russia (in Russian).

[17] Barkley L R C, Baskin K A, Locke S J, Vinquist M R (1989) Absolute rate constants for lipid peroxidation and inhibition in model biomembranes, *Can. J.Chem.* 68: 2258-2269.

[18] Burton GW, Doba T, Gabe EJ, Hughes L, Lee FL, Prasad L, Ingold KU (1985) Antioxidants in biological molecules. 4. Maximizing the antioxidant activity of phenols. *J. Am. Oil Chem. Soc.* 107: 7053-7065.

[19] Small RD, Scaiano JC, Patterson LK.(1979) radical process in lipids:\a laser photolysis study of t-butoxy radical reactivity toward fatty acids. *Photochem. Photobiol.* 29:49-51.

[20] Kaouadji MN, Jore D, Ferradini C, Patterson LK (1987) Radiolytic scanning of vitamin E/vitamin C oxidation reduction mechanisms. *Bioelectrochem. Bioenerg.* 18: 59-70.

[21] Roginsky VA (1990) Kinetics of polyunsaturated fatty acid esters oxidation inhibited by substituted phenols. Kinet. Catal. 31: 546-552.

[22] Kortenska-Kancheva VD, Belyakov VA (2005) Simulation of lipid oxidation kinetics in various mechanism of hydroperoxides decomposition, *Riv. Ital. delle Sost. Grasse* 82: 177-185.

[23] Belyakov VA, Kortenska VD, Rafikova VS, Yanishlieva NV (1992a) Kinetics of the inhibited oxidation of model lipid systems. Role of fatty alcohols. Kinetics and Catalysis, 33: 617-622;

[24] Belyakov VA, Kortenska VD, Rafikova VS, Yanishlieva NV (1992b) Mechanism of lipid autoxidation in presence of fatty alcohols and partial glycerols. Kinetics and Catalysis 33: 756-771.

[25] Kortenska VD, Yanishlieva NV, Roginskii VA (1991) Kinetics of inhibited oxidation of lipids in the presence of 1-Octadecanol and 1-Palmitoylglycerol, *J. Am. Oil Chem. Soc.* 68: 888-890.

[26] Kortenska VD, Yanishlieva NV (1995) Effect of the phenol antioxidant type on the kinetics and mechanism of inhibited lipid oxidation in the presence of fatty alcohols, *J. Sci. Food Agric.* 68:117-126.

[27] Kortenska VD, Yanishlieva NV, Kasaikina OT, Totzeva IR, Boneva MI, Rusina IF (2002) Phenol antioxidant efficiency in presence of lipid hydroxy compounds in various lipid systems, *European J. Lipid Science and Technology* 104: 513-519.

[28] Burlakova E B (2007) Bioantioxidants. Molecular Cell Biophysics. Russ.Chem. J. 51:3-12 (in Russian).

[29] Pobedimskiy DG, Burlakova EB (1993) Mechanism of Antioxidant Action in Living Organisms in Atmospheric Oxidation and Antioxidants. Ed. J. Scott. 2: 223-235.

[30] Kancheva VD (2009) Phenolic antioxidants – radical scavenging and chain breaking activities. Comparative study. Eur. J. Lipid Sci. Technol. 111: 1072-1089.

[31] Kancheva VD (2010) New bioantioxidants strategy. T3D-2010 International Symposium on Trends in Drug DSiscivery and Development 5th-8th January 2010 Delhi (India), Book of AbstractsIL-50.

[32] Denisov ET, Denisova TG: Handbook of antioxidants. Bond dissociation energies, rate constasnts, activation energies and enthalpies of reactions. Second Ed. CRS Press, New York, Washington, D.C. 2001.

[33] Shahidi F: Natural antioxidants, a review in: Natural antioxidants. Chemistry, Health effects and applications, Shahidi, F. (ed) AOCS, Press, Champaign, Illinois, USA, 1999, 1-11.

[34] Vermeris W, Nikolson R. Phenolic compound biochemistry. Springer 2006, The Nederlands, pp 267.

[35] Frankel E N (1998) Lipid oxidation, The Oily Press, Dundee (Scotland) 1998, pp. 303.

[36] Kamal-Eldin A, Makinen M, Lampi A-M (2003) The Challenging Contribution of Hydroperoxides to the Lipid Oxidation Mechanism, in Lipid Oxidation Pathways (Kamal-Eldin, A., ed.), Chapter 1, AOCS Press, Champaign, II pp 1-36.

[37] Vasilev RF, Kancheva VD, Fedorova GF, Batovska DI, Trofimov AV (2010) Antioxidant activity of chalcones. The chemiluminescence determination of the reactivity and quantum – chemical calculation of the energies and structures of reagents and intermediates. Kinetics and Catalysis 51: 507-515.

[38] Kancheva VD, Boranova PV, Nechev JT, Manolov II (2010) Structure-activity relationship of new 4-hydroxy-bis-coumarins as radical scavengers and chain-breaking antioxidants. Biochimie 92:1138-1146.

[39] Kancheva VD, Saso L, Angelova SA, Foti MC, Slavova-Kazakova A, Draquino C, Enchev E, Firuzi O, Nechev J (2012) Antiradical and antioxidant activities of new bioantioxidants. Biochemie 94: 403-415.

[40] Kancheva VD, Saso L, Boranov PV, Khan A, Saron M …., Parmar V S (2010a) Structure-activity relationship of dihydroxy-4methyl-coumarins as powerful antioxidants. Correlation between experimental & theoretical data and synergistic effect. Biochemie 92: 1089-1100.

[41] Kancheva VD, Taskova R, Totseva I, Handjieva N (2007) Antioxidant activity of extracts, fractions and flavonoid constituents from Carthamus lanatus L., Riv. Ital. delle Sost. Grasse, 84: 77-86.

[42] Kancheva VD, Dinchev D, Tsimidou M, Kostova I, Nenadis N (2007) Antioxidant properties of *Tribulus Terrestris* from Bulgaria and radical scavenging activity of its flavonoid components, Riv. Ital. delle Sost. Grasse 54: 10-19.

[43] Amic D., Davidovic-Amic D, Beslo D, Trrinajstic N (2003) Structure-radical scavenging activity relationships of flavonoids, Croatia Chemica Acta, CCACAA, 76: 55-61.

[44] Kortenska VD, Yanishlieva NV, Kasaikina OT, Totzeva IR, Boneva MI, Rusina IF (2002) Phenol antioxidant efficiency in presence of lipid hydroxy compounds in various lipid systems, European J. Lipid Science and Technology 104: 513-519.

[45] Kasaikina O T, Golyavin A A, Krugovov D A, Kartasheva Z S, Pisarenko L M (2010) Micellar catalysis in the lipid oxidation *Mosc. Univ. Chem. Bull.* (Engl. Transl.) 65: 206-209.

[46] Kasaikina OT, Kortenska VD, Kartasheva ZS (1999) Hydrocarbon and lipid oxidation in microheterogeneous systems formed by surfactants or nanodispersed Al2O3, SiO2, and TiO2, *Coll. Surf. A: Physicochem. Eng. Asp.*149: 29–38.

[47] Kasaikina OT, Kartasheva ZS, Pisarenko LM (2008) Effect of surfactants on the liquid phase oxidation of hydrocarbon and lipids. *Rus. J. Gen. Chem.* (Engl. Transl.)78:1533–1544.

[48] Kasaikina O T, Kancheva V D, Maximova T V, Kartasheva S Z, Vedutenko VV, Yanishlieva NV, Kondratovich VG, Totseva IR (2006) Catalytic Effect of Amphiphilic Components on the Lipid Oxidation and Lipid Hydroperoxide Decomposition. *Oxid. Commun.* 29:574-584.

[49] Trunova NA, Kartasheva ZS, Maksimova TV, Bogdanova YuG (2007) Cumene hydroperoxide decomposition in the system of direct and reverse micelles formed by cationic surfactants. *Colloid. J.* (Engl. Transl.) 69(5): 655–659.

[50] Krugovov DA, Pisarenko LM, Kondratovich BG, Shchegolikhin AN, Kasaikina OT (2009) Catalysis of limonene oxidation by cationic surfactants in combination with transition metal acetylacetonates, Petr. Chem. (Engl. Transl.)49:120–126.

[51] Pisarenko L M, Krugovov D A, Kasaikina O T (2004) . Influence of Cationic Surfactants on the Limonene Oxidation, *Russ. Chem. Bull., Int. Ed.* 53:2210-13.

[52] Maksimova TV, Sirota TV, Koverzanova EV, Kashkai AM (2001) Effect of Surfactants on the Ethylbenzene Oxidation. 4. Catalysis of Oxidation by CTAB in Combination with $Co(acac)_2$, *Petr. Chem. (Engl. Transl.)* 41(5): 389-395

[53] Bakunin VN, Popova ZV, Oganesova EYu , Parenago (2001) The change of hydrocarbon medium structure in the course of liquid phase oxidation, *Petr. Chem. (Engl. Transl.)* 41: 41-46.

[54] Yoshimoto M, Miyazaki Y, Umemoto A, Walde P, Kuboi R, Nakao K (2007) Phosphatidylcholine vesicle-mediated decomposition of hydrogen peroxide, *Langmuir* 23: 9416-9422.

[55] Gregor Cevc. *Phospholipids Handbook,* Marvel Dekker Inc., New York, 1993.

[56] Xu L, Davis TA, Porter NA (2009) Rate Constants for Peroxidation of Polyunsaturated Fatty Acids and Sterols in Solution and in Liposomes, *J. Am. Chem. Soc.* 131: 13037–13044.

[57] Wang X, Ushio H, Ohshima T (2003). Distributions of Hydroperoxide Positional Isomers Generated by Oxidation of 1-Palmitoyl-2-arachidonoyl-sn-glycero-3-phosphocholine in Liposome and in Methanol Solution, *J. Lipids* 38: 65-72.

[58] Gupta R, Muralidhara HS, Davis HT (2001) Structure and phase behavior of phospholipids-based micelles in nonaqueous media, *Langmuir* 17: 5176-5183.

[59] Mengele EA, Kartasheva ZS, Plashchina IG, Kasaikina O.T. (2008) Specific features of lecithin oxidation in organic solvents, *Colloid J.* 70: 753-758.

[60] Mengele E.A., Plashchina I.G., Kasaikina O.T. (2011) Kinetics of lecithin oxidation in liposome water solutions, *Colloid J.* 73: 701-707.

[61] Barclay LRC, MacNeil J M, VanKessel JA, Forrest BJ, Porter NA, Lehman LS, Smith IKJ, Ellington JC (1984) Autoxidation and aggregation of phospholipids in organic solvents *J. Am. Chem. Soc.* 106: 6740-6747.

[62] Burton GW, Doba T, Gabe EJ, Hughes L, Lee FL, Prasad L, Ingold KU (1985) Autoxidation of biological molecules. 4. Maximizing the antioxidant activity of phenols, *J. Am. Chem. Soc.* 107:7053-7065

[63] Roginsky V (2003) Chain breaking antioxidant activity of natural polyphenols as determined during the chain oxidation of methyl linoleate in Triton X-100 micelles, *Arch. Biochem. Biophysics* 414: 261-267.

[64] Roginsky V, Tikhonov I. (2010) Natural polyphenols as chain-breaking antioxidants during lipid peroxidation, *Chem. Phys. Lipids* 163: 127133.

[65] Nakamura T., Maeda H.A. (1991) Simple assay for lipid hydroperoxides based on triphenylphosphine oxidation and high performance liquid chromatography, *Lipids* 26: 765-768.

[66] Barclay LRC, Ingold KU (1981) Autoxidation of Biological Molecule. 2. The Autoxidation of a Model Membrane. A Comparison of the Autoxidation of Egg Lecithin Phospha-tidylcholine in Water and in Chlorobenzene *J. Am. Chem. Soc.* 103: 6478-6485.

[67] Roginsky V, Lissi EA (2005) Review of methods to determine chain-breaking antioxidant activity in food *Food Chem* 92: 235-254.

[68] Green, S. (1956) Mechanism of the catalytic oxidation of adrenaline by ferritin. *J. Biol. Chem.* 220: 237—255.

Chapter 3

Lipid Peroxidation End-Products as a Key of Oxidative Stress: Effect of Antioxidant on Their Production and Transfer of Free Radicals

Hanaa Ali Hassan Mostafa Abd El-Aal

Additional information is available at the end of the chapter

http://dx.doi.org/10.5772/45944

1. Introduction

1.1. Oxidative stress

The term oxidative stress; is a state of unbalanced tissue oxidation refers to a condition in which cells are subjected to excessive levels of molecular oxygen or its chemical derivatives called reactive oxygen species (ROS).Under physiological conditions, the molecular oxygen undergoes a series of reactions that ultimately lead to the generation of superoxide anion (O_2-), hydrogen peroxide (H_2O_2) and H_2O. Peroxynitrite (OONO-), hypochlorus acid (HOCl), the hydroxyl radical (OH.), reactive aldehydes, lipid peroxides and nitrogen oxides are considered among the other oxidants that have relevance to vascular biology.

Oxygen is the primary oxidant in metabolic reactions designed to obtain energy from the oxidation of a variety of organic molecules. Oxidative stress results from the metabolic reactions that use oxygen, and it has been defined as a disturbance in the equilibrium status of pro-oxidant/anti-oxidant systems in intact cells. This definition of oxidative stress implies that cells have intact pro-oxidant/anti-oxidant systems that continuously generate and detoxify oxidants during normal aerobic metabolism. When additional oxidative events occur, the pro-oxidant systems outbalance the anti-oxidant, potentially producing oxidative damage to lipids, proteins, carbohydrates, and nucleic acids, ultimately leading to cell death in severe oxidative stress. Mild, chronic oxidative stress may alter the anti-oxidant systems by inducing or repressing proteins that participate in these systems, and by depleting cellular stores of anti-oxidant materials such as glutathione and vitamin E (Laval, 1996). Free radicals and other reactive species are thought to play an important role oxidative stress resulting in many human diseases. Establishing their precise role requires the ability to measure them and the oxidative damage that they cause (Halliwell and Whiteman, 2004).

Oxidative stress is involved in the process of aging (Kregel and Zhang 2007) and various chronic diseases such as atherosclerosis (Fearon and Faux 2009), diabetes (Ceriello and Motz, 2004) and eye disease (Li et al. 2009), whereas fruit and vegetable diets rich in antioxidants such as polyphenols, vitamin C, and carotenoids are correlated with a reduced risk of such chronic diseases (Dherani et al. 2008). An excessive amount of reactive oxygen/nitrogen species (ROS/RNS) leading to an imbalance between antioxidants and oxidants can cause oxidative damage in vulnerable targets such as unsaturated fatty acyl chains in membranes, thiol groups in proteins, and nucleic acid bases in DNA (Ceconi et al. 2003). Several assays to measure " total " antioxidant capacity of biological systems have been developed to investigate the involvement of oxidative stress in pathological conditions or to evaluate the functional bioavailability of dietary antioxidants. Conventional assays to determine antioxidant capacity primarily measure the antioxidant capacity in the aqueous compartment of plasma. Consequently, water - soluble antioxidants such as ascorbic acid, uric acid, and protein thiols mainly influence these assays, whereas fat - soluble antioxidants such as tocopherols and carotenoids show little inf uence over the many results. However, there are new approaches to define the total antioxidant capacity of plasma, which reflect the antioxidant network between water - and fat - soluble antioxidants. Revelation of the mechanism of action of antioxidants and their true antioxidant potential can lead to identifying proper strategies to optimize the antioxidant defense systems in the body.

1.2. Measurement of oxidative damage

A basic approach to study oxidative stress would be to measure some products such as (i) free radicals; (ii) radical-mediated damages on lipids, proteins or DNA molecules; and iii) antioxidant enzymatic activity or concentration.

1.2.1. Free radicals

Free radicals are reactive compounds that are naturally produced in the human body. They can exert positive effects (e.g. on the immune system) or negative effects (e.g. lipids, proteins or DNA oxidation). Free radicals are normally present in the body in minute concentrations. Biochemical processes naturally lead to the formation of free radicals, and under normal circumstances the body can keep them in check. If there is excessive free radical formation, however, damage to cells and tissue can occur (Wilson, 1997). Free radicals are toxic molecules, may be derived from oxygen, which are persistently produced and incessantly attack and damage molecules within cells; most frequently, this damage is measured as peroxidized lipid products, protein carbonyl, and DNA breakage or fragmentation. Collectively, the process of free radical damage to molecules is referred to as oxidative stress (Reiter et al., 1997).To limit these harmful effects, an organism requires complex protection – the antioxidant system. This system consists of antioxidant enzymes (catalase, glutathione peroxidase, superoxide dismutase) and non-enzymatic antioxidants (e.g. vitamin E [tocopherol], vitamin A [retinol], vitamin C [ascorbic acid], glutathione and uric acid). An imbalance between free radical production and antioxidant defence leads to an oxidative stress state, which may be involved in aging processes and even in some pathology (e.g. cancer and Parkinson's disease).

1.2.2. Formation of free radicals

Normally, bonds don't split in a way that leaves a molecule with an odd, unpaired electron. But when weak bonds split, free radicals are formed. Free radicals are very unstable and react quickly with other compounds, trying to capture the needed electron to gain stability. Generally, free radicals attack the nearest stable molecule, gaining its electron. When the "attacked" molecule loses its electron, it becomes a free radical itself, beginning a chain reaction. Once the process is started, it can cascade, finally resulting in the disruption of a living cell. Some free radicals arise normally during metabolism. Sometimes the body's immune system's cells purposefully create them to neutralize viruses and bacteria. However, environmental factors such as pollution, radiation, and toxins can also spawn free radicals. Normally, the body can handle free radicals, but if antioxidants are unavailable, or if the free-radical production becomes excessive, damage can occur. Of particular importance is that free radical damage accumulates with age (Packer, 1994).

1.2.3. Sources of free radicals

Free radicals have two principle sources: endogenous sources and exogenous sources. Endogenous sources of free radicals include those that are generated intracellularly, acting within the cell, and those that are formed within the cell, but are released into the surrounding area. These intracellular free radicals result from auto-oxidation and consequent inactivation of small molecules such as reduced thiols and flavins. They may also occur as a result of the activity of certain oxidases, lipoxygenases, cyclo-oxygenases, dehydrogenases and peroxidases. Electron transfer from metals such as iron to oxygen-containing molecules can also initiate free radical reactions paradoxically; antioxidants may also produce free radicals (Weir et al., 1996). A wide range of free radical molecular species are endogenous. The singlet oxygen is not a free radical but is nevertheless a reactive oxygen species and capable of causing tissue damage (Zebger et al., 2004). Exogenous sources of free radicals are environmental sources. Environmental sources of free radicals include exposure to ionizing radiation (from industry, sun exposure, cosmic rays, and medical X-rays), ozone and nitrous oxide (primarily from automobile exhaust), heavy metals (such as mercury, cadmium, and lead), cigarette smoke (both active and passive), alcohol, unsaturated fat, and other chemicals and compounds from food, water, and air. The exogenous sources of free radicals resulting from ionizing radiation play a major role in free radical production. The energy transferred into water from ionizing particles ionizes the water molecule. The water ions produced dissociate yielding free radicals (Valencia and Moran, 2004).

There are two enzymes including Aldehyd oxidase (AO) and xanthine oxidase (XO), they have a very close evolutionary relationship, based on the recent cloning of the gens and they show a high degree of amino acid sequence homology (Terao et al., 2000). They have been suggested to be relevant to the pathophysiology of a number of clinical disorders (Wright et al., 1995). **Aldehyd oxidase (AO)** commonly exists in vertebrates. Although the liver is the main site for aldehyde oxidase this enzyme has also been reported in kidney, lung, muscle, spleen, stomach, heart and brain (Beedham, 2002). The enzyme in liver of various species catalyzes the oxidation of a number of aldehydes and nitrogenous and also catalyzes the metabolism of physiological compounds such as retinaldehyde and monoamine neurotransimeters (Huang and Ichikawa, 1994). Reduction of oxygen during substrate turnover, leads to the formation of superoxide anion and hydrogen peroxide as ROS. This capacity has attracted attention to the possible role of aldehyde oxidase as a source of ROS. In vivo, it seems that aldehyde oxidase together with cytochrome P450 are quantitative, the most important cellular sources for ROS (Al-Omar et al., 2004). Additionally, the most likely sources of free radicals are **xanthine oxidase (XO)** (McCord, 1985). This enzyme is high particularly in liver and intestine. Although XO generates ROS and evidence has been presented for its role in the development of ischaemic intestinal, hepatic and renal damage (Cohen, 1992). It may also contribute to the development of lung and myocardial reperfusion injury after ischaemic episodes.

1.2.4. Production of free radicals

Free radicals are produced in a number of ways in biological systems (Halliwell and Whiteman, 2004):

a. Exposure to ionizing radiation is a major cause of free radical production. When irradiated water is ionized, and electron is removed from the molecule, leaving behind an ionized water molecule. The damaging species resulting from the radiolysis of water are the free radicals •H and •OH and hydrated electrons. They are highly reactive and have a lifetime on the order of 10 -9 to 10 -11 seconds. The hydroxyl radical is extremely reactive and is carcinogenic. Since water presents the largest number of target molecules in a cell, most of the energy transfer goes on in water when a cell is irradiated, rather then the solute consisting of protein, carbohydrate, nucleic acid, and bioinorganic molecules. Oxygen is an excellent electron acceptor and can combine with the hydrogen radical to form a peroxyl radical. Hydrogen peroxide is toxic and when present in sufficient quantities can interfere with normal cellular metabolism.
b. Enzymes and transport molecules also generate free radicals as a normal consequence of their catalytic function.
c. Auto-oxidation reactions produce free radicals from the spontaneous oxidation of biological molecules involved in non-enzymatic electron transfers.
d. Physical exercise also increases oxidative stress and causes disruptions of the homeostasis. Training can have positive or negative effects on oxidative stress depending on training load, training specificity and the basal level of training. Moreover, oxidative stress seems to be involved in muscular fatigue and may lead to overtraining.

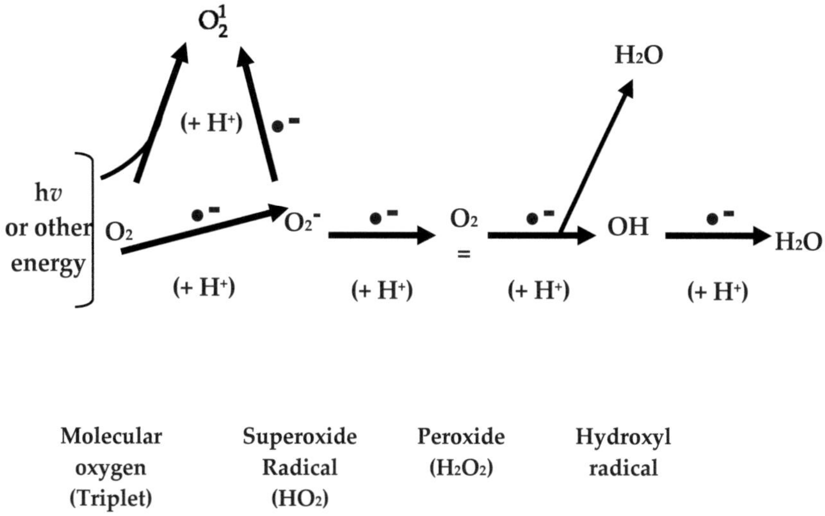

Figure 1. The active oxygen system. Molecular oxygen is reduced to water in four single-electron steps. Reduction of non-radical forms of oxygen is a "forbidden" process and thus usually involves spin-orbit coupling by a heavy metal or a halide or excitation to singlet state. An example is Fenton's reaction, the reduction of peroxide to water and hydroxyl radical by ferrous iron. Hydroxyl radical is one of the most powerful oxidizing agents known.

2. Oxidative damage to lipids (Lipid peroxidation)

The peroxidation of lipids is basically damaging because the formation of lipid peroxidation products leads to spread of free radicals reactions. The important role of lipids in cellular components emphasizes the significance of understanding the mechanisms and consequences of lipid peroxidation in biological systems. Polyunsaturated fatty acids (PUFAs) serve as excellent substrates for lipid peroxidation because of the presence of active bis-allylic methylene groups. The carbon-hydrogen bonds on these activated methylene units have lower bond dissociation energies, making these hydrogen atoms more easily abstracted in radical reactions (Davies et al., 1981). The susceptibility of a particular PUFA toward peroxidation increases with an increase in the number of unsaturated sites in the lipid chain (Nagaoka et al., 1990).

Lipid hydroperoxides are non-radical intermediates derived from unsaturated fatty acids, phospholipids, glycolipids, cholesterol esters and cholesterol itself. Their formation occur in enzymatic or non-enzymatic reactions involving activated chemical species known as "reactive oxygen species" (ROS) which are responsible for toxic effects in the body via various tissue damages. These ROS include among others hydroxyl radicals, lipid oxyl or peroxyl radicals, singlet oxygen, and peroxinitrite formed from nitrogen oxide (NO), all these groups of atoms behave as a unit and are now named "free radical". These chemical forms are defined as any species capable of independent existence that contains one or more unpaired electrons (those which occupy an atomic or molecular orbital by themselves). They are formed either by the loss of a single electron from a non-radical or by the gain of a single electron by a non-radical. They can easily be formed when a covalent bond is broken if one electron from each of the pair shared remains with each atom, this mechanism is known as homolytic fission. In water, this process generates the most reactive species, hydroxyl radical OH.. Chemists know well that combustion which is able at high temperature to rupture C-C, C-H or C-O bonds is a free-radical process. The opposite of this mechanism is the heterolytic fission in which, after a covalent break, one atom receives both electrons (this gives a negative charge) while the other remains with a positive charge.Lipid peroxidation leads to the breakdown of lipids and to the formation of a wide array of primary oxidation products such as conju- dienes or lipid hydroperoxides, and secondary products including MDA, F2-isoprostane or expired pentane, ethane or hexane. Measurement of conjugated dienes is interesting because it detects molecular reorganisation of poly- unsaturated fatty acids during the initial phase lipid peroxidation. Lipid hydroperoxide is another marker of the initial.(reaction of FR and is a specific marker of cellular damage. Other products are often used to measure oxida- stress but have the disadvantage of being secon- dary oxidation products. One of them, MDA, is during fatty acid auto-oxidation. This sub- is most commonly measured by its reaction with thiobarbituric acid, which generates thiobarbi- turic acid reactive substances (TBARS). Although of the results.MDA overestimation), this method is accepted as a the general marker of lipid peroxidation but results are subject to caution (Sies, 1997). In addition, some studcable ies tend to show that MDA is not an adapted method ucts used for such methods.

The peroxidation of lipids involves three distinct steps: initiation, propagation and termination. The initiation phase of lipid peroxidation may proceed by the reaction of an activated oxygen species such as singlet oxygen (1O2), O2-, or HO· with a lipid substrate or by the breakdown of preexisting lipid hydroperoxides by transition metals. In the former case, peroxidation occurs by abstraction of a hydrogen atom from a methylene carbon in the lipid substrate (LH) to generate a highly reactive carbon-centered lipid radical (L·) (Kelly et al., 1998). In the propagation phase of lipid peroxidation, molecular oxygen adds rapidly to L at a diffusion controlled rate to produce the lipid peroxyl radical (LOO·). The peroxyl radical can abstract a hydrogen atom from a number of in vivo sources, such as DNA and proteins, to form the primary oxidation product, a lipid hydroperoxide (LOOH). Alternatively, antioxidants such as α -tocopherol (α -TOH) can act as excellent hydrogen atom donors , generating LOOH and the relatively inert α-tocopherol phenoxyl radical (α -TO·). In the absence of antioxidants or other inhibitors, LOO· can abstract a hydrogen from another lipid molecule (LH), producing another highly reactive carbon centered radical (L·), which then propagates the radical chain as presented in Figure 2 (Waldeck and Stocke , 1996). The lipid hydroperoxide (ROOH) is unstable in the presence of iron or other metal catalysts because ROOH will participate in a Fenton reaction leading to the formation of reactive alkoxy radicals. Therefore, in the presence of irron, the chain reactions are not only propagated but amplified. Among the degradation products of ROOH are aldehydes, such as malondialdehyde, and hydrocarbons, such as ethane and ethylene, which are commonly measured end products of lipid peroxidation (Sener et al., 2004).

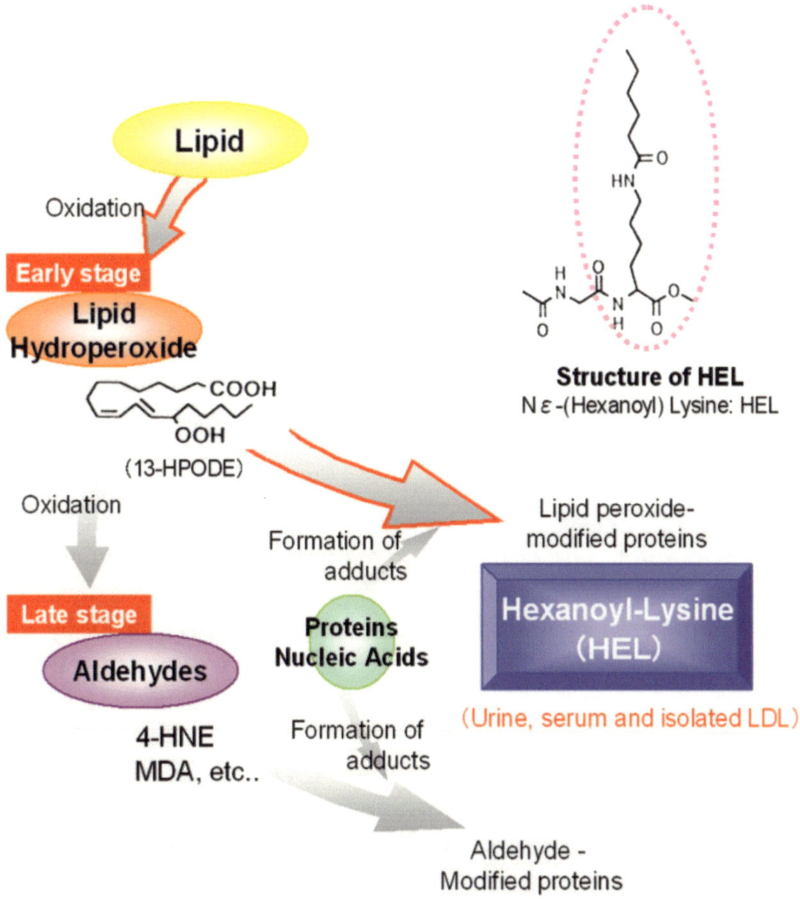

Abbreviations: NRP, nonradical product; LOOH, lipid hydroperoxide; α-TOH, α -tocopherol; α -TO·, α -TOH radical; LH, lipid substrate; LOO·, lipid peroxyl radical. Adapted from Waldeck and Stocke(1996)

Figure 2. Overview of lipid peroxidation.

During peroxidation pathway via reactive intermediates, several end products are formed such as aldehyde [malondialdehyde + 4-hydroxynonenal], pentane and ethane, 2,3 transconjugated diens, isoprostains and chlesteroloxides. The biological activities of MDA and other aldehydes include cross-linking with DNA and proteins, which alters the function/activity of these molecules. MDA + 4HNE have shown tissue toxicity. MDA can react with amino and thiol groups, the aldehydes are more diffusible than free radicals, which means damage is exported to distance sites. Aldehydes are quickly removed from cells as several enzymes control their metabolism (Ustinova and Riabinin 2003).

3. Antioxidants

To minimize the negative effects of ROS generated by any pro-oxidant, endogenous defensive mechanisms called antioxidant defense (AD) system, which utilizes enzymatic and non-enzymatic mechanisms. Antioxidants are naturally occurring substances that combat oxidative damage in biological entities. An antioxidant achieves this by slowing or preventing the oxidation process that can damage cells in the body. This it does by getting oxidized itself in place of the cells. Thus an antioxidant can also be termed as a reducing agent. Antioxidants are considered as important in the fight against the damage that can be done by free radicals produced due to oxidative stress. Although the human body has its own defenses against oxidative stress, these become weak with age or in the case of an illness. Although, antioxidants are sold in various forms as dietary supplements there is no clinching clinical evidence in favor of antioxidants as beneficial in maintaining health and preventing disease. However, there is a lot of anecdotal evidence that those who partake of antioxidant-rich food are better protected against problems such as heart disease, macular degeneration, diabetes, and cancer. Antioxidants are either hydrophilic or hydrophobic. Water soluble or hydrophilic antioxidants are active in the blood plasma while the water insoluble antioxidants protect the cell membranes. How do antioxidants work? Antioxidants work by bringing under control the rogue and unstable oxygen molecules that have an odd number of electrons. These oxygen molecules known as free radicals are highly reactive; they attack cells, DNA, and protein thereby accelerating the aging process. The antioxidants work in harmony and the efficacy of one antioxidant depends upon the availability and concentration of another. Essentially, antioxidants work by donating an electron to the unstable free radical. This stabilizes the free radical and converts it into a harmless compound that may safely be removed from the body. Antioxidants are segregated into two classes based on their mode of operating. They can either be chain-breaking or preventive. Chain-breaking antioxidants such as vitamins E and C halt the process of radical formation by stabilizing free radical molecules so that the chain-like process of radical formation is arrested. Preventive antioxidants such as superoxide dismutase and catalase prevent chain initiations by scavenging for initiator radicals and stabilizing them. They also stabilize transition metal radicals like iron and copper. These metals work as catalysts in the production of free radicals. Antioxidants and their various forms. Antioxidants are chiefly available to us through vitamins, enzymes, and minerals. Vitamin E is actually a group of eight tocopherols. Alpha-tocopherol is the most widely available tocopherol and also the most potent in terms of its effect on the body. Vitamin E is fat-soluble and protects cell membranes that are mainly composed of fatty acids. Vitamin C or ascorbic acid is water-soluble and it scavenges for free radicals that are present in aqueous environments within the human body. Beta carotene is also water soluble and is particularly effective in tackling free radicals in areas of low oxygen concentration. Selenium, manganese and zinc are trace elements that are important components of several antioxidant enzymes such as superoxide dismutase (SOD), catalase (CAT) and glutathione peroxidase. Enzymes work as both primary and secondary antioxidants and help repair oxidized DNA and target lipids that are oxidized. Other substances that are now being considered for their antioxidant properties include uric acid and phytochemicals found in plants.

Fruits and vegetables that have been identified as sources of powerful antioxidants help people counter the risk of heart ailments and different types of cancers. However, there is a possibility that these benefits obtained from fruits and vegetables could be a result of not just antioxidants but a mix that includes flavonoids as well. Although, clinical trials have not put forth conclusive evidence in favor of antioxidants as being helpful to our health the vast number of observational studies and anecdotal evidence offers a very strong suggestion that antioxidants are indeed of much use in keeping the body healthy. It is only a matter of time before scientists unravel the exact mechanism that governs the working of antioxidants in the body. Most nutritionists agree that the best source of antioxidants is natural food. One should try and avoid supplements if possible. It is also important to keep in mind that a high dosage of antioxidant supplements can have a detrimental effect on the body. Excessive vitamin E can lead to blood hemorrhage. Vitamin C in large amounts can cause diarrhea and also atherosclerosis. High amounts of selenium can cause hair loss and rashes on the skin. Antioxidants are thought to protect the body against the destructive effects of free radicals. Antioxidants neutralize free radicals by donating one of their own electrons, ending the electron- gain reaction. The antioxidant nutrients themselves don't become free radicals by donating an electron because they are stable in either form. They act as scavengers, helping to prevent cell and tissue damage that could lead to cellular damage and disease (Reiter, 2003; Reiter et al., 2004).

A rangeor of antioxidants are active in the body including enzymatic and non-enzymatic. All of them can to redox status. Redox status is directly linked and be intracellular or extracellular antioxidants . The body produces several enzymes, including superoxide dismutase (SOD), catalase (CAT), glutathione peroxidase (GSHPX), and glutathione reductase (GR) that neutralize many types of free radicals. Supplements of these enzymes are available for oral administration. However, their absorption is probably minimal at best. Supplementing with the "building blocks" the body requires to make SOD, catalase, and glutathione peroxidase may be more effective. These building block nutrients include the minerals manganese, zinc, and copper for SOD and selenium for GSHPX. While the nonenzymatic defense consists of substances of low molecular weight such as reduced glutathione vitamin C, vitamin E, beta-carotene, lutein, lycopene, vitamin B2, coenzyme Q10, and cysteine (an amino acid). Herbs, such as bilberry, turmeric (curcumin), grape seed or pine bark extracts, and ginkgo can also provide powerful antioxidant protection for the body. Melatonin is a hormone secreted by pineal gland and proves to be powerful antioxidant and free radical scavenger (Yang et al.,2002 and Koc et al., 2003).

3.1. Enzymatic antioxidants

Antioxidant enzymes include superoxide dismutase (SOD), catalase (CAT), glutathione peroxidase (GSHPX), and glutathione reductase (GR). Non-en-zymatic antioxidants include a variety of FR quenchers such as vitamin A (retinol), vitamin C - (ascorbic acid), vitamin E (tocopherol), flavonoids, thiols (including glutathione [GSH], ubidecarenone uric acid, bilirubin, ferritin) and micronutrients (iron, copper, zinc, selenium, mangawhich which act as enzymatic cofactors. The antioxidant system efficiency depends on nutritional ineccentric (vitamins and micronutrients) and on endoge nous antioxidant enzyme production, which can be modified by exercise, training, nutrition and agdative Moreover, the antioxidant system efficiency is important in sport physiology because exercise increases the production of FR.

3.1.1. Superoxide dismutase (SOD, EC 1.15.1.1)

Superoxide dismutase (SOD) is the major defence upon superoxide radicals and is the first defence line against oxidative stress. SOD represents a group of enzymes that catalyse the dismutation of O2– and the formation of H2O2. SOD is an enzyme (EC 1.15.1.1) discovered by McCord and Fridovich, which plays an important role in the defense mechanism of biological cells exposed to oxygen (McCord and Fridovich 1969). SOD catalyzes the dismutation of superoxide anion radical (O 2 • –) into an oxygen molecule and a hydrogen peroxide. This reaction is recognized as an antioxidant system that protects cells from superoxide toxicity. There are several types of SOD, depending on the type of metal ion. Three major isoforms of mammalian SOD have been identif ed with different tissue distributions (Zelko 2002). Cu/Zn - SOD (SOD1) exists in the cytoplasm, lysosomes, and nuclear compartments of mammalian cells (Bannister and et al., 1987; Zelko et al., 2002). In humans, the liver has a relatively high amount and activity of SOD1 (Nozik-Grayck et al., 2005). Human SOD1 is a homodimer containing one copper ion and one zinc ion in each 16 - kDa subunit which consists of 153 amino acids. The copper ion is held by interaction with imidazolate ligands of the histidine residues in SOD1 in the enzymatic active site. The zinc ion (Zn 2 +) contributes to the stabilization of the enzyme (Johnson and Giulivi , 2005).

3.1.2. Catalase (CAT, EC 1.11.1.6)

Catalse (CAT) is one of the major antioxidant enzymes (Scandalios et al., 1997). It is one of the first enzymes to be purified and crystallized and has gained a lot of attention in recent years because of its link to cancer, diabetes and aging in humans and animals (Preston et al., 2001). It is present in every cell and in particular in cell structures that use oxygen in order to detoxify toxic substances and produce H2O2. Catalase converts H2O2 into water andoxygen (Greenwald, 1990 ;Yasminch and Theologides, 1993). Catalase can also use H2O2 in order to detoxify some toxic substances via a peroxidase reaction (Mayo et al., 2003). There are many evidences that the changes of catalase activity as well as the mechanisms of its regulation are essential in the response to stress situations which catalyzes the dismutation of H2O2, forming O2 and H2O resulting good protection the cells from the toxic effects of hydrogen peroxide (Brioukhanov and Netrusor, 2004).

3.1.3. Glutathione peroxidase (GPX, EC 1.11.1.9)

GPX was discovered in 1957 ny Mills. It exists in cell cytosol and mitochondria and has the ability to transform H2O2 into waterThis reaction uses GSH and transforms it into oxidised glutathione (GSSG). GPX and CAT have the same action upon H2O2, debut GPX is more efficient with high ROS concentra-tion and CAT has an important action with lower H2O2 concentration. GPx is a glycoprotein containing a single selenocysteine residue at the active center of each subunit. To protect biological organisms from oxidative damage, GPx catalyses the reduction of hydrogen peroxide and lipid hydroperoxides to water and their corresponding alcohols, respectively, as follows (Antunes et al., 2002): ROOH + 2GSH→ROH +GSSG + H2O2 where reduced monomeric glutathione (GSH) is essential as a hydrogen donor, and GSH is oxidized to glutathione disulf de (GSSG). There are five main mammalian isozymes, which vary in the structure (amino acid sequence and subunit), tissue distribution (liver, kidney,erythrocyte, blood plasma, among others), location (cytoplasm, intestine, extracellular f uid), and substrate specif city (hydrogen peroxide and lipid hydroperoxides) (Dudek et al., 2002). GXP is a selenium dependent enzyme that is ubiquitously expressed and protects cells against oxidative damage by reducing hydrogen peroxide and a wide range of organic peroxides with reducing glutathione (Arthur, 2000). It has been suggested that GPX has anti - inflammatory activity in the cardiovascular system. An increase in cytosolic GPx is linked to a lower risk of cardiovascular disease (Blankenberg et al. 2003).

3.1.4. Glutathione reductase (GR, EC 1.6.4.2)

Glutathione Reductase (GR) is a key enzyme of glutathione metabolism and is widespread in all tissues and blood cells. It a flavin enzyme involved in the defense of the erythrocyte against hemolysis. This enzyme catalyses reduction of oxidized glutathione (GSSG) to reduced glutathione (GSH) in the presence of NADPH and maintains a high intracellular GSH/GSSG ratio of about 500 in red blood cells (Kondo et al., 1980). GR is important not only for the maintaining thr required GSH level but also for reducing protein thiols to their native state.This enzyme is conserved between all kingdoms. In bacteria, yeasts and animals, one GR gene is found, however in plant genomes two GR genes are enclosed. Under normal conditions, GSH and GR are involved in the detoxification of H2O2 generated in the light by the Mehler reaction in chloroplasts. Disturbances GH level have been correlated with oxidative stress induced by various factors including toxicity, pollutants, inflammation and different diseases particularly red blood cell defects.

3.1.5. Glutathione-S-transferase (GST, EC 2.5.1.18)

Glutathione-S-Transferase (GST) catalyzes the conjugation with glutathione of a number of electrophilic xenobiotics, including several carcinogens, mutagens and anticancer drugs (Hayes and Pulford, 1995). These electrophiles are made less reactive by conjugation with glutathione and the conjugates are thought to be less toxic to the cell. Consiquently, GSTs are believed to play an important role in the defense of cells against these zenobiotic toxins. Several antineoplastic drugs particularly the reactive electrophilic alkylating agents, can form conjugates with glutathione both spontaneously and in GST-catalyzed reactions (Awasthi et al., 1996). Morever, Some studies reported that there are a good association between cellular resistance to some anticancers drugs and expression of particular isozymes of GST (Hayes and Pulford, 1995).

3.2.1. Glutathione (GSH)

Gultathione (GSH) is a small molecule found in almost every cell (Anderson, 1997). It is the smallest intrecellular thiol (SH) molecule. Its high electron-donating capacity (high negative redox potential) combined with high intracellular concentration generate great reducing power (Kidd, 1997). Glutathione is a cysteine-containing peptide found in most forms of aerobic life. It is not required in the diet and is instead synthesized in cells from its constituent amino acids (Meister and Anderson, 1983; Meister, 1988). It can react non-enzymatically with ROS and GSH peroxidase catalyses the destruction of hydrogen peroxide and hydroperoxides resulting in its oxidation to the disulphide form (GSSG). Firstly, glutathione is the major antioxidant produced by the cell protecting it from free radicals as oxygen radicals which are highly reactive substances can damage or destroy key cell components. Its antioxidant properties result from the thiol group in its cysteine moiety is a reducing agent and can be reversibly oxidized and reduced. In cells, glutathione is maintained in the reduced form by the enzyme glutathione reductase and in turn reduces other metabolites and enzyme systems, such as ascorbate in the glutathione-ascorbate cycle, glutathione peroxidases and glutaredoxins, as well as reacting directly with oxidants. Due to its high concentration and its central role in maintaining the cell's redox state, glutathione is one of the most important cellular antioxidants. In some organisms glutathione is replaced by other thiols, such as by mycothiol in the Actinomycetes, bacillithiol in some Gram-positive bacteria, or by trypanothione in the Kinetoplastids (Meister and Larsson, 1995). Secondly, GSH is a very important detoxifing agent, enabling the body to get rid of undesirable toxins and pollutants. It forms a soluble compound with the toxin that can then be excreted through the urine or the gut. The liver and kidneys contain high levels of GSH as they have the greatest exposure to toxins. The lung are also rich in glutathione partly for the same reason. Thirdly, GSH plays a crucial role in maintaining a normal balance between oxidation and anti-oxidation. This in turn regulates many of the cell's vital functions such as the synthesis and repair of DNA, the synthesis of proteins and the activation, maintaining the essential thiol status of protein, immune function, regulate nitric oxide homeostasis, modulate the activity of neurotransmitter receptors and regulation of enzymes (Oja et al., 2000; Hogg, 2002). The lower level of GSH is related to different physiological and biochemical disturbances.

3.2.2. Ascorbic acid

Ascorbic acid or "vitamin C" is a monosaccharide oxidation-reduction (redox) catalyst found in both animals and plants (Peake , 2003). As one of the enzymes needed to make ascorbic acid has been lost by mutation during primate evolution, humans must obtain it from the diet; it is therefore a vitamin. Most other animals are able to produce this compound in their bodies and do not require it in their diets. Ascorbic acid is required for the conversion of the procollagen to collagen by oxidizing proline residues to hydroxyproline. In other cells, it is maintained in its reduced form by reaction with glutathione, which can be catalysed by protein disulfide isomerase and glutaredoxins. Ascorbic acid is redox catalyst which can reduce, and thereby neutralize, reactive oxygen species such as hydrogen peroxide. In addition to its direct antioxidant effects, ascorbic acid is also a substrate for the redox enzyme ascorbate peroxidase, a function that is particularly important in stress resistance in plants. Ascorbic acid is present at high levels in all parts of plants and can reach concentrations of 20 millimolar in chloroplasts. It act as a marked antioxidant that help in the treatement of different diseases such as cancer and cardiovascular (Coulter et al., 2006 ; Cook et al., 2007).

3.2.3. Melatonin

Melatonin (N-acetyl-5-methoxytryptamine), is synthesized from serotonin in the pineal gland which contains all the enzymes necessary for the methoxylation and acetylation reactions. Melatonin is released in mammals during the dark-phase of the circadian cycle, and declines with age (Tan et al., 2001). It is able to reduce the free radical formation which follows the interaction between transition metal ions and amyliod-beta peptide (Zatta et al., 2003). As a free radical scavenger melatonin exhibits several important properties: It has both lipophilic and hydrophilic and it passes all bio-barriers, e.g. blood brain barrier and placenta (Wakatsuki et al., 1999).

Reiter (1995) reported that melatonin seems to be more effective than other antioxidants (e.g. mannitol, glutathione and vitamin E) in protecting against oxidative damage. Thus, it may provide protection against diseases that cause degenerative or proliferative changes by shielding macromolecules, particularly DNA from such injuries. Besides its direct free radical scavenging action, melatonin functions as an indirect antioxidant by stimulating the activities of antioxidiative enzymes in addition to protecting against lipid peroxidation (Undeger et al., 2004).

Melatonin has been found to be a direct free radical scavenger and an indirect antioxidant that, may have an active role in protection against genetic damage due to endogenously produced free radicals and it may be of use in reducing damage from physical and chemical mutagens and carcinogens that generate free radicals (Bandyopadhyay et al., 2000).

3.2.4. Tocopherols and tocotrienols (vitamin E)

Vitamin E is the collective name for a set of eight related tocopherols and tocotrienols, which are fat-soluble vitamins with antioxidant properties (Roberts et al., 2007). Of these, α-tocopherol has been most studied as it has the highest bioavailability, with the body preferentially absorbing and metabolising this form. It has been claimed that the α-tocopherol form is the most important lipid-soluble antioxidant, and that it protects membranes from oxidation by reacting with lipid radicals produced in the lipid peroxidation chain reaction. This removes the free radical intermediates and prevents the propagation reaction from continuing (Cook et al., 2007). This reaction produces oxidised α-tocopheroxyl radicals that can be recycled back to the active reduced form through reduction by other antioxidants, such as ascorbate, retinol or ubiquinol.This is in line with findings showing that α-tocopherol, but not water-soluble antioxidants, efficiently protects glutathione peroxidase 4 (GPX4)-deficient cells from cell death. GPx4 is the only known enzyme that efficiently reduces lipid-hydroperoxides within biological membranes. However, the roles and importance of the various forms of vitamin E are presently unclear, and it has even been suggested that the most important function of α-tocopherol is as a signaling molecule, with this molecule having no significant role in antioxidant metabolism. The functions of the other forms of vitamin E are even less well-understood, although γ-tocopherol is a nucleophile that may react with electrophilic mutagens, and tocotrienols may be important in protecting neurons from damage. However it has a protective action against different diseases including cancer (Coulter et al., 2006).

3.3. Total antioxidant capacity

Epidemiologic studies have demonstrated an inverse association between consumption of fruits and vegetables and morbidity and mortality from degenerative diseases. The antioxidant content of fruits and vegetables may contribute to the protection they offer from disease. Because plant foods contain many different classes and types of antioxidants, knowledge of their total antioxidant capacity (TAC), which is the cumulative capacity of food components to scavenge free radicals, would be useful for epidemiologic purposes. To accomplish this, a variety of foods commonly consumed in Italy, including 34 vegetables, 30 fruits, 34 beverages and 6 vegetable oils, were analyzed using three different assays, i.e., Trolox equivalent antioxidant capacity (TEAC), total radical-trapping antioxidant parameter (TRAP) and ferric reducing-antioxidant power (FRAP). These assays, based on different chemical mechanisms, were selected to take into account the wide variety and range of action of antioxidant compounds present in actual foods. Among vegetables, spinach had the highest antioxidant capacity in the TEAC and FRAP assays followed by peppers, whereas asparagus had the greatest antioxidant capacity in the TRAP assay. Among fruits, the highest antioxidant activities were found in berries (i.e., blackberry, redcurrant and raspberry) regardless of the assay used. Among beverages, coffee had the greatest TAC, regardless of the method of preparation or analysis, followed by citrus juices, which exhibited the highest value among soft beverages. Finally, of the oils, soybean oil had the highest antioxidant capacity, followed by extra virgin olive oil, whereas peanut oil was less effective. Such data, coupled with an appropriate questionnaire to estimate antioxidant intake, will allow the investigation of the relation between dietary antioxidants and oxidative stress-induced diseases (Pellegrini et al., 2003 ; Puchau et al., 2009 ; Dilis and Trichopoulou, 2010).

4. Nutritional therapy with natural antioxidants

Antioxidants have been the focus of research on the relationship between The role of dietary factors in protecting against the change from native to oxidized LDL has received considerable attention. An overview of epidemiological research suggests that individuals with the highest intakes of antioxidant vitamins, whether through diet or supplements, tend to lower of various disease. Research examining the effects of a diet rich in fruits and vegetables on disease has been carried out using several types of study.There is strong scientific evidence to support an increase in intakes of vegetables and fruit in the prevention of disease. Further research is required to clarify which particular components of fruit and vegetables are responsible for their protective effects. Numerous epidemiological studies have indicated that diets rich in fruits and vegetables are correlated with a reduced risk of chronic diseases (German, 1999; Benzie, 2003; Hassan, 2005; Hassan and Yosef, 2009; Hassan et al., 2010). It is probable that antioxidants, present in the fruits and vegetables such as polyphenols, carotenoids, and vitamin C, prevent damage from harmful reactive oxygen species, which either are continuously produced in the body during normal cellular functioning or are derived from exogenous sources (Gate et al. 1999). The possible protective effect of antioxidants in fruits and vegetables against ROS has led people to consume antioxidant supplements against chronic diseases.

5. Nutritional factors as natural antioxidants agents in alleviating the oxidative stress induced by environmental pollutents: Some experimental studies for the author

5.1. Mitigating effects of antioxidant properties of black berry juice on sodium fluoride induced hepatotoxicity and oxidative stress in rats

Fluorosis is a serious public health problem in many parts of the world. As in the case of many chronic degenerative diseases, increased production of reactive oxygen species has been considered to play an important role, even in the pathogenesis of chronic fluoride toxicity. Black berry is closely linked to its protective properties against free radical attack. Therefore, the aim of this study was to demonstrate the role of black berry juice (BBJ) in decreasing the hepatotoxicity and oxidative stress of sodium fluoride)NaF). Results showed that NaF caused elevation in liver TBARS and nitric oxide (NO), and reduction in superoxide dismutase (SOD), catalase (CAT), total antioxidant capacity (TAC) and glutathione (GSH.(Plasma transaminases (AST and ALT), creatine kinase (CK), lactate dehydrogenase (LDH), total lipids)TL), cholesterol, triglycerides (TG), and low density lipoprotein–cholesterol (LDL–c) were increased, while high density lipoprotein–cholesterol (HDL–c) was decreased. On the other hand, BBJ reduced NaF-induced TBARS, NO, TL, cholesterol, TG, LDL–c, AST, ALT, CK and LD. Moreover, it ameliorated NaFinduced decrease in SOD, CAT, GSH, TAC and HDL–c. Therefore, the present results revealed that BBJ has a protective effect against NaF-induced hepatotoxicity by antagonizing the free radicals generation and enhancement of the antioxidant defence mechanisms. (Hassan and Yousef, 2009).

5.2. Evaluation of free radical-scavenging and antioxidant properties of black berry against fluoride toxicity in rats

Oxidative damage to cellular components such as lipids and cell membranes by free radicals and other reactive oxygen species is believed to be associated with the development of degenerative diseases. Fluoride intoxication is associated with oxidative stress and altered anti-oxidant defense mechanism. So the present study was extended to investigate black berry anti-oxidant capacity towards superoxide anion radicals, hydroxyl radicals and nitrite in different organs of fluoride-intoxicated rats. The data indicated that sodium fluoride (10.3 mg/kg bw) administration induced oxidative stress as evidenced by elevated levels of lipid peroxidation and nitric oxide in red blood cells, kidney, testis and brain tissues. Moreover, significantly decreased glutathione level, total anti-oxidant capacity and superoxide dismutase activitywere observed in the examined tissues. On the other hand, the induced oxidative stress and the alterations in anti-oxidant system were normalized by the oral administration of black berry juice (1.6 g/ kg bw). Therefore it can be concluded that black berry administration could minimize the toxic effects of fluoride indicating its free radical-scavenging and potent anti-oxidant activities. (Hassan and Fattoh, 2010)

5.3. Garlic oil as a modulating agent for oxidative stress and neurotoxicity induced by sodium nitrite in male albino rats

In the present study, we investigated the neurobiochemical alterations and oxidative stress induced by food preservative; sodium nitrite (NaNO2) as well as the role of the garlic oil in amelioration of the neurotoxicity in male albino rats. Serum and brain homogenates of the rats received NaNO2 (80 mg/kg body weight) for 3 months exhibited significant decrease in acetylcholine esterase (AChE) activity as well as the levels of phospholipids, total protein and the endogenous antioxidant system (glutathione; GSH and superoxide dismutase; SOD). In contrast, lactic dehydrogenase (LDH) activity, brain thiobarbituric acid reactive substances (TBARS) and nitric oxide (NO) levels were significantly increased. On the other hand, the oral administration of garlic oil (5 ml/kg body weight) daily for 3 months significantly improved the neurobiochemical disorders and inhibited the oxidative stress induced by NaNO2 ingestion. So, this study reveals the neural toxic effects of NaNO2 by exerting oxidative stress and retrograde the endogenous antioxidant system. However, garlic oil has a promising role in attenuating the obtained hazard effects of sodium nitrite by its high antioxidant properties which may eventually be related with the preservation of SOD activity and primary mitochondrial role against nitrite-induced neurotoxicity in rats . (Hassan et al., 2010).

5.4. *In vivo* evidence of hepato-and-reno-protective effect of garlic oil against sodium nitrite-induced oxidative stress

Sodium nitrite (NaNO2), a food color fixative and preservative, contributes to carcinogenesis. We investigated the protective role of garlic oil against NaNO2-induced abnormalities in metabolic biochemical parameters and oxidative status in male albino rats. NaNO2 treatment for a period of three months induced a significant increase in serum levels of glucose, aspartate aminotransferase (AST), alanine aminotransferase (ALT), alkaline phosphatase (ALP), bilirubin, urea and creatinine as well as hepatic AST and ALT.

However, significant decrease was recorded in liver ALP activity, glycogen content, and renal urea and creatinine levels. In parallel, a significant increase in lipid peroxidation, and a decrease in glutathione content and catalase activity were observed in the liver and the kidney. However, garlic oil supplementation showed a remarkable amelioration of these abnormalities. Our data indicate that garlic is a phytoantioxidant with powerful chemopreventive properties against chemically-induced oxidative stress (Hassan et al., 2009).

5.5. Ameliorating effect of chicory (Cichorium intybus L)-supplemented diet against nitrosamine precursors-induced liver injury and oxidative stress in male rats

The current study was carried out to elucidate the modulating effect of chicory (*Cichorium intybus L.)*-supplemente diet against nitrosamnine-induced oxidative stress and hepatotoxicity in male rats. Rats were divided into four groups and treated for 8 weeks as follow: group 1 served as control; group 2 fed on chicory-supplemented diet (10% w/w); group 3 received simultaneously nitrosamine precursors [sodium nitrite (0.05% in drinking water) plus chlorpromazine (1.7 mg/kg body weight)] and group 4 received nitrosamine precursors and fed on chicory-supplemented diet. The obtained results revealed that rats received nitrosamine precursors showed a significant increase in liver TBARS and total lipids, total cholesterol, bilirubin, and enzymes activity (AST, ALT, ALP and GGT) in both serum and liver. While a significant decrease in the levels of GSH, GSH-Rx, SOD, catalase, total protein and albumin was recorded. On the other hand, chicory-supplemented diet succeeded to modulate these observed abnormalities resulting from nitrosamine compounds as indicated by the reduction of TBARS and the pronounced improvement of the investigated biochemical and antioxidant parameters. So, it could be concluded that chicory has a promising role and it worth to be considered as a natural substance for ameliorating the oxidative stress and hepatic injury induced by nitrosamine compounds **(Hassan and Yousef, 2010).**

6. Summary

6.1. Oxidative stress

The term oxidative stress refers to a condition in which cells are subjected to excessive levels of molecular oxygen or its chemical derivatives called reactive oxygen species (ROS).Under physiological conditions, the molecular oxygen undergoes a series of reactions that ultimately lead to the generation of superoxide anion (O_2-), hydrogen peroxide (H_2O_2) and H_2O. Peroxynitrite (OONO-), hypochlorus acid (HOCl), the hydroxyl radical (OH.), reactive aldehydes, lipid peroxides and nitrogen oxides are considered among the other oxidants that have relevance to vascular biology.

Oxygen is the primary oxidant in metabolic reactions designed to obtain energy from the oxidation of a variety of organic molecules. Oxidative stress results from the metabolic reactions that use oxygen, and it has been defined as a disturbance in the equilibrium status of pro-oxidant/anti-oxidant systems in intact cells. This definition of oxidative stress implies that cells have intact pro-oxidant/anti-oxidant systems that continuously generate and detoxify oxidants during normal aerobic metabolism. When additional oxidative events occur, the pro-oxidant systems outbalance the anti-oxidant, potentially producing oxidative damage to lipids, proteins, carbohydrates, and nucleic acids, ultimately leading to cell death in severe oxidative stress. Mild, chronic oxidative stress may alter the anti-oxidant systems by inducing or repressing proteins that participate in these systems, and by depleting cellular stores of anti-oxidant materials such as glutathione and vitamin E (Laval, 1996). Free radicals and other reactive species are thought to play an important role oxidative stress resulting in many human diseases. Establishing their precise role requires the ability to measure them and the oxidative damage that they cause (Halliwell and Whiteman, 2004).

A basic approach to study oxidative stress would be to measure some products such as (i) free radicals; (ii) radical-mediated damages on lipids, proteins or DNA molecules; and iii) antioxidant enzymatic activity or concentration.

6.2. Free radicals

Free radicals are reactive compounds that are naturally produced in the human body. They can exert positive effects (e.g. on the immune system) or negative effects (e.g. lipids, proteins or DNA oxidation). Free radicals are normally present in the body in minute concentrations. Biochemical processes naturally lead to the formation of free radicals, and under normal circumstances the body can keep them in check. If there is excessive free radical formation, however, damage to cells and tissue can occur (Wilson, 1997). Free radicals are toxic molecules, may be derived from oxygen, which are persistently produced and incessantly attack and damage molecules within cells; most frequently, this damage is measured as peroxidized lipid products, protein carbonyl, and DNA breakage or fragmentation. Collectively, the process of free radical damage to molecules is referred to as oxidative stress (Reiter et al., 1997).To limit these harmful effects, an organism requires complex protection – the antioxidant system. This system consists of antioxidant enzymes (catalase, glutathione peroxidase, superoxide dismutase) and non-enzymatic antioxidants (e.g. vitamin E [tocopherol], vitamin A [retinol], vitamin C [ascorbic acid], glutathione and uric acid). An imbalance between free radical production and antioxidant defence leads to an oxidative stress state, which may be involved in aging processes and even in some pathology (e.g. cancer and Parkinson's disease).

6.3. Oxidative damage to lipids (Lipid peroxidation)

The peroxidation of lipids involves three distinct steps: initiation, propagation and termination. The initiation reaction between an unsaturated fatty acid and the hydroxyl radical involves the abstraction of a H atom from the methylvinyl group on the fatty acid. The remaining carbon-centred radical, forms a resonance structure sharing this unpaired electron among carbons 9 to 13. In the propagation reactions, this resonance structure reacts with triplet oxygen, which is a biradical having two unpaired electrons and therefore reacts readily with other radicals. This reaction forms a peroxy radical. The peroxy radical then abstracts a H atom from a second fatty acid forming a lipid hydroperoxide and leaving another carbon centered free radical that can participate in a second H abstraction. Therefore, once one hydroxyl radical initiates the peroxidation reaction by abstracting a single H atom, it creates a carbon radical product that is capable of reacting with ground state oxygen in a chain reaction. The role of the hydroxyl radical is analogous to a "spark" that starts a fire. The basis for the hydroxyl radical's extreme reactivity in lipid systems is that at very low concentrations it initiates a chain reaction involving triplet oxygen, the most abundant form of oxygen in the cell (Benderitter et al., 2003).

The lipid hydroperoxide (ROOH) is unstable in the presence of irron or other metal catalysts because ROOH will participate in a Fenton reaction leading to the formation of reactive alkoxy radicals. Therefore, in the presence of irron, the chain reactions are not only propagated but amplified. Among the degradation products of ROOH are aldehydes, such as malondialdehyde, and hydrocarbons, such as ethane and ethylene, which are commonly measured end products of lipid peroxidation (Sener et al., 2004). During peroxidation pathway via reactive intermediates, several end products are formed such as aldehyde [malondialdehyde + 4-hydroxynonenal], pentane and ethane, 2,3 transconjugated diens, isoprostains and chlesteroloxides. The biological activities of MDA and other aldehydes include cross-linking with DNA and proteins, which alters the function/activity of these molecules. MDA + 4HNE have shown tissue toxicity. MDA can react with amino and thiol groups, the aldehydes are more diffusible than free radicals, which means damage is exported to distance sites. Aldehydes are quickly removed from cells as several enzymes control their metabolism (Ustinova and Riabinin 2003).

6.4. Antioxidants

Antioxidants are thought to protect the body against the destructive effects of free radicals. Antioxidants neutralize free radicals by donating one of their own electrons, ending the electron- gain reaction. The antioxidant nutrients themselves don't become free radicals by donating an electron because they are stable in either form. They act as scavengers, helping to prevent cell and tissue damage that could lead to cellular damage and disease (Reiter, 2003).

The body produces several enzymes, including superoxide dismutase (SOD), catalase (CAT), and glutathione peroxidase (GSHPX), that neutralize many types of free radicals. Supplements of these enzymes are available for oral administration. However, their absorption is probably minimal at best. Supplementing with the "building blocks" the body requires to make SOD, catalase, and glutathione peroxidase may be more effective. These building block nutrients include the minerals manganese, zinc, and copper for SOD and selenium for GSHPX.

In addition to enzymes, many vitamins, minerals and hormones act as antioxidants in their own right, such as vitamin C, vitamin E, beta-carotene, lutein, lycopene, vitamin B2, coenzyme Q10, and cysteine (an amino acid). Herbs, such as bilberry, turmeric (curcumin), grape seed or pine bark extracts, and ginkgo can also provide powerful antioxidant protection for the body. Melatonin is a hormone secreted by pineal gland and proves to be powerful antioxidant and free radical scavenger (Yang et al.,2002 and Koc et al., 2003).

6.5. Nutritional therapy with natural antioxidants

Antioxidants have been the focus of research on the relationship between The role of dietary factors in protecting against the change from native to oxidized LDL has received considerable attention. An overview of epidemiological research suggests that individuals with the highest intakes of antioxidant vitamins, whether through diet or supplements, tend to experience 20–40% lower risk of coronary heart disease (CHD) than those with the lowest intake or blood levels (diet and disease. Research examining the effects of a diet rich in fruits and vegetables on disease has been carried out using several types of study.

There is strong scientific evidence to support an increase in intakes of vegetables and fruit in the prevention of disease. Further research is required to clarify which particular components of fruit and vegetables are responsible for their protective effects.

Numerous epidemiological studies have indicated that diets rich in fruits and vegetables are correlated with a reduced risk of chronic diseases (Banerjee and Maulik, 2002 ; Sesso et al., 2003). It is probable that antioxidants, present in the fruits and vegetables such as polyphenols, carotenoids, and vitamin C, prevent damage from harmful reactive oxygen species, which either are continuously produced in the body during normal cellular functioning or are derived from exogenous sources (Gate et al. 1999). The possible protective effect of antioxidants in fruits and vegetables against ROS has led people to consume antioxidant supplements against chronic diseases.

6.6. Some experimental studies for the author about food factors as a nutritional antioxidants agents in alleviating the oxidative stress induced by environmental pollutents

1. In vivo evidence of hepato-and-reno-protective effect of garlic oil against sodium nitrite-induced oxidative stress Int. Int. J. Biol. Sci. (5)3: 249-255. (Hassan et al., 2009)
2. Mitigating effects of antioxidant properties of black berry juice on sodium fluoride induced hepatotoxicity and oxidative stress in rats. Food Chem.Toxicol., 47 2332–2337. (Hassan and Yousef, 2010).
3. Evaluation of free radical-scavenging and antioxidant properties of black berry against fluoride toxicity in rats. Food Chem.Toxicol., 48: 1999-2004. (Hassan and Fattoh, 2010).
4. Garlic oil as a modulating agent for oxidative stress and neurotoxicity induced by sodium nitrite in male albino rats. Food Chem. Toxicol., 48: 1980-1985. (Hassan et al., 2010)
5. Ameliorating effect of chicory (Cichorium intybus L)-supplemented diet against nitrosamine precursors-induced liver injury and oxidative stress in male rats. Food Chem. Toxicol., 48: 2163-2169. (Hassan and Yousef, 2010).

Author details

Hanaa Ali Hassan Mostafa Abd El-Aal
Zoology Department, Faculty of Science, Mansoura University, Mansoura, Egypt

7. References

Antunes, F., Derick, H. and Cadenas, E. (2002). Relative contributions of heart mitochondria glutathione peroxidase and catalase to H2O2 detoxification in in vivo conditions. Free Radic. Biol. Med., 33 (9): 1260-1267

Al-Omar, M. A., Beedham, C. and al-Sarra, I. (2004). Pathological roles of reactive oxygen species and their defence mechsnisms. Saudi Pharm. J., 12:1-18.

Arthur, J. R. (2000). The glutathione peroxidase. Cell Mol. Life Sci., 57:1825 1835.

Awasthi, S., Bajpai, K. K., Piper, J. T., Singhal, S. S., Ballaatore, A., Seifert, W. E., Awasthi. Y. C. and Ansari, G. A. S. (1996): Drug metabolism. Dispos. 24: 371-374.

Bandyopadhyay, D., Biswas, K., Bandyopadhyay, U., Reiter, R. J. and Banerjee, R. K. (2000). Melatonin protects against stress-induced gastric lesions by scavenging the hydroxyl radical. J. Pineal Res., 29: 143-151.

Banerjee, S. K. and Maulik, s. K. (2002): Effect of garlic on cardiovascular disorders. J. Nutr.,1:4-36.

Bannister, J., Bannister, W. and Rotilio, G. (1987). Aspects of the structure, function, and applications of superoxide dismutase". CRC Crit. Rev. Biochem. 22 (2): 111–180.

Benderitter, M.; Vincent-Genod, L.; Pouget, J. P. and Voisin, P. (2003): The cell membrane as a biosensor of oxidative stress induced by radiation exposure: a multiparameter investigation. Radiat. Res., 159(4): 471-483.

Benzie, I. (2003). Evolution of dietary antioxidants. Comp. Biochem. Phys., 136 (1): 113–126.

Blankenberg, S., Rupprecht, H. J., Bickel, C., Torzewski, M., Hafner, G., Tiret, L., Smieja, M., Cambien, F., Meyer, J., and Lackner, K.J. (2003). Glutathione peroxidase 1 activity and cardiovascular events in patients with coronary artery disease atherogene investigators. Engl. J. Med., 349:1605-1613

Brioukhanov, A; L; and Netrusoe, a. I. (2004): Catalase and superoxide dismutase, distribution, properties and physiological role in cells of strict anaerobes. Biochem. 69:949-962.

Ceconi, C. , Boraso, A. , Cargnoni, A. and Ferrari, R., (2003). Oxidative stress in cardiovascular disease:myth or fact? Arch. Biochem. Biophys., 420: 217 – 221 .

Ceriello, A. and Motz, E. (2004). Is oxidative stress the pathogenic mechanism underlying insulin resistance, diabetes, and cardiovascular disease? The common soil hypothesis revisited .Arterioscler. Thromb. Vasc. Biol., 24 : 816 – 823 .

Cohen, P. J. (1992): Allopurinol administered prior hepatic ischaemic in the rat prevents chemiluminescence following restoration of circulation. Can. J. Anaesth., 39: 1090-1093.

Cook, NR., Albert, CM. and Gaziano, JM. (2007). A randomized factorial trial of vitamins C and E and beta carotene in the secondary prevention of cardiovascular events in women: results from the Women's Antioxidant Cardiovascular Study. Arch. Int. Med., 167 (15): 1610–1618.

Coulter, I., Hardy, M., Morton, S., Hilton, L., Tu, W., Valentine, D. and Shekelle, P. (2006). Antioxidants vitamin C and vitamin e for the prevention and treatment of cancer. J. f general int. medici.: official j. Soc. Res. Educ. in Primary Care Int. Medici., 21 (7): 735–744

Davies, AG., Griller, D., Ingold, KU., Lindsay, DA. and Walton, JC. (1981). An electron spin resonance study of pentadienyl and related radicals: homolytic fission of cyclobut-2-enylmethyl radicals. J. Chem. Soc. Perkin. Trans., II:633-641.

Dherani, M. , Murthy, GV., Gupta, SK., Young, IS. and Maraini, G, Dherani, M, Murthy, GV, Gupta, SK, Young, IS, Maraini, G, Camparini, M, Price, GM, John, N, Chakravarthy, U, Fletcher, AE. (2008). Blood levels of vitamin C,carotenoids and retinol are inversely associated with cataract in a North Indian population .Invest. Ophthalmol. Vis. Sci., 49: 3328 – 3335 .

Dilis, V. and Trichopoulou, A. (2010). Antioxidant intakes and food sources in greek adults J. Nutr., 140: 1274-1279.

Dudek, H., Farbiszewski, R., Michno, T., Łebkowski, W. J. and Kozłowski, A. (2002). Activity of glutathione peroxidase (GSH-Px), glutathione reductase (GSSG-R) and superoxide dismutase in the brain tumors 55, 5-6: 252-256.

Fearon, IM. and Faux, SP. (2009) . Oxidative stress and cardiovascular disease: novel tools give (free)radical insight . J. Mol. Cell Cardiol., 47 : 372 – 381 .

Gate L, Paul J, Ba GN, Tew KD and Tapiero H. (1999): Oxidative stress induced in pathologies: the role of antioxidants. Biomed Pharmacother,53: 169 -180.

German, JB. (1999). Food processing and lipid oxidation". Advances in experimental medicine and biology. Advanc. Experim. Medici. Biol., 459: 23–50.

Greenwald, R. A. (1990). Superoxide dismutase and catalase as therapeutic agents for human diseases. A critical review. Free Radic. Boil. Med., 8(2): 210-219.

Halliwell, B. and Whiteman, M. (2004): Measuring reactive species and oxidative damage in vivo and in cell culture: how should you do it and what do the results mean?. Br. J. Pharmacol.., 142(2):231-255.

Hassan, H. A. (2005). The ameliorating effect of black grape juice on fluoride induced testicular toxicity in adult albino rats. Egypt. J. Zool., 44:507–525.

Hassan, H. A ; El-Agmy Sh. M; Gaur, R. L; Fernando, A. Raj, M., HG and Ouhtit A. (2009): In vivo evidence of hepato-and-reno-protective effect of garlic oil against sodium nitrite-induced oxidative stress Int. Int. J. Biol. Sci. (5)3: 249-255.

Hassan, H. A and A. M. Fattoh (2010): Evaluation of free radical-scavenging and antioxidant properties of black berry against fluoride toxicity in rats. Food Chem.Toxicol.,48:1999-2004.

Hassan H. A. Hafez H. S. and Zeghebar F. E. (2010): Garlic oil as a modulating agent for oxidative stress and neurotoxicity induced by sodium nitrite in male albino rats. Food Chem. Toxicol., 48: 1980-1985.

Hassan, H. A and Yousef M. I. (2009): Mitigating effects of antioxidant properties of black berry juice on sodium fluoride induced hepatotoxicity and oxidative stress in rats. Food Chem.Toxicol., 47 2332–2337

Hassan, H. A and Yousef, M. I. (2010): Ameliorating effect of chicory (Cichorium intybus L)-supplemented diet against nitrosamine precursors-induced liver injury and oxidative stress in male rats. Food Chem. Toxicol., 48: 2163-2169.

Hayes, J. D. and Pulford, D. J. (1995): The glutathione S transferase supergene family: regulation of GST and the contribution of the isoenzymes to cancer chemoprotectiion and drug resistance. Crit. Rev. Biochem. Mol. Biol., 30: 445-600.

Hogg, N. (2002): The biochemistry and physiology of S nitrosothiols. Ann. Rev. Pharmacol. Toxicol., 42:585-600.

Huang, D. Y. and Ichikawa, Y. (1994): Two different enzymes are primary responsible for retinoic acid synthesis in rabbit liver cystosol. Biochem. Biophys. Res. Commin., 205: 1278-1283.

Johnson, F. and Giulivi, C. (2005). Superoxide dismutases and their impact upon human health. Mol. Aspects Med. 26 (4–5): 340–352.

Kelly, S. A., Havrilla, C. M., Brady, T. C., Abramo, K. H. and Levin, E. D. (1998). Review of oxidative stress in toxicology: Established mammalian and emerging piscine model systems. Environ. Health Perspect., 106:375-384.

Kidd, P. M. (1997): Glutathione systemic protectant against oxidative and free radical damage. Altern. Med. Res., 1-155-176.

Koc, M.,Taysi, S.,Buyukokuroglu, M. E. and Bakan, N. (2003). The effect of melatonin against oxidative damage during total- body irradiation in rats. Radiat. Res., 160: 251-255.

Kondo, T., Dale, O.L. and Beutler, E. (1980): Glutathione transport by inside-out vesicles from human erythrocytes. Proc. Nat. Acad. Sci. Biochem., 77: 6359-6362.

Kregel, KC. and Zhang, HJ . (2007) . An integrated view of oxidative stress in aging: basic mechanisms,functional effects, and pathological considerations . Am. J. Physiol. Regul. Integr. Comp. Physiol .,292: R18 – 36 .

Laval, J. (1996): Role of DNA repair enzymes in the cellular resistance to oxidative stress. Pathol. Biol., 44: 14-24

Mayo, J. C., Tan, D. X., Sainz, R. M., Lopez-Burillo, S. and Reiter, R, J. (2003). Oxidative damage to Catalase induced by peroxyl radicals: Functional protection by melatonin and other antioxidants. Free Radic. Res., 37: 543-553.

McCord, J. M. (1985): Oxygen-derived free radicals in postischemic tissue injury. N. Eng. J. Med., 312: 159-163.

Meister, A. (1988). Glutathione metabolism and its selective modification. J. Biol. Chem., 263(33):17205-17208.

Meister, A. and Anderson, M. N. (1983). Glutathione. Ann. Rev. Biochem., 52: 611-660.

Meister, A. and Larsson, A. (1995). Glutathione synthetase deficiency and other disorders of the gamma-glutamyl cycle. In: Scriver, C. R.; Kinzler, K. W. ; Valle, D., et al., eds. The

Metabolic and molecular bases of inherited diseases. New York: Mc Graw- Hill, 1461-1477.

Nagaoka, S., Okauchi, Y., Urano, S., Nagashima, U. and Mukai, K. (1990). Kinetic and ab initio study of the prooxidant effect of vitamin E: hydrogen abstraction from fatty acid esters and egg yolk lecithin. J. Am. Chem. Soc., 112:8921-8924.

Nozik-Grayck, E., Suliman, H. and Piantadosi, C. (2005). Extracellular superoxide dismutase. Int. J. Biochem. Cell Biol., 37 (12): 2466–2471.

Oja, S. S., Janaky, R., Varga, V. and Saranasaari, P. (2000): Modulation of glutamate receptor functions by glutathione. Neurochem. Int., 37: 299-306.

Packer, L. (1994). Methods in Enzymology; Oxygen Radicals in Biological Systems, Part C, Academic Press, 1994.

Peake, J. (2003). Vitamin C: Effects of exercise and requirements with training. Int. J. Sport Nutr. Exerc. Metab., 13 (2): 125–151.

Pellegrini, N. Serafini, M., Colombi, B., Del Rio, D., Salvatore, S., Bianchi, M. and Brighenti, F. (2003): Total antioxidant capacity of plant foods, beverages and oils consumed in italy assessed by three different in vitro assays. J. Nutr. 133:2812-2819.

Preston, T.J., Muller. W.J. and Singh, G. (2001): Scavenging of extracellular H2O2 by catalase inhibits the proliferation of HER-2/Neu-transformed Rat-1 fibroblasts through the induction of a stress response. J. Biological Chem., 276: 9558-9564.

Puchau, B., Ángeles, M., Zulet, de Echávarri, A. G., Hermsdorff, H. H. M., and Martínez, J. A., (2009). Dietary total antioxidant capacity: A novel indicator of diet quality in healthy young adults J. Am. Coll. Nutr., 28: 648-656.

Reiter, R. J. (1995). Oxygen radical detoxification processes during aging: the functional importance of melatonin. Aging (Milano), 7: 340-351.

Reiter, R. J. (2003). Melatonin: clinical relevance. Best Pract. Res. Clinic. Endocr. Metabol., 17 (2):273–285.

Reiter, R. J.; Carneiro, R. C. and Oh, C. S. (1997): Melatonin in relation to cellular antioxidiative defense mechanisms. Horm. Metab. Res., 29: 363-372.

Reiter, R. J.,Tan, D. X., Gitto, E., Sainz, R. M., Mayo, J.C., Leon, J., Manchester, L. C.,Vijayalaxm, Kilic, E. and Kilic, U. (2004). Pharmacological utility of melatonin in reducing oxidative cellular and molecular damage. Pol. J. Pharmacol.,56(2):159-170.

Roberts, LJ., Oates, JA. and Linton, MF. (2007). The relationship between dose of vitamin E and suppression of oxidative stress in humans. Free Radic. Biol. Med., 43 (10): 1388–1393.

Scandalios, J. G., Guan, l. and Polidoros, A. (1997): Catalase in plants.: Gene structure, properties, regulation and expression In J. G. Scandalios Ed., Oxidative stress and the molecular biology of antioxidant defenses Pp 343-406.

Sener, G.; Paskaloglu, K.; Toklu, H.; Kapucu, C.; Ayanoglu-Dulger, G.; Kacmaz, A. and Sakarcan, A. (2004): Melatonin ameliorates chronic renal failure-induced oxidative organ damage in rats. J Pineal Res., 36(4):232-241.

Sesso, H.D.; Liu, S.; Gaziano, J.M. and Buring, J. E. (2003): Dietary Lycopene, Tomato-Based Food Products and Cardiovascular Disease in women. J. Nutr., 133: 2336-2341.

Sies, H. (1997). Oxidative stress: Oxidants and antioxidants. Exp. Physiol., 82(2): 291-295.

Tan, D. X., Manchester, L. C.,Burkardt, S., Sainz, R. M., Mayo, J. C., Kohen, R., Shohami, E., Huo, Y. S., Hardeland, R. and Reiter, R. J. (2001). N-acetyl-N-formyl-5-methoxyknuramine, a biogenic amine and melatonin metabolite, functions as a potent antioxidant. FASEB. J., 15: 2294-2296.

Terao, M., Kurosaki, M., Saltini, G., Demontis, S., Marini, M., Salmona, M. and Garattini, E. (2000): Cloning of the cDNAs coding for two novel molybdoflavoproteins showing high similarity with aldehyde oxidase and xanthine oxidoreductase. J. Biol. Chem., 275: 30690-30700.

Undeger, U., Giray, B., Zorlu, A. F., Oge, K. and Bacaran, N. (2004). Protective effects of melatonin on the ionizing radiation induced DNA damage in the rat brain. Exp. Toxicol. Pathol., 55: 379-384.

Ustinova, A. A. and Riabinin, V. E. (2003): Effect of chronic gamma-irradiation on lipid peroxidation in CBA mouse blood serum. Radiat. Biol. Radioecol., 43(4):459-463.

Wakatsuki, A.,Okatani, Y., Izumiya, C. and Lkenoue, N. (1999). Melatonin protect against ischemia and reperfusion- induced oxidative lipid and DNA damage in fetal rat brain. J. Pineal Res., 26: 147-152.

Waldeck, A. R. and Stocker, R. (1996). Radical-initiated lipid peroxidation in low density lipoprotein. Insights obtained from kinetic modeling. Chem. Res. Toxicol., 9: 954-964.

Weir, E.K., Archer, S.L. and Reeves, J.T. (1996). Nitric Oxide and Radicals in the Pulmonary Vasculature, Futura Publishing, Armonk, N.Y.

Wilson, J. (1997): Antioxidant defense of the brain: a role for astrocytes (review). Can. J. phys. Pharmacol., 75: 1149 – 1163.

Wright, R. M., Vaitaitis, G. M., Weigel, L.K., Repine, T. B., McManaman, J. L. and Repine, J.E. (1995): Identification of the candidate ALS2 gene at chromosome 2q33 as a human aldehyde oxidase gene. Redox Report 1:313-321.

Yang, J.; Lam, E. W.; Hammad, H. M.; Oberley, T. D. and Oberley, L. W. (2002): Antioxidant enzyme levels in oral squamous cell carcinoma and normal human oral epithelium. J. Oral Pathol. Med., 31: 71-77.

Yasminch, W. and Theologides, A. (1993). Catalase as a removing scavenger of hydrogen peroxide: a hypothesis. J. Lab. Clin. Med., 122(1): 110-114.

Zatta, P.,Tognon, G. and Carampin, P. (2003). Melatonin prevent free radical formation due to the interaction between beta-amyloid peptides and metal ions [Al(III), Zn(II), Cu(II), Mn(II), Fe(II)]. J. Pineal. Res., 35: 98-103.

Zebger, I.,Snyder, J. W., Andersen, L. K., Poulsen, L., Gao, Z., Lambert, J. D., Kristiansen, U. and Ogilby, P. R. (2004). Direct optical detection of singlet oxygen from a single cell. Photochem. Photobiol., 79(4): 319-322.

Zelko, I., Mariani, T. and Folz, R. (2002). Superoxide dismutase multigene family: a comparison of the CuZn-SOD (SOD1), Mn-SOD (SOD2), and EC-SOD (SOD3) gene structures, evolution, and expression. Free Radic. Biol. Med., 33 (3): 337–349.

Chapter 4

Iron Overload and Lipid Peroxidation in Biological Systems

Paula M. González, Natacha E. Piloni and Susana Puntarulo

Additional information is available at the end of the chapter

http://dx.doi.org/10.5772/46181

1. Introduction

Fe is an essential element for the growth and well-being of almost all living organisms, except for some strains of lactobacillus, where the role of Fe may be assumed by another metal [1]. It is involved in many biological functions since by varying the ligands to which it is coordinated, Fe has access to a wide range of redox potentials and can participate in many electron transfer reactions, spanning the standard redox potential range. It is also involved in O_2 transport, activation, and detoxification, in N_2 fixation and in several of the reactions of photosynthesis [2]. However, there are problems in the physiological management of Fe, since in spite of its overall abundance, usable Fe is in short supply because at physiological pH under oxidizing conditions, Fe is extremely insoluble. Anytime Fe exceeds the metabolic needs of the cell it may form a low molecular weight pool, referred to as the labile iron pool (LIP), which catalyzed the conversion of normal by-products of cell respiration, like superoxide anion (O_2^-) and hydrogen peroxide (H_2O_2), into highly damaging hydroxyl radical (•OH) through the Fenton reaction (reaction 1) or by the Fe^{2+} catalyzed Haber-Weiss reaction (reaction 2), or into equally aggressive ferryl ions or oxygen-bridged Fe^{2+}/Fe^{3+} complexes. Fe^{3+} can be reduced either by O_2^- (reaction 3) or by ascorbate leading to further radical production.

$$Fe^{2+} + H_2O_2 \xrightarrow[(Fe)]{} Fe^{3+} + HO^- + \bullet OH \quad (1)$$

$$O_2^- + H_2O_2 \Rightarrow O_2 + HO^- + \bullet OH \quad (2)$$

$$Fe^{3+} + O_2^- \Rightarrow Fe^{2+} + O_2 \quad (3)$$

Defense against the toxic effect of Fe and O_2 mixtures is provided by two specialized Fe-binding proteins: the extracellular transferrin (Tf) and the intracellular ferritin (Ft). Both retain Fe in the form of Fe^{3+} which unless mobilized will not be able to efficiently catalyze the production of free radicals. Fe is stored mainly intracellularly, where its potentially damaging effects are greatest.

The marine ecosystem can be seen as an integrative system with many factors that interact with the biota. Natural variables such as temperature, winds, precipitations, tide flows, currents, human activities, affect metal deposition into the sea. Once metals become bioavailable, they can enter the food web starting with the primary producers, and also in heterotrophic organisms at the bottom of the marine food chain, such as benthic filter feeders. Metals follow a bioaccumulation process inside the animals, depending on the animal's detoxification capacities and on exogenous Fe availability.

In plants, Fe concentrations increased during seed maturation, and by immunodetection experiments it was indicated that Ft concentration of seeds also increased with maturity, containing up to 1800 atoms of Fe per molecule [3]. This seed Fe could be stored for future use during seedling growth, as has been proposed by Hyde et al. [4], avoiding toxicity. Over growth, the oxidative stress depends upon a wide array of factors related to an enhanced radical production due to several metabolic pathways activated during the initial water uptake, including mitochondrial O_2 consumption. On the other hand, excess Fe effects seem to be limited mostly to the hydrophobic domain of the cell following different profile than during physiological development.

In the last decade or so, important advances have been made in the knowledge of conditions that involve Fe-overload in humans. Those conditions would include short term processes, as organ or tissue ischemia-reperfusion and local inflammation, as well as progressive pathologies essentially affecting the central nervous system. In the first case, the de-compartamentalization of Fe would lead to the expansion of the LIP and the increase of the oxidative damage. In the second case, it has been described an increase in Fe levels in the substantia nigra of Parkinsonian brains [5], Hallervorden-Spatz syndrome [6] and in mitochondria of Friedrich´s ataxia cerebella [7]. Hereditary hemochromatosis is a very common genetic defect in the Caucasian population, with an autosomal recessive inheritance. It is characterized by inappropriately increased Fe absorption from the duodena and upper intestine, with consequent deposition in various parenchymal organs, notably the liver, pancreas, heart, pituitary gland and skin [8]. Fe overload is characterized by the presence of several clinical manifestations such as: increased susceptibility to infections, hepatic dysfunction, tumors, joint diseases, myocardiopathy, and endocrine alterations. Fe overload has been also observed (a) if dietary Fe is excessive, such as in the severe Bantu siderosis, reported in the Bantu tribe of Africa who drink acidic beer out of Fe pots, (b) in other inherited diseases, such as congenital atransferrinemia (lacking circulating Tf), and (c) during the medical treatment of thalassemia. Moreover, clinical and epidemiologic observations indicated that increased Fe storage status is a risk factor in several diseases such as porphyria cutanea tarda and sudden infant death syndrome, among others.

Oxidative damage to lipids had been studied over several decades, and it had been characterized in terms of the nature of the oxidant, the type of lipid, and the severity of the oxidation. Many stable products are formed during the process and accordingly, the assays developed to assess these products to evaluate lipid peroxidation include many techniques. The most currently used assay is the determination of malondialdehyde (MDA) formation with the thiobarbituric acid reactive substances test (TBARS). However, electron paramagnetic resonance (EPR) spectroscopy has shown the capacity of detecting, in the presence of exogenous traps, the presence of the lipid radical formed during peroxidation, by yielding unique and stable products. EPR, also known as Electron Spin Resonance (ESR) is at present the only analytical approach that permits the direct detection of free radicals. This technique reports on the magnetic properties of unpaired electrons and their molecular environment [9].

This chapter will be dedicated to overview the Fe-related alterations in oxidative metabolism in photosynthetic and non-photosynthetic organisms after experimental exposure to excess Fe employing different protocols of administration. Data assessing lipid peroxidation post-treatment both, as TBARS generation and/or EPR detection of lipid radicals, are reviewed in a wide range of biological systems.

2. Fe overload in aquatic organisms

Fe content in the upper earth's crust is around 6% [10]. The Fe concentration in sediments influences the Fe concentration in the associated surrounding seawater. However, the concentration of dissolved Fe (defined as Fe that can diffuse through a membrane of less than 0.45 μm) in open-oceanic waters is extremely low (< 56 ng/l) [11]. Natural parameters that augment the Fe levels in coastal and central oceanic areas are: aeolian deposition of dust, river discharge, washout of dust particles in the atmosphere by rainfall, ground water discharge, glacial melting, volcanic sediments, coastal erosion and up-welling of Fe-rich deep waters over hydrothermal vents [12]. Human activities also have a great impact on Fe levels, especially around coastal areas. Chemical and mining industries, disposal of waste metal, ports, aeolian deposition of atmospheric dust from polluted areas, are some of the human activities bringing Fe and other metals to the marine ecosystem. Therefore waters from different regions may have different Fe concentrations. Fe was recognized as a bioactive element [13] and a deficiency in Fe had been suggested to limit primary productivity in some ocean regions [14,15]. Fe uptake is strictly required for phytoplankton development since the photosynthetic apparatus contains numerous loci for Fe. Moreover, it was pointed out that it is critical to avoid Fe overload in water with low organic matter content under aquarium conditions to prevent Fe-dependent toxicity [16].

Over a decade ago, Estévez et al. [17] studied the effect of *in vivo* Fe supplementation to the green algae *Chlorella vulgaris* in terms of the establishment of oxidative stress conditions. Growth under laboratory conditions increased with Fe availability up to 90 μM with increases in biomass, suggesting that Fe supply at concentrations lower than 90 μM could be considered limiting for algal growth. However, Kolber et al. [18] pointed out that in their field experiments in the equatorial Pacific, 2 days following Fe enrichment, photosynthetic energy conversion efficiency began to decline. It was also indicated that some algal cultures showed deleterious effects if exceeding an Fe threshold (14-28 μM) in unpolluted freshwater [16]. Between 90 and 200 μM Fe in *C. vulgaris* cultures, there was no effect on growth with increased Fe additions and further increases on Fe availability led to a drastic decrease in the growth of the cultures (Table 1). The increase of Fe at the intracellular level showed a linear dependence with the concentration of added Fe below 200 μM Fe, however concentrations between 200 and 500 μM Fe added to the medium led to a less active increase in intracellular Fe (Table 1), suggesting an intracellular control for Fe uptake. Thus, the data presented by Estévez et al. [17] under laboratory conditions suggested the possibility that excess Fe could be responsible for the decrease in *C. vulgaris* growth by inducing oxidative stress. Accordingly, when *C. vulgaris* cells were incubated with an EPR-spin trapping for lipid radicals (α-(4-pyridyl 1-oxide)-*N*-*t*-butyl nitrone, POBN), a POBN-spin adduct was observed. The spin adduct EPR spectra exhibit hyperfine splitting that were characteristic for POBN/lipid radicals, a_N = 15.56 G and a_H = 2.79 G, possibly generated from membrane lipids as a result of β-scission of lipid-alkoxyl radicals [19,20]. Quantification of lipid radical EPR signals in algal cells indicated that Fe supplementation significantly increased radical content in the membranes supplemented with the higher Fe dose, as compared to cells supplemented with 90 μM Fe (Table 1). These results indicate that lipid peroxidation was increased by Fe availability. In this context, even though an increased content of antioxidants has been detected in *C. vulgaris* cells exposed to increased Fe, the damaging potential of Fe excess in the cell did not seem to be efficiently controlled by the activity of the antioxidants [17].

Fe added (μM)	Chlorophyll content (μM)	Intracellular Fe (nmol (10^7 cell)$^{-1}$)	Lipid radicals (pmol (10^7 cell)$^{-1}$)
0	1.2	4 ± 1	nd
50	4.5*	18 ± 2*	nd
90	7.0*	28 ± 3*	6 ± 2
200	6.8*	62 ± 3*	nd
300	6.0*	65 ± 7*	nd
500	1.2	85 ± 10*	36 ± 9**

[1]Taken from [17]. nd stands for not determinated.
C. vulgaris cultures were supplemented with up to 500 μM Fe (EDTA:Fe, 2:1). Development was followed measuring chlorophyll content and each experimental value represents chlorophyll content of the cultures after 12 days of development. Intracellular Fe content as a function of the Fe addition to the incubation medium was spectrophotometrically measured. Data are expressed as means ± SE of 4-6 independent experiments, with two replicates in each experiment. Lipid radicals were detected and quantify by EPR.
*significantly different from value without Fe added, $p \leq 0.05$. ANOVA.
**significantly different from value in the presence of 90 μM Fe added, $p \leq 0.05$. ANOVA.

Table 1. Fe supplementation effect on *C. vulgaris* culture after 12 days of development[1]

It has been postulated that if as a result of ozone loss, UV-B flux at the surface of the earth increases, negative impacts on biological organisms will be inevitable since UV-B radiation causes a multitude of physiological and biochemical changes in photosynthetic organisms, probably related to oxidative stress [21,22]. Estévez el at. [17] exposed to 30 kJ/m² UV-B *C.*

vulgaris cells grown at up to 500 μM Fe. They observed that either 50 or 90 μM Fe did not alter significantly cell morphology. However, 30 kJ/m^2 UV-B exposure of algal cultures grown at 500 μM Fe affected cellular internal structure and there were no signs of cellular division. Exposure of *C. vulgaris* cells to 30 kJ/m^2 UV-B during lag phase did not significantly affect the content of lipid radicals in log phase of development under conditions of standard supplementation of Fe (90 μM) (Figure 1). This parameter was significantly increased by the addition of 500 μM Fe during development of the cultures in the absence of UV-B irradiation. Exposure of the cultures grown at 500 μM Fe to 30 kJ/m^2 UV-B during log phase led to a further increase in the content of lipid radicals in the membranes. In conclusion, even though exposure of *C. vulgaris* cells to UV-B under Fe standard concentration did not lead to cellular oxidative alterations, increase in Fe availability (500 μM Fe) was responsible for a substantial increase in lipid deterioration in the membranes by oxidative stress. These data strongly suggest that oxidative stress triggered by an excess content of Fe could affect cellular growth and have a negative biogeoimpact to phytoplankton when exposed to other environmental conditions.

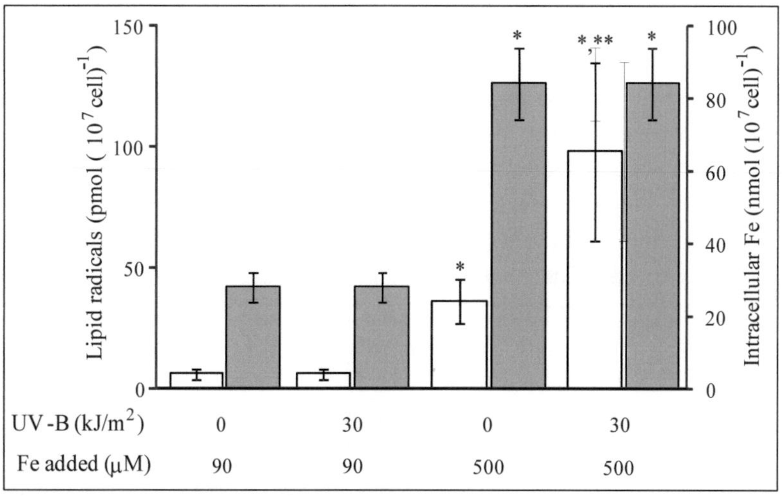

Algal cultures (2×10^5 cells ml^{-1}) were added with Fe starting day 0 of growth, and were assayed on day 12 (log phase). Data are expressed as means ± SE of 4-6 independent experiments, with 2 replicates in each experiment.

*significantly different from values for cells grown in the presence of 90 μM Fe without exposure to UV-B, $p \leq 0.05$. ANOVA.

**significantly different from values for cells added with 500 μM Fe without exposure to UV-B, $p \leq 0.05$. ANOVA. (Statview SE+, v 1.03, Abacus Concepts Inc, Berkeley, CA).

Lipid radicals were detected and quantified by EPR and intracellular Fe by the use of acid solutions to digest the cells which were measured spectrophotometrically after reduction with thioglycolic acid followed by the addition of bathophenanthroline.

Figure 1. Effect of Fe addition on UV-B-dependent lipid radical (□) and intracellular Fe content (■) in algae cells. Taken from [17].

Marine animals incorporate Fe bound to inorganic particles or to organic matter during food ingestion. Further, dissolved Fe is absorbed over the respiratory surfaces and mantle tissue in filter-feeding molluscs. The extrapallial water around these tissues is constantly exchanged with the surrounding seawater. Marine invertebrates are less tolerant of metal accumulation than vertebrates and can be affected at lower metal concentrations. Bivalves are widely used as sentinel organisms in marine pollution monitoring programs, due to their sessile and filtering habits, and their ability to bioaccumulate organic pollutants and metals in their tissues [23]. The exposure of marine molluscs to metals has been shown to induce oxidative stress through the formation of reactive O_2 species (ROS) and reactive nitrogen species (RNS), leading to lipid peroxidation. Bivalves have also been used as models for the study of the effect of Fe supplementation. Viarengo et al. [24] treated the mussel *Mytilus galloprovincialis* with 300-600 μg Fe/l (as $FeCl_3$) and observed a significantly Fe accumulation in the digestive gland (DG) (190 ± 25, 394 ± 131 and 412 ± 146 μg Fe/l in 0, 300 and 600 μg Fe/l supplemented animals, respectively). The TBARS content was measured in animals treated with 600 μg Fe/l, and a significant increase was observed among control and treated mussels. Lately, Alves de Almeida et al. [25] exposed mussels from *Perna perna* species to 500 μg/l Fe (as $FeSO_4$) and it was reported that mussels exposed to Fe for 12, 24 and 72 h presented increased phospholipid hydroperoxide glutathione peroxidase (PHGPx) activity, and no differences in MDA levels. However, at 120 h of Fe exposure both, MDA and PHGPx were significantly higher than control. Such increased MDA levels agree with previous findings by Viarengo et al. [24]. The negative correlation observed between PHGPx activity and MDA levels after Fe exposure, supports an interpretation that PHGPx protects tissues from lipid peroxidation. Thus, the exposure of mussels to Fe along with a concomitant increase in •OH formation would be involved in the modulation of PHGPx activity, however the precise mechanism remains unclear. Also, exposure of mussels to 500 μg/l of Fe caused no changes in other antioxidant enzymes such as glutathione *S*-transferase and glutathione peroxidase. These data suggest that PHGPx have a role in the susceptibility of DG of mussels against lipid peroxidation, and that exposure to transition metals such as Fe could lead mussels to stimulate PHGPx in order to prevent lipid peroxidation. Thus, the authors postulated that the evaluation of MDA levels in parallel with antioxidant defenses, such as PHGPx, could be considered as a potential new biomarker of toxicity associated with contaminant exposure in marine organisms.

Recently, González et al. [26] investigated the oxidative effects produced by the *in vivo* Fe exposure of the bivalve *Mya arenaria.* The soft shell clams were collected on an intertidal sand flat near Bremerhaven, Germany, and the bivalves were placed in small aquaria containing 500 μM Fe (EDTA:Fe, 2:1). Exposure to 500 μM Fe in natural seawater resulted in a significant increase in DG total Fe content (Table 2). After 2 days of exposure to Fe, TBARS content showed a significant increase by approximately 3.8-fold as compared to control values. This increase was followed by a decrease to control values at treatment day 7 and afterwards TBARS concentration increased constantly until day 17 (Table 2). The LIP in DG tissue increased on day 7 of exposure to high dissolved Fe concentration. By day 9, the LIP increase was accompanied by a significant induction of the oxidative stress signals, ROS and ascorbyl radical content and correlated with the final increase of TBARS content in tissues. Once the LIP has increased, the catalytically active Fe is able to efficiently catalyze Fenton [27,28] and Haber-Weiss reactions [29,30] and consistently and drastically accelerated accumulation of TBARS. Contrary, oxidative stress effects measured on day 2 of treatment cannot be attributed to a significant increase of the LIP, since neither total Fe content nor the LIP were enhanced over the initial values in the 0 day exposure group. However, the H_2O_2 scavenging antioxidant, catalase (CAT), increased after 2 days of treatment compared to controls (day 0) but the activity went back to control level on day 7 of exposure. Catalase activity was, however, increased again on day 9 of exposure compared to controls [26]. It was postulated that the initial phase of elevated oxidative stress, occurring before significant Fe accumulation could be attributed to indirect effects under the experimental exposure conditions. Metabolic rates were not measured, but it is possible that Fe exposure triggers an initial stress response including accelerated respiration as the animals are pumping to rid themselves of the inflowing Fe enriched seawater. H_2O_2 is a good candidate for triggering cellular responses since it is a stable species [27]. H_2O_2 diffuses freely into the tissue and leads the oxidative stress, and further increases causes oxidative damage, assessed as TBARS content. H_2O_2 induced oxidative stress may have triggered the endogenous antioxidant system in such a manner that by day 7 of exposure to Fe excess the TBARS content was reduced to the starting values. Even though the superoxide dismutase (SOD) activity was not changed, induction of other protective mechanisms, such as metallothioneins, might act as effective transient control of heavy metal effects during the initial phase of exposure [24,25].

Time (days)	Total Fe content (ng Fe/mg FW)	TBARS (pmol/mg FW)	LIP (ng LIP/mg FW)
0	39 ± 4	57 ± 8	3.8 ± 0.4
2	48 ± 8	218 ± 14***	5.3 ± 1.3
7	42 ± 6	75 ± 13	7.2 ± 0.3*
9	66 ± 4**	157 ± 14***	14.2 ± 1.1*
17	106 ± 3**	226 ± 20***	10.4 ± 0.7*

[2]Taken from [26].
*significantly different from the value at day 0 with $p < 0.05$,
**$p < 0.01$ and
***$p < 0.001$. ANOVA.
Experimental bivalves were placed in small aquaria containing 13 l (1 l/animal) of natural seawater of 23-26‰ at 10°C, and 500 μM Fe (Fe:EDTA, 1:2).

Table 2. Fe supplementation effect on lipid peroxidation in *Mya arenaria*[2]

Other studies evaluate the impact of nutritional Fe on Fe level and concentrations of MDA in tissues. Baker et al. [31] analyzed the Fe in the diet of the African catfish, *Clurims gariepinus*. This fish model is of particular relevance when considering that *C. gariqinus* is typically cultured in earth-ponds, and these may be high in dissolved Fe content. Additionally, catfish may consume mud-burrowing organisms to supplement their diet, with incidental associated silt consumption, and therefore further metal loading. After 5 weeks of feeding the animals with a diet supplemented with Fe (6354.4 mg Fe/kg), the total

Fe content was measured in muscle, liver and blood-plasma and no significant differences with control animals were found, suggesting the possibility of efficient regulation of Fe status by the fish. MDA determination in tissues revealed that there was significantly more MDA in livers and hearts of fish fed high Fe diets than in controls, and no significant difference was found in skeletal-muscle. Values of MDA concentration were higher in Fe-stressed liver tissue comparative to other tissues, possible because hepatic tissue is lipid-rich making the liver a target organ for lipid peroxidation. The relative lack of response in skeletal muscle may have resulted from decreased abundance of polyunsaturated fatty acids within this tissue, and these findings are consistent with those of Desjardins et al. [32].

All together these data show that Fe in aquatic ecosystems could be a major stressor having a main role in lipid peroxidation not only in unicellular species, such as algae, but also in higher organisms, such as invertebrates and vertebrates. These kind on analyses should be performed before consider ecological strategies which may involved Fe fertilization in seawater [33-35], to increase primary production in the oceans as an answer to global temperature increments. These actions may drastically modify marine communities in ocean layers triggering oxidative reactions, which should be properly considered due to the fact that Fe may be profitable or unfavorable, depending of its usefulness as a micronutrient or as a catalyzer of free radical reactions.

3. Fe overload in soybean seeds

Plants have developed several mechanisms to maintain fairly constant internal concentrations of mineral nutrients over a wide range of external concentrations. To avoid Fe-dependent oxidative cellular damage, Fe^{2+} is either incorporated into the mineral core of Ft [36] which is located exclusively in the plastids [37] or reoxidized by O_2 and chelated by organic acids [38]. Bienfait et al. [39] reported that plants grown on Fe-EDTA formed a substantial pool of free space Fe in the roots and that Fe could be mobilized upon Fe-free growth in order to be transferred to the leaves. During growth in water culture at pH 5 to 6, a free space pool of 500 to 1000 nmol/g FW was formed in roots of bean grown in the presence of Fe-EDTA 20 μM and a pool of 20 to 50 nmol/g FW in roots without Fe supplementation. Like Ft in the cell, the free space Fe^{3+} precipitate is not only an immobile result of a defensive action against an excessive Fe supply; the plant may also use it as storage form of Fe that can be mobilized [39]. Even more, Caro and Puntarulo [40] indicated that O_2 radical generation depends on total Fe content, however it could mostly reflect Fe content in the free space. In soybean, Fe^{3+} reduction is an obligatory step in Fe uptake, and this is probably true for all strategy I plants [41]. Both total Fe content and the *in vitro* rate of Fe reduction were higher in roots grown in the presence of exogenously added Fe (up to 500 μM) than in roots grown in absence of supplemented Fe (Table 3). However, no visual differences (e.g. evidence of damage) between any of the roots or growth (assessed as the fresh weight of the roots, 0.21 ± 0.01 g/root) have been observed at the studied range of Fe supplementation. Total Fe content in soybean roots exposed to 50 and 500 μM Fe-EDTA, was higher than in roots grown in absence of supplemented Fe

(Table 3) and lipid oxidation, assessed as the content of TBARS, were not significantly affected by Fe supplementation up to 500 μM, to the incubation medium (Table 3). However, Fe supplementation to the roots did affect α-tocopherol content that was significantly decreased in the homogenates and the microsomes isolated from roots supplemented with Fe, as compared with values in roots developed in absence of Fe [40]. These data suggest that *in vivo* Fe supplementation could increase O_2 radical generation in soybean roots that was adequately control.

	No added Fe	500 μM added Fe	Ref
Soybean roots			
Total Fe content (μg/g FW)	0.07 ± 0.01	0.15 ± 0.02*	[40]
Fe-EDTA reduction rate (nmol/min/mg prot)	1.4 ± 0.1	3.1 ± 0.6*	[40]
TBARS (nmol MDA eq/mg)	5.7 ± 0.7	5.7 ± 0.7	[40]
Soybean embryonic axes			
Total Fe content (nmol/mg DW)	1.3 ± 0.2	3.9 ± 0.8*	[42]
Fe-EDTA reduction rate (nmol/min/mg DW)	15 ± 1	22 ± 2	[42]
Ft (μg Ft/ g DW)	34 ± 11	27 ± 10	[42]
Ft Fe content (Fe atoms/molec Ft)	1054 ± 111	494 ± 103*	[42]
LIP (pmol/mg DW)	50 ± 10	310 ± 50*	[42]
TBARS (nmol MDA eq/mg)	0.4 ± 0.1	0.3 ± 0.1	[75]

*significantly different from values without Fe addition, $p \leq 0.05$. ANOVA.

Table 3. Fe supplementation effects in soybean after 24 h of incubation

Robello et al. [42] reported that total Fe content in soybean embryonic axes exposed to 500 uM Fe-EDTA was higher than in axes grown in absence of supplemented Fe after 24 h of incubation. However, neither Fe reduction rate nor growth assessed, either as the fresh weight or the dry weight of the embryonic axes, were significantly affected by Fe supplementation to the incubation medium. Membrane integrity was no affected by the supplementation with 50 and 500 μM Fe:EDTA (1:2) since electrolyte leakage at 24 h and 48 h of imbibition was not significantly different from electrolyte leakage found in non-supplemented Fe axes (15.3 ± 0.7 and 8.0 ± 0.3%, after 24 h and 48 h of incubation with 500 μM Fe, as compared to 12.4 ± 0.4 and 8.6 ± 0.6%, after 24 h and 48 h of incubation in the absence of added Fe, respectively). Moreover, as it was previously reported in soybean roots [43], Fe accumulation was not followed by Ft accumulation in soybean embryonic axes upon growth. Without any significant change in the content of Ft in the embryonic axes incubated for 24 h upon Fe supplementation, a 53% decrease in the Fe content per molecule of Ft was observed in the presence of 500 μM Fe (Table 3). These data differed from previous observations showing Fe induction of Ft synthesis and accumulation in soybean [44], however, the nature of the model employed by Lescure et al. [44], cells in suspension grown heterotropically, could alter the kinetic of the response. In this regard, it should not be discarded that a transient increase in Ft content could occur under these experimental conditions before 24 h of imbibition. The observed rapid decrease in Fe content per molecule of Ft, as compared to non-added Fe conditions, could reflect an early loosing of Ft molecules altered by free radicals, or a reduction of its capacity of binding Fe, or both. The increase in the protein sensitivity to proteases would lead to an early degradation, as compared to axes grown in a non-added Fe medium. The increased rate of ROS generation could be due to the significant increase in the LIP under conditions of Fe supplementation. However, it is important to point out that the substantial increase in the total Fe content in axes grown in the presence of 500 μM Fe for 24 h, as compared to seeds grown in non-added Fe medium, could not be allocated as the measured increase in the LIP that would represent only the 10% of the increase in the total Fe content. Besides the LIP critical importance as initiator of free radical reactions and the decisive requirement of keeping Fe concentration as low as possible to minimize cellular deterioration, the role of other soluble and insoluble Fe-storage proteins, the formation and contribution of Fe-nitrosyl complexes, glutathione, nitric oxide, etc. should be considered among other non-protein agents, as possible candidates to handle Fe transport and storage under stress conditions since TBARS content was not significantly affected in Fe overloaded soybean embryonic axes (Table 3). Beside the apoplastic space [45], Lanquar et al. [46] identified the vacuole as a major compartment for Fe storage in plant seeds and showed that retrieval of the Fe stored in vacuoles is an essential step for successful germination in a wide range of environments.

On the other hand, recently Simontacchi et al. [47] summarized assays performed to characterize lipid radical-dependent oxidation in photosynthetic organisms where EPR was successfully employed to evaluate not only lipidperoxidation but also to analyze the relative scavenging capacity of plant extracts, the effects of both, natural environmental challenges and oxidative stress situations, in several model and biological systems. Further studies should be oriented in this direction to explore the critical effect of Fe overload on radical-dependent pathways that play a major role in plant metabolism.

4. Fe overload in mammals

Fe overload in mammals has been often associated with injury, fibrosis, and cirrhosis in the liver followed by cardiac disease, endocrine abnormalities, arthropathy, osteoporosis and skin pigmentation [48]. Several mechanisms has been proposed whereby excess hepatic Fe causes cellular injury, but Fe-induced peroxidative injury to phospholipids of organelle membranes is a potential unifying mechanisms underlying the major theories of cellular injury in Fe overload [49]. With progressively increasing Fe deposition, the capacity to maintain Fe in storage forms is exceeded resulting in a transient increase in the hepatic LIP [50]. Moreover, Fe-catalyzed generation of ROS has been implicated in the pathogenesis of many disorders including atherosclerosis [51,52], cancer [53], ischaemia reperfusion injury [54,55] besides in Fe overload [56], such as haemochromatosis [57].

Several experimental models of Fe overload have been developed. In the dietary model used by Dabbagh et al. [58] rats were fed for 10 weeks a chow diet enriched with 3% (w/w) reduced pentacarbonyl Fe (a 99%, w/w, pure form of elemental Fe). Dietary Fe overload resulted in significant increases in hepatic Fe levels; with no difference in Fe content in serum (Table 4). Lipid peroxidation was assessed by measuring TBARS and F2-isoprostanes. The latter are a series of prostaglandin-F2-like compounds derived from the free-radical-catalyzed, non-enzymic peroxidation of arachidonic acid [59] and the *in vivo* levels of F2-isoprostanes have been shown to increase dramatically in acute hepatotoxicity [60]. Direct evidence for moderately increased lipid peroxidation products in liver was reported after dietary Fe overload. In addition to hepatic oxidative damage, Fe overload also caused changes in the plasma lipid profile. These data suggest that in this rat model of Fe overload, oxidative stress is associated with depletion of endogenous antioxidants in plasma and liver, and although no conclusive evidence for lipid peroxidation in plasma was found, hepatic F2-isoprostane levels were significantly increased in treated rats.

Experimental Fe overload in rats using dietary supplementation with carbonyl Fe is a well established model, where Fe deposition results mainly in the hepatocytes in a periportal distribution, as observed in idiopathic hemochromatosis [48]. Galleano and Puntarulo [61] used the dietary carbonyl-Fe model carried out on male Wistar rats that were fed during 6 weeks with either a) control chow diet, or b) control chow diet supplemented with 2.5% (w/w) carbonyl-Fe. Both, Fe and TBARS content, were increased in liver (Table 4). However, mild dietary Fe overload increased Fe content in plasma but did not lead to a significant increase in TBARS probably because Fe content after dietary Fe supplementation was increased less dramatically in plasma than in liver (88% and 15-fold, respectively), suggesting that plasma mechanisms for sequestering catalytically active Fe were fully operative (Table 4). Under these conditions, TBARS content in plasma does not seem to be a good indicator of oxidative stress conditions in the liver, and more sensitive techniques should be used in plasma to assess Fe-dependent oxidative stress.

Cockell et al. [62] used sucrose-based modified AIN-93G diets formulated to differ in Fe (35 mg/kg and 1500 mg/kg for control and Fe overloaded diets). Weanling male Long-Evans rats were fed these diets for 4 weeks and killed. Fe content was measured in plasma and liver. No differences in plasma between control and treated groups were found, meanwhile a significantly increase in liver between control and treated groups was observed. Since TBARS content in livers was significantly increased in Fe overloaded animals, hepatic Fe concentrations in this study were correlated positively with increases in TBARS. However, Fischer et al. [63] showed that Fe overloaded diets did not significantly alter other oxidative stress indices, such as DNA double-strand breaks or NF-κB activation despite observed increases in hepatic lipid peroxidation.

	Fe content		TBARS	
	control	Fe-overload	control	Fe-overload
Pentacarbonyl Fe, diet 3% (w/w)				
Liver [58]	104 ± 15(a)	1391 ± 242*(a)	-	-
Plasma [58]	134 ± 55(c)	124 ± 46(c)	nd	nd
Carbonyl Fe, diet 2.5% (w/w)				
Liver [61]	69 ± 16(a)	1091 ± 178*(a)	0.45 ± 0.05(b)	0.58 ± 0.01*(b)
Plasma [61]	179 ± 43(c)	336 ± 57*(c)	0.6 ± 0.1(d)	0.6 ± 0.2(d)
Sucrose-basemodified AIN-93G, diet 1500 mg/kg				
Liver[62]	218 ± 46(e)	895 ± 376**(e)	0.54 ± 0.07(b)	0.78 ± 0.19**(b)
Plasma [62]	2.72 ± 1.74(i)	3.82 ± 1.21(i)	-	-
Fe-dextran, ip 500 mg/kg				
Liver [68]	257 ± 11(e)	1837 ± 205*(e)	40 ± 1(f)	110 ± 30*(f)
Plasma [70]	126 ± 20(g)	1538 ± 158*(g)	0.7 ± 0.1(h)	2.7 ± 0.1*(h)
Kidney [49]	14 ± 3(e)	113 ± 15*(e)	29 ± 2(f)	37 ± 3*(f)

Letters indicate the units for each parameter as follows: (a) μg Fe/g FW; (b) nmol/mg prot; (c) μg/dl; (d) nmol/ml; (e) μg Fe/g DW; (f) pmol/min/mg prot; (g) μg Fe/dl; (h) nmol/l; (i) mg/l.
*significantly different from control values $p < 0.01$,
**$p < 0.001$, ANOVA.
nd stands for not-detectable.

Table 4. Fe effects in different organs and plasma employing several models of Fe overload

Fe-dextran treatment seems as a good model for the study of Fe toxicity resembling the pathological and clinical consequences of acute Fe overload in humans [48]. Fe supplied as Fe-dextran, is initially taken up by Kupffer cells, and when their storage capacity is exceeded the metal is accumulated by parenchymal cells producing a mild Fe overload. The increased Fe content alters the Kupffer cell functional status by inducing a progressive increase in macrophage-dependent respiration at earlier times after treatment. The effect is sensitive to macrophage inactivation by $GdCl_3$ pretreatment, decreases the respiratory response of the Kupffer cell to particle stimulation, plays a role in the development of liver injury, and seems to condition the impairment of hepatic respiration observed at later times after Fe overload [64]. Other pathological situations that increase oxidative conditions in the cell, could enhance Fe-dependent damage. As an example, hyperthyrodism increases the susceptibility of the liver to the toxic effects of Fe, which seems to be related to the development of a severe oxidative stress status in the tissue, thus contributing to the concomitant liver injury and impairment of Kupffer cell phagocytosis and particle-induced respiratory burst activity [65]. It was also shown that acute Fe overload was responsible for oxidative stress in rat testes with a concurrent decrease of antioxidant content [66,67]. The oxidative stress has been developed using Fe-dextran intra peritoneal (ip) administration as 500 mg/kg body weight and killed after 20 h.

Spontaneous organ chemiluminescence (CL) reflects the rate of lipid peroxidation reactions through the detection of the steady-state level of excited species and is considered to be an useful technique to evaluate oxidative stress *in vivo*. Galleano and Puntarulo [68] reported an association between Fe content and light emission in rats exposed to Fe-dextran after 2-6 h. Presumably, with progressively increasing Fe deposition, the capacity of maintaining Fe in storage forms is exceeded resulting in a transient increase in the hepatic LIP. However, at longer times (20 h) the significant increase in cytosolic Fe is limited, and CL goes back to control values. Moreover, cytochrome P_{450} inactivation is an early event and precedes other enzyme inactivation [68]. Data included in Table 4 show that liver Fe content was increased by 7-fold after 8 h of Fe-dextran administration, and TBARS generation rate was enhanced by 3-fold (6 h after ip) suggesting that liver is deeply affected by acute Fe-overload.

Mammalian red blood cells are particularly susceptible to oxidative damage because (i) being an O_2 carrier, they are exposed uninterruptedly to high O_2 tension, (ii) they have no capacity to repair their damaged components, and (iii) the haemoglobin is susceptible to autoxidation and their membrane components to lipid peroxidation. Red blood cells, however, are protected by a variety of antioxidant systems which are capable of preventing most of the adverse effects of oxidative stress, under normal conditions [69]. Galleano and Puntarulo [70] reported, employing the ip Fe-dextran model of Fe overload, that 20 h after Fe-dextran injection Fe concentration in plasma of treated rats showed approximately 12-fold increase, and TBARS content in plasma showed a 285% increase as compared to control values (Table 4). On the other hand, *in vitro* studies showed that Fe can stimulate the peroxidation of erythrocytes membrane lipids. Since red blood cells from Fe overloaded rats are continuously being exposed to an increase Fe content, no differences in TBARS content were detected in red blood cells from control rats as compared to erythrocytes from Fe overloaded rats, suggesting high resistance to oxidative stress of these cells.

Galleano et al. [71] also employed this model to comparatively studying Fe overload in kidney. Fe content in whole kidney was 8-fold increased (Table 4), and 5-fold increased in kidney mitochondria (16 ± 5 to 78 ± 1 nmol/mg prot for control and treated animals, respectively). Even thought TBARS content showed no significant differences after Fe administration, in Fe-treated rats TBARS production rate by kidney homogenates was higher in treated animals than in kidneys from control rats (Table 4). The authors suggested that Fe-dextran treatment does not affect kidney integrity, even though increases in lipid peroxidation rate occurs. α-tocopherol, one of the most efficient antioxidant in the hydrophobic phase, appeared to be effective in controlling Fe-dextran dependent damage in kidney.

Brain tissue is thought to be very sensitive to oxidative stress. Neurons are enriched in mitochondria and possess a very high aerobic metabolism, which makes these tissues susceptible to ROS-dependent damage than other organs. Moreover, low levels of some antioxidant enzymes, high contents of polyunsaturated fatty acids in brain membranes, and high Fe content may combine their effects to make the brain a preferential target for oxidative stress-related degeneration [72]. Maaroufi et al. [73] developed a chronic Fe overload model consisting in a daily 3 mg Fe/kg administrated in adult rats during 5 days. These treatments resulted, 16 days after treatment, in a significant Fe accumulation in the hippocampus, cerebellum, and basal ganglia. Lately, Maaroufi et al. [74] studied rats which received daily one ip injection of 3 mg $FeSO_4$/kg dissolved in sodium chloride 0.9% (or vehicle) during 21 consecutive days, and this accumulation was correlated to behavioral deficits. No increase levels of the TBARS content in different brain structures were observed in any brain region investigated. This observation suggested that chronic Fe administration had induced adaptive responses involving stimulation of the antioxidant defenses since, both SOD and CAT activities, were increased after treatment.

Thus, different forms and quantities of Fe administrated to rats, supplemented either as diets or ip, lead to an increase in Fe content in several tissues and plasma. This Fe increase seems to be associated with an increase in lipid peroxidation. The underlying mechanisms of tissue damage are unclear, but they probably depend on the Fe administration protocol. Even though lipid damage was observed in many cases after Fe overload, antioxidant capacity seems to play a crucial role in controlling the impairment mechanisms.

5. Concluding remarks

Fe metabolism is very complex since Fe is both, an essential element and a toxic compound that has to be carefully kept under a regulated concentration in a living cell. Toxic Fe activity is due to its ability of catalyzing free radical reactions. The most efficient Fe fraction to act as a free radical promoter is that forming the LIP. LIP content is the resultant of multiple dynamic equilibrium between the Fe incorporated to the cell, utilized and intracellularly stored. We have briefly reviewed the role of Fe on the oxidative damage to lipid membranes employing both *in vitro* and *in vivo* models of Fe overload in several biological systems. Much progress has still to be made in order to understand the nature and function of the LIP, the mechanisms of the Fe-catalyzed reactions *in vivo*, the contribution of Fe to oxidative stress and disease, and the development of appropriate chemotherapeutic strategies. Thus, alterations in Fe metabolism should be carefully analyzed before evaluating cellular responses to either damaging agents or xenobiotics of biomedical or ecological impact since Fe is a double-faced element that can be either good or bad to the cell, depending on whether it serves as a micronutrient or as a catalyst of free radical reactions.

Since a tight metabolic organization is required to successfully face oxidative external conditions in invertebrates, anthropogenic contamination with Fe could be toxic for animals that are adapted to their natural environment. As it could be understood from the data presented here, it is strongly suggested that natural habitats should be strictly preserved even though absolute Fe content did not seem to reach critical values to avoid cellular deterioration.

Mobilization of Fe stored in plant seeds is an essential step for germination in a wide range of environments. The analysis of these aspects would provide information that could be the key to understand Fe nutrition in plants, and will allow the designing and engineering of crop plants requiring minimal fertilizer input, contributing to a more ecological agricultural practice under optimal and sub-optimal environmental conditions avoiding reaching Fe overload conditions that would jeopardize successful plant development.

Moreover, therapeutic strategies should be designed to chelate either Fe from the LIP or Fe loosely bound to Ft to avoid Fe-related oxidative damage. Focus in chemical-related aspects of the Fe-chelator complexes should help to fulfill the new drugs designing expectances to control Fe toxicity in humans that through promoting lipid peroxidation could severely affect human health.

Author details

Paula M. González, Natacha E. Piloni and Susana Puntarulo*

Physical Chemistry-PRALIB, School of Pharmacy and Biochemistry, University of Buenos Aires, Junín, Buenos Aires, Argentina

Acknowledgement

This study was supported by grants from the University of Buenos Aires and CONICET. S.P. is career investigator from CONICET, and P.M.G. and N.P. are fellows from CONICET.

6. References

[1] Harrison PM, Arosio P (1996) The ferritins: molecular properties, iron storage function and cellular regulation. Biochim. Biophys. Acta 1275: 161-203.

[2] Crichton RR, Ward RJ (1992) Iron metabolism. New perspectives in view. Biochemistry 31: 11255-11264.

[3] Laulhère JP, Lescure AM, Briat JF (1988) Purification and characterization of ferritins from maize, pea, and soyabean seeds. J. Biol. Chem. 263: 10289-10294.

[4] Hyde BB, Hodge AJ, Kahn A, Birnstiel ML (1963) Studies on phytoferritin: I. identification and localization. J. Ultrastruct. Res. 9: 248-258.

[5] Dexter DT, Wells FR, Agid F, Agid Y, Lees AJ, Jenner P, Marsden D (1987) Increased nigral iron content in postmortem parkinsonian brain. Lancet 8569: 1219-1220.

[6] Ponka P (2004) Hereditary causes of disturbed iron homeostasis in the central nervous system. Ann. N.Y. Acad. Sci. 1012: 267-281.

[7] Chaston TB, Richardson DR (2003) Iron chelators for the treatment of iron overload disease: relationship between structure, redox activity, and toxicity. Am. J. Hematol. 73: 200-10.

[8] Limdi JK, Crampton JR (2004) Hereditary haemochromatosis. QJM 97: 315-324.

[9] Tarpey MM, Wink DA, Grisham MB (2004) Methods for detection of reactive metabolites of oxygen and nitrogen: in vitro and in vivo considerations. Am. J. Physiol. Regul. Integr. Comp. Physiol. 286R: 431-444.

[10] Wedepohl KH (1995) The composition of the continental crust. Geochim. Cosmochim. Acta 59 (7): 1217-1232.

---

* Corresponding Author

[11] Rue EL, Bruland KW (1995) Complexation of iron(III) by natural organic ligands in the Central North Pacific as determined by a new competitive ligand equilibration/adsorptive cathodic stripping voltammetric method. Mar Cehm. 50: 117-138.

[12] Watson AJ (2001) Iron limitation in the oceans. In: Turner DR, Hunter KA, editors. The biogeochemistry of iron in seawater. New York: Wiley and Sons. pp 85-121.

[13] Bruland, KW, Donat JR, Hutchins DA (1991) Interactive influences of bioactive metals on biological production in oceanic waters. Limnol. Oceanogr. 36: 1555-1577.

[14] Martin JH, Fitzwater SE, Gordon RM (1990) Iron deficiency limits phytoplankton growth in Antarctic waters. Global Biogeochem. Cycles 4: 5-12.

[15] Martin JH, Fitzwater SE, Gordon RM, Hunter CN, Tanner SJ (1993) Iron, primary productivity and carbon nitrogen flux studies during the JGOFS North Atlantic Bloom Experiment. Deep-Sea Res. 40: 115-134.

[16] Brand LE, Sunda WG, Guillard RRL (1983) Limitation of phytoplankton reproductive rates by zinc, manganese and iron. Limnol. Oceanogr. 28: 1182-1198.

[17] Estévez MS, Malanga G, Puntarulo S (2001) Iron-dependent oxidative stress in *Chlorella vulgaris*. Plant Sci. 161: 9-17.

[18] Kolber ZS, Barber R, Coale KH, Fitzwater SE, Greene RM, Johnson KS, Lindley S, Falkowski PG (1994) Iron limitation of phytoplankton photosynthesis in the equatorial Pacific Ocean. Nature 371: 145-149.

[19] North JA, Specto AA, Buettner GR (1992) Detection of lipid radicals by electron paramagnetic resonance spin trapping using intact cells enriched with polyunsaturared fatty acids. J. Biol. Chem. 267: 5743-5746.

[20] Jurkiewicz BA, Buettner GR (1994) Ultraviolet light-induced free radical formation in skin: an electron paramagnetic resonance study. Photochem. Photobiol. 59: 1-4.

[21] Malanga G, Puntarulo S (1995) Oxidative stress and antioxidant content in *Chlorella vulgaris* after exposure to ultraviolet-B radiation. Plant Physiol. 94: 672-679.

[22] Kozak RG, Malanga G, Caro A, Puntarulo S (1997) Ascorbate free radical content in photosynthetic organisms after exposure to ultraviolet-B. Recent Res. Devel. Plant Physiol. 1: 233-239.

[23] Goldberg ED (1975) The mussel watch: a first step in global marine monitoring. Mar. Poll. Bull. 6: 111-132.

[24] Viarengo A, Burlando B, Cavaletto M, Marchi B, Ponzano E, Blasco J (1999) Role of metallothionein agaist oxidative stress in the mussel *Mytilus galloprovincialis*. Am. J. Physiol. Regul. Integr. Comp. Physiol. 277: 1612-1619.

[25] Alves de Almeida E, Dias Bainy AC, de Melo Loureiro AP, Martinez GR, Miyamoto S, Onuki J, Barbosa LF, Machado Garcia CC, Manso Prado F, Ronsein GE, Sigolo CA, Barbosa Brochini C, Gracioso Martins AM, Gennari de Medeirosa MH, Di Mascio P (2007) Oxidative stress in *Perna perna* and other bivalves as indicators of environmental stress in the Brazilian marine environment: Antioxidants, lipid peroxidation and DNA damage. Comp. Biochem. Physiol. 146(4)A: 588-600.

[26] González PM, Abele D, Puntarulo S (2010) Exposure to excess of iron in vivo affects oxidative status in the bivalve *Mya arenaria*. Comp. Biochem. Physiol. C 152: 167-174.

[27] Boveris A (1998) Biochemistry of free radicals: from electrons to tissues. Medicina 54: 350-356.

[28] Pierre JL, Fontecave M (1999) Iron and activated oxygen species in biology: the basic chemistry. Biometals 12: 195-199.

[29] Rauen U, Petrat F, Tongju L, De Groot H (2000) Hypothermia injury/cold induced apoptosis -evidence an increase of chelatable iron causing oxidative injury in spite of low O_2/H_2O_2 formation. FASEB J. 14: 1953-1964.

[30] Livingstone DR (2001) Contaminant-stimulated reactive oxygen species production and oxidative damage in aquatic organisms. Mar. Pollut. Bull. 42: 656-666.

[31] Baker RTM, Martin P, Davies SJ (1997) Ingestion of sub-lethal levels of iron sulphate by African catfish affects growth and tissue lipid peroxidation. Aquat. Toxicol. 40: 51-61.

[32] Desjardins LM, Hicks BD, Hilton JW (1987) Iron catalysed oxidation of trout diets and its elect on the growth and physiological response of rainbow trout. Fish Physiol. Biochem. 3: 173-182.

[33] Martin JH, Coale KH, Johnson KS, Fitzwater SE, Gordon RM, Tanner SJ, Hunter CN, Elrod VA, Nowicki JL, Coley TL, Barber RT, Lindley S, Watson AJ, Van Scoy K, Law CS, Liddicoat MI, Ling R, Stanton T, Stockel J, Collins C, Anderson A, Bidigare R, Ondrusek M, Latasa M, Millero FJ, Lee K, Yao W, Zhang JZ, Friederich G, Sakamoto C, Chavez F, Buck K, Kolber Z, Greene R, Falkowski P, Chisholm SW, Hoge F, Swift R, Yungel J, Turner S, Nightingale P, Hatton A, Liss P, Tindale NW (1994) Testing the iron hypothesis in ecosystems of the equatorial Pacific Ocean. Nature 371: 123-129.

[34] Gervais F, Riebesell U, Gorbunov MY (2002) Changes in primary productivity and chlorophyll a in response to iron fertilization in the Southern Polar Frontal Zone. Limnol. Oceanogr. 47(5): 1324-1335.

[35] Blain S, Quéguiner B, Armand L, Belviso S, Bombled B, Bopp L, Bowie A, Brunet C, Brussaard C, Carlotti F, Christaki U, Corbière A, Durand I, Ebersbach F, Fuda J-L, Garcia N, Gerringa L, Griffiths B, Guigue C, Guillerm C, Jacquet S, Jeandel C, Laan P, Lefèvre D, Lo Monaco C, Malits A, Mosseri J, Obernosterer I, Park Y-H, Picheral M, Pondaven P, Remenyi T, Sandroni V, Sarthou G, Savoye N, Scouarnec L, Souhaut M, Thuiller D, Timmermans K, Trull T, Uitz J, van Beek P, Veldhuis M, Vincent D, Viollier E, Vong L, Wagener T (2007) Effect of natural iron fertilization on carbon sequestration in the Southern Ocean. Nature 446: 1070-1074.

[36] Laulhère JP, Laboure AM, Briat JF (1990) Photoreduction and incorporation of iron into ferritins. Biochem. J. 269: 79-84.

[37] Seckbach, J. (1968) Studies on the deposition of plant ferritin as influenced by iron supply to iron-deficient beans. J. Ultrastruct. Res. 22: 413-423.

[38] Cataldo DA, McFadden KM, Garland TR, Wildung RE (1988) Organic constituents and complexation of nickel (II), iron (III), cadmium (II), and plutonium (IV) in soybean xylem exudates. Plant Physiol. 86: 734-739.

[39] Bienfait HF, Van Den Briel W, Mesland-Mul NT (1985) Free space iron pools in roots. Generation and mobilization. Plant Physiol. 78: 596-600.

[40] Caro A, Puntarulo S (1996) Effect of *in vivo* iron supplementation on oxygen radical production by soybean roots. Biochim. Biophys. Acta 1291: 245-251.

[41] Schaedle M, Basshamb JA (1977) Chloroplast Glutathione Reductase. Plant Physiol. 59(5): 1011-1012.

[42] Robello E, Galatro A, Puntarulo S (2007) Iron role in oxidative metabolism of soybean axes upon growth. Effect of iron overload. Plant Sci. 172: 939-947.

[43] Lobréaux S, Briat JF (1991) Ferritin accumulation and degradation in different organs of pea (*Pisum sativum*) during development. Biochem. J. 274: 601-606.

[44] Lescure AM, Massenet O, Briat JF (1990) Purification and characterization of an iron induced ferritin from soybean (*Glycine max*) cell suspensions. Biochem. J. 272: 147-150.

[45] Briat JF, Lobréaux S (1997) Iron transport and storage in plants. Trends Plant Sci. 2: 187-193.

[46] Lanquar V, Lelièvre F, Bolte S, Hamès C, Alcon C, Neumann D, Vansuyt G, Curie C, Schröder A, Krämer U, Barbie-Brygoo H, Thomine S (2005) Mobilization of vacuolar iron by AtNRAMP3 and AtNRAMP4 is essential for seed germination on low iron. EMBO J. 24: 4041-4051.

[47] Simontacchi M, Buet A, Puntarulo S (2011) The use of electron paramagnetic resonance (EPR) in the study of oxidative damage to lipids in plants. In: Catalá A editor. Lipid Peroxidation: Biological Implications. Kerala, India: Res. Signpost Transworld Res. Network. pp 141-160.

[48] Puntarulo S (2005) Iron, oxidative stress and human health. Molec. Asp. Med. 26: 299-312.

[49] Galleano M, Puntarulo S (1994) Mild iron overload effect on rat liver nuclei. Toxicology 93: 125-134.

[50] Galleano M, Simontacchi M, Puntarulo S (2004) Nitric oxide and iron. Effect of iron overload on nitric oxide production in endotoxemia. Molec. Asp. Med. 25: 141-154.

[51] Heinecke JW, Rosen H, Chait A (1984) Iron and copper promote modification of low density lipoprotein by human arterial smooth muscle cells in culture. J. Clin. Invest. 74: 1890-1894.

[52] Salonen JT, Nyyssonen K, Korpela H, Tuomilehto J, Seppanen R and Salonen R (1992) High stored iron levels are associated with excess risk of myocardial infarction in eastern Finnish men. Circulation 86: 803-811.

[53] Loeb LA, James EA, Waltersdorph AM, Klebanoff SJ (1988) Mutagenesis by the autoxidation of iron with isolated DNA. Proc. Natl. Acad. Sci. U.S.A. 85: 3918-3922.

[54] Aust SD, White BC (1985) Iron chelation prevents tissue injury following ischemia. Free Radic. Biol. Med. 1: 1-17.

[55] Katoh S, Toyama J, Kodama I, Akita T, Abe T (1992) Deferoxamine, an iron chelator, reduces myocardial injury and free radical generation in isolated neonatal rabbit hearts subjected to global ischaemia-reperfusion. J Mol Cell Cardiol. 24(11): 1267-1275.

[56] Burkitt M J and Mason R P (1991) Direct evidence for in vivo hydroxyl-radical generation in experimental iron overload: an ESR spin-trapping investigation. Proc. Natl. Acad. Sci. U.S.A. 88: 8440-8444.

[57] Bacon BR, Britton RS (1990) The pathology of hepatic iron overload: a free radical mediated process? Hepatology 11: 127-137.

[58] Dabbagh AJ, Mannion T, Lynch SM, Frei B (1994) The effect of iron overload on rat plasma and liver oxidant status in vivo. Biochem. J. 300: 799-803.
[59] Morrow JD, Harris TM, Roberts LJ (1990) Noncyclooxygenase oxidative formation of a series of novel prostaglandins: analytical ramifications for measurement of eicosanoids. Anal. Biochem. 184: 1-10.
[60] Morrow JD, Awad JA, Kato T, Takahashi K, Badr KF, Roberts LJ II , Burk RF (1992) Formation of novel non-cyclooxygenase-derived prostanoids (F2-isoprostanes) in carbon tetrachloride hepatotoxicity. An animal model of lipid peroxidation. J. Clin. Invest. 90: 2502-2507.
[61] Galleano M, Puntarulo S (1997) Dietary alpha-tocopherol supplementation on antioxidant defenses after in vivo iron overload in rats. Toxicology 124(1): 73-81.
[62] Cockell KA, Wotherspoon ATL, Belonje B, Fritz ME, Madère R, Hidiroglou N, Plouffe LJ, Nimal Ratnayake WM, Kubow S (2005) Limited effects of combined dietary copper deficiency/iron overload on oxidative stress parameters in rat liver and plasma. J. Nutr. Biochem. 16(12): 750-756.
[63] Fischer JG, Glauert HP, Yin T, Sweeney-Reeves ML, Larmonier N, Black MC (2002) Moderate iron overload enhances lipid peroxidation in livers of rats, but does not affect NF-kappaB activation induced by the peroxisome proliferator, Wy-14,643. J. Nutr. 132: 2525-2531.
[64] Tapia G, Troncoso P, Galleano M, Fernandez V, Puntarulo S, Videla LA (1998) Time course study of the influence of acute iron overload on Kupffer cell functioning and hepatotoxicity assessed in the isolated perfused rat liver. Hepatology 27: 1311-1316.
[65] Boisier X, Schon M, Sepulveda A, Cornejo P, Bosco C, Carrion Y, Galleano M, Tapia G, Puntarulo S, Fernandez V, Videla LA (1999) Derangement of Kuppfer cell functioning and hepatotoxicity in hyperthyroid rats subjected to acute iron overload. Redox Rep. 4: 243-250.
[66] Lucesoli F, Fraga CG (1995) Oxidative damage to lipids and DNA concurrent with decrease of antioxidants in rat testes after acute iron intoxication. Arch. Biochem. Biophys. 316: 567-571.
[67] Lucesoli F, Caliguri M, Roberti MF, Perazzo JC, Fraga CG (1999) Dose-dependent increase of oxidative damage in the testes of rats subjected to acute iron overload. Arch. Biochem. Biophys. 372: 37-43.
[68] Galleano M, Puntarulo S (1992) Hepatic chemiluminiscence and lipid peroxidation in mild iron overload. Toxicol. 76: 27-38.
[69] Chow CK (1992) Oxidative damage in the red cells of vitamin E-deficient rats. Free Radic. Res. Commun. 16: 247-258.
[70] Galleano M, Puntarulo S (1995) Role of antioxidants on the erythrocytes resistance to lipid peroxidation after acute iron overload in rats. Biochim. Biophys. Acta 1271: 321-326.
[71] Galleano M, Farre SM, Turrens JF, Puntarulo S (1994) Resistance of rat kidney mitochondrial membranes to oxidation induced by acute iron overload. Toxicology 88:141-149.

[72] Halliwell B (2006) Oxidative stress and neurodegeneration: where are we now? J. Neurochem. 97(6): 1634-1658.

[73] Maaroufi K, Ammari M, Jeljeli M, Roy V, Sakly M, Abdelmelek H (2009) Impairment of emotional behavior and spatial learning in adult Wistar rats by ferrous sulfate. Physiol. Behav. 96(2): 343-349.

[74] Maaroufi K, Save E, Poucet Sakly B, Abdelmelek H, Had-Aissouni L (2011) Oxidativestress and prevention of the adaptive response to chronic iron overload in the brain of young adult rats exposed to a 150 kilohertz electromagnetic field. Neuroscience 186: 39-47.

[75] Caro A, Puntarulo S (1995) Effect of iron-stress on antioxidant content of soybean embryonic axes. Plant Physiol. (Life Sci. Adv.) 14: 131-136.

Section 2

Evaluation of Lipid Peroxidation Processes

Chapter 5

Trends in the Evaluation of Lipid Peroxidation Processes

Mihaela Ilie and Denisa Margină

Additional information is available at the end of the chapter

http://dx.doi.org/10.5772/46075

1. Introduction

Oxidative stress occurs as a result of imbalance between the antioxidant and prooxidant systems acting at certain points in metabolic processes, in favor of the last. The oxidative stress, defined by H Sies following extensive research performed between 1981 and 1993, is the outcome of intense generation of reactive oxygen species (ROS), which are not counteracted by endogenous antioxidant molecules [Sies, 1985]. Current knowledge links many types of pathologies to oxidative damage; among them, most cited are atherosclerosis, diabetes mellitus, neurodegenerative disorders, cancers, rheumatic diseases, autoimmune disorders, etc. Figure 1, sometimes referred to as "oxidative stress wheel", presents the most important diseases in which oxidative stress is involved resulting in biochemical lessions

Free radicals are chemical species containing unpaired electrons, which can increase the reactivity of atoms or molecules. Free radicals are highly reactive and unstable, due to their impaired electrons; they can react locally, accepting or donating electrons, in order to become more stable. The reaction between a radical and a non-radical compound generally leads to the propagation of the radical chain reaction, and to an increasing generation of new free radicals. During biochemical processes that normally take place in living cells, many types of free radicals are generated: oxygen-, sulfur-, bromide- and chloride- centered species [Halliwell & Gutteridge, 2007]. The most common reported cellular free radicals are singlet oxygen ($^1\Sigma g^+O_2$), hydroxyl ($OH\cdot$), superoxide ($O_2^-\cdot$) and nitric monoxide ($NO\cdot$). Also, some other molecules like hydrogen peroxide (H_2O_2) and peroxynitrite ($ONOO^-$) (which are not free radicals from the chemical point of view, having all-paired electrons) are reported to generate free radicals in living organisms through various chemical reactions [Halliwell, 2006].

In this context, it is extremely important to evaluate the extent and rate of the lipid peroxidation process using different methods and experimental models, ranging from quantitative assay of lipoperoxides end products to the evaluation of changes in certain metabolic processes under the influence of pro-oxidative or antioxidative known substances. The present article aims at reviewing different techniques, methods and experimental models for the evaluation of lipid peroxidation that can be used in clinical research and in basic biochemical research as well. Simple, rapid, cost effective, and more elaborated, expensive methods are critically evaluated, presenting the advantages and limitations of each one. A special emphasis is given to fluorescent methods, which our team is frequently using to evaluate the lipid peroxidation processes.

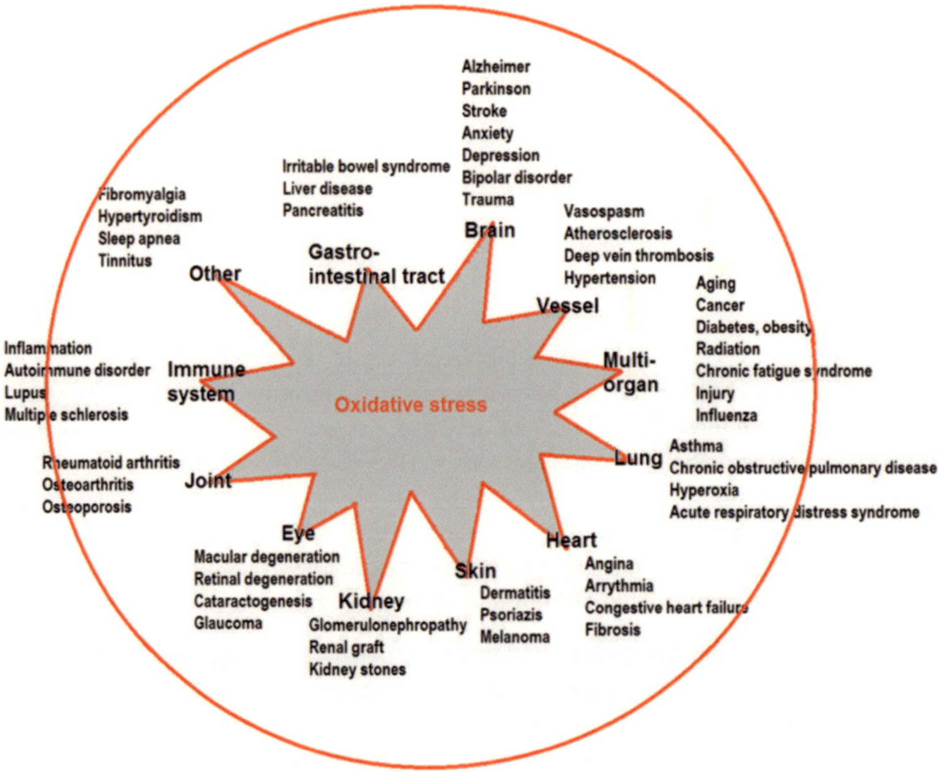

Figure 1. Implication of oxidative stress in pathology ("oxidative stress wheel")

2. Oxidative stress, ROS and implication in metabolic procesess

2.1. Oxygen centered reactive species (ROS)

Reactive oxigen species is a generic term that includes both oxygen radicals and certain non-radicals that are oxidizing agents and/or are easily converted into radicals, such as H_2O_2, ozone (O_3), singlet oxygen (Δg^1O_2), peroxynitrite, hypochlorous acid (HOCl), etc. [Halliwell, 2006].

ROS are generated as a result of oxygen action on nutrients or on physiological components in living organisms.

The sources for oxidative stress are either endogenous (abnormal mithocondria and peroxisomes function, lipoxygenase, NADPH-oxidase, cytochrome P450 activity), endogenous antioxidant systems dysfunction (low amount of non-enzymatic antioxidants such as gluthatione, vitamins A, C and E, reduced enzimatic activity) or exogenous agents (ultraviolet or ionizing radiation, toxins, chemotherapy, bacteria, etc.) (Figure 2).

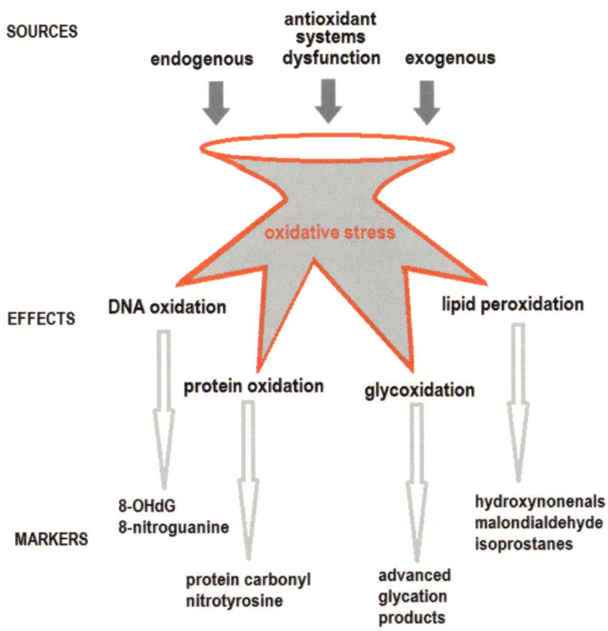

Figure 2. Sources, effects and main markers of oxidative stress

ROS can induce many damaging cellular processes, such as DNA oxidative lesions, loss of membrane integrity due to lipid peroxidation, protein and functional carbohydrate structural changes, etc. All these structural and functional changes have direct clinical consequences, leading to the acceleration of the general aging process, but also to some pathological phenomena, associated with the increase of the capillary permeability, impairment of the blood cell function, etc. ROS lesions are frequently associated with aging [Dröge & Schipper, 2007; Griffiths et al., 2011], atheroclerosis [Hulsmans & Holvoet, 2010], cardio-vascular disease [Dikalov & Nazarewicz, 2012; Puddu et al., 2009], type I or type II diabetes mellitus [Cai et al., 2004], autoimmune disorders, neurodegenerative disorders such as Parkinson [Yoritaka et al., 1996] or Alzheimer's disease [Sayre et al., 1997; Takeda et al., 2000], inflammatory diseases such as reumatoid arthritis [Griffiths et al., 2011] or different types of cancers [Lenaz, 2012; Li et al., 2009; Manda et al., 2009].

In order to counteract the damaging action of the physiologically generated ROS, the living organisms developed efficient antioxidant systems [Christofidou-Solomidou & Muzykantov 2006; Halliwell, 2006; Sies, 1997; Veskoukis et al., 2012]. Endogenous antioxidants in the human body act through different types of mechanisms:

- reducing the ROS generation – through chelating the metal ions (ferum, copper, etc.) by specific or non-specific proteins (ferritine, transferrine, albumin), so that these ions can no longer participate to redox reactions [Aruoma et al., 1989; Elroy-Stein et al., 1986; Freinbichler, 2011; Halliwell & Gutteridge, 1984; Velayutham, 2011]
- stopping the ROS formation chain reaction – generally via antioxidants with small molecular mass, such as reduced glutathion (GSH), vitamins E and C, uric acid, etc. [Gordon, 2012; Halliwell, 2006]
- scavenging ROS with antioxidant enzymes such as superoxyde dismutase (SOD), catalase (CAT), glutathion-peroxidase (GPx), glutathion-reductatse (GR), etc. [Halliwell, 2006; Sies, 1997]
- reparing lesions caused by ROS via specific enzymes such as endonucleases, peroxidases, lipases, etc. [Sies, 1997].

All these antioxidant systems act differently, depending on their structure and properties, their hydrophilic or lipophilic character, and also depending on their localization (intracellular or extracellular, in cell or organelles membrane, in the cytoplasm, etc.). All the aforementioned systems act sinergically and form a network which protects living cells from the destructive action of ROS (Figure 3).

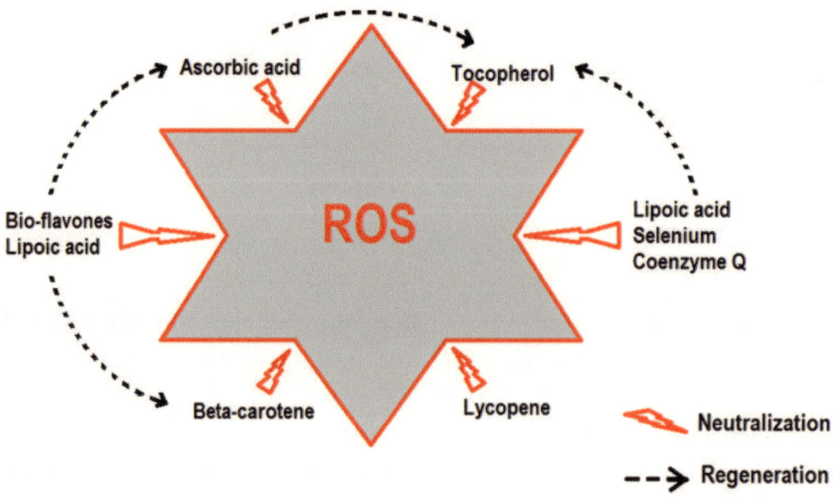

Figure 3. ROS neutralization by several biomolecules

2.2. Nitrogen centered radical species (RNS)

After the discovery of the physiological role of endogenously produced nitric oxide (NO), the capacity of this bio-molecule to react with other cellular components (such as proteins and lipids), specific nitrosative chemical changes have emerged as a key signaling mechanism in cell physiology. Several studies reported the involvement of excess generation of NO and its adducts in the etiology of multiple disease states, including insulin resistance and diabetes, atherosclerosis or Alzheimer's disease [Duplain et al., 2008; Parastatidis et al., 2007; Uehara, 2007; Yasukawa T et al., 2005; White et al., 2010].

The nitrosative modifications of proteins take two main forms: either S-nitrosylation of cysteine thiols or nitration of tyrosine residues. Both chemical processes may arise from protein interactions with NO or with secondary intermediates of NO, otherwise termed reactive nitrogen species (RNS) [White et al., 2010]. One of the very important members of the RNS group is represented by peroxynitrite ($ONOO^-$), produced from the reaction of NO with the superoxide anion (O^{2-}), which is considered as one of the major cellular nitrating agents [Hogg, 2002; White et al., 2010]. Other nitrating agents are the nitrosonium cation (derived from the action of myeloperoxidase), produced from the reaction of nitrite with hydrogen peroxide and nitroso-peroxocarbonate, which results from the reaction of carbon dioxide with peroxynitrite. Lipid peroxyl radicals have been recently shown to promote tyrosine nitration by inducing tyrosine oxidation and also by reacting with NO_2^- to produce · NO_2 [Bartesaghi et al., 2010; Denicola et al., 1996; Lang et al., 2000].

2.3. Oxidative stress and lipid peroxidation

Among the targets of ROS and RNS, lipids are basically the most vulnerable, as their peroxidation products can result in further propagation of free radical reactions [Halliwell & Chirico, 1992]. The brain is a high oxygen-consuming organ and the nervous cell has also the greatest lipid-to-protein ratio; besides, the brain has a relatively week protection systems against ROS generation, therefore it is particularly vulnerable to oxidative stress. The age-related increase in oxidative brain damage results in intense generation of lipid peroxidation products, protein oxidation, oxidative modifications in nuclear and mitochondrial DNA [Grimsrud et al., 2008].

Polyunsaturated fatty acids (PUFAs) and their metabolites have many physiological roles such as energy generation, direct involvement in cellular and sub-cellular membrane structure and function, implication in cell signaling processes and in the regulation of gene expression as well. They constitute the main target of ROS action in the lipid peroxidation reactions.

The general process of lipid peroxidation consists of three stages: initiation, propagation, and termination [Auroma et al., 1989; Leopold & Loscalzo, 2009]. Many species can be responsible for the initiation of the chain reaction the radicals: hydroxyl, alkoxyl, peroxyl, superoxide or peroxynitrite. As a consequence, a free radical atracts a proton from a carbon of a fatty acyl side chain leaving the remaining carbon radical accessible to molecular oxygen to form a lipid peroxyl radical. This is also highly reactive and the chain reaction is propagated further. As a result, PUFA molecules are transformed into conjugated dienes, peroxy radicals and hydroperoxides, which will undergo a cleavage mainly to aldehydes. More than 20 lipoperoxidation end-products were identified [Niki, 2009]; among the components of PUFAs oxidative degradation products, the most frequently mentioned were acrolein, malondialdehyde (MDA), 4-hydroxyalkenals and isoprostanes [Esterbauer et al., 1991; Leopold & Loscalzo, 2009].

RNS also play a major role in lipid biology, by two major pathways: targeting some enzymes (COX-2 and cytochrome P-450) and thus influencing bioactive lipid synthesis and interacting with unsaturated fatty acids (such as oleate, linoleate and arachidonate) and generating novel nitro-fatty acids. Nitrated lipids, such as nitroalkenes, may undergo aqueous decay and release independent of thiols, isomerize to a nitrite ester with N-O bond cleavage, or generate an enol group and ·NO [Leopold & Loscalzo, 2009]. The nitro-fatty acids have distinct bioactivities from their precursor lipids [Baker et al., 2009; Freeman et al., 2008; Kim et al., 2005; Lee et al., 2008; White et al., 2010]. Studies identified a high number of nitro-fatty acid species and proved an elevated formation of RNS in hydrophobic environments, such as the lipid bilayer, suggesting that lipids might constitute candidates for nitrosative signal transduction [Jain et al., 2008; Moller et al., 2005; Moller et al., 2007; Thomas et al., 2001].

Nitroalkenes may also participate in reactions with cysteine and histidine residues in proteins and with the thiolate anion of glutathione (GSH) to initiate reversible modification(s) of proteins. Thiyl radicals may also initiate lipid peroxidation by extraction of a hydrogen atom from bis-allylic methylene groups of fatty acids generating pentadienyl radicals. These radicals, in turn, may react with oxygen to generate peroxyl radicals [Leopold & Loscalzo, 2009].

Lipid hydroperoxides (LOOHs) are intermediates of PUFAs lipid peroxidation and can be also found as minor constituents of cell membranes; these compounds are also final products of prostaglandin and leukotriene biosynthesis, and can be decomposed by transition metals to form alkoxyl and lipoperoxyl radicals. Furthermore, biomolecules such as proteins or amino-lipids, can be covalently modified by lipid decomposition products (i.e. by forming Schiff bases with aldehydes or/and by activating membrane-bound enzymes). In consequence, lipid peroxidation may alter the arrangement of proteins in bilayers and thereby interfere with their physiological role in the membrane function.

2.4. Pathological involvement of the lipid peroxidation process

Many lipid peroxidation products, either full chain or chain-shortened, have been reported to be harmful or to have pro-infl ammatory effects [Birukov 2006; Niki 2009; Salomon 2005].

Lipid peroxidation increases the permeability of cellular membranes, resulting in cell death.

The lipid peroxidation process located at the cell membrane level may lead to loss of integrity and viability, and also to altered cell signaling and finally to tissue dysfunction; the oxidation of plasma lipoproteins is probably a major contributor to the formation of lipid peroxidation products and is widely thought to be involved in atherosclerosis. Malondialdehyde and 4-hydroxy-2-nonenal (HNE), the well known products of lipid peroxidation, react with a variety of biomolecules, such as proteins, lipids and nucleic acids, and they are thought to contribute to the pathogenesis of human chronic diseases [Breusing et al., 2010].

A clear example of the pathological role of PUFAs peroxidation is the evolution of atherosclerotic lesions to cardio-vascular disease. Until 1970, it was considered that dyslipidemia was the main factor initiating the atherosclerotic lesions. Later on, researchers emphasized the involvement of inflammatory processes, growth factors, smooth muscle cells proliferation, as well as viruses, bacteria or tumor phenomenon in the atherosclerosis, besides lipids and lipoproteins.

RNS interaction with different types of proteins is directly involved in physio-pathological processes. Several studies proved that key enzymes involved in glycolysis, β-oxidation, the tricarboxylic acid cycle and electron transport chain are targets of tyrosine nitration or S-nitrosylation. Modifications by nitrosylation that reduce the activity or function of important tricarboxylic acid cycle and electron transport proteins have the potential to slow substrate oxidation and probably lead to the build up of metabolic intermediates (particularly lipids) that could impair signaling pathways to reduce insulin action. Also, key insulin-signaling intermediaries are *S*-nitrosylated, and this could constitute a potential mechanism of insulin resistance [Chouchani et al., 2010].

3. Evaluation of the end products of lipid peroxidation

Malonidialdehyde (MDA) is one of the most cited lipoperoxidation product originating from PUFAs. Several generation mechanisms have been proposed for MDA. Pryor & Stanley (1975) considered them as bicyclic endoperoxydes coming up from nonvolatiles MDA precursors, similar to prostaglandins. This mechanism was confirmed in 1983 by Frankel & Nef. Two other mechanisms were postulated by Esterbauer (1991) and consist in the successive generation of peroxydes and β-cleavage of the lipid chain or as a reaction of acroleine radical with a hydroxyl moiety (exemplification for arachidonic acid in Figure 4, adapted from Esterbauer, 1991). Hecker & Ulrich (1989) consider that MDA can be generated *in* vivo by means of enzymatic processes linked to prostaglandins.

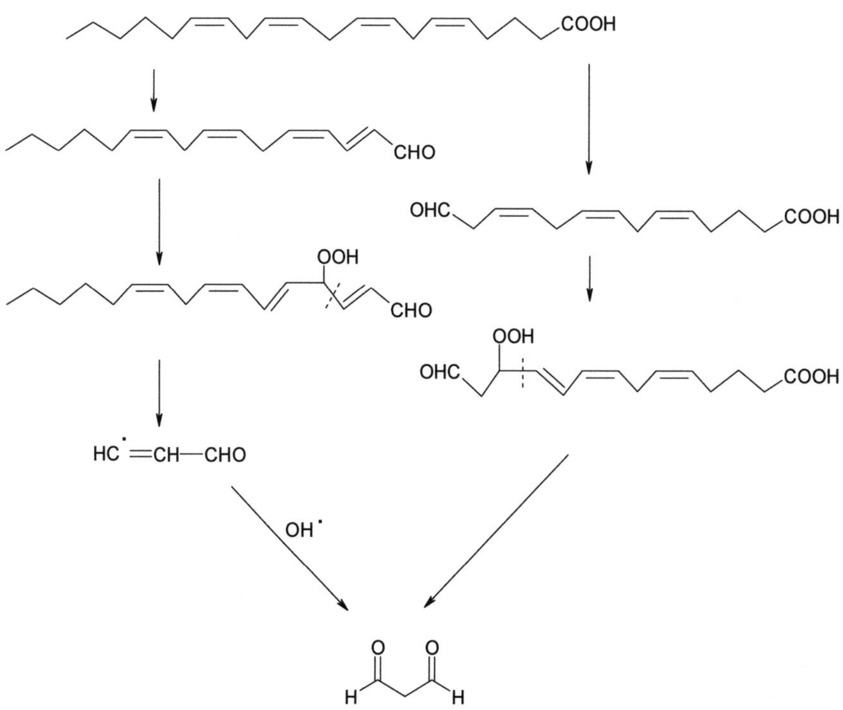

Figure 4. Generation of MDA as proposed by Esterbauer (1991)

Because the evaluation of the end products of lipid peroxidation at the tissue level is considered of maximum importance in both clinical and toxicological research, several methods to assay MDA have been proposed [Del Rio, 2005]. Since 1948, the most used method to assay lipoperoxidation end products is based on the studies of Bernheim et al., consisting in MDA condensation with the thiobarbituric acid (TBA) leading to a red complex which can be quantified by visible absorption spectrophotometry (in the range 500-600 nm, depending on the procedure used) or fluorescence spectroscopy (Figure 5). As TBA reacts also with several other aldehydes commonly present in the biological sample, the agents reacting with TBA are frequently denoted as thiobarbituric acid reacting species (TBARS). It must be said that TBARs and MDA are a rather imprecise measure of the lipid peroxidation process, since many substances that are present in human biological fluids can also react with TBA.

Moreover the reaction conditions (heating) can lead to the degradation of other molecules in the sample, increasing the amount of MDA that is available for the reaction with TBA. Therefore this assay usually gives an overestimation of free radical damage [Cherubini et al., 2005].

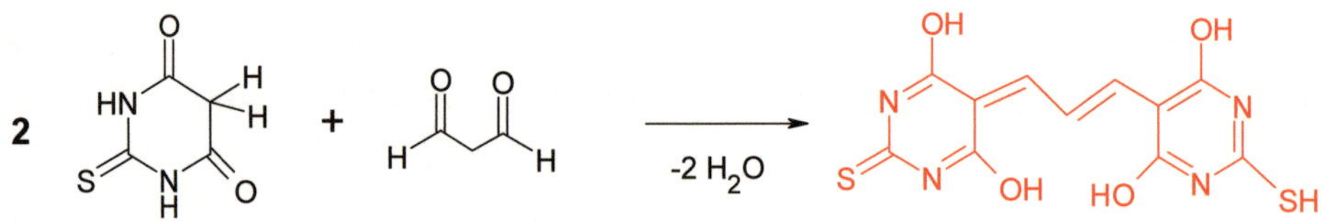

Figure 5. Mechanism of reaction for TBARS quantification

The authors generally avoid interferences by using different methods: Kwiecieñ et al., 2002, used BHT (butylated hydroxytoluene) to prevent further oxidation of the sample; deproteinisation was performed with trichloracetic acid [Cassini et al., 1986], forced the lipoperoxidation and stopped the reaction with SDS and acetic acid [Gautam et al., 2010; Ohkawa et al., 1979], added EDTA and refered the results to standards prepared from tetramethoxypropane [Houglum et al., 1990] or used the standard addition method [Sprinteroiu et al., 2010].

The specificity of the measurement is improved by HPLC to separate the MDA-TBA adduct from interfering chromogens [Agarwal & Chase, 2002; Del Rio et al., 2003; Lykkersfeldt, 2001; Templar et al., 1999].

The principles of TBARS assay is so popular, that a few companies even developed kits for clinical research to assay MDA spectrophotometrically from biological samples.

Apart from the above mentioned method, other methods were applied for the quantitative assay of lipoperoxidation end-products: direct HPLC [Karatas et al., 2002], capillary electrophoresis [Wilson et al., 1997], RP-HPLC, derivatisation with 2,4-diphenylhydrazine [Sim et al., 2003], pre-column derivatisation with diaminonaphtalene at acidic pH (for protein bound MDA) or alkaline pH (for non protein-bound MDA), followed by HPLC-UV analysis [Stegens, 2001], GC-MS analysis following derivatisation with phenylhydrazine [Cighetti et al., 2002], GC-ECD-MS after derivatisation with 2,4,6-trichlorophenylhydrazine [Stalikas & Konidari, 2001].

4. Evaluation of the antioxidant status

For the evaluation of the antioxidant status in biological samples several markers (either enzymatic or non-enzymatic) can be used. Generally, the results obtained from the evaluation of antioxidant status markers should be correlated with certain peroxidative parameters, in order to be able to draw conclusions from the experiments.

Among the enzymatic markers of the antioxidant defense mechanism of the biological samples, literature cites some specific enzymes, such as catalase, superoxide-dismutase, glutathione-reductase, glutathione-peroxidase, etc. Each of these enzymes can be assayed using specific kits.

Commonly used are also non-enzyimatic markers, such as reduced glutathione, some vitamins (ascorbic acid, tocopherols, carotenoides).

A series of commercial kits for measuring the antioxidant status in biological samples (blood, serum, but also food products) are also available - the so-called *total antioxidant status* kits. The assays can be colorimetric (one example is the kit using as a chromogen 2,2′-azino-di-[3-ethylbenzthiazole sulfonate], which reacts with methmyoglobin and hydrogen peroxide to give a coloured cation), chemiluminometric (using the reaction of luminol with hydrogen peroxyde), or physico-chemical (potentiometric).

These methods allow the evaluation of the total antioxidant capacity in the biological samples, thus accounting both for enzymatic and for non-enzymatic bio-molecules.

Our group developed a method enabling a distinct evaluation for the biological samples antioxidant capacity as resulting exclusively from redox hydrophilic biomolecules. The method is based on potentiometric evaluation of the status of oxidant and reducing species in samples of human serum. We used a micro Pt/AgCl combination redox electrode, with an internal reference, and a Tistand 727 Potentiometer (Metrohm AG, Switzerland). The baseline apparent redox potential of the human serum (ARP0) was measured; than a mild prooxidant chemical system (quinhydrone) was added to the biological samples. After incubating at 25°C for 1h respectively 3 hours, two final apparent redox potentials were recorded (ARPf). Quinhydrone mimics the prooxidant conditions developing *in vivo* and consumes the reducing species leading to an increase of the apparent redox potential in time. This dynamic recording of the data allowed the calculation of a difference between the final value of the ARP (ARPf) and the initial one (ARP0), thus defining the redox stability index (RSI). This parameter illustrates the serum sample capacity to counteract the prooxidant agent. The lower the RSI, the higher the activity of the hydrosoluble antioxidant protective systems in the serum [Margina et al., 2009].

5. Monitoring the induced-peroxidation process

J. Goldstein, M. Brown, and D, Steinberg emphasized more than a decade ago, that low density lipoproteins can be chemically modified, loosing their ability to be recognized by the classical LDL receptors [Brown & Jessup, 1999; Steinberg, 2009]. The oxidation of LDL does not take place in the blood stream, but inside the intimae, after lipoprotein complexes crossing through the endothelium. Therefore, in the oxidation process intracellular as well as extra cellular components are involved. This change of the LDL is realized mainly by oxidation with the free radicals that appear in large amounts in hypertension, diabetes mellitus, as a consequence of smoking, or in viral and bacterial infections, but LDL may be also modified by glycation (in type II diabetes mellitus), by association with proteoglycans or by incorporation in immune complexes [Ross, 1999].

Oxidised LDL particles (LDLox) have been proved to have different proatherogenic effects that can be predominantly attributed to their lipid components. Mainly, the uptake of LDLox by macrophages is enhanced through the scavenger receptor. Products generated from the decomposition of peroxidized lipids, such as aldehydes, modify the apolipoprotein B-100 (apoB-100) structure to a more electronegative form able to interact with the macrophage scavenger receptor. The modified LDL particles are taken up by scavenger receptors on the macrophages instead of the classical LDL receptors. This process is not regulated by feed back inhibition and allows the excessive build-up of cholesterol inside the cells, leading to their transformation into foam cells that are involved in the initiation and progression of atherosclerotic lesion [Parthasarathy et al., 1999; Shamir, 1996].

Alpha-tocopherol is an antioxidant from the LDL structure, and is the first one degraded during the radical attack on these lipoproteins. When this antioxidant protection is exhausted, PUFA are changed into lipid hydroperoxides. There is a great variability between subjects regarding the amount of PUFA and of antioxidants from LDL, which explains the variability concerning the susceptibility to oxidation of LDL particles. There are also a lot of other factors that influence *in vitro* evaluation of LDLox: some endogenous compounds, diet, some medicines, and probably genetic factors. Therefore, the assay of the *in vitro* LDL susceptibility to lipid peroxidation constitutes an important marker in the evaluation of atherogenic models/patients [Parthasarathy et al., 1999; Shamir, 1996].

The susceptibility of lipoprotein particles to lipid peroxidation can be assessed, after the isolation of LDL, either by treatment with copper salts, with mixtures of ferric compounds and ascorbic acid, or other prooxidant systems.

In order to evaluate this susceptibility to oxidation Esterbauer et al. proposed in 1989 a method based on the variation of absorbance of conjugated dienes in time at 234 nm. These dienes are relatively stable products resulted by a rearrangement of the double bonds from the PUFA molecules after the radical hydrogen abstraction. The increase in absorption at 234 nm is due to the formation of conjugated dienes during the peroxidation of polyunsaturated fatty acids. Absorbance at 234 nm shows an initial slower increase (lag phase) as antioxidants are destroyed and then increases more rapidly (propagation phase) reaching a plateau phase at which the absorbance is maximal as the rate of formation of dienes approaches their rate of decomposition [Esterbauer et al., 1990].

Another method for the assay of lipoprotein susceptibility to lipid peroxidation is based on the ability of lipid hydroperoxides to convert iodide (I^-) to iodine (I_2), which will than react with the iodide excess and form I_3^- that absorbs at 365 nm. There is a direct stoechiometric relationship between the amount of organic peroxides that resulted from the reaction and the concentration of I_3^- [Steinberg D., 1990, Esterbauer H. 1993]

One of the symplest methods for the assay of the susceptibility of LDL particles to lipid peroxidation is based on the selective precipitation of serum LDL with heparin at isoelectrical point (pH=5.4), method which is cheaper and quicker than the one based on ultracentrifugation. Personal results proved that the susceptibility of LDL to lipid peroxidation is correlated with the fasting plasma glucose level as well as with the lipid level [Margina et al., 2004].

The same type of assay can be used in order to evaluate the susceptibility to induced peroxidation for other kinds of biological samples (red blood cells, sub-cellular fractions such as mitochondria, or even tissue homogenates).

Previously published results [Margina D et al., 2011] proved that, for patients diagnosed with central obesity (BMI>30Kg/m2), adipose tissue susceptibility to lipid peroxidation correlated significantly with the total cholesterol (TC) level and with the LDL level. The susceptibility of adipose tissue to lipid peroxidation was assessed on white adipose tissue harvested from the abdominal area, homogenated in NaOH 0.015M, followed by TBARS evaluation. This parameter reflects the tendency to accumulation of free radicals in the adipose tissue of obese patients. In the same study, we pointed out that patients with impaired lipid profile (TC>220 mg/dl, LDL>150 mg/dl) had a significantly higher susceptibility of the adipose tissue to lipid peroxidation (p=0.036), associated with the decrease of the adiponectin level (Figure 6), compared to obese patients with physiologic lipid profile (TC<220mg/dl, LDL<150mg/dl).

Literature data also mention the assay of circulating LDLox, using different ELISA methods; one of these methods uses antibodies against a conformational epitope in the apolipoprotein B-100 (apoB-100) moiety of LDL that is generated as a consequence of substitution of at least 60 lysine residues of apoB-100 with aldehydes. This number of substituted lysines corresponds to the minimal number required for scavenger-mediated uptake of ox-LDL. The substituting aldehydes can be produced by peroxidation of lipids of LDL, leading to the generation of ox-LDL [Holvoet et al., 2006]. Another method might be the electrophoretic separation of LDL and LDLox particles from serum samples; studies proved that the electrophoretic mobility of oxidized LDL particles is increased compared to that of standard LDL [Lougheed et al., 1996].

Besides biochemical determination, noninvasive, real-time monitoring of lipid peroxidation using fluorescent probes has also been developed. The assays can be performed either on living cells (for example using cis-parinaric acid, fluoresceinated phosphoethanolamine, undecylamine-fluorescein, diphenyl-1-pyrenylphosphine – DPPP or other fluorescent

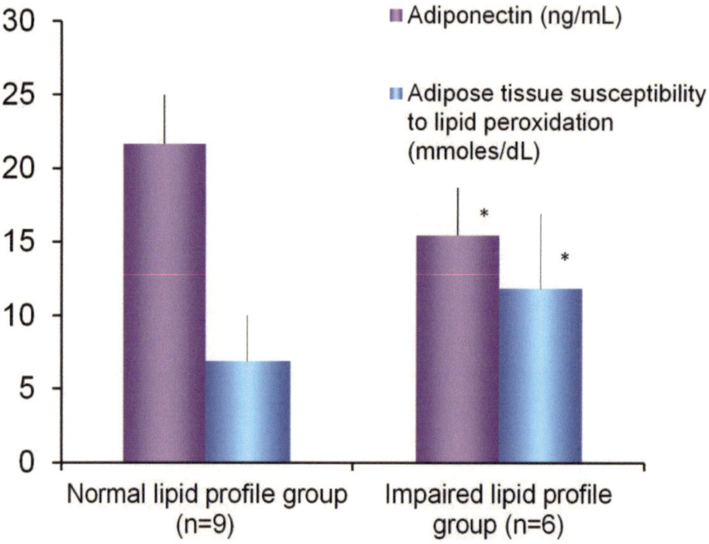

Figure 6. Obese patients with impaired plasma lipid profile (TC>220mg/dl, LDL>150mg/dl) are characterized by significantly different levels for the susceptibility of adipose tissue to induced lipid peroxidation as well as adiponectin level, compared to obese patients with normal plasma lipid profile (TC<220mg/dl, LDL<150mg/dl); * p<0.05 for cardio-vascular group compared to normal lipid profile group)

markers) or non-living samples (liposomes, tissue homogenates, plasma, serum, etc). In the cases of experiments that are performed on living cells, common limitation of use of fluorescent probes is that the probes often are cytotoxic or affect physiological activities of the cell [Drummen et al., 2004; Margina et al., 2012, Takahashi M. et al., 2001]. DPPP stoichiometrically reduces biologically generated hydroperoxides (such as fatty acid hydroperoxides, and triacylglycerol hydroperoxides) to their corresponding alcohols, and is transformed consequently into its oxide. DPPP is essentially non-fluorescent until oxidized to a phosphine-oxide by peroxides. Due to its solubility in lipids, DPPP intercalates into the membrane leaflets and reacts with lipid hydroperoxides, thus allowing the evaluation of peroxide formation in the membranes of live cells [Kawai et al.,, 2007]. Due to these chemical properties, the probe can be used in order to evaluate the extent of lipid peroxidation of biological materials such as cell membranes [Akasaka et al., 1993; Ohshima et al., 1996; Takahashi et al., 2001]. Because DPPP molecules are incorporated into the cell membranes, hydroperoxides located in the membrane are supposed to preferably react with DPPP.

Diphenyl-1-pyrenylphosphine (DPPP) is a synthetic compound with high reactivity against hydroperoxides, which has been used as a sensitive fluorescent probe for hydroperoxide analysis for HPLC methods. H_2O_2, which is least lipid-soluble, does not induce the peroxidation reaction of DPPP located in cell membranes. Although H_2O_2 is highly permeable to the membrane, it may not stay within the membrane long enough to react with DPPP effectively. Experimental lipid peroxidation can be induced by 10μM cumene hydroperoxide (CuOOH) which generates an effective reaction with DPPP in the membrane [Gomes et al., 2005; Takahashi et al., 2001].

We proved using DPPP in cell models (U937 human macrophage cell line as well as Jurkat lymphocytic cells) that the increase of certain polyphenol concentration (quercetin or epigallocatechin gallate) induces a decrease of the CuOOH induced lipid peroxidation of cell membranes. Fluorescence signals for the cells labeled with 5 μM DPPP are presented in Figure 7 [Margina et al, 2012].

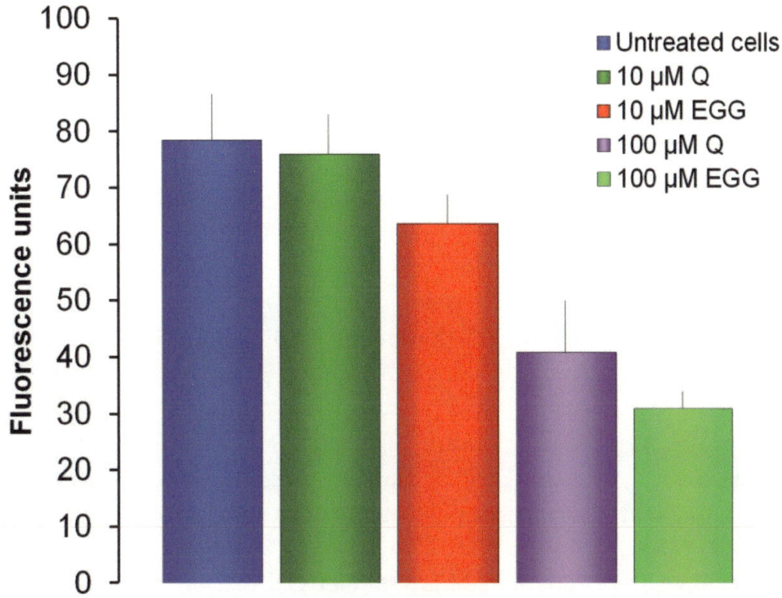

Figure 7. DPPP-oxide generation from Jurkat cell membrane before and after 20 minutes exposure to 10μM and 100 μM quercetin (Q) and epigallocatechin gallate (EGG) respectively; the fluorescence intensity was registered after 5 minutes of incubation of DPPP-labeled cells with CuOOH (for the induction of the lipid peroxidation)

6. Conclusions

Oxidative stress is among the most claimed causes of disease, as by its very definition indicates an abnormal biochemical function of the body. Among the targets of the oxidative stress, lipids are favorites, susceptible to structural changes that can decisively influence their normal function, and also generating hydroxyperoxides that further propagate the lipoperoxidation process.

Lipid peroxidation has been intensively studied in connection with normal and pathological metabolic processes; one of the main purposes was the understanding of toxicity triggered by lipoperoxidation end-products. That is why direct or indirect quantification of these products (TBARS, MDA, hydroxynonenals, prostaglandins, DNA and protein-adducts of the former, etc.) remains of interest for traditional and nowadays methods. The process of lipoperoxidation is often monitored in dynamics, even on living cells, using various techniques.

It is also of a great interest to seek for efficient antioxidants to prevent the excess lipoperoxidation, therefore one component of such kind of studies consist in finding out efficient markers for the oxidative stress and the antioxidant status. This can be fulfilled if a right and comprehensive understanding of the lipoperoxidation process is achieved. But this is still a faraway target.

Author details

Mihaela Ilie and Denisa Margină
Carol Davila University of Medicine and Pharmacy, Faculty of Pharmacy, Bucharest, Romania

Acknowledgments

The work was performed under the CNCSIS grant PD 132/30.07.2010 (PD 29/2010).

7. References

Agarwal R. & Chase S.D. (2002) Rapid, fluorimetric-liquid chromatographic determination of malondialdehyde in biological samples. *J Chromatogr B*, Vol. 775, No. 1, 121-126, ISSN 1570-0232.

Akasaka K., Ohrui H., Meguro H. & Tamura M. (1993). Determination of triacylglycerol and cholesterol ester hydroperoxides in human plasma by high-performance liquid chromatography with fluorometric postcolumn detection. *J. Chromatogr. B* Vol 617, No2, 205–211, ISSN 1570- 0232

Aruoma O.I., Halliwell B., Laughton M.J., Quinlan G.J. & Gutteridge J.M.C. (1989). The mechanism of initiation of lipid peroxidation. Evidence against a requirement for an iron(II)-iron(III) complex. *Biochem. J.* Vol. 258, No. 2, 617-620, ISSN 0264-6021

Baker P.R., Schopfer F.J., O'Donnell V.B. & Freeman B.A. (2009). Convergence of nitric oxide and lipid signaling: anti-inflammatory nitro-fatty acids. *Free Radic Biol Med* Vol 46, No 8, 989–1003, ISSN 0891-5849.

Bartesaghi S., Wenzel J., Trujillo M., Lopez M., Joseph J., Kalyanaraman B. & Radi R. (2010) Lipid peroxyl radicals mediate tyrosine dimerization and nitration in membranes. *Chem Res Toxicol* Vol 23, No 4, 821–835, ISSN 1520-5010.

Benkhai H., Lemanski S., Below H., Heiden J.U., Below E., Lademann J.,Bornewasser M., Balz T., Chudaske C. & Kramer A. (2010). Can physical stress be measured in urine using the parameter antioxidative potential? *GMS Krankenhaushyg Interdiszip.* Vol.5 No 2, Doc03.DOI: 10.3205/dgkh000146 ISSN 1863-5245

Bernheim F., Bernheim M.L.C. & Wilbur K.M. (1948). The reaction between thiobarbituric acid and the oxidation products of certain lipids. *J Biol Chem*, Vol. 174, 257-264, ISSN 0021-9258

Birukov KG. (2006) Oxidized lipids: the two faces of vascular inflammation. *Curr Atheroscler Rep* Vol 8, No 3, 223 – 231, ISSN 1534-6242.

Brown A.J. & Jessup W. (1999) Oxysterols and atherosclerosis. *Atherosclerosis* Vol. 142, 1–28, ISSN 0021-9150.

Cai W., He J.C., Zhu L., Peppa M., Lu C., Uribarri J. & Vlassara H. (2004). High Levels of Dietary Advanced Glycation End Products Transform Low-Density Lipoprotein Into a Potent Redox-Sensitive Mitogen-Activated Protein Kinase Stimulant in Diabetic Patients *Circulation*, Vol. 110, No. 3, 285-291, ISSN: 0009-7322.

Casini A.F, Ferrali M., Pompella A., Maellaro E. & Comporti M. (1986) Lipid Peroxidation and Cellular Damage in Extrahepatic Tissues of Bromobenzene-Intoxicated Mice. *Am J Pathol*, Vol. 123, 520-531, ISSN 0002-9440.

Cherubini A., Ruggiero C., Polidori M.C. & Mecocci P. (2005) Potential markers of oxidative stress in stroke. *Free Radic Biol Med* Vol. 39, No. 7, 841 – 852, ISSN 0891-5849

Chouchani E.T., Hurd T.R., Nadtochiy S.M., Brookes P.S., Fearnley I.M., Lilley K.S., Smith R.A. & Murphy MP. (2010) Identification of *S*-nitrosated mitochondrial proteins by *S*-nitrosothiol difference in gel electrophoresis (SNO-DIGE): implications for the regulation of mitochondrial function by reversible *S*-nitrosation. *Biochem J* Vol 430, No 1, 49–59, ISSN 0264-6021

Christofidou-Solomidou M. & Muzykantov V.R. (2006). Antioxidant Strategies in Respiratory Medicine, *Treat Respir Med* Vol. 5, No 1, 47-78, ISSN 1176-3450

Cighetti G., Allevi P., Anastasia L., Bortone L. & Paroni R. (2002). Use of methyl malondialdehyde as an internal standard for malondialdehyde detection: validation by isotope-dilution gas chromatography-mass spectrometry. *Clin Chem* Vol. 48, No. 12, 2266-2269, ISSN 1530-8561

Del Rio D., Pellegrini N., Colombi B., Bianchi M., Serafini M., Torta F, Tegoni M. & Musci M. (2003). Rapid fluorimetric method to detect total plasma malondialdehyde with mild derivatization conditions. *Clin Chem*, Vol. 49, No. 4, 690-692, ISSN 1530-8561

Denicola A., Freeman B.A., Trujillo M. & Radi R. (1996) Peroxynitrite reaction with carbon dioxide/bicarbonate: kinetics and influence on peroxynitrite mediated oxidations. *Arch Biochem Biophys* Vol 333, No 1, 49–58, ISSN 0003-9861.

Dikalov S.I. & Nazarewicz R.R. (2012). Angiotensin II-induced production of mitochondrial ROS: Potential mechanisms and relevance for cardiovascular disease. *Antioxid Redox Signal*. Mar 24. [Epub ahead of print], ISSN 1557-7716

Dröge W. & Schipper H.M. (2007). Oxidative stress and aberrant signaling in aging and cognitive decline. *Aging Cell* Vol. 6, No. 3, 361–370, ISSN 1474-9718.

Drummen G.P., Makkinje M., Verkleij A.J., Op den Kamp J.A., Post J.A. (2004) Attenuation of lipid peroxidation by antioxidants in rat-1 fibroblasts: comparison of the lipid peroxidation reporter molecules cis-parinaric acid and C11-BODIPY(581/591) in a biological setting. *Biochim Biophys Acta*. Vol. 1636, No. 2-3, 136-150, 0006-3002 .

Duplain H., Sartori C., Dessen P., Jayet P.Y., Schwab M., Bloch J., Nicod P. & Scherrer U. (2008) Stimulation of peroxynitrite catalysis improves insulin sensitivity in high fat diet-fed mice. *J Physiol* Vol 586, No 16, 4011–4016, ISSN 1469-7793

Eiserich J.P., Hristova M., Cross C.E., Jones A.D., Freeman B.A., Halliwell B. & van der Vliet A. (1998) Formation of nitric oxide-derived inflammatory oxidants by myeloperoxidase in neutrophils. *Nature* Vol 391, No 6665, 393–397, ISSN 1476-4687.

Elroy-Stein O., Bernstein Y. & Groner Y. (1986). Overproduction of human Cu/Zn-superoxide dismutase in transfected cells: extenuation of paraquat-mediated cytotoxicity and enhancement of lipid peroxidation. *EMBO J* Vol.5 No.3, 615-622, ISSN 0261-4189

Esterbauer H. (1993). Cytotoxicity and genotoxicity of lipid-oxydation products, *Am. J. Clin. Nutr.*, Vol 57, No 5, 779-785. ISSN 0002-9165.

Esterbauer H., Dieber-Rothender M., Waeg G., Striegl G. & Jurgens, G. (1990). Biochemical, structural, and functional properties of oxidized low-density lipoproteins. *Chem. Res. Toxicol.*, Vol. 3, No. 2, 77-92, ISSN 0893-228X .

Esterbauer H., Schaur R.J, & Zollner H. (1991). Chemistry and biochemistry of 4-hydroxynonenal, malonaldehyde and related aldehydes. *Free Radic Biol Med,* Vol. 11, No. 1, 81-128, ISSN 0891-5849.

Frankel E.N. & Neff W.E. (1983) Formation of malonaldehyde from lipid oxidation products, *Biochim Biophys Acta* Vol. 754, No. 3, 264-270, ISSN 0006-3002

Freeman B.A., Baker P.R., Schopfer F.J., Woodcock S.R., Napolitano A. & d'Ischia M. (2008) Nitro-fatty acid formation and signaling. *J Biol Chem* Vol 28, No 23, 15515–15519, ISSN 1083-351X.

Freinbichler W., Colivicchi M.A., Stefanini C., Bianchi L., Ballini C., Misini B., Weinberger P., Linert W., Vareŝlija D., Tipton K.F. & Della Corte L. 2011. Highly reactive oxygen species: detection, formation, and possible functions. *Cell Mol Life Sci.* Vol. 68 No.12, 2067-79, ISSN 1420-682X.

Gautam N., Das S., Mahapatra S.K., Chakraborty S.P., Kundu P.K. & Roy S. (2010). Age associated oxidative damage in lymphocytes. *Oxidative Medicine and Cellular Longevity* Vol. 3, No.4, 275-282, ISSN 1942-0900.

Gomes A., Fernandes E. & Lima J.L.F.C. (2005). Fluorescence probes used for detection of reactive oxygen species. *J. Biochem. Biophys. Methods* Vol 65, No 2-3, 45–80, ISSN 0165-022X

Gordon M.H. (2012). Significance of Dietary Antioxidants for Health. *Int. J. Mol. Sci.*, Vol. 13, No. , 173-179, ISSN 1422-0067.

Griffiths H.R., Dunston C.R., Bennett S.J., Grant M.M., Phillips D.C. & Kitas G.D. 2011. Free radicals and redox signalling in T-cells during chronic inflammation and ageing. *Biochem Soc Trans.* Vol. 39, No. 5, 1273-1278, ISSN 0300-5127.

Grimsrud P.A., Xie H., Griffin T.J & Bernlohr D.A. (2008). Oxidative Stress and Covalent Modification of Protein with Bioactive Aldehydes. *J Biol Chem* Vol. 283, No. 32, 21837–21841, ISSN 0021-9258.

Halliwell B. & Chirico S. (1993). Lipid peroxidation: its mechanism, measurement, and significance. *J Clin Nutr* Vol 57(suppl), No. 7, 155-255, ISSN 1938-3207.

Halliwell B. & Gutteridge J.M.C. (1984). Oxygen toxicity, oxygen radicals, transition metals and disease. *Biochem. J.* Vol. 219, No. 1, 1-14, ISSN 0264-6021

Halliwell B. (1992) Reactive oxygen species and the central nervous system. *J. Neurochem.* Vol. 59, No. 5, 1609-1623, ISSN 0022-3042.

Halliwell B.& Gutteridge J.M.C. (2007). *Free radicals in Biology and Medicine*, 4-th edition Oxford University Press, ISBN 978-0-19-856869-8, New York.

Halliwell, B. (2006). Reactive Species and Antioxidants. Redox Biology Is a Fundamental Theme of Aerobic Life. *Plant Physiology*, Vol. 141, No. 2, 312–322, ISSN 0981-9428.

Hecker M. & Ullrich V. (1989) On the mechanism of prostacyclin and thromboxane A2 biosynthesis. *J Biol Chem* Vol. 264, No. 1, 141-150, ISSN 1083-351X.

Hogg N. (2002). The biochemistry and physiology of S-nitrosothiols. *Annu Rev Pharmacol Toxicol* Vol 42, 585–600, ISSN 0362-1642.

Holvoet P., Macy E.,Landeloos M., Jones D., Nancy J.S.,Van de Werf F. & Tracy R.P. (2006). Analytical Performance and Diagnostic Accuracy of Immunometric Assays for the Measurement of Circulating Oxidized LDL. *Clinical Chemistry* Vol 52, No 4, 760-764, ISSN 1530-8561.

Houglum K., Filip M., Witztum J.L. & Chojkier M. (1990). Malondialdehyde and 4-Hydroxynonenal Protein Adducts in Plasma and Liver of Rats with Iron Overload. *J. Clin. Invest.* Vol. 86, 1991-1998, ISSN 0021-9738.

Hulsmans M. & Holvoet P. (2010). The vicious circle between oxidative stress and inflammation in atherosclerosis. *J Cell Mol Med* Vol. 14, No. 1-2, 70–78, ISSN 1422-0067.

Jain K., Siddam A., Marathi A., Roy U., Falck J.R. & Balazy M. (2008) The mechanism of oleic acid nitration by $\cdot NO_2$. *Free Radic Biol Med,* Vol 45, No 3, 269–283, ISSN 0891-5849.

Karatas F., Karatepe M. & Baysar A. (2002) Determination of free malondialdehyde in human serum by high-performance liquid chromatography, *Anal Biochem,* Vol 311, 76-79, ISSN 1096-0309.

Kawai Y., Miyoshi M., Moon J.H. & Terao J. (2007). Detection of cholesteryl ester hydroperoxide isomers using gas chromatography–mass spectrometry combined with thin-layer chromatography blotting. *Anal. Biochem.* Vol 360, No 1, 130–137, ISSN 1096-0309

Kim S.F., Huri D.A. & Snyder S.H. (2005) Inducible nitric oxide synthase binds, S-nitrosylates, and activates cyclooxygenase-2. *Science* Vol 310, No 5756, 1966–1970, ISSN 0036-8075.

Kwiecień S., Brzozowski T., Konturek P.Ch. and Konturek S.J. (2002). The role of reactive oxygen species in action of nitric oxide-donors on stress-induced gastric mucosal lesions. *Journal of Physiology and Pharmacology,* Vol. 53, No. 4, 761 -773, ISSN 0867-5910.

Lang J.D. Jr, Chumley P., Eiserich J.P., Estevez A., Bamberg T., Adhami A., Crow J. & Freeman B.A. (2000) Hypercapnia induces injury to alveolar epithelial cells via a nitric oxide-dependent pathway. *Am J Physiol Lung Cell Mol Physiol* Vol 279, No 5, L994–L1002, ISSN 1522-1504.

Lenaz G. (2012). Mitochondria and reactive oxygen species. Which role in physiology and pathology? *Adv Exp Med Biol.* Vol. 942, 93-136, ISSN 0065-2598.

Leopold J.A., & Loscalzo L. (2009). Oxidative Risk for Atherothrombotic Cardiovascular Disease. *Free Radic Biol Med.* Vol. 47, No. 12, 1673–1706, ISSN 0891-5849.

Liu Y & Schubert D.R. (2009). The specificity of neuroprotection by antioxidants. *J Biomed Sci* Vol. 16, 98-112, ISSN 1021-7770.

Lougheed M. & Steinbrecher U.P. (1996) Mechanism of Uptake of Copper-oxidized Low Density Lipoprotein in Macrophages Is Dependent on Its Extent of Oxidation, *J Biol Chem* Vol 271, No 20, 11798–11805, ISSN 1083-351X

Lykkersfeld, J. (2001). Determination of Malondialdehyde as Dithiobarbituric Acid Adduct in Biological samples by HPLC with fluorescence detection: Comparison with

Ultraviolet-Visible Spectrophotometry. *Clin Chem*, Vol. 47, No. 9, 1725-1727, ISSN 1530-8561.

Manda G., Nechifor M.T. & Neagu T.M. (2009). Reactive Oxygen Species, Cancer and Anti-Cancer Therapies. *Current Chemical Biology* Vol. 3, No. 1, 342-366, ISSN 1872-3136

Margina D, Gradinaru D., Panaite C., Cimponeriu D., Vladica M., Danciulescu R., Mitrea N. (2011) The association of adipose tissue markers for redox imbalance and the cardio-vascular risk in obese patients, *HealthMED* Vol. 5, No 1, 194-199, ISSN 1840-2291.

Margina D., Gradinaru D. & Mitrea N. (2004). Clinical study regarding the atherogenic properties of the oxidatively modified low-density lipoproteins. *Archives of the Balkan Medical Union*, Vol 39, No. 2, 78-82, ISSN 0041-6940

Margina D., Gradinaru D., Mitrea N. (2009) Development of a potentiometric method for the evaluation of redox status in human serum, *Revue Roumaine de Chimie,* Vol 54, No 1, 45–48, ISSN 0035-3930

Moller M., Botti H., Batthyany C., Rubbo H., Radi R. & Denicola A. (2005) Direct measurement of nitric oxide and oxygen partitioning into liposomes and low density lipoprotein. *J Biol Chem* Vol 280, No 10, 8850–8854, ISSN 1083-351X.

Moller M.N., Li Q., Lancaster J.R.Jr & Denicola A. (2007) Acceleration of nitric oxide autoxidation and nitrosation by membranes. *IUBMB Life*, Vol 59, No 4-5, 243–248, ISSN 1521-6551.

Niki E. (2009). Lipid peroxidation: physiological levels and dual biological effects. *Free Radic Biol Med*; Vol 47, No 5, 469 – 484. ISSN 0891-5849.

Ohkawa H., Ohishi N. & Yagi K. (1979). Assay for lipid peroxides in animal tissues by thiobarbituric acid reaction. *Anal Biochem* Vol. 95, No. 2, 351-358, ISSN1096-0309.

Ohshima T., Hopia A., German J.B. & Frankel E.N. (1996). Determination of hydroperoxides and structures by high-performance liquid chromatography with post-column detection with diphenyl-1-pyrenylphosphine. *Lipids* Vol 31, No 10, 1091–1096, ISSN: 1558-9307

Parastatidis I., Thomson L., Fries D.M., Moore R.E., Tohyama J., Fu X., Hazen S.L., Heijnen H.F., Dennehy M.K., Liebler D.C., Rader D.J. & Ischiropoulos H. (2007) Increased protein nitration burden in the atherosclerotic lesions and plasma of apolipoprotein A-I deficient mice. *Circ Res* Vol 101, No 4, 368–376, ISSN 0009-7330.

Parthasarathy S., Santanam N., Ramachandran S. & Meilhac O (1999) Oxidants and antioxidants in atherogenesis: an appraisal, *J. Lipid Res.* Vol 40, No 12, 2143–2157. ISSN 1539-7262.

Pryor W.A. & Stanley J.P. (1975). A suggested mechanism for the production of malonaldehyde during the autoxidation of polyunsaturated fatty acids. Nonenzymatic production of prostaglandin endoperoxides during autoxidation. *J Org Chem* Vol. 40, No. 24, 3615-3617, ISSN 1099-0690.

Puddu P., Puddu G.M., Cravero E., De Pascalis S & Muscari A. (2009). The emerging role of cardiovascular risk factor-induced mitochondrial dysfunction in atherogenesis *J. Biomed Sci* Vol. 16, 112, ISSN 1021-7770.

Rizzo M.A. & Piston D.W. (2003). Regulation of beta cell glucokinase by S-nitrosylation and association with nitric oxide synthase. *J Cell Biol* Vol 161, No 2, 243–248, ISSN 1540-8140.

Ross R. (1999) Atherosclerosis–an Inflammatory Disease, *NEJM,* Vol 340, No 2, 115-126, ISSN 1533-4406.

Salomon RG. (2005). Isolevuglandins, oxidatively truncated phospholipids, and atherosclerosis. *Ann NY Acad Sci* Vol 1043, 327 – 342, ISSN 1749- 6632.

Sayre L.M., Zelasko D.A., Harris P.L., Perry G., Salomon R.G. & Smith MA. (1997). 4-Hydroxynonenal-derived advanced lipid peroxidation end products are increased in Alzheimer's disease. *J. Neurochem.* Vol. 68, No. 5, 2092–2097, ISSN 0022-3042.

Shamir R., Johnson W., Morlock-Fitzpatrick K, Zolfaghari R., Ling Li, Mas E., Lombardo D., Morel D.W. & Fisher E.A. (1996). Pancreatic Carboxyl Ester Lipase: A Circulating Enzyme That Modifies Normal and Oxidized Lipoproteins In Vitro, *J Clin Invest* Vol 97, No 7, 1696-1704, ISSN 1365- 2362.

Sies H. (1997). Oxidative stress: oxidants and antioxidants. *Experim Phys* Vol. 82, No. 2, 291-295, ISSN 0144-8757

Sies, H. (1985). Oxidative stress: introductory remarks. In: *Oxidative Stress*. H. Sies, (Ed.). Academic Press, pp (1–7), ISBN 978-0-12-642760-8, London

Sim A.S., Salonikas C., Naidoo D. & Wilcken D.E. (2003). Improved method for plasma malondialdehyde measurement by high-performance liquid chromatography using methyl malondialdehyde as an internal standard. *J Chromatogr B* Vol. 785, No. 2, 337-344, ISSN 1570- 0232

Sprinteroiu M., Balalau D., Gutu C.M., Ilie M. & Petraru C. (2010). Quantitative analysis of malondialdehyde in normal human plasma using fluorescence and the standard addition method, *Farmacia* Vol. 58, No. 4, 509-514, ISSN 0014-8237

Stalikas C.D. & Konidari C.N. (2001). Analysis of malondialdehyde in biological matrices by capillary gas chromatography with electron-capture detection and mass spectrometry. *Anal Biochem* Vol. 290, No. 1, 108-115, ISSN 1096-0309

Steghens J.P., van Kappel A.L., Denis I. & Collombel C. (2001). Diaminonaphthalene, a new highly specific reagent for HPLC-UV measurement of total and free malondialdehyde in human plasma or serum. *Free Radic Biol Med,* Vol. 31, No. 2, 242-249, ISSN 0891-5849

Steinberg D. (1990). Arterial metabolism of lipoproteins in relation to atherogenesis. *Ann N Y Acad Sci,* Vol 598, 125-135, ISSN 1749-6632.

Steinberg, D. (2009). The LDL modification hypothesis of atherogenesis: an update. *J. Lipid Res.* Vol. 50 (S), S376–S381, ISSN 0022-2275.

Takahashi M., Shibata M. & Niki E. (2001). Estimation of lipid peroxidation of live cells using a fluorescent probe, Diphenyl-1-pyrenylphosphine. *Free Rad. Biol. Med.* Vol 31 No 2, 164-174, ISSN 1873-4596

Takeda A., Smith M.A., Avila J., Nunomura A., Siedlak S.L., Zhu X., Perry G., Sayre L.M. (2000). In Alzheimer's disease, heme oxygenase is coincident with Alz50, an epitope of tau induced by 4-hydroxy-2- nonenal modification. *J. Neurochem.* Vol. 75, No. 3, 1234–1241, ISSN 0022-3042.

Templar J., Kon S.P., Milligan T.P., Newman D.J. & Raftery M.J. (1999). Increased plasma malondialdehyde levels in glomerular disease as determined by a fully validated HPLC method. *Nephrol Dial Transplant,* Vol. 14, No. 4, 946-951, ISSN 0931-0509.

Thomas D.D., Liu X., Kantrow S.P. & Lancaster J.R. Jr. (2001) The biological lifetime of nitric oxide: implications for the perivascular dynamics of NO and O2. *Proc Natl Acad Sci USA* Vol 98, No 1, 355–360, ISSN 1091-6490.

Uehara T. (2007) Accumulation of misfolded protein through nitrosative stress linked to neurodegenerative disorders. *Antioxid Redox Signal* Vol 9, No 5, 597–601, ISSN: 1557-7716.

Velayutham M., Hemann C. & Zweier J.L. (2011). Removal of H_2O_2 and generation of superoxide radical: role of cytochrome c and NADH. *Free Radic Biol Med*. Vol. 51, No. 1, 160-170, ISSN 0891-5849.

Veskoukis A.S., Tsatsakis A.M. & Kouretas D. (2012). Dietary oxidative stress and antioxidant defense with an emphasis on plant extract administration. *Cell Stress and Chaperones* Vol. 17, No. 1, 11–21, ISSN 1355-8145

White P.J., Charbonneau A., Cooney G.J. & Marette A. (2010). Nitrosative modifications of protein and lipid signaling molecules by reactive nitrogen species, *Am J Physiol Endocrinol Metab,* Vol 299, No 6, E868-E878, ISSN: 0193-1849.

Wilson D.W., Metz H.N., Graver L.M. & Rao P.S. (1997). Direct method for quantification of free malondialdehyde with high-performance capillary electrophoresis in biological samples. *Clin Chem.* Vol. 43, No. 10, 1982-1984, ISSN 1530-8561

Yasukawa T., Tokunaga E., Ota H., Sugita H., Martyn J.A. & Kaneki M. (2005) S-nitrosylation-dependent inactivation of Akt/protein kinase B in insulin resistance. *J Biol Chem* Vol 280, No 9, 7511–7518, ISSN 1083-351X

Yoritaka A., Hattori N., Uchida K., Tanaka M., Stadtman E.R & Mizuno Y. (1996). Immunohistochemical detection of 4-hydroxynonenal protein adducts in Parkinson disease. *PNAS* Vol. 93, No. 7, 2696–2701, ISSN 1091-6490.

Chapter 6

Automation of Methods for Determination of Lipid Peroxidation

Jiri Sochor, Branislav Ruttkay-Nedecky, Petr Babula, Vojtech Adam, Jaromir Hubalek and Rene Kizek

Additional information is available at the end of the chapter

http://dx.doi.org/10.5772/45945

1. Introduction

Free radicals are atoms or molecules having one (or rarely more) free electron(s). These compounds may attack most of the (bio)molecules in organisms, which leads to the oxidative stress, which belongs to the causes of pathological processes in organisms [1-6]. Oxidative stress occurs in a situation, when the imbalance between the production of free radicals and effectiveness of antioxidant defence system occurs in a healthy organism. Determination of antioxidant activity or eventually markers directly connected with this variable is one way how to monitor the damage of organisms by these compounds [7-14]. The negative effect of free oxygen radicals consists in the lipid peroxidation. This type of peroxidation is a chemical process, in which unsaturated fatty acids of lipids are damaged by free radicals and oxygen under lipoperoxides formation. Lipoperoxides are unstable and decompose to form a wide range of compounds including reactive carbonyl compounds, especially certain aldehydes (malondialdehyde (MDA), 4-hydroxy-2-nonenal (4-HNE)) [15-22] that damage cells by the binding the free amino groups of amino acids of proteins. Consequently, the proteins' aggregates become less susceptible to proteolytic degradation [23-25]. In tissues, the accumulation of age pigment spots appears. In addition, free radicals effects are connected with a formation of atherosclerotic lesions. In body fluids (blood, urine) the increased levels of peroxidation end-products (MDA, 4-HNE, isoprostanes) are present [26,27]. The lipid peroxidation by free radicals occurs in three stages: initiation, propagation and termination [2,26]. Reaction (1) represents initiation, in which a fatty acid molecule of lipid is attacked by free radicals leading to a detachment of the hydrogen atom under fatty acid radical formation. In its structure, a rearrangement of the double bond to form conjugated diene occurs. This diene structure subsequently reacts with oxygen molecule to form a lipoperoxyl radical, which leads to the initiation of the second phase called propagation (2). In another part of the promotion, lipoperoxyl radical further reacts with another molecule of fatty acid, from which a hydrogen atom is detached under formation of lipid hydroperoxide from original molecule (3). After pairing of all radicals, the last stage of the reaction called termination occurs. In addition to the above-mentioned chemical non-enzymatic peroxidation, enzymatic lipid peroxidation that is catalysed by the enzymes cyclooxygenase and lipoxygenase takes place. [26,28]. Both enzymes are involved in the formation of eicosanoids, which represent a group of biologically active lipid compounds derived from unsaturated fatty acids containing 20 carbon atoms. Cyclooxygenase is involved in the genesis of prostaglandins [29].

$$(1)\ LH + R^{\bullet} \rightarrow L^{\bullet} + RH\ (2)\ L^{\bullet} + O_2 \rightarrow LOO^{\bullet}\ (3)\ LOO^{\bullet} + L'H \rightarrow L'^{\bullet} + LOOH$$

Scheme 1. The scheme of lipid peroxidation. Initiation (1), the first part of the propagation (2), the second part of propagation (3).

For the monitoring of lipid peroxidation, spectrophotometric [30,31], chromatographic [32] and immunochemical [33] methods can be used. The analysis itself may be based on the analysis of the primary products of lipid peroxidation as conjugated dienes [34] and lipid hydroperoxides [35], or secondary products, such as malondialdehyde [36], alkanes [37] or isoprostanes [32,38-40]. Chromatographic methods represent the special group of methods, which are mostly based on the decrease of unsaturated fatty acids' concentration [41]. The scope of this review was to summarize the photometric analyses of lipid peroxidation. Less common method - FOX (ferrous oxidation in xylenol orange) was suggested to be automated.

1.1. Spectrophotometric methods in lipid peroxidation analysis

Spectrophotometric methods for the analysis of lipid peroxidation (see Table 1) are well reproducible and low cost. They usually consist of several steps that can be automated without much difficulty. Determination of conjugated dienes and TBARS belong to the one of the oldest and mostlz used methods for their rapidity and simplicity. On the other hand, they are criticized for their non-specificity [42,43]. Lipid hydroperoxides may be determined by the iodometric method and FOX test [44].

Determined analyte	Method	Type of analysed sample	Reference
Conjugated dienes	The structures of conjugated dienes absorb in the UV spectrum of 230-235 nm	Serum lipoproteins, tissue lipids	[34,45]
TBARS/MDA	TBA complex with MDA, Measurement at 532 nm	plasma, urine, tissues (liver), Cell lysates	[36,46-53]
Lipid hydroperoxides	Iodometric method	plasma, plant tissues	[44,54]
	FOX test	plasma, serum lipoproteins, both animal and plant tissues	[35,44,55,56]

Table 1. Summary of spectrophotometric methods used in lipid peroxidation determination.
FOX –ferrous oxidation in xylenol orange, MDA – malondialdehyde, TBARS – thiobarbituric acid reactive substances

1.2. Conjugated dienes

The structures of conjugated dienes (Fig. 1) with alternating double and single bonds between carbon atoms (-C=C-C=C-) absorb wavelengths of 230-235 nm in the UV region. Therefore, it is possible to use UV absorption spectrometry for their determination [41,42]. The method is used for determination of a non-specific lipid peroxidation caused by free radicals in biological samples, and is successfully used in the study of peroxidation in isolated lipoprotein fractions (LDL lipoproteins) [45]. However, its use in the direct analysis of plasma is controversial because of the presence of interfering substances, such as heme proteins, purines or pyrimidines in the UV region measurement [42,57].

Figure 1. Structural formula of conjugated diene arising from the fatty acids by the free radicals effects during lipid peroxidation.

Increased sensitivity of the method can be achieved by an extraction of lipids into organic solvents in combination HPLC with UV detection [34,58]. However, the result of application the method to lipid extracts from human body fluids after HPLC separation was surprising, because the majority of pre-treated lipid fraction absorbs at wavelengths typical for conjugated dienes consisting of conjugated linoleic acid isomer (*cis*-9, *trans*-11-octadecadienoic acid) [59]. The main sources of conjugated isomer of linoleic acid (CLA) are dairy products and ruminant meat, especially beef [60]. They come into human serum and tissues probably from the diet [61], but can be also produced by bacteria [62,63]. Therefore, formation of large amounts of CLA by free radicals seems unlikely. In addition, the presence of CLA was not detected in the plasma of animals suffering from oxidative stress. *In vivo* induction of lipid peroxidation in rats treated with phenylhydrazin trichlorbrommethan did not cause an increase of CLA plasma values [64]. In the case of the use this method, it is necessary to take into account the above-mentioned shortcomings in the analysis of biological fluids or tissues.

1.3. TBARS, TBA-MDA adducts

TBARS (TBA-MDA) (**T**hio**b**arbituric **A**cid **R**eactive **S**ubstances) is the most widely used method for determination of lipid peroxidation method, especially due to its simplicity and cheapness. As the name of this method implies, it is based on the ability of malondialdehyde, which is one of the secondary products of lipid peroxidation, to react with thiobarbituric acid (TBA) [65]. The principle of this method consists in the reaction of MDA with thiobarbituric acid in acidic conditions and at a higher temperature to form a pink MDA-$(TBA)_2$ complex (Fig. 2), which can be quantified spectrophotometrically at 532 nm [17,66-70]. TBARS method measures the amount of MDA generated during lipid peroxidation, however, other aldehydes generated during lipid peroxidation, which also absorb at 532 nm, may react with TBA [71]. The results of the assay are expressed in μmol of MDA equivalents. TBARS method can be also used in the case of defined membrane systems, such as microsomes and liposomes, but its application in biological fluids and tissue extracts appears to be problematic [72-74]. The first problem is based on the fact that MDA can be formed by the decomposition of lipid peroxides under heating of the sample with TBA. This decomposition is accelerated by traces of iron in the reagents and is inhibited by the use of chelating agents [42]. At the decomposition of lipid peroxides in the analysis, the originating radicals can amplify the entire process and the amount of MDA could be overestimated [74]. To prevent the decomposition of lipid peroxides during the analysis, inhibitor of the lipid peroxidation called butylated hydroxytoluene is added to the sample [42]. One of the other problems of the TBARS method application has been found in the analysis of biological fluids. In this case, some substances, such as bile pigments and glycoproteins provide a false positive reaction with TBA [71,75]. Unspecificity TBARS test problems can be partially overcome by the using of HPLC techniques for the separation of "authentic", original MDA-$(TBA)_2$ adduct from other chromogens absorbing at 532 nm [76]. Nevertheless, this approach cannot solve all problems. In addition, next molecules, such as aldehydes originated from lipid peroxidation, can form with TBA a original MDA-TBA_2 adduct, which has been demonstrated in the deoxyribose [77]. Using of different techniques in the determination of lipid peroxides in plasma or serum of healthy people (spectrophotometric versus HPLC method) leads to significantly different results. When using spectrophotometric techniques, the content of TBARS in plasma (serum) reached values from 0.9 to 42.7 $\mu mol \cdot L^{-1}$ of MDA equivalents, when HPLC technique was used, the content of TBARS in human plasma (serum) reached values of 0.6 – 1.4 $\mu mol \cdot L^{-1}$ of MDA equivalents [78-84]. This was probably caused by the using different methods for modifying the preparation of plasma (serum) sample. Method for the non-specific index of lipid peroxidation determination in isolated purified lipid fractions seems to be most useful [42].

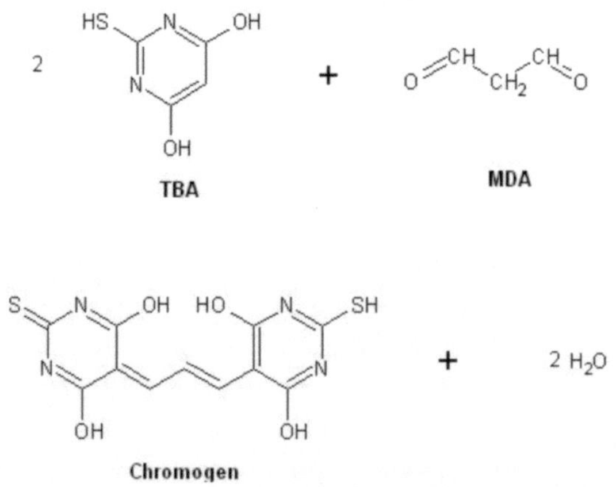

Figure 2. Chromophore produced by a condensation of MDA with TBA

1.4. Lipid hydroperoxides

1.4.1. Iodometric method

Iodometric method for lipid hydroperoxides determination is one of the oldest methods and is still used to determine lipid peroxide number [42,85]. Principle of this method is based on the ability of lipid hydroperoxides to oxidize iodide (I^-) to iodine (I_2), which further reacts with unreacted iodide (I^-) to triiodide anion (I_3^-) [86] and can be determined spectrophotometrically at 290 or 360 nm [87]. Modification of the iodometric method using commercially available reagent used for the determination of cholesterol can also be used to determine lipid (hydro)peroxides spectrophotometrically at 365 nm [54]. The method can be applied to extracts of biological samples without present the oxidizing agents. The possible interfering factors are especially the presence of oxygen, hydrogen peroxide and protein peroxides, which are able to oxidize iodide. Oxygen interference can be avoided by the using the anaerobic cuvettes and cadmium ions, which form a complex with unreacted iodide [86]. Values of lipid hydroperoxides in human plasma determined by iodometry are about 4 $\mu mol.L^{-1}$ [88,89].

1.4.2. Ferrous oxidation in xylenol orange

Total hydroperoxides can be determined using the oxidation of ferrous ions in the test with xylenol orange (FOX). The principle of the FOX method is based on the oxidation of ferrous ions to ferric by the hydroperoxide activity in the acidic environment [90-94]. The exact mechanism of the sequence of radical reactions is not known, but the mechanism has been designed by Gupta et al. [95] and is shown in reactions 1-4 (equation 2) [96]. The increase in the concentration of ferric ion is then detected using xylenol orange (Fig. 3), which forms a blue-violet complex with ferric ion (equation 2, reaction 5) with an absorption maximum at 560 nm [35]. However, the experimentally determined stoichiometry of 3 moles of Fe^{3+}-xylenol orange produced from 1 mol of peroxide [96,97] cannot be explained by the mechanism proposed by Gupta [95].

$$(1)\ Fe^{2+} + LOOH + H^+ \rightarrow Fe^{3+} + H_2O + LO^{\bullet}$$

$$(2)\ LO^{\bullet} + \text{xylenol orange} + H^+ \rightarrow LOH + \text{xylenol orange}^{\bullet}$$

$$(3)\ \text{Xylenol orange}^{\bullet} + Fe^{2+} \rightarrow \text{xylenol orange} + Fe^{3+}$$

$$(4)\ LO^{\bullet} + Fe^{2+} + H^+ \rightarrow Fe^{3+} + LOH$$

$$(5)\ Fe^{3+} + \text{xylenol orange} \rightarrow \text{blue} - \text{violet complex}\ (560\ \text{nm})$$

Scheme 2. Equation of mechanism sequence of radical reactions.

Gay et al. [90] have found during comparison of the reactions of different peroxides with FOX reagents that the stoichiometry of the reaction ranged from 2.2 (H_2O_2) to 5.3 moles (Cu-OOH, *t*-BuOOH) Fe^{3+}-xylenol orange (Fe-XO) generated from 1 mol of peroxide, which was observed due to determination of molar absorption coefficients of Fe-XO complexes. Therefore, it is possible to compare only the results of FOX method analyses, in which the same type of peroxide was used in calibration. Hydrogen peroxide (H_2O_2) and Cumene hydroperoxide (Cu-OOH) are the most often peroxides used to calibrate the FOX method.

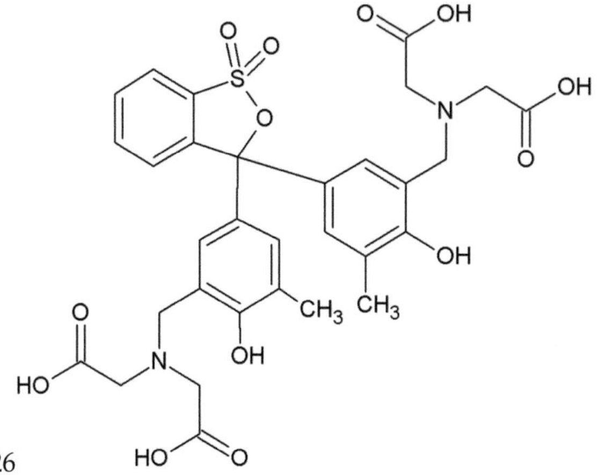

26

Figure 3. Structural formula of xylenol orange

The literature describes two versions of the FOX method called FOX1 and FOX2.. - FOX1 method can be used for the hydroperoxides determination in water phase and FOX2 method is suitable for the hydroperoxides of the lipid phase [30,35,98]. In the FOX1 method, chemicals used for a preparation of reagents (ferrous salt and sulphuric acid) are dissolved in water, whereas in FOX2 method methanol (90 % *v*/*v*) is the solvent [35]. FOX methods are not specific to hydroperoxides, the presence of oxidizing agent(s) in sample leads to the oxidization of ferrous ions to ferric ions. In the case of FOX2, the specificity of the method is achieved by the first FOX2 test performance in the presence of triphenylphosphine (TPP), which selectively reduces hydroperoxides to alcohols. The result of this test is used as a blank. After it, the FOX test without triphenylphosphine is performed and after deduction of blank values, we get the real value of lipid hydroperoxides. Improved specificity of the method using triphenylphosphine was later achieved also in FOX1 test [99]. Peroxidation chain reactions, which might occur during the analysis, are prevented by the addition of butylated hydroxytoluene prevented into the FOX1 agent. Plasma samples collected using ethylenediaminetetraacetic acid (EDTA) or diethylenetriaminepentaacetic acid pentasodium salt abbreviated as DETAPAC (anticoagulants or iron chelating agents) cannot be used due to interference with FOX reagents [30]. FOX1 method has been automated [100].

Measurement of lipid peroxidation in (blood) plasma

Banerjee et al. [99] enhanced sensitivity of FOX1 method by the addition of sorbitol into the FOX1 reagent in accordance with Wolff [98], and concurrently by the stabilization of pH of reagents at the values of 1.7 - 1.8. Improved specificity of method was obtained using triphenylphosphine and butylated hydroxytoluene. A comparison of both FOX1 and FOX2 methods on plasma samples of healthy individuals and diabetic patients was performed, where modified FOX1 method was more sensitive compared to the FOX2 method. Another advantage of the FOX1 method was based on the skip the centrifugation step that is necessary in FOX2 method. Nourooz-zadeh et al. [55] determined total lipid hydroperoxides in plasma by the use the FOX2 method and subsequently monitored content of lipid hydroperoxides in individual lipoprotein fractions (VLDL, LDL and HDL fractions). Content of total lipid hydroperoxides in plasma was 3.50±2.05 μmol/L. The highest rate of hydroperoxides (67 %) was detected in LDL lipoprotein fractions. Södergren et al. [101] studied the impact of the storage of samples at low temperatures on the total lipid hydroperoxide content by the use the FOX2 method. They were focused on possible reduction of total lipid hydroperoxides content during the storage of samples under these conditions. Researchers found that storage of samples for 6 weeks at -70 °C leads to the 23 % average reduction of hydroperoxides content. The finding that the content of lipid hydroperoxides in short-term stored plasma samples (6 weeks) did not differ from the content of lipid hydroperoxides in the long-term stored samples (60 weeks) was interesting too.

Measurement of lipid peroxidation in animal tissues

Hermes-Lima et al. [96] proposed and elaborated methodology for application of FOX1 test in determination of lipid hydroperoxides in animal tissue extracts. They used methanol extracts of kidney, liver and heart from adult mice (*Mus musculus* Linnaeus), brain and lungs from adult Wistar rats (*Rattus norvegicus* Berkenhout var. *alba*), liver and adipose tissues from adult golden-mantled ground squirrels (*Spermophilus lateralis* Say), and liver and muscle tissues from adult red-eared slider turtles (*Trachemis scripta elegans* Wied-Neuwied). The highest values of lipid hydroperoxide content were detected in mice organs. The contents of lipid peroxides in animal tissues measured by the FOX1 method well correlated with results obtained by the TBARS. Grau et al. [102] adapted the FOX2 method for the determination of lipid hydroperoxides in raw and cooked dark chicken meat. Chickens were fed by a diet with different contents of α-tocopherol and fats from different sources. They determined the absolute values of lipid hydroperoxides in different experimental groups of chickens. Eymard et al. [56] modified the FOX1 method used by Hermes-Lima et al. [96] for the determination of lipid hydroperoxides in small pelagic fish. They used methanol extracts of ground tissues of the Atlantic horse mackerel (*Trachurus trachurus* Linnaeus). The original FOX1 reagent was replaced by the FOX2 reagent used by Wolff et al. [98] with the increased content of methanol to increase a solubility of extracts.

Measurement of lipid peroxidation in plant tissues

De Long et al. [44] applied the FOX2 method in the determination of hydroperoxides in plant tissues. They used ethanol extracts of pericarp of avocado (*Persea americana* P. Mill.), periderm of potatoes (*Solanum tuberosum* L.), leaves of red cabbage (*Brassica oleracea* convar. *capitata* var. *rubra* DC. Ranost), leaves of spinach (*Spinacia oleracea* L.), pericarp of the European Pear (*Pyrus communis* L.) and fruits of red pepper (*Capsicum annuum* L.) for analyses. The effect of UV radiation on lipid peroxidation was monitored. Parts of plants were exposed to UV radiation for 10-12 days prior the extraction due to induction of lipid peroxidation in plants. Lipid hydroperoxides were determined by the FOX2, the TBARS and the iodometric methods. UV radiation induced an increase in lipid peroxidation values in all samples of different plant tissues determined by the FOX method. The good correlation was found between the FOX and iodometric methods. However, the iodometric method had limitations in the determination of the low concentrations of lipid hydroperoxides. Similar results were obtained by the use the TBARS method. Griffiths et al. [103] applied the FOX2 method in determination of lipid peroxides in different types of plant tissues. They analysed plant tissues, such as extracts of bean hypocotyls (*Phaseolus* sp.) and microsomes, potato leaves (*Solanum tuberosum* L.), flowers of alstromeria (*Alstroemeria* spp.), broccoli (*Brassica oleracea* var. *italica* Plenck) and cells of green algae (*Chlamydomonas* sp.). Lipid hydroperoxide levels ranged from 26 to 602 nmol.g^{-1} of FW. The highest content of lipid hydroperoxides was detected in broccoli and green alga cells in their study.

2. Experimental section

2.1. Instruments

For dilution of stock solutions of standards an epMotion 5075 (Eppendorf, Germany) automated pipetting system was used (Fig. 4). The pipetting provides a robotic arm with adapters (TS 50, TS 300 and TS 1000) and Gripper (TG-T). The empty microtubes are placed in the position B3 (Fig. 4) in adapter Ep0.5/1.5/2 ml. Module Reservoir is located in the position B1, where stock solutions are available. The device is controlled by the epMotion control panel. The tips are located in the A4 (ePtips 50), A3 (ePtips 300) and A2 (ePtips 1000) positions. For preparation of the standards tips of sizes 300 μl and 1000 μl (Eppendorf – Germany) were used. For determination of antioxidant activity, a BS-400 automated spectrophotometer (Mindray, China) was used. It is composed of cuvette space tempered to 37±1 °C, reagent space with a carousel for reagents (tempered to 4±1 °C), sample space with a carousel for preparation of samples and an optical detector. Transfer of samples and reagents is provided by robotic arm equipped with a dosing needle (error of dosage up to 3 % of volume). Cuvette content is mixed by an automatic mixer including a stirrer immediately after addition of reagents or samples. Contamination is reduced due to its rinsing system, including rinsing of the dosing needle as well as the stirrer by MilliQ water. For detection itself, the following range of wave lengths can be used as 340, 380, 412, 450, 505, 546, 570, 605, 660, 700, 740 and 800 nm. In addition, a SPECOL 210 two beam UV-VIS spectrophotometer (Analytik Jena AG, Germany) with cooled semiconductor detector for measurement within range from 190 to 1,100 nm with control by an external PC with the programme WinASPECT was used as the manual instrument in this study. Laboratory scales (Sartorius, Germany) and pipettes (Eppendorf Research, Germany) were used.

2.2. Chemicals

Xylenol orange disodium salt, iron D-gluconate dihydrate, glycerol, *tert*-butylhydroperoxide (t-BHP) 70% in water, sodium chloride, sulphuric acid, formic acid and water ACS reagent were purchased from Sigma Aldrich (USA).

2.3. Preparation of reagents and standards

FOX1 reagents were prepared according Arab et al. [100]. The general acidic reagent (acidic reagent A) final concentrations were 0.9 % NaCl, 40 mM H_2SO_4, 20 mM formic acid and 1.37 M glycerol in ACS water. The pH of the reagent was adjusted to the value of 1.35. The reagent R1 contained 167 μM xylenol orange disodium salt, which was dissolved in acidic reagent A. The reagent R2 contained 833 μM iron D-gluconate dehydrate, which was also dissolved in acidic reagent A. Standards were prepared from the 70% water solution of *tert*-butylhydroperoxide, which was diluted by ACS water to the 20 mM pre-stock solution. From the pre-stock solution, five stock solutions: and 0.2, 3.9, 62.5, 375 and 1,000 μM were prepared daily by dilutions of pre-stock solution with 0.9 % NaCl. For further preparation of 20 standards from five stock solutions, an automated pipetting system epMotion 5075 was used to minimalize possible pipetting errors. The standards had following concentrations: 0.06, 0.12, 0.24, 0.48, 0.97, 1.9, 3.9, 7.8, 15.6, 31.2, 46.8, 62.5, 93.7, 125, 187, 250, 375, 500, 750 and 1000 μM. These standards were used for the preparation of calibration curves in both manual and automatic measurements.

2.4. Working procedure for manual spectrophotometric determination

A volume of 720 μl of the reagent R1 (167 μM xylenol orange in acidic reagent) was pipetted into plastic cuvettes. Subsequently, a volume of 100 μl of the sample was added. Absorbance was measured at λ = 591 nm. After it, a volume of 180 μl of the reagent R2 (833 μM iron D-gluconate in acidic reagent A) was pipetted to a reaction mixture and after 6 minutes of the incubation, absorbance was measured. Final value is calculated from the absorbance value of the mixture of the reagent R1 with sample and from the absorbance value after 6 minutes of incubation of the mixture with the reagent R2. The final concentrations in the cuvette of xylenol orange (R1) and iron D-gluconate (R2) were 120 and 150 μM, respectively.

2.5. Working procedure for automated spectrophotometric determination

A volume of 180 μL of the solution R1 (167 μM xylenol orange in acidic reagent) was pipetted into a plastic cuvette with subsequent addition of a 25 μL of sample. This mixture was incubated for 4.5 minutes. Subsequently, 45 μL of solution R2 (833 μM iron D-gluconate in acidic reagent) was added and the solution was incubated for next 6 minutes. Absorbance was measured at λ = 570 nm. Final value is calculated from the absorbance value of the mixture of reagent R1 with sample before the addition of the reagent 2 and from the absorbance value after 6 minutes of incubation of the mixture with the reagent 2. The final concentrations in the cuvette of xylenol orange (R1) and iron D-gluconate (R2) were 120 and 150 μM, respectively.

3. Results and discussion

Spectrophotometric methods for determination of lipid peroxidation have a relatively simple procedure of a measurement. In addition, they are relatively low-cost with easy applicability and they do not require specialized equipment or personnel. To maintain the sustainability of these methods, it is necessary to introduce these methods to automated operation, which has not been yet satisfactorily solved. Analyses of samples performed due to intensive work of personnel, which is expensive, slow, and, in addition, the human factor is responsible for a high percentage of errors. Requirement for laboratories, in which a large number of samples is analysed per day, consists in relatively simple and easy to apply method. Our aim was to automate the pre-analytical and analytical phase of the FOX1 method. For specification and comparison of this method, the method based on the use the manual spectrophotometer was also carried out.

3.1. Pre-analytical phase

Pre-analytical processing of biological samples in the laboratory is a necessary and important part of laboratory work. It represents a wide range of manual, often stereotyped operations that do not require special knowledge and skills, but require maintenance of the standard procedure(s) and prevent the possibility of errors connected with this analytical phase. Pre-analytical laboratory process is destined to automation and robotics. Automation and robotics of the pre-analytical phase brings many benefits and advantages to laboratory. It reduces the number of errors, the time necessary for sample manipulation, and the response time. It significantly increases the productivity, cost savings connected with productivity, and minimizes the exposure of personnel with biological material [104].

For automation of pre-analytical phase, the epMotion 5075 automated pipetting system was used. Stock solutions of *tert*-butylhydroperoxide (*t*-BHP) at the concentrations of 1000, 375, 62.5, 3.9 and 0.2 μM prepared in 0.9 % NaCl solution were applied into five vials. Sixth vial contained diluting solution (0.9% NaCl). Twenty empty Eppendorf tubes (1.5 ml) were placed into the metal holder. Scheme of the preparation of standards is shown in Table 2. Pipetting robot first pipetted different volumes of diluting solution (0.9% NaCl) into vials and after it, different volumes of stock solutions of various concentrations of *t*-BHP were pipetted. When pipetting the stock solution into the dilution buffer in micro test tube, robot three times mixed the solution by a pipetting.

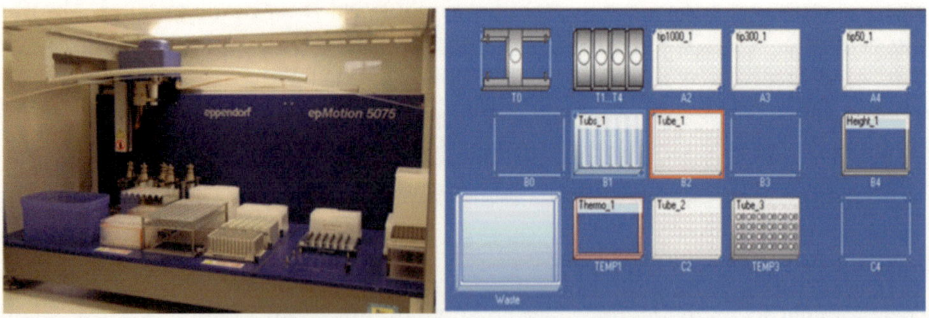

Figure 4. epMotion 5075 automated pipetting system from frontal part.

Tube nb.	Final concentration t-BHP (μM)	Pipetting volume (μl)					
		solution 0.9% NaCl	solution 1 1000 μM t-BHP	solution 2 375 μM t-BHP	solution 3 62.5 μM t-BHP	solution 4 3.906 μM t-BHP	solution 5 0.244 μM t-BHP
1	1000	-	1000	-	-	-	-
2	750.0	250	750	-	-	-	-
3	500.0	500	500	-	-	-	-
4	375.0	-	-	1000	-	-	-
5	250.0	750	250	-	-	-	-
6	187.5	500	-	500	-	-	-
7	125.0	875	125	-	-	-	-
8	93.75	750	-	250	-	-	-
9	62.50	-	-	-	1000	-	-
10	46.87	875	-	125	-	-	-
11	31.25	500	-	-	500	-	-
12	15.62	750	-	-	250	-	-
13	7.812	875	-	-	125	-	-
14	3.906	-	-	-	-	1000	-
15	1.953	500	-	-	-	500	-
16	0.977	750	-	-	-	250	-
17	0.488	875	-	-	-	125	-
18	0.244	-	-	-	-	-	1000
19	0.122	500	-	-	-	-	500
20	0.061	750	-	-	-	-	250

Table 2. Volume of the solution in the preparation of standards using epMotion 5075 automated pipetting system.

Using the epMotion 5075 automated pipetting system, work time of 20 minutes was saved (time, when laboratory staff was not needed). The only time-demanding operation consisted in replenishment of vials and initiation of the program. Potential errors that arise due to human activity were avoided. Accuracy of a pipetting was verified by weighing, the average error was approximately 1.8 %.

3.2. Analytical phase

Our goal was to introduce the FOX1 method to an automated operation and improve both analysis itself and conditions of analysis. The experiment was carried out using *tert*-butylhydroperoxide standard prepared at the concentrations from 0.06 to 1000 μM. Furthermore, the spectral curves of generated chromatic complexes were observed and the concentration dependence on temperature and time were determined. In addition, reaction kinetics during the reaction was established.

3.2.1. Monitoring the spectral courses at different concentrations and times

Spectral changes in the *t*-BHP concentration range from 0.06 to 1000 μM (Figures 5A and 4B) were observed. Two peaks at the wavelengths of 444 and 591 nm were detected in the formed complex at the recommended temperature of interaction of 37 °C.

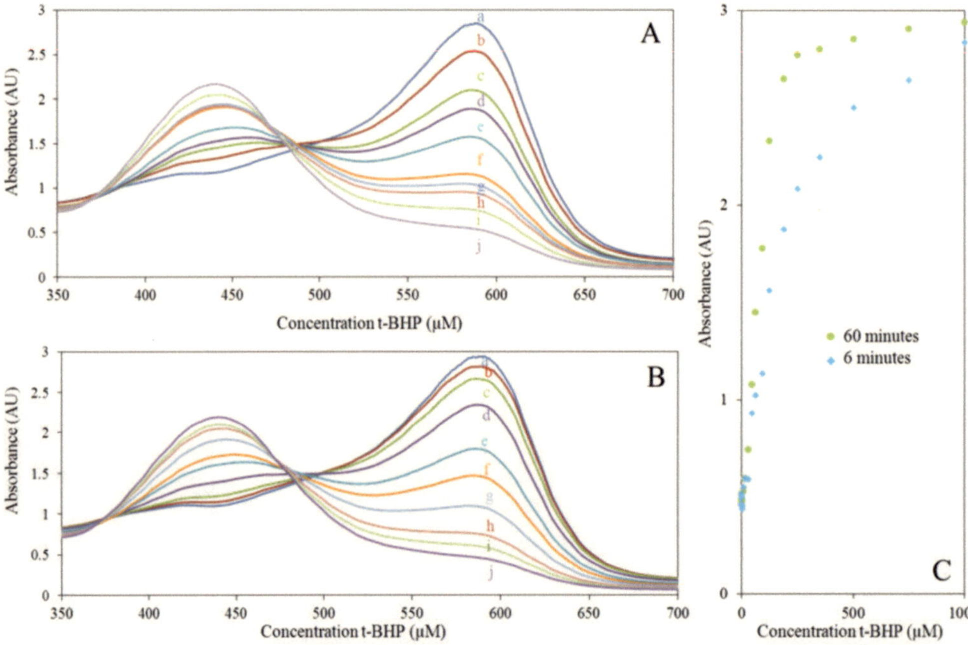

Figure 5. Courses of spectra of *t*-BHP in the concentrations from 0.06 to 1000 μM - **a)** 1000, **b)** 500, **c)** 250, **d)** 125, **e)** 62.50, **f)** 31., **g)** 7.8, **h)** 1.9, **i)** 0.4, **j)** 0.06 in the time of 6 **(A)** and 60 **(B)** minutes. **(C)** Comparison of values of absorption maximum at the wavelength of 591 nm and a time period of 6 and 60 minutes. The courses were measured in the interval form 350 to 700 nm using the SPECORD 210 apparatus. All analyses were carried out in triplicates.

Absorption maximum at low concentrations (up to the concentration of 0.122 μM) was at 444 nm, and with the increasing concentrations (higher than 0.122 μM) the absorption maximum was sifted and observed at 591 nm. Interaction of sample and reagents proceeded in six minutes, after this time, absorbance could be measured and the final value of lipid peroxidation calculated. We wanted to determine the changes in the absorbance during one hour. Comparison of absorbance values at the time of 6 and 60 min at λ = 591 nm is shown in Figure 5C. Absorbance values during the monitoring decreased for about 13 % on an average. When interlaying the trends points in the linear concentration part from 0.12 to 125 μM, the determination factor decreased from 0.996 (for the 6-minute reaction time) to 0.987 (for the 60-minute reaction time). This fact can be explained by unequal reaction kinetics during the analysis (see the reaction kinetics, Chapter 3.2.3) and oxidation of the sample during the analysis.

3.2.2. Monitoring the reaction under different temperature conditions

Dependences of representative concentration (62.5 µM) on the temperature conditions (17, 27, 37 and 47 °C) and the time from 0 to 30 minutes and absorption maximum of 591 nm is shown in Figure 6. The absorbance increased with the increasing temperature; after 6 minutes of reaction, the difference of absorbance value between the lowest (17 °C) and the highest (47 °C) temperature was about 0.64 AU. In other words, the value of absorbance at 47 °C was higher for 71 % compared to the absorbance determined at 17 °C. The highest values of absorbance and concurrently the most prominent difference was detected at 47 °C, therefore, this temperature was the most suitable for our purposes. On the other hand, this temperature may lead to degradation of biological samples. Due to this fact, the temperature of interaction of 37 °C was selected for further analyses.

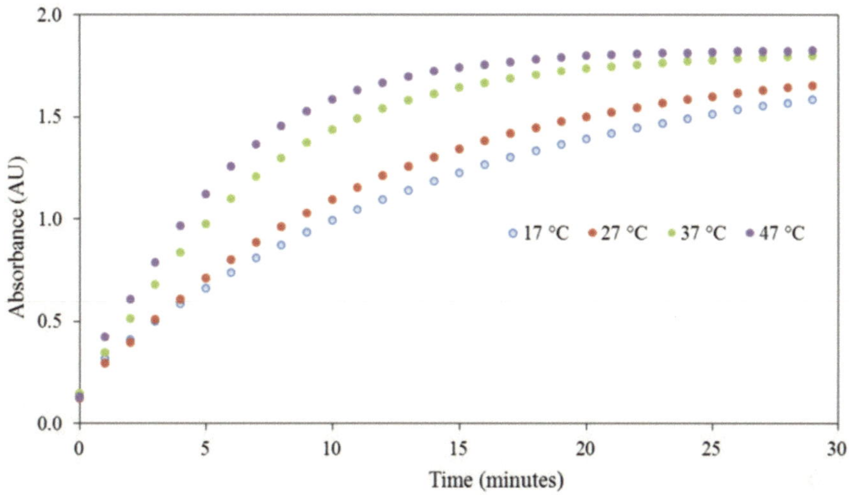

Figure 6. Dependences of representative concentration (62.5 µM) of applied *t*-BHP on temperature conditions (17, 27, 37 and 47 °C) and the time of interaction. Detected at 591 nm, interval of record is 1 minute, interval period 0 - 30 minutes. All analyses were carried out in triplicates.

3.2.3. Determination of reaction kinetics

Reaction kinetics at the temperature of 37 °C in the shortest time intervals in all concentrations (0.06 – 1000 µM) was monitored. Automated analyser BS-400 was used for this purpose. All samples could be studied at all once. This is not possible using the manual spectrophotometer, thus, use the automated analyser represents one of the most important steps in the analysis automation.

The curves were used for the calculating the reaction rate constants indicating the course and conception of the impact of the effect of *t*-BHP concentration on the reaction rate. The constant was calculated as the change in the absorbance per time unit (second, minute) according to the equation $x = A/t$, where x is the rate constant, A the value of absorbance after 6 minutes and t time for which the rate constant was related (second, minute). The effect of each of concentrations on the change in absorbance value was determined.

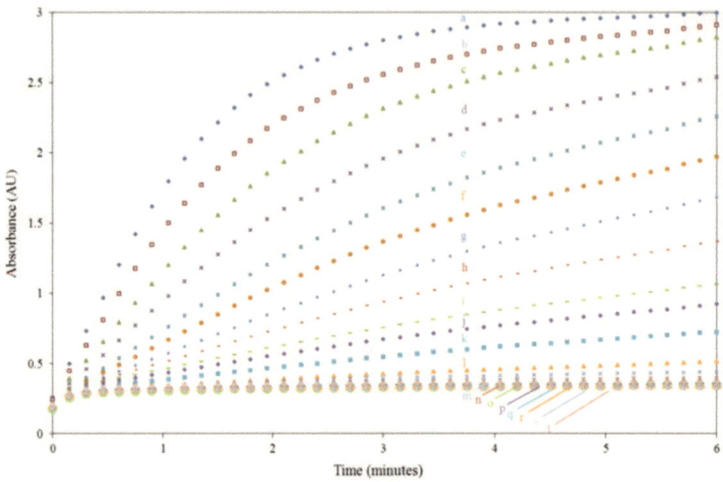

Figure 7. Monitoring of reaction curves of *t*-TBH in the concentrations from 0.06 to 1000 µM - **a)** 1000, **b)** 750, **c)** 500, **d)** 375, **e)** 250, **f)** 187, **g)** 125 **h)** 94, **i)** 63, **j)** 47., **k)** 31, **l)** 15.6, **m)** 7.8, **n)** 3.9, **o)** 1.9, **p)** 0.9, **q)** 0.4, **r)** 0.2, **s)** 0.1, and **t)** 0.06 µM in the time interval from 0 to 6 minutes. All analyses were carried out in triplicates.

Concentration	Logarithmic equation	Change in absorbance per second	Change in absorbance per minute	Change in abs. per minute recalculated to 1 µM t-BHP
1000	y = 3.7532ln(x) - 5.899	0.02304	1.383	0.0013
750.0	y = 3.6495ln(x) - 5.544	0.02211	1.345	0.0017
500.0	y = 3.4895ln(x) - 5.241	0.02168	1.301	0.0028
375.0	y = 3.1895ln(x) - 4.872	0.01987	1.258	0.0036
250.0	y = 3.2076ln(x) - 4.677	0.01853	1.112	0.0044
187.5	y = 2.7574ln(x) - 4.375	0.01534	0.924	0.0052
125.0	y = 2.2477ln(x) - 3.945	0.01298	0.779	0.0062
93.75	y = 1.7316ln(x) - 2.968	0.01000	0.600	0.0060
62.50	y = 1.2213ln(x) - 1.998	0.00705	0.423	0.0068
46.87	y = 1.0049ln(x) - 1.596	0.00580	0.348	0.0070
31.25	y = 0.7102ln(x) - 1.054	0.00410	0.246	0.0079
15.62	y = 0.3846ln(x) - 0.445	0.00222	0.133	0.0085
7.812	y = 0.2525ln(x) - 0.183	0.00146	0.088	0.0112
3.906	y = 0.1765ln(x) - 0.037	0.00102	0.061	0.0157
1.953	y = 0.1303ln(x) + 0.033	0.00075	0.045	0.0231
0.976	y = 0.1177ln(x) + 0.073	0.00068	0.041	0.0418
0.488	y = 0.1031ln(x) + 0.089	0.00060	0.036	0.0732
0.244	y = 0.0965ln(x) + 0.101	0.00057	0.034	0.1370
0.122	y = 0.0926ln(x) + 0.105	0.00055	0.033	0.2629
0.061	y = 0.0957ln(x) + 0.131	0.00053	0.032	0.5434

Table 3. Mathematical formularization of the course of reaction curves for *t*-TBH in the concentration range from 0.06 to 1000 µM by the use the logarithmic equation. Reaction rate constant is expressed as a

change in absorbance per second, and per minute. In addition, change in absorbance per minute recalculated to 1 μM t-BHP is introduced.

3.2.4. Dependence on concentration

By the using manual spectrophotometer and automated analyser, the dependence of *t*-TBH concentration (0.06 – 1000 μM) on the changes of coloured complex was determined. The calibration curves were calculated from final values.

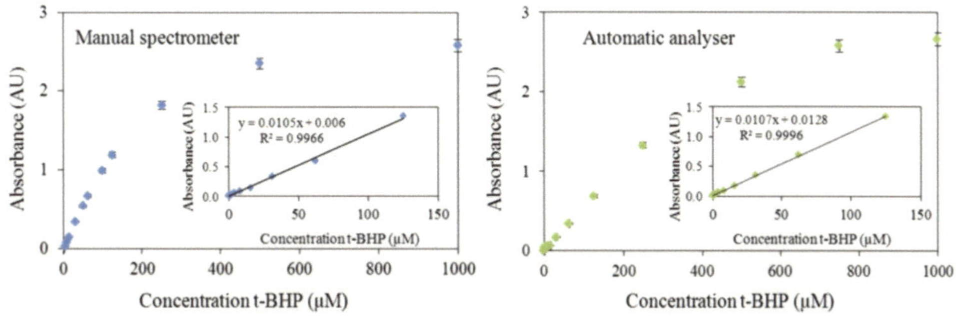

Figure 8. Dependence of absorbance on applied *t*-BHP concentration measured by manual spectrophotometer SPECOL 210 and automated analyser BS-400. All analyses were carried out in triplicates. For other experimental detail, see Fig. 7.

The analysis of 60 samples (20 samples in a standard three repetitions) took using the BS-400 automated analyser only 24 minutes. The analysis of 60 samples including delays for the pipetting, mixing and displacement of samples using the manual spectrophotometer took about 7 hours (6 minutes per sample + one minute of delay, 60 × 6 minutes of sample analysis). By using the fully automated analyser, results were obtained in more than 17 times less time compared to manual spectrophotometer. Shortening of the time of analysis contributes especially to higher quality of results due to reduction of possibility of chemical modification including degradation of the measured samples. This fact resulted in the preparation of calibration curves, where the determination factor for the calibration curve obtained using the automatic analyser was $R^2 = 0.9996$, while the determination factor for the results from manual spectrophotometer was $R^2 = 0.9966$. In addition, a limit of detection (LOD) and limit of quantification (LOQ) were determined. In the case of both automated and manual analyses, the LOD was determined as LOD = 0.06 μM of *t*-BHP, limit of quantification (LOQ) was also determined as LOQ = 0.2 μM of *t*-BHP (see Table 3). All measurements of all concentrations of *t*-BHP (concentration range from 0.06 to 1000 μM) were carried out in 3 repetitions and repeatability (RSD) was determined. In the case of automated method, the repeatability was RSD = 2.6 % compared to manual spectrophotometer, where RSD = 3.8 %.

Technical development is responsible for a tendency to increase the speed of analysis and analytical process itself. Automatic analysers allow analysing more samples at the same time, reducing the time required to analyse one sample and errors caused by incorrect pipetting and manipulation with sample, and generally provide higher data quality compared to manual analysis. Due to automation, the risk of sample confusion is significantly reduced. In addition, the whole process is much faster, the consumption of reagents and demands of personnel staff are reduced. The aim of automation is to eliminate stereotypical incompetent operation, eliminate the possibility of error, and to accelerate operations under significant increase of capacity while maintaining the precise performance of all necessary operations. The disadvantage, however, consists in still high acquisition costs and the need for compete service [105,106].

Apparatus	Wavelength (nm)	LOD	LOQ	Measuring range (μM)	Calibration equation	Confidence coefficient (R^2)	RSD	Time analysis of 60 samples (min)
SPECOL	591	0.06	0.2	0.012 - 125	y=0.0105x +0.006	0.9969	3.8	420
BS-400	570	0.06	0.2	0.012 - 125	y=0.0107x +0.0128	0.9996	2.6	24

Table 4. Analytic parameters for the FOX1 method for *t*-BHP analysis using manual SPECOL and automated BS-400 analysers.

4. Conclusion

This chapter brought a comprehensive overview of photometric methods used in the study of lipid peroxidation. Main attention was devoted to the detection of lipid peroxidation by using the less common FOX1 method. The proposal to automation the pre-analytical and analytical phases of the sample was introduced. In addition, conditions and parameters influencing the photometric reaction were studied and described. The comparison of results obtained using the manual and automated apparatuses (manual/automated operation) is introduced and discussed.

Author details

Jiri Sochor, Branislav Ruttkay-Nedecky, Vojtech Adam, Jaromir Hubalek and Rene Kizek*

Department of Chemistry and Biochemistry, Faculty of Agronomy, Mendel University in Brno, Brno, Czech Republic, European Union

Department of Microelectronics, Faculty of Electrical Engineering and Communication, Brno University of Technology, Brno, Czech Republic, European Union

Petr Babula

Department of Natural Drugs, Faculty of Pharmacy, University of Veterinary and Pharmaceutical Sciences Brno, Brno, Czech Republic, European Union

* Corresponding Author

Acknowledgement

This project was supported by SIX research centre CZ.1.05/2.1.00/03.0072. The authors wish to express their thanks to Lukáš Nejdl for excellent technical assistance.

5. References

[1] Grinna LS (1977) Age-related changes in lipids of microsomal and mitochondrial-membranes of rat-liver and kidney Mech. Ageing Dev. 6: 197-205.

[2] De Zwart LL, Meerman JHN, Commandeur JNM, Vermeulen NPE (1999) Biomarkers of free radical damage applications in experimental animals and in humans. Free Radic. Biol. Med. 26: 202-226.

[3] Knight JA (1995) Diseases related to oxygen-derived free-radicals. Ann. Clin. Lab. Sci. 25: 111-121.

[4] Gutteridge JMC (1993) Free-radicals in disease processes - A compilation of cause and consequence. Free Rad. Res. Commun. 19: 141-158.

[5] Evans PH (1993) Free-radicals in brain metabolism and pathology. Br. Med. Bull. 49: 577-587.

[6] Adams JD, Odunze IN (1991) Oxygen free-radicals and Parkinsons-disease. Free Radic. Biol. Med. 10: 161-169.

[7] Pohanka M, Novotny L, Zdarova-Karasova J, Bandouchova H, Zemek F, Hrabinova M, Misik J, Kuca K, Bajgar J, Zitka O, Cernei N, Kizek R, Pikula J (2011) Asoxime (HI-6) impact on dogs after one and tenfold therapeutic doses: Assessment of adverse effects, distribution, and oxidative stress. Environ. Toxicol. Pharmacol. 32: 75-81.

[8] Diopan V, Babula P, Shestivska V, Adam V, Zemlicka M, Dvorska M, Hubalek J, Trnkova L, Havel L, Kizek R (2008) Electrochemical and spectrometric study of antioxidant activity of pomiferin, isopomiferin, osajin and catalposide. J. Pharm. Biomed. Anal. 48: 127-133.

[9] Babula P, Kohoutkova V, Opatrilova R, Dankova I, Masarik M, Kizek R (2010) Pharmaceutical importance of zinc and metallothionein in cell signalling. Chim. Oggi-Chem. Today. 28: 18-21.

[10] Jurikova T, Rop O, Mlcek J, Sochor J, Balla S, Szekeres L, Hegedusova A, Hubalek J, Adam V, Kizek R (2012) Phenolic Profile of Edible Honeysuckle Berries (Genus Lonicera) and Their Biological Effects. Molecules. 17: 61-79.

[11] Sochor J, Skutkova H, Babula P, Zitka O, Cernei N, Rop O, Krska B, Adam V, Provaznik I, Kizek R (2011) Mathematical evaluation of the amino acid and polyphenol content and antioxidant activities of fruits from different apricot cultivars. Molecules. 16: 7428-7457.

[12] Sochor J, Zitka O, Skutkova H, Pavlik D, Babula P, Krska B, Horna A, Adam V, Provaznik I, Kizek R (2010) Content of phenolic compounds and antioxidant capacity in fruits of apricot genotypes. Molecules. 15: 6285-6305.

[13] Rop O, Sochor J, Jurikova T, Zitka O, Skutkova H, Mlcek J, Salas P, Krska B, Babula P, Adam V, Kramarova D, Beklova M, Provaznik I, Kizek R (2011) Effect of five different

stages of ripening on chemical compounds in Medlar (Mespilus germanica L.). Molecules. 16: 74-91.

[14] Rop O, Jurikova T, Sochor J, Mlcek J, Kramarova D (2011) Antioxidant capacity, scavenging radical activity and selected chemical composition of native apple cultivars from Central Europe. J. Food Qual. 34: 187-194.

[15] Balestrieri ML, Dicitore A, Benevento R, Di Maio M, Santoriello A, Canonico S, Giordano A, Stiuso P (2012) Interplay between membrane lipid peroxidation, transglutaminase activity, and Cyclooxygenase 2 expression in the tissue adjoining to breast cancer. J. Cell. Physiol. 227: 1577-1582.

[16] Cai F, Dupertuis YM, Pichard C (2012) Role of polyunsaturated fatty acids and lipid peroxidation on colorectal cancer risk and treatments. Curr. Opin. Clin. Nutr. Metab. Care. 15: 99-106.

[17] Zalejska-Fiolka J, Wielkoszynski T, Kasperczyk S, Kasperczyk A, Birkner E (2012) Effects of Oxidized Cooking Oil and alpha-Lipoic Acid on Blood Antioxidants: Enzyme Activities and Lipid Peroxidation in Rats Fed a High-Fat Diet. Biol. Trace Elem. Res. 145: 217-221.

[18] Flohr L, Fuzinatto CF, Melegari SP, Matias WG (2012) Effects of exposure to soluble fraction of industrial solid waste on lipid peroxidation and DNA methylation in erythrocytes of Oreochromis niloticus, as assessed by quantification of MDA and m(5)dC rates. Ecotox. Environ. Safe. 76: 63-70.

[19] Janowska B, Kurpios-Piec D, Prorok P, Szparecki G, Komisarski M, Kowalczyk P, Janion C, Tudek B (2012) Role of damage-specific DNA polymerases in M13 phage mutagenesis induced by a major lipid peroxidation product trans-4-hydroxy-2-nonenal. Mutat. Res.-Fundam. Mol. Mech. Mutagen. 729: 41-51.

[20] Demir E, Kaya B, Soriano C, Creus A, Marcos R (2011) Genotoxic analysis of four lipid-peroxidation products in the mouse lymphoma assay. Mutat. Res. Genet. Toxicol. Environ. Mutagen. 726: 98-103.

[21] Guo LL, Chen ZY, Cox BE, Gragg S, Zhang YQ, Amarnath V, van Lenten B, Epand R, Davies SS (2011) Lipid Peroxidation Generates Aldehyde-Modified Phosphatidylethanolamines That Induce Inflammation. Free Radic. Biol. Med. 51: S103-S104.

[22] Guo J, Prokai L (2011) To tag or not to tag: A comparative evaluation of immunoaffinity-labeling and tandem mass spectrometry for the identification and localization of posttranslational protein carbonylation by 4-hydroxy-2-nonenal, an end-product of lipid peroxidation. J. Proteomics. 74: 2360-2369.

[23] Esterbauer H, Schaur RJ, Zollner H (1991) Chemistry and biochemistry of 4-hydroxynonenal, malonaldehyde and related aldehydes. Free Radic. Biol. Med. 11: 81-128.

[24] Benedetti A, Comporti M, Esterbauer H (1980) Identification of 4-hydroxynoneal as a cyto-toxic product originating from the peroxidation of liver microsomal lipids. Biochim. Biophys. Acta. 620: 281-296.

[25] Esterbauer H, Comporti M, Benedetti A (1980) Biochemical effects of 4-hydroxyalkenals, in particular 4-hydroxynonenal produced by microsomal lipid-peroxidation. J. Am. Oil Chem. Soc. 57: A144-A145.

[26] Gutteridge JMC (1995) Lipid-peroxidation and antioxidants as biomarkers of tissue-damage Clin. Chem. 41: 1819-1828.

[27] Tanrikulu AC, Abakay A, Evliyaoglu O, Palanci Y (2011) Coenzyme Q10, Copper, Zinc, and Lipid Peroxidation Levels in Serum of Patients with Chronic Obstructive Pulmonary Disease. Biol. Trace Elem. Res. 143: 659-667.

[28] Viita H, Pacholska A, Ahmad F, Tietavainen J, Naarala J, Hyvarinen A, Wirth T, Yla-Herttuala S (2012) 15-Lipoxygenase-1 Induces Lipid Peroxidation and Apoptosis, and Improves Survival in Rat Malignant Glioma. In Vivo. 26: 1-8.

[29] Kuehl FA, Egan RW (1980) Prostaglandins, arachidonic-acid, and inflammation. Science. 210: 978-984.

[30] Nourooz-Zadeh J Ferrous ion oxidation in presence of xylenol orange for detection of lipid hydroperoxides in plasma, in: L. Packer (Ed.), Oxidants and Antioxidants, Pt B, 1999, pp. 58-62.

[31] Rop O, Mlcek J, Jurikova T, Valsikova M, Sochor J, Reznicek V, Kramarova D (2010) Phenolic content, antioxidant capacity, radical oxygen species scavenging and lipid peroxidation inhibiting activities of extracts of five black chokeberry (Aronia melanocarpa (Michx.) Elliot) cultivars. J. Med. Plants Res. 4: 2431-2437.

[32] Nouroozzadeh J, Gopaul NK, Barrow S, Mallet AI, Anggard EE (1995) Analysis of F-2-isoprostanes as indicators of nonenzymatic lipid-peroxidation in-vivo by gas-chromatography mass-spectrometry - Development of a solid-phase extraction procedure. J. Chromatogr. B-Biomed. App. 667: 199-208.

[33] Bachi A, Zuccato E, Baraldi M, Fanelli R, Chiabrando C (1996) Measurement of urinary 8-epi-prostaglandin F-2 alpha, a novel index of lipid peroxidation in vivo, by immunoaffinity extraction gas chromatography mass spectrometry. Basal levels in smokers and nonsmokers. Free Radic. Biol. Med. 20: 619-624.

[34] Iversen SA, Cawood P, Dormandy TL (1985) A method for the measurement of a diene-conjugated derivative of linoleic-acid, 18-2(9,11), in serum phospholipid, and possible origins. Ann. Clin. Biochem. 22: 137-140.

[35] Nouroozzadeh J, Tajaddinisarmadi J, Wolff SP (1994) Measurement of plasma hydroperoxide concentrations by the ferrous oxidation xylenol orange assay in conjunction with triphenylphosphine. Anal. Biochem. 220: 403-409.

[36] Draper HH, Squires EJ, Mahmoodi H, Wu J, Agarwal S, Hadley M (1993) A comparative-evaluation of thiobarbituric acid methods for the determination of malondialdehyde in biological-materials. Free Radic. Biol. Med. 15: 353-363.

[37] Burk RF, Ludden TM (1989) Exhaled alkanes as indexes of invivo lipid-peroxidation. Biochem. Pharmacol. 38: 1029-1032.

[38] Nikolaidis MG, Kyparos A, Vrabas IS (2011) F(2)-isoprostane formation, measurement and interpretation: The role of exercise. Prog. Lipid Res. 50: 89-103.

[39] Janicka M, Kot-Wasik A, Kot J, Namiesnik J (2010) Isoprostanes-Biomarkers of Lipid Peroxidation: Their Utility in Evaluating Oxidative Stress and Analysis. Int. J. Mol. Sci. 11: 4631-4659.

[40] Soffler C, Campbell VL, Hassel DM (2010) Measurement of urinary F(2)-isoprostanes as markers of in vivo lipid peroxidation: a comparison of enzyme immunoassays with gas chromatography mass spectrometry in domestic animal species. J. Vet. Diagn. Invest. 22: 200-209.

[41] Gutteridge JMC, Halliwell B (1990) The measurement and mechanism of lipid-peroxidation in biological-systems. Trends Biochem.Sci. 15: 129-135.

[42] Halliwell B, Chirico S (1993) Lipid-peroxidation - its mechanism, measurement, and significance Am. J. Clin. Nutr. 57: S715-S725.

[43] Devasagayam TPA, Boloor KK, Ramasarma T (2003) Methods for estimating lipid peroxidation: An analysis of merits and demerits. Indian J. Biochem. Biophys. 40: 300-308.

[44] DeLong JM, Prange RK, Hodges DM, Forney CF, Bishop MC, Quilliam M (2002) Using a modified ferrous oxidation-xylenol orange (FOX) assay for detection of lipid hydroperoxides in plant tissue. J. Agric. Food Chem. 50: 248-254.

[45] Esterbauer H, Striegl G, Puhl H, Rotheneder M (1989) Continuous monitoring of invitro oxidation of human low-density lipoprotein. Free Rad. Res. Commun. 6: 67-75.

[46] Yagi K (1984) Assay for blood-plasma or serum Methods Enzymol. 105: 328-331.

[47] Hong R, Kang TY, Michels CA, Gadura N (2012) Membrane Lipid Peroxidation in Copper Alloy-Mediated Contact Killing of Escherichia coli. Appl. Environ. Microbiol. 78: 1776-1784.

[48] Martins DB, Mazzanti CM, Franca RT, Pagnoncelli M, Costa MM, de Souza EM, Goncalves J, Spanevello R, Schmatz R, da Costa P, Mazzanti A, Beckmann DV, Cecim MD, Schetinger MR, Lopes STD (2012) 17-beta estradiol in the acetylcholinesterase activity and lipid peroxidation in the brain and blood of ovariectomized adult and middle-aged rats. Life Sciences. 90: 351-359.

[49] Kwok CT, van de Merwe JP, Chiu JMY, Wu RSS (2012) Antioxidant responses and lipid peroxidation in gills and hepatopancreas of the mussel Perna viridis upon exposure to the red-tide organism Chattonella marina and hydrogen peroxide. Harmful Algae. 13: 40-46.

[50] Ahmad MK, Syma S, Mahmood R (2011) Cr(VI) Induces Lipid Peroxidation, Protein Oxidation and Alters the Activities of Antioxidant Enzymes in Human Erythrocytes. Biol. Trace Elem. Res. 144: 426-435.

[51] Naziroglu M, Akkus S, Celik H (2011) Levels of lipid peroxidation and antioxidant vitamins in plasma and erythrocytes of patients with ankylosing spondylitis. Clin. Biochem. 44: 1412-1415.

[52] Pimentel VC, Pinheiro FV, Kaefer M, Moresco RN, Moretto MB (2011) Assessment of uric acid and lipid peroxidation in serum and urine after hypoxia-ischemia neonatal in rats. Neurol. Sci. 32: 59-65.

[53] Tonin AA, Thome GR, Calgaroto N, Baldissarelli J, Azevedo MI, Escobar TP, dos Santos LG, Da Silva AS, Badke MRT, Schetinger MR, Mazzanti CM, Lopes STD (2011) Lipid

Peroxidation and Antioxidant Enzymes Activity of Wistar Rats Experimentally Infected with Leptospira interrogans. Acta Sci. Vet. 39: 966-976.

[54] Elsaadani M, Esterbauer H, Elsayed M, Goher M, Nassar AY, Jurgens G (1989) A spectrophotometric assay for lipid peroxides in serum-lipoproteins using a commercially available reagent. J. Lipid Res. 30: 627-630.

[55] NouroozZadeh J, TajaddiniSarmadi J, Ling KLE, Wolff SP (1996) Low-density lipoprotein is the major carrier of lipid hydroperoxides in plasma - Relevance to determination of total plasma lipid hydroperoxide concentrations. Biochem. J. 313: 781-786.

[56] Eymard S, Genot C (2003) A modified xylenol orange method to evaluate formation of lipid hydroperoxides during storage and processing of small pelagic fish. Eur. J. Lipid Sci. Technol. 105: 497-501.

[57] Corongiu FP, Banni S (1994) Detection of conjugated dienes by 2nd-derivative ultraviolet spectrophotometry. Oxygen Radicals in Biological Systems, Pt C. 233: 303-310.

[58] Banni S, Day BW, Evans RW, Corongiu FP, Lombardi B (1995) Detection of conjugated diene isomers of linoleic-acid in liver lipids of rats fed a choline-devoid diet indicates that the diet does not cause lipoperoxidation. J. Nutr. Biochem. 6: 281-289.

[59] Dormandy TL, Wickens DG (1987) The experimental and clinical pathology of diene conjugation. Chem. Phys. Lipids. 45: 353-364.

[60] Chin SF, Storkson JM, Liu W, Pariza MW (1991) Dietary sources of the anticarcinogen CLA (conjugated dienoic derivatives of linoleic-acid). Faseb J. 5: A1444-A1444.

[61] Britton M, Fong C, Wickens D, Yudkin J (1992) Diet as a source of phospholipid esterified 9,11-octadecadienoic acid in humans. Clin. Sci. 83: 97-101.

[62] Fairbank J, Ridgway L, Griffin J, Wickens D, Singer A, Dormandy TL (1988) Octadeca-9-11-dienoic acid in diagnosis of cervical intraepithelial neoplasia. Lancet. 2: 329-329.

[63] Jack CIA, Ridgway E, Jackson MJ, Hind CRK (1994) Serum octadeca-9,11 dienoic acid - an assay of free-radical activity or a result of bacterial production. Clin. Chim. Acta. 224: 139-146.

[64] Thompson S, Smith MT (1985) Measurement of the diene conjugated form of linoleic-acid in plasma by high-performance liquid-chromatography - A questionable noninvasive assay of free-radical activity Chem.-Biol. Interact. 55: 357-366.

[65] Janero DR (1990) Malondialdehyde and thiobarbituric acid-reactivity as diagnostic indexes of lipid-peroxidation and peroxidative tissue-injury. Free Radic. Biol. Med. 9: 515-540.

[66] Yu TC, Sinnhuber RO (1957) 2-Thiobarbituric acid method for the measurement of rancidity in fishery products. Food Technol. 11: 104-108.

[67] Mihara M, Uchiyama M (1981) Evaluation of thiobarbituric acid (TBA) value as an index of lipid-peroxidation in CCl4-intoxicated rat-liver. Yakugaku Zasshi-J. Pharm. Soc. Jpn. 101: 221-226.

[68] Mihara M, Uchiyama M (1983) Properties of thiobarbituric acid-reactive materials obtained from lipid peroxide and tissue homogenate. Chem. Pharm. Bull. 31: 605-611.

[69] Mohebbi-Fani M, Mirzaei A, Nazifi S, Shabbooie Z (2012) Changes of vitamins A, E, and C and lipid peroxidation status of breeding and pregnant sheep during dry seasons on medium-to-low quality forages. Trop. Anim. Health Prod. 44: 259-265.

[70] Keles H, Ince S, Kucukkurt I, Tatli, II, Akkol EK, Kahraman C, Demirel HH (2012) The effects of Feijoa sellowiana fruits on the antioxidant defense system, lipid peroxidation, and tissue morphology in rats. Pharm. Biol. 50: 318-325.

[71] Kosugi H, Kato T, Kikugawa K (1987) Formation of yellow, orange, and red pigments in the reaction of alk-2-enals with 2-thiobarbituric acid. Anal. Biochem. 165: 456-464.

[72] Frankel EN (1991) Recent advances in lipid oxidation. J. Sci. Food Agric. 54: 495-511.

[73] Bonnestaourel D, Guerin MC, Torreilles J (1992) Is malonaldehyde a valuable indicator of lipid-peroxidation Biochem. Pharmacol. 44: 985-988.

[74] Lapenna D, Cuccurullo F (1993) TBA test and free MDA assay in evaluation of lipid-peroxidation and oxidative stress in tissue systems Am. J. Physiol. 265: H1030-H1031.

[75] Gutteridge JMC, Tickner TR (1978) Thiobarbituric acid reactivity of bile-pigments. Biochem. Med. 19: 127-132.

[76] Largilliere C, Melancon SB (1988) Free malondialdehyde determination in human-plasma by high-performance liquid-chromatography. Anal. Biochem. 170: 123-126.

[77] Halliwell B, Gutteridge JMC, Aruoma OI (1987) The deoxyribose method - a simple test-tube assay for determination of rate constants for reactions of hydroxyl radicals. Anal. Biochem. 165: 215-219.

[78] Wade CR, Vanrij AM (1989) Plasma malondialdehyde, lipid peroxides, and the thiobarbituric acid reaction. Clin. Chem. 35: 336-336.

[79] Wade CR, Jackson PG, Vanrij AM (1985) Quantitation of malondialdehyde (MDA) in plasma, by ion-pairing reverse phase high-performance liquid-chromatography. Biochem. Med. 33: 291-296.

[80] Wade CR, Vanrij AM (1988) Plasma thiobarbituric acid reactivity - Reaction conditions and the role of iron, antioxidants and lipid peroxy-radicals on the quantitation of plasma-lipid peroxides. Life Sciences. 43: 1085-1093.

[81] Wong SHY, Knight JA, Hopfer SM, Zaharia O, Leach CN, Sunderman FW (1987) Lipoperoxides in plasma as measured by liquid-chromatographic separation of malondialdehyde thiobarbituric acid adduct. Clin. Chem. 33: 214-220.

[82] Santos MT, Valles J, Aznar J, Vilches J (1980) Determination of plasma malondialdehyde-like material and its clinical-application in stroke patients. J. Clin. Pathol. 33: 973-976.

[83] Mezes M, Bartosiewicz G (1983) Investigations on vitamin-E and lipid peroxide status in rheumatic diseases. Clin. Rheumatol. 2: 259-263.

[84] Kedziora J, Bartosz G, Gromadzinska J, Sklodowska M, Wesowicz W, Scianowski J (1986) Lipid peroxides in blood-plasma and enzymatic antioxidative defense of erythrocytes in Downs-syndrome. Clin. Chim. Acta. 154: 191-194.

[85] Takagi T, Mitsuno Y, Masumura M (1978) Determination of peroxide value by colorimetric iodine method with protection of iodide as cadmium complex. Lipids. 13: 147-151.

[86] Jessup W, Dean RT, Gebicki JM (1994) Iodometric determination of hydroperoxides in lipids and proteins. Oxygen Radicals in Biological Systems, Pt C. 233: 289-303.
[87] Hicks M, Gebicki JM (1979) Spectrophotometric method for the determination of lipid hydroperoxides. Anal. Biochem. 99: 249-253.
[88] Cramer GL, Miller JF, Pendleton RB, Lands WEM (1991) Iodometric measurement of lipid hydroperoxides in human plasma. Anal. Biochem. 193: 204-211.
[89] Chajes V, Sattler W, Stultschnig M, Kostner GM (1996) Photometric evaluation of lipid peroxidation products in human plasma and copper oxidized low density lipoproteins: Correlation of different oxidation parameters. Atherosclerosis. 121: 193-203.
[90] Gay C, Collins J, Gebicki JM (1999) Hydroperoxide assay with the ferric-xylenol orange complex. Anal. Biochem. 273: 149-155.
[91] Griffiths G, Leverentz M, Silkowski H, Gill N, Sanchez-Serrano JJ (2000) Lipid hydroperoxides in plants. Biochem. Soc. Trans. 28: 837-839.
[92] Nouroozzadeh J, Tajaddinisarmadi J, Birlouezaragon I, Wolff SP (1995) Measurement of hydroperoxides in edible oils using the ferrous oxidation in xylenol orange assay J. Agric. Food Chem. 43: 17-21.
[93] Burat KM, Bozkurt O (1996) Improvement of calibration curve for determining peroxide values of food lipids by the modified ferrous oxidation-xylenol orange method. J. AOAC Int. 79: 995-997.
[94] Bou R, Codony R, Tres A, Decker EA, Guardicila F (2008) Determination of hydroperoxides in foods and biological samples by the ferrous oxidation-xylenol orange method: A review of the factors that influence the method's performance. Anal. Biochem. 377: 1-15.
[95] Gupta BL (1973) Microdetermination techniques for H2O2 in irradiated solutions. Microchem J. 18: 363-374.
[96] Hermeslima M, Willmore WG, Storey KB (1995) Quantification of lipid-peroxidation in tissue-extracts based on Fe(III)xylenol orange complex-formation. Free Radic. Biol. Med. 19: 271-280.
[97] Jiang ZY, Hunt JV, Wolff SP (1992) Ferrous ion oxidation in the presence of xylenol orange for detection of lipid hydroperoxide in low-density-lipoprotein Anal. Biochem. 202: 384-389.
[98] Wolff SP Ferrous ion oxidation in presence of ferric ion indicator xylenol orange for measurement of hydroperoxides, in: L. Packer (Ed.), Oxygen Radicals in Biological Systems, Pt C, 1994, pp. 182-189.
[99] Banerjee D, Madhusoodanan UK, Sharanabasappa M, Ghosh S, Jacob J (2003) Measurement of plasma hydroperoxide concentration by FOX-1 assay in conjunction with triphenylphosphine. Clin. Chim. Acta. 337: 147-152.
[100] Arab K, Steghens JP (2004) Plasma lipid hydroperoxides measurement by an automated xylenol orange method. Anal. Biochem. 325: 158-163.
[101] Sodergren E, Nourooz-Zadeh J, Berglund L, Vessby B (1998) Re-evaluation of the ferrous oxidation in xylenol orange assay for the measurement of plasma lipid hydroperoxides. J. Biochem. Biophys. Methods. 37: 137-146.

[102] Grau A, Codony R, Rafecas M, Barroeta AC, Guardiola F (2000) Lipid hydroperoxide determination in dark chicken meat through a ferrous oxidation-xylenol orange method. J. Agric. Food Chem. 48: 4136-4143.

[103] Griffiths G, Leverentz M, Silkowski H, Gill N, Sanchez-Serrano JJ (2000) Lipid hydroperoxide levels in plant tissues. J. Exp. Bot. 51: 1363-1370.

[104] Huska D, Adam V, Babula P, Trnkova L, Hubalek J, Zehnalek J, Havel L, Kizek R (2011) Microfluidic robotic device coupled with electrochemical sensor field for handling of paramagnetic micro-particles as a tool for determination of plant mRNA. Microchim. Acta. 173: 189-197.

[105] Sochor J, Salas P, Zehnalek J, Krska B, Adam V, Havel L, Kizek R (2010) An assay for spectrometric determination of antioxidant activity of a biological extract. Lis. Cukrov. Repar. 126: 416-417.

[106] Sochor J, Ryvolova M, Krystofova O, Salas P, Hubalek J, Adam V, Trnkova L, Havel L, Beklova M, Zehnalek J, Provaznik I, Kizek R (2010) Fully automated spectrometric protocols for determination of antioxidant activity: Advantages and disadvantages. Molecules. 15: 8618-8640.

Chapter 7

Liposomes as a Tool to Study Lipid Peroxidation

Biljana Kaurinovic and Mira Popovic

Additional information is available at the end of the chapter

http://dx.doi.org/10.5772/46020

1. Introduction

Lipid peroxidation is used as a marker of cellular oxidative stress and contributes to the oxidative damage that occurs as a result of xenobiotics metabolism, inflammatory processes, ischemia, reperfusion injuries and chronic diseases such as atherosclerosis and cancer [1,2].

Cell membrane lipids (phospholipids, glycolipids and cholesterol) are the most common substrates of oxidative attack. Once initiated reaction autocatalytic continues, it has pro-gradient flow, and the ultimate consequence is the structural-functional changes of the substrate. Lipid peroxidation is one of the best studied processes of cell damage under conditions of oxidative stress [3-5]. In 1960s Hochstein et al. [6] found that the initiation of lipid peroxidation require the presence of iron ions. From that moment the mechanism of lipid peroxidation process has been studied in many *in vitro* systems. However, accurate and precise mechanism is still not fully understood. Peroxidation in liposomes is usually studied after adding iron ions (Fe^{2+} plus ascorbic acid). Although the mechanism is not fully understood, it is known that redox chemistry of iron plays an important role in the occurrence and the rate of lipid peroxidation. Many studies have shown that the iron-dependent lipid peroxidation in systems comprised initially of Fe^{2+} and liposomes requires Fe^{2+} oxidation. In their research work, Minotti and Aust [7] assumed that the complex is formed between Fe^{2+} and Fe^{3+} ions could be initiator of iron-dependet lipid peroxidation. However, the existence of this complex has never been proven. In contrast, Aruoma et al. [8] argue against the participation of a Fe^{2+}-Fe^{3+}-O_2 complex, or a critical 1:1 ratio of Fe^{2+} to Fe^{3+}, in the initiation of lipid peroxidation in liposomes. Study of Tang et al. [9] showed that whether adding 100 or 150 mM Fe^{2+} initially or adding 100 mM Fe^{2+} initially and then 50 mM Fe^{2+} later at various times during the latent period in the liposomal system, the concentration of the remaining Fe^{2+} at the end of the latent period was almost the same every time.

Since lipid peroxidation causes oxidative damage to cell membranes and all other systems that contain lipids, in investigation of total antioxidative activity of plant extracts it is necessary to investigate their effects on lipid peroxidation. However, the impact of various natural products (isolated compounds and extracts) on the intensity of lipid peroxidation is studied in a number of substrate (linoleic acid, liposomes, various fatty oils, liver homogenates or hepatocytes isolated from it). Some substrates (liposomes and linolenic acids) are used more frequently than others mainly because of the simpler ways of performing the method. Also, due to the complex composition, examining the process of lipid peroxidation in fatty oils and liver homogenates makes research more difficult.

2. Liposomes as a model system

Liposomes are microscopic structure consisting of the one or more lipid bilayer enclosing the same number of water compartments. First, they were produced in Great Britain in 1961 by Alex D. Bangham while he was studying blood clotting. It was discovered that when phospholipids were combined with water they immediately formed a sphere. This is due to the fact that one end of each molecule is water soluble, while the oposite end is water insoluble. Water-soluble medications added to the water were trapped inside the aggregation of the hydrophobic ends; fat-soluble medications were incorporated into phospholipids layer and then – an important delivery system was born! Generally, such a structure formed polar lipids (such as phospholipids) [10]. Liposomes could be characterized as particles, similar to the structure and composition of cell membrane (Figure 1.). They occur in nature and could be artificially prepared [11].

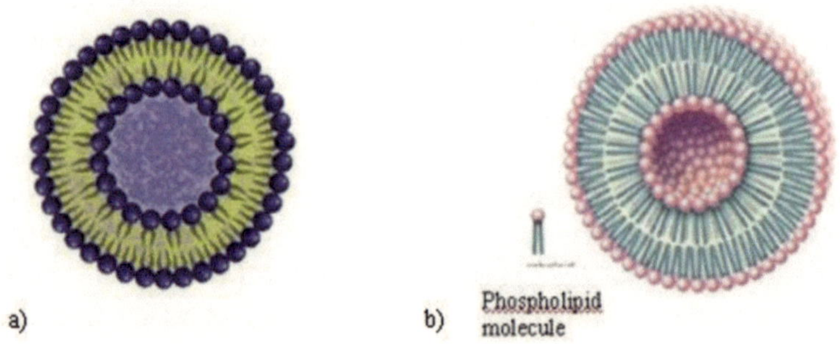

Figure 1. Example of a) empty liposome; b) liposome (2007 Encyclopadeia Britannica, Inc.)

The behaviour of liposomes in physical and biological systems is governed by the factors such as physical size, membrane permeability, percent entrapped solutes, chemical composition (estimation of phospholipids, phospholipids oxidation, and analysis of cholesterol), and quantity and purity of the starting material. Therefore, liposomes are characterized for physical attributes: shape, size, and its distribution; percentage drug capture; entrapped volume; lameliarity; percentage drug release. Based on the structure and size, we distinguish between different types of liposomes: Multilamellar Vesicles (MLV, size >0.5μm), Oligolamellar Vesicles (OLV, size 0.1-1μm), Unilamellar Vesicles (UV, all size ranges), Multivesicular Vesicle (MVV/MV, size >1μm). Unilamellar Vesicles are further divided into Small Unilamellar Vesicles (SUV, size 20-50nm), Medium Unilamellar Vesicles (MUV, size 50-100nm), Large Unilamellar Vesicles (LUV, size >100nm) and Giant Unilamellar Vesicles (GUV) (Figure 2.).

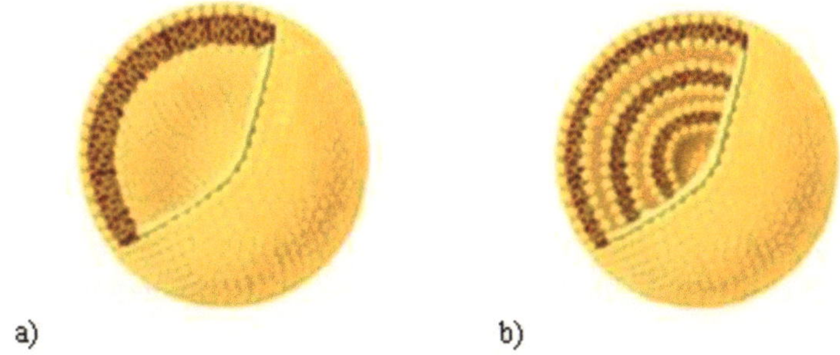

Figure 2. Example of a) unilamellar liposome; b) multilamellar liposome

Based on composition and applications, liposomes are divided into conventional liposomes (CL), fusogenic liposomes, pH sensitive liposomes, cationic liposomes, long circulatory (stealth) liposomes (LCL) and immuno-liposomes [12]. It is very difficult to measure dirrectly the phospholipid concentration, since dried lipids can often contain considerable quantities of residual solvent. Because of that, the method most widely used is an indirect one in which the phosphate content of the sample is first measured. The phospholipid concentration is measured using two methods - Bartlett and Stewart. In the Bartlett method the phospholipid phosphorous in the sample is first hydrolyzed to inorganic phosphate. This is converted to phospho-molybdic acid by the addition of ammonium molybdate and phospho-molybdic acid is quantitatively reduced to a blue colored compound by amino-naphthyl-sulfonic acid. The intensity of the blue color is measured spectrophotometrically and is compared with the curve of standards to give phosphorous and hence phospholipid content. This method is very sensitive. The problem is that test is easily upset by trace contamination with inorganic phosphate. In the other test, Stewart test, the phospholipid forms a complex with ammonium ferrothiocyanate in organic solution. The advantage of this method is that the presence of inorganic phosphate does not interfere with the test.

Until recently, liposomes are used as inert particles, carries of active principles, mostly for cosmetic purposes [13]. Today liposomes are used as very useful models, reagents and tools in various scientific disciplines, including biophysics (properties of cell membranes and channels), chemistry (catalysis, energy conversion, photosynthesis), biochemistry (the function of membrane proteins) and biology (excretion, cellular functions, transports and signaling, the transfer of genes and their functions). Liposomal formulation of several active molecules are currently in pre-clinical and clinical trials in different fields, with promising results. Two of the key problems in drug therapy (biodistribution throught the body and targeting to specific receptors) can be overcome by using liposomal formulations – liposomes protect encapsulated molecules from degradation and can passively target tissues or organs that have a discontinuous endothelium, such as liver, spleen, and bone marrow [14]. Comercial use of liposome was based on their colloidal, chemical and surface and microcapsuled proporties. These products include dosage formes of drugs (anti-cancer and antifugal agents, vaccines), cosmetic formulation (skin care products, shampoos), diagnostic products, a variety of applications in the food chemistry, as well as oral nutrient transport (liposomal vitamins, minerals and plants extracts for oral use). Liposome stability is an importrant aspect that must be met to be able to apply. By selecting the optimal value and size, pH and ionic strenght and the addition of complexing agents, liquid liposomical formulations could be stable for years.

Liposomal models have helped us to better understand the structure and dynamics of natural biomembrane systems. The concepts of structure and function of biomembranes, such as membrane fluidity, phase transition, the movement of lipids and proteins, triggering prosesses that affect metal ions or pH, have been very developed in this way. Modulatind effect of internal molecules (such as cholesterol) and insight into the mechanisms of membrane permeabillity for non-electrolytes and ions, are obtained by testing the model membranes. Liposomes that contains proteins as a components of membrane (reconstructes liposomes) were used in testing lipid-protein interactions in biological membranes, in examining the activities of active components such as membrane ionophores, anesthetics and divalent cations and mechanisms of antybody-antigen interactions [10].

Liposomes are very good models because they show the selectivity of the membrane to ions, osmotic swelling and response of range to agents that speed up or slow down the loss of ions and molecules from the particles in a way that at least, qualitatively mimic their activity in the natural membrane systems. Liposomes have also been successfully applied to „exclude the role" of membranes lipids and other components in biomembranes interact with the physical or chemical agents. Nevertheless, liposomal systems are useful because they allow the manipulation of membrane lipid composition, pH, temperature, content of different compounds in a limited way and provide the ability to determine the individual effect of the investigation product [15].

An example of the advantages of liposome in investigation of lipid peroxidation is that the influence of free radicals can be explored in the absence of chemical systems that produce free radicals, which may affect the test reaction. It is also possible to control the chemical composition of the liposome. This is particularly useful for the determination of lipid peroxidation induced by different systems for the generation of free radicals, and monitoring the overall effect of the combined system, or synergistic effects of combined systems that may arise. In addition to this it is possible to determine the antioxidant activity of tested compounds and determine which system works the best, by simple monitoring of lipid peroxidation. The tests used to determine the power of antioxidants to exert suppression of lipid peroxidation based on an assessment of the strength of oxidation of lipid substrate in the presence or absence of potential antioxidant molecules of plant extracts. There are four different strategies for assessing antioxidant capacity of molecules to the lipid substrate. They include the determination of oxygen consumed, the loss of substrate and formation of primary and secondary oxidation products [16]. The first method for determining the degree of lipid peroxidation, which includes the determination of oxygen consumed, based on following of initiation phase and its extension in the absence of antioxidants. The second method is based on measuring the loss of substrate in systems such as samples of food or biological samples and is very complicated, because they are full of potential oxiable substrates that are difficult to identify and characterize. The third method is based on monitoring the formation of primary oxidation products. It is a method that is well adapted to study such complex model systems and often involves the spectrophotometric determination of hydroperoxide, the dominant primary products of lipid peroxidation. Monitoring of secondary products of oxidation is the most commonly used method for the study of lipid model systems and lipid isolated from their natural environment. Both the *in vitro* and *in vivo* conditions are very often used TBA (thiobarbituric acid) test for detection of MDA (malondialdehyde), a secondary product of oxidation. This test is based on the reaction between TBA and MDA, which produces red chromophore with maximum of absorbance at 532 nm. This reaction is widely used and is performed by means of determination of many oxiable substrates (free fatty acids, LDL, body fluids). However, this method has some drawbacks. One is that the MDA is formed from free fatty acids which contain at least three double bonds. The next disadvantage is that the TBA is not specific for MDA because it can react with other aldehydes, such as occurs brown color that comes from the reactions of decomposition of sugar, amino acids, proteins and nucleic acids. Finally, MDA is not generated during the oxidation of many lipids and is often less important secondary oxidation product, and therefore not representative enough for the individual measurements. However, the TBA test was held for examination of lipid peroxidation, due to the simplicity of the method.

Despite all these advantage, liposomal systems remain different from the natural cellular systems. For this reason, regardless of the results obtained by testing the liposomes, they could not be reproduced on the natural membrane system, but they can provide useful information.

Therefore, the liposomes are still mostly used as a model system of biological membrane for testing the LP, especially when testing extracts and essential oils from plants on the intensity of LP. These studies are important because free radical oxidation of lipid components of food is a major strategic problem of food producers. The degree of oxidation of fatty acids and their esters in foods depends on the chemical structure of fatty acids, food processing technology, the temperature at which food is stored or prepared for eating and the presence of antioxidants. Synthetic antioxidants are widely used in many foods to retard undesirable changes as a result of oxidation. Chemicals, like tert-butyl-4-hydroxyanisole (BHA) and tert-butyl hydroxytoluene (BHT), can be used as antimicrobial and antioxidants agents. However, the use of some of these chemicals is restricted in several countries, as they may be dangerous to human health [17]. Therefore, the search for new natural antioxidant sources has been greatly intensified. For this reason, there is a growing interest in the studies of natural additives as potential antioxidants. The antioxidant properties of many herbs and spices are reported to be effective in retarding the process of lipid peroxidation in oils and fatty foods and have gained the interest of many research groups. A number of studies on the antioxidant activities of various aromatic plants have been reported over the last 20 years [18,19]. Their aroma is associated with essential oils, complex mixtures of volatile compounds, dominated by mono- and sesquiterpenes. It is known that essential oils exhibit significant biological and pharmacological activities such as anti-inflammatory, antimicrobial, spasmolytic, stimulant effect on the CNS and the like. New research shows that they possess significant antitumor activity [20], and act as inhibitors of growth of breast tumors [21]. It was confirmed that essential oils of some aromatic plants possess a high antioxidant potential [22]. Widely used in the food industry to improve the flavor of foods.

In addition to essential oils, aromatic plants and characterized by the presence of plant phenolic compounds, primarily phenylpropanoids and coumarins which are proven to have multiple pharmacological activities. Studies of these secondary biomolecules have become intensified when some commercial synthetic antioxidants found to be expressed toxic, mutagenic and carcinogenic activities [23]. In addition, it was found that excessive production of oxygen radicals in the body initiates oxidation and degradation of polyunsaturated fatty acids. It is known that free radicals attack the highly unsaturated fatty acid of membrane system and induce lipid peroxidation, which is a key process in many pathological conditions, and one of the reactions caused by oxidative stress. Particularly vulnerable are the biological membrane lipids in the spinal cord and brain because they contain high oxiable polyunsaturated fatty acids. These features facilitate the formation of oxygen radicals involved in the processes of aging, Alzheimer's and Parkinson's disease, ischemic damage, arthritis, myocardial infarction, arteriosclerosis and cancer. Phenolic antioxidants "stop" oxygen free radicals and free radicals formed from the substrate by giving hydrogen atom or an electron. Some flavonoids have strong inhibitory effect on lipid peroxidation processes. This action is based on their ability to chelate transition metal ions, thereby preventing the formation of radicals (initiators of LP), caught radicals initiators of LP (ROS), scavenge lipid-alkoxyl and lipid-peroxyl radicals and regenerate α-tocopherol by reduction of α-tocopheryl radicals. Flavonoids have the following characteristics: 3 ', 4'-dihydroxy group in ring B, or 4-keto and 3-hydroxy group in C ring, or 4'-keto group in C ring and 5-hydroxy group in A ring have the metal chelated properties (Figure 3.).

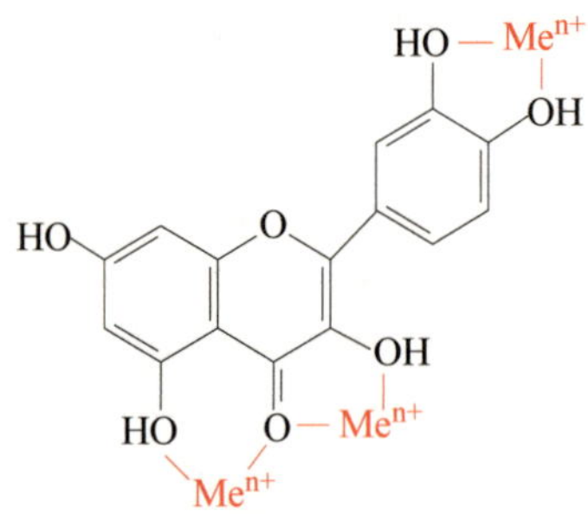

Figure 3. Possible places on flavonoids for chelating the transition metal ions in the process of lipid peroxidation.

Different metals have different binding affinity of the flavonoids [24]. Thus, for example, iron has the highest binding affinity for 3-OH group of ring C, then catechol group ring B and at the end of 5-OH group of ring A, while the copper ions bind to the first ring catechol group B [25]. Solubility of flavonoids in the lipid phase and the ability to penetrate the lipid membrane is small, since flavonoids in nature are mostly in the form of polar glycosides. Numerous tests of the inhibitory effects of flavonoids on lipid peroxidation were carried out on models of cell membranes. Based on these studies, it is assumed that quercetin and other flavonoids probably located on the surface membrane could easily capture radicals from the aqueous phase and thus prevent the initiation of LP. Thus located, flavonoids faster capture radicals initiators LP than α-tocopherol, which is located within phospholipid bilayer and that the switch is a typical chain reaction. Prevention of initial attacks radicals from the aqueous phase to membrane phospholipids is essential in the antioxidant protection of biomembranes because free radicals are constantly generated in the aqueous phase of cellular and sub cellular structure [25,26].

In the present chapter, lipid peroxidation in a liposomal system was initiated by Fe^{2+}-ascorbic acid system and the effects of four different Lamiaceae species (*Melittis melissophyllum, Marrubium peregrinum, Ocimum basilicum,* and *Origanum vulgare*) extracts and essential oils were investigated. Particular attention was paid to the chemical composition of extracts and essential oils and their capability to reduce lipid peroxidation. The plant leaves were dried in air and ground in a mixer. Finely powdered material (200 g) was macerated three times in 70% methanol (MeOH) with 4 L during a 24-h period. The macerates were collected, filtered, and evaporated to dryness under vacuum. The residues were dissolved in water and successively extracted with four solvents of increasing polarity: ether (Et_2O), chloroform ($CHCl_3$), ethyl acetate (EtOAc), and *n*-butanol (*n*-BuOH). The extraction was carried out until a colorless extract was obtained. The residue was the aqueous extract. All of five extracts (Et_2O, $CHCl_3$, EtOAc, *n*-BuOH, and H_2O) were evaporated to dryness and then dissolved in 50% ethanol to make 10% (w=v) solutions. Both, these and the diluted solutions, were further used for examination. Essential oil was made when air-dried plant material was submitted to hydrodistillation according to Eur. Pharm. 4 [27], using *n*-hexane as a collecting solvent. The solvent was removed under vacuum. The oils were dried over anhydrous sodium sulphate and kept at +4 ºC. The inhibition of LP was determined by measuring the formation of secondary components (malondialdehyde) of the oxidative process, using liposomes as an oxidizable substrate [28-30]. However, because the thiobarbituric acid test is not specific for MDA, other non-lipid substances present in plant extracts, or peroxidation products other then malondialdehyde, could react positevely with TBA. These interfering compounds distort the results and therefore all the final results of investigated extracts have been corrected using the absorbances of the investigated extracts after the TBA-test (without liposomes) [31]. The commercial preparation of liposomes 'PRO-LIPO S' (Lucas-Meyer) pH = 5–7 was used as a model system of biological membranes. The liposomes, 225–250 nm in diameter, were obtained by dissolving the commercial preparation in demineralized water (1:10), in an ultrasonic bath.

3. Lamiaceae (Labiatae) family

The Lamiaceae family is one of the largest and most distinctive families of flowering plants, with about 220 genera and almost 4000 species worldwide [32]. Lamiaceae are best known for the essential oils common to many members of the family [33]. The family was established by De Jussieu in 1789 as the order Labiatae. This was the original family name, so given because the flowers typically have petals fused into an upper lip and a lower lip, the flower thus having an open mouth. Although this is still considered an acceptable alternative name, most botanists now use the name "Lamiaceae" in referring to this family. The main centre of diversity is the Mediterranean region to central Asia. Members are found in tropical and temperature regions [34]. All Lamiaceae are aromatic plants. The essential oil contains mainly monoterpenes, sesquiterpenes and phenylpropanoid compounds. Also, the plant species of Lamiaceae have been shown as rich sources of phenolic compounds mostly flavonoids and phenolic acids.

3.1. Balm (*Melittis melissophyllum* L.)

The name *melittis* of the genus derives from a Greek words *Melissa* or *Melitta,* meaning "honey bee" and refers to the properties of flowers of attracting these insects. The name *melissophyllum* of the species simply means "with leaves similar to melissa". This is a tall plant which likes shady places and is ideal for a sunny woodland edge or scrubby border, where it will be attractive to bees and other insects. Bastard balm is a strongly aromatic plant that smells like fresh mowed grass and has erect hairy stems. It blooms white with a large pinkish purple blotch on the lower lip. The flowers are hermaphrodite and get pollinated by bees and moths. It has oval, bluntly-toothed, leaves in opposite pairs up the stems. Bastard balm is a herb native to the Mediterranean region.

Main flavonoids in balm are glycosides of apigenin and luteolin. However, presence of some other flavonoids as kaempherol, quercetin (Figure 4.) and ramnocitrin have been also reported [35].

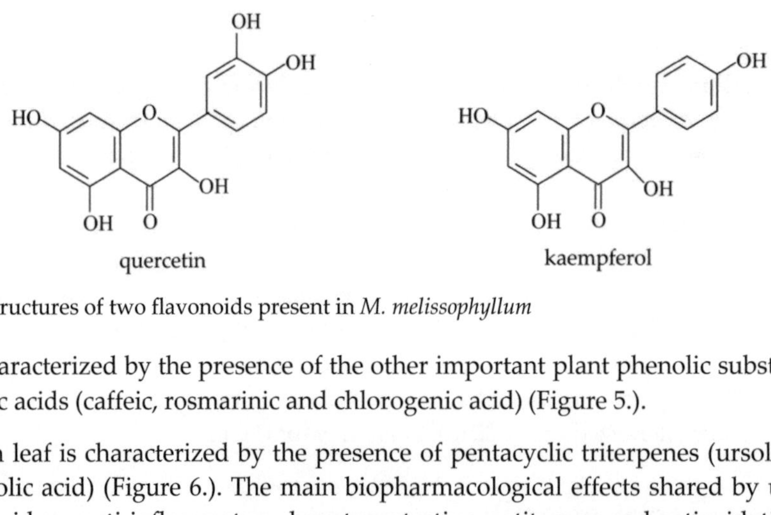

Figure 4. Structures of two flavonoids present in *M. melissophyllum*

Balm is characterized by the presence of the other important plant phenolic substances such as phenolic acids (caffeic, rosmarinic and chlorogenic acid) (Figure 5.).

Also, balm leaf is characterized by the presence of pentacyclic triterpenes (ursolic, pomolic and oleanolic acid) (Figure 6.). The main biopharmacological effects shared by ursolic and oleanolic acid are anti-inflammatory, hepatoprotective, antitumor, and antioxidative [36-39].

Essential oil is present in all parts of the plant. The largest amount of oil obtained from aerial parts of plants, harvested in late summer. Balm leaves contain no more than 0.13% of essential oil which is of complex and variable composition. Among the more than 50 compounds identified to date, citronellal (dominantly the (*R*) enantiomer), β-caryophyllene, ◎-caryophyllene oxide, germacrene-D, nerol, geranial, citronellol, and geraniol amount to about 70% of the oil (Figure 7.) [40]. The composition is similar to that of lemongrass, but balm oil can be identified by its typical pattern of chiral compounds; for example, almost enantiomerically pure (*R*)-(+◎)-methyl citronellate is a good indicator of true balm oil. For distinguish between two oils there is used the carbon isotopic ratio (IRMS-*isotope ratio mass spectrometry)* [41]. The essential oil exhibits spasmolitic action and acts as a muscle relaxant, sedative, narcotic, antibacterial, and antifungal [42,43].

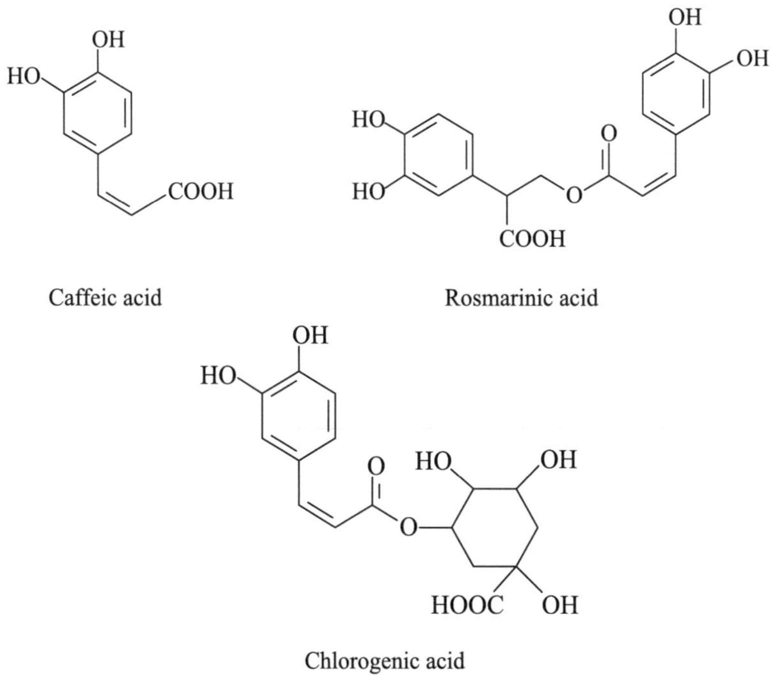

Figure 5. Structures of phenolic acids in *M. Melissophyllum*

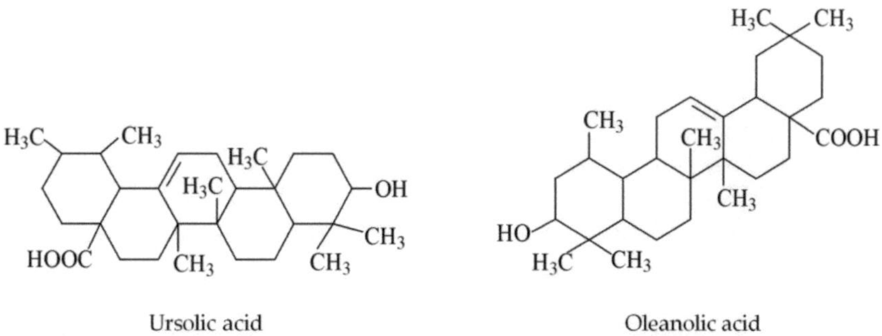

Figure 6. Structures of triterpenoids compounds present in *M. melissophyllum* leaves.

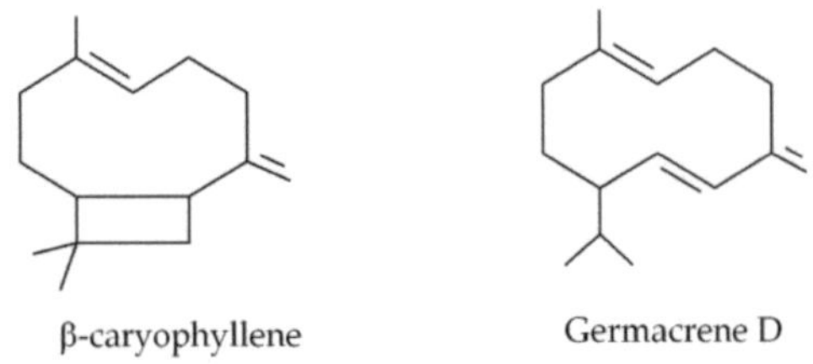

Figure 7. Sesquiterpenes in *M.melissophyllum* leaves

Beneficial effects of plants introduced by ancient Greeks and Romans.There is overlap with the use of plants in folk medicine and science. In relation to its complex composition it has multiple medicinal effects. Its herb has wide applications in the folk medicine. Due to the soothing action balm leaves enters into the composition of tea for calming, which is recommended for hysteria and neuralgia. Balm leaves mixed with bitter herbs are a great tool to enhance appetite. Various preparations containing extract or essential oil balm leaves are used as an addition to baths against rheumatism. In the folk medicine of Belarus alcoholic extract is drunk for stomach ulcer and duodenum, to calm the pain in the stomach, intestines, the liver, heart, and women's diseases. Terpenes found in essential oil of balm leaves, have a relaxing and antiviral effects. Eugenol calms muscle spasms and destroy bacteria [44]. It is also used as a carminative and sedative. Recent results indicate that the balm extract acts as depressants and have sedative effect on central nervous system of mice [45]. In the folk medicine of central Italy inflorescences of this plant, called "Erba Lupa", were used under infusion as antispasmodic, against insomnia and eyes inflammations [46,47].

Our research on balm was recently extended to the comprehensive *in vitro* and *in vivo* studies of antioxidant properties of balm essential oil and extracts measuring their capability to reduce lipid peroxidation in liposomes and effect on some enzymes of antioxidant defense systems [48]. Investigation of balm essential oil showed that with increasing concentration of essential oil reduces the intensity of lipid peroxidation compared to hexane-control (Table 1). Also, only the most diluted solution of essential oil of balm (0.213 and 0.535 μg/mL) has a weaker protective effect than the synthetic antioxidant BHT. The capability to reduce lipid peroxidation of essential oil was dose-dependent. This high inhibitory effect of balm essential oil was found to be in correlation with the content of monoterpene alcohols and ketones.

	Concentration (μg/mL)					
	BHT	**0.213**	**0.535**	**1.065**	**1.598**	**2.130**
LP	26.15	13.05	21.03	39.62	46.09	56.81

Table 1. Inhibition of LP (%) in Fe^{2+}/ascorbate system of induction by essential oil of balm leaves and BHT (as a positive control) in the TBA assay.

The protective effects on lipid peroxidation of balm extracts have been evaluated using the Fe^{2+}/ascorbate system of induction, by the TBA-assay (Table 2.). In general, all of the examined extracts (except *n*-BuOH extract) expressed strong antioxidant capacity and ability to reduce lipid peroxidation in liposomes. The largest inhibitory activity was exhibited by EtOAc and H_2O extracts because the 5% solutions show better protective effect than BHT. All extracts of the highest concentrations (10%) exhibited a better inhibitory effect than BHT. Protective activity of these extracts and its components towards Fe^{2+}-dependent LP of liposomes can be explained by present of phenolic acids and flavonoids and their influence on antioxidative capacity of ascorbic acid, which doesn`t show a strong antioxidative effect in lipid phase, but different phenolic compounds can result increase of its antioxidant activity [49].

	Extracts					
Concentration	**BHT**	**Et_2O**	**$CHCl_3$**	**EtOAc**	***n*-BuOH**	**H_2O**
1%	26.15	17.52	16.15	22.87	-10.59	24.24
5%	26.15	20.82	20.59	27.88	-13.19	39.36
10%	26.15	26.40	28.40	38.47	-18.52	41.32

Table 2. Inhibition of LP (%) in Fe^{2+}/ascorbate system of induction by extracts of balm leaves and BHT (as a positive control) in the TBA assay.

It is known that quercetin, like many other flavonoids, prevents oxidation of LDL cholesterol, and its anti-inflammatory activity comes from inhibition of the enzyme lipooxigenase and inhibition of inflammatory mediators [50]. Kaempferol acts synergistically with quercetin to reduce the proliferation of malignant cells, and treatments are a combination of quercetin and kaempherol efficient than their single use [51]. It is, also, known that rutin has strong antioxidant effects, as well as a feature to built chelates with metal ions (e.g. iron) and reduces the Fenton reaction in which the resulting harmful oxygen radicals. It is supposed to stabilize vitamin C. If rutin is taken together with vitamin C, increases the activity of ascorbic acid [52]. In addition, HPLC-DAD analysis showed that the aqueous extract, in large quantities, present phenolic acids (rosmarinic, chlorogenic and caffeic acid), which are known antioxidants. It was determined that rosmarinic acid has stronger antioxidant activity than vitamin E. Rosmarinic acid prevents cell damage caused by free radicals and reduce the risk of cancer and atherosclerosis. In contrast to the histamines, rosmarinic acid prevents activation of the immune system cells that cause swelling and fluid collection. Also, it is known that the caffeic acid by far surpassing other antioxidants because it reduces the production of α-toxin for more than 95% [35]. Furthermore, it can be supposed that the reduction process of lipid peroxidation is caused, besides flavonoids, also by triterpenoids acids (especially ursolic, oleanolic, and pomolic acid) since non-polar extracts (Et_2O and $CHCl_3$) also exhibited high antioxidant potential [39]. The *n*-BuOH extract shows a prooxidative effect that is increased by increasing concentration of added extract. It can be supposed that compounds with polar groups were extracted by *n*-BuOH, and are present in high concentration in the extract. It is notable that molecules which show antioxidant activity, when they are present in high concentration, might behave as prooxidants [53], so *n*-BuOH extract of balm leaves probably have this kind of activity. The antioxidant activities of all extracts of balm leaves were dose dependent.

The represented antioxidant activity results show that extracts of examined plant species, especially EtOAc and H_2O extracts are efficient in the protection of tissues and cells from oxidative stress. Anyway, according to variations in regard to antioxidant activity of tested by different *in vitro* models, there are also requiste *in vivo* test that would confirm the capability of extracts to reduce the lipid peroxidation. *In vivo* tests are also necessary because a lot of plant phenols are biotransformed during their active metabolism. *In vivo* effects are evaluated on LP in the mice liver (Table 3.) and blood hemolysate (Table 4.) after treatment with examined balm extracts, or in combination with carbon tetrachloride (CCl_4).

Parameter	Control	Et_2O	$CHCl_3$	EtOAc	*n*-BuOH	H_2O
LP	7.19±0.23	7.36±0.21	7.91±0.19	6.71±0.16	7.12±0.23	6.19±0.27
LP + CCl_4	8.91±0.29	7.12±0.21	7.06±0.24	6.92±0.17	6.98±0.24	6.81±0.24

Table 3. Effect of extracts of balm leaves on intensity of lipid peroxidation (nmol malondialdehyde/mg of proteines) in liver homogenate before and after treatment with CCl_4

As compared with control, intensity of LP is statistically significant reduced during the treatment with ethylacetate and water extracts of balm leaves. The result derived by treatment with ethylacetate and water extracts is in according with amounts got *in vitro* experiment. Using $CHCl_3$ extract leads to a significant increase of LP intensity, whereas the other two extracts had no effect on this parameter. All extracts of balm leaves combine with CCl_4 have showed a statistically significant decrease of LP intensity, and this behavior of the extract probably results from the presence of secondary biomolecules like flavonoids and phenolic acids. Handa *et al.* [54] determined that secondary biomolecules such as flavonoids, xanthones and tannins in combination with CCl_4 have protective effects on liver. Phenolic components present in balm leaves (rutin, luteolin, kaempherol) are known as strong inhibitors of CCl_4-induced LP [55]. Flavonoids could affect the initiation phase of lipid peroxidation, where they influence the metabolism of CCl_4, they scavenge the free radicals, or they decrease the microsomal enzyme systems that are claimed for CCl_4 metabolism [56]. In continuation of this process, flavonoids can scavenge lipoperoxides and their radicals or they can act as chelating agents for Fe^{2+} ion, and in this way can stop Fenton reactions [57]. Furthermore, Afanas'ev et al. [28] found that quercetin and rutin exhibited a high inhibitory effect on the Fe^{2+}-induced liposomal LPx and NADPH-dependent CCl_4-induction LPx in liver microsomes. Luteolin, one of the main active component in the balm, is responsible for the inhibitory effect on the former reaction. In addition to the above-mentioned mechanism (chelate formation with Fe^{2+}) it is possible that these compounds (of flavonoid type) act as scavengers of OH radicals, whereby they are transformed in the corresponding radical form which is stabilized by resonance. On the basis of these results, it can be concluded that all of extracts of balm leaves showed protection effect in relation to the CCl_4-induced lipid peroxidation.

Similar results were obtained during examining the effects of extracts of bastard balm on LP in blood hemolysate in mice (Table 4.). Three extracts, $CHCl_3$, EtOAc and H_2O, induced a significant decrease of LP intensity, while Et_2O and *n*-BuOH ones decreased the level of this enzyme insignificantly.

Parameter	Control	Et_2O	$CHCl_3$	EtOAc	*n*-BuOH	H_2O
LP	4.81±0.24	4.59±0.28	3.78±0.17	2.96±0.13	4.74±0.19	4.07±0.24
LP + CCl_4	5.11±0.24	5.31±0.17	4.92±0.21	3.02±0.24	5.17±0.25	2.98±0.12

Table 4. Effect of extracts of balm leaves on intensity of lipid peroxidation (nmol malondialdehyde/mL erythrocytes) in blood hemolysate before and after treatment with CCl_4

The LP value showed a statistically insignificant increase with CCl_4-treated animals compared with the untreated ones. A clear protective effect was seen in experimental animals administered H_2O extract and CCl_4 compared with untreated animals. Furthermore, EtOAc extract also significantly decreased the activity of LP, while Et_2O, $CHCl_3$ and *n*-BuOH extracts did not change notably the levels of lipid peroxidation. These results suggest that these two extracts (EtOAc and H_2O) had a protective effect. According to the literature data [58], the reduction of the serum LP might be the result of antioxidant activity of several classes of plant phenolic constituents, such as cinnamic acids (ferulic, caffeic, and chlorogenic), flavonoids and biflavonoids, 1,3,6,7-tetrahydroxyxynthones, and acylphoroglycinols such as hyperforin and adhyperforin. Cock and Samman [59] showed that quercetin and rutin and their glycosides show strong inhibitory effect in respect of LP. The observed differences in the action of particular balm extracts are probably due to the different contents of flavonoids, but the potential protective effects of some other groups of compounds can not be ruled out.

3.2. Horehound (*Marrubium peregrinum* L.)

Marrubium genus includes about 40 species. Species of this genus growing in dry pastures, abandoned the places along the roads in central and southern Europe, but also in North Africa, in parts of Asia and the Americas. Horehound is a perennial plant with a rectangular stem, branched in the upper part. Rhizomes of this species are ligneous, leaves oblong, flowers grouped in loose inflorescence [60]. A common plant blooms from July to September and harvested in that period. It has a bitter and pungent taste and smell. It is the drug of Herba *Marrubii albi*. This plant doesn`t require special conditions for growth.

In previous phytochemical investigations on *M. peregrinum*, different groups of chemicals were isolated: flavones (apigenin and luteolin) [61] (Figure 8.), flavonols (kaempferol) [62], glycosylated flavonoids (quercetin-3-O-β-D-rutinoside, naringenin-7-O-β-D-glucoside, kaempferol-3-O-β-D-rutinoside, quercetin-3-O-β-D-glucoside) [63], caffeic acid derivatives [64], and four diterpenoids (peregrinin, peregrinol, marrubiin and premarrubiin) [65]. T. Hennebelle et al. [66] have established the presence of acteoside, forsythoside B, arenarioside and terniflorine (apigenin-7-*O*-[6″-E-p-coumaroyl]β-D-glucopyranoside) in the MeOH extract of *M. peregrinum*.

Marrubium peregrinum essential oil yield between 0.02-0.07% [67]. Dominant monoterpenes are: α-pinene, sabinene, limonene, camphene and α-terpinolene. In a Greek sample, β-phellandrene, epi-bicyclosesquiphellandrene and bicyclogermacrene proved to be the major compounds [68], whereas the essential oil of a sample from Central Europe was rich in β-caryophyllene and its oxide, bicyclogermacrene and germacrene D [69]. The main sesquiterpene compounds are Z- and E- β-farnesene (~12%), β-caryophyllene (~8.5%), heksahidrofarnesil acetone (~ 6.5%), spathullenol (~5%) i germacrene D (~4.5%) (Figure 9.) [68].

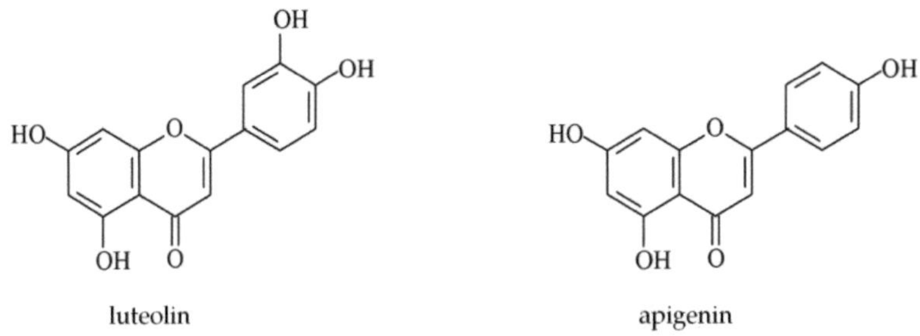

Figure 8. Figure 8. Structures of two main flavonoids in *M. peregrinum*.

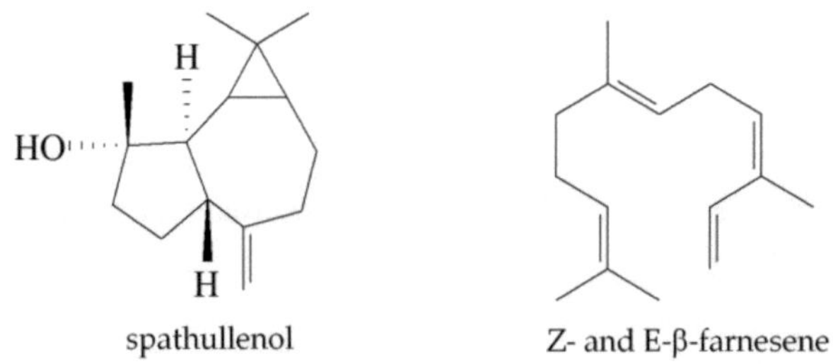

Figure 9. Main constituents of *M. peregrinum* essential oil

Some species of *Marrubium* are used in traditional and modern medicine. Many studies have shown various activities in this genus, such as hypoglycemic effect, anti-schistosoma, antioxidant, calcium channel blocker and hypotensive activity [70]. As a medicinal plant, *M. peregrinum* have been employed against vascular diseases (antihypertensive, antispasmolitic) [61].

In our comprehensive study of chemical and biochemical investigation of *M. peregrinum* from three different locations (Backo Gradiste-Rimski Sanac (No 1.); Novi Knezevac (No 2.) and Senta (No 3.)), we have identified more than 40 compounds in essential oil of *M. peregrinum* (44 for *M. peregrinum* from Senta locality, 42 for *M. peregrinum* from Novi Knezevac locality and 41 for *M. peregrinum* from Rimski Sanac locality, representing 96.15%, 87.60% and 83.66% of the total oil contents, respectively), in which dominant compounds were β-caryophyllene (13.20-17.99%), bicyclogermacrene (6.42-9.80%) and germacrene-D (6.79-9.05%). Besides sesquiterpene hydrocarbons, oxygenated sesquiterpenes, spathulenol (3.76-5.78%) and caryophyllene oxide (3.73-4.78%) are also present in relevant quantities. However, we must point out that the amounts of these components in essential oil from different localities are very different. Essential oil obtain from plant collected in Senta is the richest of sesquiterpene hydrocarbons (62.71%), while oxygenated sesquiterpenes are most represented (11.84%) in essential oil from plants collected in the Rimski Sanac area.

	Concentration (µg/mL)					
	BHT	**0.213**	**0.535**	**1.065**	**1.598**	**2.130**
***M. peregrinum* (No 1.)**	26.15	12.24	26.00	35.36	44.26	56.18
***M. peregrinum* (No 2.)**	26.15	19.15	20.05	39.37	52.14	61.18
***M. peregrinum* (No 3.)**	26.15	21.17	37.02	55.81	65.16	71.32

Table 5. Inhibition of LP (%) in Fe^{2+}/ascorbate system of induction by essential oil of *M. peregrinum* from three different location, and BHT (as a positive control) in the TBA assay.

Also, our study showed that all of the examined essential oils express strong antioxidant activity and capability to reduce lipid peroxidation (Table 5.). The largest inhibitory activity was exhibited by essential oil from plant collected at Senta locality (No. 3.). Solution of all concentrations, except the most diluted (0.213 µg/mL), have exhibited a stronger protective effect (from 37.02 to 71.32% of inhibition of LP) than BHT (26.15%). The other two essential oils (from Rimski sanac and Novi Knezevac), at higher concentration (from 1.065 to 2.130 µg/mL), have also exhibited more intense protective effect than BHT [71].

The effect of crude MeOH extracts of *M. peregrinum* was preliminarily determined from the three localities. There were taken three concentrations of MeOH extracts (1, 5, and 10% extracts). All of the examined extracts expressed stronger antioxidant capacity as compared to the 50% solution of MeOH. In particular, the largest inhibitory activity was established by the MeOH extracts of *M. peregrinum* collected from Senta locality. Also, the best results were obtained using solutions of the highest concentrations [72]. Because of all this there was carried out successive extractions of *M. peregrinum* from all three localities, and for further work 10% extracts are prepared. Successive extraction was performed as the extraction of antioxidant substances of different chemical structure, was achieved using solvents of different polarity. Numerous investigations of qualitative composition of plant extracts revealed the presence of high concentration of phenols in the extracts obtained using polar solvents [73]. The extracts that perform the highest antioxidant activity have the highest concentration of phenols. Phenols are very important plant constituents because of their scavenging ability on free radicals due to their antioxidant action [74]. The examination of capability to reduce intensity of lipid peroxidation of plant extracts from *M. peregrinum* showed different values (Table 6.). The first two extracts (Et_2O and $CHCl_3$) obtained from plants from all three locality are exhibited weaker protective effect than BHT, while the other three extracts (EtOAc, *n*-BuOH and H_2O) showed better protective properties than synthetic antioxidant.The largest inhibitory activity, again, was exhibited by the EtOAc and H_2O extracts of *M. peregrinum* collected from Senta locality.

Obtained results can be related to the experiments in which the total amount of flavonoids was determined, which show that EtOAc and H_2O extracts from Senta locality contains the largest amounts of total flavonoids, namely of luteolin, either being present as free or in the form of its glucosides. The suggested mechanism of flavonoid antioxidative action is as follows: the double bond in position 2, 3 is conjugated with C_4-carbonyl group, and free OH groups (C_5, C_3 and C_7) can form chelates with ions of d-elements. Once formed, complex with Fe^{2+} ion prevents formation of $OH^{\bullet}$ radicals in Fenton's reaction [59]. Also, luteolin is thought to play an important role in the human body as an antioxidant, a free radical scavenger, an agent in the prevention of inflammation, a promoter of carbohydrate metabolism, and an immune system modulator. These characteristics of luteolin are also believed to play an important part in the prevention of cancer. Multiple research experiments describe luteolin as a biochemical agent that can dramatically reduce inflammation and the symptoms of septic shock [75]. Furthermore, it is well known that some other flavonoids isolated from *M. peregrinum* possess certain biological and pharmacological activity. For example, apigenin, one of the flavonoids present in *M. peregrinum*, was shown to express strong antioxidant effects, increasing the activities of antioxidant enzymes and, related to that, decreasing the oxidative damage to tissues [61].

	Extracts					
	BHT	**Et_2O**	**$CHCl_3$**	**EtOAc**	***n*-BuOH**	**H_2O**
***M. peregrinum* (No 1.)**	26.15	9.38	14.22	29.41	27.39	32.35
***M. peregrinum* (No 2.)**	26.15	14.27	21.19	29.54	26.83	37.55
***M. peregrinum* (No 3.)**	26.15	17.11	23.52	38.83	28.73	41.18

Table 6. Inhibition of LP (%) in Fe^{2+}/ascorbate system of induction by extracts of *M. peregrinum* and BHT (as a positive control) in the TBA assay.

3.3. Basil (*Ocimum basilicum* L.)

Basil is originally native to India and other tropical regions of Asia, having been cultivated there for more then 5.000 years. Ocimum genus includes about 150 species [76]. There are many varieties of *Ocimum basilicum*, as well as several related species or species hybrids also called basil. These varieties differ in morphological and general structure, and also in the content and composition of essential oil. The chemotype is determined by chemical composition of essential oil and it is basic for chemotaxonomy within the genus Ocimum and species *Ocimum basilicum* [77].

The word market has several types of essential oils that differ in chemical structure, composition and fragrance. The dominant compounds of basil essential oil occur in two different biochemical pathways: phenylpropanoids (methyl chavicole, eugenol, methyl eugenol, and methyl cinnamate) through shicimic acid, and terpenoids (linalool and geraniol) through mevalonic acid. Based on chemical content, basils can be divided into four groups: European (French) *O. basilicum* (contains lower amounts of phenols); Exotic (contains methyl chavicol (40-80%)); Reunion and Javanean. European type of essential oil is the finest quality, has the finest fragrance and the highest price in the market. Other components that can be found in higher concentrations in this type of oil are: linalool, methyl chavicol (estragole) (Figure 10.), 1,8-cineole, eugenol, geraniol, germacrene D, α-terpinolene, β-caryophyllene, ocimene, sabinene, thujone, and γ-terpinene [78].

Among phenolic constituents flavonoids and their glucosides are dominant. The major flavonoids are: quercetin, kaempferol, apigenin, luteolin and rutin. Quercetin-3-Odiglucoside and kaempferol-3-O-β-rutinoside have been also identified. Beside, basil is rich in triterpenoid acids (ursolic and oleanolic), cinnamic acid (caffeic and rosmarinic), vitamin C and β-carotene, as well with calcium, copper, magnesium, sodium and potassium [79].

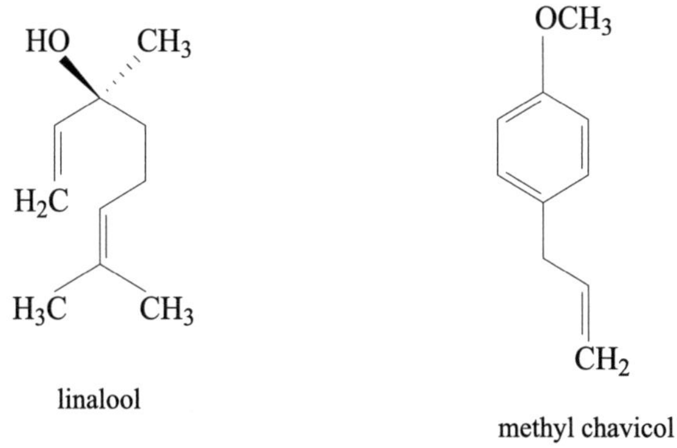

Figure 10. Main constituents of *O. basilicum* essential oil

Basilici herba has been used in traditional and homeopathic medicine to treat number of diseases. Essential oil (*Basilici aetheroleum*) extracted from fresh leaves and flowers can be used as aroma additives in foods, pharmaceuticals, and cosmetics [80]. Traditionally, basil has been used as a medicinal plant in the treatment of headaches, coughs, diarrhea, constipation, warts, worms, and kidney malfunction. Major aroma compounds from volatile extracts of basil present anti-oxidative activity [81]. Among the many studies to determine the antioxidant activities of basil, most have focused mainly on the antioxidant activities of crude extracts, using methanol, acetone, or water as a solvent [82,83].

	Concentration (μg/mL)					
	BHT	**0.213**	**0.535**	**1.065**	**1.598**	**2.130**
LP	26.15	24.12	35.17	48.41	64.13	79.14

Table 7. Inhibition of LP (%) in Fe^{2+}/ascorbate system of induction by essential oil of basil leaves and BHT (as a positive control) in the TBA assay.

In our investigation, the examined essential oil expressed strong antioxidant activity (Table 7.). Solutions of all concentrations, except the most diluted (0.213 μg/mL), have exhibited a stronger protective effect (from 35.17 to 79.14% of inhibition of LP) than BHT (26.15%). The largest inhibitory activity was achieved by using the solution of the highest concentration. For the inhibition of LP, the most responsible compounds were the oxygenated phenolic monoterpens (methyl chavicole) and the mixture of mono- and sesquiterpene hydrocarbons. These findings are in correlation with the earlier published data on the antioxidant activities of the investigated essential oil and selected oil components [84,85].

	Extracts					
Concentration	BHT	Et_2O	$CHCl_3$	EtOAc	*n*-BuOH	H_2O
1%	26.15	-0.87	-0.86	37.42	26.31	31.74
5%	26.15	-0.94	-0.89	38.91	27.06	35.29
10%	26.15	-1.01	-1.04	41.56	28.83	36.54

Table 8. Inhibition of LP (%) in Fe^{2+}/ascorbate system of induction by extracts of basil leaves and BHT (as a positive control) in the TBA assay.

The data presented in Table 8. show that the last three extracts of *O. basilicum* (EtOAc, *n*-BuOH and H_2O) reduced the intensity of lipid peroxidation, while the first two extracts (Et_2O and $CHCl_3$) increased the intensity of LP, but statistically insignificant. The largest inhibitory activity was exhibited by ethyl acetate extract. High inhibitory effect of this three extracts can be related to the presence of the amount of total phenolic compounds and content of total flavonoids in the extracts, because a considerably content of total phenolic compounds and total flavonoids was determined in EtOAc and H_2O extract of *O. basilicum*. Preliminary 2D-TLC (Two Dimensional - Thin Layer Chromatography) analysis showed that the dominant flavonoid in the EtOAc extract of *O. basilicum* is derivative of quercetin. It is known that quercetin shows high antioxidant activity because of present OH groups in position 3' ring B (includes a 3', 4'-dihydroxy group). In the same experiment we established the presence of caffeic acid and its derivatives in the H_2O extract, which has two hydroxyl groups in ortho position. This was confirmed once again that the antioxidant capacity depends not only on quantity, but also depends on of the type of phenols and flavonoids present in the extracts. However, in extracts of *O. basilicum* higher content of total phenols and flavonoids from the EtOAc extract had an H_2O extract. From all this it can be assumed that the polarity of flavonoid components affects their ability to inhibit the process of LP. Specifically, in this test we used liposomes as a model system of biological membranes, and the least polar flavonoids present in the EtOAc extract could help to approach the scene and engage in the process of defense from the LP, compared to more polar compounds that are found in H_2O extract. A little less of total flavonoids was determined in *n*-BuOH extracts, while the smallest quantity of these compounds was found in Et_2O and $CHCl_3$ extracts. Differences in the amount of total phenolic compounds and flavonoid content between extracts can be explained by different number of secretory structures in various plant tissues [86]. Furthermore, the obtained results could be related to the protective role of phenolics, especially the flavonoid aglycones, in plants collected on the outskirts of big cities. One of the functions of these biomolecules, which are produced in response to ecological stress factors like pollution, is to serve as UV-B filters in plants [87]. It was established that flavonoids act as mighty scavengers of free radicals [88]. Different flavonoids inhibit LP *in vitro* and the most pronounced effect is exhibited by quercetin whose presence is found in extracts of *O. basilicum* using 2D-TLC [89]. More investigation is required to explain the enhanced production of phenolics in certain geographic areas [90]. Also, from the presented results we can conclude that the increase in concentration of the extracts does not significantly affect the inhibition of lipid peroxidation.

3.4. Oregano (*Origanum vulgare* L.)

Origanum is one of the most variable genera of Lamiaceae family. Originates from Europe, but is now cultivated throughout the world including USA, India and South America. This is an extremely variable species with several subspecies and named cultivars grown for ornamental, culinary and medicinal uses. Oregano is a bushy, semi-woody sub-shrub with upright or spreading stems and branches. Some varieties grow in mound like mats, spreading by underground stems (called rhizomes), and others with a more upright habit. The aromatic leaves are oval-shaped. Oregano will grow in a pH range between 6.0 (mildly acid) and 9.0 (strongly alkaline) with a preferred range between 6.0 and 8.0. The flowers are purple, 3–4 mm long, produced in erect spikes.

As the other three Lamiaceae species oregano is characterized by the presence of essential oil, flavonoids, phenolic acids (caffeic, chlorogenic and rosmarinic), triterpenoid acids (oleanolic and ursolic) and tannins. The oregano essential oil yield between 0.35-0.55% [91]. According to Arnold et al. [92], the content of essential oil in *Origanum ssp.* may come up even to 8.8%. Essential oils obtained from different parts of plant have a similar chemical profile. The dominant components are oxygenated phenolic monoterpenes thymol and carvacrol (Figure 11.), as well as sabinene, linalool, terpine-4-ol, α-pinene, caryophyllene, caryophyllene-oxide and 1,8-cineole.

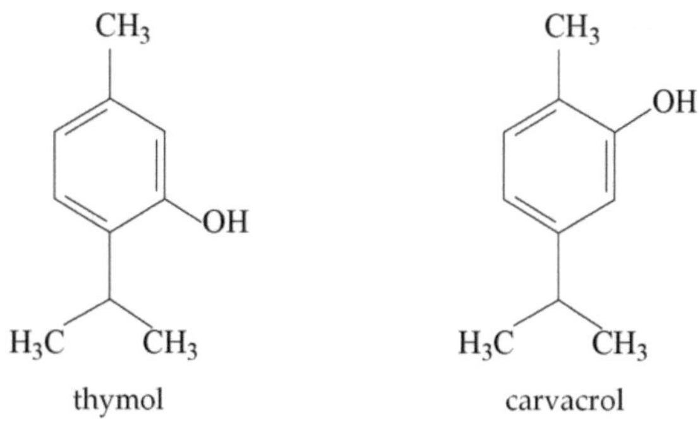

Figure 11. Oxygenated phenolic monoterpens from *O. vulgare* essential oil

According to Duke [93], flavonoids are found in the leaves and whole plant, mostly as kaempferol, quercetin, apigenin, luteolin and rutin. Beside, oregano is rich in apigenin-7-O-β-D-glucoside and luteolin-7-O-β-D-glucuronide. In oregano flavanon naringenin and flavanon glucoside (naringin), have also been identified (Figure 12.).

Most of the healing properties are attributed to the essential oil and flavonoids. It has been widely used in agricultural, pharmaceutical and cosmetic industries as a culinary herb, flavoring substances in food products, alcoholic beverages and perfumery for its spicy fragrance [94]. Regarding the nonvolatile components, the extracts of oregano have the most effective antioxidant activity among aromatic herbs [95]. Oregano family, is widely known as possessing therapeutic properties (diaphoretic, carminative, antispasmodic, antiseptic, tonic) being used in traditional medicine systems in many countries. Different groups of researchers [96,97] studied oregano alcohol extracts. The antioxidant effect of the mentioned extracts is generally due to the presence of rosmarinic and caffeic acid [98].

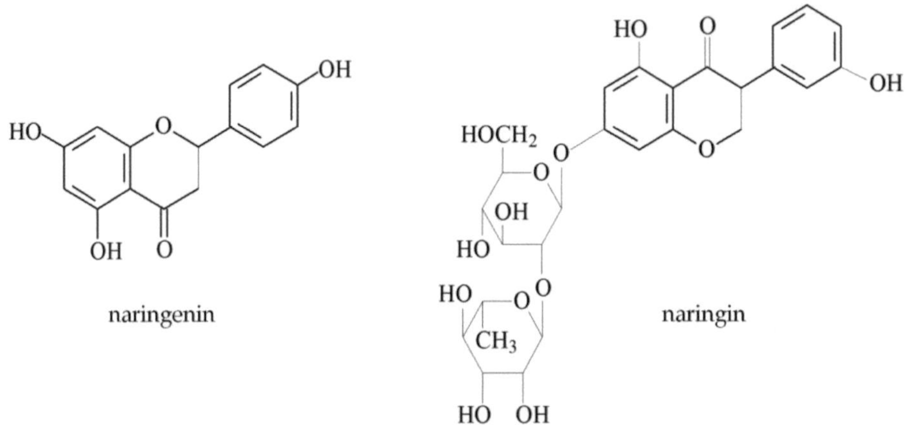

Figure 12. Structures of main flavonoids of *O. vulgare*

	Concentration (μg/mL)					
	BHT	**0.213**	**0.535**	**1.065**	**1.598**	**2.130**
LP	26.15	17.31	24.35	37.17	49.58	51.13

Table 9. Inhibition of LP (%) in Fe^{2+}/ascorbate system of induction by essential oil of oregano leaves and BHT (as a positive control) in the TBA assay.

Our tests showed that only concentrated solutions of essential oil exhibit a greater ability to inhibit LP in liposomes of synthetic antioxidant BHT. The antioxidant activities were dose dependent, but it is noticeable that the values obtained using two most concentrated solution of essential oils (1.598 and 2.130 μg/mL) are very close (49.58 and 51.13% of inhibition of LP). For the inhibition of LP, the most responsible compounds were the oxygenated phenolic monoterpens (thymol and carvacrol) and the mixture of mono- and sesquiterpene hydrocarbons [98].

	Extracts					
Concentration	**BHT**	**Et_2O**	**$CHCl_3$**	**EtOAc**	***n*-BuOH**	**H_2O**
1%	26.15	-0.46	-0.92	24.17	11.31	13.58
5%	26.15	-0.77	-0.94	26.04	14.57	16.49
10%	26.15	-0.91	-0.97	30.28	19.78	23.24

Table 10. Inhibition of LP (%) in Fe^{2+}/ascorbate system of induction by extracts of oregano leaves and BHT (as a positive control) in the TBA assay.

The data presented in Table 10. show that the last three extracts of *O. vulgare* (EtOAc, *n*-BuOH and H_2O) reduced the intensity of lipid peroxidation while the first two extracts (Et_2O and $CHCl_3$) have prooxidative effect (but not statistically significant). The largest inhibitory activity was exhibited by ethyl acetate extract. High inhibitory effect of this extract and its components towards Fe^{2+}-dependent LP of liposomes can be related to the presence of flavonoids in the extract. It was established that flavonoids that antiradical potential of flavonoids are the most pronounced towards OH, peroxy- and alkoxy radicals, which are formed in the process of lipid peroxidation [99]. Also, these results are consistent with 2D-TLC analysis which showed that the dominant component of the EtOAc extract was kaempferol monoglycoside, while the H_2O extract contains multiple kaempferol diglycosides. From the literature it is known that additional glycosylation reduces the antioxidant activity and capability to reduce lipid peroxidation [100].The antioxidant and prooxidant activities of all extracts of oregano leaves were dose dependent.

4. Conclusions

It was found that excessive production of oxygen radicals in the body initiates oxidation and degradation of polyunsaturated fatty acids. It is known that free radicals attack the highly unsaturated fatty acid of membrane system and induce lipid peroxidation. Since lipid peroxidation causes oxidative damage to cell membranes and all other systems that contain lipids, in any investigation of total antioxidative activity of extracts and essential oils it is necessary to investigate their effects on lipid peroxidation. Some substrates (for example liposomes) are used more frequently than others, mainly because of the simplicity of the methods involved. In this way we get very useful information to direct future research. The results of our *in vitro* assays of examined four different Lamiaceae species extracts expressed significant protective effects on LP, which was found to be correlated to different compounds. It can be concluded that ethyl acetate and water proved to be the best solvent for extraction of plant material. Also, a very strong protective activity of the EtOAc and H_2O extracts in lipid peroxidation processes was recorded, which means that they may have a protective role in oxidative stress. Experimental results indicate that the essential oil of *M. peregrinum* collected from the Senta locality (No.3) exhibited the strongest inhibitory effect on lipid peroxidation. Furthermore, the present chapter on the chemistry and biological activity of four well known Lamiaceae species explicitly prove that these plants may be an important sources of pharmalogically active substances, and thus can be used in the preparation of various herbal medicine.

Author details

Biljana Kaurinovic and Mira Popovic
Department of Chemistry, Biochemistry and Environmental Protection, University of Novi Sad, Novi Sad, Republic of Serbia

Acknowledgement

This work was supported by the Ministry of Science and Environmental Protection of the Republic of Serbia (Project No. 172058) and by the Provincial Secretariat for Science and Technological Development, Autonomous Province of Vojvodina, Republic of Serbia.

5. References

[1] Dargel R (1992) Lipid Peroxidation – A Common Pathogenic Mechanism? Exp. toxicol. pathol. 44: 169-181.

[2] Ursini F, Maiorino M, Sevanian A (1991) Membrane Hydroperoxides. In: Sies H, editor. Oxidative Stress: Oxidants and Antioxidants. London: Academic Press. pp. 319-336.

[3] Kagan VE (1988) Lipid Peroxidation in Biomembranes. Boca Raton: CRC Press. 1 p.

[4] Farber JL, Kyle ME, Coleman JB (1990) Mechanisms of Cell Injury by Activated Oxygen Species. Lab. invest. 62: 670-679.

[5] Halliwell B, Gutteridge JMC. (1984) Lipid Peroxidation, Oxygen Radicals, Cell Damage and Antioxidant Therapy. Lancet. 23: 1396-1397.

[6] Holchestein P, Nordenbrand K, Ernster L (1964) Evidence for the Involvement of Iron in the ADP-Activated Peroxidation of Lipids in Microsomes and Mitochondria. Biochem. biophys. res. commun. 14: 323-328.

[7] Minotti G, Aust SD (1987) The Role of Iron in the Initiation of Lipid Peroxidation. Chem. phys. lipids. 44: 191-208.

[8] Aruoma OI, Halliwell B, Laughton MJ, Quinlan J, Gutteridge JMC (1989) The Mechanism of Initiation of Lipid Peorxidation. Evidence Against a Requirement for an Iron(II)-Iron(III) Complex. Biochem. j. 258: 617-620.

[9] Tang L, Zhang Y, Qian Z, Shen X (2000) The Mechanism of Fe^{2+}-Initiated Lipid Peroxidation in Liposomes: The Dual Function of Ferrous Ions, the Roles of the Pre-Existing Lipid Peroxides and the Lipid Peroxyl Radical. Biochem. j. 352: 27-36.

[10] Chatterjee SN, Agarwal S (1988) Liposomes as Membrane Model for Study of Lipid Peroxidation. Free rad. biol. med. 4: 51-72.

[11] Bombardelli E, Cristoni A, Morazzobi P (1994) PHYTOSOME® in Functional Cosmetics. Fitoterapia. 65: 397-401.

[12] http://www.pharmaxchange.info

[13] Foldvari M (2000) Non-Invasive Administration of Drugs Through the Skin. Challenges in Delivery System Design. PSTT. 3: 417-425.

[14] Immordino ML, Dosio F, Cattel L (2006) Stealth Liposomes: Review of the Basic Science, Rationale, and Clinical Applications, Existing and Potential. IJN 1(3): 297-315.

[15] Kujundžić S (2002) Biochemical Investigaion of Plant Species from the Apiaceae Family (in Serbian). Master Thesis. Novi Sad: PMF. 48 p.

[16] Laguerre M, Lecomte J, Villeneuve P (2007) Evaluation of the Ability of Antioxidants to Counteract Lipid Oxidation: Existing Methods, New Trends and Challenges. Prog. lipid res. 46: 244-282.

[17] Safer AM, Al-Nughamish AJ (1999) Hepatotoxicity Induced by the Antioxidant Food Additive Butylated Hydroxytoluene (BHT) in Rats: An Electron Microscopical Study. Histol. histopathol. 197: 391-406.

[18] Brraco U, Loliger J., Viret J (1981) Production and Use of Natural Antioxidants. JAOCS. 58: 686-690.

[19] Lagouri V, Blekas G, Tsimidou M, Kokkini S, Boskou D (1993) Composition and Antioxidant Activity of Essential Oil from Oregano Plants Grown in Greece. Z. lebensmitt.- unters. forsch. 197: 20-23.

[20] Crowell PL, Ayoubi S, Burke YD (1996) Antitumorigenic Effects of Limopnene and Perillyl Alcohol Against Pancreatic and Breast Cancer. Adv. exp. med. biol. 401: 131-136.
[21] Carlson J, Borg AK, Unelius R, Shoshan MC, Wilking N, Ringborg U, Linder S (1996) Inhibition of Tumor Cell Growth by Monoterpenes *in Vitro*: Evidence of a Ras-Independent Mechanism of Action. Anti-cancer drugs. 74(4): 422-429.
[22] Ruberto G, Baratta M.T, Deans SG, Dorman HJD (2000) Antioxidant and Antimicrobal Activity of *Foeniculum vulgare* and *Crithmum maritimum* Essential Oils. Planta med. 66: 687-693.
[23] Namiki M (1990) Antioxidants/Antimutagens in Food. Food sci. nutr. 29: 273-300.
[24] Rice-Evans CA, Miller NJ, Paganga G (1996) Structure-Antioxidant Activity Relationships of Flavonoids and Phenolic Acids. Free radic. biol. med. 20: 933-956.
[25] Cotelle N (2001) Role of Flavonoids in Oxidative Stress. Cur. topics med. chem. 1: 569-590.
[26] Rice-Evans CA, Miller NJ, Paganga G (1996) Antioxidant Properties of Phenolic Compounds. Trends plant sci. 2: 152-159.
[27] 1. & 2. European Directorate for the Quality of Medicines (2002) European Pharmacopeia, 4th edition. Council of Europe. France: Strasbourg Cedex. pp. 183-184.
[28] Afanas'ev A, Dorozhko I, Brodski V, Kostyk A, Potapovich I (1989) Chelating and Free Radical Scavenging Mechanism of Inhibitory Action of Rutin and Quercetin in Lipid Peroxidation. Biochem. pharmacol. 38: 1763-1769.
[29] Fukuzawa K, Seko T, Minami K, Terao J (1993) Dynamics of Iron-Ascorbate-Induced Lipid Peroxidation in Charged and Uncharged Phospholipid Vesicles. Lipids. 28: 497-503.
[30] Buege AJ, Aust DS (1988) Methods in Enzymology. New York: Academic Press. 302 p.
[31] Janero DR (1990) Malondialdehyde and Thiobarbituric Acid-Reactivity as Diagnostic Indices of Lipid Peroxidation and Peroxidative Tissue Injury. Free rad. biol. med. 9: 515-540.
[32] Naghibi F, Mosaddegh M, Motamed MS, Ghorbani A (2005) Labiatae Family in Folk Medicine in Iran: From Ethnobotany to Pharmacology. IJPR. 2: 63-79.
[33] Hammer KA, Carson CF, Riley TV (1999) Antimicrobial Activity of Essential Oils and Other Plant Extracts. J. appl. microbiol. 86(6): 986-990.
[34] Cantino PD, Harley RM, Wagstaff SJ (1992) Genera of Labiatae: Status and Classification. In: Harley RM, Reynolds T, editors. Advances in Labiate Science. London: Royal Botanic Gardens. pp. 511-522.
[35] Kaurinović B (2008) Antioxidant Activities of *Laurus nobilis* L. and *Melittis melissophyllum* L. Extracts (in Serbian). PhD Thesis. Novi Sad: PMF. 68 p.
[36] Honda T, Rounds BV, Bore L, Finlay HJ, Favaloro FG, Suh N, Wang Y, Sporn MB, Gribble GW (2000) Synthetic Oleanane and Ursane Triterpenoids with Modified Rings A and C: A Series of Highly Active Inhibitors of Nitric Oxide Production in Mouse Macrophages. J. med. chem. 43: 4233-4246.
[37] Kuzuhara H, Nishiyama S, Minova N, Sasaki K, Omoto S (2000) Protective Effects of Soyasapogenol a on Liver Injury Mediated by Immune Response in a Concavalin A-Induced Hepatitis Model. EJP. 391: 175-181.
[38] Shibata SS (2001) Chemistry and Cancer Preventing Activities of Ginseng Saponins and Some Related Triterpenoid Compounds. J. Korean med. sci. 16: 28-37.
[39] Zhang Z, Chang Q, Zhu M, Huang Y, Ho WK, Chen Z (2001) Characterization of Antioxidants Present in Hawthorn Fruits. J. nutr. biochem. 12: 144-152.
[40] Velasco-Negueruela A, Sanz J, Perez-Alonso MJ, Pala-Paul J (2004) The Volatile Components of the Aerial Parts of *Melittis melissophyllum* L. Bot. complutensis. 28: 133-136.

[41] http://www.ang.kfunigraz.ac.at/~katzer/engl.html

[42] Skrzypczak-Pietraszek E, Hensel A (2000) Polysaccharides from *Melittis melissophyllum* L. Herb and Callus. Pharmazie. 55: 768-771.

[43] Skrzypczak, E.; Skrzypczak, L. (1993) The Tissue Culture and Chemical Analysis of *Melittis melissophyllum* L. Acta hort. 330: 263-267.

[44] http://www.umm.edu/altmed.html

[45] http://www.hort.purdue.edu/newcrop/med-aro.html

[46] Guarrera PM (2005) Traditional Phytotherapy in Central Italy (Marche, Abruzzo, and Latium). Fitoterapia. 76: 1–25.

[47] Maggi F, Bílek T, Lucarini D, Papa F, Sagratini G, Vittori S (2009) *M. melissophyllum* L. subsp. *melissophyllum* (Lamiaceae) from Central Italy: A New Source of a Mushroom-Like Flavour. Food chem. 113: 216-221.

[48] Kaurinovic B, Popovic M, Vlaisavljevic S, Raseta M (2011) Antioxidant Activities of *Melittis melissophyllum* L. (Lamiaceae). Molecules. 16: 3152-3167.

[49] Doba T, Burton GW, Ingold KU (1985) Antioxidant and Co-Oxidant Activity of Vitamin C. The Effects of Vitamin C, Either Alone or in the Presence of Vitamin E or a Water-Soluble Vitamin E Analogue, Upon the Peroxidation of Aqueous Multimalleral Phospholipid Liposomes. Biochim. biophys. acta. 835: 298-303.

[50] http://www.phytochemicals.info/phytochemicals/quercetin.php

[51] Acland ML, van de Waarsenburg S, Jones R (2005) Synergistic Antiproliferative Action of the Flavonols Quercetin and Kaempferol in Cultured Human Cancer Cell Lines. In vivo. 19(1): 69-76.

[52] http://www.phytochemicals.info/phytochemicals/rutin.php

[53] Decker EA (1997) Phenolics: Prooxidants or Antioxidants? Nutr. rev. 55: 396-407.

[54] Handa SS, Sharma A, Chakraborti KK (1986) Natural Products in Plants as Liver Protecting Drugs. Fitoterapia. 57: 307-310.

[55] Cholbi MR, Paya M, Alcaraz MJ (1991) Inhibitory Effect of Phenolic Compounds on CCl_4-Induced Microsomal Lipid Peroxidation. Experimentia. 47: 195-198.

[56] Sousa RL, Marletta MA (1985) Inhibition of Cytochrome P-450 Activity in Rat Liver Microsomes by the Natural Occuring Flavonoid, Quercetin. Arch. biochem. biophys. 240: 345-348.

[57] Rekka E, Kourounakis PN (1991) Effect of Hydroxyethyl Rutoside and Related Compounds on Lipid Peroxidation and Free Radical Scavening Activity. Some Structural Aspects. J. pharmac. pharmacol. 43: 486-490.

[58] Larson RA (1988) The Antioxidant of Higher Plants. Phytochemistry. 27: 969-978.

[59] Cock C, Samman S (1996) Flavonoids Chemistry Metabolism, Cardioprotectioeffects and Dietary Sources. Nutrit. biochem. 7: 66-76.

[60] Stanković SM (2011) Total Phenolic Content, Flavonoid Concentration and Antioxidant Activity of *Marrubium peregrinum* L. Extracts. Kragujevac j. sci. 33: 63-72.

[61 Vlaisavljevic S (2007) Antioxidant Activities of *Marrubium peregrinum* L. Essential Oil and Extracts. Master Thesis. Novi Sad: PMF. 36 p.

[62] Nagy M, Gergel D, Grancai D, Novomesky P, Ubik K (1996) Antilipoperoxidative Activity of Some Phenolic Constituents from *Marrubium peregrinum* L. Farmaceuticky-Obzor. 65: 283-285.

[63] Sahpaz S, Hennebelle T, Bailleul F (2002) Marruboside, a New Phenylethanoid Glycoside from *Marrubium vulgare* L. Nat. prod. lett. 16: 195-199.
[64] Gruenwald J, Brendler T, Jaenicke C (2000) PDRc for Herbal Medicines. NewYork: Medical Economics Co. 271 p.
[65] Salei LA, Popa DP, Lazurēvskii GV (1967) Diterpenoids from *Marrubium peregrinum* L. Chem. nat. comp. 2: 200-201.
[66] Hennebelle T, Sahpaz S, Skaltsounis AL, Bailleul F (2007) Phenolic Compounds and Diterpenoids from *Marrubium peregrinum*. Biochem. systemat. ecol. 35: 624-626.
[67] Demirci B, Baser HC, Kirimer N (2004) Composition of the Essential Oil of *Marrubium bourgaei* ssp. Caricum P.H. Davis. J. essent. oil res. 16: 133-134.
[68] Lazari DM, Skaltsa HD, Constantinidis T (1999) Essential Oils of *Marrubium velutinum* Sm. and *Marrubium peregrinum* L., Growing Wild in Greece. Flavour frag. j. 14: 290-292.
[69] Nagy M, Svajdlenka E (1998) Comparison of Essential Oils from *Marrubium vulgare* L. and *M. peregrinum* L. J. essent. oils res. 10: 585-587.
[70] Khanavi M, Ghasemian L, Motlagh EH, Hadjiakhoondi A, Shafiee A (2005) Chemical Composition of the Essential Oils of *Marrubium parviflorum* Fisch. & C.A. Mey. and *Marrubium vulgare* L. from Iran. Flavour frag. j. 20: 324-326.
[71] Kaurinovic B, Vlaisavljevic S, Popovic M, Vastag Dj, Djurendic-Brenesel M (2010) Antioxidant Properties of *Marrubium peregrinum* L. (Lamiaceae) Essential Oil. Molecules. 15: 5943-5955.
[72] Kaurinovic B, Popovic M, Vlaisavljevic S, Zlinska J, Trivic S (2011) *In Vitro* Effect of *Marrubium Peregrinum* L. (Lamiaceae) Leaves Extracts. FEB. 20(12): 3152-3157.
[73] Čanadanović-Brunet J, Ćetković G, Đilas S, Tumbas V, Bogdanović G, Mandić A, Markov S, Cvetković D, Čanadanović V (2008) Radical Scavenging, Antibacterial, and Antiproliferative Activities of *Melissa officinalis* L. Extracts. J. med. food. 11: 133-143.
[74] Tosun M, Ercisli S, Sengul M, Ozer H, Polat T (2009) Antioxidant Properties and Total Phenolic Content of Eight *Salvia* Species from Turkey. Biol. res. 41: 175-181.
[75] Mann J (1992) Secondary Metabolism, 2nd edition. Oxford: Oxford University Press. 279 p.
[76] Jelačić ČS, Beatović VD, Prodanović SA, Tasić RS, Moravčević ŽĐ, Vujošević MA, Vučković MS (2011) Chemical Composition of the Essential Oil of Basil (*Ocimum basilicum* L. Lamiacea). Chem. ind. 65(4): 465-471.
[77] Grayer RI, Kite GC, Goldstone J, Bryan SE, Paton A, Putievsky E (1996) Intraspecific Taxonomy and Essential Oil Chemotypes in Sweet Basil, *Ocimum basilicum*. Phytochemistry. 43: 1033-1039.
[78] Marotti M, Piccaglia E, Giovanelli E (1996) Differences in Essential Oil Composition of Basil (*Ocimum basilicum* L.) Italian Cultivars Related to Morphological Characteristics. J. agric. food chem. 44: 3926-3929.
[79] http://www.ars-grin.gov/cgi-bin/duke/farmacy2.pl
[80] Telci I, Elmastas M, Sahin A (2009) Chemıcal Composition and Antioxidant Activity of *Ocimum minimum* Essential Oils. Chem. nat. comp. 45: 568-571.
[81] Soran ML, Cobzac-Codruta S, Varodi C, Lung I, Surducan E, Surducan V (2009) The Extraction and Chromatographic Determination of the Essentials Oils from *Ocimum basilicum* L. by Different Techniques. J. phys. conf. ser. 182: 012016.

[82] Gülçin I, Elmastaş M, Aboul-Enein YH (2007) Determination of Antioxidant and Radical Scavenging Activity of Basil (*Ocimum basilicum* L. Family Lamiaceae) Assayed by Different Methodologies. Phytother. res. 21: 354-361.

[83] Lee SJ, Umano K, Shibamoto T, Lee KG (2005) Identification of Volatile Components in Basil (*Ocimum basilicum* L.) and Thyme Leaves (*Thymus vulgaris* L.) and Their Antioxidant Properties. Food chem. 91: 131-137.

[84] Ruberto G, Baratta MT (2000) Antioxidant Activity of Selected Essential Oil Components in Two Lipid Model Systems. Food chem. 68: 167-174.

[85] Lewinsohn E, Ziv-Raz I, Dudai N, Tadmor Y, Lastochkin E, Larkov O, Chaimovitsh D, Ravid U, Putievsky E, Pichersky E, Shoham Y (2000) Biosynthesis of Estragole and Methyl Eugenol in Sweet Basil (*Ocimum basilicum* L.). Developmental and Chemotypic Association of Allaphenol O-Methyltransferase Activities. Plant sci. 160: 27-35.

[86] Tucakov J (1997) Healing With Herbs (Phytotherapy) (in Serbian). Beograd: Rad. 34 p.

[87] Kootstra A (1994) Protection from UV-B-Induced DNA Damage by Flavonoids. Plant mol. biol. 26: 771-774.

[88] Robak J, Gryglewski J (1988) Flavonoids are Scavengers of Superoxide Anion. Biochem. pharmacol. 37: 837-841.

[89] Kaurinovic B, Popovic M, Vlaisavljevic S, Trivic S (2011) Antioxidant Capacity of *Ocimum basilicum* L. and *Origanum vulgare* L. Extracts. Molecules. 16: 7401-7416.

[90] Lakic N, Mimica-Dukic N, Isak J, Božin B (2010) Antioxidant Properties of *Galium verum* L. (Rubiaceae) Extracts. Cent. eur. j. biol. 5: 331-337.

[91] W'glarz Z, Osidska E, Geszprych A, Przybyb J (2006) Intraspecific Variability of Wild Marjoram (*Origanum vulgare* L.) Naturally Occurring in Poland. Rev. bras. pl. med. botucatu. 8: 23-26.

[92] Arnold N, Bellomaria B, Valentini G (2000) Composition of the Essential Oil of Different Species of *Origanum* in the Eastern Mediterrenean. JEOR. 12: 192-196.

[93] Duke J (1997) The Green Pharmacy, the Ultimate Compendium of Natural Remedies from the World's Foremost Authority on Healing and Herbs. New York: St. Martin's Press.

[94] Souza EL, Stamford TLM, Lima EO, Trajano VN (2007) Effectiveness of *Origanum vulgare* L. Essential Oil to Inhibit the Growth of Food Spoiling Yeasts. Food Control. 18: 409-413.

[95] Vekiari SA, Oreopoulou V, Tzia C, Thomopoulos CD (1993) Oregano Flavonoids as Lipid Antioxidants. J. am. oil chem. soc. 70: 483-487.

[96] Chevolleau S, Mallet JF, Ucciani E, Gamisans J, Gruber M (1992) Effects of Rosemary Extracts and Major Constituents on Lipid Oxidation and Soybean Lipoxygenase Activity. J. am. oil chem. soc. 69: 1269-1271.

[97] Kramer RE (1985) Antioxidants in Clove. J. am. oil chem. soc. 62: 111-113.

[98] Milos M, Mastelic J, Jerkovic I (2000) Chemical Composition and Antioxidant Effect of Glycosidically Bound Volatile Compounds from Oregano (*Origanum vulgare* L. ssp. *Hirtum*). Food chem. 71: 79-83.

[99] Sichel G, Corsaro C, Scalia M, Dibilio J, Bonomo R (1991) *In Vitro* Scavenger Activity of Some Flavonoids and Melanins Against O_2. Free rad. biol. med. 11: 1-8.

[100] Rice-Evans C, Miller NJ, Paganga G (1997) Antioxidant Properties of Phenolic Compounds. Trends plant sci. 2: 152-159.

Chapter 8

Liposomes as a Tool to Study Lipid Peroxidation in Retina

Natalia Fagali and Angel Catalá

Additional information is available at the end of the chapter

http://dx.doi.org/10.5772/46074

1. Introduction

In living organisms, the oxidative stress is associated with several physio- pathological affections (e.g. atherosclerosis, cancer, aging, neurodegenerative diseases). The oxidative stress is generally initiated by generation of reactive oxygen (ROS) and nitrogen species (RNS) (Halliwell & Gutteridge, 1990). ROS are continuously formed during cellular metabolism and are removed by antioxidants defences. ROS from endogenous and exogenous sources results in continuous and accumulative oxidative damage to cellular components and alters many cellular functions. The most vulnerable molecules to oxidative damage are proteins, lipids and DNA (Kohen & Nyska, 2002; Catalá, 2009, 2011a, 2011b).

In mammalian retina, free radicals and lipoperoxides seem to play important roles in the evolution of different retinopathies including glaucoma, cataractogenesis, diabetic retinopathy, ocular inflammation and retinal degeneration (Ueda et al., 1996; De La Paz & Anderson, 1992). Because of free radicals production induces the lipid peroxyl radical formation, known as secondary free radicals products; this chain reaction of lipid peroxidation can damage the retina, especially the membranes that play important roles in visual function (Catalá, 2006). The retina is the neurosensorial tissue of the eye. It is very rich in membranes and therefore in polyunsaturated fatty acids (PUFAs) such as docosahexenoic acid (22:6 n-3), that are quite vulnerable to lipid peroxidation. Also, the human retina is a well oxygenated tissue. High-energy short-wavelength visible light promotes the formation of ROS which can initiate lipid peroxidation in the macula and elsewhere. The macular carotenoids are thought to combat light-induced damage mediated by ROS by absorbing the most damaging incoming wavelengths of light prior to the formation of ROS and by chemically quenching ROS once they are formed.

Although peroxidation in model membranes may be very different from peroxidation in biological membranes, the results obtained in model membranes may be used to progress our understanding of subjects that cannot be studied in biological membranes. Nevertheless, in spite of the relative simplicity of peroxidation of liposomal lipids model, these reactions are still relatively complex because they depend in a complex fashion on liposome type, reaction initiator and reaction medium (Fagali & Catalá, 2009). This complexity is the most likely cause of the apparent contradictions of literature results.

Biological membranes are complex systems. In view of this complexity and in order to avoid collateral effects that may arise during lipid peroxidation process of whole retinal membranes, we have attempted to gain understanding of the mechanisms responsible for peroxidation in a simple model system, made by dispersing retinal lipids in the form of liposomes.

This chapter describes a very useful method to prepare liposomes with natural phospholipids and the necessary methodology to follow the lipid peroxidation of these liposomes.

2. Materials and methods

2.1. Materials

Chloroform, methanol, trizma base, butylated hydroxytoluene (BHT), NaCl, $FeSO_4$ heptahydrate and 2-thiobarbituric acid (TBA) were purchased from Sigma Chemical Co. Suitable plastic lab ware was used throughout this study to avoid effects of adventitious metals. Other reagents were of the highest quality commercially available. All solutions were prepared using distilled water treated with a Millipore Q system.

2.2. Isolation of bovine retina

Eyes were enucleated at slaughter (Frigorífico Gorina), transported in ice to laboratory where retinas were taken out within 1–2 h. Under red light and with all tubes and solutions in ice buckets, corneas were excised; lenses and vitreous were subsequently removed. Eye cups were inverted and retinas were carefully peeled from the eyes. Retinas were briefly homogenized in 0.15 M NaCl (1 ml/retina) 120 s (20on-20off) at 4 ºC in an Ultraturrax X25 homogenizer at 7000 rpm.

2.3. Lipid extraction

Total lipids were extracted from retinal homogenates with chloroform/methanol (2:1 v/v) (Folch et al, 1957) at 4 ºC (sample:Folch = 1:5). A volume of water corresponding to 20 % of total volume was added. This mixture was shaken and kept in rest in cold to allow phases separation. Chloroformic phase was kept at -22 °C.

2.4. Preparation of liposomes made of retinal lipids

Total lipids obtained from retinal homogenates dissolved in chloroform were evaporated under nitrogen until constant weight and submitted to vacuum to remove traces of chloroform. Resultant films were dispersed at room temperature in a saline solution (0.15 M NaCl). Dispersed lipids were mixed to homogeneity using a vortex-mixer to obtain non-sonicated liposomes (NSL). Sonicated liposomes (SL) were prepared by sonication of NSL under nitrogen and ice cooling (Huang, 1969), using a Sonics vibra cell, probe-sonicator Model VCX 750 (750 W, 20kHz) at 75% of maximal output. Preparation of liposomes required about 2.5 min of sonication to reach apparently minimal optical density values.

2.5. Determination of liposomes size by Dynamic Light-Scattering (DLS)

The time correlation $G(q,t)$ of the light-scattering intensity was measured at 90° with a goniometer, ALV/CGS-5022F, with a multiple-τ digital correlator, ALV-5000/EPP, covering a 10^{-6}-10^{3} s time range. The light source was a helium/neon laser with a wavelength of 633 nm operating at 22 mW. Each correlation function was analyzed by the well known cumulant fit yielding the apparent mean diffusion coefficient and the distribution δD of this value (Koppel, 1972). The measurements were carried out with 80 μl of SL and NSL (lipid concentration= 2 mg/ml) in water, 0.15 M NaCl and Tris-HCl buffer 20 mM (final volume= 2 mL).

2.6. Measurements of lipid peroxidation by detection of conjugated dienes and trienes production

In order to determine conjugated dienes and trienes production, absorption spectra were recorded by means of a Shimadzu UV-1800 spectrophotometer, in the range 200 to 300 nm, at 22 °C, with 1 cm path length quartz cell. Liposomes made of retinal lipids (80 μl, 2 g/l of lipids) were diluted to 2 ml with water, 0.15 M NaCl or 20 mM Tris-HCl pH 7.4, and oxidation was initiated by the addition of $FeSO_4$ (final concentration = 25 μM). Lipid peroxidation was assessed continuously by measuring the increase in absorbance at 234 nm (formation of conjugated dienes) and 270 nm (formation of conjugated trienes) taken at 1 min intervals. Oxidation rates were determined as the slope of a regression line drawn through linear range of absorbance versus time curve. Lag times were determined as time corresponding to intersection of oxidation rate regression line with a regression line drawn through initial phase of oxidation (Sargis & Subbaiah, 2003).

2.7. Measurements of lipid peroxidation by detection of Thiobarbituric Reactive Substances (TBARS)

During Fe^{2+}- initiated reactions, extent of liposomal lipid peroxidation was assessed using a TBA assay. In this procedure, 850 μL of TBA (0.375% w/v TBA, 0.25 N HCl) were added to aliquots of 150 μl of reaction mixture containing BHT (0.1 % w/v in ethanol) to prevent possible peroxidation of liposomes during incubation. The aliquots were taken at different intervals of time. Samples were heated for 30 min at 75 °C. Absorbance was measured at 532 nm for determination of aldehydic breakdown products of lipid peroxidation.

2.8. Preparation of Fatty Acids Methyl Esters (FAME)

Lipids from retina, liposomes or liposomes exposed to peroxidation initiated by Fe^{2+}, in absence or presence of BHT, were extracted according to the method of Folch et al (1957). A similar reaction mixture to that used in the analysis of conjugated dienes but scaled up 7.5 times was used to analyze the fatty acid composition of the samples

After one hour of incubation of liposomes with or without Fe^{2+} in the presence or absence of BHT, the samples were mixed with 15 ml of chloroform:methanol (2:1 v/v) containing 0.01 % BHT to stop the reaction. The mixture was stirred, gassed with nitrogen and kept in refrigerator overnight to achieve separation of phases. The lower chloroform phase was filtered through paper filter containing anhydrous sodium sulphate. The solvent was evaporated to dryness under nitrogen. Dry lipids of retina and/or liposomes were transmethylated with 300 µl of 1.3 M BF_3 in methanol at 65°C during 180 min. After incubation 1 ml of 0.15 M NaCl was added and the fatty acid methyl esters were extracted with 1 ml of hexane. This phase was injected onto the chromatograph.

2.9. Gas chromatography – Mass spectrometry analyses

GC–MS analyses were done using a Perkin Elmer Clarus 560D MS - gas chromatograph equipped with a mass selective detector with quadrupole analyzer and photomultiplier detector and a split/splitless injector. In the gas chromatographic system, a Elite 5MS (Perkin Elmer) capillary column (30 m, 0.25 mm ID, 0.25 µm df) was used. Column temperature was programmed from 130 to 250 °C at a rate of 5 °C/min and 250 °C for 6 min. Injector temperature was set to 260 ºC and inlet temperature was kept at 250 °C. Split injections were performed with a 10:1 split ratio. Helium carrier gas was used at a constant flow rate of 1 ml/min. In the mass spectrometer, electron ionization (EI+) mass spectra was recorded at 70 eV ionization energy, in full scan mode (50-400) unit mass range. The ionization source temperature was set at 180 °C. The fatty acid composition of the lipid extracts was determined by comparing their methyl derivatives mass fragmentation patterns with those of mass spectra from the NIST databases.

3. Results

3.1. Size of sonicated and non-sonicated liposomes made of retinal lipids in different aqueous media

Average hydrodynamic radii of liposomes determined by DLS studies are presented in Table 1. We noted that NSL display a multimodal size distribution when analyzed by inverse Laplace transform (CONTIN), a result that is compatible with the high polydispersity index (PI > 0.4) from cumulants fit. Thus, hydrodynamic radii for NSL, at room temperature in different aqueous media, cover a broad range with intensity weighted maxima centered between 190 and 320 nm. On the other hand, results for liposomes formed by sonication gave, through cumulant method, hydrodynamic radii in the order of 76.4-83.3

nm, showing as expected significant influence of sonication on size and distribution. It is clear that NSL possessed higher hydrodynamic radii than SL. Either NSL or SL in water were slightly smaller than that in 0.15 M NaCl and Tris-buffer.

Type of liposome	Aqueous media	Hydrodynamic radii (nm)	Polydispersity index
Sonicated	Water	76.4	0.31
	0.15 M NaCl	83.3	0.27
	20 mM Tris–HCl	83.3	0.27
Non-sonicated	Water	190–225	0.44
	0.15 M NaCl	260–320	0.45
	20 mM Tris–HCl	200–240	0.43

Table 1. Summary of values obtained by dynamic light scattering of SL and NSL made of retinal lipids in different aqueous media. Hydrodynamic radii values are the average of at least three representative determinations in each media.

3.2. Evolution of UV spectra as a function of time for Fe^{2+} initiated lipid peroxidation of SL and NSL in different aqueous media

Figure 1 shows evolution of UV spectra as a function of time, for Fe^{2+} initiated lipid peroxidation of SL and NSL, in different aqueous media. This figure showed increases in UV absorption with a maximum at 234 nm and at 270 nm, due to conjugated dienes and trienes respectively, and a decrease of absorbance at 200-215 nm, due to loss of methylene interrupted double bonds (unoxidized lipids). When lipid peroxidation was carried out in water or 0.15 M NaCl decreases at 200-215 nm were more notorious than in reactions carried out in Tris-buffer.

3.3. Conjugated dienes, trienes and TBARS are excellent markers of lipid peroxidation of liposomes made of retinal lipids

Figure 2 shows changes in TBARS production and variation of absorbance at 234 nm and 270 nm as a function of time.

When SL were peroxidized in water (**Figure 2A**) a lag phase of 30 min, followed by a fast rate, was observed in TBARS production. Absorbance final value at 532 nm reached was 0.24. Increase of absorbance at 234 nm showed a small lag phase followed by a fast initial phase until 40 min, since then speed of reaction became slighter. This behaviour was also observed in measured absorbance at 270 nm, but all absorbance values were lower than that at 234 nm in the range of time studied.

Lipid peroxidation of SL in 0.15 M NaCl (**Figure 2B**). showed an immediate and fast production of TBARS without lag phase, reaching a final value ($Abs_F \approx$ 0.23) similar to that obtained in water. The absorbance at 234 nm increased with an initial speed greater than

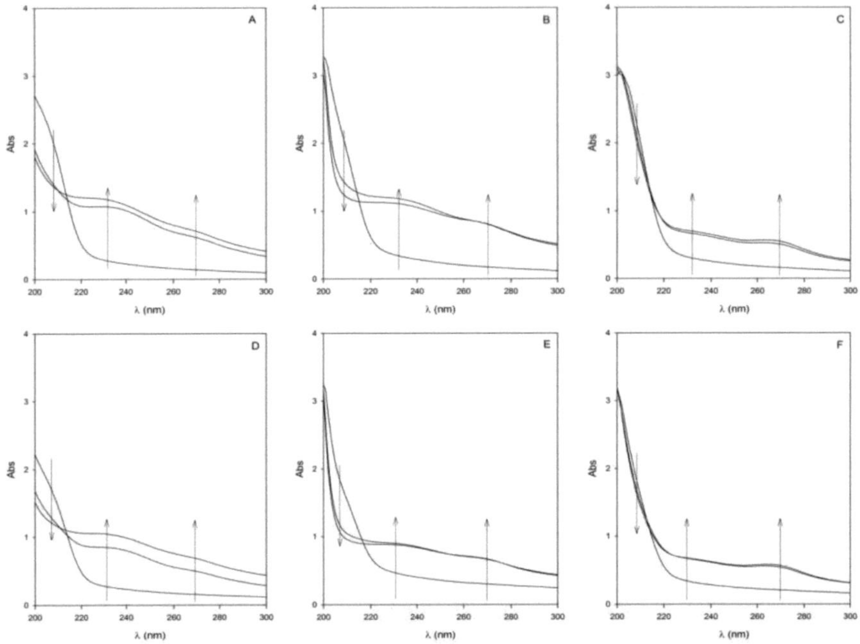

Figure 1. Time evolution (0, 90 and 180 min) of UV spectra of liposomes peroxidized with Fe^{2+} as an initiator of the reaction. SL in A) water, B) 0.15 M NaCl, C) buffer Tris. NSL in D) water and) 0.15 M NaCl, F) buffer Tris

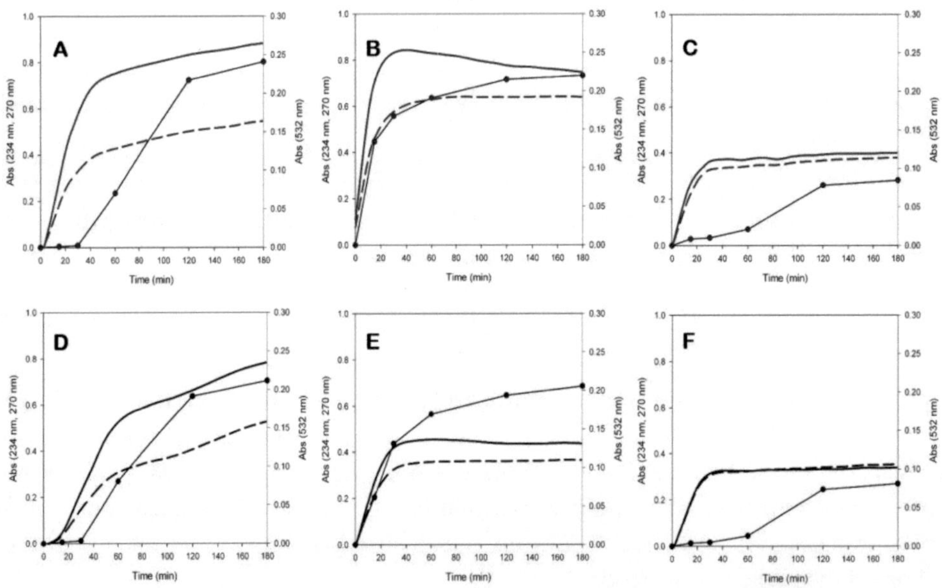

Figure 2. TBARS production (—•—) and variation of absorbance at 234 nm (—) and 270 nm (---) as a function of time, during Fe^{2+}-catalyzed lipid peroxidation of SL (top) and NSL (bottom). TBARS were determined at 0, 15, 30, 60, 120 and 180 min after addition of Fe^{2+}. Aqueous media where reactions were carried out: A, D: water; B, E: 0.15 M NaCl; C, F: 20 mM Tris-HCl pH 7.4.

that observed in water, became the highest to 30 minutes and, then, diminished slowly. The absorbance at 270 nm increased with an initial speed greater than that observed in water, became the highest around the 30 min and then remained constant.

Lipid peroxidation of SL in Tris- buffer (**Figure 2C**) showed the largest lag phase and the lowest final value of absorbance (Abs_f= 0.09) for TBARS formation. The initial speed of TBARS production was also the lowest. Initial speed of reaction observed, by increase of absorbance at 234 nm, was lower than that measured on water and 0.15 M NaCl. Absorbance reached the maximum at 30 minutes and then remained constant. Conjugated trienes production was very similar to that of conjugated dienes.

Lipid peroxidation of NSL in water (**Figure 2D**) showed a lag phase of 30 min for the TBARS production and a final value of absorbance of 0.21. Changes of absorbance at 234 nm displayed a lag of 16 min, increased quickly from this time to 60 min and since then continued increasing with lower speed. Values of absorbance at 270 nm were below than those observed at 234 nm, although the behavior was similar.

Lipid peroxidation of NSL in 0.15 M NaCl (**Figure 2E**) showed an initial speed of TBARS production greater than that observed in water, but with a very similar final value (Abs_f= 0.21). Changes in absorbance at 270 nm and 234 nm showed greater initial speeds than the corresponding ones in water. These speeds stayed until 30 minutes and since then, absorbance values did not change. Final values of absorbance in 0.15 M NaCl were smaller than the water ones.

Lipid peroxidation of NSL in Tris-buffer (**Figure 2F**) showed the greatest lag phase (60 min) in TBARS production and the smallest initial reaction rate. The final value was 0.08, a result much smaller than those obtained in water and 0.15 M NaCl. Values of change of absorbance determined at 270 nm and 234 nm were practically the same. Initial speeds were similar to those obtained in water and slower to those observed in 0.15 M NaCl. The reached final values were below to those obtained in water and 0.15 M NaCl.

SL were more susceptible to lipid peroxidation than NSL both in water as in 0.15 M NaCl. Nevertheless, both types of liposomes were equally peroxidized in Tris-buffer.

3.4. Fatty acid composition of retinal lipids and liposomes made of these retinal lipids

Figure 3 shows the fatty acid composition (area %) of retinal lipids and of liposomes made of these retinal lipids (SL-Fe, control). This table also compares fatty acid profiles of control with liposomes incubated with Fe^{2+} for 1 h, in absence and in presence of BHT. Retinal lipids show a high percent (25.8 ± 0.6 %) of docosahexaenoic acid (22:6 n-3), characteristic of this tissue. The retina has approximately 40 percent of PUFAs and 60 percent of saturated and monounsaturated fatty acids. SL prepared with these lipids show a decrease of 22:6 n-3. The PUFAs diminished significantly after incubation with Fe^{2+}. This produce a relative increase of saturated and monounsaturated fatty acids. 5 μM BHT protected PUFAs avoiding lipid peroxidation effects and the fatty acid profile there was not significant differences with control.

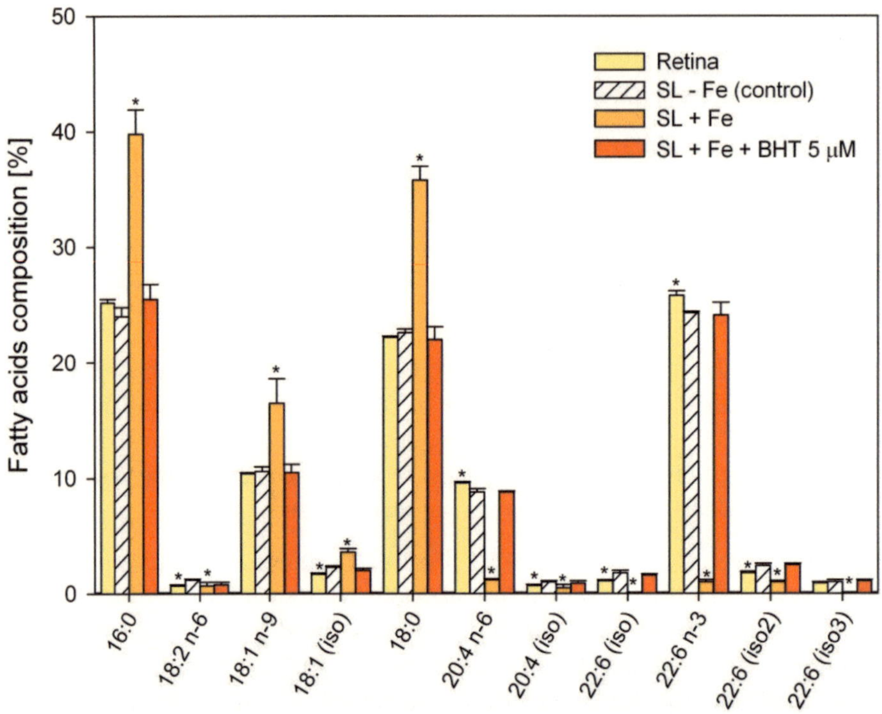

Figure 3. Fatty acid composition (area %) of retinal lipids, liposomes made of retinal lipids (SL -Fe, control), liposomes incubated with Fe^{2+} (SL + Fe) and liposomes incubated with Fe^{2+} in the presence of BHT. Results are expressed as $\tilde{x} \pm SD$. $\tilde{x}$: Average of area % of 3 assays, SD: standard deviation. Significant differences analyzed by ANOVA with control are marked with (*).

4. Conclusion

In summary, the presented results are indicative that liposomes made of retinal lipids by their structural similarities with the biomembranes constitute a very useful analytical system and can mimic the cellular membranes, providing additional information to that obtained with the whole retina. In addition, SL prepared with phospholipids obtained from selected tissues should be used in order to measure lipid peroxidation and the effect of different antioxidants. Additionally, we presented some simple techniques of many possibles that can be applied to study the lipid peroxidation process, different reaction initiators and the antioxidant effect of new compounds.

Abbreviations

16:0: palmitic acid, 18:1 n-9: oleic acid, 18:2 n-6: linoleic acid, 20:4 n-6: arachidonic acid, 22:6 n-3: docosahexaenoic acid, BHT: butylated hydroxitoluene, GC-MS: gas chromatography–mass spectrometry, PUFAs: polyunsaturated fatty acids, TBARS: thiobarbituric acid reactive substances, RNS: reactive nitrogen species, ROS: reactive oxygen species, SL: sonicated liposomes.

Author details

Natalia Fagali* and Angel Catalá
Instituto de Investigaciones Fisicoquímicas Teóricas y Aplicadas, (INIFTA-CCT La Plata-CONICET), Facultad de Ciencias Exactas, Universidad Nacional de La Plata. Casilla de Correo 16, La Plata, Argentina

Acknowledgement

The financial support of Consejo Nacional de Investigaciones Científicas y Técnicas (CONICET), Argentina, Grant PIP-0157, is gratefully acknowledged. The authors thank at the Departamento de Física, Facultad de Ciencias Exactas, Universidad Nacional de La Plata, Argentina for providing light scattering measurements facilities. Special thanks to Prof. J.L. Alessandrini for fitting experimental data.

5. References

Halliwell, B. and Gutteridge, J.M. (1990). Role of free radicals and catalytic metal ions in humandisease: an overview, *Methods Enzymol.* 186, 1–85

Huang, C. (1969). Studies on phosphatidylcholine vesicles, formation and physical characteristics. *Biochemistry* 8, 344–352.

Kohen, R., Nyska, A. (2002). Oxidation of biological systems: oxidative stress phenomena, antioxidants, redox reactions, and methods for their quantification. *Toxicol. Pathol.* 30, 620–650.

Catalá, A. (2009). Lipid peroxidation of membrane phospholipids generates hydroxy-alkenals and oxidized phospholipids active in physiological and/or pathological conditions, *Chem Phys Lipids*, 157, 1-11.

Catalá, A. (2011a). In: Catalá, A. (Ed.), Lipid Peroxidation: Biological Implications, Transworld Research Network Kerala, India, pp. 1–173.

Catalá, A. (2011b). Lipid peroxidation of membrane phospholipids in the vertebrate retina, *Front. Biosci.* (Schol Ed). 3, 52–60, Review.

Fagali, N., Catalá, A. (2009) Fe^{2+} and Fe^{3+} initiated peroxidation of sonicated and non-sonicated liposomes made of retinal lipids in different aqueous media, *Chem Phys Lipids*, 159, 88-94.

Folch, J., Lees, N., Sloane Stanley, G.A. (1957). A simple method for the isolation and purification of total lipids from animal tissues. *J Biol Chem*, 226, 497–509.

Koppel, D.E. (1972). Analysis ofmacromolecular polydispersity in intensity correlation spectroscopy: the methods of cumulants, *J Chem Phys*, 57, 4814–4820.

Sargis, R.M., Subbaiah, P.V. (2003). Trans unsaturated fatty acids are less oxidizable than cis unsaturated fatty acids and protect endogenous lipids from oxidation in lipoproteins and lipid bilayers. *Biochemistry*, 42, 11533–11543.

---

* Corresponding Author

Ueda, T., Ueda, T., Armstrong, D. (1996). Preventive effect of natural and synthetic antioxidants on lipid peroxidation in the mammalian eye. *Ophthalmic Res*, 28, 184–192.

Section 3

Lipid Peroxidation in Vegetables, Oils, Plants and Meats

Chapter 9

The Effect of Plant Secondary Metabolites on Lipid Peroxidation and Eicosanoid Pathway

Neda Mimica-Dukić, Nataša Simin, Emilija Svirčev,
Dejan Orčić, Ivana Beara, Marija Lesjak and Biljana Božin

Additional information is available at the end of the chapter

http://dx.doi.org/10.5772/48193

1. Introduction

Inflammation, free radical damage and oxidative stress have become major health issues in recent years and the subject of plenty of research. These processes are implicated in cancer [1], cardiovascular diseases [2], multiple sclerosis [3], diabetes mellitus [4], Alzheimer's and Parkinson's diseases [5], rheumatoid arthritis [6], premature aging [7] and almost any other degenerative condition. Reactive oxygen species (ROS), which are involved in these physiological functional changes, are often either by-products of the normal cellular processes or are formed by action of exogenous factors - xenobiotics, ionizing radiation, stress, pathogens etc. Overproduction of ROS leads to oxidative stress, with biomolecules, including lipids, proteins and nucleic acids undergoing oxidative alterations.

1.1. Lipid peroxidation

Lipid peroxidation (LP) – oxidative degradation of polyunsaturated fatty acids caused by ROS – is responsible for degradation of membrane lipids resulting in cell damage and formation of many toxic products. LP is a free radical chain reaction where three major steps - the initiation, propagation and termination – can be recognised (Figure 1.).

In initiation phase, highly reactive hydroxyl radical, formed in Fenton reaction, abstracts hydrogen atom in α position relative to the polyunsaturated fatty acid double bond. This results in the formation of fatty acid radical, highly unstable, short-lived intermediate that stabilises by abstracting hydrogen from another chemical species, or reacts with triplet oxygen to generate different radical species, including fatty acid peroxyl radical. In the termination step peroxyl radicals transform into nonradical compounds – hydrocarbons, aldehydes, alcohols, volatile ketones and lipid polymers, some of which are harmful (Figure 1.) [8, 9].

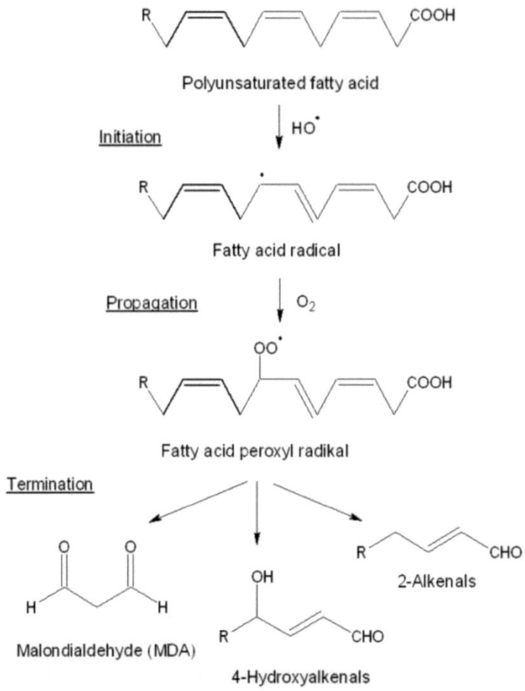

Figure 1. Steps in lipid peroxidation process

Organism uses a number of endogenous antioxidants, such as glutathione, α-lipoic acid, coenzyme Q10, bilirubin and antioxidant enzymes (glutathione peroxidase, catalase, superoxide dismutase), to protect itself from oxidative stress. When they are insufficient, it becomes necessary to introduce exogenous antioxidants. Most of these compounds are primarily taken into the body by food and are predominantly of herbal origin (phenolics, carotenoids, terpenoids and vitamins – ascorbic acid and tocopherol). Herbal antioxidants exhibit their activity through a wide variety of mechanisms, such as inhibition of oxidising enzymes, chelation of transition metals, transfer of hydrogen or single electron to radicals, singlet oxygen deactivation, or enzymatic detoxification of ROS [8]. They can stop LP process in either initiation or propagation step. Herbal antioxidants have become subjects of growing interest and targets of numerous scientific research. However, enormous number of plant species is still waiting to be investigated in this manner and explored as potential medical drugs or dietary supplements.

1.2. Eicosanoid pathway

In addition to direct detrimental effects on biomolecules and cellular structures, lipid peroxidation is also involved in biosynthesis of eicosanoids – arachidonic acid metabolites serving as inflammatory mediators. Arachidonic acid, released from phospholipids by action of phospholipase A, can be converted into these products by three different pathways: cyclooxygenase, leading to the formation of prostanoids (prostaglandins and thromboxanes), lipoxygenase, where leukotrienes and certain mono-, di- and trihydroxy acids are synthesised, and epoxygenase pathway, which includes cytochrome P-450 and gives epoxides as final products [10].

Cyclooxygenase (COX), key enzyme in cyclooxygenase pathway (Figure 2.), exists in three forms, COX-1, COX-2, and recently discovered COX-3. Despite the differences in structure, localisation and regulation, reactions catalysed by COX isoforms follow the same mechanism. All of them transform arachidonic acid into prostanoids: prostaglandins (PGH_2, PGE_2, $PGF_{2\alpha}$), thromboxanes (TXA_2, TXB_2), and 12(*S*)-hydroxyheptadeca-5*Z*,8*E*,10*E*-trienoic acid (12(*S*)-HHT) as a co-product.

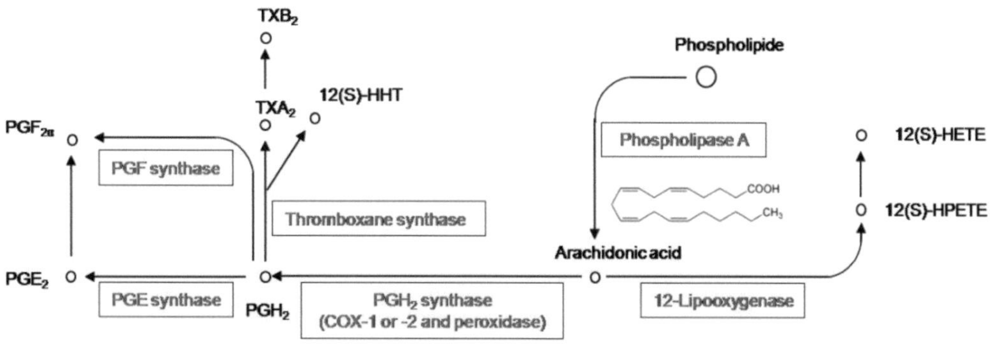

Figure 2. COX and 12-LOX branches of eicosanoid pathway

COX-catalysed transformation of arachidonic acid starts with tyrosyl radical generation through Tyr^{385} oxidation by heme in COX active site. Formed tyrosyl radical abstracts hydrogen from C-13 of arachidonic acid. In subsequent steps, free electron migration, reaction with oxygen yielding peroxyl radical, and cyclisation reactions give PGG_2 which is converted to PGH_2 through the action of peroxidase (Px) (Figure 3.) [11].

COX-1 is expressed constitutively in different tissues, blood monocytes and platelets, and is involved in normal cellular homeostasis. In contrast, COX-2 may be induced by a series of pro-inflammatory stimuli and its role in the progress of inflammation, fever and pain has been known [12]. Furthermore, COX-2 has been targeted in many cancers including: colon cancer, colorectal cancer, breast cancer, gliomas, prostate cancer, esophageal carcinoma, pancreatic cancer, lung carcinoma, gastric carcinoma, ovarian cancer, Kaposi's sarcoma and melanoma [13].

In lipoxygenase branch of arachidonic acid metabolism, there are three types of lipoxygenases, termed 5-, 12- and 15-lipoxygenase. In 12-lipooxigenase (12-LOX) pathway, the first reaction is abstraction of hydrogen from C-10 of arachidonic acid, which includes reduction of Fe^{3+} to Fe^{2+} in enzyme active site. This results in the formation of arachidonic acid radical, which than reacts with oxygen and generates 12-hydroperoxyeicosatetraenoic acid (12-HPETE). Finally, formation of 12(*S*)-hydroxy-(5*Z*,8*Z*,10*E*,14*Z*)-eicosatetraenoic acid (12-HETE) is catalysed by glutathione peroxidase (GPx), whereby glutathione (GSH) is oxidised to GS-SG (Figure 3.) [14, 15].

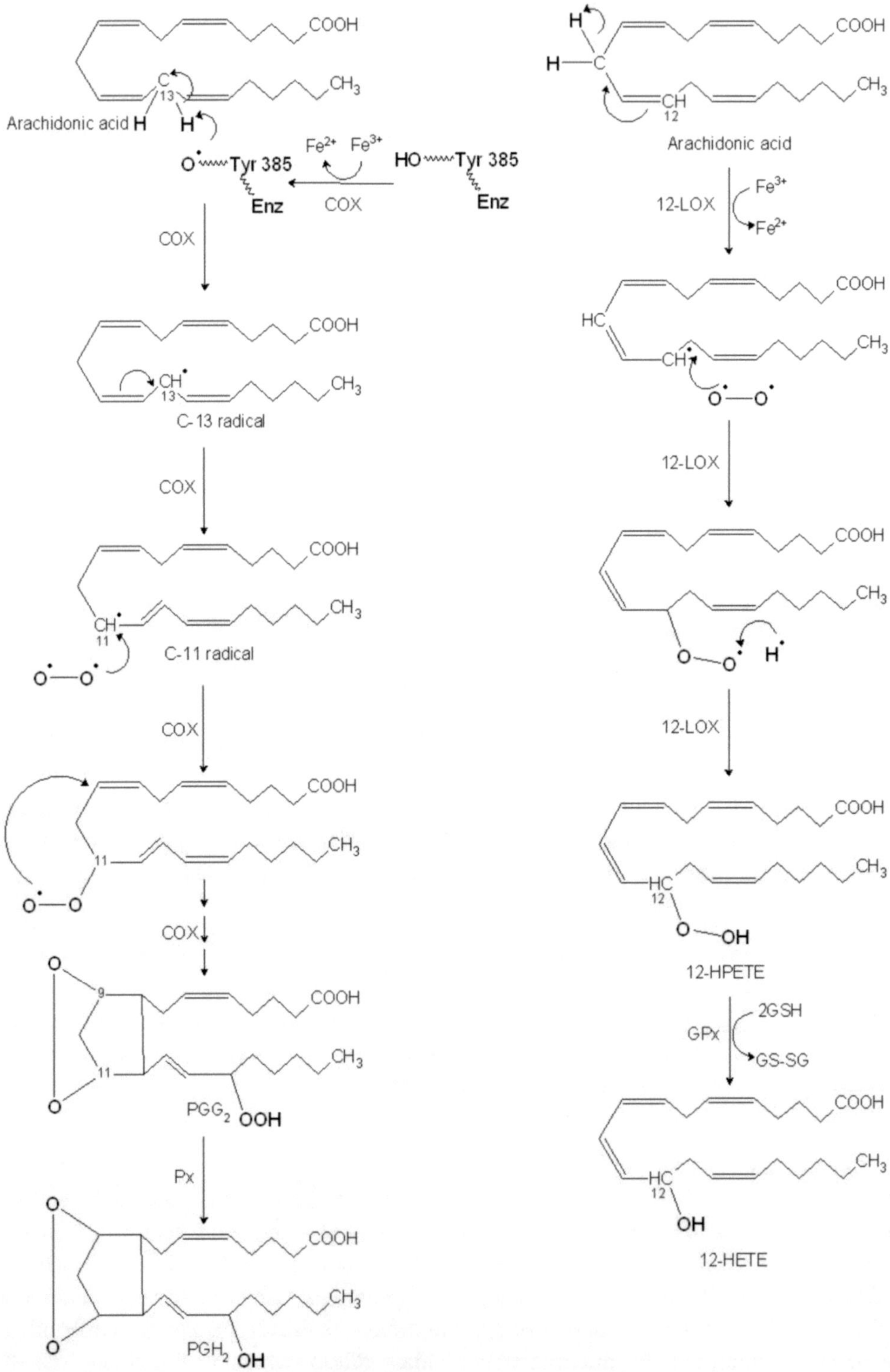

Figure 3. Mechanism of COX and 12-LOX-catalysed arachidonic acid transformation.

12-HETE is implicated in regulation of platelet aggregation, angiogenesis, as well as the progression of several human diseases like various cancers, rheumatoid arthritis and psoriasis [13, 16, 17]. Also, 12-HETE is known to take part in the metastatic cascade as a crucial intracellular signalling molecule which activates protein kinase C and mediates the biological functions of growth factors and cytokines.

Nowadays, there is a great need for a new anti-inflammatory compounds with minimal side effects. Natural products, especially phenolics frequently consumed in a diet of plant origin, are known to have great anti-inflammatory potential considering inhibition of COX and LOX enzymes [18, 19]. Thus, screening of plants for COX/LOX inhibitory activity, followed by effect-guided fractionation, can be a useful tool for discovering new secondary biomolecules with anti-inflammatory potential. Although plant species widely used in traditional medicine may be a good starting point, there is also a vast number of currently unexplored species of unknown composition and activity. Therefore, this study included the poorly investigated plant species classified into four families – Alliaceae, Cupressaceae, Plantaginaceae and Polygonaceae – wild growing in Serbia, that are part of our continuing research [20-29]. Most of these species are used in diet or used in folk medicine for healing various disorders.

Considering the fact that lipid peroxidation is involved in an inflammation process, the aim of this study was to compare antioxidant activity (more specifically, the ability to inhibit lipid peroxidation) of the selected plant extracts with their ability to inhibit production of particular arachidonic acid metabolites and to correlate these activities with the phenolic and flavonoid content.

2. Methods

2.1. Plant material and extract preparation

Plant material used in this study was collected from different locations in Serbia during the period between 2008–2010, during a flowering phase or as ripe specimens. Species belonging to four families: Alliaceae (genus Allium – *A. flavum* L., *A. carinatum* L., *A. melanantherum* Panč., *A. pallens* L., *A. rhodopeum* Velen., *A. paniculatum* L.), Cupressaceae (genus *Juniperus* L. – *J. communis* L., *J. sibirica* Burgsdorf., *J. foetidissima* Willd.), Plantaginaceae (genus *Plantago* L. – *P. major* L., *P. maritima* L., *P. media* L., *P. lanceolata* L., *P. altisima* L.) and Polygonaceae (genus *Rumex* L. – *R. patientia* L., *R. crispus* L., *R. obtusifolius* L.) were investigated. The voucher specimens were deposited in the Herbarium of the Department of Biology and Ecology (BUNS Herbarium), University of Novi Sad Faculty of Sciences.

Air-dried and grounded plant material (30 g) (whole plants of *Allium* sp., needles and cones of *Juniperus* sp., aerial parts of *Plantago* sp. and herbs and rhizomes of *Rumex* sp.) was extracted by maceration with 80% aqueous methanol (8 mL per 1 g of drug) during 72 h at room temperature. After filtration, the solvent was evaporated to dryness under reduced pressure. All raw extracts except those of *Allium* sp. and of *Rumex* sp. rhizomes were resuspended in hot distilled water (to a final concentration of approx. 1 g/mL), washed exhaustively with petroleum ether (fraction 40–60 °C) to remove nonpolar pigments, and evaporated to dryness under vacuum.

Dried extracts were dissolved in 80% aqueous methanol and DMSO for evaluation of the antioxidant and anti-inflammatory activity, respectively, to obtain 300 mg/mL stock solutions.

2.2. Determination of total phenolic content

Total phenolic content was determined according to method of Singleton et al. [30], modified for 96-well microplates. Gallic acid was used as a standard for calibration curve construction. Thirty microliters of each extract or standard solution was added to 150 μL of 0.1 mol/L Folin-Ciocalteu reagent and after 10 min mixed with 120 μL of sodium carbonate (7.5%). The same mixture, with solvent instead of extract, was used as a blank. Absorbance at 760 nm was read after 2 h. The phenolics concentration was determined by using the calibration curve of gallic acid. The total phenolics value was expressed as milligrams of gallic acid equivalents per gram of dry weight (dw).

2.3. Determination of total flavonoid content

The aluminium chloride spectrophotometric method described by Chang et al. [31] and modified for 96-well microplates, was used for determination of of the total flavonoid content. Quercetin was used as a standard to prepare a calibration curve. The reaction mixture was comprised of 30 μL of the extract or standard solution, 90 μL of methanol, 6 μL of 10% aluminium chloride (substituted with distilled water in blank probe), 6 μL of 1 mol/L potassium acetate and 170 μL of distilled water. Absorbance at 415 nm was measured after 30 min. Flavonoid content was calculated according to the standard calibration curve and are expressed in milligrams of quercetin equivalents per gram of dw.

2.4. Lipid peroxidation

There are several methods for measuring the ability of plant extracts to inhibit lipid peroxidation [8]. Some of the methods are based on monitoring of malondialdehyde (MDA), a degradation product of polyunsaturated fatty acids peroxidation. Possible ways for quantification of MDA are GC-FID after derivatisation, HPLC with DAD or fluorimetric detector and spectrophotometric method. The latter is based on the reaction of MDA with thiobarbituric acid (TBA) and it is commonly used both in *in vitro* and *in vivo* studies. Formation of the red-coloured MDA-TBA adduct is measured at 532 nm. Different substrates can be used in this test: lecithin liposomes, free fatty acids, LDL and body fluids [8, 32]. Also, a few different initiators of LP, such as ionizing radiation, chemical agents – metal ions, free radicals and metalloproteins may be used [8].

In this study, we used spectrophotometric TBA assay [28, 33] for evaluation the ability of extracts to inhibit LP. Linseed oil, used as a source of polyunsaturated fatty acids (69.7 % linolenic, 13.5 % linoleic acid, as determined by GC-MS), was obtained from linseed by

Soxhlet extraction. Oil was added to 0.067 mol/L phosphate buffer, pH 7.4, in the presence of 0.25% Tween-80 to obtain a 0.035% suspension and sonicated for 1 hour. This suspension (3.0 mL) was mixed with 20 μL of $FeSO_4$ (4.58 mmol/L), 20 μL of ascorbic acid (87 μmol/L), and 20 μL of extract (or solvent in control); 3.0 mL of phosphate buffer and 20 μL of extract were added in the blank probe. After incubation at 37 °C for 1 hour, 0.2 mL 3.72% EDTA was added to all samples followed by 2 mL of an aqueous mixture containing TBA (3.75 mg/mL), $HClO_4$ (1.3%), and trichloroacetic acid (0.15 g/mL). Reaction mixtures were heated at 100 °C for 15 min, cooled, centrifuged at 1600 g for 15 min, and absorbance was measured at 532 nm. All samples and control were made in triplicate. IC_{50} values were determined from inhibition vs. concentration plots.

2.5. Anti-inflammatory activity

There are a great number of different *in vitro* methods used for estimation of inhibitory activity of COX and LOX enzymes including a number of commercial kits. Some assays, as a source of enzymatic activity, include native or recombinant enzymes, animals or human origin, while others use different cell lines that express desirable activities. Arachidonic acid is often added in reaction medium and in some cases is radio labelled. Induction of inflammatory respond also differs, and in most cases is performed by bacterial lipopolysaccharide, various cytokines and tumour necrosis factor. Different techniques are used for detection of enzymatic activity, such as different chromatographic techniques (TLC, HPLC-UV), as well as EIA [19, 34].

In this study, COX-1 and 12-LOX inhibitory activity was investigated using *ex vivo* assay according to modified method of Safayhi *et al.* [29, 35]. Intact cells (human platelets) were used as a source of COX-1 and 12-LOX enzymes. Arachidonic acid metabolites (12-HHT and 12-HETE) were determined by use of LC–MS/MS technique [29].

An aliquot of human platelet concentrate (viable but outdated for medical use) containing $4 \cdot 10^8$ cells was suspended in buffer (0.137 mol/L NaCl, 2.7 mmol/L KCl, 2.0 mmol/L KH_2PO_4, 5.0 mmol/L Na_2HPO_4 and 5.0 mmol/L glucose, pH 7.2) to obtain final volume of 2 mL. This mixture was slowly stirred at 37 °C for 5 min. Subsequently, 0.1 mL of extracts or standard compounds solutions in DMSO (concentration ranging from 10.0 to 200.0, 0.156 to 5.0 and 0.01 to 0.6 mg/ml for extracts, quercetin and aspirin, respectively) and 0.1 mL of calcimycin (Calcium Ionophore A23187, 125 μmol/L in DMSO) were added and incubated for 2 min at 37 °C, with moderate shaking. The exact volume of extract in control and calcimycin in blank probe were substituted with solvent (DMSO). Thereafter, 0.3 mL of $CaCl_2$ aqueous solution (16.7 mmol/L), substituted with water in blank probe, was added and the mixture was incubated for further 5 min at 37 °C with shaking. Acidification with cold 1% aqueous formic acid (5.8 mL) to pH 3 terminated the reaction. If gel formation occurred, vortexing was applied before mixing with the acid. Prostaglandin B_2 (50 μL of 6 μg/mL solution in DMSO) was added as internal standard, and extraction of products was done with mixture of chloroform and methanol (1:1, 8.0 mL) with vigorous vortexing for 15 min. After centrifugation at 7012 × g for 15 min at 4 °C, organic layer was separated, evaporated to dryness, dissolved in methanol (0.5 mL), filtered and used for further LC–MS/MS analysis. All samples and control were made in triplicate.

Test for estimation of the anti-inflammatory activity, applied in our research, has a lot of advantages. Firstly, the advantage is avoidance of the undesirable *in vivo* tests on experimental animals. Even though the exact anti-inflammatory activity can be validated only through *in vivo* tests, creating *in vitro* assays, in which physiological conditions similar to *in vivo* assays are used, can provide valuable information about inhibitory potential of the compounds tested. Platelets are a suitable cell system for testing inhibitory activity, because they can provide physiological cell conditions and possibility to examine the inhibition of both enzymes at the same time. Secondly, for determination of the formed metabolites as a measure of level of inhibition of COX and LOX activity, LC-MS/MS technique was used. LC-MS/MS provided a highly sensitive and specific detection of desirable metabolites within a short analysis time [29].

2.6. Statistical analysis

Percent of lipid peroxidation inhibition achieved by different concentration of extracts was calculated by the following equation: $I(\%) = (A_0 - A)/A_0 \times 100$, where A_0 was the absorbance of the control reaction and A was the absorbance of the examined samples, corrected for the value of blank probe. Percent of COX-1 and 12-LOX inhibition achieved by different concentrations of extracts was calculated by the following equation: $I(\%) = 100 \times (R_0 - R)/R_0$, where R_0 and R were response ratios (metabolite peak area divided by internal standard peak area) in the control reaction and in the examined samples, respectively. Both R and R_0 were corrected for the value of blank probe. For both assays, corresponding inhibition-concentration curves were drawn using Origin software, version 8.0 (Origin Labs) and IC_{50} values (concentration of extract that inhibited lipid peroxidation and COX-1 and 12-LOX metabolites formation by 50%) were determined. All of the results were expressed as mean ± SD of three replicates. Correlation analyses were done using Statistica software version 6 (StatSoft). Concentrations of total phenolics and the total flavonoids were used as independent variables, while inhibitory activities towards LP and 12-HETE synthesis (expressed as $1/IC_{50}$) were used as dependent variables. Due to a wide range of values, log-log plots were applied. Pearson's correlation coefficients were calculated.

3. Results

In this study, we tested the effect of 17 taxa from four families (*Alliaceae, Cupressaceae, Plantaginaceae* and *Polygonaceae*) on lipid peroxidation and metabolism of arachidonic acid. The total phenolic and flavonoid contents were determined in the plant extracts, as well. All results are presented in Figures 4, 5, 6 and 9.

Among all tested samples, herb and rhizome extracts of *Rumex* species were the most potent inhibitors of LP (0.009-0.047 mg/mL). Slightly lower activity was shown by herb extracts of *Plantago* species (0.025-0.178 mg/mL), while extracts of cones and needles of *Juniperus* species, as well as the whole plant extracts of *Allium* species expressed a much lower activity (0.117-0.887 and 0.68-1.986 mg/mL, respectively).

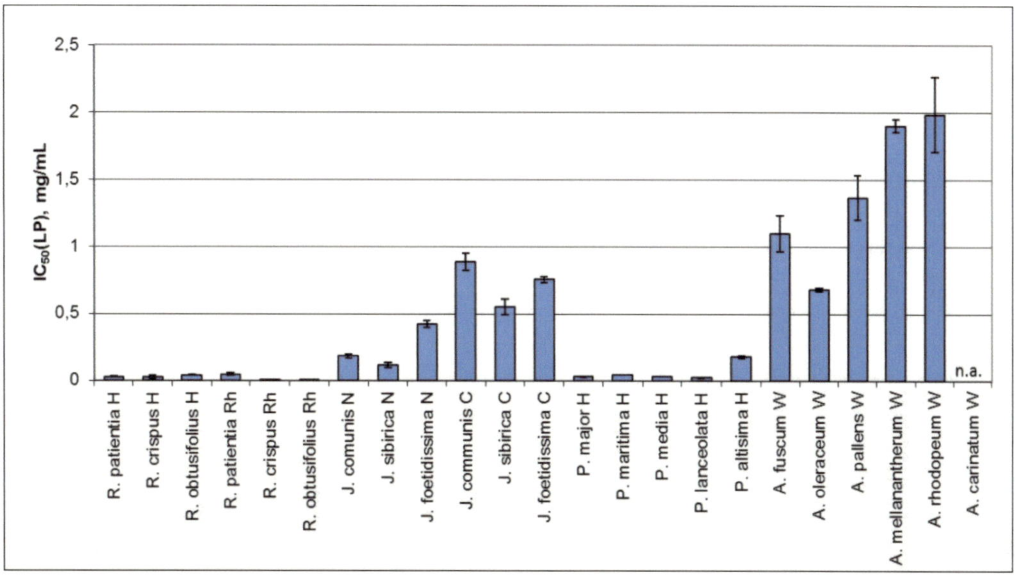

Legend: H – herb, Rh – rhizome, C – cones, N – needles, W – whole plant, n.a. - not achieved.

Figure 4. Results of lipid peroxidation inhibition assay.

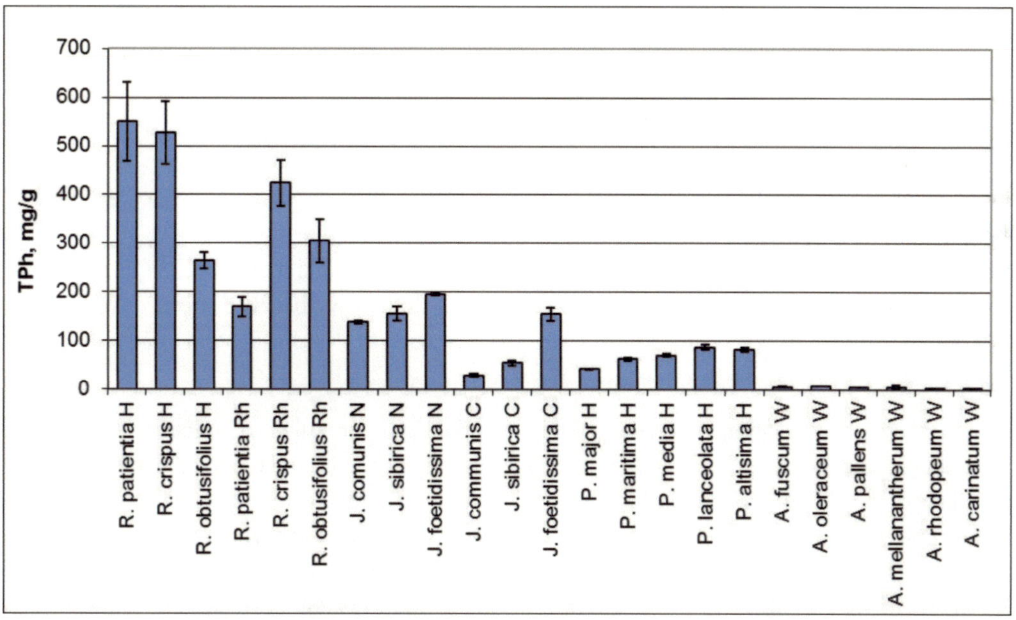

Legend: H – herb, Rh – rhizome, C – cones, N – needles, W – whole plant.

Figure 5. Results of total phenolic content assay (expressed as mg eq. gallic acid / 1 g dw).

Total phenolic content in examined genera decreased in following order: *Rumex, Juniperus, Plantago* and *Allium*, with *R. patentia* and *R. crispus* extracts being by far the richest in phenolics (550 and 527 mg eq. gallic acid per 1 g dw, respectively). Regarding the content of the total flavonoids, significant intrageneric variations were observed, hence it was not possible to point out a genus with the highest content. Flavonoids were the most abundant in *J. foetidissima* needles extract (60 mg quercetin eq per 1 g dw), and scarcest in *A. rhodopeum* (0.2 mg quercetin eq per 1 g dw).

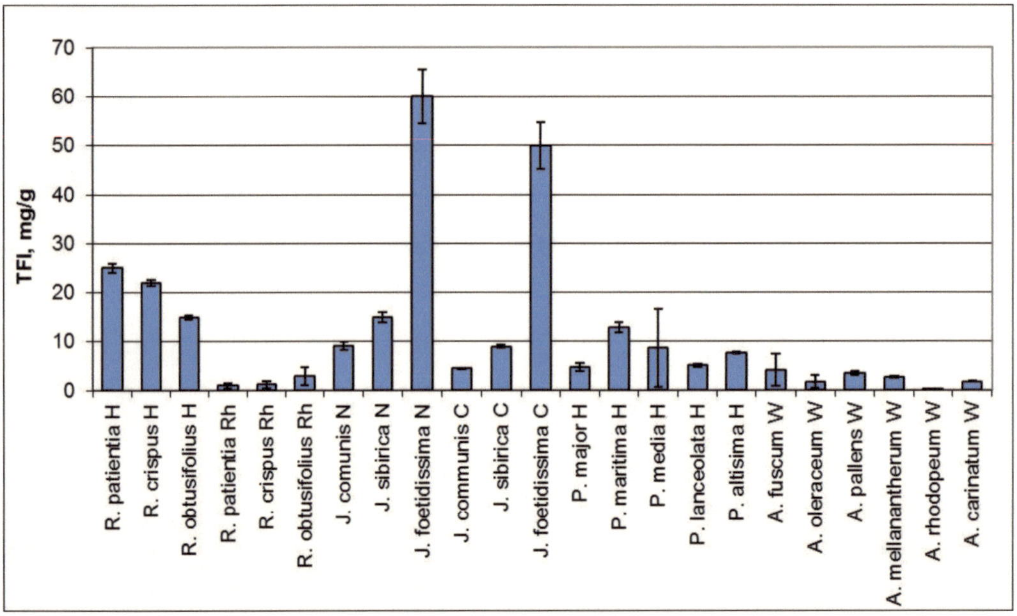

Legend: H – herb, Rh – rhizome, C – cones, N – needles, W – whole plant.

Figure 6. Results of total flavonoid content assay (expressed as mg eq. quercetin / 1 g dw).

Plant phenolics present in extracts can counteract lipid peroxidation in either initiation or propagation step. Possible mechanisms are chelation of transition metals (Fe^{2+}), reduction of Fe^{3+} and neutralisation of lipid peroxidation radical intermediates by transfer of hydrogen or single electron [8]. An attempt was made to correlate the observed lipid peroxidation inhibitory activity (given as $1/IC_{50}$) with the content of the total phenolics and total flavonoids. The corresponding plots are given in Figure 7. and Figure 8.

A high degree of correlation was observed between the lipid peroxidation inhibitory activity and the total phenolic content ($r = 0.7713$), while correlation with the total flavonoids content was not established ($r = 0.1410$). Thus, flavonoids do not contribute to the total observed activity to a significant extent. The lower activity of flavonoids, compared to other phenolics, can be explained through Porter's polar paradox. Namely, flavonoids present in the extracts investigated were predominantly in glycosylated form and thus highly polar. It is demonstrated that compounds of lower polarity are more effective in polar reaction media, such as oil-in-water emulsion used in our experiments, since they exhibit their activity at oil-water interface [36]. In addition, hydrogen atom donation ability of flavonoids decrease with glycosylation, especially if the most active hydroxyl groups (C-3, C-3′ or C-4′) are occupied by carbohydrate moiety [37, 38].

Regarding the eicosanoid pathway, *Rumex* and *Plantago* species showed, on average, very high 12-LOX inhibitory effect, with IC_{50} in range 0.75–3.59 mg/mL (Figure 9.). While

Juniperus and *Allium* also exhibited dose-dependent inhibition of 12-HETE production, their IC_{50} values were higher, ranging from 1.45 mg/mL to 11.04 mg/mL.

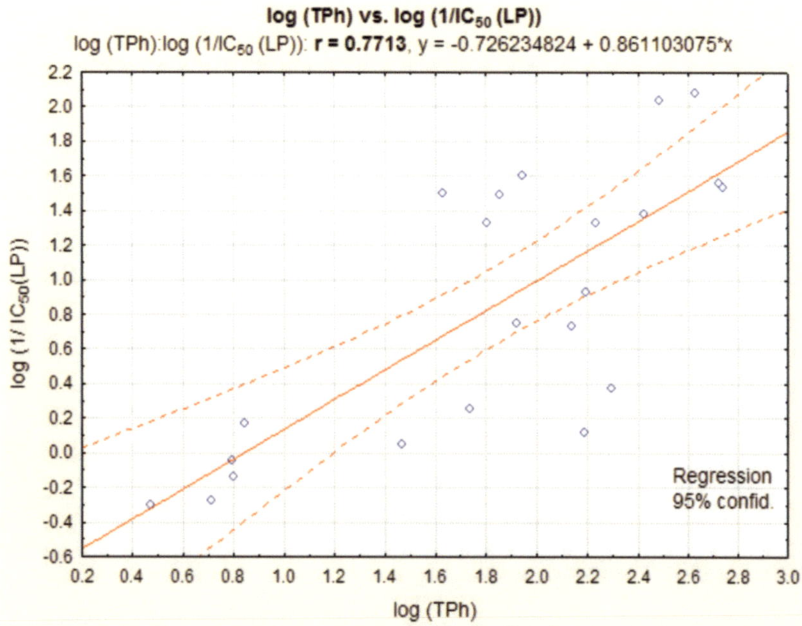

Figure 7. Correlation between total phenolic content and ability of extracts to inhibit lipid peroxidation

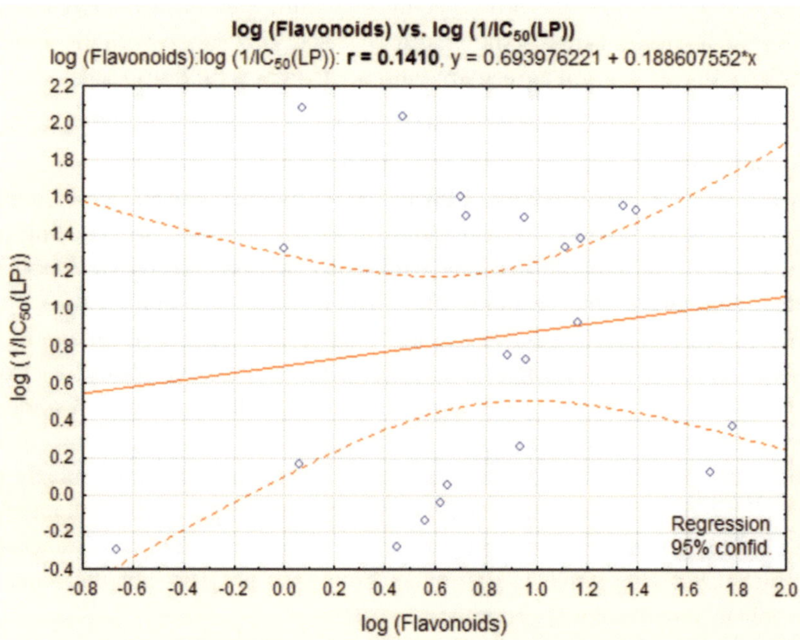

Figure 8. Correlation between total flavonoids content and ability of extracts to inhibit lipid peroxidation

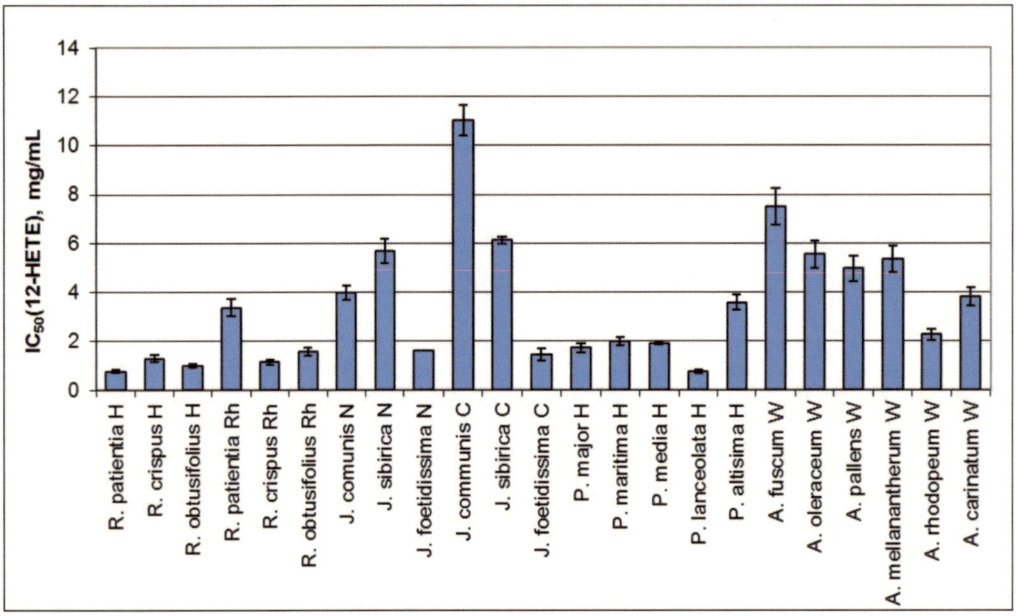

Legend: H – herb, Rh – rhizome, C – cones, N – needles, W – whole plant.

Figure 9. Results of 12-LOX inhibition assay.

Moreover, certain inhibitory activity of the examined extracts towards COX-1 was also confirmed in the anti-inflammatory assay applied in this research. Determined IC_{50} values ranged from 0.34 to 8.00 mg/mL, with no genus exhibiting significantly higher activity than others. However, the results are not shown due to lack of correlation with the total phenolics and flavonoid content. This is in agreement with our previous findings [27].

The exact mechanism of COX and LOX enzymes inhibition by natural products is still not fully elucidated. However, bearing in mind their structures and chemical properties, several mechanisms could be speculated. Both pathways of arachidonic acid metabolism include free radical reactions (Figure 3.). Due to their radical scavenging activity or reducing properties, many plant natural products can interfere with reactions catalysed by COX and LOX [39]. They can neutralise radical intermediates, thus terminating the reaction. In addition, they can reduce Fe^{3+} ion that is a part of active site of both enzymes and is necessary for initiation reaction. Some natural products, including acetylenes, can bind covalently to enzymes and inhibit them irreversibly [19]. Also, in inhibition of LOX, isoprenyl moiety of some phenolics and terpenoid backbone structure could play an important role. Prenylated phenolics are usually more hydrophobic than conventional ones. Terpenoids are also mostly non-polar compounds. These characteristics suggest an easy penetration through the cell membrane and their good 12-lipoxygenase inhibitory properties [19]. Finally, the COX or LOX activity can be decreased by suppressing their transcription by phenolics [18], although this effect is observable only in long-duration experiments.

To identify the compound class responsible for the observed 12-LOX-inhibitory activity, $1/IC_{50}$ was correlated with the content of the total phenolics and flavonoids. The corresponding plots are given in Figure 10. and Figure 11. As with the lipid peroxidation, the correlation between the content of the total phenolics and 12-LOX-inhibitory activity was also observed, although slightly weaker, with Pearson's correlation coefficient r = 0.6037. At the same time, no relationship was found between the total flavonoids content and 12-HETE production (r = 0.2825). Bearing in mind a good correlation of 12-LOX inhibition with the phenolic content, and the lack thereof with the flavonoid content, it is possible that only small phenolic molecules, but not the voluminous flavonoid glycosides, can enter 12-LOX active site and exhibit the inhibitory effect there. Thus, the observed differences in the total phenolics and the total flavonoids effects can at least partially be explained by steric hindrance.

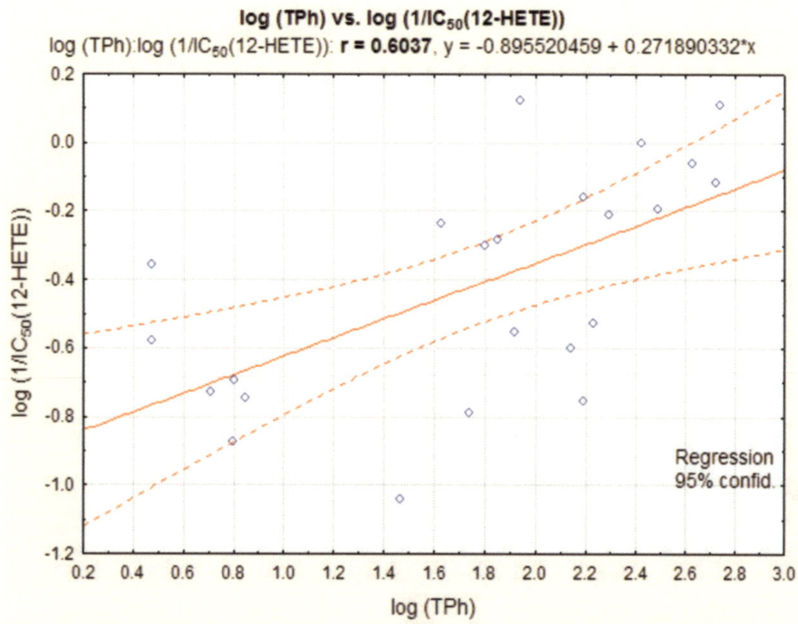

Figure 10. Correlation between total phenolic content and 12-LOX inhibitory activity

The differences in COX and LOX inhibition (LOX inhibition being correlated with phenolic content) can be attributed to the differences in reaction mechanism and active site three-dimensional structure. Namely, both mechanisms include abstraction of hydrogen from arachidonic acid leading to formation of radical species. However, hydrogen acceptor in COX is tyrosyl radical (formed through tyrosine oxidation by Fe^{3+}) while in LOX, electron is transferred directly to Fe^{3+}. Phenolics from plant extracts could reduce Fe^{3+} to Fe^{2+}, thus inactivating both enzymes. However, the presence of tyrosyl residue in COX active site could provide steric protection and prevent phenolics from approaching the Fe^{3+} ion.

Finally, the correlation between the ability of extracts to counteract lipid peroxidation and inhibit production of 12-HETE is shown in Figure 12. High correlation coefficient (r = 0.6819) suggests that extracts with a high inhibitory effect on LP, also represent potent inhibitors of 12-LOX pathway. Methods for examination of anti-inflammatory activity are labour-intensive, expensive and sometimes involve ethical issues due to the usage of laboratory animals. Therefore, *in vitro* measurement of lipid peroxidation inhibition and the total phenolic content could be a useful tool for screening of plant extracts with a potential anti-inflammatory activity.

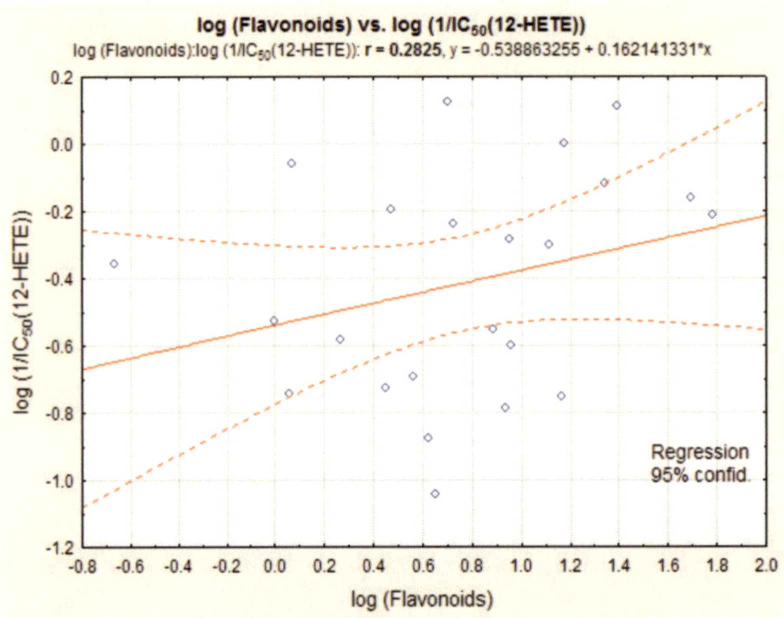

Figure 11. Correlation between total flavonoid content and 12-LOX inhibitory activity

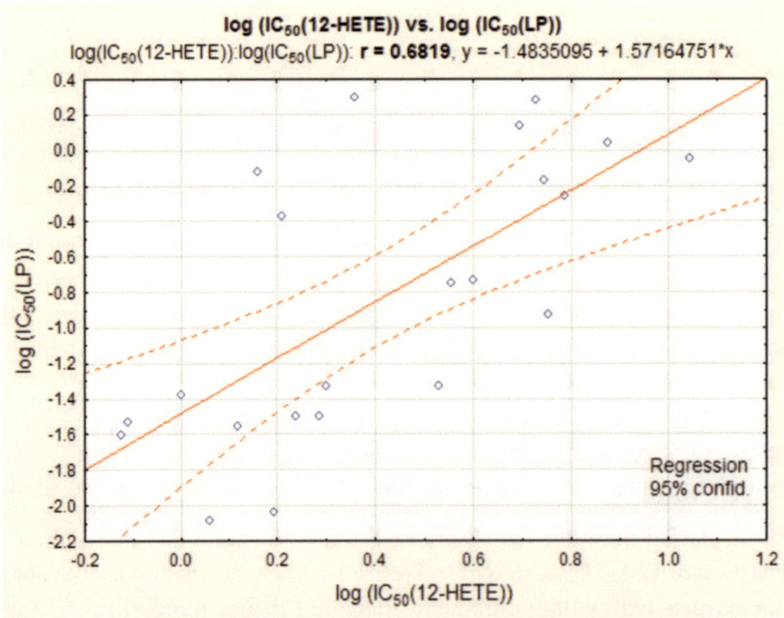

Figure 12. Correlation between the ability of extracts to inhibit 12-HETE production and lipid peroxidation

To summarise, the most of the examined species expressed high lipid peroxidation and anti-inflammatory activity, especially *Rumex* and *Plantago* species. High inhibitory activity of these species towards 12-LOX pathway makes them good candidates for further research taking into account role of this enzyme in cancer development. A good correlation was found between total phenolic content of 23 investigated plant extracts and their ability to inhibit lipid peroxidation and 12-LOX pathway. Consequently, LP inhibitory activity was also highly correlated with 12-LOX inhibition. Thus it can serve as indicator for preliminary selection of plant extracts further to be tested for anti-inflammatory activity by expensive and time-consuming methods.

Author details

Neda Mimica-Dukić, Nataša Simin, Emilija Svirčev, Dejan Orčić, Ivana Beara and Marija Lesjak
Department of Chemistry, Biochemistry and Environmental Protection, University of Novi Sad, Faculty of Sciences, Novi Sad, Republic of Serbia

Biljana Božin
Department of Pharmacy, Faculty of Medicine, University of Novi Sad, Novi Sad, Republic of Serbia

Acknowledgement

The Ministry of Education and Science, Republic of Serbia (Grant No. 172058) funded this research. We wish to thank to Goran Anačkov, PhD, for the providing voucher specimens, and Gordana Vlahović, MSc, for language corrections. Platelet concentrate was kindly provided by The Institute for Blood Transfusion of Vojvodina, Novi Sad, Republic of Serbia.

4. References

[1] Kawanishi S., Hiraku Y., Murata M., Oikawa S. The role of metals in site-specific DNA damage with reference to carcinogenesis. Free Radical Biology & Medicine 2002;32(9) 822-832.

[2] Manach C., Mazur A., Scalbert A. Polyphenols and prevention of cardiovascular diseases. Current Opinion in Lipidology 2005;16(1) 77-84.

[3] Gonsette RE. Review: Oxidative stress and excitotoxicity: a therapeutic issue in multiple sclerosis? Multiple Sclerosis 2008;14(1) 22-34.

[4] Maritim AC., Sanders RA., Watkins JB. Diabetes, Oxidative Stress, and Antioxidants: A Review. Journal of Biochemical and Molecular Toxicology 2003;17(1) 24-38.

[5] Ramassamy C. Emerging role of polyphenolic compounds in the treatment of neurodegenerative diseases: a review of their intracellular targets. European Journal of Pharmacology 2006;545(1) 51-64.

[6] Wruck CJ., Fragoulis A., Gurzynski A., Brandenburg L., Kan YW., Chan K., Hassenpflug J., Freitag-Wolf S., Varoga D., Lippross S., Pufe T. Role of oxidative stress

in rheumatoid arthritis: insights from the Nrf2-knockout mice. Annals of the Rheumatic Diseases 2010;doi:10.1136/ard.2010.132720

[7] Alvarado C., Alvarez P., Jimenez L., De la Fuente M. Oxidative stress in leukocytes from young prematurely aging mice is reversed by supplementation with biscuits rich in antioxidants. Developmental & Comparative Immunology 2006;30(12) 1168-1180.

[8] Laguerre M., Lecomte J., Villeneuve P. Evaluation of the ability of antioxidants to counteract lipid oxidation: Existing methods, new trends and challenges. Progress in Lipid Research 2007;46(5) 244-282.

[9] Gordon MH. The mechanism of antioxidant action in vitro. In: Hudson BJF. (ed.) Food antioxidants. Amsterdam: Elsevier; 1990. p1-18.

[10] Smith WL. The eicosanoids and their biochemical mechanisms of action. Biochemical Journal 1989;259(2) 315-324.

[11] Marnett LJ., Rowlinson SW., Goodwin DC., Kalgutkar AS., Lanzo CA. Arachidonic acid oxygenation by COX-1 and COX-2. Mechanisms of catalysis and inhibition. The Journal of Biological Chemistry 1999;274(33) 22903-22906.

[12] Hawkey CJ. COX-2 inhibitors. The Lancet 1999;353(9149) 307-314.

[13] Greene ER., Huang S., Serhan CN., Panigrahy D. Regulation of inflammation in cancer by eicosanoids. Prostaglandins and Other Lipid Mediators 2011;96(1-4): 27-36.

[14] Yamamoto S., Suzuki H., Ueda N., Takahashi Y., Zoshimoto T. Mammalian Lipoxygenases. In: Curtis-Prior PB. (ed.) The Eicosanoids. Cambridge: Wiley; 2004. p53-61.

[15] Müller-Decker K. Cyclooxygenases. In: Marks F., Fürstenberger G. (ed.) Prostaglandins, leukotrienes, and other eicosanoids: from biogenesis to clinical applications. Weinheim: Wiley-VCH; 1999. p65-83.

[16] Liagre B., Vergne P., Rigaud M., Beneytout JL. Expression of arachidonate platelet-type 12-lipoxygenase in human rheumatoid arthritis type B synoviocytes. FEBS Letters 1997;414(1) 159-164.

[17] Müller K. 5-lipoxygenase and 12-lipoxygenase: attractive targets for the development of novel antipsoriatic drugs. Archiv der Pharmazie. 1994;327(1) 3-19.

[18] Lee JH., Kim GH. Evaluation of Antioxidant and Inhibitory Activities for Different Subclasses Flavonoids on Enzymes for Rheumatoid Arthritis. Journal of Food Science. 2010;75(7) H212-H217.

[19] Schneider I., Bucar F. Lipoxygenase Inhibitors from Natural Plant Sources. Part 2: Medicinal Plants with Inhibitory Activity on Arachidonate 12-lipoxygenase, 15-lipoxygenase and Leukotriene Receptor Antagonists. Phytotherapy Research. 2005;19(4) 263-272.

[20] Beara IN., Lesjak MM., Orčić DZ., Simin NĐ., Četojević-Simin DD., Božin BN., Mimica-Dukić NM. Comparative analysis of phenolic profile, antioxidant, anti-inflammatory and cytotoxic activity of two closely-related Plantain species: *Plantago altissima* L. and *Plantago lanceolata* L. LWT - Food Science and Technology. 2012;47(1) 64-70.

[21] Lesjak MM., Beara IN., Orčić DZ., Anačkov GT., Balog KJ., Francišković MM., Mimica-Dukić NM. *Juniperus sibirica* Burgsdorf. as a novel source of antioxidant and anti-inflammatory agents. Food Chemistry. 2011;124(3) 850-856.

[22] Orčić DZ., Mimica-Dukić NM., Franciškovič MM., Petrović SS., Jovin ED. Antioxidant activity relationship of phenolic compounds in *Hypericum perforatum* L. Chemistry Central Journal. 2011;5(1) 34-41.

[23] Beara I. Phytochemical screening and evaluation of antioxidant and anti-inflammatory potential of secondary metaboliotes of *Plantago* L. species. PhD thesis. University of Novi Sad; 2010.

[24] Orčić D. Scandiceae tribe (Apiaceae Lindley 1836, subfam. Apioideae) species – potential resources of biologically and pharmacologically active secondary biomolecules. PhD thesis. University of Novi Sad; 2010.

[25] Mimica-Dukić NM., Simin NĐ., Cvejić JM., Jovin ED., Orčić DZ., Božin BN. Phenolic compounds in field horsetail (*Equisetum arvense* L.) as natural antioxidants. Molecules. 2008;13(7) 1455-1464.

[26] Božin BN., Mimica-Dukić NM., Simin NĐ., Anačkov GT. Characterization of the volatile composition of essential oils of some lamiaceae spices and the antimicrobial and antioxidant activities of the entire oils. Journal of Agricultural and Food Chemistry. 2006;54(5) 1822-1828.

[27] Lesjak M. Biopotential and chemical characterization of extracts and essential oils of species from *Juniperus* L. genus (Cupressaceae). PhD thesis. University of Novi Sad; 2011.

[28] Beara I., Lesjak M., Jovin E., Balog K., Anačkov G., Orčić D., Mimica-Dukić N. Plantain (*Plantago* L.) Species as novel sources of flavonoid antioxidants. Journal of Agricultural and Food Chemistry. 2009;57(19) 9268-9273.

[29] Beara I., Orčić D., Lesjak M., Mimica-Dukić N., Peković B., Popović M. Liquid chromatography/tandem mass spectrometry study of anti-inflammatory activity of Plantain (*Plantago* L.) species. Journal of Pharmaceutical and Biomedical Analysis. 2010;52(5):701-706.

[30] Singleton VL., Orthofer R., Ramuela-Raventos RM. Analysis of total phenols and other oxidation substrates and antioxidants by means of Folin-Ciocalteu reagent. Methods in Enzymology. 1999;299 152-178.

[31] Chang CC., Yang MH., Wen HM. Chern JC. Estimation of total flavonoid content in propolis by two complementary colorimetric methods. Journal of Food and Drug Analysis. 2002;10(3) 178-182.

[32] Mavi A., Lawrence GD., Kordalia A., Yildirim A. Inhibition of iron-fructose-phosphate-induced lipid peroxidation in lecithin liposome and linoleic acid emulsion systems by some edible plants. Journal of Food Biochemistry. 2011;35(3) 833-844.

[33] Aust SD. Lipid peroxidation. In: Greenwald RA. (ed.) Handbook of methods for oxygen radical research. Boca Raton: CRC Press, 1985. p203-207.

[34] Pairet M., Ryn van J. Experimental models used to investigate the differential inhibition of cyclooxygenase-1 and cyclooxygenase-2 by non-steroidal anti-inflammatory drugs. Inflammation Research. 1998;47(2) S93-101.

[35] Safayhi H., Mack T., Sabieraj J., Anazodo MI., Subramanian LR., Ammon HPT. Boswellic acids: novel, specific, nonredox inhibitors of 5-lipoxygenase. Journal of Pharmacology and Experimental Therapeutics. 1992;261(3) 1143-1146.

[36] Porter WL. Paradoxical behavior of antioxidants in food and biological systems. Toxicology and Industrial Health. 1993;9(1-2) 93-122.

[37] Rice-Evans CA., Miller NJ., Paganga G. Structure-antioxidant activity relationships of flavonoids and phenolic acids. Free Radical Biology and Medicine. 1996;20(7) 933-56.

[38] Rice-Evans CA., Miller NJ., Paganga G. Antioxidant properties of phenolic compounds. Trends in Plant Science. 1997;2(4) 152-159.

[39] Hostettmann K. Assays for immunomodulation and effects on mediators of inflammation. In: Dey PM., Harborne JB. (ed.) Methods in Plant Biochemistry vol. 6, London: Academic Press Limited; 1991. p207-211.

Chapter 10

Repeatedly Heated Vegetable Oils and Lipid Peroxidation

Kamsiah Jaarin and Yusof Kamisah

Additional information is available at the end of the chapter

http://dx.doi.org/10.5772/46076

1. Introduction

Deep frying is the most common and one of the oldest methods of food preparation worldwide. It involves heat and mass transfer. To reduce the expenses, the oils tend to be used repeatedly for frying. When heated repeatedly, changes in physical appearance of the oil will occur such as increased viscosity and darkening in colour [1], which may alter the fatty acid composition of the oil. Heating causes the oil to undergo a series of chemical reactions like oxidation, hydrolysis and polymerization [2]. During this process, many oxidative products such as hydroperoxide and aldehydes are produced, which can be absorbed into the fried food [3].

Palm and soy oil are the most commonly used vegetable oils in the household and industry in Malaysia for deep frying purposes. Both palm and soy oils are rich in tocopherols [4-5]. In addition to the tocopherols, palm oil also contains an abundant amount of tocotrienols. The latter form of vitamin E was consistently shown to possess better antioxidant activity than the former form [6]. The soy oil has bigger proportion of polyunsaturated fatty acid compared to the palm oil. Whereas in the palm oil, the major fatty acids present are the monounsaturated and saturated fatty acids [7].

A survey conducted in Kuala Lumpur recently had revealed that majority of the respondents admitted using repeatedly heated cooking oil [8]. The public level of awareness regarding such usage is influenced by the socioeconomic status. Respondents with higher income and education level had higher level of awareness [9]. Chronic consumption of repeatedly heated vegetable oils could be detrimental to health. It was shown to demonstrate genotoxic and preneoplastic change in the rat liver [10]. It also impaired fluid and glucose intestinal absorption in rats [11]. In rats given alcohol plus heated sunflower, an apparent liver damage as well as increased cholesterol level was observed [12]. Soriguer et al. [13] found an independent positive association between the risk of hypertension and intake of heated cooking oil. These accumulating data suggest chronic intake of heated cooking oils increases the risk of cancer and cardiovascular diseases.

The incidence of cardiovascular disease is higher in men compared to women of similar age [14]. However, after menopause, the risk increases in women due to deficiency of estrogen level. The estrogen was shown to be cardioprotective in ovariectomized female rats [15]. It also possesses antioxidant activity which can lower the risk of lipid peroxidation [16]. Ovariectomized female rats have been used widely as a postmenopausal model [17-18].

In this study, we used two models, normal male rats and ovariectomized female rats to compare the effects of prolonged consumption of repeatedly heated palm oil and soy oil on in vivo plasma lipid peroxidation content. The oxidative stability of these repeatedly heated palm oil and soy oil were also compared.

2. Materials and methods

2.1. Animals, materials and chemicals

Male and female Sprague Dawley rats (180-200 gram) were obtained from the Laboratory Animal Resource Unit of Universiti Kebangsaan Malaysia. They were kept in polyethylene cages in a well ventilated room at room temperature. Food and water were provided *ad libitum*. Palm oil (Lam Soon Edible Oil, Malaysia) and soy oil (Yee Lee Edible Oil, Malaysia) were used in this study. Sweet potatoes were purchased from the same source at a local market. All chemicals and enzymes were obtained from Sigma-Aldrich (St. Louis, MO, USA), unless otherwise stated.

Ethical approval regarding the experimental procedure and humane animal handling in the study was obtained from the Universiti Kebangsaan Malaysia Animal Ethical Commitee and Medical Research Ethics Committee.

2.2. Frying procedure

The oils were heated according to the method of Owu et al. [19]. A kilogram of sweet potato slices were fried in a stainless steel wok containing two and half litres of palm oil or soy oil for 10 minutes at 180°C. Upon completion of the frying process, once heated oil was obtained. The process was repeated four times to obtain five times heated oil with a cooling interval of at least five hours. The food quantity was proportionately adjusted with the amount of vegetable oil left. No fresh oil was added between the frying processes to make up for the loss due to uptake by the frying materials. After being heated, small quantity of the oils was extracted for the peroxide value, fatty acid composition and vitamin E content measurements.

2.3. Diet preparation

Diets enriched with heated palm or soy oils were prepared by mixing 850 g grinded standard mouse pellet with 150 g heated oil. While to prepare high cholesterol (2%) diets fortified with heated palm or soy oils, 150 g palm or soy oil were mixed with 20 g cholesterol

(MP Biomedical Inc., Australia) with 830 g grinded standard mouse pellet. The pellets were then remolded and dried in an oven at 80°C overnight.

2.4. Effects of heated oils in male rats

After one week of acclimatization period, forty-two male Sprague Dawley rats were randomly divided into seven groups. The first group was given standard mouse pellet (control). The second, third and fourth groups were given fresh, once and five times heated palm oil diets, while the fifth, sixth and the last groups were given fresh, once and five times heated soy oil diets, respectively for four months. Blood was sampled before and at the end of treatment duration.

2.5. Effects of heated oils in ovariectomized female rats fed high cholesterol diet

Forty-two female Sprague Dawley rats were allowed to acclimatize for a week before the treatment was started and were ovariectomized ahead of the study. They were randomly divided into seven groups. Group 1 was fed 2% cholesterol diet (control), while groups 2, 3 and 4 were respectively fed 2% cholesterol diet added with fresh, once and five times heated palm oil. Groups 5, 6 and 7 were respectively given 2% cholesterol diet added with fresh, once and five times heated soy oil. The treatment duration was four months, after which the rats were sacrificed and blood samples were taken. Blood sample was also taken prior to the treatment.

2.6. Peroxide content measurement

Measurement of peroxide values of the heated oils was done according to the American Oil Chemists' Society (AOCS) Official Methods Cd 8-53 [20]. Briefly, five grams of the oil sample were added with 30 ml of acetic acid-chloroform (3:2) in a flask. The flask was then swirled before the addition of 0.5 ml saturated potassium iodide. The solution was swirled again for a minute. An amount of 30 ml distilled water and a few drops of starch solution (10%) were added. The solution was titrated against 0.01 N sodium thiosulphate solution which was priorly standardised using potassium dichromate and potassium iodide, until blue colour disappeared. The peroxide value in the oils was calculated as the difference in the volume of sodium thiosulphate solution (ml) used for samples and blank, divided by its normality. The values were expressed in miliequivalents of peroxide per kilogram of the sample.

2.7. Fatty acid composition measurement

Fatty acid composition in the fresh and heated oils was analysed using gas chromatography (GC-17A, Shimadzu, Japan), which consisted of a flame ionisation detector, a BPX 70 capillary column (30 m x 0.25 mm x 0.25 μm), programmable injector temperature, set at 250°C and detector temperature, set at 280°C. The oil samples (100 μl) were first transesterified to fatty acid methyl ester using 1 ml of 1 M sodium methoxide in 1 ml hexane before injected into the gas chromatographic system. The injection volume was 1 μl. Nitrogen at a flow rate of 0.40 ml/min was used as carrier gas in the analysis. Identification of fatty acid methyl ester peaks was carried out by comparing their retention times with their authentic standards. The fatty acid composition was expressed as the percentage of the total fatty acids.

2.8. Vitamin E content measurement

The vitamin E content in the oil samples (20 μl sample) was analysed using an analytical high performance liquid chromatography (HPLC) using a programmable fluorescence detector at excitation 295 nm and emission 330 nm (Hewlett Packard HP1100, USA). The chromatographic system consisted of an isocratic pump and the stationary phase was a 150 mm silica normal phase column (YMC 5U) with an internal diameter 6 mm. The mobile phase was 0.5% isopropanol in hexane at a flow rate of 1 ml/min. The oil samples were injected directly into the HPLC system without any processing, after being cooled from the heating process. The vitamin E standard was obtained from the Malaysian Palm Oil Board (Bangi, Malaysia). Vitamin E measurement in the oils was done for six samples (n=6) for each of the three corresponding groups; fresh, once and five times heated. The vitamin E content in the oils was expressed as part per million (ppm). The estimated percentage of difference compared to the respective fresh oils was also calculated.

2.9. Plasma lipid peroxidation measurement

Lipid peroxidation content in the plasma measured as thiobarbituric acid reactive substance (TBARS) was determined following a method described by Ledwozyw et al. [21] with some modification. Briefly, 2.5 ml trichloroacetic acid (1.22 M in 0.6 M HCl) was used to acidify 0.5 ml plasma and incubated at room temperature for 15 minutes. Next, 1.5 ml of 0.67% thiobarbituric acid (in 0.05 M NaOH) was added. The samples were then incubated at 100°C for 30 minutes. After being cooled, the lipid peroxide content was extracted by the addition of 4 ml butanol using vigorous shaking. Later, the samples fluorescence unit was read at 515 nm (excitation wavelength) and 553 nm (emission wavelength) using a spectrofluorometer (Shimadzu RF500, Japan). 1,1,3,3-Tetraethoxypropane was used as the standard. The unit of the plasma lipid peroxidation was nmol malondialdehyde/mg protein. The results were shown as the difference percentage of post-treatment content compared to the pretreatment content.

The protein content in the plasma was carried out according to the method of Lowry et al. [22], using bovine serum albumin as the standard. A plasma sample (0.5 ml) was added with 5 ml mixture of sodium carbonate (2%), sodium potassium tartrate (2%) and copper sulphate (1%) solution at a ratio of 100 : 1 : 1, prior to incubation at room temperature for 15 minutes. Subsequently, 0.5 ml Folin-Ciocalteau phenol reagent (0.5 N) was added and then was left to stand at room temperature for 35 minutes. The absorbance of the samples was read at 700 nm with a spectrophotometer (Shimadzu UV-160A, Japan).

2.10. Statistical analysis

The results were expressed as the means ± standard error of mean (SEM). Normality of the data was analysed using Kolmogorov-Smirnov test. For normally distributed data, they were then analysed using one-way analysis of variance (ANOVA) followed by Tukey's HSD post-hoc test. While for not normally distributed data, the differences among the groups were determined using Kruskal-Wallis H and Mann-Whitney U test. Values of $P<0.05$ were considered statistically significant. All statistical analyses were performed using SPSS version 13.0 (SPSS Inc., Chicago, IL, USA).

3. Results

3.1. Peroxide value in the oils

The peroxide values measured in the oils are shown in Figure 1. The values in the once and five times heated palm and soy oils were significantly elevated compared to the fresh oil respectively. In five times heated oils, the values were also significantly higher than the once heated oils. Fresh and five times heated soy oils had bigger peroxide values compared to those of palm oils. According to the American Oil Chemists' Society (AOCS) [23], only five times heated soy oil had peroxide value exceeded the maximum allowable peroxide value for edible oils (> 10 meq/kg).

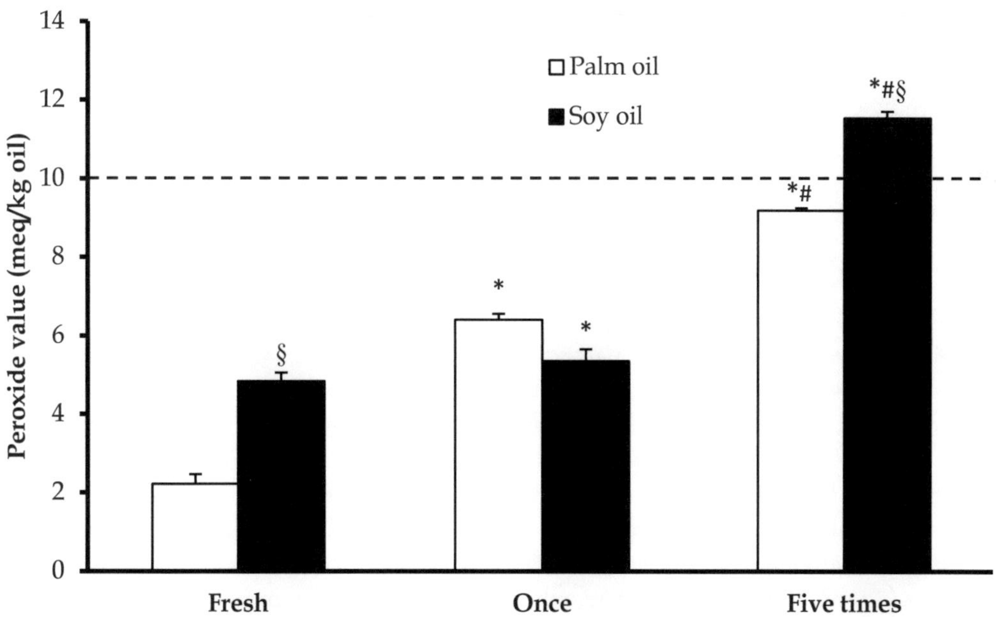

Figure 1. The peroxide values (meq/kg) in fresh and heated palm and soy oils. Bars represent mean ± SEM (n=6). *Significantly different from the fresh oils respectively ($P<0.05$). #Significantly different from the once heated oils respectively ($P<0.05$). §Significantly different from the palm oil groups respectively ($P<0.05$). Dashed horizontal line indicates maximum allowable peroxide value for edible oils according to the American Oil Chemists' Society (AOCS) [23].

3.2. Fatty acid composition of the oils

The fatty acid composition in both palm and soy oils is tabulated in Table 1. All the main components of fatty acid were present in the oils regardless of the frequency of heating. Once and five times heated palm oils had similar percentages of saturated, monounsaturated and polyunsaturated fatty acids composition compared to the fresh palm oil. In heated soy oil, generally the fatty acid composition was somewhat similar to the fresh oil. However, its polyunsaturated fatty acid percentage seemed to be lower than once heated and fresh oils. Overall, heating did not affect saturated and monounsaturated and polyunsaturated fatty acids components of palm oil. However, the repeated heating reduced the percentage of polyunsaturated and monounsaturated fatty acids, and increased saturated fatty acids components of the soy oil.

	Saturated	Monounsaturated	Polyunsaturated
Palm oil			
Fresh	42.87	48.94	8.18
Once heated	42.64	49.24	8.52
Five times heated	43.25	48.21	7.97
Soy oil			
Fresh	16.69	25.00	52.48
Once heated	17.14	26.10	51.78
Five times heated	18.10	24.21	41.72

Table 1. The percentage of saturated, monounsaturated and polyunsaturated fatty acids in the fresh, once and five times heated palm and soy oils.

3.3. Vitamin E content in the oils

The vitamin E isoforms, namely α-, γ- and δ-tocopherols, α-, γ- and δ-tocotrienols content in the oils is tabulated in Table 2. In palm oil, only α-tocopherol, α-, γ- and δ-tocotrienols were present. Whilst in soy oil, only tocopherol isoforms (α, γ and δ) were detected but none of tocotrienols. Fresh palm oil had larger content of total vitamin E compared to fresh soy oil. Once and five times heated palm oils had significantly lower content of all vitamin E isoforms than the fresh palm oil. In five times heated palm oil, all the isoforms content were also significantly reduced compared once heated palm oil. For soy oil, once and five times heating decreased the α-tocopherol content significantly compared to the fresh oil. The contents of γ- and δ-tocopherols were only reduced significantly in five times heated soy oil compared to once heated and fresh soy oils.

The difference in vitamin E content relative to fresh palm oil is diagrammatically shown in Figure 2. The relative reductions in α-tocopherol and all tocotrienol isoforms as well as total contents were greater in five times heated palm oil than once heated palm oil. The relative reduction was the least seen in the δ-tocotrienol content. In once heated soy oil, a big relative

	Tocopherols (T) (ppm)			Tocotrienols (T3) (ppm)			Total (ppm)
	αT	γT	δT	αT3	γT3	δT3	
Palm oil							
Fresh	178.3 ± 3.7	ND	ND	188.5 ± 16.3	260.8 ± 19.7	69.8 ± 2.2	697.4
Once heated	79.0 ± 8.1*	ND	ND	80.2 ± 14.6*	193.2 ± 21.4*	59.8 ± 10.4*	412.2
Five times heated	3.3 ± 3.3*#	ND	ND	3.8 ± 3.8*#	31.8 ± 20.8*#	35.8 ± 11.6*#	74.7
Soy oil							
Fresh	66.2 ± 1.4	247.2 ± 15.0	122.7 ± 0.7	ND	ND	ND	436.1
Once heated	25.3 ± 1.7*	254.7 ± 20.1	127.7 ± 0.7	ND	ND	ND	407.7
Five times heated	11.5 ± 2.0*	97.8 ± 18.0*#	80.5 ± 4.9*#	ND	ND	ND	189.8

Table 2. Composition of tocopherols and tocotrienols (ppm) in the fresh, once and five times heated palm and soy oils. Values are mean ± SEM (n=6). *Significantly different from the fresh oils respectively (P<0.05), #Significantly different from once heated oil respectively (P<0.05). ND, not detectable.

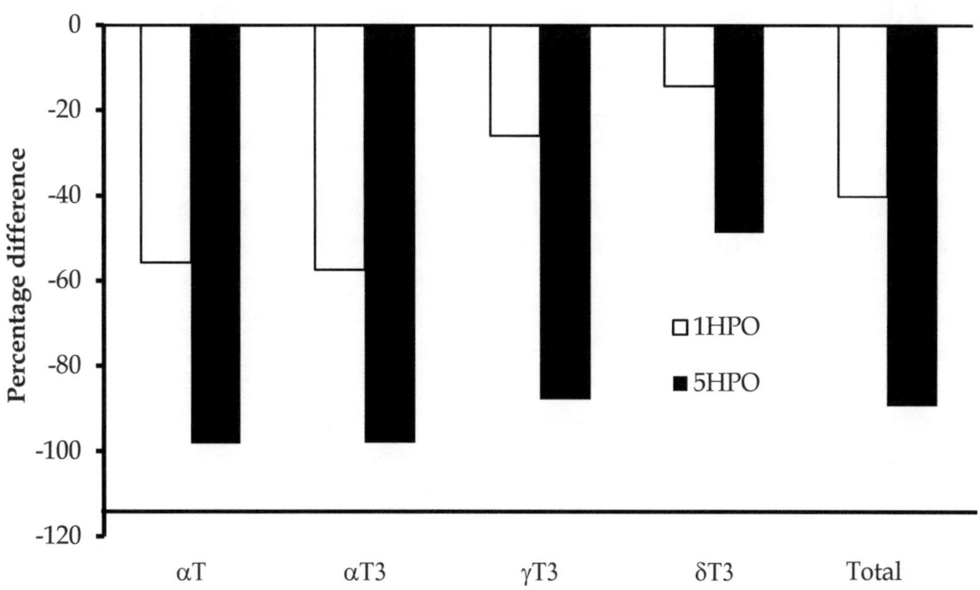

Figure 2. The percentage difference in α-tocopherol (αT), α-tocotrienol (αT3), γ-tocotrienol (γT3) and δ-tocotrienol (δT3) contents in heated palm oils (once and five times) in comparison to fresh palm oil.

decrease was seen in α-tocopherol content, whereas other isoforms content were not much affected. However, there was only a slight relative decrease in total vitamin E content. In five times heated soy oil, the relative reductions were seen in all tocopherol isoforms (α, γ and δ isoforms) (Figure 3). However, the least relative reduction was noted in the δ-tocopherol content. It thus appeared that δ-tocopherol and δ-tocotrienol were more resistant to heat compared to α and γ isoforms.

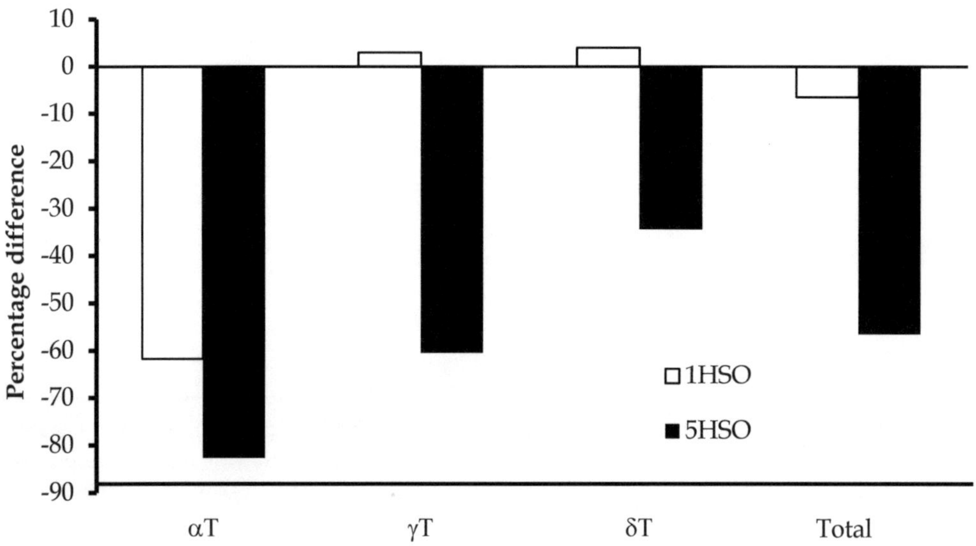

Figure 3. The percentage difference in α-tocopherol (αT), γ-tocopherol (γT) and δ-tocopherol (δT) contents in heated soy oils (once and five times) in comparison to the fresh soy oil.

3.4. Plasma lipid peroxidation in male rats

Relative plasma lipid peroxidation, measured as TBARS was increased significantly in male rats that were given diet containing 15% once and five times heated palm or soy oil for 4 months compared to the control and fresh oils, respectively (Figure 4). The five times heated groups also had significantly higher TBARS content than the once heated groups, respectively. In the once and five times heated groups, the TBARS content was significantly elevated in the soy oil-fed group compared to the palm oil-fed group, respectively. Both fresh palm and soy oil groups had significantly lower relative plasma TBARS content than the control.

3.5. Plasma lipid peroxidation in ovariectomized female rats

Ovariectomized female rats that were fed once and five times heated palm and soy oils in addition to 2% cholesterol for four months had significantly elevated relative plasma TBARS compared to both groups that were given either control diet or fresh oil respectively. The rats that ingested five times heated oils had significantly higher plasma TBARS than the

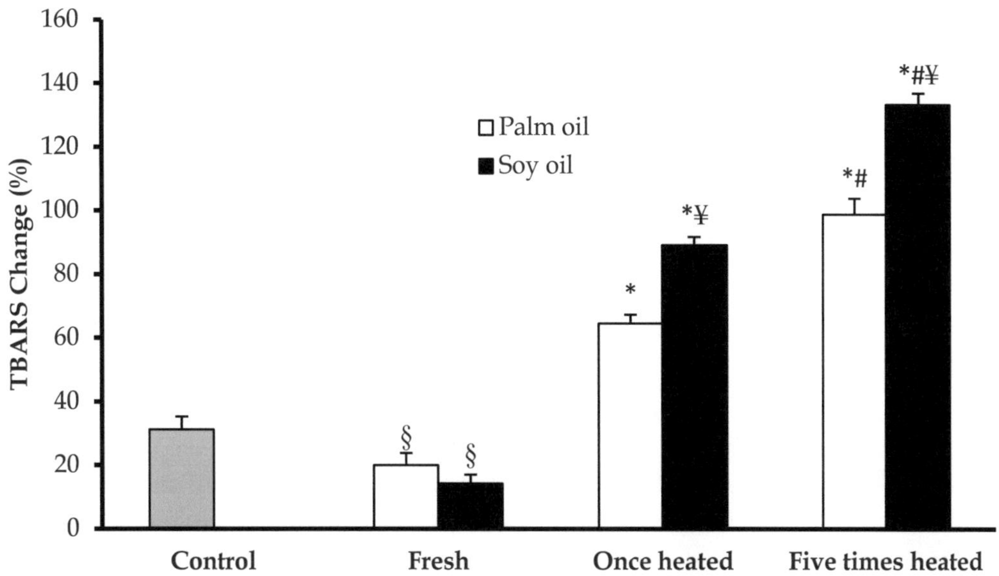

Figure 4. The change percentage of thiobarbituric acid reactive substance (TBARS), a lipid peroxidation product in male rats that were fed 15% once or five times heated (w/w) palm or soy oils for 4 months. Bars represent mean ± SEM (n=6). *Significantly different from the control and fresh oil respectively ($P<0.05$). #Significantly different from once heated soy oil ($P<0.05$). ¥Significantly different from the heated palm oil, respectively ($P<0.05$). §Significantly different from the control ($P<0.05$).

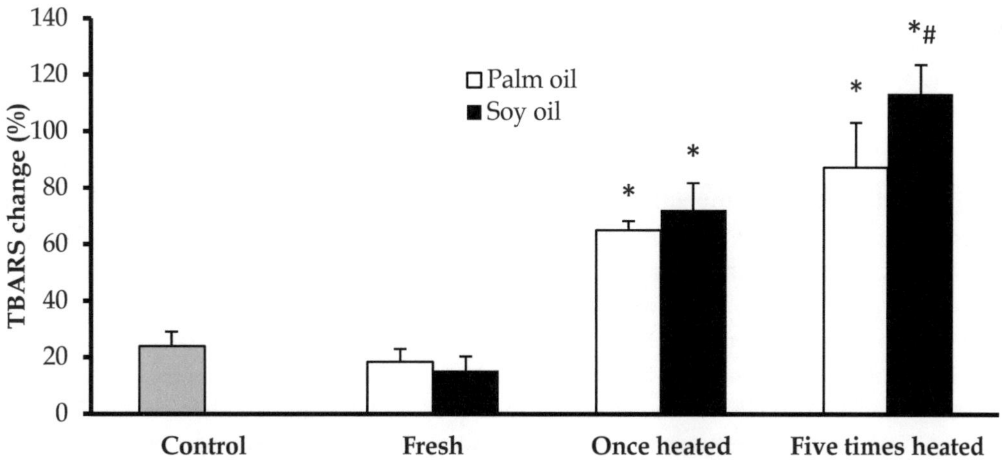

Figure 5. The percentage of change of lipid peroxidation product measured as thiobarbituric acid reactive substance (TBARS) in ovariectomised female rats that were fed 2% cholesterol together with 15% once or five times heated (w/w) palm or soy oils for 4 months. Bars represent mean ± SEM (n=6). *Significantly different from the control and fresh oil respectively ($P<0.05$). #Significantly different from once heated soy oil ($P<0.05$).

once heated groups respectively. The plasma TBARS in palm oil-fed groups were similar to the soy oil-fed groups except in five times heated groups which the palm oil group had a lower plasma TBARS ($P<0.05$). The plasma TBARS of the fresh oil groups was not different from that of the control.

4. Discussion

Repeatedly heated cooking oil is often used interchangeably with thermoxidized or recycled cooking oil. Repeated use of this oil has become a common practice due to low level of awareness among the public about the bad effect of this practice [9]. Nowadays, the consumption of deep-fried food has gained popularity which may cause increased risk of obesity [24].

During frying, food is immersed in hot oil at a high temperature of 150 °C to 190 °C. The heat and mass transfer of oil, food and air that occurs during deep frying produces the unique and desirable quality of fried foods [2]. It was shown in the present study that the peroxide values were increased with the increasing frequency of heating in both types of oil. Increased values indicate increased lipid peroxidation byproduct content, mainly the peroxides that were formed in the oil during heating process. The extent of oxidation in the oils was affected by the number of frying. Other than the peroxides, there are other oxidized components that are formed during oil heating such as oxidative dimers and oxidized triacylglycerols [25]. Only five times heated soy oil exceeded the upper limit of peroxide value set by the American Oil Chemists' Society (AOCS) that is 10 meq/kg oil [23]. However, if the Food Sanitation Law of Japan guideline (peroxide value ≤ 30 meq/kg oil) is used instead, all fresh and heated oils can be considered safe for human ingestion. However, a recent study done by Awney [26] has demonstrated that in male rats fed thermally oxidized soy oil with peroxide value of 14 meq/kg, had significantly increased lipid peroxidation content in various organs such as liver, kidney, testes and brain. Chronic intake of repeatedly heated cooking could be harmful to health and increase risk of many diseases including hypertension [13] and cancer [10].

During heating at high temperature, several complex chemical reactions take place in the oil. It is dependent on the temperature, duration of heating, type of frying materials, type of oils, presence of antioxidant and prooxidant as well as the amount of oxygen [2,27]. Repeatedly heated oils at high temperature will undergo chemical conversion of fatty acid configuration from cis to trans isomer [28]. Diets containing trans fatty acid could be detrimental to cardiovascular health because it was reported that this fatty acid isomer could induce inflammation of the blood vessels and decrease its nitric oxide production [29-30]. The polymer and polar compounds content are also increased more than 37% and 47% respectively when the oil is used to fry [31]. The repeatedly heating would reduce the quality of cooking oil by darkening its color and changing the smell as well as the taste [1].

We found that the fresh and repeatedly heated soy oil had greater peroxide value than palm oil. The peroxide value is often used as an indicator for oxidative stability or the extent of degradation for fats and oils [32]. Therefore, greater values found in soy oil suggest that the soy oil were more susceptible to oxidative modification than palm oil. Vegetable oils rich in polyunsaturated fatty acid are more prone to oxidation compared to the oils which are rich in monounsaturated fatty acids [33]. Fats are usually oxidized by prooxidants or free radicals at the unsaturated bonds of the fatty acids, which are abundantly present in the polyunsaturated fatty acids such as linoleic (C18:2) and linolenic (C18:3) acids. The bigger the numbers of unsaturated bonds, the more prone the fatty acids are to oxidation. Intense heating of oils decreases unsaturation of the fatty acids [34].

The fresh soy oil contained about five times more polyunsaturated fatty acid compared to the palm oil. It seemed that five times heating had reduced about 10% of the polyunsaturated fatty acid content in the soy oil. The content of monounsaturated fatty acid in the fresh palm oil was higher than that of the fresh soy oil. Palm oil had a quite balanced ratio of saturated and unsaturated fatty acids, whereas more than 70% of soy oil fatty acid was unsaturated (polyunsaturated and monounsaturated). This unique fatty acid composition of palm oil renders its stability against oxidative insult. The fatty acid composition of heated soy oil [26] and palm oil [35] was similarly obtained in previous studies.

The examples of the monounsaturated are oleic (C18:1) and palmitoleic (C16:1), whereas palmitic (C16:0) and stearic (C18:0) are saturated fatty acids [26]. Soy oil majorly contains linoleic (54%) and oleic (23%) acids. It also contains linolenic (8%), palmitic (11%) and stearic (4%) acids [26]. While palm oil has a bigger proportion of oleic (46%) and palmitic acids (36%), a smaller amount of linoleic (12%) and linolenic (< 1%) acids, and similar proportion of stearic acid (4.5%) compared to soy oil [36]. However in the present study, the individual fatty acid composition in the oil was not determined.

Most antioxidants are heat labile. Vitamin E is not an exception. When subjected to heat at high temperature, we found that all tocotrienol isoforms (α, γ and δ) as well as α-tocopherol in the palm oil were lost. It became more prominent when the oil was heated five times. The fresh palm oil contained almost 700 ppm of total vitamin E, but once and five times heating had further reduced it to barely 400 and 75 ppm, respectively. γ-Tocotrienol content was the highest in fresh and once heated palm oil, followed by α-tocotrienol, α-tocopherol and the least being the δ-tocotrienol. Once heating caused almost 50% reduction in α-tocopherol and α-tocotrienol contents in palm oil. While the loss of other isoforms (γ- and δ-tocotrienols) were at smaller percentages. In five times heated palm oil, the percentage loss of all vitamin E components were almost 100% except δ-tocotrienol (almost 50%), compared to the fresh palm oil. It seemed that amongst the tocotrienol isoforms, δ-tocotrienol had the highest resistance to heat.

Different from palm oil, the soy oil exclusively contained α-, γ- and δ-tocopherols. γ-Tocopherol was the largest constituent in the soy oil, followed by δ-tocopherol and lastly α-tocopherol. The total content of vitamin E in fresh soy oil was about 60% of that of fresh palm oil. Compared to tocopherols, tocotrienols are less widespread in plants [37]. Other than palm oil, tocotrienols are found abundantly in barley [38], rice bran [39], grape seed oil [40], rye, oat, maize and wheat germ [41]. Tocopherols on the other hand, are distributed at a wider range of food and vegetable oils such as soybean, corn, sunflower and cottonseed oils [40,42].

Once heating did not significantly affect γ- and δ-tocopherols contents in the soy oil, but the α-tocopherol content was apparently reduced. However in five times heated, all tocopherol isoforms content were significantly decreased. When the percentage difference relative to the fresh soy oil calculated, the most prominent reduction was seen in α-tocopherol content after once heating, followed by the γ-tocopherol and δ-tocopherol contents. Quite a similar finding was reported by Rennick & Warner [43] that the α-tocopherol content of soy oil was significantly decreased when was heated for 5 hours and was further decreased after heated for 10 hours. The loss of the tocopherol content was accompanied by an increasing appearance of α-tocopherolquinones.

From the findings obtained, it seemed that the loss of tocotrienol was more than the same isoform of tocopherol. The vitamin E chemical structure is consisted of a chromanol ring and a 15-carbon phytyl side chain. Tocotrienol differs from tocopherol by the presence of three unsaturated bonds in its phytyl chain [44]. The presence of unsaturations in the tocotrienol phytyl chain may explain the susceptibility of the compound to repeated heating. It was also noted that the α isoform of both tocopherol and tocotrienol was the mostly susceptible to oxidative loss in both palm and soy oils. The second sensitive isoform was the γ isoform (both γ-tocopherol and tocotrienol), and the least sensitive being the δ isoform. The α isoform has three methylated groups, the γ isoform has two while the δ isoform has only one methylated group on the chromanol ring [45]. However Miyagawa et al. [46] had reported that the decomposition rate of γ-tocopherol was the fastest, followed by δ- and α-tocopherol when a mixture of soy and rapeseed oil was used to deep fry potato. The discrepancy could be due to the different type of oils used in their study. It can be postulated that the degree and position of the methylation on the chromanol ring also determine the susceptibility of the vitamer to oxidative loss in addition to the degree of saturation of the phytyl chain.

The higher vitamin E content in the palm oil than soy oil might contribute to lesser peroxide value in the former oil. Both tocopherols and tocotrienols possess antioxidant property, with the latter having greater antioxidant property [6]. Vitamin E could effectively protect the fatty acids in the oil from oxidation. During oil heating, the vitamin E is consumed by scavenging the lipid free radicals which are derived from the oxidation of unsaturated fatty acids in the oils. α-Tocopherol addition to frying oil increased the stability and resistance of polyunsaturated fatty acid against oxidation [47]. Inclusion of antioxidants lemon seed extract and tert-butylhydroquinone was demonstrated to retard lipid oxidation and contribute to the α-tocopherol retention in the soy oil heated at high temperatures for several hours [48].

Male rats that were fed heated oils had elevated plasma lipid peroxidation content. The elevation was more prominent with the five times heated oils. The increased peroxide value of the heated oils may be associated with the significant increase in plasma lipid peroxidation. Chronic ingestion of heated oil was shown to cause an elevation of blood pressure [49] and necrotic cardiac changes [50]. This would disrupt the endogenous antioxidant defense in our body in order to overcome the overwhelming of oxidative stress. Linoleyl, peroxy, and alkoxy radicals were reported to be produced in heated oils. These radicals act on the fatty acids of the oil producing oxidized products via hemolytic β-scission [25]. Dietaries fresh palm and soy oil were shown to attenuate oxidative stress and augment antioxidant enzymes activities in rat models [51-53]. Their findings are in agreement with the present study where a significant reduction in plasma lipid peroxidation was observed in the fresh oil-fed male rats. This positive effect was attributed to the rich antioxidant content of the oils.

Heating diminished the vitamin E content and rendered the oils to lose their beneficial effects. Chronic ingestion of heated soy oil diet was shown to induce the increase in hepatic lipid peroxidation in rats [54]. They also observed a reduction in serum and hepatic vitamin E content in rats that were fed heated oil containing diet. An evidence of hepatic damage was also reported with a significant elevation of aspartate and alanine transaminases, the marker enzymes for liver function in rats that were given combination of heated soy and rapeseed oils [55]. The administration of dietary heated soy oil affected the activities of antioxidant enzymes in various rat organs. It increased superoxide dismutase in the liver and brain, while increment in glutathione reductase was seen in the liver and kidneys [26]. Heated palm oil also increased catalase, glutathione peroxidase and glutathione S-transferase in the rat liver [35]. It shows dietary heated oil contained prooxidant substances that would evoke the body primary defenses.

Plasma lipid peroxidation was increased significantly in the rats that were given heated soy oil (once and five times heated) when compared to the heated palm oil-fed groups. This finding was suggestive of the better effect of the palm oil in terms of oxidative stability when exposed to extreme heat. High composition of saturated fats in palm oil confers it to withstand thermal oxidative changes, in addition to its rich content of tocotrienols. The better effect of heated palm oil compared to heated soy oil was reported elsewhere [56].

Ovariectomized female rats were often used as an experimental postmenopausal model. Ovariectomy results in an estrogen deficiency state. The estrogen provides protection to the body against oxidative stress [16]. In the present study, dietaries once heated and five times heated palm and soy oil for four months had significantly increased the plasma lipid peroxidation in ovariectomized female rats. The result is in agreement with other studies which also reported that in ovariectomized-induced estrogen deficiency, an increase in oxidative stress accompanied by a reduction in antioxidant status was seen [57-58]. A study by Sánchez-Rodríguez [59] had recently found that menopause could be one of the main risk factors for oxidative stress. Therefore, this finding suggests that chronic ingestion of repeatedly heated vegetable oils is detrimental to health and could further aggravate the increase in oxidative stress in postmenopausal women. Shuid et al. [56] also had demonstrated that fresh and once heated palm and soy oil diets for six months reduced the risk of osteoporosis in ovariectomized rats, a positive effect that was not seen in rats that were fed five times heated oils.

The patterns of plasma lipid peroxidation increase in both male and ovariectomized female rats were somewhat similar. However, the plasma lipid peroxidation of the fresh oil-fed groups in the ovariectomized female rats that were similar to the control group, different from the one observed in the male rats. The addition of dietary 2% cholesterol and the ovariectomy status of the female rats made them more susceptible to oxidative insult. This may explain why there was no significant reduction in plasma TBARS seen in the female rats fed fresh oils compared to the control. The plasma lipid peroxidation was significantly elevated in the five times heated soy oil group compared to the same heating frequency of palm oil. The difference was also attributed to the higher component of polyunsaturated fatty acids present in the soy oil, which are prone to oxidation.

5. Conclusion

Our findings suggest that it is recommended not to heat cooking oil more than once in view of its possible deleterious effect on health. The use of palm oil possibly has better effect on health due to its stability against oxidative insult.

Author details

Kamsiah Jaarin and Yusof Kamisah
Department of Pharmacology, Faculty of Medicine,Universiti Kebangsaan Malaysia, Malaysia

Acknowledgement

The study was funded by grants from Universiti Kebangsaan Malaysia (UKM-GUP-SK-08-21-299) and Malaysian Ministry of Science, Technology and Innovations (IRPA 06-02-02-0050-EA242). The authors also would like to thank Prof. Dr Jumat Salimon, Ms Xin-Fang Leong and Mr. Chun-Yi Ng for their technical help as well as Ms Jurika Sharaton Abdul Wahed for editorial assistance.

6. References

[1] Rani AKS, Reddy SY, Chetana R (2010) Quality changes in trans and trans free fats/oils and products during frying. Eur. Food Res. Technol. 230(6): 803–811.
[2] Choe E, Min DB (2007) Chemistry of deep-fat frying oils. J. Food Sci. 72(5):R77-R86.
[3] Choe, E. & Min, D. B (2006) Chemistry and reactions of reactive oxygen species in foods. Crit. Rev. Food Sci. Nutr. 46(1): 1-22.
[4] Kamisah Y, Adam A, Wan Ngah WZ, Gapor MT, Azizah O, Marzuki A (2005) Chronic intake of red palm olein and palm olein produce beneficial effects on plasma lipid profile in rats. Pakistan J. Nutr. 4(2): 89-96
[5] Clemente TE, Cahoon EB (2009) Soybean oil: genetic approaches for modification of functionality and total content. Plant Physiol. 151(3): 1030-1040.
[6] Bardhan J, Chakraborty R, Raychaudhuri U (2011). The 21st century form of vitamin E - Tocotrienol. Curr. Pharm. Des. 17(21): 2196-2205.

[7] Edem DO (2002) Palm oil: biochemical, physiological, nutritional, hematological, and toxicological aspects: a review. Plant Foods Hum. Nutr. 57(3-4): 319-341.

[8] Azman, SM Shahrul, XS Chan, AP Noorhazliza, M Khairunnisak, MF Nur Azlina, HMS Qodriyah, Y Kamisah, K Jaarin (2012) Level of knowledge, attitude and practice of night market food outlet operators in Kuala Lumpur regarding the usage of repeatedly heated cooking oil. Med. J. Malays. 67(1): 97-107.

[9] Abdullah A, Shahrul S, Chan XS, Noorhazliza AP, Khairunnisak M, Qodriyah HMS, Kamisah Y, Nur Azlina MF, Kamsiah J (2010) Level of awareness amongst the general public regarding usage of repeatedly heated cooking oil in Kuala Lumpur, Malaysia. Inter. Med. J. 17(4): 310-311.

[10] Srivastava S, Singh M, George J, Bhui K, Murari Saxena A, Shukla Y (2010) Genotoxic and carcinogenic risks associated with the dietary consumption of repeatedly heated coconut oil. Br. J. Nutr. 104(9): 1343-1352.

[11] Obembe AO, Owu DU, Okwari OO, Antai AB, Osim EE (2011) Intestinal Fluid and Glucose Transport in Wistar Rats following Chronic Consumption of Fresh or Oxidised Palm Oil Diet. ISRN Gastroenterol. 2011: 972838.

[12] Latha P, Chaitanya D, Rukkumani R (2010) Protective effect of Phyllanthus niruri on alcohol and heated sunflower oil induced hyperlipidemia in Wistar rats. Toxicol Mech. Methods. 20(8): 498-503.

[13] Soriguer F, Rojo-Martínez G, Dobarganes MC, García Almeida JM, Esteva I, Beltrán M, Ruiz De Adana MS, Tinahones F, Gómez-Zumaquero JM, García-Fuentes E, González-Romero S (2003) Hypertension is related to the degradation of dietary frying oils. Am. J. Clin. Nutr. 78(6): 1092-1097.

[14] Rossouw JE (2002) Hormones, genetic factors and gender differences in cardiovascular disease. Cardiovasc. Res. 53: 550-557.

[15] Lu H, Meléndez GC, Levick SP, Janicki JS (2012) Prevention of adverse cardiac remodeling to volume overload in female rats is the result of an estrogen-altered mast cell phenotype. Am. J. Physiol. Heart Circ. Physiol. 302(3): H811-H817.

[16] Abbas AM, Elsamanoudy AZ (2011) Effects of 17β-estradiol and antioxidant administration on oxidative stress and insulin resistance in ovariectomized rats. Can. J. Physiol. Pharmacol. 89(7): 497-504.

[17] Shuid AN, Mohamad S, Muhammad N, Fadzilah FM, Mokhtar SA, Mohamed N, Soelaiman IN (2011) Effects of α-tocopherol on the early phase of osteoporotic fracture healing. J. Orthop. Res. 29(11): 1732-1738.

[18] Mvondo MA, Njamen D, Fomum ST, Wandji J, Vollmer G (2011) A postmenopause-like model of ovariectomized Wistar rats to identify active principles of Erythrina lysistemon (Fabaceae). Fitoterapia. 82(7): 939-949.

[19] Owu DU, Osim EE, Ebong PE (1998) Serum liver enzymes profile of Wistar rats following chronic consumption of fresh or oxidized palm oil diets. Acta Trop. 69(1): 65-73.

[20] AOCS (2003) *Official methods and recommended practices of the American Oil Chemists' Society*. 4th ed. AOCS Press, Champaign , Illinois.

[21] Ledwozyw A, Michalak J, Stepian A, Kadziolka A (1986) A relationship between plasma triglycerides, cholesterol, total lipids and lipid peroxidation products during human atherosclerosis. Clin. Chim. Acta. 155: 275-285.

[22] Lowry OH, Rosebrough NJ, Farr AL, Randall RJ (1951) Protein measurement with Folin Phenol reagent. J. Biol. Chem. 193: 265-275.

[23] Matthaus B (2006) Utilization of high-oleic rapeseed oil for deep-fat frying of French fries compared to other commonly used edible oils. Eur. J. Lipid Sci. Tech. 108(3): 200-211.

[24] Sayon-Orea C, Bes-Rastrollo M, Basterra-Gortari FJ, Beunza JJ, Guallar-Castillon P, de la Fuente-Arrillaga C, Martinez-Gonzalez MA (2012) Consumption of fried foods and weight gain in a Mediterranean cohort: The SUN project. Nutr. Metab. Cardiovasc. Dis. (in press)

[25] Picariello G, Paduano A, Sacchi R, Addeo F (2009) Maldi-tof mass spectrometry profiling of polar and nonpolar fractions in heated vegetable oils. J. Agric. Food Chem. 57(12): 5391-5400.

[26] Awney HA (2011) The effects of Bifidobacteria on the lipid profile and oxidative stress biomarkers of male rats fed thermally oxidized soybean oil. Biomarkers. 16(5): 445-452.

[27] Leskova E, Kubikova J, Kovacikova E, Kosicka M, Pobruska J, Holcikova K (2006) Vitamin losses: Retention during heat treatment and continual changes expressed by mathematical models. J. Food Comp. Anal. 19: 252–276.

[28] Velasco J, Marmesat S, Bordeaux O, Márquez-Ruiz G, Dobarganes C (2004) Formation and evolution of monoepoxy fatty acids in thermoxidized olive and sunflower oils and quantitation in used frying oils from restaurants and fried-food outlets. J. Agr. Food Chem. 52(14): 4438-4443.

[29] Iwata NG, Pham M, Rizzo NO, Cheng AM, Maloney E, Kim F (2011) Trans fatty acids induce vascular inflammation and reduce vascular nitric oxide production in endothelial cells. PLoS One. 6(12): e29600.

[30] Bryk D, Zapolska-Downar D, Malecki M, Hajdukiewicz K, Sitkiewicz D (2011) Trans fatty acids induce a proinflammatory response in endothelial cells through ROS-dependent nuclear factor-κB activation. J. Physiol. Pharmacol. 62(2): 229-238.

[31] Saka S, Aouacheri W, Abdennour C (2002) The capacity of glutathione reductase in cell protection from the toxic effect of heated oils. Biochimie 84(7): 661-665.

[32] Karoui IJ, Dhifi W, Jemia MB, Marzouk B (2011) Thermal stability of corn oil flavoured with Thymus capitatus under heating and deep-frying conditions. J. Sci. Food Agr. 91(5): 927-933.

[33] Kochhar SP, Henry CJ (2009) Oxidative stability and shelf-life evaluation of selected culinary oils. Inter. J. Food Sci. Nutr. 60(7 Suppl): 289-296.

[34] Moya Moreno MC, Mendoza Olivares D, Amézquita López FJ, Gimeno Adelantado JV, Bosch Reig F. Analytical evaluation of polyunsaturated fatty acids degradation during thermal oxidation of edible oils by Fourier transform infrared spectroscopy. Talanta. 1999 Sep 13;50(2):269-75.

[35] Purushothama S, Ramachandran HD, Narasimhamurthy K, Raina PL (2003) Long-term feeding effects of heated and fried oils on hepatic antioxidant enzymes, absorption and excretion of fat in rats. Mol. Cell Biochem. 247(1-2): 95-99.

[36] Vara Prasad SS, Jeya Kumar SS, Kumar PU, Qadri SS, Vajreswari A (2010) Dietary fatty acid composition alters 11β-hydroxysteroid dehydrogenase type 1 gene expression in rat retroperitoneal white adipose tissue. Lipids Health Dis. 9: 111.

[37] Horvath G, Wessjohann L, Bigirimana J, Jansen M, Guisez Y, Caubergs R, Horemans N (2006) Differential distribution of tocopherols and tocotrienols in photosynthetic and non-photosynthetic tissues. *Phytochemistry*. 67: 1185–1195.

[38] Shen Y, Lebold K, Lansky EP, Traber MG, Nevo E (2011) 'Tocol-omic' diversity in wild barley, short communication. Chem. Biodivers. 8(12): 2322-2330.

[39] Huang SH, Ng LT (2011) Quantification of tocopherols, tocotrienols, and γ-oryzanol contents and their distribution in some commercial rice varieties in Taiwan. J. Agric. Food Chem. 59(20): 11150-11159.

[40] Hassanein MM, Abedel-Razek AG (2009) Chromatographic quantitation of some bioactive minor components in oils of wheat germ and grape seeds produced as by-products. J. Oleo Sci. 58(5): 227-233.

[41] Milagros Delgado-Zamarreño M, Bustamante-Rangel M, Sierra-Manzano S, Verdugo-Jara M, Carabias-Martínez R (2009) Simultaneous extraction of tocotrienols and tocopherols from cereals using pressurized liquid extraction prior to LC determination. J. Sep. Sci. 32(9): 1430-1436.

[42] Traber MG (2007) Vitamin E regulatory mechanisms. *Annu. Rev. Nutr.* 27: 347–362.

[43] Rennick KA, Warner K (2006) Effect of elevated temperature on development of tocopherolquinones in oils. J. Agric. Food Chem. 54(6): 2188-2192.

[44] Sen CK, Rink C, Khanna S (2010) Palm oil-derived natural vitamin E alpha-tocotrienol in brain health and disease. J. Am. Coll. Nutr. 29(3 Suppl): 314S-323S.

[45] Aggarwal BB, Sundaram C, Prasad S, Kannappan R (2010) Tocotrienols, the vitamin E of the 21st century: its potential against cancer and other chronic diseases. Biochem. Pharmacol. 80(11): 1613-1631.

[46] Miyagawa K, Hirai K, Takezoe R (1991) Tocopherol and fluorescence levels in deepfrying oil and their measurement for oil assessment. J. Am. Oil Chem. Soc. 68: 163–166.

[47] Quiles JL, Ramírez-Tortosa MC, Ibáñez S, Alfonso González J, Duthie GG, Huertas JR, Mataix J (1999) Vitamin E supplementation increases the stability and the in vivo antioxidant capacity of refined olive oil. Free Radic. Res. 31 Suppl: S129-S135.

[48] Luzia DM, Jorge N (2009) Oxidative stability and alpha-tocopherol retention in soybean oil with lemon seed extract (Citrus limon) under thermoxidation. Nat. Prod. Commun. 4(11): 1553-1556.

[49] Leong X, Najib MNM, Das S, Mustafa MR, Jaarin K (2009) Intake of repeatedly heated palm oil causes elevation in blood pressure with impaired vasorelaxation in rats. Tohoku J. Exp. Med. 219: 71-78.

[50] Leong XF, Aishah A, Aini UN, Das S, Jaarin K (2008) Heated palm oil causes rise in blood pressure and cardiac changes in heart muscle in experimental rats. Arch. Med. Res. 39: 567-572.

[51] Bayorh MA, Abukhalaf IK, Ganafa AA (2005) Effect of palm oil on blood pressure, endothelial function and oxidative stress. Asia Pac. J. Clin. Nutr. 14(4): 325-339.

[52] Narang D, Sood S, Thomas M, Dinda AK, Maulik SK (2005) Dietary palm olein oil augments cardiac antioxidant enzymes and protects against isoproterenol-induced myocardial necrosis in rats. J. Pharm. Pharmacol. 57(11): 1445-1451.

[53] Hassan HA, Abdel-Wahhab MA (2012) Effect of soybean oil on atherogenic metabolic risks associated with estrogen deficiency in ovariectomized rats : Dietary soybean oil modulate atherogenic risks in overiectomized rats. J. Physiol. Biochem. (in press).

[54] Corcos-Benedetti P, Di Felice M, Gentili V, Tagliamonte B, Tomassi G (1990) Influence of dietary thermally oxidized soybean oil on the oxidative status of rats of different ages. Ann. Nutr. Metab. 34(4): 221-231.

[55] Totani N, Burenjargal M, Yawata M, Ojiri Y (2008) Chemical properties and cytotoxicity of thermally oxidized oil. J. Oleo. Sci. 57(3): 153-160.

[56] Shuid AN, Chuan LH, Mohamed N, Jaarin K, Fong YS, Soelaiman IN (2007) Recycled palm oil is better than soy oil in maintaining bone properties in a menopausal syndrome model of ovariectomized rat. Asia Pac J Clin Nutr. 16(3): 393-402.

[57] Huang YH, Zhang QH (2010) Genistein reduced the neural apoptosis in the brain of ovariectomised rats by modulating mitochondrial oxidative stress. Br. J. Nutr. 104(9): 1297-1303.

[58] Topçuoglu A, Uzun H, Balci H, Karakus M, Coban I, Altug T, Aydin S, Topçuoglu D, Cakatay U (2009) Effects of estrogens on oxidative protein damage in plasma and tissues in ovariectomised rats. Clin. Invest. Med. 32(2): E133-143.

[59] Sánchez-Rodríguez MA, Zacarías-Flores M, Arronte-Rosales A, Correa-Muñoz E, Mendoza-Núñez VM (2012) Menopause as risk factor for oxidative stress. Menopause. 19(3): 361-367.

Chapter 11

Effects of Hydroperoxide in Lipid Peroxidation on Dough Fermentation

Toshiyuki Toyosaki

Additional information is available at the end of the chapter

http://dx.doi.org/10.5772/39187

1. Introduction

The oxidation of lipids in foods is responsible for the formation of off-flavors and chemical compounds that may be detrimental to health; it is a well known problem in the food chemistry and biochemistry fields (Ames et al., 1994; Gardnaer, 1996; Grosch, 1987; Pokorny, 1999; Shewfelt & Del Rosario, 2000; Mercier & Gelinas, 2001; Toyosaki & Sakane, 2002; Toyosaki & Koketsu, 2004). Currently, the various lipid peroxides produced by such lipid peroxidation are treated only as a nuisance. However, among longstanding traditional foods there are foods with fine flavors that are brought out by inducing lipid peroxidation; such foods include fine, thin noodles and certain dried foods. Thus, lipid peroxides produced by lipid peroxidation can also be advantageous. The properties of foods can be improved by better use of the properties of lipid peroxides. The current work provided an interesting finding: when lipoxygenase was added during the fermentation of bread dough, the fermentation of dough was promoted. During this event, hydroperoxides produced by lipid peroxidation triggered the promotion of fermentation and promoted the fermentation of dough. The phenomenon by which hydroperoxides are produced by lipid peroxidation and promote the fermentation of bread dough is decidedly not beneficial when assessed from a nutritional standpoint, but this phenomenon is extremely desirable when assessed from a food science standpoint. The objective of the current study was to investigate the bread dough fermentation-promoting action of hydroperoxides produced by lipid peroxidation in bread dough and the mechanism that is involved.

2. Methods

Preparation of dough

The wheat flour (strong flour) that was used to make adjustments in the bread dough was the type that is readily commercially available. The lipid added was linoleic acid (more than

95% pure), which was added at a 3% level. Other ingredients used to make the bread were all commercially available. Lipoxygenase was added at the end of bread dough adjustment and underwent primary fermentation in an incubator at 37°C with 75-80% humidity. After fermentation, gas was released; after a bench time of 10 min, the dough underwent final fermentation for 90 min and was then baked for 12 min at 200°C.

Preparation of the model system of gluten and linoleic acid

By mixing a fixed amount of commercially available gluten and linoleic acid (more than 95% pure) of 3% level, which served as the test sample, a model system was created. This sample underwent fermentation by lipoxygenase induction.

Measurement of the rate of dough expansion

To determine the rate of dough expansion with fermentation, a fixed amount of dough was placed in a graduated cylinder and fermented in an incubator (temperature 30°C, humidity 75%). The rate of expansion over a fixed period of time was then measured.

Measurement of hydroperoxide

Hydroperoxide concentration was calculated in terms of 2',7'-dichlorofluorescein (DCF). To de-emulsify, the 5.0 ml samples were centrifuged (10,000 x g, 30 min). The linoleic acid of the supernatant was then measured to determine the hydroperoxide level using the method of Cathcart et al. (1984). First, 1.0 ml of a 1.0 mM solution of DCF in ethanol and 2.0 ml of 0.01N NaOH were mixed and stirred for 30 min before being neutralized with 10 ml of 25 mM phosphate buffer (pH 7.2). Then 2.0 ml of the neutralized DCF solution were added to a solution of hematin (10 mM) in 25 mM phosphate buffer (pH 7.2; 0.01 mg DCF/ml); subsequently, 2.0 ml of this hematin-DCF solution and 10 ml of the linoleic acid sample were mixed and left at 50°C for 50 min, before fluorometry treatment (excitation. 400 nm; emission. 470 nm) to measure DCF. This method measures hydroperoxide with more sensitivity than the iron rhodanide method that is usually used.

Diethylaminoethyl (DEAE) column chromatography

The extracted dough was separated by Tris-HCl buffer (pH 8.0). The extracted sample fractions were separated by DEAE-cellulose (DE52, Whatman, Ltd., Tokyo, Japan) column chromatography as follows. A DEAE-cellulose column (3.8 x 54 cm) was equilibrated with 50 mM Tris-HCl buffer (pH 8.0), washed with the same buffer, and developed in a linear gradient made with 300 ml of this buffer and 300 ml of the same buffer containing 0.6 M NaCl. The flow rate was 30 ml/hr, and 3.0 ml fractions were collected.

Measurement of the amount of protein

The amount of protein was measured using the Lowry method (1951).

SDS-polyacrylamide gel electrophoresis (PAGE)

The measurement was done according to the method of Laemmli (1979). Electrophoresis was performed using the Mini-Protean II Electrophoresis Cell (Bio-Rad Laboratories, Inc.,

Tokyo, Japan) at 18 mA/gel with Ready Gel J of differing gel concentrations. After electrophoresis, the gels were stained using Coomassie brilliant blue-R250. In addition, automated electrophoresis (Phast System; Pharmacia LKB, Biotechnology AB, Uppsala, Sweden) equipment was used.

Statistical analysis

Analysis of variance (ANOVA) was performed and mean comparisons were obtained by Duncan's multiple range test (Steel & Torrie, 1980). Significance was established at $P<0.05$.

3. Results and discussion

Lipid peroxidation during fermentation and accompanying changes in the rate of expansion

After 3% linoleic acid was added to the other bread ingredients, the doughs with and without lipoxygenase were individually mixed with a mixer for a fixed time. Next, the dough underwent primary fermentation in an incubator for 90 min, and the amount of hydroperoxide produced over this time period was measured. These results are shown in Fig. 1. Lipid peroxidation by lipoxygenase induction increased as the fermentation time progressed. However, the overall amount of hydroperoxide produced in the lipoxygenase-free dough tended not to increase. The rate of expansion during this time is shown in Fig. 2.

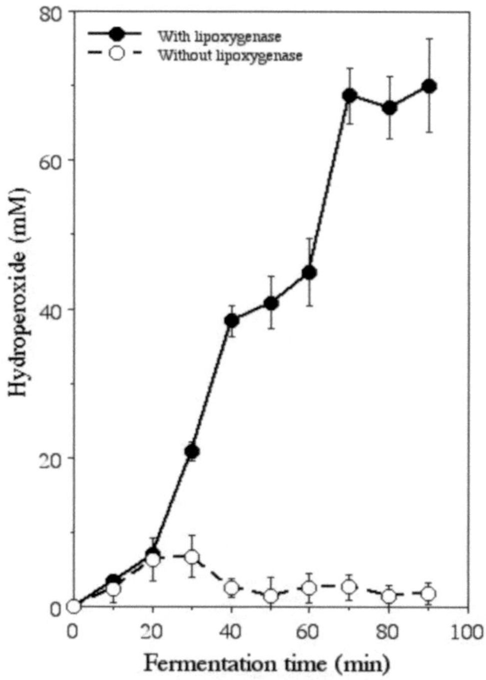

Figure 1. Changes in the amount of hydroperoxide produced with the fermentation of dough. Each value represents the mean ± standard error in triplicate.

For up to 40 min, dough with lipoxygenase expanded rapidly after the start of fermentation, but after this time the expansion tended to decrease abruptly. In contrast, lipoxygenase-free dough reached its maximum rate of expansion in 30 min from the start of fermentation, and this tended to decrease gradually afterwards. Further, in the dough without lipoxygenase, changes in the rate of expansion per unit time were smaller than in the dough with lipoxygenase. Changes brought about by this phenomenon are quite likely due to the effect that hydroperoxide, which is produced by lipid peroxidation, has on the fermentation stage.

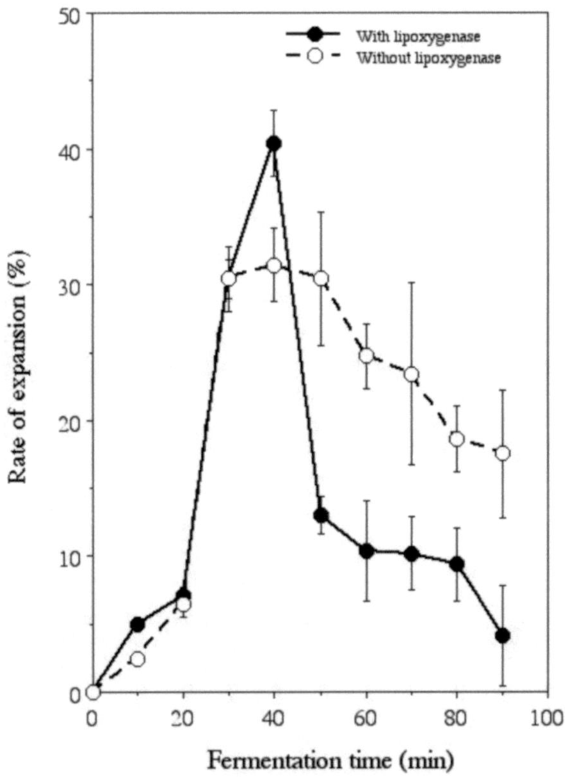

Figure 2. Changes in the rate of expansion with the fermentation of dough. Each value represents the mean ± standard error in triplicate.

Relationship between the rate of expansion and hydroperoxide

A model system of gluten and linoleic acid was created, and the involvement of the hydroperoxide that was produced in the fermentation of dough was examined. These results are shown in Fig. 3. The rate of dough expansion was affected by hydroperoxide concentration, and in this experiment the rate of expansion reached its maximum at a hydroperoxide concentration of 30-40 mM. Based on these results, the fermentation of dough was influenced by the concentrations of hydroperoxide that were produced.

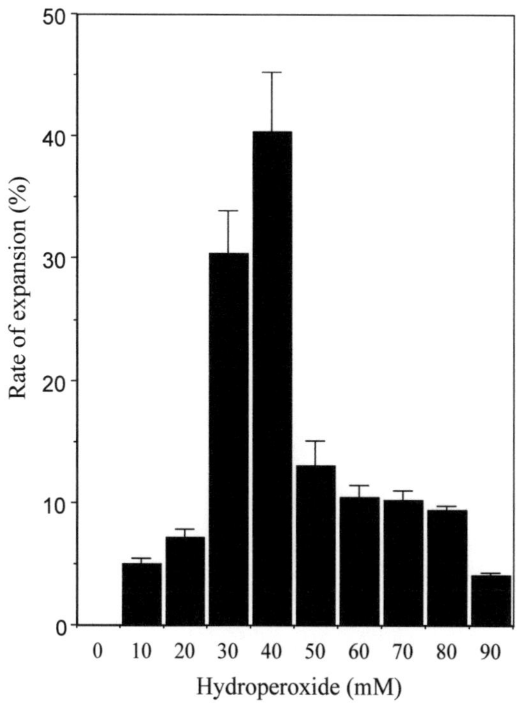

Figure 3. Changes in the rate of dough expansion with hydroperoxides. Each value represents the mean ± standard error in triplicate.

Effects of yeast and gluten during fermentation

The effects of the yeast on dough fermentation were also studied. The comparison of the dough with and without lipoxygenase is shown in Fig. 4. Both doughs with ≤ 2.0% yeast had similar rates of expansion that tended to increase with fermentation time. There were almost no changes in the rate of expansion with yeast concentrations of ≥ 2.5%; in fact, the rate of expansion tended to decrease. However, the rate of expansion of dough with lipoxygenase tended to increase more than the rate of the dough without lipoxygenase. Since a detailed study of the relationship between yeast and hydroperoxide produced was not done in this experiment, further study is needed.

Next, changes in the amount of hydroperoxide were studied; these results are shown in Fig. 5. For the dough with lipoxygenase, the amount of hydroperoxide reached its maximum when the yeast concentration was 1%; as the yeast concentration increased, the amount of hydroperoxide that was produced tended to decrease. Comparing these results with those in Fig. 4 indicates that there is a relationship between the amount of hydroperoxide produced and the yeast concentration; the specifics of this relationship need to be further investigated. Lipoxygenase-free dough produced almost no hydroperoxide. However, based on the results in Fig. 4, the production of hydroperoxide may not be the sole factor involved in the fermentation of dough. Thus, the hydroperoxide that is produced may be synergistically involved in the mechanism of yeast fermentation.

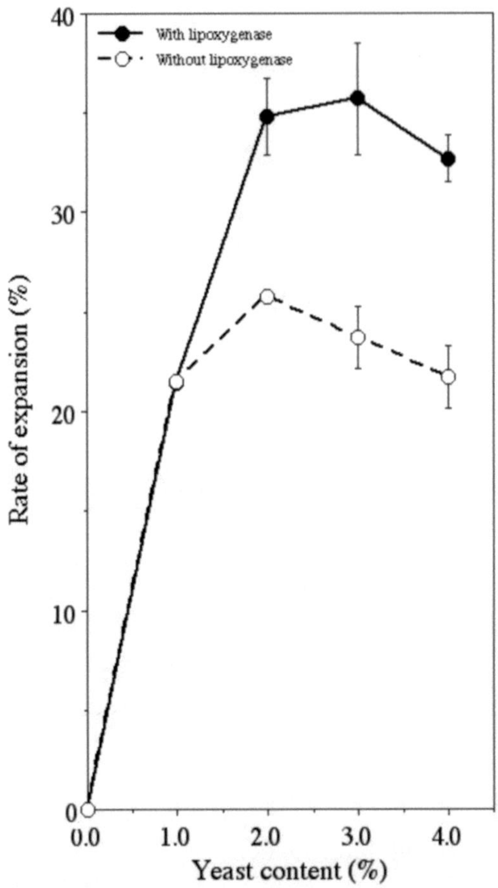

Figure 4. Effect of yeast contents on the rate of dough expansion. Each value represents the mean ± standard error in triplicate.

Next, the effect of differences in gluten content on dough fermentation was studied. The results are shown in Fig. 6. The expansion of dough began at a gluten content of 40%, and the rate of expansion reached its maximum at a gluten content of 60%. Beyond this concentration, the rate of expansion tended to gradually decrease. Dough with lipoxygenase had a rate of expansion of about 35% at a gluten content of 60%, while lipoxygenase-free dough had a rate of expansion of 15%. Thus, the presence of hydroperoxide had an effect on the rate of expansion; the hydroperoxide that was produced promoted fermentation.

The relationship between gluten and hydroperoxide

Hydroperoxide is involved in the fermentation of dough in a facilitatory manner, and, as a result, the rate of dough expansion is increased. Consequently, this phenomenon has a positive effect on dough. To study the effect of the hydroperoxide that is produced during dough fermentation on gluten, the gluten was separated and purified after the completion of fermentation using affinity chromatography, so that, ultimately, the gluten fraction was obtained. This gluten fraction was subjected to SDS–gel electrophoresis, and the relationship between gluten and hydroperoxide in the fermentation stage was studied; the results are shown in Fig. 7. In the dough without lipoxygenase, there were almost no changes in the molecular weight of gluten during 100 min of fermentation time. In contrast, in the dough with lipoxygenase, changes in the molecular weight of gluten were seen with fermentation, and formation of gluten polymers was noted with fermentation. This phenomenon was caused by hydroperoxide that was produced, which acted on the gluten and may have induced denaturation. A comparison of the results shown in Figure 4 and 5 shows that the gluten network was tightened, because the hydroperoxide that was produced by the addition of lipoxygenase denatured the gluten and, subsequently, increased dough expansion.

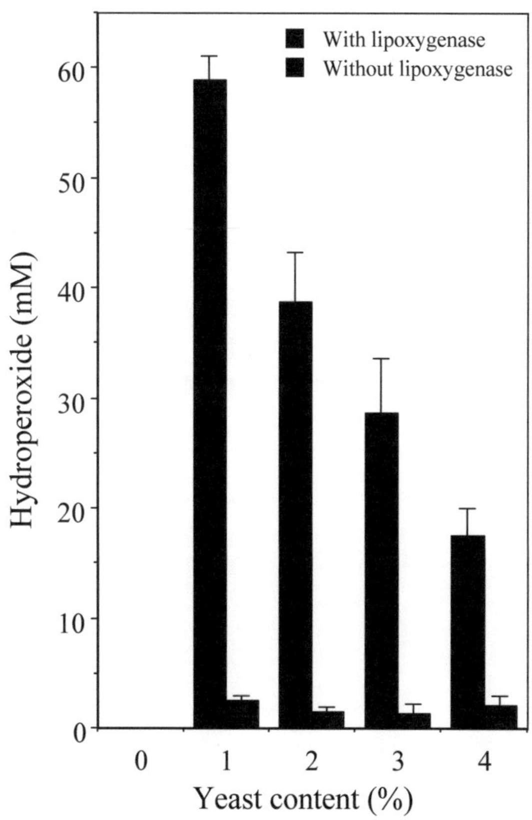

Figure 5. Effect of yeast contents on the amount of hydroperoxide produced in dough. Each value represents the mean ± standard error in triplicate.

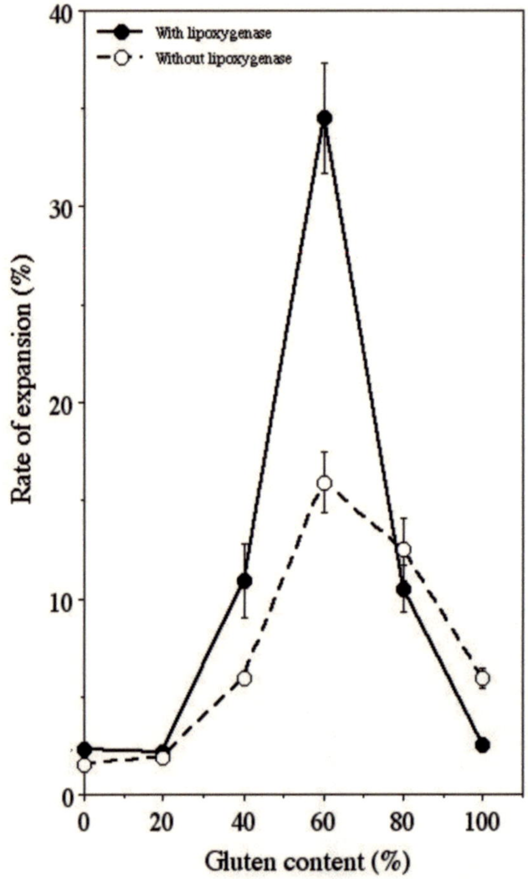

Figure 6. Effect of gluten content on the rate of dough expansion. Each value represents the mean ± standard error in triplicate.

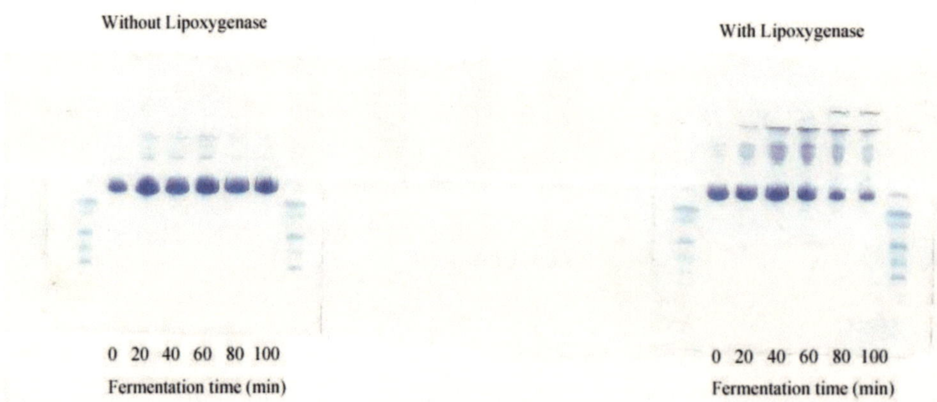

Figure 7. Changes in the molecular weight of gluten when dough was fermented for 100 min.

The mechanism by which hydroperoxide acts to promote fermentation

The various experimental results that were obtained were comprehensively analyzed to determine the mechanism of action by which hydroperoxide acts to promote fermentation; this is shown in Fig. 8. During gluten formation, gluten is formed when gliadin and glutenin form a network structure. When gluten is crosslinked in the presence of hydroperoxide, the molecules themselves form macromolecules. As a result, expansion is promoted by the uptake of large amounts of carbon dioxide gas produced during dough fermentation. This phenomenon is ultimately advantageous when baking dough, and it improves the bread's texture. When very little lipid peroxidation is induced, the unoxidized linoleic acid has no interaction with gluten, and, as a result, gluten crosslinking does not occur, which results in baked bread with a poor texture.

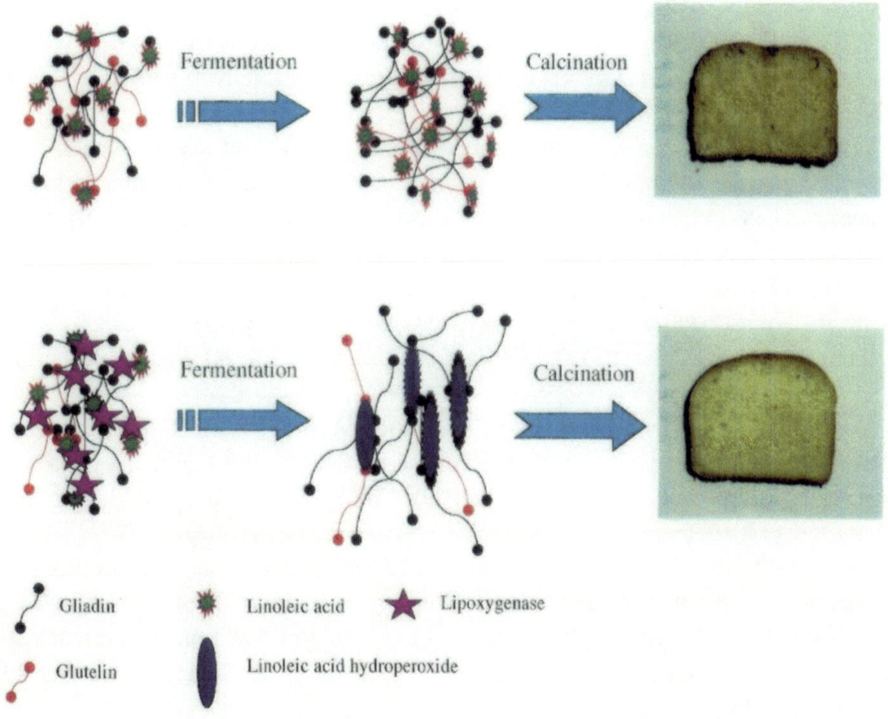

Figure 8. The mechanism by which hydroperoxides accelerate fermentation.

4. Conclusions

The current research demonstrated that well-fermented dough can be produced by the induction of lipid peroxidation when fermenting dough. The induction of lipid peroxidation was achieved in the current study by using lipoxygenase induction, but a similar phenomenon should also occur with lipid peroxidation induced by other methods. This phenomenon is advantageous when baking bread and can be used to enhance the quality of baked bread. Based on the results of these tests of physical properties, further detailed study is needed of the effect of lipid peroxidation on the flavor of baked bread.

Author details

Toshiyuki Toyosaki
Department of Foods and Nutrition, Koran Women's Junior College, Fukuoka, Japan

5. References

[1] Ames, B.M., Shigena, K., & Hagen, T.M. (1994). Oxidants, antioxidants and the degenerative diseases of aging. *Proceeding National Academy Science,* 90, 7915-7922.
[2] Cathcart, R., Schwiers, E., & Ames, N. (1984). Detection of picomole levels of lipid hydroperoxides using a dichloro-fluorescein fluorescent assay. In L. Packer (Ed), *Methods in Enzymology,* vol. 105 (p. 352). New York; Academic Press Inc.
[3] Gardnaer, H.W. (1996). Lipoxygenase as a versatile biocatalyst. *Journal of American Oil Chemical Society,* 73, 1347-1350.
[4] Grosch, W. (1987). Reactions of hydroperoxides-products of low molecular weight. In H.W.-S Chan (Ed), *Autoxidation of Unsaturated Lipids.* London; Academic Press Inc.
[5] Laemmli, U.K. (1979). Cleavage of structural proteins during the assembly of the head of bacteriophage T4. *Nature,* 227, 680-685.
[6] Lowry, O.H., Rosebrough, N.J., Farr A.L., & Randall, R.J. (1951). Protein measurement with the Folin-Phenol reagent. *Journal of Biological Chemistry,* 193, 265-275.
[7] Mercier, M., & Gelinas, P. (2001). Effect of lipid oxidation on dough bleaching. *Cereal Chemistry,* 78, 36-38.
[8] Pokorny, J. (1999). *Handbook of Food Preservation, Antioxidants in Food Preservation.* New York; Marcel Dekker.
[9] Shewfelt, R.L., & Del Rosario, B.A. (2000). The role of lipid peroxidation in storage disorders of fresh fruits and vegetables. *Horticulture Science,* 35, 575-579.
[10] Steel, R.G.D., & Torrie, J.H.(1980). *Principles and procedures of statistics: a biometrical approach* (2nd ed). New York: McGraw-Hill.
[11] Toyosaki, T., & Sakane, Y. (2002). Antioxidant effect of NaCl on the aqueous solution, emulsified and enzymic lipid peroxidation. *Bulletin of the Society of Sea Water Science, Japan,* 56, 10-16.
[12] Toyosaki, T., & Koketsu, M. (2004). Oxidative stability of silky fowl eggs. Comparison with hen eggs. *Journal of Agriculture and Food Chemistry,* 52, 1328-1330.

Chapter 12

Modification of Fatty Acid Composition in Meat Through Diet: Effect on Lipid Peroxidation and Relationship to Nutritional Quality – A Review

Gema Nieto and Gaspar Ros

Additional information is available at the end of the chapter

http://dx.doi.org/10.5772/51114

1. Introduction

The use of nutritional strategies to improve quality of food products from livestock is a new approach that emerges at the interface of food science and animal science. These strategies have emphasized in the alteration of nutritional profile, for example increasing the content of polyunsaturated fatty acid (*PUFA*), and in the improvement of the oxidative stability, such as supplementation of animal with natural antioxidants to minimize pigment and lipid oxidation in meat.

The interest in the modification of fatty acid of meat is due to that fatty acid composition plays an important role in the definition of meat quality because it is related to differences in sensory attributes and in the nutritional value for human consumption [1]. Meat is a major source of fat in the diet, especially of saturated fatty acids (*SFA*), which have been implicated in diseases, especially in developed countries, such as cardiovascular diseases and some types of cancer.

One of the key goals of nutritional research focuses on establishing clear relationships between components of diet and chronic diseases, considering that nutrients could provide beneficial health results. The incidence of these diseases in humans is associated with the amount and the type of fat consumed in the diet. Diets high in *SFA* contribute to increase LDL-cholesterol level, which is positively related to the occurrence of heart diseases. However, some monounsaturated fatty acids (*MUFA*) and *PUFA*, in particular long-chain n-3 *PUFA* have favourable effects on human health.

In recent years, consumers' pressure to reduce the composition and quality of fat in meat has led to attempts to modify meat by dietary strategies. Where as in recent years consumers have been advised to limit their intake of saturated fats and to reach a ratio of *PUFA*:*SFA* greater than 4 and the type of polyunsaturated fatty acid is now being emphasized and a higher ratio of n-3: n-6 fatty acids is advocated [1]. There is also now concern about the consumption of unsaturated fatty acids that are formed during high-temperature hydrogenation of oils for use in food products: the *trans*-unsaturated fatty acids in which the double bonds are in the *trans*-stereometric position.

Nutritional approaches to improve the oxidative stability of muscle foods are often more effective than direct addition of food ingredients since the antioxidants are preferentially deposited where it is most needed. In addition, diet often represents the only technology available to alter the oxidative stability of intact muscle foods, where utilization of exogenous antioxidants additives is difficult if not impossible. Since product composition is altered biologically, nutritional alteration of muscle composition is more label-friendly since no additive declarations are required.

Among the strategies used, meat and meat products can be modified by adding ingredients considered beneficial for health where the ingredients are able to eliminate or reduce components that are considered harmful. In this sense, several studies have shown that animal diet can strongly influence the fatty acid composition of meat. Scerra et al. [2] showed that feeding ewes with pasture increases the *PUFA* content of intramuscular fat of the lamb infant compared with diets consisting of concentrate. Nieto et al. [3] showed that feeding Segureña ewes with thyme increases the *PUFA* content of intramuscular fat of the lamb meat compared with control diets. Similarly, Elmore et al. [4] showed that feeding lambs with diets rich in fish oil can modify the fatty acid profile of meat (increasing the level of *PUFA*). Moreover Bas et al. [5] used linseed diet and Ponnampalam et al. [6] used fish oil, in order to increase the content of long-chain n-3 fatty acids in lamb meat.

The variation of fatty acid compositions has profound effects on meat quality, because fatty acid composition determines the firmness/oiliness of adipose tissue and the oxidative stability of muscle, which in turn affects flavour and muscle colour. It is well known that high *PUFA* levels may produce alterations in meat flavour due to their susceptibility to oxidation and the production of unpleasant volatile components during cooking [7]. Therefore, it's important to study the implications of the modification of fatty acid in the quality of the meat and the lipid stability, for that it would be interesting the use of liposomes to study the lipid oxidation.

Since liposomes mimic cellular structures [8], the feasibility to protect lipid membranes in the presence of natural antioxidants can be investigated in model systems prior to administration trough feeding. Such previous experiments are particularly interesting for meat industry as they furnish preliminary insights with respect to lipid oxidation at relatively short timescales [9].

2. Lipid digestion in ruminants and non-ruminants

It is well known that lipid digestion is different in ruminant and non-ruminant and that the nature of lipid digestion by the animal has an important effect on the transfer of fatty acids from the diet into the animal product. In case of non-ruminant, the principal site of digestion of dietary lipid is the small intestine, where the pancreatic lipase breaks the triacylglycerols down to mainly 2-monoacylglycerols and free fatty acids and the formation of micelles aids absorption, with lipid uptake mediated by the lipoprotein lipase enzyme, which is widely distributed throughout the body. Therefore dietary fatty acids in the non-ruminant are absorbed unchanged before incorporation into the tissue lipids. Dietary lipid sources have a direct and generally predictable effect on the fatty acid composition of pig and poultry products and the supply of unsaturated fatty acids (*UFA*) to tissues may be simply increased by increasing their proportion in the diet [10].

However, digestion and metabolism of ingested lipids in the rumen results in the exit of mainly long-chain, saturated fatty acids from the rumen. The rumen microorganisms in the ruminant digestive system have a major impact on the composition of fatty acids leaving the rumen for absorption in the small intestine. Microbial enzymes are responsible for the isomerisation and hydrolysis of dietary lipid and the conversion of *UFA* to various partially and fully saturated derivatives, including stearic acid ($C_{18:0}$). Although linoleic ($C_{18:2}$ n-6) and linolenic ($C_{18:3}$ n-3) acids are the main *UFA* in the diet of ruminants, the processes within the rumen ensure that the major fatty acid leaving the rumen is $C_{18:0}$. The intestinal absorption coefficient of individual fatty acids is higher in ruminants than nonruminants, ranging from 80% for *SFA* to 92% for *PUFA* in conventional low fat diets. Therefore, the higher absorption efficiency of *SFA* by ruminants has been attributed to the greater capacity of the bile salt and lysophospholipid micellar system to solubilise fatty acids, as well as the acid conditions within the duodenum and jejunum (pH 3.0–6.0).

3. Fatty acid in meat

Taking into accounts that fat is currently an unpopular constituent of meat and however contributes to meat quality and is important to the nutritional value of meat. This section considers the fatty acid composition in different species and the roles of the fat in meat quality.

Doing a brief introduction of the importance of fatty acids, firstly we will highlight the essential unsaturated fatty acids, linoleic ($C_{18:2}$), linolenic ($C_{18:3}$) and arachidonic ($C_{20:4}$). They are necessary constituents of mitochondria and cell walls. These fatty acids are specials, because contrary to the production from saturated sources, the body can not produce any of the fatty acid mentioned above, unless one of them is available in the diet. Oleic, linoleic and linolenic acids each belong to a different family of compounds in which unsaturation occurs at the n–9, the n–6 and n–3 carbon atoms, respectively, in the hydrocarbon chain numbering from the methyl carbon (n). They are thus referred to as the ω –9, ω –6 and ω –3 series. Linoleic acid is abundant in vegetable oils and at about 20 times the concentration found in meat; and linolenic acid is present in leafy plant tissues [11].

Doing a comparative data between the content of *PUFA* in the muscular tissue of the beef, lamb and pork (Table 1), it is clear that linoleic acid ($C_{18:2}$) is markedly greater in the lean meat of pigs than in that of either the beef or lamb.

	C18:2	C18:3	C20:4	C22:5	C22:6
Beef	2.0	1.3	1.0	Tr.	-
Lamb	2.5	2.5	-	Tr.	-
Pork	7.4	0.9	Tr.	Tr.	1.0

Table 1. Polyunsaturated fatty acids and cholesterol in lean meat (as % total fatty acids)

In addition, the Table 2 shows the study of Enser at al. [12], who obtained 50 samples of beef sirloin steaks, pork chops, and lamb chops and determined the fatty acid profile of the muscle portions of these retail meat cuts. In the same way that Table 1, the most notable difference among the ruminant species and pork was the fivefold greater concentration of linoleic in pork and significantly greater proportions of $C_{20:3}$, $C_{20:4}$, and $C_{22:6}$, and $C_{14:0}$. For example, pork have a proportions of linoleic acid ($C_{18:2}$ n-6): 302 mg/100g of loin muscle, while beef and lamb contains 89 and 25 mg/100g, respectively. The reason of this is because linoleic acid is derived entirely from the diet. It passes through the pig's stomach unchanged and is then absorbed into the blood stream in the small intestine and incorporated from there into tissues. When linoleic acid is ingested, they are metabolized by animal liver to produce two families of long chain polyunsaturated fatty acids which are specific to animals, respectively, the n-6 and n-3 series.

Fatty acid	Pork	Beef	Lamb
$C_{12:0}$ (lauric)	2.6	2.9	13.8
$C_{14:0}$ (myristic)	30	103	155
$C_{16:0}$ (palmitic)	526	962	1101
$C_{18:0}$ (stearic)	278	507	898
$C_{18:1}$ (trans)	-	104	231
$C_{18:1}$ (oleic)	759	1395	1625
$C_{18:2}$ n-6 (linoleic)	302	89	125
$C_{18:3}$ n-3 (α-linolenic)	21	26	66
$C_{20:3}$ n-6 (lauric)	7	7	2
$C_{20:4}$ n-6 (arachidonic)	46	22	29
$C_{20:5}$ n-3 (eicosopentaenoic)	6	10	21
$C_{22:5}$ n-3 (docosopentaenoic)	13	16	24
$C_{22:6}$ n-3 (docosohexaenoic)	8	2	7
Total	2255	3835	4934
P:S	0.58	0.11	0.15
n-6:n-3	7.22	2.11	1.32

Source: Enser et al. [12]

Table 2. Fatty acid Content (mg/100g) of loin muscle in steaks or chops.

However, in ruminants, linoleic acid ($C_{18:2}$ n–6) and α-linolenic acid ($C_{18:3}$ n-3) which are at present in many concentrate feed ingredients, are degraded into monounsaturated (*MUFA*) and saturated fatty acids (*SFA*) in the rumen by microbial biohydrogenation (70–95% and 85-100%, respectively) and only a small proportion, around 10% of dietary consumption, is available for incorporation into tissue lipids. By that reason, beef and lamb contain lower content of linoleic acid, compared with pork meat. Muscle also contains significant proportions of long chain (C20-22) *PUFAS* which are formed from $C_{18:2}$ n–6 and $C_{18:3}$ n–3 by the action of Δ5 and Δ 6 desaturase and elongase enzymes. Important products are arachidonic acid ($C_{20:4}$ n-6) and eicosapentaenoic acid (EPA, $C_{20:5}$ n-3).

Taking into accounts that in ruminants, rumen microorganisms hydrogenate a substantial proportion of *PUFA* diet, resulting in high levels of *SFA* for deposition in muscle tissue, lamb or beef meat contain a low relationship between fatty acids *PUFA* and *SFA* (ratio P/S), which increases the risk of cardiovascular problems and other diseases.

The consequences of a greater incorporation of $C_{18:2}$ n-6 into pig muscle fatty acids compared with ruminants produces higher levels of $C_{20:4}$ n-6 by synthesis and the net result is a higher ratio of n-6:n-3 *PUFA* compared with the ruminants. If the nutritional advice is for ratios <4.0, the present value of 7 on pig muscle is unbalanced relative to that of the ruminants (1.32 in lamb and 2.11 in beef). In addition, another ratio is the ratio of all *PUFA* to *SFA* (P:S). The ratio P/S in a normal diet is 0.4 [13] and in lamb meat is 0.15 and in beef 0.11, while in pork is 0.58.

For all these reasons, there is an increase interesting in research intended to modify the fatty acid composition in meat, especially reducing the concentration of *SFA* and increasing *PUFA*.

4. Dietary modification of fatty acid in meat

Doing the comparison between ruminants and non-ruminants, the fatty acid composition of stored lipids of the ruminant is relatively unwilling to changes in the fatty acid profile of ingested lipids. This logic has been the basis for expression of the concept that ruminant fats are more saturated than those of non-ruminants. Although effects of ruminal biohydrogenation on ruminant tissue fatty acid profiles are generalized, numerous researchers have demonstrated that ruminant lipids can be manipulated by dietary means to contain a higher proportion of unsaturated fatty acids. The next paragraphs will show the different strategies to modify the fatty acids of ruminants and nonruminants.

4.1. Altering quality of muscle from monogastric

Monogastric farm animals are worldwide a main source of high-quality products with a high content of highly available protein, minerals, and vitamins. Pigs and chickens are the main monogastric farm animals [1].

To altering the quality of muscle of monogastric, it's necessary to know the digestion of nutrients in these animals. Anaerobic microorganisms are able to hydrogenate unsaturated fatty acids (*UFA*) preferably polyunsaturated fatty acids (*PUFA*). During these processes they build trans-fatty acid as well as conjugated fatty. While in ruminants these fatty acids are absorbed to a great extent, monogastric animals excrete most of them with the feces as they are produced in the lower parts of the digestive tract.

4.2. Altering quality of muscle from ruminants

Meat from ruminants is a major source of essential nutrients (amino acids, iron, zinc and vitamins from the B group). Meat from ruminants (huge diversity of breeding systems and pieces) is characterised by great variations in fats, quantitatively and qualitatively. Some saturated ($C_{14:0}$ and $C_{16:0}$) and monounsaturated trans fatty acids are not recommended for human consumption and it is possible to reduce their concentrations in meats by increasing the proportions of polyunsaturated fatty acids absorbed by the animals from their diets. To achieve this goal, fatty acids must be protected against hydrogenation in the rumen. Dietary intake of *PUFA* from the n-3 series and especially from the n-6 series by the animals favour the production of conjugated linoleic acid by the rumen bacteria. Some of these fats, such as CLA (conjugated linoleic acid), could be beneficial to human health. CLA is important in the prevention of specific cancers and in the treatment of obesity that has been demonstrated in animal models and, at least partly, in humans.

Alteration of quality in food products from ruminants requires knowledge of the nutritional and metabolic principles that influence product composition. Ruminants are unique among mammals due to their pregastric fermentation. Microflora and microfauna present in the ruminant forestomach dramatically modify the ingested nutrients and consequently have a large impact on the metabolism and composition of the muscle and milk [1].

5. Fatty acid sources

It's important to take into accounts several factors to choice the ingredient and the form by which it is included in the feed: (a) the cost and availability; (b) the impact of the ingredients and its fatty acid composition on feed digestibility; (3) the influence of consumers and retailers regarding the introduction of ingredients into the food chain and (4) animal feed regulations regarding permitted supplements.

Recognizing that fatty acids are readily absorbed from the diet and incorporated into tissue fat, producers have attempted to improve the nutritional quality of meat by incorporating various sources of n-3-*PUFA* [7, 14]. The main dietary sources of n-3 fatty acids fed to pig are vegetable oils [15-20], fish oil/fish meal [21,22] and forage [23]. Novel oil sources such as chia seed, marine algae, lupin and camelina have been investigated as lipid sources in animal feeds.

5.1. Lipid sources for ruminants

5.1.1. Forages

Several studies have shown that ruminants consuming fresh pasture have higher content of UFA in their meat that those receiving a cereal-based concentrate diet.

Grass is a good source of n–3 *PUFA* although there can be variation due to maturity and variety. Grass lipids contain high proportions of the unsaturated linolenic acid ($C_{18:3}$ *n–3*). Other studies have suggested that the (n–6)/(n–3) ratio in phospolipids may be useful to discriminate grass-fed from grain-fed lambs [1, 24].

Therefore, pasture-raised animals have higher proportions of linolenic acid in their fat than stallfed animals [25].

French et al. [15] compared the effect of offering grazed grass, grass silage and concentrates on the fatty acid composition of intramuscular fat in steers. Similar low intramuscular fat contents (<4.5 g/100 g muscle) were determined in meat from all diets offered, hence a possible confounding effect due to differences in the amount of fat deposited was avoided. Decreasing the proportion of concentrate in the ration effectively increased the proportion of grass intake and resulted in a linear increase in *PUFA*:*SFA* ratio (P<0.01) and a linear decrease in the concentration of SFA (P<0.001). The highest concentration of *PUFA* in the intramuscular fat was found in those animals that had consumed grass only (22 kg grazed grass). Grass and grass silage had a much greater proportion of α-linolenic acid than the concentrates, although levels of linoleic acid were similar. The content of linoleic acid in the intramuscular fat was not significantly different between treatments but concentrations of α-linolenic acid and total conjugated linoleic acid were significantly higher for grass-fed steers than for steers offered grass silage and/or concentrates.

Moreover, Nuernberg et al. [26] showed that the concentration of lauric acid was higher in subcutaneous fat and muscle of lambs fed on pasture compared to lambs fed concentrate. Similar results were found by Demirel et al. [27], who studied the fatty acids of lamb meat from two breeds fed different forage: concentrate ratio. And Scerra et al. [2], who showed that lamb meat derived from pasture-fed ewes had a lower levels of lauric and palmitic acid (compared with diets with concentrate) that are though to be a public health risk

Sañudo et al. [28] studied British lambs compared with lambs fed grass fed grain, the result showed that a higher percentage of linolenic acid in the meat of grass-fed lambs, as result of the introduction of this natural antioxidant.

Realinia et al. [29] studied thirty Hereford steers that were finished either on pasture or concentrate to determine dietary and antioxidant treatment effects on fatty acid composition and quality of beef. These authors reported that the percentages of $C_{14:0}$, $C_{16:0}$, and $C_{18:1}$ fatty acids were higher (P<0.01) in the intramuscular fat of concentrate-fed steers, whereas pasture-fed cattle showed greater (P<0.01) proportions of $C_{18:0}$, $C_{18:2}$, $C_{18:3}$, $C_{20:4}$, $C_{20:5}$, and $C_{22:5}$. Total conjugated linoleic acid (CLA) was higher (P<0.01) for pasture- than concentrate-fed cattle. Therefore, these authors reported that finishing cattle on pasture enhanced the unsaturated fatty acid profile of intramuscular fat in beef including CLA and omega-3 fatty acids. Results from this study suggest that the negative image of beef attributed to its highly saturated nature may be overcome by enhancing the fatty acid profile of intramuscular fat in beef through pasture feeding from a human health perspective.

5.1.2. Oilseeds

Many studies to manipulate the fatty acid composition of meat using whole oilseeds have been conducted. For example, the effect of the physical form of linseed offered on the fatty acid composition of meat has been reported by several workers: Raes et al. [30] reported that the replacement of whole soyabean with extruded linseed or crushed linseed in the finishing diet of Belgian Blue young bulls increased α-linolenic acid. Mach et al. [31] reported that whole canola seed (α-linolenic acid content 10.6 g/100 g total FA) or whole linseed (α-linolenic acid content 54.2 g/100 g total FA), at three lipid levels (50, 80 and 110 g/kg DM) to 54 Holstein bulls increased linearly with lipid level the concentration of n–3 *PUFA* in the longissimus dorsi muscle.

Elmore et al. [4] reported that the feeding of lamb with diets rich in fat and oils (fish oils, kelp and flax seed) increased the level of polyunsaturated fatty acids. Similarly, Nute et al. (2007) studied the oxidative stability and quality of fresh meat from lambs fed different levels of n-3 *PUFA* from linseed oil, fish oil, a supplement produced from flax seed (PLS), seed sunflower and soybean meal, seaweed, and combinations of these different oils. They reported that the fatty acid composition of semimembranosus muscle phospholipids was affected by diet.

The rabbit meat was also used in several studies with the objective of fatty acid modification. As the study of Kouba et al. [32], who studied rabbits fed with a diet containing 30 g of extruded linseed/kg. Feeding the linseed diet increased ($P < 0.005$) the content of 18:2n-3 in muscles, perirenal fat, and raw and cooked meat. The long chain n-3 polyunsaturated fatty acid (*PUFA*) contents were also increased ($P < 0.01$) in the meat. The linseed diet produced a decrease in the n-6/n-3 ratio. These authors highlights that the inclusion of linseed in rabbit diets is a valid method of improving the nutritional value of rabbit meat.

5.1.3. Marine algae

Marine algae are an alternative to fish oil as a dietary source of n–3 long chain PUFA (LCPUFA), eicosapentaenoic acid (EPA) and docosahexaenoic acid (DHA).

In the study of Cooper [33], the marine algae were included in the sheep diet not only as a source of DHA (fish oil/algae diet supplied 15 g/100 g total FA as DHA) but also because it had been previously shown to undergo a lower level of biohydrogenation than fish oil [33]. In another study of the same author [34] studied the manipulation of the n-3 PUFA fatty acid content of muscle and adipose tissue lamb was studied. For that fifty lambs, with an initial live weight of 29 ± ◎2.1 kg, were allocated to one of five concentrate-based diets formulated to have a similar fatty acid content (60 g/kg DM), but containing either linseed oil (high in 18:3*n*◎3); fish oil (high in 20:5*n*◎3 and 22:6*n*◎3); protected linseed and soybean (PLS; high in 18:2*n*◎6 and 18:3*n*◎3); fish oil and marine algae (fish/algae; high in 20:5*n*◎3 and 22:6*n*◎3); or PLS and algae (PLS/algae; high in 18:3*n*◎3 and 22:6*n*◎3). Lambs fed either diet containing marine algae contained the highest ($P < 0.05$) percentage of 22:6*n*◎3 in the phospholipid (mean of 5.2%), 2.8-fold higher than in sheep fed the fish oil diet.

A more limited number of studies have looked into the effects of dietary supplementation with DHA-rich marine algae on the fatty acid composition of muscle tissue of rabbits [35], lambs [4] and pigs [36, 37].

5.2. Lipid sources for non-ruminants

The cereal-based diet commonly offered to poultry and pigs supplies mainly n−6 *PUFA* and a small amount of n−3 *PUFA*. This is reflected in the fatty acid composition of the animal product. Dietary modification of poultry meat, eggs or pork to increase the n−3 *PUFA* content requires a supply of n−3 *PUFA* from the diet.

The actual strategies to non-ruminants are focused on assessing the effect of offering terrestrial versus marine sources of n−3 PUFA and the subsequent implications for product quality.

5.2.1. Vegetable oils

Enrichment of poultry diets with plant oils has been shown to have different impacts on abdominal fat and the site of fatty acid deposition depending on the *SFA*, *MUFA* and *PUFA* content of the oil [38].

Crespo & Esteve-García [38] studied broiler chickens fed with a basal diet supplemented for 20 days before slaughter with 10% inclusion of linseed oil, sunflower oil and olive oil. As expected with non-ruminant, the fatty acid profile of the deposited fat in the broiler carcase reflected the dietary fat source. The supplementation with olive oil resulting in the highest proportion of $C_{18:1}$, sunflower oil supplementation resulting in the highest proportion of linoleic acid (51.1 g/100 g total body FA), while linseed oil contributed the highest amount of n−3 *PUFA* and most favourable n−6:n−3 ratio in the carcase fat.

In addition, Lu et al. [39] investigated the effects of soybean oil and linseed oil on the fatty acid compositions of pork. The three dietary treatments were: (a) no oil supplement; (b) 3% soybean oil supplement; (c) 3% linseed oil supplement. Dietary linseed oil and soybean oil significantly increased the contents of $C_{18:3}$ and $C_{18:2}$ in the neutral lipids and phospholipids in both longissimus muscle and biceps brachii muscle, respectively.

5.2.2. *Linseed and fish oil*

The n–3 long chain polyunsaturated fatty acids can be incorporated into non-ruminant products from dietary fish oil. The transfer of these fatty acids was found to be influenced by time and duration of feeding and the presence of other oil supplements. Haak et al. [40] offered to pigs a basal diet composed of barley, wheat and soyabean meal ad libitum alone or supplemented with 1.2% linseed or fish oil during: the whole fattening period; the first fattening phase (weeks 1–8) only; or the second fattening phase (6 weeks or 9 weeks, until slaughter at 100 kg). Haak et al. [40] reported that incorporation of α-linolenic acid into the longissimus thoracis muscle was similar (1.24 g/100 g total FA) when linseed was offered throughout the fattening period or only during the second phase. When fish oil was offered during either of the fattening phases, only the proportion of DHA incorporated was affected, being greater when fish oil was offered during the second fattening phase ($P<0.05$). Incorporation of EPA (Eicosapentaenoic acid) and DHA (Docosahexaenoic acid) following fish oil supplementation for the whole fattening period was 1.37 and 1.02 g/100 g total FA in the longissimus thoracis muscle, representing a six-fold increase compared to the basal diet and a three- and five-fold increase respectively, compared to the linseed diet. In agreement with other animal work [34], Haak et al. [40] concluded that a direct dietary source of DHA was required to increase DHA in animal muscle and that levels in pork could not be substantially influenced by dietary supply of precursors.

5.2.3. *Marine algae*

It's well known that microalgae are the original source of DHA in the marine food chain [41], dried marine algae have also been included in animal feeds to improve the DHA level of foods of animal origin. Studies has mainly focused on the quality of eggs [42, 43] and chicken meat [44]. A more limited number of studies have looked into the effects of dietary supplementation with DHA-rich marine algae on the fatty acid composition of muscle tissue of pigs [36, 37].

6. Effects of fatty acid modification on the nutritional value of meat

There is a growing consumers resistance to the incorporation of additives into foods, especially where the additives are of synthetic origin, even when they have a nutritional or health advantage. Dietary supplementation of the growing animal provides a unique method of manipulating the content of some micronutrients and other nonnutrient bioactive compounds in meat, with a view to improving the nutrient intake of consumers or improving their overall health.

Research on heart disease in humans has tended to implicate high intakes of saturated fat and cholesterol as contributory factors with a possible protective effect of polyunsaturated fat and a neutral effect of monounsaturated fat [45]. Overall, the advice to consumers has been to control the level of energy consumed as fat to under 35% and in particular, to limit saturated fats to 10%of energy intake [13]. It is also recommended that the proportion of short- and medium-chain saturated fatty acids be reduced and that intake of n-6 fatty acids be reduced relative to n-3 [45].

The nutritional properties of meat are largely related to its fat content and its fatty acid composition. In this sense, long-chain n-3 fatty acids, such as $C_{20:5}$ n-3 and $C_{22:6}$ n-3 have beneficial health effects, such as reduction in the thrombotic tendency of blood, associated with lower coronary heart disease in humans [46]. In addition, the role of dietary fat in human health is further complicated by the differing biological activity of some fatty acids when present at different stereospecific positions in triacylglycerols [47].

To avoid possible health dangers from the consumption of the meat of ruminants, a greater degree of unsaturation could be introduced into their fats. One example of the modification of fatty acid in meat resulting in an improvement of human health is the study of Diaz et al. [48]. These authors studied the fatty acid content and sensory characteristics of meat from light lambs fed three diets supplemented with different sources of n-3 fatty acids (fish oil, linseed and linseed plus microalgae) and a control diet during refrigerated storage. The meat from lambs fed linseed diets had the highest levels of $C_{18:3}$ n-3,while animals fed fish oil, had the highest long-chain n-3 polyunsaturated fatty acids (*PUFA*). Thus, 100 g of meat from lamb fed the fish oil diet provided 183 mg of long-chain n-3 *PUFA*, representing 40% of the daily recommended intake. The levels of n-3, n-6 and long-chain n-3 *PUFA* decreased during a 7-day storage period. These authors reported that consumption of 100 g of lamb muscle from lambs fed control diet would provide about 5% of the daily recommended intake for long chain n-3 fatty acids (500 mg per day, according to EFSA [49]. In case of linseed plus microalgae and linseed diets, would provide nearly 10% of the daily recommended intake for long-chain n-3 fatty acids. The greatest supply of n-3 *PUFA* and long-chain n-3 fatty acids would come from lambs fed fish oil diet, which would provide about 34% of the daily recommended intake for long-chain n-3 fatty acids. Moreover, the highest *PUFA/SFA* ratio was found in lambs fed linseed and fish oil diets, which was close to the recommended value (0.35). The lowest value was observed in lambs fed control diet. During storage, the total content of *PUFA*, including n-3 *PUFA*, n-6 *PUFA* and the long-chain n-3 *PUFA*, decreased. Thus, meat from lambs fed fish oil could supply close to 40% of the daily recommended intake for long-chain n-3 fatty acids on day 0. On day 7 this meat supplies almost 31% and, therefore, this could be considered a reduction in the nutritional value of the meat. This decrease could be a consequence of oxidation changes, since *PUFA* are more prone to oxidation than *MUFA* or *SFA*; the meat from the supplemented groups and especially from animals fed fish diet, are more prone to oxidation than the control diet (with lower content in *PUFA* and long-chain n-3 *PUFA*). Therefore, the importance of the influence of the modification of fatty acid profile on the lipid peroxidation should be studied.

7. Quality of *PUFA* enriched animal products and relation with lipid peroxidation

One of the main factors limiting the quality of meat and meat products is lipid oxidation. Lipid oxidation results in rancid odour and flavour, sometimes referred to as warmed-over flavour. Fatty acids are oxidised into aldehydes, alkanes, alcohols and ketones by chemical (auto-oxidation) or enzymatic (β-oxidation) reactions. In this sense, rancid aroma is apparently due to the dominance of alkanal (hexanal, nonanal) or certain alcohols (1-penten-3-ol, 1-octen-3-ol). The reason is due to the first step of lipid oxidation, which involves the removal of hydrogen from a methylene carbon in the fatty acid. This becomes easier as the number of double bonds in the fatty acid increases, which is why polyunsaturated fatty acids are particularly susceptible to oxidation. Therefore, increasing the degree of unsaturation of muscle membranes reduces the oxidative stability of the muscle. In addition, the relative oxidation rates of fatty acids containing 1, 2, 3, 4, 5 or 6 double bonds are 0.025, 1, 2, 4, 6 and 8, respectively [50].

It is very interesting to correlate the fatty acid profile of the meat with the development of off-odour and off-flavour in order to understand the susceptibility of oxidative damage in the meat. For example, If the ratio of *PUFA* to *SFA* is higher in the meat, this softer fat is more susceptible to oxidative damage, and this may cause difficulties for the retailers who are increasingly turning toward centralized butchery and modified atmosphere packaging, both of which lead to meats being exposed to higher levels of oxygen for a longer period of time prior to retail.

There are few studies that examine the effect of an enrichment of the diet in n-3 *PUFA* and the oxidation potential of muscle. Some of these studies are made in rabbit meat [51-52, 32] and lamb meat [53]. While Kouba et al. [32], reported that the enriched Longissimus dorsi did not exhibit a lower oxidative stability, Castellini et al. [51] and Dal Bosco et al. [52] found that feeding a n-3 *PUFA* enriched diet lowered significantly TBARS level in, raw meat and Longissimus dorsi. One plausible explanation could be that these authors only supplemented the experimental diet with a high amount of vitamin E, which led to an increase of vitamin E level in tissues of rabbits fed this diet, as already described by Oriani et al. [54], and it is well known that the susceptibility of lipids to oxidation can be reduced by vitamin E, as described by Lin et al. [55] in poultry and Monahan et al. [56] in pigs.

The oxidative processes in living animals are dependent on the endocrine and enzymatic activities in the tissues. There is some evidence that differences between species and breeds of animals exists. However, a high individual variation has also to be assumed. Oxidative processes can occur at many different stages in animal nutrition. During digestion the nutrients are soluble and therefore can be easily oxidized. Extrinsic influences on the oxidative processes mainly derive from the composition of the feedstuffs and feed additives. Therefore, feed should be protected against oxidative damage already during storage.

With increased content of polyunsaturated fatty acids (*PUFA*) a higher oxidation rate in the feed, in the digest as well as in the intermediate metabolism, occurs. But feedstuffs can also contain antioxidants like vitamins, carotenoids, or phenols, or prooxidative compounds like some trace elements. To improve the oxidative stability of the feed, antioxidative additives are often used as supplements to the diets.

Notwithstanding the beneficial attributes of polyunsaturated fatty acids, it should be noted that lipid oxidation products are believed to adversely affect the health of cells. Fortunately muscular tissue contains several enzymes that protect cells against such change, the most important of which is glutathione peroxidase [57].

To avoid the lipid oxidation tendency shown in meat rich-PUFA, Díaz et al. [48], recommended the inclusion of antioxidants in the diet of lambs, in order to avoid the negative impact on the flavour and to prevent fatty acids from oxidation of these on lamb meat enriched in n-3 fatty acids. Therefore, the inclusion of antioxidants with the incorporation of the ingredients responsible of the fatty acid modification through the feed could be an interesting strategy to prevent oxidation of the meat. Similarly, it has been shown that some fatty acids (such as conjugated linoleic acid) can exert antioxidant activity in meat by reducing lipid oxidation [58]. In a previous study made by our group, the effectiveness of thyme leaves diet (during pregnancy and lactation of ewes) to improving the lamb meat lipid stability was attributed to the antioxidant effect of the phenolic compounds present in the thyme leaf. These bioactive compounds in the leaves may interfere with the propagation reaction of lipid oxidation, besides inhibiting the enzymatic systems involved in initiation reactions [59]. It has been shown that diet with natural antioxidants interferes with the metabolism of fatty acids in ruminants [60]. Taking into accounts another studies using plants of the family Labiatae in the diet, Youdim and Deans [61] showed that a dietary supply of thyme oil or thymol to ageing rats showed a beneficial effect on the antioxidative enzymes superoxide dismutase and glutathione peroxidase, as well as on the polyunsaturated fatty acid composition in various tissues. Animals receiving these supplements had higher concentrations of polyunsaturated fatty acids in phospholipids of the brain compared to the untreated controls. Similarly, Lee et al. [62] showed that the pattern of fatty acids of the abdominal fat of chicken was also altered by oregano oil and dietary carvacrol lowered plasma triglycerides. In animals for food productions, such effects are of importance for product quality: these supplement may improve the dietary value and lead to a better oxidative stability and longer shelf-life of fat, and meat [63].

7.1. Liposomes

Oxidative stress leads to oxidation of low-density lipoproteins (LDL), which plays a key role in the pathogenesis of atherosclerosis, which is the primary cause of coronary heart disease [64]. The nutritional manipulations of the fatty acid composition of meats increase the susceptibility of their lipids to peroxidation; because as have been explained in the previous sections, *PUFA* are known to act as substrates initiating the oxidative process in meat. In this sense, much attention has been paid to the use of the natural antioxidants, since potentially these components may reduce the level of oxidative stress in the feed.

Several methods have been described in the literature for assessing antioxidant activity. These include radical scavenging assays, ferric reducing assay, or inhibition of the oxidation of oils, emulsions, low-density lipoproteins (LDL), or liposomes. The use of LDL is an interesting method of assessing antioxidant properties relevant to human nutrition, since these systems allow investigation of the protection of a substrate by an antioxidant in a model biological membrane or a lipoprotein. Assessment of the activity of mixtures of lipid-soluble and water-soluble antioxidants in liposomes has clear advantages over other commonly used methods. The liposome system allows the lipid-soluble components to be present in the lipid phase without the presence of a cosolvent, while the water soluble antioxidants can be added to the aqueous phase of the liposome [8].

The liposome system also allows study the synergy between different antioxidants ingredients used in the manufacture of feed, as tocopherols or other water-soluble antioxidants to be demonstrated [65- 66], whereas synergy is not normally observed if these components are present in homogeneous solution.

Therefore, the use of a liposome system is an interesting strategy for a preliminary assessment of the antioxidant activity of ingredients used in the manufacture of feedstuffs.

This was the objective of a previous study [9] where the use of liposomes as biological membrane models to evaluate the potential of natural antioxidants as inhibitors of lipid peroxidation was described. For that, the antioxidative effects of by-products from manufacturing of essential oils, i.e., distilled rosemary leaf residues (*DRL*), distilled thyme leaf residues (*DTL*), and the combined antioxidative effects of *DRL* or *DTL* with α-tocopherol (*TOH*), ascorbic acid (*AA*), and quercetin (*QC*) on peroxidation of L-α-phosphatidylcholine liposomes as initiated by hydrophilic azo-initiators, were investigated. The results showed that the extracts from *DRL* and *DTL* all had an obvious antioxidative effect as evidenced by a lag phase for the formation of phosphatidylcholine-derived conjugated dienes. Combination of *TOH* or *QC* with *DRL* and *DTL*, respectively, showed synergism in prolonging of the lag phase. Distilled leaves of rosemary and thyme were found to be a rich source of antioxidants as shown by the inhibition of the formation of conjugated dienes in a liposome system. Based on this study, it can be concluded that rosemary and thyme residues, as by-products from distillation of essential oils, are a readily accessible source of natural antioxidants, which possibly provides a good alternative to using synthetic antioxidants in the protection of foods and meat products in particular.

After this study, it was reported that both distilled leaves (rosemary and thyme) were readily accessible source of natural antioxidants in animal feedstuffs, these by-products were added to the feed of pregnant ewes [67- 71]. As shown previously with the liposomes model system study, the meat of lambs from ewes fed with distilled rosemary and thyme leaf had lower levels of lipid oxidation and these additives were considered a good alternative to using synthetic antioxidant in animal diets.

8. Conclusions

This review suggest that the negative image of meat attributed to its highly saturated nature may be overcome by enhancing the fatty acid profile of intramuscular fat through feeding from a human health perspective. Increasing the n-3 *PUFA* content of animal feedstuffs can be a promising and sustainable way to improve the nutritional value of meat, without forcing consumers to change their eating habits.

It's well known that although dietary *PUFA* improves meat nutritional qualities, such meats are more susceptible to lipid oxidation during processing. Therefore, there is a need to study the differences in oxidative stability of the muscles in order to understand the effect of dietary on lipid peroxidation. For that the use of liposomes is an interesting strategy to study the lipid peroxidation in model system as preliminary studies (prior the administration of fatty acid sources through feeding).

When all of these considerations are taken into account, the possibility of preserving the nutritional qualities of processed meat rich in *PUFA* by an original dietary antioxidant strategy is recommended, in order to prevent the lipid peroxidation and the decrease of overall liking of meat.

Author details

Gema Nieto and Gaspar Ros
Department of Food Technology, Nutrition and Food Science,
Faculty of Veterinary Sciences, University of Murcia, Murcia, Spain

Acknowledgement

We thank the University of Murcia for the postdoctoral contract of Gema Nieto.

9. References

[1] Wood JD, Richardson RI, Nute GR, Fisher AV, Campo MM, Kasapidou E, Sheard PR, Enser, M.(2003). Effects of fatty acids on meat quality: A review. Meat Sci.66: 21–32.
[2] Scerra M, Caparra P, Foti F, GalofaroV, Sinatra MC, & Scerra V (2007) Influence of ewe feeding systems on fatty acid composition of suckling lambs. Meat Sci. 76: 390-394.
[3] Nieto G, Bañon S, Garrido MD (2012a) Incorporation of thyme leaves in the diet of pregnant and lactating ewes: Effect on the fatty acid profile of lamb. Small Ruminant Res. 105: 140-147.
[4] Elmore JS, Cooper SL, Enser M, Mottram DS, Sinclair LA, Wilkinson, RG, Wood JD (2005) Dietary manipulation of fatty acid composition in lamb meat and its effect on the volatile aroma compounds of grilled lamb. Meat Sci. 69: 233-242.
[5] Bas, P.; Berthelot, E.; Pottier, E. Normand, J. (2007) Effect of level of linseed on fatty acid composition of muscles and adipose tissues of lambs with emphasis on trans fatty acids. Meat Sci. 77: 678-688.
[6] Ponnampalam EN, Trout GR, Sinclair AJ, Egan AR, Leury BJ (2001) Comparison of the color stability and lipid oxidative stability of fresh and vacuum packaged lamb muscle containing elevated omega-3 and omega-6 fatty acid levels from dietary manipulation. Meat Sci. 58: 151–161.
[7] Wood JD, Enser M, Fisher AV, Nute GR, Richardson RI, Sheard PR. (1999) Manipulating meat quality and composition. Proceedings of the Nutrition Society. 58: 363– 370.
[8] Roberts WG, Gordon MH (2003) Determination of the total antioxidant activity of fruits and vegetables by a liposome assay. J. Agr. Food Chem. 51 (5): 1486–1493
[9] Nieto G, Huvaere K, Skibsted LH (2011a) Antioxidant activity of rosemary and thyme by-products and synergism with added antioxidant in a liposome system. Eur. Food Res.Technol. 233: 11-18.

[10] Woods VB, Fearon AM (2009) Dietary sources of unsaturated fatty acids for animals and their transfer into meat, milk and eggs: A review. Livest. Sci. 126:1– 20.

[11] Decker EA, Faustman C, Lope-Bote C.J (2000) Antioxidants in muscle foods. Nutritional strategies to improve quality. Wiley-Interscience. 465 p.

[12] Enser MS, Hallet K, Hewitt B, Fursey GAJ, Wood JD (1996) Fatty acid content and composition of English beef, lamb and pork retail. Meat Sci. 42: 443-456.

[13] Department of Health (1994) Nutritional aspects of cardiovascular disease (Report on health and social subjects no. 46). H.M. Stationery Office, London.

[14] Raes K, Haak L, Balcaen A, Claeys E, Demeyer D, De Smet S (2004) Effect of linseed feeding at similar linoleic acid levels on the fatty acid composition of double-muscled Belgian Blue young bulls. Meat Sci. 66 (2): 307–315.

[15] French P, Stanton C, Lawless F, O'Riordan EG, Monahan FJ, Caffrey PJ, Moloney AP (2000) Fatty acid composition, including conjugated linoleic acid, of intramuscular fat from steers offered grazed grass, grass silage or concentrate-based diets. J. Anim. Sci.78: (11): 2849–2855.

[16] Loor JJ, Hoover WH, Miller-Webster TK, Herbein JH, Polan CE (2003) Biohydrogenation of unsaturated fatty acids in continuous culture fermenters during digestion of orchard grass or red clover with three levels of ground corn supplementation. J. Anim. Sci. 81: 1611–1627.

[17] Romans JR, Johnson RC, Wulf DM, Libal GW, Costello WJ (1995) Effects of ground flaxseed in swine diets on pig performance and on physical and sensory characteristics and omega-3 fatty acids content of pork: I. Dietary level of flaxseed. J. Anim. Sci. 73: 1982– 1986.

[18] Leskanich CO, Matthews KR, Warkup CC, Noble RC, Hazzledine M (1997) The effect of dietary oil containing (n-3) fatty acids on the fatty acid, physicochemical and organoleptic characteristics of pig meat and fat. J. Anim. Sci. 75: 673– 683.

[19] Enser M, Richardson RI, Wood JD, Gill BP, Shaeard PR (2000) Feeding linseed to increase the n-3 PUFA of pork: fatty acid composition of muscle, adipose tissue, liver and sausage. Meat Sci. 55: 201–212.

[20] Kouba M, Enser M, Whittington FM, Nue GR, Wood JD (2003) Effect of a high & linolenic acid diet on lipogenic enzyme activities, fatty acid composition, and meat quality in the growing pig. J. Anim. Sci. 81: 1967– 1979.

[21] Øverland M, Taugbl O, Haug A, Sundstl E (1996) Effect of fish oil on growth performance, carcass characteristics, sensory parameters, and fatty acid composition in pigs. Acta Agric. Scand., A. Anim. Sci. 46: 11 –17.

[22] Kjos NP, Skrede A, Øverland M (1999) Effect of dietary fish silage and fish fat on growth performance and sensory quality of growing-finishing pigs. Can. J. Anim. Sci. 79: 139–147.

[23] Nilzen V, Babol J, Dutta PC, Lundeheim N, Enfalt AC, Lundstrom K (2001) Free-range rearing of pigs with access to pasture grazing-effect on fatty acid composition and lipid oxidation products. Meat Sci. 58: 267–275.

[24] Aurousseau B, Bauchart D, Calichon E, Micol D, Priolo A (2004) Effect of grass or concentrate feeding systems and rate of growth on triglyceride and phospholipid and their fatty acids in the M. *Longissimus thoracis* of lambs. Meat Sci. 66: 531–541.

[25] Wood JD, Enser M (1997) Factors influencing fatty acids in meat and the role of antioxidants in improving meat quality. British J. Nutrition. 78: S49–S60.

[26] Nuernberg K, Fischer A, Nuernberg G., Ender, K., & Dannenberger, D. (2008). Meat quality and fatty acid composition of lipids in muscle and fatty tissue of Skudde lambs fed grass versus concentrate. Small Ruminant Res. 74: 279-283.

[27] Demirel G, Ozpinar H, Nazli B, Keser O (2006) Fatty acids of lamb meat from two breeds fed different forage: concentrate ratio. Meat Sci. 72: 229–235.

[28] Sañudo C, Enser M, Campo MM, Nute GR, Maria G, Sierra I, Wood J D (2000) Fatty acid composition and sensory characteristics of lamb carcasses from Britain and Spain. Meat Sci. 54: 339–346.

[29] Realinia CE, Ducketta SK, Britob GW, Dalla Rizzab M, De Mattos D (2004) Effect of pasture vs. concentrate feeding with or without antioxidants on carcass characteristics, fatty acid composition, and quality of Uruguayan beef. Meat Sci. 66: 567–577.

[30] Raes, K., De Smet, S., & Demeyer, D. (2004). Effect of dietary fatty acids on incorporation of long chain polyunsaturated fatty acids and conjugated linoleic acid in lamb, beef and pork meat: a review. Anim. Feed Sci. Tech. 113: 199–221.

[31] Mach N, Devant M, Díaz I, Font-Furnols M, Oliver MA, García JA, Bach A (2006) Increasing the amount of n–3 fatty acids in meat from young Holstein bulls through nutrition. J. Anim. Sci. 84: 3039–3048.

[32] Kouba M, Benatmane F, Blochet JE, Mourot J (2008) Effect of a linseed diet on lipid oxidation, fatty acid composition of muscle, perirenal fat, and raw and cooked rabbit meat. Meat Sci. 77 (80): 829-834.

[33] Cooper SL (2002). Dietary manipulation of the fatty acid Composition of sheep Meat. Ph.D. Diss. Open University, Milton Keynes, UK.

[34] Cooper SL, Sinclair LA, Wilkinson RG, Hallett KG, Enser M, Wood, JD (2004) Manipulation of the n–3 polyunsaturated fatty acid content of muscle and adipose tissue in lambs. J. Anim. Sci. 82: 1461–1470.

[35] Tassinari M, Mordenti AL, Testi S, Zotti A (2002) Esperienze sulla possibilita` di arricchire con la dieta la carne di coniglio di LCPUFA n-3. Prog. Nutr. 4: 119– 124.

[36] Marriott NG, Garret JE, Sims MD, Abrul JR (2002) Composition of pigs fed a diet with docosahexaenoic acid. J. Muscle Foods. 13: 265– 277.

[37] Sardi L, Martelli G, Lambertini L, Parisini P, Mordenti A (2006) Effects of a dietary supplement of DHA-rich marine algae on Italian heavy pig production parameters. Livest. Sci. 103: 95–103.

[38] Crespo N, Esteve-García E (2002) Nutrient and fatty acid deposition in broilers fed different dietary fatty acid profiles. Poultry Sci. 81: 1533–1542.

[39] Lu P, Zhang LY, D Yin JD, Everts AKR, Li DF (2008) Effects of soybean oil and linseed oil on fatty acid compositions of muscle lipids and cooked pork flavour. Meat Sci. 80: 910–918.

[40] Haak L, De Smeet A, Fremaut D, Van Walleghem K, Raes K (2008) Fatty acid profile and oxidative stability of pork influence by duration and time of dietary linseed or fish oil supplementation. J. Anim. Sci. 86:1418–1425.

[41] Abril R, Garret J, Zeller SG, Sande WJ, Mast RW (2003) Safety assessment of DHA-rich microalgae from Schizochytrium sp.: Part V. Target animal safety/toxicity in growing swine. Regulatory Toxicology and Pharmacology. 37:73– 82.

[42] Van Elswyk, M.E., (1997). Comparison of n–3 fatty acid sources in laying hen rations for improvement of whole egg nutritional quality. A review. *British Journal and Nutrition*, Vol. 78 (Suppl. 1), 61–69.

[43] Meluzzi A, Sirri F, Tallarico N, Franchini A (2001) Effect of different vegetable lipid sources on the fatty acid composition of egg yolk and on hen performance. Arch. Geflugelkd. 65: 207– 213.

[44] Sirri F, Minelli G, Iaffaldano N, Tallarico N, Franchini A (2003) Oxidative stability and quality traits of n-3 PUFA enriched chicken meat. Italian J. Anim. Sci. 2 (1): 450–452.

[45] Gibney MJ (1993) Fat in animal products: Facts and perceptions. In Safety and Quality of Food from Animals, J.D. Wood and T.L.J. Lawrence (eds.), p.p.57-61. British Society of Animal Production Occasional Publication No. 17, BSAP, Edinburgh.

[46] EFSA (European Food Safety Authority) (2010a). Scientific opinion on dietary reference values for fats, including saturated fatty acids, polyunsaturated fatty acids, monounsaturated fatty acids, trans fatty acids, and cholesterol. EFSA Journal, 8(1461), 1–107.

[47] Decker EA (1996) The role of stereospecific saturated fatty acid positions on lipid nutrition. Nutrition Review. 54: 108–110.

[48] Díaz MT, Cañeque V, Sánchez S, Lauzurica C, Pérez C, Fernández C, Àlvarez I, De la Fuente J (2011) Nutritional and sensory aspects of light lamb meat enriched in n-3 fatty acids during refrigerated storage. Food Chem.124: 147–155.

[49] EFSA (European Food Safety Authority) (2010b). Outcome of the public consultation on the draft opinion of the scientific panel on dietetic products, nutrition, and allergies (NDA) on dietary reference values for fats, including saturated fatty acids, polyunsaturated fatty acids, monounsaturated fatty acids, trans fatty acids, and cholesterol. EFSA ON–1507, pp. 1–23.

[50] Horwitt MK (1986) Interpretations of requirements of thiamine, riboflavin, niacin-tryptophan and vitamin E plus comments on balance studies and vitamin B-6. The American Journal of Clinical Nutrition, 44: 5–973.

[51] Castellini C, Dal Bosco A, Bernardini M, Cyril HW (1998) Effect of dietary vitamin E on the oxidative stability of raw and cooked rabbit meat. Meat Sci. 50: 153–161.

[52] Dal Bosco A, Castellini C, Bianchi L, Mugnai C (2004) Effect of dietary alinolenic acid and vitamin E on the fatty acid composition, storage stability and sensory traits of rabbit meat. Meat Sci. 66: 407–413.

[53] Berthelot, V., Bas, P., Pottier, E., & Normand, J. (2012). The effect of maternal linseed supplementation and/or lamb linseed supplementation on muscle and subcutaneous adipose tissue fatty acid composition of indoor lambs. Meat Science, 90, 548–557.

[54] Oriani G, Salvatori G, Pastorelli G, Pantaelo L, Ritieni A, Corino C (2001) Oxidative status of plasma and muscle in rabbits supplemented with dietary vitamin E. J. Nutritional Biochem. 12: 138–143.

[55] Lin, C. F., Asghar, A., Gray, J. I., Buckley, D. J., Booren, A. M., Crackel, R. L., et al. (1989). Effects of oxidised dietary oil and antioxidant supplementation on broiler growth and meat stability. *British Poultry Science*, Vol. 30, p.p. 855–864.

[56] Monahan FJ, Buckley DJ, Morrissey PA, Lynch PB, Gray JI (1992) Influence of dietary fat and a-tocopherol supplementation on lipid oxidation in pork. Meat Sci. 31: 229–241.

[57] Halliwell B, Aeschbach R, Loeliger J, Auroma OI (1995) The characterization of antioxidants. Food Chem. Toxicol. 33: 601–617.

[58] Joo ST, Lee JI, Ha YL, Park GB (2002) Effects of dietary conjugated linoleic acid on fatty acid composition, lipid oxidation, color, and water-holding capacity of pork loin. J. Anim. Sci. 80: 108–112.

[59] You KM, Jong HG, Kim HP (1999) Inhibition of cyclooxygenase/ lipoxygenase from human platelets by polyhydroxylated/ methoxylated flavonoids isolated from medicinal plants. Arch. Pharm. Res. 22 (1): 18–24.

[60] Vasta V, Pennisi M, Lanza D, Barbagallo M, Bella A, Priolo A (2007) Intramuscular fatty acid composition of lambs given a tanniniferous diet with or without polyethylene glycol supplementation. Meat Sci. 76: 739–745.

[61] Youdim KA, Deans SG (2000) Effect of thyme oil and thymol dietary supplementation on the antioxidant status and fatty acid composition of the ageing rat brain. British J. Nutrition. 83: 878.

[62] Lee, K.W., Everts, H., Kappert, H.J., Frehner, M., Losa, R., & Beynen, A.C. (2003). Effects of dietary essential oil components on growth performance, digestive enzymes and lipid metabolism in female broiler chickens. *British Poultry Science*, Vol. 44, p.p. 450.

[63] Govaris A, Botsoglou E, Florou-Paneri P, Moulas A, Papageorgiou G (2005) Dietary supplementation of oregano essential oil and α-tocopheryl acetate on microbial growth and lipid oxidation of turkey breast fillets during storage. Int. J. Poultry Sci. 4: 969.

[64] Reaven PD, Witztum JL (1996) Oxidized low-density lipoproteins in atherogenesis: Role of dietary modification. Annu. Rev. Nutr. 16: 51-71.

[65] Niki, E., Noguchi, N., Tsuchihashi, H., & Gotoh, N. (1995). Interaction among vitamin C, vitamin E, and beta-carotene. Am. J. Clin. Nutr. 62: S1322-S1326.

[66] Livrea MA, Tesoriere L, D'Arpa D, Morreale M (1997) Reaction of melatonin with lipoperoxyl radicals in phospholipid bilayers. Free Radic. Biol. Med. 23: 706-711.

[67] Nieto G, Bañon S, Garrido MD (2010a) Effect on lamb meat quality of including thyme (*Thymus zygis* ssp. gracilis) leaves in ewes' diet. Meat Sci. 85: 82-88.

[68] Nieto G, Díaz P, Bañon S, Garrido MD (2010b) Dietary administration of ewes diets with a distillate from rosemary leaves (*Rosmarinus officinalis*): influence on lamb meat quality. Meat Sci. 84: 23-29.

[69] Nieto G, Bañon S, Garrido MD (2011b) Effect of supplementing ewes' diet with thyme (*Thymus zygis* ssp. gracilis) leaves on the lipid oxidation of cooked lamb meat. Food Chem. 125(4): 1147-1152.

[70] Nieto G, Estrada M, Jordán MJ, Garrido MD, Bañon S (2011c) Effects in ewe diet of rosemary by-product on lipid oxidation and the eating quality of cooked lamb under retail display. Food Chem. 124(4): 1423-1429.

[71] Nieto G, Bañon S, Garrido MD (2012b) Administration of distillate thyme leaves into the diet of Segureña ewes: effect on lamb meat quality. *Animal,* In press.

Section 4

Lipid Peroxidation in Health and Disease

Chapter 13

Lipid Peroxidation After Ionizing Irradiation Leads to Apoptosis and Autophagy

Juliann G. Kiang, Risaku Fukumoto and Nikolai V. Gorbunov

Additional information is available at the end of the chapter

http://dx.doi.org/10.5772/48189

1. Introduction

A living cell is a dynamic biological system composed primarily of nucleic acids, carbohydrates, lipids, and proteins that structurally and functionally interact with many other molecules--organic and inorganic--to carry out normal cell metabolism. Exposure of a cell to radiation can both directly and indirectly alter molecules within the cell to affect cell viability. Radiation energy absorbed by tissues and fluids is dissipated by the radiolysis of water molecules and biomolecules [1-3]. These reactions result in redox-reactive products such as hydroxyl radical (HO*), hydrogen peroxide (H_2O_2), hydrated electron (e^-_{aq}), and an array of biomolecule-derived carbon-, oxygen-, sulfur-, and nitrogen-centered radicals (i.e., RC*, RO*, RS*, and RN*) that can in turn lead to the formation of organic peroxides and superoxide anion radicals (O_2*-) in the presence of molecular oxygen [3, 4].

While the strongly electrophilic HO* has the capacity to damage molecules like polypeptides, amino acids, and polyunsaturated fatty acids (PUFAs) directly, the alterations caused by peroxide and superoxide radicals are usually produced indirectly via Fenton-type reactions [1-3, 5]. It is the interaction of these radiation-induced free radicals with important biomolecules within the cell that is the basis of the cellular sensitivity to radiation.

Free radical reactions generated after short-term radiation exposure are often quickly terminated by antiradical/antioxidant redox cycles. However, in certain cases and certain cellular environments, free radicals can initiate self-propagating chain reactions that can magnify the effects of the initial oxidations induced by radiation, leading to major disruptions that affect basic cell function [4, 6-9]. This especially true in cellular structures rich in poly-unsaturated fatty acids (PUFAs), such as cellular membranes, where radiation exposure can induce lipid peroxidation chain reactions that trigger reactions both within and beyond of the membrane [see 10-13]. Although the chemical mechanisms by which radiation induces the formation of redox-reactive products in cells are fairly well-understood, many of the mechanisms by which they impact specific cellular processes to produce radiation injury are only beginning to be elucidated.

Lipid peroxidation is a process in which free radicals remove electrons from lipids, producing reactive intermediates that can undergo further reaction. Cellular membranes, because of their high lipid content, are especially susceptible to damage. Because lipid peroxidation reactions can alter the structure and function of critical membrane lipids, they can lead to cell injury and cell death.

Lipid peroxidation reactions take place in three steps. The first step is initiation, which produces a fatty acid radical. In polyunsaturated fatty acids, methylene groups next to carbon-carbon double bonds possess especially reactive hydrogen atoms. Lipid peroxidation is most commonly initiated when reactive oxygen species (ROS) such as OH· and HO_2 interact with a reactive methylene hydrogen atom to produce water and a fatty acid radical:

$$\text{PUFA} + \text{OH}\cdot \rightarrow \text{PUFA}^* + H_2O \tag{1}$$

The second step is propagation. Molecular oxygen reacts with the unstable lipid radical to produce a lipid peroxyl radical. This radical is also unstable; it reacts readily with an unsaturated fatty acid to regenerate a new fatty acid radical as well as a lipid peroxide. The generation of the new fatty acid radical in this step reinitiates the cycle. For this reason, this series of reactions is referred to as a lipid peroxidation chain reaction.

$$\text{PUFA}^* + O_2 \rightarrow \text{PUFAOO}^* \tag{2}$$

$$\text{PUFAOO}^* \rightarrow \text{Fenton} \rightarrow \text{HO-PUFA}^* + OH^- \tag{3}$$

The third step is termination. As the chain reaction continues, an increasing concentration of lipid radicals are produced, thereby increasing the probability two lipid radicals will react with each other, which can produce a non-radical species. This constitutes a chain-breaking step, which in combination with the activity of natural radical scavenging molecules in cellular systems ultimately quells the chain reaction.

There are four types of radiation capable of ionizing, and thus damaging, target molecules via mechanisms such as lipid peroxidation: alpha particles, beta particles, gamma rays, and neutrons. Gamma radiation has been shown to increase lipid peroxidation in a variety of biological systems. After gamma radiation exposure, levels of the lipid peroxidation indicator MDA have been shown to increase in brain [14], liver [15-19], lens [20], serum [21], and skeletal muscle [22] of rats as well as in bacteria [23].

The possibility of exposure to radiation doses significant enough to cause lipid peroxidation leading to tissue injury is more than a hypothetical hazard. It is estimated that more than 50% of cancer patients receive radiotherapy at some point during the course of their disease, and these exposures can injure normal tissues [24, 25]. Potentially harmful radiation exposures after a nuclear power plant accident are also possible, either as a plant worker or a citizen who lives in or moves through fallout areas. Such exposures are unlikely; however, as the Fukushima, Japan, reactor incident showed, the threat is still very real. The threat of general exposure to radiation via a nuclear or radionuclide-based terrorist device is unfortunately also a real-world scenario. In order to provide public health protection in such cases, it is important to understand more about how radiation affects cells and tissues and learn how to ameliorate radiation injury.

Living organisms have evolved a variety of free radical-scavenging molecules to help protect the cell membrane from damage. Endogenous antioxidants include the enzymes superoxide dismutase (SOD), catalase, and peroxidase. Other antioxidants such as exogenously derived vitamin E can also play a role.

Free radical-mediated lipid peroxidation is harmful not only because damaged lipids disrupt membrane structure and function, but also because the process produces potentially mutagenic and carcinogenic byproducts [26]. One such product is the highly reactive carbonyl compound, malondialdehyde (MDA), which can react with deoxyadenosine and deoxyguanosine in DNA to form DNA adducts, primarily pyrimido[1,2-a]purin-10(3H)-one (M_1G) [23, 26]. M_1G toxicity has been demonstrated in experiments with glutathione peroxidase 4 knockout mice, which have a diminished capacity to protect themselves from lipid peroxidation and thus M_1G toxicity; mice with this lethal phenotype do not survive past embryonic day 8 [27].

Lipid peroxidation reactions can occur at the both the cell membrane and mitochondria membranes, and either can subsequently trigger cell death through apoptosis and/or autophagy [28]. Apoptosis is typically executed by caspases, which are cysteine aspartic acid-specific proteases that cleave an amino acid sequence-motif located N-terminal to a specific aspartic acid residue. Caspases can be broadly divided into two functional subgroups: (1) those activated during apoptosis (caspases -2, -3, -6, -7, -8, -9, and -10) ; and (2) those implicated in the processing of proinflammatory cytokines during responses (caspases -1, -4, and -5) [29-31].

The apoptosis process can follow two pathways, extrinsic and intrinsic. The extrinsic pathway involves the activation of pro-caspase-8 by external, typically molecular signals such as FAS ligand binding to FAS or TNF binding to TNF receptors on the cell membrane. Through a series of steps pro-caspase-8 becomes activated caspase-8, which then acts on the mitochondrial membrane either directly or via the activation of caspase-3. Subsequent steps then follow those described for the intrinsic pathway. The intrinsic pathway involves mitochondria directly. When mitochondria are under stress, cytochrome c is released from the mitochondria. The released cytochrome c conjugates with Apf-1and caspase-9 tin the cytoplasm to form apoptosomes, which in turn activate caspase-3 and -7. Activated caspase-3 then activates caspase-2, -6, -8, and -10. It should be noted that caspase-independent apoptosis pathways also exist, such as the intrinsic, apoptosis inducible factor (AIF) pathway and the intrinsic, mitochondria-derived endonuclease G-related pathway [29-31].

Autophagy is a catabolic process involving the bulk degradation of cellular constituents in lysosomes [32]. Autophagy under normal conditions is a cytoprotective process involved in tissue remodeling, recovery, and rejuvenation. Autophagy dynamics in mammalian cells are well-described in several recent reviews [33-37]. The autophagic pathway is complex. To date there are over thirty genes identified in mammalian cells as regulators of various steps of autophagy. The misregulation of the autolysosomal pathway during autophagy can eventually cause cell death either by triggering apoptosis in apoptosis-sensitive cells or as a result of destructive self-digestion [38]. Light chain 3 (LC3) is a protein involved in the formation of autophagosomes in mammalian cells that serves as a biomarker for occurrence of autophagy [13].

Given the widespread potential for cellular damage from lipid peroxidation after exposure to ionizing radiation, we hypothesize that ionizing radiation-induced lipid peroxidation leads to caspase-mediated apoptotic cell death and LC-3-mediated autophagic cell death. The objective of this current chapter is to provide evidence of this hypothesis. We used human Jurkat T cultured cells and mouse ileum to investigate the relationship between lipid peroxidation and cell death both *in vitro* and *in vivo.*

2. Experimental procedures and technical approach

2.1. Cell culture

Human Jurkat T cells (American Type Cell Collection, Rockville, MD, USA) were grown in 75 cm^2 tissue culture flasks (Costar, Cambridge, MA, USA) containing RPMI 1640 medium supplemented with 0.03% glutamine, 4.5 g/L glucose, 25 mM HEPES, 10% fetal bovine serum, penicillin (50 μg/mL), and streptomycin (50 U/mL) (Gibco/BRL, Gaithersburg, MD, USA). Cells were incubated in a 5% CO2 atmosphere at 37 ^{o}C and fed every 3-4 d.

2.2. Animal

CD2F1 male mice (25-30 g) were purchased from Harlan Laboratories (Indianapolis, IN). All mice were randomly assigned to experimental groups. Eight mice were housed per filter-topped polycarbonate cage (MicroIsolator) in conventional holding rooms. Rooms were provided 20 changes per hour of 100% fresh air, conditioned to 72 ± 2 ^{o}F and a relative humidity of 50 ± 20%. Mice were maintained on a 12-h light/dark, full-spectrum light cycle with no twilight. Research was conducted in a facility accredited by the Association for Assessment and Accreditation of Laboratory Animal Care-International (AAALAC-I). All procedures involving animals were reviewed and approved by the Armed Forces Radiobiology Research Institute (AFRRI) Institutional Animal Care and Use Committee.

2.3. Radiation exposure

2.3.1. Cell culture

Radiation exposures were conducted using AFRRI's ^{60}Co source. Cells suspended in growth medium were placed in 6-well plates (5 x 10^6 cells/ml; 2 ml per well) and exposed to ^{60}Co gamma-radiation at various total doses using a dose rate of 0.6 Gy/min. Cells were then returned to the incubator in a 5% CO_2 atmosphere at 37 ^{o}C for the specified time.

2.3.2. Animals

For survival experiments, mice (n=16 per group) received 9.25 Gy (equivalent to $LD_{90/30}$) of total-body ^{60}Co gamma-photon radiation administered at a dose rate of 0.6 Gy/min. Sham-treated mice were handled identically but received no radiation. After treatment, mice were returned to their original cages and survival was monitored for 30 d. Body weight and facial dropsy were assessed. Mean survival times (ST_{50}) were observed. The moribund mice found during the observation period were euthanized in accordance with recommendations [39, 40] and guidelines [41]. For mechanistic experiments, mice also received 9.25 Gy (n=6 per group). At specified time points after irradiation, mice were euthanized in accordance with recommendations [39, 40] and guidelines [41]. Interested tissues were harvested 1 and 7 d after irradiation and stored at -80°C until use for biochemical assays and western blots. Ileum was also prepared for immunofluorescence assessment.

2.4. Western blots

To investigate amounts of caspase-3 and LC3 proteins, ileum was minced, mixed in 100 μL Na^+ Hanks' solution containing protease inhibitors, sonicated, and centrifuged at 8000 x g for 10 min. The supernatant was collected and total protein was determined with Bio-Rad reagent (Bio-Rad, Richmond, CA, USA). Aliquots containing 20 μg of protein in tris buffer (pH=6.8) containing 1% sodium dodecyl sulfate (SDS) and 1% 2-mercaptoethanol were resolved on SDS-polyacrylamide slab gels (Novex precast 4-20 % gel; Invitrogen, Grand Island, NY, USA). After electrophoresis, proteins were blotted onto a PDVF nitrocellulose membrane (type NC, 0.45 μm; Invitrogen), using a Novex blotting apparatus and the manufacturer's protocol. After blocking the nitrocellulose membrane by incubation in tris-buffered saline-0.5% tween20 (TBST) containing 3% nonfat dried milk for 90 min at room temperature, the blot was incubated for 60 min at room temperature with monoclonal antibodies directed against caspase-3 and LC3 at a concentration of 1 μg/ml in TBST - 3% dry milk. The blot was then washed 3 times (10 min each) with TBST before incubating the blot for 60 min at room temperature with a 1000X dilution of species-specific IgG peroxidase conjugate (Santa Cruz Biotechnology) in TBST. The blot was washed 6 times (5 min each) in TBST before detection of peroxidase activity using the Enhanced Chemiluminescence Plus kit (Amersham Life Science Inc., Arlington Heights, IL, USA). IgG levels were not altered by radiation; we therefore used IgG as a control for protein loading. Protein bands of interest were quantitated densitometrically and normalized to IgG.

2.5. Nitric oxide measurements

Nitric oxide (NO) production was measured under acidic conditions as nitrite, using a commercial kit (Biomedical Research Service, School of Medicine and Biomedical Sciences, State University of New York at Buffalo, NY, USA; www.bmrservice.com).

2.6. Lipid peroxidation measurements

Malondialdehyde (MDA), a lipid peroxidation end product, was measured colorimetrically using a commercial lipid peroxidation assay kit (CalBiochem, San Diego, CA).

2.7. Detection and analysis of caspase-3/7 activity by confocal microscopy

The Magic Red® Caspase Detection kit (MP Biomedicals; Solon, OH, USA) was used for the detection of caspase-3/7 activity, following the manufacturer's protocol. Briefly, about 2 x 10^5 cells were stained in the presence of up to 300 μl of OPTI-MEM I medium (Invitrogen). Cells were seeded onto #1 borosilicate glass slides with 4-well chambers (Fisher Science Education, Hanover Park, IL). An LSM 5 PASCAL Zeiss laser scanning confocal microscope (Carl Zeiss MicroImaging; Thornwood, NY, USA) with a 100×/1.3 NA Plan Apochromat oil objective was used to scan the signals. Each resulting image was provided with a simultaneous scan of differential interference contrast (DIC).

2.8. Immunofluorescence staining and image analysis

Small intestine specimens (5 per each of animal groups) collected at necropsy were processed for the immunofluorescence analysis and analyzed using fluorescence confocal microscopy [42]. Donkey normal serum and antibody were diluted in phosphate buffered saline (PBS) containing 0.5% BSA and 0.15% glycine. Any nonspecific binding was blocked by incubating the samples with purified donkey normal serum (Santa Cruz Biotechnology, Inc., Santa Cruz, CA, USA) diluted 1:20. The primary antibodies were raised against CD15 (mouse monoclonal biotin conjugated IgM from eBioscience), MAP LC3and AD4 (vendors indicated above). This was followed by incubation with secondary fluorochrome-conjugated antibody and/or streptavidin-AlexaFluor 610 conjugate (Molecular Probes, Inc., Eugene, OR, USA), and with Heochst 33342 (Molecular Probes) diluted 1:3000. The secondary antibodies used were AlexaFluor 488 and AlexaFluor 594 conjugated donkey IgG (Molecular Probes Inc.) Negative controls for nonspecific binding included normal goat serum without primary antibody or with secondary antibody alone. Five confocal fluorescence and DIC images of crypts (per specimen) were captured with Zeiss LSM 7100 microscope. Immunofluorescence image analysis was conducted as described previously [43]. The index of spatial correlation (r) of proteins was determined by multiple pixel analysis for pairwise signal interaction of green and red channels. Paneth cell identification was conducted by i) their spatial localization in crypts; ii) presence of immunoreactivity to CD15 and AD4, which are specific to this epithelial phenotype; iii) spatial appearance of the immunoreactivity to CD15, AD4, and the FISH reactivity to AD4 mRNA (see above) in morphologically identified Paneth cells.

2.9. Solutions

Na^+ Hanks' solution contained in mM: 145 NaCl, 4.5 KCl, 1 .3 $MgCl_2$, 1.6 $CaCl_2$, and 10 HEPES (pH 7.40 at 24 °C). Na^+ Hanks' stop buffer contained in mM: 50 tris-HCl, 1% NP-40, 0 .25% Na^+-deoxycholate, 150 NaCl, 1 EDTA, 1 phenylmethanesulfonyl fluoride, 1 Na_3VO_4, 1 NaF, along with aprotinin, leupeptin, and pepstatin (10 μg/mL each). Na^+ Hanks' wash buffer contained in mM: 1 EDTA, 1 phenylmethanesulfonyl fluoride, 1 DTT, 1 Na_3VO_4, 1 NaF, along with aprotinin, leupeptin, and pepstatin (10 μg/mL each).

2.10. Statistical analysis

Results represent the mean ± s.e.m. One-way ANOVA, Studentized-range test, Bonferroni's inequality, and Student's t-test were used for comparison of groups with 5% as a significant level.

3. Response of human T cells to irradiation

3.1. Gamma radiation increased lipid peroxidation production

Cells were irradiated with 2, 4, 6, or 8 Gy; the lipid peroxidation marker MDA was measured in these cells at 4, 24, 48, and 72 h postirradiation. Figure 1 shows MDA levels increased in a radiation dose- and postirradiation time-dependent manner in cells receiving 2, 4, and 6 Gy (Fig. 1A-C). Cells receiving 8 Gy (Fig. 1D) had lipid peroxidation levels above the baseline at all times tested. But in cells receiving 6 Gy and 8 Gy, by 24 h postirradiation the levels were significantly lower than those observed in cells receiving 4 Gy (Fig. 1C vs. 1B and Fig. 1D vs. 1B). This observation may be a reflection of the drop in cell viability previously observed in cells after doses greater than 4 Gy.

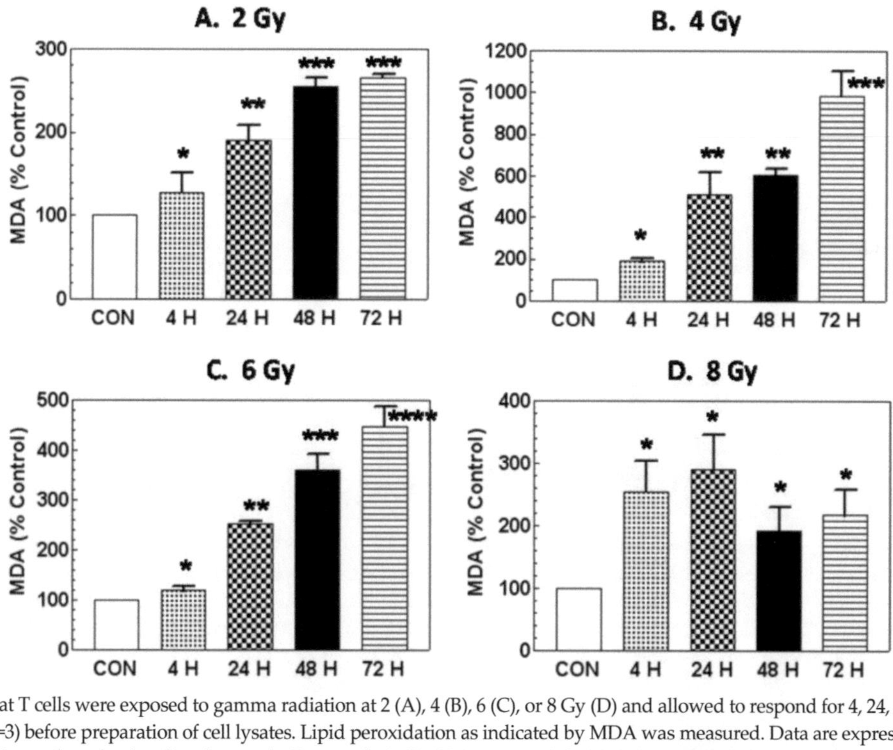

Jurkat T cells were exposed to gamma radiation at 2 (A), 4 (B), 6 (C), or 8 Gy (D) and allowed to respond for 4, 24, 48, or 72 h (n=3) before preparation of cell lysates. Lipid peroxidation as indicated by MDA was measured. Data are expressed relative to that of unirradiated controls. For panels A: *P<0.05 vs. control, 24 H, 48 H, and 72 H; **P<0.05 vs. control, 4 H, 48 H, and 72 H; ***P<0.05 vs. control, 4 H, and 24 H. For Panel B: *P<0.05 vs. control, 24 H, 48 H, and 72 H; **P<0.05 vs. control, 4 H, and 72 H; ***P<0.05 vs. control, 4 H, 24 H, and 48 H. For panels C: *P<0.05 vs. control, 24 H, 48 H, and 72 H; **P<0.05 vs. control, 4 H, 48 H, and 72 H; ***P<0.05 vs. control, 4 H, 24 H, and 72 H; ****P<0.05 vs. control, 4 H, 24 H, and 48 H. For panel D: *P<0.05 vs. control, determined by one way ANOVA and studentized-range test. CON: unirradiated controls

Figure 1. Gamma radiation increased lipid peroxidation in T cells.

3.2. Gamma radiation increased NO production

Because NO is known to react with O_{2}- to form peroxynitrite (ONOO-) that can oncrease lipid peroxidation [23], we measured NO production levels in irradiated Jurkat cells. Figure 2 shows NO production in irradiated cells. By 4 h postirradiation cells receiving 2, 4, or 6 Gy showed NO levels statistically lower than non-irradiated controls, but cells receiving 8 Gy showed an increase. By 24 and 48 h, all irradiated cells exhibited increased NO production, generally in a radiation dose-dependent manner. By 72 h postirradiation NO level returned to baseline in cells receiving 4, or 8 Gy, while NO levels in cells receiving 2 or 6 Gy remained above the baseline and had dropped below baseline, respectively.

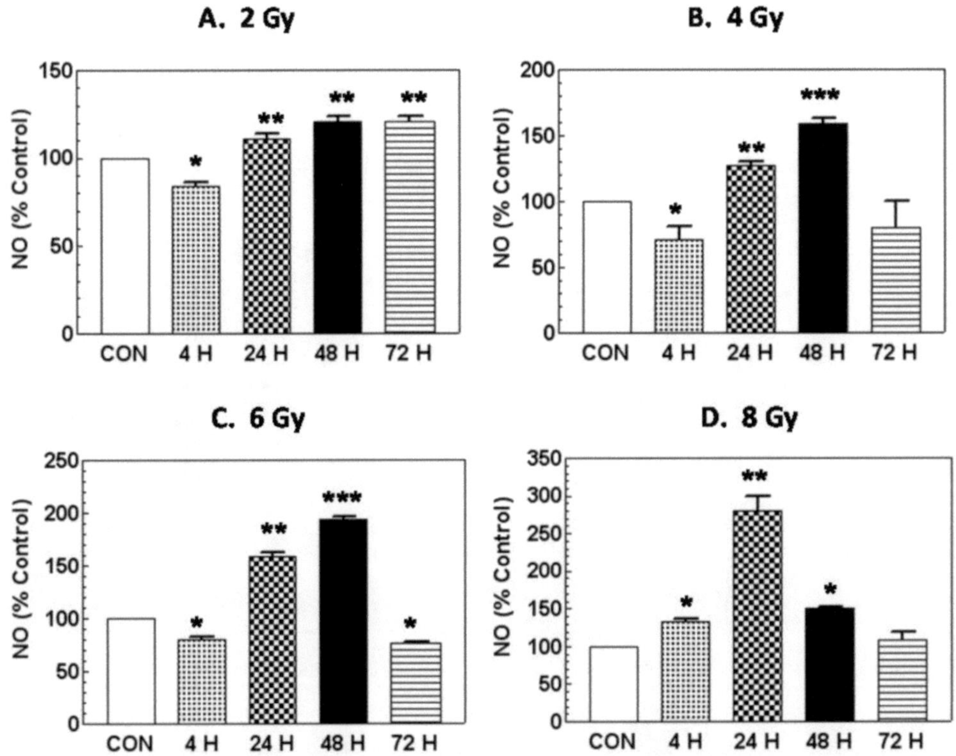

Jurkat T cells were exposed to gamma radiation at 2 (A), 4 (B), 6 (C), or 8 Gy (D) and allowed to respond for 4, 24, 48, or 72 h (n=3) before preparation of cell lysates. NO production was measured. Data are expressed relative to that of unirradiated controls. For panel A: *P<0.05 vs. control, 24 H, 48 H, and 72 H; **P<0.05 vs. control and 4 H. For Panel B: *P<0.05 vs. control, 24 H, and 48 H; **P<0.05 vs. control, 4 H, 48 H, and 72 H; ***P<0.05 vs. control, 4 H, 24 H, and 72 H. For panel C: *P<0.05 vs. control, 24 H, and 48 H; **P<0.05 vs. control, 4 H, 48 H, and 72 H. ***P<0.05 vs. control, 4 H, 24 H, and 72 H. For panel D: *P<0.05 vs. control, 24 H, and 72 H; **P<0.05 vs. control, 4 H, 48 H, and 72 H, determined by one way ANOVA and studentized-range test. CON: unirradiated controls

Figure 2. Gamma radiation increased NO production in T cells.

3.3. Gamma radiation increased apoptosis

Lipid peroxidation occurring within the cell and mitochondrial membranes can trigger apoptosis [28]. Because increased caspase-3 and -7 are indicators of cells undergoing apoptosis [16-18], we measured caspase-3 and -7 in irradiated Jurkat cells. As shown in Fig. 3, cells receiving 8 Gy displayed significantly increased immunofluorescence of caspase-3/-7, compared to non-irradiated cells. The percentages of non-irradiated cells and irradiated cells presenting the caspase-3/-7 immunofluorescence were 23 and 85 %, respectively, suggesting that gamma irradiation increases apoptosis.

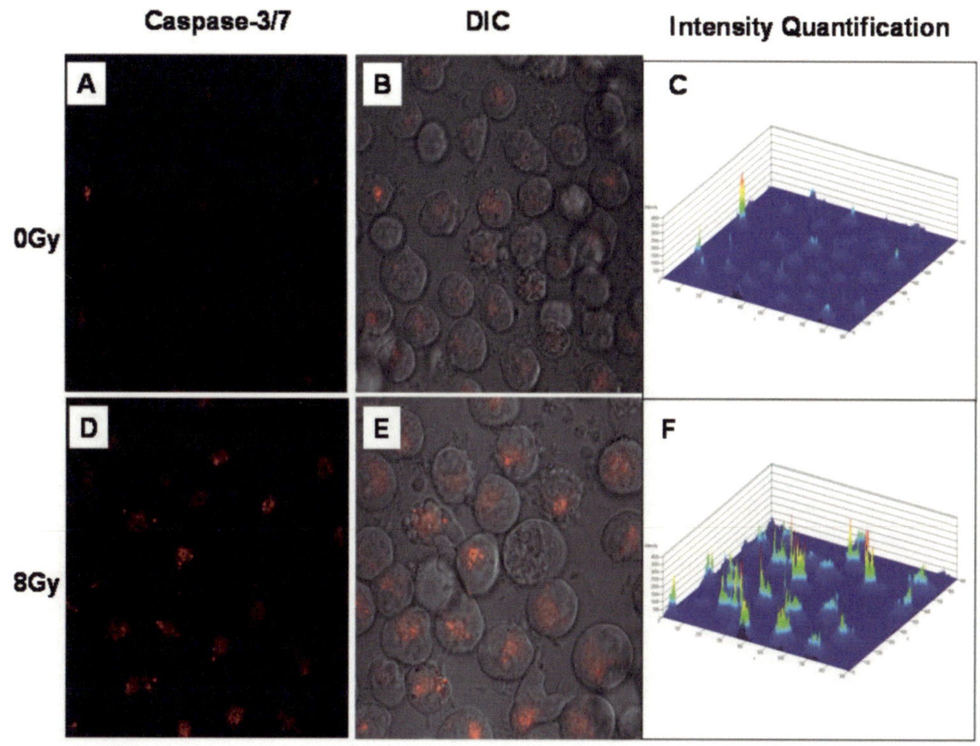

Freshly isolated human Jurkat T cells were counted, seeded, and allowed to grow for 24h. Cells were then irradiated at 8 Gy and returned to incubator. Live cells were studied 24 h postirradiation for internal caspase-3/7 activities by Magic Red® staining.

Figure 3. Gamma radiation increased apoptosis in T cells.

4. Response of mice to irradiation

The organs most sensitive to radiation are the hematopoietic, lymphoid, gastrointestinal, reproductive, vascular, and cutaneous systems [44]. We used mice irradiated with 9.25 Gy ^{60}Co-gamma photons to determine if events similar to those observed *in vitro* also occurred *in vivo*. Thirty-day survival, body weight, facial dropsy, and ileum were also assessed.

4.1. Gamma radiation decreased mouse survival and body weight

Irradiated mice first demonstrated mortality 9 d after irradiation and showed no survival 23 d after irradiation. The mean survival time (ST_{50}) was 12 d. All non-irradiated mice survived

(Fig. 4A). Irradiated mice demonstrated significant body weight loss 10 d after irradiation and lost over 20% by the end of the experiment; non-irradiated mice gained weight daily. The rate of weight change was -0.6 g/d for irradiated mice and 0.17 g/d for non-irradiated mice. The observations are in agreement with those obtained in previous experiments using irradiated B6D2F1/J mice [45]

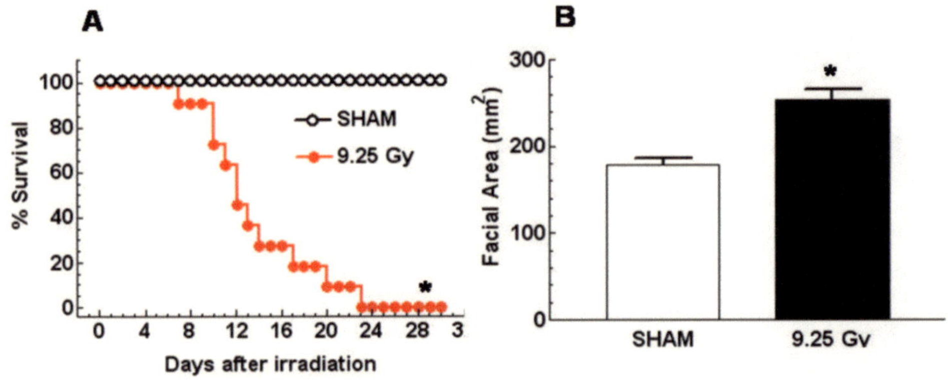

Mice (n=16 per group) received 9.25 Gy. (A) Radiation significantly reduced 30-d survival after irradiation. *P<0.01 vs. sham group, determined by one-way ANOVA and Student's t-test. (B) Radiation induced facial dropsy. Facial dropsy was assessed by measuring increase in facial area. Facial area was approximated by multiplying width between outer edges of ears and distance from ear-to-ear midpoint to tip of snout and then dividing by 2 (area of a triangle). Facial dropsy in each mouse was calculated approximately 10 d after 9.25 Gy (9-12 d, point of maximal swelling) and averaged. *P<0.05 vs. sham group, determined by Student's t-test.

Figure 4. Gamma radiation decreased survival and increased facial dropsy.

4.2. Gamma radiation induced facial dropsy

Irradiated mice began to show facial dropsy around 10 d postirradiation (Fig. 4B). The overall increase in dropsy was approximately 43%. Facial dropsy did not occur in previous experiments using irradiated B6D2F1/J mice [44] or in humans [46], suggesting the response is mouse strain- and species-specific.

4.3. Gamma radiation induced lipid peroxidation and NO production

Ileum lysate was obtained for MDA measurement from mice exposed to 9.25 Gy. As shown in Fig. 5, irradiation increased MDA levels (Fig. 5A) and NO production (Fig. 5B). The results were consistent with *in vitro* findings using human T cells.

4.4. Gamma radiation induced increases in caspase-3 and LC3

To measure apoptosis in ileum from irradiated mice, we performed an immunoblot analysis of ileum caspase-3, a biomarker for apoptosis [29-31]. Radiation induced an approximate 2.5-fold increase in caspase-3 (Fig. 6A-B). The result was consistent with the *in vitro* findings with Human T cells.

To measure autophagy in ileum from irradiated mice, we performed an immunoblot analysis of ileum LC3 protein, a biomarker for Autophagy [13]. Radiation induced an approximate 3.5-fold increase in LC3 (Fig. 6A and C).

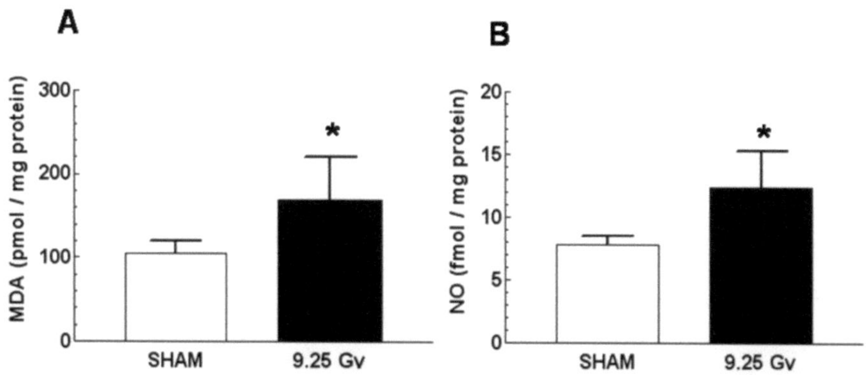

Mice (n=6 per group) received 9.25 Gy. Ileum tissues were collected 1 d postirradiation and cell lysates prepared. (A) Lipid peroxidation as indicated by MDA was measured. (B) NO production was measured. *P<0.05 vs. sham group, determined by Student's t-test.

Figure 5. Gamma increased lipid peroxidation and NO production in ileum.

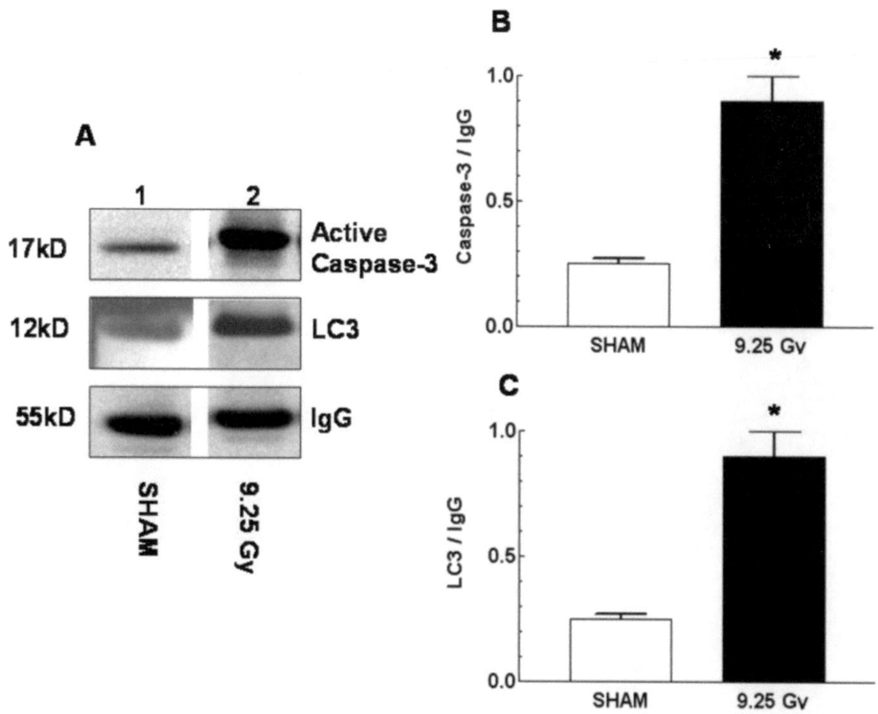

Mice (n=6 per group) received 9.25 Gy. Ileum tissues were collected 1 d postirradiation and cell lysates prepared. (A) Representative Western blots. (B) Caspase-3 and LC3 were quantitated densitometrically and normalized to IgG. *P<0.05 vs. sham, determined by Student's t-test.

Figure 6. Gamma radiation increased caspase-3 and LC3 in ileum.

4.5. Gamma radiation induced apoptosis in mouse ileal villi

The immunofluoresence images in Fig. 7 show that there was little immunofluorescence present in ileal villi of non-irradiated mice (Fig. A and C), whereas there was a significant increase in immunofluorescence in ileal villi of irradiated mice (Fig. B and D). It is known that ileal epithelial cells regularly slough off after undergoing apoptosis and are then replenished within 7 d [47]. It is not clear if irradiation accelerated the rate of apoptosis, but ionizing radiation-induced increases in apoptosis were observed in these studies (Fig. B vs. A). Similar results were observed in earlier studies using irradiated B6D2F1/J mice [48].

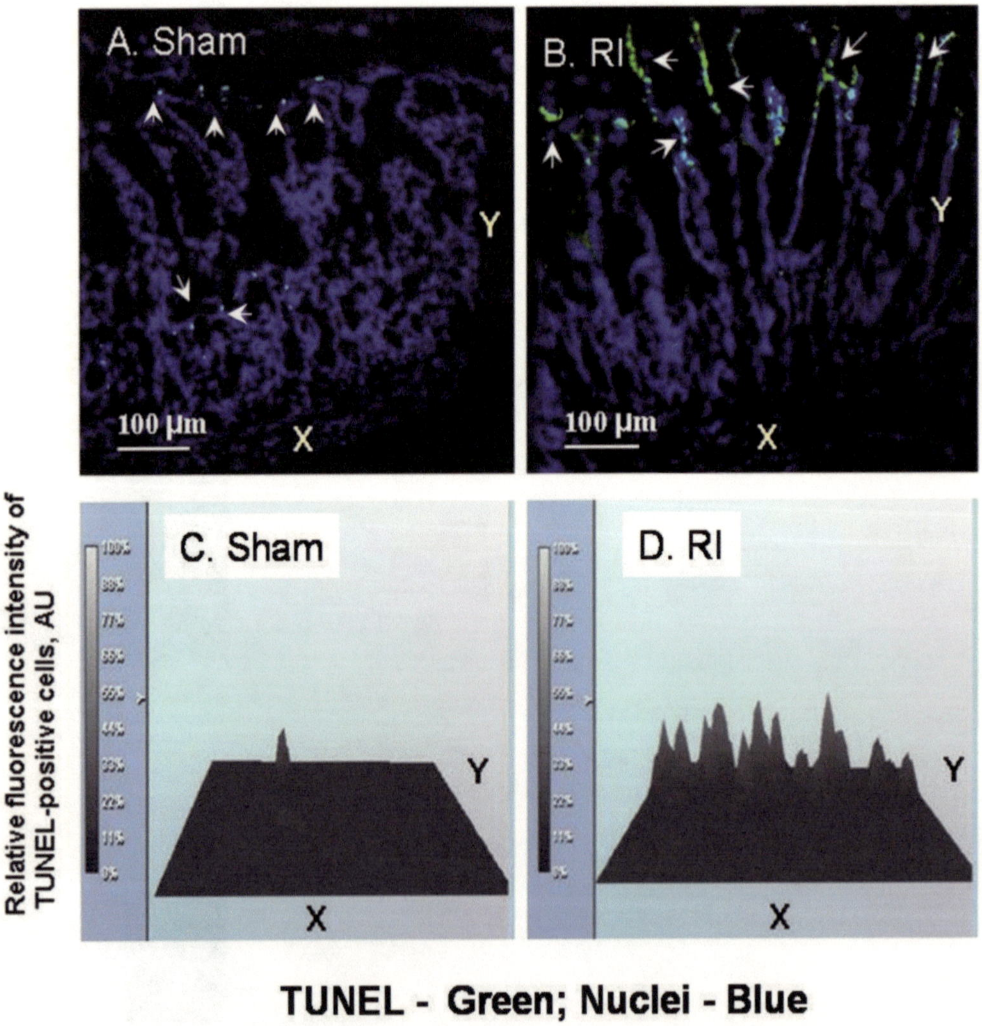

Mice (n=6 per group) received 9.25 Gy. Ileum tissues were collected 7 d postirradiation and slide preparations were stained using TUNEL assay to detect apoptosis. (A-B) Confocal microscopy fluorescent images: green fluorescence indicates apoptotic cells; blue Hoechst 33342 fluorescence indicates all nuclei. (C-D) Quantitation of fluorescence intensities.

Figure 7. Gamma radiation induced apoptosis in ileum.

4.6. Gamma radiation induced autophagy in mouse ileal crypts

In ileum, stem cells anchored in the crypts gives rise to proliferating progenitor cells that exit the cell cycle as they migrate and differentiate into 4 different cell types, including enterocytes, goblet cells, enteroendocrine cells, and Paneth cells [see reviews 49, 50]. Paneth cells produce defensins to protect against bacterial entry via the gut barrier [49, 50]. However, stem cells have also been shown to inhibit bacterial entry [51]. The immunofluoresence images in Fig. 8 show that there was little immunofluorescence present in ileal crypts of non-irradiated mice (A and C), whereas there was significant increase in immunofluorescence in ileal crypts of irradiated mice (B and D). Autophagy occurred in both stem cells and Paneth cells (using CD-15 as a biomarker). Ionizing radiation-induced increases in autophagy (B vs. A) have also been observed in B6D2F1/J mice [44].

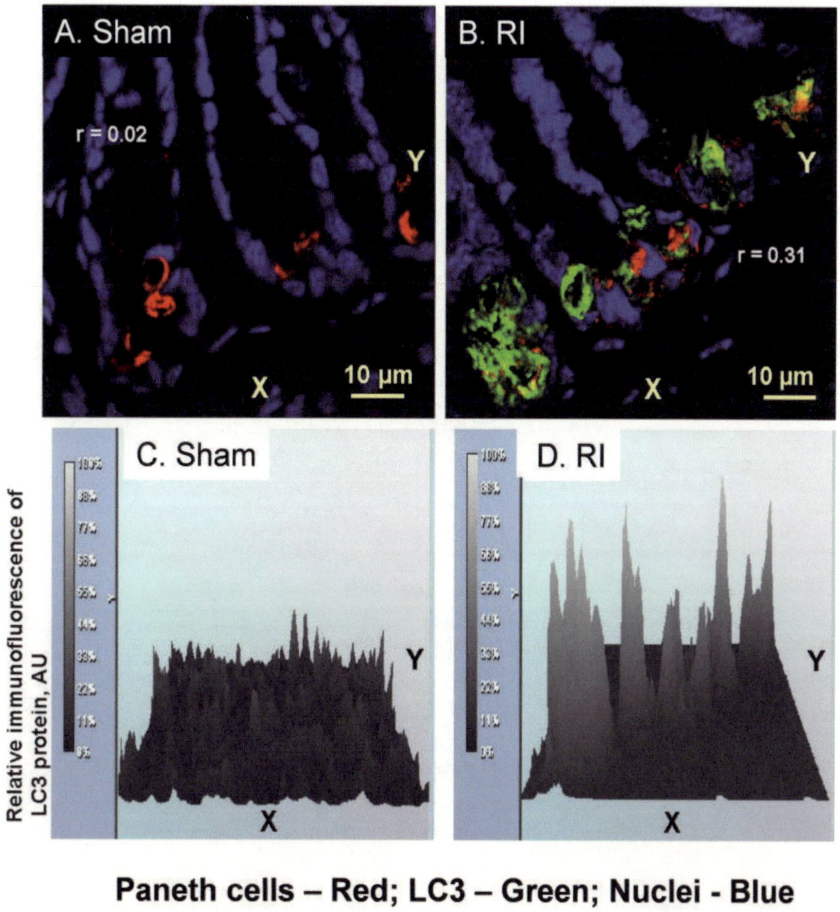

Mice (n=6 per group) received 9.25 Gy irradiation. Ileum tissues were collected 7 d postirradiation and slide preparations were stained with anti-LC3 antibody to detect autophagy indicator LC-3. (A-B) Confocal microscopy fluorescent images: green fluorescence indicates anti-LC3 antibody; blue Hoechst 33342 fluorescence indicates all nuclei. (C-D) Quantitation of fluorescence intensities.

Figure 8. Gamma radiation induced autophagy in ileum.

5. Conclusion

The complexity of the cellular response to ionizing radiation complicates efforts to design approaches to treat or prevent injury resulting from ionizing radiation. Ionizing radiation activates many signal transduction pathways [44, 52], including the one involving lipid peroxidation. It is known that Bcl-2 (an anti-apoptotic protein) decreases lipid peroxidation [53]. Although in cultured cells, silencing of the iNOS gene by iNOS siRNA inhibits lipid peroxidation and the subsequent production of apoptosis-related proteins [12], ionizing radiation increases lipid peroxidation even in iNOS knockout mice (Lu and Kiang, unpublished data). This suggests that radiation-induced activation of iNOS pathway is not all for its occurrence. Nevertheless, total body ionizing irradiation causes lipid peroxidation, caspase-3 activation, LC3 increases, and apoptosis and autophagy in ileum, suggesting the ileal cell death may contribute to ionizing radiation-induced mortality and body weight loss (Fig. 4). The mechanism underlying radiation-induced facial dropsy is unclear and warrants further investigation.

Figure 9 shows a schematic representation of lipid peroxidation-mediated induction of apoptosis and autophagy after exposure to gamma-radiation. Mitochondrial membrane lipid peroxidation is induced either directly by radiation or indirectly by increased NO production as a result of radiation-induced iNOS upregulation. Lipid peroxidation in the mitochondrial membrane leads to release of cytochrome c into the cytoplasm. Cytochrome c complexes with Apf-1 and seven molecules of caspase-9 to form an apoptosome. Apoptosomes then activate caspase-3 and -7. Active caspase-3 subsequently activates other caspases in the cytoplasm, which leads to apoptosis. It is not known what role plasma membrane lipid peroxidation plays in the apoptosis and autophagy processes. Lipid peroxidation also increases LC3 to cause autophagy, a process whose poorly understood molecular mechanisms warrant further investigation.

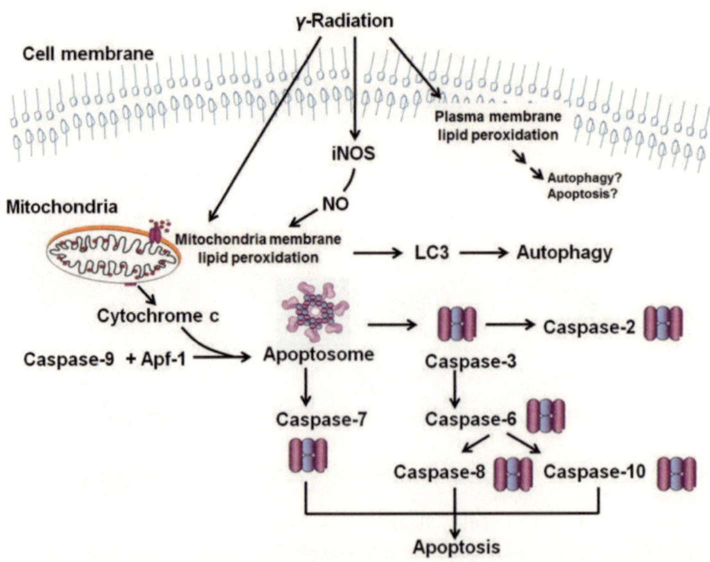

Figure 9. Schematic representation of lipid peroxidation-mediated induction of apoptosis and autophagy after exposure to gamma-radiation. iNOS: inducible nitric oxide synthase; NO: nitric oxide; LC3: light chain 3; Apf-1: ATP-dependent proteolysis factor 1

6. Perspective

Cell death resulting from ionizing radiation is a common scenario in the clinical practice of medicine and probably occurs in virtually all organ systems. Its hallmarks are relatively consistent across species and probably organ systems. Blockade of lipid peroxidation could be a useful approach to prevent radiation injury. In the murine model [14, 17-19, 21], radiation-induced lipid peroxidation occurs in fetal brain cultures [14], liver [17-19], and serum [21] as well as in the ileum, as reported in this chapter. A wide range of agents has been shown to effectively inhibit radiation-induced lipid peroxidation, including N-acetylcysteine [15], melatonin [16], vitamin E [20, 22], MnSOD-Plasmid Liposomes [14], leaf extract of Moringa oleifera [17], cinnamon extract [18], green tea polyphenol [21], hesperidin [19], and 17-DMAG [Lu and Kiang, unpublished data]. Since the response to radiation involves many signal transduction pathways, a combination of drugs targeting different signaling pathways may be a useful approach to address the difficult problem of protecting from ionizing radiation injury.

Author details

Juliann G. Kiang*, Risaku Fukumoto and Nikolai V. Gorbunov
Radiation Combined Injury Program, Armed Forces Radiobiology Research Institute,
Uniformed Services University of the Health Sciences, Bethesda, Maryland,
USA

Acknowledgement

The authors thank HM1 Neil Agravante and Ms. Joan T. Smith for their technical support and Dr. David E. McClain for his scientific discussion and editorial assistance.

Grants

This work was supported by AFRRI Intramural RAB2CF (to JGK) and DTRA CBM.RAD.01.10.AR.010 (to JGK). There are no ethical and financial conflicts in the presented work.

Disclaimer

The opinions or assertions contained herein are the authors' private views and are not to be construed as official or reflecting the views of the United States Department of Defense.

7. References

[1] Jay A, LaVerne JA. OH Radicals and Oxidizing Products in the Gamma Radiolysis of Water. Radiation Research. 2000; 153:196-200.

* Corresponding Author

[2] Nakagawa K, Nishio T. Electron paramagnetic resonance investigation of sucrose irradiated with heavy ions. Radiat Res. 2000; 153:835-9.

[3] Nauser T, Koppenol WH, Gebicki JM. The kinetics of oxidation of GSH by protein radicals. Biochem J. 2005; 392(Pt 3):693-701.

[4] Gebicki JM, Nauser T, Domazou A, Steinmann D, Bounds PL, Koppenol WH. Reduction of protein radicals by GSH and ascorbate: potential biological significance. Amino Acids. 2010; 39:1131-7.

[5] Østdal H, Davies MJ, Andersen HJ. Reaction between protein radicals and other biomolecules. Free Radic Biol Med. 2002; 33:201-9.

[6] Tappel AL. Vitamin E and selenium protection from in vivo lipid peroxidation. Ann N Y Acad Sci. 1980; 355:18-31.

[7] Davies MJ, Forni LG, Willson RL. Vitamin E analogue Trolox C. E.s.r. and pulse-radiolysis studies of free-radical reactions. Biochem J. 1988; 255:513-22.

[8] Bild W, Ciobica A, Padurariu M, Bild V. The interdependence of the reactive species of oxygen, nitrogen, and carbon. J Physiol Biochem. 2012 Mar 29. [Epub ahead of print]

[9] Azzam EI, Jay-Gerin JP, Pain D. Ionizing radiation-induced metabolic oxidative stress and prolonged cell injury. Cancer Lett. 2011 Dec 17. [Epub ahead of print]

[10]. Leach JK, Van Tuyle G, Lin PS, Schmidt-Ullrich R, Mikkelsen RB. Ionizing radiation-induced, mitochondria-dependent generation of reactive oxygen/nitrogen. Cancer Res. 2001; 61:3894-901.

[11] Tyurina YY, Tyurin VA, Epperly MW, Greenberger JS, Kagan VE. Oxidative lipidomics of gamma-irradiation-induced intestinal injury. Free Radic Biol Med. 2008; 44:299-314.

[12] Kiang JG, Smith JA, and Agravante NG. Geldanamycin analog 17-DMAG inhibits iNOS and caspases in gamma irradiated human T cells. Radiat Res. 2009; 172: 321-30.

[13] Gorbunov NV, Kiang JG. Up-regulation of Autophagy in the Small Intestine Paneth Cell in Response to Total-Body γ-Irradiation. J Pathol. 2009; 219: 242-52.

[14] Bernard ME, Kim H, Rwigema JC, Epperly MW, Kelley EE, Murdoch GH, Dixon T, Wang H, Greenberger JS. Role of the esophageal vagus neural pathway in ionizing irradiation-induced seizures in nitric oxide synthase-1 homologous recombinant negative NOS1-/- mice. In Vivo. 2011; 25:861-9.

[15] Mansour HH, Hafez HF, Fahmy NM, Hanafi N. Protective effect of N-acetylcysteine against radiation induced DNA damage and hepatic toxicity in rats. Biochem Pharmacol. 2008; **75:**773-89.

[16] El-Missiry MA, Fayed TA, El-Sawy MR, El-Sayed AA. Ameliorative effect of melatonin against gamma-irradiation-induced oxidative stress and tissue injury. Ecotoxicol Environ Saf. 2007; 66:278-86.

[17] Sinha M, Das DK, Bhattacharjee S, Majumdar S, Dey S. Leaf extract of Moringa oleifera prevents ionizing radition-induced oxidative stress in mice. J Med Food. 2011; 14:1167-72.

[18] Azab KSh, Mostafa AH, Ali EM, Abdel-Aziz MA. Cinnamon extract ameliorates ionizing radiation-induced cellular injury in rats. Ecotoxicol Environ Saf. 2011; 74:2324-9.

[19] Kalpana KB, Devipriya N, Srinivasan M, Vishwanathan P, Thayalan K, Menon VP. Evaluating the radioprotective effect of hesperidin in the liver of Swiss albino mice. Eur J Pharmacol. 2011; 658:206-12.

[20] Karslioglu I, Ertekin MV, Kocer I. Protective role of intramuscularly administered vitamin E on the levels of lipid peroxidation and the activities of antioxidant enzymes in the lens of rats made cataractous with gamma radiation. Eur J Ophthalmol. 2004; 14:478-95.

[21] Hu Y, Guo DH, Liu P, Cao JJ, Wang YP, Yin J, Zhu Y, Rahman K. Bioactive components from the tea polyphenols influence on endogenous antioxidant defense system and modulate inflammatory cytokines after total-body irradiation in mice. Phytomedicine. 2011; 18:970-5.

[22] Yilmaz S, Yilmaz E. Effects of melatonin and vitamin E on oxidative-antioxidative status in rats exposed to irradiation. Toxicology 2006; 222:1-7.

[23] Zhou X, Taghizadeh K, Dedon PC. Chemical and biological evidence for base propenals as the major source of the endogenous M_1dG adduct in cellular DNA. J Biol Chem 2005; 280:25377-82.

[24] Bentzen SM. Preventing or reducing late side effects of radiation therapy: radiobiology meets molecular pathology. Nat Rev Cancer. 2006; 6:702-13.

[25] Buonanno M, de Toledo SM, Pain D, Azzam EI. Long-term consequence of radiation-induced bystander effects depend on radiation quality and dose and correlate with oxidative stress. Radiat Res. 2011; 175:405-15.

[26] Marnett LJ. Lipid peroxidation-DNA damage by malondialdehyde. Mutation Research. 1999; 424:83-95.

[27] Muller FL, Lustgarten M S, Jang Y, Richardson A, Van Remmen H. Trends in oxidative aging theories. Free Radic Biol Med. 2007; 43:477-503.

[28] Choi AO, Cho SJ, Desbarats J, Lovrić J, Maysinger D.Quantum dot-induced cell death involves Fas upregulation and lipid peroxidation in human neuroblastoma cells. J Nanobiotechnology. 2007; 5:1.

[29] Hill MM, Adrain C, Martin SJ. Portrait of a killer: The mitochondrial apoptosome emerges from the shadow. Mol Interventions. 2003; 3:19-25.

[30] Jiang X, Wang X. Cytochrome-c-mediated apoptosis. Annu Rev biochem. 2004; 73:87-106.

[31] Kiang JG, Tsen KT. Biology of hypoxia. Chin J Physiol. 2006; 49:223-33.

[32] Mizushima N, Levine B, Cuervo AM, Klionsky DJ. Autophagy fights disease through cellular self-digestion. Nature. 2008; 451:1069–75.

[33] Yang Z, Klionsky DJ. Eaten alive: a history of macroautophagy. Nat Cell Biol. 2010; 12:814-22.

[34] Klionsky DJ. The Autophagy Connection. Dev Cell. Author manuscript; available in PMC 2011 July 20.

[35] Tooze SA, Yoshimori T. The origin of the autophagosomal membrane. Nat Cell Biol. 2010; 12:831-5.

[36] Weidberg H, Shvets E, Elazar Z. Biogenesis and cargo selectivity of autophagosomes. Annu Rev Biochem. 2011; 80:125-56.

[37] Eskelinen EL. New insights into the mechanisms of macroautophagy in mammalian cells. Int Rev Cell Mol Biol. 2008; 266:207-47.

[38] Sridhar S, Botbol Y, Macian F, Cuervo AM. Autophagy and Disease: always two sidesto a problem. J Pathol. 2011 Oct 12. doi: 10.1002/path.3025. [Epub ahead of print]

[39] Montgomery CA. Oncologic and toxicologic research: Alleviation and control of pain and distress in laboratory animals. Cancer Bulletin. 1990; 42:230-37.

[40] Tomasivic SP, Coghlan LG, Gray KN, Mastromarino AJ, Travis EL. IACUC evaluation of experiments requiring death as an end point: A cancer center's recommendations. Lab Animal January/February, 31-34 (1988).

[41] American Veterinary Medical Association, Report of the AVMA Panel on Euthanasia. J Am Veterinary Medical Association. 2001; 218:669-96.

[42] Müller CA, Autenrieth IB, Peschel A. Innate defenses of the intestinal epithelial barrier. Cell Mol Life Sci. 2005; 62:1297-307.

[43] Gorbunov NV, Garrison BR, Kiang JG. Response of crypt Paneth cells in the small intestine following total-body gamma-irradiation. Int J Immunopathol Pharmacol. 2010; 23:1111-23.

[44] Kiang JG, Garrison BR, Gorbunov NV. Radiation combined injury: DNA damage, apoptosis, and autophagy. Adapt Med. 2010; 2:1-10.

[45] Kiang JG, Garrison BR, Burns TM, Zhai M, Fukumoto R, Cary LH, Elliott TB, Ledney GD. Radiation combined injury complicates biodosimetric assessment. Radiat Res. 2012; In review.

[46] http://en.wikipedia.org/wiki/Poisoning_of_Alexander_Litvinenko.

[47] AFRRI Handbook. 2009 QuickSeries Publishing, Canada.

[48] Kiang JG, Jiao W, Cary L, Mog SR, Elliott TB, Pellmar TC, Ledney GD. Wound trauma increases radiation-induced mortality by increasing iNOS, cytokine concentrations, and bacterial infections. Radiate Res. 2010; 173: 319-332.

[49] Haegebarth A, Clevers H. Wnt signaling, lgr5, and stem cells in the intestine and skin. Am J Pathol. 2009; 174:715-21.

[50] Van der Flier LG, Clevers H. Stem cells, self-renewal, and differentiation in the intestinal epithelium. Annu Rev Physiol. 2009; 71:241-60.

[51] Gorbunov NV, Garrison BR, Zhai M, McDaniel DP, Ledney GD, Elliott TB, Kiang JG. Autophagy-mediated defense response of mouse mesenchymal stromal cells (MSCs) to challenge with Escherichia coli. In: Protein Interaction / Book 1; ISBN 979-953-307-577-7. Eds.: Cai J. InTech Open Access Publisher; 2011; www.intechweb.org. pp. 23-44.

[52] Haegebarth A, Perekatt AO, Bie W, Gierut JJ, Tyner AL. Induction of protein tyrosine kinase 6 in mouse intestinal crypt epithelial cells promotes DNA damage-induced apoptosis. Gastroenterology. 2009; 137:945-54.

[53] Wilkins HM, Marquardt K, Lash LH, Linseman DA. Bcl-2 is a novel interacting partner for the 2-oxoglutarate carrier and a key regulator of mitochondrial glutathione. Free Radical Biol & Med. 2012; 52:410-19.

Chapter 14

Tissue Occurrence of Carbonyl Products of Lipid Peroxidation and Their Role in Inflammatory Disease

Maria Armida Rossi

Additional information is available at the end of the chapter

http://dx.doi.org/10.5772/51343

1. Introduction

The lipid peroxidation is a diffuse process which regards the polyunsaturated fatty acids of lipids, when they are exposed to oxygen-derived free radicals.

The process occurs when oils or foods, vegetables or meats, or other materials are exposed to air and causes their alteration at least in part through the peroxidative decomposition of the fatty acids contained in their lipids.

The lipid peroxidation does not need the action of enzymes and brings to the progressive decomposition of the unsaturated fatty acids till to the formation of carbonylic end products, aldehydes and ketones.

Oxygen-derived free radicals can be produced by the effect of sun rays on O_2, but an important source of them is the cellular metabolism, too.

The interest and the importance of the lipid peroxidation arise from the fact that the polyunsaturated fatty acids are contained in the phospholipids present in all cellular membranes; their structure and function can be strongly modified by this process.

The cellular effects of the lipid peroxidation change according to its degree. A high lipoperoxidative rate can produce serious damages to the cells and their death; on the contrary, a low degree of it allows cell survival and may modulate tissue metabolism.

2. Steps of the lipoperoxidative process

The lipid peroxidation has many steps,as shown in the Figure 1. The process is started by the attach of free radicals to poly-unsaturated fatty acids of lipids. Free radicals are chemical species which have a single, unpaired electron in an outer orbit. Their molecular configuration is unstable and so they react with the adjacent molecules to acquire a more stable configuration. The polyunsaturated fatty acids contained in the phospholipids of cell membranes are a good target for their reaction; in the attach of free radicals to the unsaturated fatty acids of lipids a methylen group near a double bond can give the electron required by the free radical to form the electon pair. So the unsaturated fatty acid has become a free radical and reacts with another molecule, starting the propagation phase which characterizes the lipoperoxidative process. In our cells the molecular oxygen is always present and can react with the lipid radical to form a lipoperoxide. This molecule has un unstable configuration too. The formed lipoperoxides react with adjacent membrane molecules, either other lipids or proteins. The reaction of a lipoperoxide with a protein molecule changes it in a reactive free radical; the so activated protein can interact with another protein to give a protein complex or interacts with lipids to form lipofuscin molecules. The presence of lipofuscin is frequent in the tissues of old people; this fact was well known by the anatomists already in the past century; the mechanism of their formation has been clarified with the discovery of the lipoperoxidative process. Beside the reaction with other molecules, the lipoperoxides can break to give more stable end products, aldehydes and ketones. These carbonylic end products of lipid peroxidation are formed above all in the microsomes where the rate of the lipid peroxidation is strong, but they can diffuse and react with various molecular targets both within the cell and outside it.

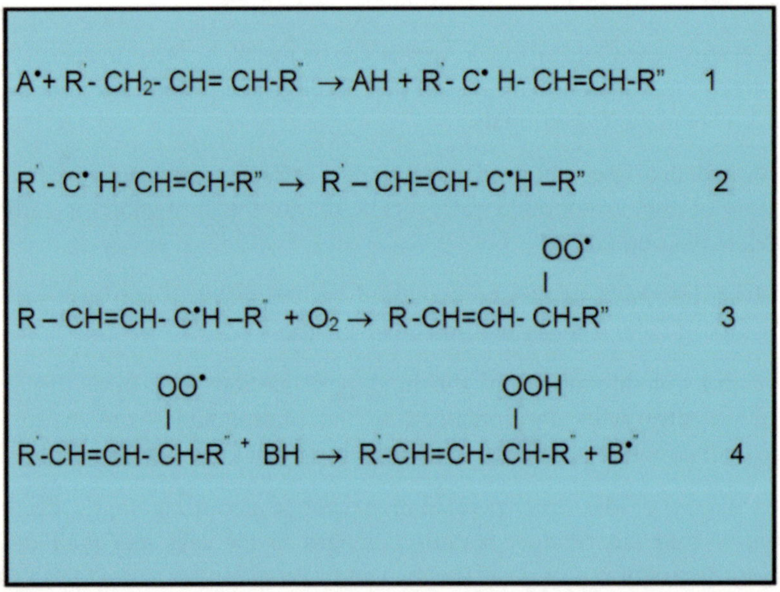

Figure 1. First steps of the lipid peroxidative process. 1) The free radical $A^{\bullet}$ abstracts an electron from a near molecule, which becomes a free radical; the target molecule is often a polyunsaturated fatty acid. 2) Its molecular configuration is unstable, so a shift of the double bond occurs. 3) This still unstable free radical binds O_2 and becomes a peroxide. 4) The peroxide captures an electron from the molecule B and forms an hydro-peroxide. Now the molecule B is a free radical. The further fate of the hydroperoxide is its fragmentation in small carbonylic compounds (not shown iin the Figure)

The first reports of the actual occurrence of the lipid peroxidation in tissues include the researches separately carried on by Comporti M et al.[1] and by Recknagel RO and Ghoshal AK. [2] to explain the liver damage induced by the rat treatment with CCl_4. Both these works used methods of investigation quite modern for those years and brought important findings to understand the structure and the functions of the different cell compartments: nucleus, mithochondria, microsomes, lysosomes. The further experimental studies on the steps and the effects of the lipid peroxidation have been deeply facilitated by Benedetti al. [3] who were able to develop a method to synthetize its carbonylic end-products, above all the aldehyde 4-hydroxy-2,3-trans-nonenal (HNE), whose chemical structure is shown in the Figure 2. HNE was shown to be produced in good amounts when the lipid peroxidation was stimulated; furthermore several experimental researches found that this aldehyde was the more cytotoxic end product of the lipoperoxidative process [4]. The first experimental works on the effects used millimolar concentrations of the aldehyde which were rather high; later the researchers found that it could display several biological effects at concentrations micromolar or less, which can be easily found in tissues even in normal conditions.

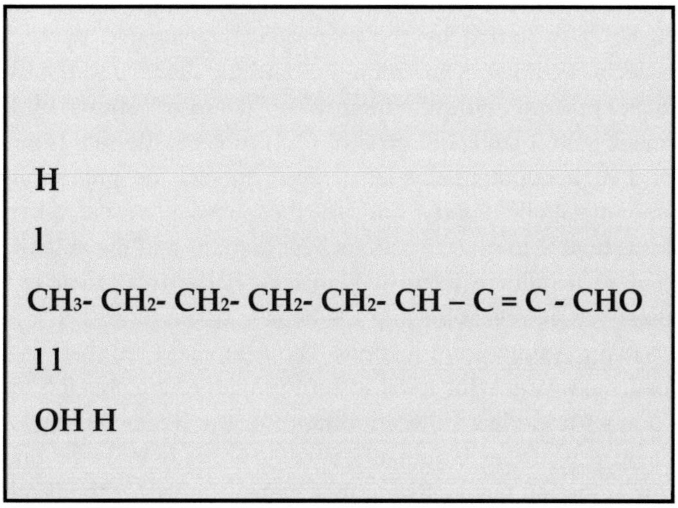

Figure 2. Structure of 4-hydroxy-2,3-trans-nonenal.

3. Lipoperoxidative effects on cell compartments

The damage to the different cell structures induced by the rat treatment with CCl_4 are similar to the alterations found in different pathological processes and are followed by similar changes in the tissue metabolism. In fact all the cell structures, mithocondria, microsomes, lysosomes, nuclei, are delimitated by membranes where the lipid peroxidation can take place and cause damage, bringing to changes in their functions. a) Effects of the lipid peroxidation on microsomes. The action of toxic compounds on cells leads to a quite rapid swelling of rough and smooth endoplasmic reticulum; the ribosomes dissociate from the rough endoplasmic reticulum and the protein synthesis decreases within less than 30 minutes [5,6]. The inhibition of microsomal glucose-6-phosphatase activity is seen very early in the action of toxic compounds; afterwords the activity of several other enzymes (hexokinase, lactate dehydrogenase, alpha and beta polymerases, 5'nucleotidase) has been found to decrease under the effect of the end products of the lipid peroxidation. However the effects of the inhibition of the protein synthesis can be seen only after different hours because the cells have a reserve of preformed proteins which can be used. The microsomes are the site of the drug metabolizing enzyme system (d.m.e.s.) which metabolizes different compounds, either endogenous components, such as different hormones, or various xenobiotics. The result of the changes induced by the d.m.e.s. on a compound can be different: the compound can be inactivated, it can change its functions or it can even acquire a toxic action. . . . CCl_4 has solvent properties in high amounts, but much smaller quantities can induce biological toxic effects through its homolitical cleavage catalysed by the d.m.e.s. In fact CCl_4 fission generates free radicals able to trigger the lipid peroxidative process, starting from microsomal membranes. In the rats intoxicated with CCl_4 an early effect is the decrease of the hepatic content of cytochrome P_{450}, which is part of the d.m.e.s., the enzyme system which metabolizes the haloalkane, generating the free radicals responsible of many of its dangerous effects. The lipid peroxidation increases strongly in the liver of rats treated with this haloalkane; it is started by the free radicals generated by CCl_4 fission in the microsomes. The decrease of the cytochrome P_{450} and the damage to the liver endoplasmic reticulum lead to an apparent and quite interesting "paradox", shown by Ugazio et al. [7], i.e. the pre-treatment with a sublethal dose of CCl_4 protects the rats from the subsequent administration of a higher, potentially lethal dose. In fact the pretrearment impairs the hepatic microsome metabolic ability and so the subsequent haloalkane dose is less metabolized and it is unable to cause a serious liver damage and the animal death. A single, non lethal dose, of CCl_4 induces fatty liver in rats; if the treatmemt is unrepeated this degenerative process can be reversible and the hepatic tissue returns to a normal anatomic aspect and to its usual physiological functions. The demonstration that the toxicity of small doses of this haloalkane was not due to its solvent properties but was the consequence of its cleavage by the d.m.e.s underlined the importance of the interaction between the various xenobiotics, foods or drugs or air inquinants, and the living organism, human or animal. A different behaviour of the cell metabolism due to genetic factors or to different other causes, such as diseases or the assumption of various substances, can higly modify the response to xenobiotics and their effects on the health. An important step in the studies on CCl_4 toxicity was the finding that the pre-treatment of rats with antioxidants (DPPD, GSH, propyl gallate) could prevent them both from liver damage and cell death, suggesting the role of an oxidative mechanism in the development of its toxic action [8,9]. b) Lipoperoxidative effects on mitochondria. The effects of the lipid peroxidation induced by toxic compounds can be seen also in mitochondria which show a swelling and a change in ATP synthesis [9,10]. The studies with the electron microscope revealed the damage to mitochondrial components. In the first phases of the mitochondrial swelling the production of ATP can increase for an easier entry of the substrates in the organelle through the more permeable membrane, then it decreases and stops. In many intoxication both the decrease of the synthesis of ATP and the damage of the plasma membrane contribute to an increase of the Ca^{2+} influx in cells, which comes before the cell necrosis. [11]. Ca^{2+} concentration is strictly checked in cells; the cytosolic free calcium is maintained at concentrations < 0.1 μM , which are much lower than the extracellular ones. This control is very important because an increase in Ca^{2*} activates several enzymes (ATPases, phospholipases, proteases, endonucleases) which can damage the same cell structures. c) Effects on lysosomes. The lysosomes are damaged by the attack of free radicals and by the onset of the lipid peroxidation in their membrane (12,13); so their lytic enzymes can be released in the cytoplasm. In the injured cells the intracellular pH tend to be acid; so the released lysosomal enzymes can be activated and destroy important cell components. The damage of the lysosomal membrane can lead to the enzymatic digestion of proteins, RNA, DNA and the cell dies by necrosis. The occurrence of the lipid peroxidation in lysosomes can also lead to the inactivation of their lytic enzymes if the lipoperoxidative rate is very high; Krohne et al. [14] have shown that the lipid peroxidaction end products, HNE and malonaldehyde (MDA), inactivated lysosomal cysteine proteases by covalent binding to their active center. d) Lipid peroxidation-induced changes in the nucleus. The cell nucleus has a membrane, like the other cell organelles; if the lipid peroxidation occurs in the nuclear membrane, it can cause serious damage. The nuclear importance is due to the presence of the DNA molecule; a damage to the DNA can lead to alterations in the codified proteins. If these changes involve important sites of the molecule, the protein can be no more functional. Some alterations in the DNA molecule are lethal, others lead to vital, but modified cells. Some changes in the DNA molecule can bring to the generation of transformed cells which show different changes in their morphology, metabolism and behaviour toward the near cells. The reaction of the different products of the lipid peroxidation with DNA has been extensively studied [15,16]; it can lead to the formation of adducts to DNA bases, which have profound mutagenic potential. The alterations of DNA molecule are believed to be important in the pathogenesis of cancer; a special attention has been given to the oncogenes and antioncogenes, which play an important role in regulating cell division.

4. Effects of lipid peroxidation in inflammation

The lipid peroxidation plays an important role in inflammation; in this process its presence is constant and its degree can reach high values.

Inflammation is the local response to any tissue damage. It is characterized by two main events: 1. changes in the blood flow in the microcirculation of the injured site. 2. recruitment of leukocytes, neutrophils and monocyte-macrophages; these cells phagocyte and destroy the agents of the tissue injury: bacteria, virus, parasites, dead cells, tissue debris. The leukocytes which gain the damaged tissue are activated by cytokines (IL-1, IL-6, TNF, MCP-1), which trigger the respiratory or phagocytic burst in them. This process is characterized by a strong increase of the consumption of oxygen, which is used to produce the superoxide anion ($O_2^{\bullet}$); its synthesis is catalysed by the NADPH (nicotinamide adenine dinucleotide phosphate) oxidase [17]. The NADPH oxidase is formed by a complex of proteins which are located both in the plasma membrane and in the cytoplasm in the resting neutrophil. When the neutrophil is activated by different stimuli (the phagocytosis itself, various cytokines), the components of the NADPH oxidase assemble on the membrane of the phagosome and the enzymatic complex can reduce oxygen to superoxide anion as shown in the following reaction:

$$NADPH + O_2 \rightarrow NADP^{+} + O_2^{\bullet}$$

Chemotactic compounds

- Microrganisms and microbic compounds
- Complement components (C3a,C5a)
- Leukotriene B_4
- Lipid peroxidation products

  4-hydroxyexenal (HEE)
  4-hydroxyoctenal (HOE)
  4-hydroxynonenal (HNE)

- Cytokines

  Interleukin-8 (specific for neutrophils)
  Monocyte chemotactic protein-1
  (specific for monocyte-macrophages)

Figure 3. List of the principal chemotactic compounds.

The superoxide anion is a free radical and so it is highly reactive.Two molecules of superoxide anion can react together and form the hydrogen peroxide (H_2O_2); this molecule has a low bactericidal power and it is also used as a disinfectant in pharmacology. In the tissues the hydrogen peroxide is used in a reaction catalysed by the myeloperoxidase (MPO) to form hypochlorite ($OCl^{\bullet}$). The microbial power of the hypochlorite is very strong; furthermore it can oxidase protein and lipids and so it can trigger the lipid peroxidation. The hydrogen peroxide can also be converted to the hydroxyl radical ($OH^{\bullet}$), a free radical with a very short lifetime; in fact it reacts with the nearest molecule to acquire a more stable configuration. Both the anion superoxide and the hydroxyl radical are able to induce the lipid peroxidation and this fact explains its steady presence in inflammation. The occurrence of the lipoperoxidative process may lead to a worsening of the tissue damage, but it also contributes to the recruitment of leukocytes, both neutrophils, and monocyte-macrophages since some lipid peroxidation end products display a chemotactic power, as shown in the Figure 2. The migration of leukocytes from blood to the inflammed tissue requires several passages [18]. Both the leukocytes and the endothelial cells need the presence of adhesion molecules on their surface to allow the leukocyte binding to the microcirculation of the damaged tissue. At first the binding is not firm and allows the leukocyte rolling on the endothelial surface; afterwards it becomes very strong and this firm adhesion is followed by the leukocyte passage outside the blood vessels to gain the site of the inflammation.

The chemotactic compounds or chemotaxins display different actions on the leukocytes. The term "chemotaxis" refers to the ability of a molecule to stimulate the oriented migration of a cell in the presence of a chemical gradient of the chemotactic compound or chemotaxin; the leukocytes have specific receptors for the different chemotaxins and move toward the site where the chemotaxins have the highest concentration. Beside this property, the chemotactic compounds display many other functions on the leukocytes: they induce the phagocyte burst, activate the adhesion molecules which are expressed on the plasmamembrane of the neutrophils, promote the synthesis of different cytokines, expecially by the macrophages.

a. Chemotactic activity of the products of the lipid peroxidation. The lipid peroxidation end product HNE has been shown to display a chemotactic power toward the polymorphonuclear leukocytes. At first this property was found by Curzio et al. (19) on rat neutrophils; the chemotactic concentrations of this aldehyde ranged from 10 μM to 0.1 μM. These doses are rather low and are devoid of any cytotoxic property. HNE chemotactic activity was initially demonstrated "in vitro" by the use of a Boyden chamber. This chamber has two compartments separated by a filter made of a mixture of cellulose esters with a pore size of 3 μ; the cells are placed in the upper chamber, the solution containing the substances to be tested in the lower one. The so mounted chamber is incubated at 37ºC for 75 min; then the chamber is removed and opened; the filter is removed, fixed in ethanol and stained with haematoxylin. The cell migration can be evaluated under the light microscope by the leading front technique. The first demonstration of HNE chemotactic power was obtained "in vitro", but afterwards it was confirmed by Schaur et al. [20] who carried on "in vivo" experimental researches. They induced an aseptic inflammation in the subcutaneous tissue of a rat leg by injecting in it some polydextrane Sephadex G-200; in control rats they inoculated Sephadex alone, while in the experimental group of rats they inoculated Sephadex together with a solution of preformed HNE. When they examinaed the histological samples obtained from the two groups of rats, they found the migration of neutrophils in both of them, but their number was much more higher around the Sephadex plus HNE. In their experimental researches the authors excluded the presence of any cytotoxic effects by the aldehyde concentrations able to stimulate the oriented migration of the neutrophils. Beside HNE, other 4-hydroxy-alkenals have been shown to display a chemotactic activity toward rat neutrophils: 4-hydroxy-2,3-hexenal(HEE) and 4-hydroxy-2,3-octenal(HOE). HOE was the most active of the lipoperoxidative end products; it could stimulate the oriented migration of neutrophils even at very low concentrations [21] between 10^{-11} and 10^{-8} M. Most chemotactic compounds can activate a phosphoinositide specific phospholipase C (PL-C) [22]; their stimulation of PL-C activity is mediated by a regulatory G protein and leads to the production of diacylglycerol and inositol-1,4,5-tris-phosphate (Ins-P_3). The diacylglycerol activates the protein kinase C and the Ins-P_3 promotes the mobilization of Ca^{++} from intracellular stores. The well known chemotaxin N-formylmethionyl-leucyl-phenylalanine (fMLP) increases the PL-C activity of neutrophils and its action is prevented by the cell pretreatment with pertussis toxin, which ADP ribosylates the alpha subunit of some G proteins. The chemotactic 4-hydroxy-alkenals formed by the lipid peroxidation have been found to activate the PL-C [23] of rat neutrophils and a good correspondence could be found between the concentrations able to increase the PL-C activity and those which regulated the cell migration. The pretreatment of neutrophils with pertussis toxin prevented the activation of PL-C by HOE, too; this finding suggested that its mechanism of action was like that of other well known chemotaxins. This discovery of the stimulation of an enzyme activity by very low doses of 4-hydroxyalkenals represented a clean change in the evaluation of the lipoperoxidative process and of the functions of its end-products. The first experimental studies on the biological effects of the lipid peroxidation supplied a lot of proofs about the inhibition of several enzymes in tissues where the lipid peroxidation rate was stimulated [5] or about the decrease of their activity in tissue homogenates or in subcellular fractions incubated in the presence of high concentrations of HNE [24].

b. Activation of the exocytosis by 4-hydroxynonenal. HNE was found to induce the exocytosis in DMSO-differentiated HL-60 cells. [25] This human promyelocitic cell line was chosen because it could be induced to differentiate toward the granulocytic cell line and therefore it represented a good in vitro model to study the mechanism of action of a chemotactic compound, like 4-hydroxynonenal. The exocytosis was valued by measuring the secretion of ß-glucuronidase, an enzyme of the azure granules, by the cells incubated in the presence of different HNE concentrations. The exocytosis was triggered by HNE doses between 10^{-8} and 10^{-6} M, which are wholly devoid of any cytotoxic power. The lack of any effect on the cell viability was checked by measuring the release of lactate dehydrogenase (LDH) in the cells incubated at 37°C for 1 hour in the presence of different HNE concentrations; the presence of HNE between 0.01 and 1.0 μM failed to induce any increase of the enzyme loss by the cells in the incubation period [25].

c. Stimulation of IL-8 release by 4-hydroxynonenal. I recently found HNE ability to change the release of the chemokine interleukin-8 (IL-8) in DMSO–differentiated HL-60 cells [26]; the aldeyde failed to modify the intracellular concentration of IL-8, but after 30 min of incubation it began to enhance the chemokine release. The increase of IL-8 level in the cell suspensions incubated in the presence of HNE was quite slow and became remarkable only after 1 h. This fact suggested that the effects shown by the aldehyde both on the chemotaxis and on the exocytosis were not mediated through the release of IL-8.

d. 4-hydroxynonenal induced synthesis of cyclooxygenase-2. The vascular reactions of inflammation are regulated by many chemicals mediators; among them the prostaglandins influence several cell functions. The prostaglandins play an important role in inflammation; above all the PGE_2 and the PGD_2 induce vasodilation and increase the permeability of post-capillary venules. These prostaglandins are produced from arachidonic acid by two cyclooxigenases, COX-1 and COX-2; the COX-1 is constitutively, while the COX-2 is inducible. The COX-2 js present in leukocytes and mastzellen and is induced by different mediators of inflammation. HNE has been shown to induce the synthesis of COX-2 [27] too; this finding underlines the importance of the lipid peroxidation role in inflammation.

5. Positive and negative actions of inflammation

Inflammation has many positive effects and it is considered a defensive response of the organism, but it is followed by negative aspects which may contribute to increase the tissue damage, as shown in Figure 3.

The leukocytes which reach a damaged tissue can remove the injurious agents. They can phagocyte and kill the microrganisms of an infectious disease; they also phagocyte the dead cells or the cell debris which are left in any damaged tissue.

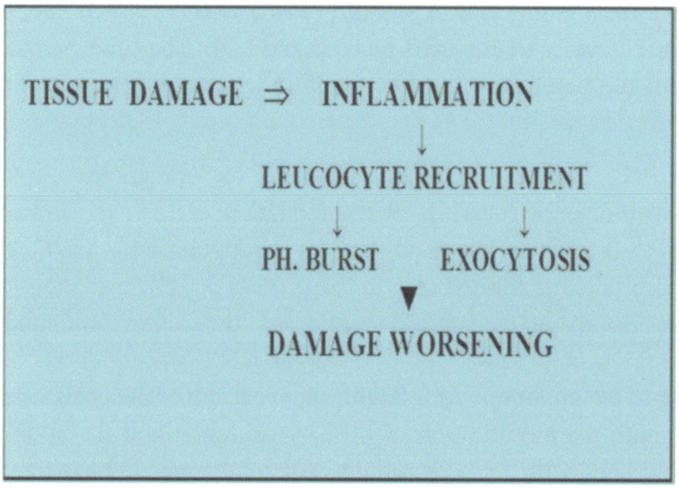

Figure 4. Main events in inflammation.

The Figure 3 underlines that the positive functions of the leukocytes are also followed by unpleasant effects, which can be caused both by the phagocytic burst and by the exocytosis. The phagocytic burst leads to the generation of free radicals and reactive oxygen species (ROS) which can diffuse outside the phagocytic cells and amplify the effects of the initial injurious agent. Moreover the induction of the lipid peroxidation by the ROS and the free radicals can worsen the tissue damage. . The exocytosis is a kind of physiological, controlled secretion of lysosomal enzymes by neutrophils and is activated by several chemotactic agent; their azurofil granules fuse with the plasma membrane and release their content in the extracellular space. In this way different lytic enzymes can diffuse in the inflammed tissue; the blood stasis which is always present in the late phases of the inflammatory process favours the lowering of pH in the damaged tissue; this fact leads to the activation of the released lysosomal enzymes and the tissue itself can be damaged.

6. Action of lipid peroxidation on atherosclerosis

Medical progress has brought good successes against many diseases in the past century. Antibiotics can win many infectious agents; the progress in surgery can correct cardiac malformations; the consequences of a vessel obstruction can be obviated by the insertion of a by pass.

In our times, when the life is becoming longer, the atherosclerosis represents a serious problem which can compromise the life of many people. The complications of the atherosclerosis are becoming the main causes of death.

Its pathological lesion is the atheroma, which is localized in the arteries of big and medium calibre: aorta, carotids, coronaries.

The atheroma is characterized by an accumulation of cholesterol and cholesterol esters both inside and outside the cells. It contains lipid-loaded cells, called foam cells. They are thought to derive from monocytes or smooth muscular cells, which have migrated in the arterial intima and have been engulfed by oxidized LDL. The lipid peroxidation plays an important role in the pathogenesis of this diffuse process through its intervent in LDL oxidation.

The modification of LDL by oxidation leads to its unregulated uptake by intimal macrophages to form foam cells[28]. In the oxidation of LDL the lipid peroxidation is stimulated and it contributes to modify their apolipoprotein. HNE, the major lipid peroxidation end product is formed also in the process of LDL oxidation and is present in the oxidized-LDL. Esterbauer et al. found that the aldehyde alone could modify the LDL. He incubated native LDL in the presence of different HNE concentrations and observed its covalent binding to the apolipoprotein B with the blockage of the epsilon-amino groups on lysine residues. Both the modification of LDL by oxidation and its modification by HNE binding were associated with an increased degradation by macrophages and a lipid loading of them.

The migration of macrophages toward the arterial intima is stimulated by chemotactic compounds, like the migration of leukocytes to a site of inflammation. The oxidized LDL have been shown to stimulate the synthesis of the monocyte chemotactic protein-1 (MCP-1) by macrophages [29]. This cytokine has a chemotactic power specific for the monocyte-macrophages; however the macrophage recruitment in the atheroma could be also favoured by the 4-hydroxyalkenals which have been found in the oxidized LDL: 4-hydrohexenal (HEE), 4-hydroxyoctenal (HOE) and 4-hydroxynonenal (HNE). These aldehydes are lipid peroxidation end products which display both a cytotoxic and a chemotactic power. They are likely to be produced by the oxidation of the LDL which have reached the intima of arteries and can contribute to the recruitment of monocyte-macrophages[30]; a direct cytotoxic effect on the foam cells of the atheroma was considered unlikely by Muller because it required higher levels of the aldehydes.

7. The lipid peroxidation role in ischemia-reperfusion.

The onset of the lipid peroxidation in a tissue requires the presence of molecular oxygen; however the ischemia can induce changes of the cell metabolism which may increase the tissue damage if the blood supply returns. This unexpected fact happens in the ischemia-reperfusion[31]. During ischemia the lack of oxygen causes the catabolism of ATP with an increased production of ipoxantine. which is an oxidable substrate for the xanthine dehydrogenase. Moreover in the ischemic tissue there is the conversion of the native xanthine dehydrogenase to a superoxide producing-oxidase; this conversion is thought to be produced by a calcium triggered protease. In the reperfusion the O_2 which reaches the tissue is transformed by the xanthine oxidase in superoxide anion; this free radical contribute to extend the tissue damage induced by the ischemia.

The tissue necrosis triggers an inflammatory process and the leukocytes which can reach the tissue in the reperfusion can worsen the tissue damage through the production of ROS and the release of lytic enzymes.

8. Conclusions

Tne lipid peroxidation can be regarded as a common process which happens in our cells. In fact low levels of the lipid peroxidation end products have been shown in tissues even in normal conditions [20]. The rate of the lipid peroxidation can be stimulated by the ROS and the free radicals which can arise also from the normal metabolism of cells.

The lipid peroxidation rate can be increased by some xenobiotics; the experimental works about the action of the haloalkane CCl_4 have been the source of the first explanations of its effects [1-3].

In any inflammation the leukocytes, above all the neutrophils and the macrophages, produce the superoxide anion and other free radicals, which can increase the lipoperoxidative rate[18]. A high degree of the lipid peroxidation is followed by some unavoidable damages to the tissue; some lesions can be reversible and can be repaired by the normal reparative process or by the aid of a pharmachological support; however a low alteration of the tissue integrity and function can follow any inflammatory event .

The aging is viewed by some authors [32] as the sum of the repeated tissue damages which occur in our life and the lipid peroxidation can take a part in them.

Another aspect of the lipid peroxidation regards its possible modulation of the normal metabolism; low concentrations of HNE, the major lipid peroxidation product, can modulate the activity of some enzymes, like the phosphoinositide-dependent phospholipase C [23] and several other enzymes [24].

Author details

Maria Armida Rossi
Department of Experimental Medicine and Oncology, University of Turin, Turin, Italy

Acknowledgement

The Author is grateful to his teacher, the Prof. Mario Umberto Dianzani (University of Turin, Turin, Italy) and to the late Prof. Hermann Esterbauer (University of Gratz, Gratz, Austria).

9. References

[1] Comporti M, Saccocci C, Dianzani MU. Effect of CCl_4 in vitro and in vivo on lipid peroxidation of rat liver homogenates and subcellular fractions. Enzymologia 1965;29(3)185-204

[2] Recknagel RO, Ghoshal AK. Lipoperoxidation of rat liver microsomal lipids induced by carbon tetrachloride. Nature 1966;210(5041) 1162-63

[3] Benedetti A, Comporti M, Esterbauer H. Identification of 4-hydroxynonenal as a cytotoxic product originating from the production of liver microsomal lipids. Biochim Biophys Acta 1980; 620(2), 281-96

[4] Esterbauer H, Schaur RJ, Zollner H. Chemistry and bio-chemistry of 4-hydroxy-nonenal, malonaldehyde and related aldehydes. Free Radic Biol Med. 1991;11(9) 81-128

[5] Glende EA, Hruszkawycz AM, Recknagel RO. Critical role of lipid peroxidation in carbon tetrachloride-induced loss of aminopyrine demethylase, cytochrome P-450 and glucose-6-phosphatase. Biochem Pharmacol. 1976;25(19) 2163-70

[6] Bidlack WR, Tappel AL. Damage to microsomal membrane by lipid peroxidation. Lipids 1973; 8(4)177-82

[7] Ugazio G, Torrielli MV, Burdino E, Sawyer BC, Slater TF. Long range effects of products of carbon tetrachloride-stimulated lipid peroxidation. Biochem Soc Trans. 1976; 4(2) 353-6

[8] Dianzani MU, Ugazio G. Lipoperoxidation after carbon tetrachloride poisoning in rats previously treated with antioxidants. Chem Biol Interact. 1973; 6(2) 67-79

[9] Christophersen BO. The inhibitory effect of reduced gluthathione on the lipid peroxidation of the microsomal fraction and mitochondria. Biochem J. 1968; 106(2) 515-22

[10] Hyams DE, Issebalbacher KJ. Prevention of fatty liver by administration of adenosine triphosphate. Nature 1964; 204,1196-7

[11] Dianzani MU, Viti I. The content and distribution of cyto-chrome c in the fatty liver of rats. Biochem J. 1955; 59(1) 141-5

[12] Yoshikawa T, Yokoe N, Takamura S, Kato A, HottaT, Matsumura N, Ikezaki M, Hosokawa K, Kondo M. Lipid peroxidation and lysosomal enzymes in D-galactosamine hepatitis and its protection by vitamin E. Gastroenterol Jpn. 1979; 14(1) 31-9

[13] Mak IT. Misra HP,,Weglicki WB. Temporal relationship of free radical-induced lipid peroxidation and loss of latent enzyme activity in highly enriched hepatic lysosomes. J Biol Chem. 1983; 258(22) 13733-7

[14] Krohne TU, Kaemmer E, Hotz EG, Kopitz J. Lipid peroxidation products reduce lysosomal protease activities in human retinal pigment. J Exp Eye Res. 2010; 90(2) 261-6

[15] Lim P, Wesenschelln GE, Holland V, Lee DH, Pfeifer GP, Rodriguez H, Termini J. Peroxyl radical mediated oxidative DNA base damage. Biochemistry 2004; 43(49) 15339-48

[16] Winczura A, Zdzalik D, Tudek B. Damage of DNA and proteins by major lipid peroxidation products in genoma stability. Free Rad Res. 2012;46(4):442-59

[17] Babior BM. NADPH oxidase. Curr Opin Immunol. 2004;(16)42-7

[18] Muller WA. Leukocyte-endothelial-cell interactions in leukocyte transmigration and the inflammatory response. Trends Immunol. 2003;24(6) 327-34

[19] Curzio M., Torrielli MV., Giroud JP., Esterbauer H., Dianzani MV. Neutrophil chemotactic responses to aldehydes. Res Commun Chem Pathol Pharmacol. 1982;36(3)463-476

[20] Schaur RJ, Dussing G, Kink E,Schauenstein E, Poscch W, Kukovetz E, Egger G. The lipid peroxidation product 4-hydroxynonenal is formed by– and is able to attract rat neutrophils in vivo. Free Rad Res.1994; 20(4) 365-373

[21] Curzio M.,Esterbauer H.Dianzani MU. Chemotactic power of 4-hydroxy-octenal. IRCS Med Sci. 1983;11,521

[22] Sandborg RR, Smolen JE Biology of disease. Early biochemical events in leukocyte activation. Lab Invest 1988,59(3) 300-320

[23] Rossi MA, Di Mauro C,Esterbauer H, Fidale F, Dianzani MU. Activation of phosphoinositide-specific phospholipase C by the chemotactic aldehydes 4-hydroxy-2,3-trans-nonenal and 4-hydroxy-2,3-trans-octenal. Cell Biochem Funct. 1994;12(4) 275-280

[24] Dianzani MU. Biochemical effects of saturated and unsaturated aldehydes. In: McBrien DCH and Slater TF (eds) Free Radicals, Lipid Peroxidation and Cancer. Academic Press, London; 1981. p.129-158,

[25] Maggiora M, Dianzani MU, Rossi MA Effect of 4-hydroxy-nonenal. A lipid peroxidation product, on exocytosis in HL-60 cells. Cell Biochem Funct. 2002;20(4)303-307

[26] Maggiora M., Rossi M.A. Changes in IL-8 release and intracellular content in DMSO-differentiated HL-60 cells after treatment with 4-hydroxynonenal. Cell Biochem. Funct. 2008; 26(5) 566-570

[27] Uchida K, Kumagai T. 4-hydroxynonenal as a COX-2 inducer. Mol Aspects Med. 2003, 24(4-5) 213-8

[28] Hoff HF,O'Neil J, Chisolm GM 3rd, Cole TB, Quehenberger O, Esterbauer H, Jürgens G. Modification of low density lipoprotein with 4-hydroxynonenal induces uptake by macrophages. Arteriosclerosis. 1989; 9(4) 538-4

[29] Wang GP, Deng ZD, Ni J, Qu ZL. Oxidized low density lipoprotein and very low density lipoprotein enhance expression of monocyte chemoattractant protein-1 in rabbit peritoneal exudate macrophages. Atherosclerosis 1997; 133(1) 31-6

[30] Muller K, Hardwick SJ, Marchant CE, Law NS, Waeg G, Esterbauer H. Carpenter KL, Mitchinson MJ. Cytotoxic and chemotactic potencies of several aldehydic components of oxidized low density lipoproteins for human monocyte-macrophage. FEBS Lett. 1996,388(2-3)165-168

[31] McCord JM, Roy RS, Schaffer SW. Free radicals and myocardial ischemia. The role of xanthina oxidase. Adv Myocardiol. 1985;5:183-9
[32] Balaban RS, Nemoto S, Finkel T. Mitochondria,oxidants and aging. Cell 2005,120(4) 483-95

Chapter 15

The Role of Physical Exercise on Lipid Peroxidation in Diabetic Complications

Yaşar Gül Özkaya

Additional information is available at the end of the chapter

http://dx.doi.org/10.5772/46182

1. Introduction

Diabetes mellitus is a group of metabolic disorder characterized by hyperglicemia and insufficiency of action or secretion of insulin. More than 346 million people worldwide have diabetes. 80 per cent of diabetes-induced deaths ocur in low- and middle-income countries. Most people with diabetes are above the age of retirement in developed countries, whereas in developing countries those most frequently affected are aged between 35 and 64 [1]. Although the etiology of this disease is not well defined, viral infections, autoimmunity, genetic and environmental factors have been implicated [2-5]. Four major types of diabetes have been defined by the American Diabetes Association (ADA): type 1 diabetes, type 2 diabetes, other spesific types of diabetes and gestational diabetes mellitus (GDM) [6].

Type 1 diabetes (T1D) usually develops in childhood and adolescence and the cause of the disease is an absolute deficiency of insulin secretion. Individuals at increased risk of developing this type of diabetes can often be identified by serological evidence of an autoimmune pathologic process occurring in the pancreatic islets and by genetic markers [6].

Type 2 diabetes (T2D) usually develops in adulthood and is related to obesity, lack of physical activity, and unhealthy diets. This is the more common type of diabetes (representing 90% of diabetic cases worldwide) and the cause is a combination of resistance to insulin action and an inadequate compensatory insulin secretory response. In Type 2 diabetes, a degree of hyperglycemia sufficient to cause pathologic and functional changes in various target tissues, but without clinical symptoms, may be present for a long period of time before diabetes is detected. During this asymptomatic period, it is possible to demonstrate an abnormality in carbohydrate metabolism by measurement of plasma glucose in the fasting state or after a challenge with an oral glucose load.

The third category "other spesific types of diabetes" includes diabetes caused by a spesific and identified underlying defect, such as genetic syndromes, acquired processes such as pancreatitis, diseases such as cystic fibrosis, exposure to certain drugs, viruses, and unknown causes. Gestational diabetes is a state of hyperglicemia which develops during pregnancy [6].

Currently, ADA recommends the use of any of the following four criteria for diagnosing diabetes: 1) glycated hemoglobin (A1c) value of 6.5% or higher, 2) fasting plasma glucose ≥ 126 $mg.dL^{-1}$ (7.0 $mmol.L^{-1}$), 3) 2-h plasma glucose ≥ 200 $mg.dL^{-1}$ (11.1 $mmol.L^{-1}$) during an oral glucose tolerance test using 75 g of glucose, and/or 4) classic symptoms of hyperglycemia (e.g., polyuria, polydipsia, and unexplained weight loss) or hyperglycemic crisis with a random plasma glucose of 200 $mg.dL^{-1}$ (11.1 $mmol.L^{-1}$) or higher. In the absence of unequivocal hyperglycemia, the first three criteria should be confirmed by repeat testing [6].

Hyperglycaemia and hyperlipidaemia are key promoters of diabetes dysmetabolism, namely, through the formation of reactive oxygen species (ROS) and advanced glycation end products (AGEs), which causes cell damage and insulin resistance [7-9]. Moreover, both of them stimulate proinflammatory cytokines, thus contributing to β-cell degradation, particularly due to apoptosis pathways [10].

Increased oxidative stress is a widely accepted participant in the development and progression of diabetes and its complications [11-13]. Diabetes is usually accompanied by increased production of free radicals [14,15] or impaired antioxidant defenses [16]. Mechanisms by which increased oxidative stress is involved in the diabetic complications are partly known, including activation of transcription factors, advanced glycated end products (AGEs) [2], and protein kinase C [17].

Modern medical care uses a vast array of lifestyle and pharmaceutical interventions aimed at preventing and controlling hyperglycemia. In addition to ensuring the adequate delivery of glucose to the tissues of the body, treatment of diabetes attempts to decrease the likelihood that the tissues of the body are harmed by hyperglycemia. The importance of protecting the body from hyperglycemia cannot be overstated; the direct and indirect effects on the human vascular tree are the major source of morbidity and mortality in both type 1 and type 2 diabetes. Generally, the injurious effects of hyperglycemia are separated into macrovascular complications (coronary artery disease, peripheral arterial disease, and stroke) and microvascular complications (diabetic nephropathy, neuropathy, and retinopathy) [11].

Physical activity (PA) and diet are cornerstones of diabetes therapy [19]. Physical activity is a multifaceted behavior of which exercise is just one component. PA is defined as "bodily movement produced by the contraction of skeletal muscle that substantially increases energy expenditure" and exercise is defined as "a subset of PA done with the intention of developing physical fitness (i.e., cardiovascular, strength, and flexibility training)." [19]. In this chapter, PA and exercise is used interchangeably.

In last decades, an impressive body of research has accumulated that demonstrates the varied benefits of regular physical activity for people with type 1 or type 2 diabetes [20]. Notably, exercise has been shown to improve glycemic control, reduce the need for insulin and oral hypoglycemic agents, and improve body weight control. Exercise has been shown to promote beneficial effects on insulin resistance, both in humans and in rodent models of T2DM [21, 22]. Moreover, exercise has myriad benefits for all people beyond those relating to diabetes alone. It can work wonders for the heart, improving the lipid profile, reducing risk for heart disease, restoring function after a heart attack, and moderating blood pressure. It helps in maintaining bone health regardless of age, it can significantly relieve depression and anxiety, and it appears to help maintain cognitive function in old age [23, 24]. A correlation between the effects of acute and chronic aerobic exercise upon oxidative stress and inflammation and the diabetic dysmetabolism has been previously described [25-27].

This chapter focuses on recent clinical and experimental studies of diabetes and exercise interventions done within the context of lipid peroxidation.

2. Lipid peroxidation and diabetic complications

2.1. Overview of lipid peroxidation and diabetic complications

Excessively high levels of free radicals cause damage to cellular proteins, membrane lipids and nucleic acids, and eventually cell death [2]. Various mechanisms have been suggested to contribute to the formation of these reactive oxygen-free radicals in diabetic state. Glucose oxidation is believed to be the main source of free radicals. In its enediol form, glucose is oxidized in a transition-metal dependent reaction to an enediol radical anion that is converted into reactive ketoaldehydes and to superoxide anion radicals. The superoxide anion radicals undergo dismutation to hydrogen peroxide, which if not degraded by catalase or glutathione peroxidase, and in the presence of transition metals, can lead to production of extremely reactive hydroxyl radicals [28, 29]. Superoxide anion radicals can also react with nitric oxide to form reactive peroxynitrite radicals [30, 31]. Hyperglycemia is also found to promote lipid peroxidation of low density lipoprotein (LDL) by a superoxide-dependent pathway resulting in the generation of free radicals [32, 33]. Another important source of free radicals in diabetes is the interaction of glucose with proteins leading to the formation of an Amadori product and then advanced glycation endproducts (AGEs) [34, 35]. These AGEs, via their receptors (RAGEs), inactivate enzymes and alter their structures and functions [36], promote free radical formation [37, 38], and quench and block antiproliferative effects of nitric oxide [39, 40]. By increasing intracellular oxidative stress, AGEs activate the transcription factor NF-κB, thus promoting up-regulation of various NF-κB controlled target genes [41]. NF-κB enhances production of nitric oxide, which is believed to be a mediator of islet beta cell damage. Considerable evidence also implicates activation of the sorbitol pathway by glucose as a component in the pathogenesis of diabetic complications, for example, in lens cataract formation or peripheral neuropathy [42-44]. Efforts to understand cataract formation have provoked various hypotheses. In the aldose reductase osmotic hypothesis, accumulation of polyols initiates lenticular osmotic changes.

In addition, oxidative stress is linked to decreased glutathione levels and depletion of NADPH levels [45, 46]. Alternatively, increased sorbitol dehydrogenase activity is associated with altered NAD+ levels, which results in protein modification by nonenzymatic glycosylation of lens proteins [47, 48]. Mechanisms linking the changes in diabetic neuropathy and induced sorbitol pathway are not well delineated. One possible mechanism, metabolic imbalances in the neural tissues, has been implicated in impaired neurotrophism [49, 50], neurotransmission changes [51, 52], Schwann cell injury [53, 54], and axonopathy [55, 56].

2.2. Overview of antioxidants

While on the one hand hyperglycemia engenders free radicals, on the other hand it also impairs the endogenous antioxidant defense system in many ways during diabetes [57]. Antioxidant defense mechanisms involve both enzymatic and nonenzymatic strategies. Common antioxidants include the vitamins A, C, and E, glutathione, and the enzymes superoxide dismutase, catalase, glutathione peroxidase, and glutathione reductase. Other antioxidants include α-lipoic acid, mixed carotenoids, coenzyme Q10, several bioflavonoids, antioxidant minerals (copper, zinc, manganese, and selenium), and the cofactors (folic acid, vitamins B1, B2, B6, B12). They work in synergy with each other and against different types of free radicals. Vitamin E suppresses the propagation of lipid peroxidation; vitamin C, with vitamin E, inhibits hydroperoxide formation; metal complexing agents, such as penicillamine, bind transition metals involved in some reactions in lipid peroxidation [58] and inhibit Fenton and Haber- Weiss-type reactions; vitamins A and E scavenge free radicals [30, 37].

Several discrepancies observed in the activities of SOD, catalase, and glutathione peroxidase in experimentally diabetic animals. Decreased levels of glutathione and elevated concentrations of thiobarbituric acid reactants are consistently observed in diabetes [59, 60]. In addition, changes in nitric oxide and glycated proteins are also seen in diabetes.

3. Biomarkers of lipid peroxidation

Since the initial discoveries of Dilliard and colleagues [61], several commercial assay kits have been made available for the measurement of oxidative stress, with many new kits emerging each year. Furthermore, the discovery and utilization of F2-isoprostanes, a prostaglandin like compound, measured via gas chromotomography mass spectrometry has emerged as a substantially more reliable and valid measure of lipid peroxidation [62]. Newly developed ELISA kits for both isoprostanes as well as protein carbonyls are also now available, proving an opportunity for a more widespread use of these biomarkers. In regards to measurement of oxidative stress, due to the high reactivity and relatively short half lives (e.g., 10^{-5}, 10^{-9} seconds for superoxide radical and hydroxyl radical, respectively) of reactive oxygen and nitrogen species (RONS), direct measurement is extremely difficult to employ. However, direct assessment of free radical production is possible via electron spin resonance spectroscopy (ESR) involving spin traps, as well as two other less common techniques such as radiolysis and laser flash photolysis [63]. ESR works by recording the energy changes that occur as unpaired electrons align in response to a magnetic field [64]. Due to the high cost of such equipment and the high degree of labor associated with each direct method, the majority of free radial research related to exercise has utilized indirect methods for the assessment of resultant oxidative stress. Indirect assessment of oxidative stress involves the measurement of the more stable molecular products formed via the reaction of RONS with certain biomolecules. Common molecular products include stable metabolites (e.g., nitrate/nitrite), and/or concentrations of oxidation target products, including lipid peroxidation end products [isoprostanes, malondialdehyde (MDA), thiobarbituric acid reactive substances (TBARS), lipid hydroperoxides (LOOH), conjugated dienes (CD), oxidized low density lipoprotein (oxLDL)], oxidized proteins [protein carbonyls (PC), individual oxidized amino acids, nitrotyrosine (NT), and nucleic acids [8-hydroxy-2-deoxyguanosine (8-OHdG), oxidized DNA bases (via the Comet Assay), strand breaks] [65]. Additionally, oxidative stress can be measured by observing alterations in the body's antioxidant defense system. This is typically done by measuring the redox changes in the major endogenous antioxidant glutathione, as well as circulating levels of vitamin E, and vitamin C. Moreover, the activity of certain antioxidant enzymes [e.g., superoxide dismutase (SOD), glutathione peroxidase (GPx), catalase (CAT), glutathione reductase (GR)] can be assessed as indicators of the oxidative stress imposed on the tissue. Numerous antioxidant capacity assays also exist and include: Trolox Equivalent Antioxidant Capacity (TEAC), Total Antioxidant Status (TAS), Ferric Reducing Ability of Plasma (FRAP), Total RadicalTrapping Antioxidant Parameter (TRAP), and Oxygen Radical Absorbance Capacity (ORAC) [66].

4. Exercise and lipid peroxidation

4.1. Lipid peroxidation and antioxidant status in acute exercise research

Numerous studies have reported an increase in several lipid peroxidation markers following both maximal [67-69] and submaximal [70, 71] exercise. In opposition to these findings, a few studies have reported no increase in lipid peroxidation despite the use of similar maximal [72-74] and submaximal [75, 76] protocols. Increased lipid peroxidation seems to be a result of increased mitochondrial oxidative enyzme activation during aerobic exercise. However, studies reporting conflictiong findings for lipid peroxidation may be partially related to the timing of sampling, in addition to the trained status of the subjects or an insufficient intensity of exercise.

In response to conditions of strenuous physical work the body's antioxidant capacity may be temporarily decreased as its components are used to quench the harmful radicals produced. It appears that the antioxidant capacity may be temporarily reduced during and immediately post exercise [77, 78], after which time levels typically increase above basal conditions during the recovery period [79, 80]. However, conflicting findings have been reported for each of the four main enzymes, with investigators noting increases in GPx [81, 82], SOD [82, 83], and CAT [70, 84, 85], as well as decreases in GPx [86], GR [81], SOD [78].

Furthermore, no change has also been reported for GPx [69], GR [87], SOD (Tauler et al., 2006), CAT [87] activity following exercise. Clearly, these results are mixed and likely depend on the time of sampling, as well as the duration and intensity of exercise, which has varied considerably across studies.

During low-intensity and duration protocols, antioxidant defenses appear sufficient to meet the RONS production, but as intensity and/or duration of exercise increases, these defenses are no longer adequate, potentially resulting in oxidative damage to surrounding tissues [67]. Other factors appear to impact the degree of antioxidant defenses present, including age, training status [81, 88], and dietary intake [80].

It has been shown that anaerobic exercise results in increased RONS production [89]. The mechanisms responsible for the exercise-induced increases in RONS have been suggested to be largely a function of radical generating enzymes (activated in response to ischemia followed by reperfusion) and/or phagocytic immune response following muscle damaging exercise. In the literature, there are fewer data on the markers of lipid peroxidation after anaerobic exercise. It currently remains to be elucidate whether increased RONS formation observed during anaerobic exercise represents a necessary or harmful event.

4.2. Lipid peroxidation and antioxidant status after exercise training

Regular physical exercise exerts numereous adaptive responses in several tissues. In the context of lipid peroxidation, repeated exposure of RONS production appears to induced to maintain the optimal health. Literature data demonstrated that regular moderate exercise is strengthening the endogenous antioxidant defense system [90, 91,92], and in some animal studies, it has decreased lipid peroxidation. On the other hand, exercise training – both endurance and interval type - appears to protects against exercise induced oxidative stress [93, 94, 95].

5. Exercise and diabetes

The therapeutic use of physical exercise for diabetes treatment has been promoted since 600 B.C. before the discovery of insulin in 1922. Some investigators highlighted the interaction between this hormone and regular physical activity, with possible beneficial results in diabetes treatment [96]. Recent guidelines provide exercise recommendations for people with diabetes based on the strong and convincing epidemiologic association of aerobic exercise with lower cardiovascular disease risk in people with diabetes. The recent 2010 ADA/American College of Sports Medicine (ACSM) exercise guidelines recommend 150 minutes of weekly aerobic exercise (i.e. brisk walking or an equivalent activity with intensity ≥ 40% VO2max); and resistance exercise of major muscle groups two to three times weekly on non-consecutive days (ACSM evidence category B, ADA B level recommendation). The ADA/ACSM guidelines also suggest adding unstructured physical activity as much as possible. Before undertaking exercise more intense than brisk walking, sedentary people with T2D should be evaluated by a physician and an exercise trainer [19].

Exercise has been shown to promote beneficial effects on insulin resistance, both in humans and in rodent models of diabetes [97, 98]. Regular physical exercise may prevent diabetes complications through beneficial effects on glycemic control, insulin sensitivity, blood pressure, lipid profile, and endothelial function. Moderate exercise training has been demonstrated to decrease the plasma glucose concentration in STZ-induced diabetic rats [99]. Hypoglycemic effect of exercise can be explained by exercise induced increase in uptake of glucose of muscle which induces increase of GLUT 4 expression and translocation from intracellular pool [100, 101]. Increase in glucose uptake seems to be related to the increased number of GLUT-4 glucose transporters, although the type of training, strain, age and sex of the animals seem to affect significantly the expression of GLUT-4 [100]. On the other hand, Etgen et al. [101] found that exercise training of normal rats results in an elevated maximal insulin-stimulated hindlimb glucose uptake. They suggested that this increase was only partially explained by an increase in total muscle GLUT-4 protein content. A recent study showed that physical training improves in vivo mitochondrial function concomitantly with increased insulin sensitivity in type 2 diabetes patients and control participants [102].

6. Exercise and chronical complications of diabetes

6.1. Exercise and cardiovascular disease (CVD)

Regular exercise has beneficial effects on glucose control and cardiovascular disease (CVD) risk factors. Exercise improves and maintains cardiorespiratory fitness, muscular strength, endurance, and body composition [103]. Exercise has a favorable effect on cardiovascular risk factors. In particular, it has specific beneficial effects on the reduction of hypertension, hyperlipidemia, and obesity and the improvement in blood lipid profile [104] even when combined with a rigorous calorie-restricted diet in obese patients with T2DM [105].

The effects of exercise training on abnormal vascular structure and function (including endothelial dysfunction and vascular distensibility) associated with diabetes are yet to be fully understood [106].

Oxidative stress has been suggested to play a role in either the primary or secondary etiology of both congestive heart failure (CHF) and coronary arter disease (CAD) [107, 108] evident by increased oxidative stress biomarkers and/or decreased antioxidant defenses at rest in diseased compared to healthy controls [109]. Increased TBARS [110, 111] and GSSG [110] have been reported following submaximal aerobic exercise in type 1 diabetic subjects. In regards to maximal exercise, direct production of RONS via electron spin resonance spectroscopy has been reported following a graded exercise testing. However, it is important to note that significance was only achieved when data for both type 1 diabetic and healthy control subjects were pooled [112]. Despite the observation of increased levels of exercise-induced oxidative stress biomarkers in studies involving type 1 diabetics, when compared to healthy individuals, the relative magnitude of increase does not differ; rather the group differences at rest are merely maintained during the post exercise period. Other investigators have reported no changes in MDA [112], total glutathione (TGSH), antioxidant enzyme activity or circulating antioxidants [111] in response to acute exercise in type 1 diabetics .

6.2. Diabetic nephropathy

Diabetic nephropathy is the most feared complication of diabetes, due to its substantial co-morbidity (need for dialysis, blindness, amputations, etc.), cost, and mortality (the annual mortality rate of diabetic patients with kidney failure on dialysis is about 25%) [113, 114]. The major determinants of kidney disease and its progression to end-stage kidney failure in diabetes are uncontrolled blood glucose, blood pressure and albuminuria [115, 116].

Diabetic nephropathy is an important complication of diabetes since it can lead to end-stage renal failure and also it is a risk factor of cardiovascular disease. The clinical problems caused by diabetic nephropathy are proteinuria and decreased renal function. Diabetic nephropathy is defined by proteinuria > 500 mg in 24 hours in the setting of diabetes, but this is preceded by lower degrees of proteinuria, or "microalbuminuria.". Microalbuminuria is defined as albumin excretion of 30–299 mg/24 hours [11]. Without intervention, diabetic patients with microalbuminuria typically progress to proteinuria and overt diabetic nephropathy.

In vitro studies indicate that hyperglycemia directly enhances oxidative stress in cultured endothelial and mesangial cells, which are targets for injury in diabetes [38, 117]. Several different antioxidants, including vitamin E (VE), vitamin C (VC), taurine, and α-lipoic acid (LA), have been reported to ameliorate renal injury in experimental diabetes [118-120]. In human diabetes, there is evidence that short-term (3 to 4 mo), high-dose (1600 to 1800 IU/d) VE supplementation reduces proteinuria in type 1 and 2 patients with overt nephropathy and decreases hyperfiltration in type 1 patients without overt nephropathy [121].

Urinary albumin excretion occurs normally after exercise [122, 123]. Post-exercise urinary albumin excretion is explained by increased glomerular capillary membrane permeability as a result of increased filtration pressure with increased filtered protein load, and decreased tubular absorption [122, 123]. In normal subjects, proteinuria is better related to exercise intensity and lactate production than to exercise duration [124], it diminishes after 1 h [125] and returns to baseline within 24 h [122]. In diabetes mellitus, the kidneys are more sensitive to the haemodynamic exercise stress [122]. Under exercise, patients with Type 1 diabetes show a partial depletion of negative charges on the glomerular capillary wall [126] that permits the increase of urinary albumin excretion [127]. Reports indicate that urinary albumin excretion increases after exercise, without correlation with glycaemic control, renal function, disease evolution or resting urinary albumin excretion [128, 129]. In contrast, post-exercise albuminuria has been found to be associated with HbA1c [127, 129].

On the other hand, recent data demonstrated that exercise might protect the diabetic renal function [130]. Kutlu et al. demonstrated that moderate exercise with combined vitamin E and C supplement was strengthen the antioxidant defense system and reduced the lipid peroxidation in STZ-induced diabetic rat kidney [131].

6.3. Diabetic retinopathy

Diabetic retinopathy may be the most common microvascular complication of diabetes. The risk of developing diabetic retinopathy or other microvascular complications of diabetes depends on both the duration and the severity of hyperglycemia. Development of diabetic retinopathy in patients with type 2 diabetes was found to be related to both severity of hyperglycemia and presence of hypertension in the U.K Prospective Diabetes Study (UKPDS) [132] and most patients with type 1 diabetes develop evidence of retinopathy within 20 years of diagnosis [133]. Retinopathy may begin to develop as early as 7 years before the diagnosis of diabetes in patients with type 2 diabetes [134]. There are several proposed pathological mechanisms by which diabetes may lead to development of retinopathy.

Oxidative stress may also play an important role in cellular injury from hyperglycemia. High glucose levels can stimulate free radical production and reactive oxygen species formation. Animal studies have suggested that treatment with antioxidants, such as vitamin E, may attenuate some vascular dysfunction associated with diabetes, but treatment with antioxidants has not yet been shown to alter the development or progression of retinopathy or other microvascular complications of diabetes [11].

6.4. Diabetic neuropathy

Diabetic polyneuropathy affects 30% of the hospital-based population and 20% of community based samples of diabetic patients [135]. There is a growing body of evidence to support the notion that oxidative stress is the biochemical trigger for nerve dysfunction. Various disturbances such as reduced endoneurial blood flow, altered electroconductive properties of the myelin sheath, impaired incorporation of acetate and glucose into the neuron cells should also be mentioned in diabetic condition. It has been shown [136] that superoxide dismutase activity is decreased in nerves from streptozotocin-induced diabetic rats. Glutathione content and glutathione peroxidase activity are also diminished in sciatic nerves from diabetic rats [137, 138]. Nerves of diabetic rats show lower amounts of vitamin E compared to control animals [139]. Lipid peroxidation products such as malondialdehydes or conjugated dienes are elevated in diabetic sciatic nerves [136, 139]. Treatment of diabetic rats with insulin or antioxidants is associated with improved nerve function [51, 140].

Sensory, visual and auditory neural conduction deficits are well documented both in diabetic animals and human studies. As an early marker of visual system deficits observed in diabetic state, visual evoked potential (VEP) latencies were measured in STZ-induced diabetic rats in our laboratories. The results of the previous studies were demonstrated that visual evoked potential (VEP) latencies were prolonged in STZ-induced diabetic rats whereas the latencies were restored by moderate physical exercise [60, 141, 142]. The VEP alterations were found to be accompanied with the increased TBARS concentration in the brain tissues of the diabetic rats.

The impact of diabetes on nervous system is complex and poorly elucidated. The brain is particularly vulnerable to oxidative damage because of its high rate of oxygen consumption, intense production of reactive radicals, and high levels of transition metals, such as iron, that catalyze the production of reactive radicals [143]. Moreover, neuronal membranes are rich in poly unsaturated fatty acids, which are a source of lipid peroxidation [37]. Free radicals are formed disproportionately in diabetes by glucose oxidation, non-enzymatic glycation of proteins, and the subsequent oxidative degradation of glycated proteins. Abnormally high levels of free radicals and the simultaneous decline of antioxidant defense mechanisms can lead to damage of cellular organelles and enzymes, increased lipid peroxidation, and development of insulin resistance. These consequences of oxidative stress can promote the development of complications of diabetes mellitus [143].

Previous experimental studies demonstrated that diabetes resulted increased lipid peroxidation and decreased antioxidant enzymes in several brain regions such as hypocampus, striatum and cerebral cortex as well as in whole brain tissue homogenates [59]. The lipids oxidation in the CNS usually demonstrates different concentrations at different regions of the brain, and it can be attributed to regional differences in the O2 consumption [144, 145].

Nervous system complications of diabetes mellitus can become one of the most debilitating complications and affect sensitive and cognitive functions that modulates memory function, resulting in significant functional impairment and dementia. Oxidative stress forms the foundation for the induction of multiple cellular pathways that can ultimately lead to both the onset and subsequent complications of DM [146]. Defects in hippocampal synaptic plasticity and transmission resulting in impairment of learning and memory is one the central nervous system complications of diabetes mellitus [147, 148]. Increasing evidence in both experimental and clinical studies suggests that oxidative stress plays a central role in the onset and subsequent complications of diabetes mellitus [149].

Physical exercise has been demonstrated to induce several neurobiological changes in the brain and to prevent diabetes-induced cognitive decline. The neurobiological changes induced by physical exercise have been demonstrated to facilitate the acquisition of a spatial memory task in rats. Exercise has also been demonstrated to increase the cognitive function both in healthy and diabetic people [150, 151, 152]. However, intense exercise has been shown to impaired the cognitive function in murine model that was prevented by vitamin C and E supplementation [153].

7. Conclusion

Literature results emphasize the beneficial role of physical exercise in the promotion of in diabetic complications probably by decreasing hyperglicemia, increasing insulin sensitivity and enhancing antioxidant status of the several systems. The type, duration and the intensity of the exercise as well as the degree of the diabetic complications should be determined before the exercise prescription in diabetic person. For future research, the effects of the different exercise protocols for maintaining the optimum health and stimulating the cellular processes for decreasing the hyperglicemia-induced complications in diabetes in children and older people remains to be explored.

Author details

Yaşar Gül Özkaya
School of Physical Education and Sports, Akdeniz University, Antalya, Turkey

8. References

[1] World Health Organization. Media Centre. Diabetes. http://www.who.int/mediacentre/factsheets/fs312/en/index.html (accessed 12 May 2008).

[2] Maritim, A. C., R. A. Sanders & J. B. Watkins. Diabetes, oxidative stress, and antioxidants: A review. Journal of Biochemical and Molecular Toxicology. 2003;17:24-38.

[3] Paik SG, Blue ML, Fleischer N, Shin SI. Diabetes susceptibility of balb/cbom mice treated with streptozotocin - inhibition by lethal irradiation and restoration by splenic lymphocytes. Diabetes. 1982;31(9):808-15.

[4] Sandler S, Andersson AK, Barbu A, Hellerstrom C, Holstad M, Karlsson E, Sandberg JO, Strandell E, Saldeen J, Sternesjo J, Tillmar L, Eizirik DL, Flodstrom M & Welsh N. Novel experimental strategies to prevent the development of type 1 diabetes mellitus. Upsala Journal of Medical Sciences. 2000;105(2):17–34.

[5] Shewad Y, Tirth S & Bhonde RR. Pancreatic islet-cell viability, functionality and oxidative status remain unaffected at pharmacological concentrations of commonly used antibiotics in vitro. Journal of Biosciences. 2001;26(3):349–355.

[6] American Diabetes Association (ADA) Position Statement (2010). Diagnosis and Classification of Diabetes Mellitus. Diabetes Care.2010;33(Supplement 1): S62-S69.

[7] de Lemos ET, Pinto Ri Oliveira J, Garrido P, Sereno J, Mascarenhas-Melo F, Pascoa Pinheiro J, Teixeira F & Reis F. (2011). Differential effects of acute (extenuating) and chronic (training) exercise on inflammation and oxidative stress status in an animal model of Type 2 diabetes mellitus. Mediators of Inflammation. 2011;253061:1-8.

[8] P′erez-Matute P, Zulet MA & Mart′ınez JA (2009). Reactive species and diabetes: Counteracting oxidative stress to improve health. Current Opinion in Pharmacology. 2009;9(6):771–779.

[9] Brunner Y, Schvartz D, Priego-Capote F, Cout′e Y & Sanchez JC (2009). Glucotoxicity and pancreatic proteomics. Journal of Proteomics.2009;71(6):576–591.

[10] Donath MY, Schumann DM, Faulenbach M, Ellingsgaard H, Perren A & Ehses JA. Islet inflammation in type 2 diabetes: from metabolic stress to therapy. Diabetes Care. 2008;31(Supplement 2):S161–164.

[11] Fowler MJ. Microvascular and Macrovascular Complications of Diabetes. Clinical Diabetes. 2008;26(2):77-82.

[12] Ballas LM, Jirousek MR, Umeda F, Nawata H, King GL. Vitamin E prevents diabetes induced abnormal retinal blood flow via the diacylglycerolprotein kinase C pathway. American Journal of Physiology. 1995;269: E239– E246.

[13] Watkins PJ. Retinopathy. BMJ.2003;326:924–926.

[14] Gross JL, de Azevedo MJ, Silveiro SP, Canani LH, Caramori ML & Zelmanovitz T. Diabetic nephropathy: diagnosis, prevention, and treatment. Diabetes Care. 2005;28:164-176.

[15] Chaturvedi N, Bandinelli S, Mangili R, Penno G, Rottiers RE.& Fuller JH. Microalbuminuria in type 1 diabetes: rates, risk factors and glycemic threshold. Kidney International. 2001;60: 219– 227.

[16] Adler AI, Stevens RJ, Manley SE, Bilous RW, Cull CA & Holman RR (2003). Development and progression of nephropathy in type 2 diabetes: the United Kingdom Prospective Diabetes Study (UKPDS 64). Kidney International. 2003;63:225–232.

[17] Kunisaki M, Bursell SE, Clermont AC, Ishii H, Ballas LM, Jirousek MR, Umeda F, Nawata H, King GL. Vitamin E prevents diabetes-induced abnormal retinal blood flow via the diacylglycerol-protein kinase C pathway. American Journal of Physiology-Endocrinology and Metabolism. 1995;269(2):E239-E46.

[18] Sigal RJ, Kenny GP, Wasserman DH, Castaneda-Sceppa C & White RD. Physical activity/exercise and type 2 diabetes: a consensus statement from the American Diabetes Association. Diabetes Care. 2006;29(6):1433-8.

[19] Colberg SR, Albright AL, Blissmer BJ, Braun B, Chasan-Taber L, Fernhall B, Regensteiner JG, Rubin RR & Sigal RJ. Exercise and Type 2 Diabetes ACSM ADA Joint Position Statement. Medicine & Science In Sports & Exercise. 2010;42(12): 2282-2303.

[20] Marrero DG. Time to get moving: helping patients with diabetes adopt exercise as part of a healthy lifestyle. Clinical Diabetes. 2005;23 (4):154-159.

[21] Hevener AL, Reichart D & Olefsky J. Exercise and thiazolidinedione therapy normalize insulin action in the obese Zucker fatty rat. Diabetes. 2000;49(12):2154– 2159.

[22] Pold R, Jensen LS, Jessen N et al. Long-term AICAR administration and exercise prevents diabetes in ZDF rats. Diabetes. 2005;54(4):928–934.

[23] Klein S, Sheard NF, Pi-Sunyer X, Daly A, Wylie-Rosett J, Kulkarni K & Clark NG. American Diabetes Association, American Association for the Study of Obesity, American Society for Clinical Nutrition: Weight management through lifestyle modification for the prevention and management of type 2 diabetes: rationale and strategies. Diabetes Care.2004;27:2067–2073.

[24] Boule N, Haddad E, Kenny G, Wells G & Sigal R. Effects of exercise on glycemic control and body mass in type 2 diabetes: a meta-analysis of controlled clinical trials. JAMA.2001;286:1218 1227.

[25] de Lemos ET, Reis F, Baptista S. et al. Exercise training is associated with improved levels of C-reactive protein and adiponectin in ZDF (type 2) diabetic rats. Medical Science Monitor. 2007;13(8):BR168–BR174.

[26] Connor TJ, Brewer C, Kelly JP, Harkin A. Acute stress suppresses pro-inflammatory cytokines TNF-alpha and IL-1 beta independent of a catecholamine-driven increase in IL-10 production. Journal of Neuroimmunology. 2005;159(1-2):119-28.

[27] Giraldo E, Garcia JJ, Hinchado MD, Ortega E. Exercise Intensity-Dependent Changes in the Inflammatory Response in Sedentary Women: Role of Neuroendocrine Parameters in the Neutrophil Phagocytic Process and the Pro-/Anti-Inflammatory Cytokine Balance. Neuroimmunomodulation. 2009;16(4):237-44.

[28] Jiang ZY, Woollard ACS, Wolff SP. Hydrogen-peroxide production during experimental protein glycation. Febs Letters. 1990;268(1):69-71.

[29] Wolff SP, Dean RT. Glucose autoxidation and protein modification - the potential role of autoxidative glycosylation in diabetes. Biochemical Journal. 1987;245(1):243-50.

[30] Halliwell B, Gutteridge JMC. Role of free-radicals and catalytic metal-ions in human-disease - an overview. Methods in Enzymology. 1990;186:1-85.

[31] Hogg N, Kalyanaraman B, Joseph J, Struck A, Parthasarathy S. Inhibition of low-density-lipoprotein oxidation by nitric-oxide - potential role in atherogenesis. Febs Letters. 1993;334(2):170-4.

[32] Tsai EC, Hirsch IB, Brunzell JD, Chait A. Reduced plasma peroxyl radical trapping capacity and increased susceptibility of LDL to oxidation in poorly controlled IDDM. Diabetes. 1994;43(8):1010-4.

[33] Kawamura M, Heinecke JW, Chait A. Pathophysiological concentrations of glucose promote oxidative modfication of low-density-lipoprotein by a superoxide-dependent pathway. Journal of Clinical Investigation. 1994;94(2):771-8.

[34] Hori O, Yan SD, Ogawa S, Kuwabara K, Matsumoto M, Stern D, et al. The receptor for advanced glycation end-products has a central role in mediating the effects of advanced glycation end-products on the development of vascular disease in diabetes mellitus. Nephrology Dialysis Transplantation. 1996;11:13-6.

[35] Mullarkey CJ, Edelstein D, Brownlee M. Free-radical generation by early glycation products - a mechanism for accelerated atherogenesis in diabetes. Biochemical and Biophysical Research Communications. 1990;173(3):932-9.

[36] McCarthy AD, Etcheverry SB, Cortizo AM. Effect of advanced glycation endproducts on the secretion of insulin-like growth factor-I and its binding proteins: role in osteoblast development. Acta Diabetologica. 2001;38(3):113-22.

[37] Baynes JW. Role of oxidative stress in development of complications in diabetes. Diabetes. 1991;40(4):405-12.

[38] Baynes JW, Thorpe SR. Role of oxidative stress in diabetic complications - A new perspective on an old paradigm. Diabetes. 1999;48(1):1-9.

[39] Vlassara H. Recent progress in advanced glycation end products and diabetic complications. Diabetes. 1997;46:S19-S25.

[40] Wautier JL, Wautier MP, Schmidt AM, Anderson GM, Hori O, Zoukourian C, et al. Advanced glycation end-products (ages) on the surface of diabetic erythrocytes bind to the vessel wall via a specific receptor inducing oxidant stress in the vasculature - a link

between surface-associated ages and diabetic complications. Proceedings of the National Academy of Sciences of the United States of America. 1994;91(16):7742-6.

[41] Mohamed AK, Bierhaus A, Schiekofer S, Tritschler H, Ziegler R, Nawroth PP. The role of oxidative stress and NF-kappa B activation in late diabetic complications. Biofactors. 1999;10(2-3):157-67.

[42] Kador PF, Kinoshita JH. Diabetic and galactosemic cataracts. Ciba Foundation Symposia. 1984;106:110-31.

[43] Greene DA, Sima AAF, Stevens MJ, Feldman EL, Lattimer SA. Complications - neuropathy, pathogenetic considerations. Diabetes Care. 1992;15(12):1902-25.

[44] Obrosova I, Faller A, Burgan J, Ostrow E, Williamson JR. Glycolytic pathway, redox state of NAD(P)-couples and energy metabolism in lens in galactose-fed rats: Effect of an aldose reductase inhibitor. Current Eye Research. 1997;16(1):34-43.

[45] Gonzalez AM, Sochor M, Hothersall JS, Mclean P. Effect of aldose reductase inhibitor (sorbinil) on integration of polyol pathway, pentose-phosphate pathway, and glycolytic route in diabetic rat lens. Diabetes. 1986;35(11):1200-5.

[46] Cheng HM, Gonzalez RG. The effect of high glucose and oxidative stress on lens metabolism, aldose reductase, and senile cataractogenesis. Metabolism-Clinical and Experimental. 1986;35(4):10-4.

[47] Yano M, Matsuda S, Bando Y, Shima K. Lens protein glycation and the subsequent degree of opacity in streptozotocin-diabetic rats. Diabetes Research and Clinical Practice. 1989;7(4):259-62.

[48] Ramalho JS, Marques C, Pereira PC & Mota MC. Role of glycation in human lens protein structure change. European Journal of Ophthalmology.1996;6(2):155–161.

[49] Mizisin AP, Bache M, distefano PS, Acheson A, Lindsay RM, Calcutt NA. BDNF attenuates functional and structural disorders in nerves of galactose-fed rats. Journal of Neuropathology and Experimental Neurology. 1997;56(12):1290-301.

[50] Hounsom L, Horrobin DF, Tritschler H, Corder R, Tomlinson DR. A lipoic acid gamma linolenic acid conjugate is effective against multiple indices of experimental diabetic neuropathy. Diabetologia. 1998;41(7):839-43.

[51] Stevens MJ, Obrosova I, Cao XH, Van Huysen C, Greene DA. Effects of DL-alpha-lipoic acid on peripheral nerve conduction, blood flow, energy metabolism, and oxidative stress in experimental diabetic neuropathy. Diabetes. 2000;49(6):1006-15.

[52] Ralevic V, Belai A, Burnstock G. Effects of streptozotocin-diabetes on sympathetic-nerve, endothelial and smooth-muscle function in the rat mesenteric arterial bed. European Journal of Pharmacology. 1995;286(2):193-9.

[53] Mizisin AP, Kalichman MW, Bache M, Dines KC, distefano PS. NT-3 attenuates functional and structural disorders in sensory nerves of galactose-fed rats. Journal of Neuropathology and Experimental Neurology. 1998;57(9):803-13.

[54] Kalichman MW, Powell HC, Mizisin AP. Reactive, degenerative, and proliferative Schwann cell responses in experimental galactose and human diabetic neuropathy. Acta Neuropathologica. 1998;95(1):47-56.

[55] Chokroverty S, Seiden D, Navidad P & Cody R. Distal axonopathy in streptozotocin diabetes in rats. Experientia. 1988;44(5):444–446.

[56] Fernyhough P, Gallagher A, Averill SA, Priestley JV, Hounsom L, Patel J, et al. Aberrant neurofilament phosphorylation in sensory neurons of rats with diabetic neuropathy. Diabetes. 1999;48(4):881-9.

[57] Saxena AK, Srivastava P, Kale RK, Baquer NZ. Impaired antioxidant status in diabetic rat-liver - effect of vanadate. Biochemical Pharmacology. 1993;45(3):539-42.

[58] Feher J, Cosmos G & Vereckei A. Free Radical Reactions in Medicine. Berlin:Springer-Verlag;1987.

[59] Ozkaya YG, Agar A, Yargicoglu P, Hacioglu G, Bilmen-Sarikcioglu S, Ozen I, et al. The effect of exercise on brain antioxidant status of diabetic rats. Diabetes & Metabolism. 2002;28(5):377-84.

[60] Ozkaya YG, Hacioglu G, Kucukatay V, Yargicoglu P, Agar A. The Effect of Chronic N(G)-Nitro-L-arginine Methyl Ester (L-NAME) Administration on Visual Evoked Potentials and Oxidative Stress in Streptozotocin Induced Diabetic Rats. Journal of Neurological Sciences-Turkish. 2011;28(2):132-41.

[61] Dillard CJ, Litov RE, Savin WM, Dumelin EE, Tappel AL. Effects of exercise, vitamin-E, and ozone on pulmonary-function and lipid peroxidation. Journal of Applied Physiology. 1978;45(6):927-32.

[62] Morrow JD, Roberts LJ. Mass spectrometric quantification of F-2-isoprostanes in biological fluids and tissues as measure of oxidant stress. Oxidants and Antioxidants, Pt B. 1999;300:3-12.

[63] Knight JA. Free Radicals, antioxidants, aging, and disease. Washington: American Association for Clinical Chemistry Press, 1999.

[64] Halliwell B, Cross CE. Oxygen-derived species - their relation to human-disease and environmental-stress. Environmental Health Perspectives. 1994;102:5-12.

[65] Dalle-Donne I, Rossi R, Colombo R, Giustarini D, Milzani A. Biomarkers of oxidative damage in human disease. Clinical Chemistry. 2006;52(4):601-23.

[66] Fisher-Wellman K & Bloomer RJ. Acute exercise and oxidative stress: a 30 year history. Dynamic Medicine. 2009;8(1):1-8.

[67] Knez WL, Jenkins DG, Coombes JS. Oxidative stress in half and full ironman triathletes. Medicine and Science in Sports and Exercise. 2007;39(2):283-8.

[68] Steinberg JG, Ba A, Bregeon F, Delliaux S, Jammes Y. Cytokine and oxidative responses to maximal cycling exercise in sedentary subjects. Medicine and Science in Sports and Exercise. 2007;39(6):964-8.

[69] Miyazaki H, Oh-ishi S, Ookawara T, Kizaki T, Toshinai K, Ha S, et al. Strenuous endurance training in humans reduces oxidative stress following exhausting exercise. European Journal of Applied Physiology. 2001;84(1-2):1-6.

[70] Nikolaidis MG, Kyparos A, Hadziioannou M, Panou N, Samaras L, Jamurtas AZ, et al. Acute exercise markedly increases blood oxidative stress in boys and girls. Applied

Physiology Nutrition and Metabolism-Physiologie Appliquee Nutrition Et Metabolisme. 2007;32(2):197-205.

[71] Meijer EP, Goris AHC, van Dongen JLJ, Bast A, Westerterp KR. Exercise-induced oxidative stress in older adults as a function of habitual activity level. Journal of the American Geriatrics Society. 2002;50(2):349-53.

[72] Rahnama N, Gaeini AA, Hamedinia MR. Oxidative stress responses in physical education students during 8 weeks aerobic training. Journal of Sports Medicine and Physical Fitness. 2007;47(1):119-23.

[73] Silvestro A, Scopacasa F, Oliva G, de Cristofaro T, Iuliano L, Brevetti G. Vitamin C prevents endothelial dysfunction induced by acute exercise in patients with intermittent claudication. Atherosclerosis. 2002;165(2):277-83.

[74] Gaeini AA, Rahnama N, Hamedinia MR. Effects of vitamin E supplementation on oxidative stress at rest and after exercise to exhaustion in athletic students. Journal of Sports Medicine and Physical Fitness. 2006;46(3):458-61.

[75] Morillas-Ruiz J, Zafrilla P, Almar M, Cuevas M, Lopez FJ, Abellan P, et al. The effects of an antioxidant-supplemented beverage on exercise-induced oxidative stress: results from a placebo-controlled double-blind study in cyclists. European Journal of Applied Physiology. 2005;95(5-6):543-9.

[76] Goldfarb AH, Patrick SW, Bryer S, You TJ. Vitamin C supplementation affects oxidative-stress blood markers in response to a 30-minute run at 75% VO2max. International Journal of Sport Nutrition and Exercise Metabolism. 2005;15(3):279-90.

[77] Di Massimo C, Scarpelli P, Tozzi-Ciancarelli MG. Possible involvement of oxidative stress in exercise-mediated platelet activation. Clinical Hemorheology and Microcirculation. 2004;30(3-4):313-6.

[78] Tozzi-Ciancarelli MG, Penco M, Di Massimo C. Influence of acute exercise on human platelet responsiveness: possible involvement of exercise-induced oxidative stress. European Journal of Applied Physiology. 2002;86(3):266-72.

[79] Alessio HM, Hagerman AE, Fulkerson BK, Ambrose J, Rice RE, Wiley RL. Generation of reactive oxygen species after exhaustive aerobic and isometric exercise. Medicine and Science in Sports and Exercise. 2000;32(9):1576-81.

[80] Watson TA, Callister R, Taylor RD, Sibbritt DW, macdonald-Wicks LK, Garg ML. Antioxidant restriction and oxidative stress in short-duration exhaustive exercise. Medicine and Science in Sports and Exercise. 2005;37(1):63-71.

[81] Elosua R, Molina L, Fito M, Arquer A, Sanchez-Quesada JL, Covas MI, et al. Response of oxidative stress biomarkers to a 16-week aerobic physical activity program, and to acute physical activity, in healthy young men and women. Atherosclerosis. 2008;197(2):967.

[82] Buczynski A, Kedziora J, Tkaczewski W, Wachowicz B. Effect of submaximal physical exercise on antioxıdative protection of human blood-platelets. International Journal of Sports Medicine. 1991;12(1):52-4.

[83] Chen MF, Hsu HC, Lee YT. Effects of acute exercise on the changes of lipid profiles and peroxides, prostanoids, and platelet activation in hypercholesterolemic patients before and after treatment. Prostaglandins. 1994;48(3):157-74.

[84] Michailidis Y, Jamurtas AZ, Nikolaidis MG, Fatouros IG, Koutedakis Y, Papassotiriou I, et al. Sampling time is crucial for measurement of aerobic exercise-induced oxidative stress. Medicine and Science in Sports and Exercise. 2007;39(7):1107-13.

[85] Vider J, Lehtmaa J, Kullisaar T, Vihalemm T, Zilmer K, Kairane C, Landor A, Karu T &, Zilmer M. Acute immune response in respect to exercise-induced oxidative stress. Pathophysiology.2001;7(4):263-270.

[86] Akova B, Surmen-Gur E, Gur H, Dirican M, Sarandol E, Kucukoglu S. Exercise-induced oxidative stress and muscle performance in healthy women: role of vitamin E supplementation and endogenous oestradiol. European Journal of Applied Physiology. 2001;84(1-2):141-7.

[87] Tauler P, Aguilo A, Gimeno I, Fuentespina E, Tur JA, Pons A. Response of blood cell antioxidant enzyme defences to antioxidant diet supplementation and to intense exercise. European Journal of Nutrition. 2006;45(4):187-95.

[88] Fatouros IG, Jamurtas AZ, Villiotou V, Pouliopoulou S, Fotinakis P, Taxildaris K, et al. Oxidative stress responses in older men during endurance training and detraining. Medicine and Science in Sports and Exercise. 2004;36(12):2065-72.

[89] Guzel NA, Hazar S, Erbas D. Effects of different resistance exercise protocols on nitric oxide, lipid peroxidation and creatine kinase activity in sedentary males. Journal of Sports Science and Medicine. 2007;6(4):417-22.

[90] Chung HY, Cesari M, Anton S, Marzetti E, Giovannini S, Seo AY, et al. Molecular inflammation: Underpinnings of aging and age-related diseases. Ageing Research Reviews. 2009;8(1):18-30.

[91] Radak Z, Chung HY, Koltai E, Taylor AW & Goto S. Exercise, oxidative stress and hormesis. Ageing Research Reviews. 2008;7(1):34-42.

[92] Gocmen AY, Ozkaya YG, Yazar H, Agar A, Gunaydin I & Gumuslu S. (2011b). Effect of exercise on paraoxonase-1 activity and lipid peroxidation in diabetes. Bozok Medical Journal. 2011;1(2):13-21.

[93] Atalay M, Marnila P, Lilius EM, Hanninen O, Sen CK. Glutathione-dependent modulation of exhausting exercise induced changes in neutrophil function of rats. European Journal of Applied Physiology and Occupational Physiology. 1996;74(4):342-7.

[94] Atalay M, Seene T, Hanninen O, Sen CK. Skeletal muscle and heart antioxidant defences in response to sprint training. Acta Physiologica Scandinavica. 1996;158(2):129-34.

[95] Powers SK, Ji LL, Leeuwenburgh C. Exercise training-induced alterations in skeletal muscle antioxidant capacity: a brief review. Medicine and Science in Sports and Exercise. 1999;31(7):987-97.

[96] Steppel TH & Horton ES. (2005). Exercise in patients with diabetes mellitus. In: Kahn CR, Weir GC, King GL, Jacobson AM, Moses AC, Smith RJ, Editors: Joslins' Diabetes Mellitus. Philadelphia: Lippincott Williams and Wilkins, 2005. pp: 649-658.

[97] Hughes VA, Fiatrone MA, Fielding RA, Kahn BB, Ferrara CM, Shepherd P, Fisher EC, Wolfe RR, Elahi D & Evans WJ. Exercise increases muscle GLUT-4 levels and insulin action in subjects with impaired glucose tolerance. American Journal of Physiology.1993;264:E855–E862.

[98] Cox JH, Cortright RN, Dohm GL, Houmard JA. Effect of aging on response to exercise training in humans: skeletal muscle GLUT-4 and insulin sensitivity. Journal of Applied Physiology. 1999;86(6):2019-25.

[99] Giacca A, Groenewoud Y, Tsui E, McClean P, Zinman B. Glucose production, utilization, and cycling in response to moderate exercise in obese subjects with type 2 diabetes and mild hyperglycemia. Diabetes. 1998 Nov;47(11):1763-70.

[100] Kainulainen H, Komulainen J, Joost HG, Vihko V. Dissociation of the effects of training on oxidative-metabolism, glucose-utilization and GLUT4 levels in skeletal-muscle of streptozotocin-diabetic rats. Pflugers Archiv-European Journal of Physiology. 1994;427(5-6):444-9.

[101] Etgen GJ, Brozinick JT, Kang HY, Ivy JL. Effects of exercise training on skeletal-muscle glucose-uptake and transport. American Journal of Physiology. 1993;264(3):C727-C33.

[102] Wohaieb SA, Godin DV. Alterations in free-radical tissue-defense mechanisms in streptozocin-induced diabetes in rat - effects of insulin-treatment. Diabetes. 1987;36(9):1014-8.

[103] Marwick TH, Hordern MD, Miller T, Chyun DA, Bertoni AG, Blumenthal RS, et al. Exercise Training for Type 2 Diabetes Mellitus Impact on Cardiovascular Risk A Scientific Statement From the American Heart Association. Circulation. 2009;119(25):3244-62.

[104] Wilmore JH, Green JS, Stanforth PR, Gagnon J, Rankinen T, Leon AS, et al. Relationship of changes in maximal and submaximal aerobic fitness to changes in cardiovascular disease and non-insulin-dependent diabetes mellitus risk factors with endurance training: The Heritage Family Study. Metabolism-Clinical and Experimental. 2001;50(11):1255-63.

[105] Pi-Sunyer, X., Blackburn, G., Brancati, F.L., et al. (2007). Look AHEAD Research Group. Reduction in weight and cardiovascular disease risk factors in individuals with type 2 diabetes: one-year results of the Look AHEAD trial. Diabetes Care. Volume 30, pp:1374–1383.

[106] Simpson SH, Corabian P, Jacobs P, Johnson JA. The cost of major comorbidity in people with diabetes mellitus. Canadian Medical Association Journal. 2003;168(13):1661-7.

[107] Fisher-Wellman K, Bell HK, Bloomer RJ. Oxidative stress and antioxidant defense mechanisms linked to exercise during cardiopulmonary and metabolic disorders. Oxidative Medicine and Cellular Longevity. 2009;2(1):43-51.

[108] Soccio M, Toniato E, Evangelista V, Carluccio M, De Caterina R. Oxidative stress and cardiovascular risk: the role of vascular NAD(P)H oxidase and its genetic variants. European Journal of Clinical Investigation. 2005;35(5):305-14.

[109] Tsutsui H. Oxidative stress in heart failure: The role of mitochondria. Internal Medicine. 2001;40(12):1177-82.

[110] Laaksonen DE, Uusitupa M, Atalay M, Hanninen O, Niskanen L, Sen CK. Increased resting and exercise-induced oxidative stress in young IDDM men. Diabetes Care. 1996;19(6):569-74.

[111] Atalay M, Laaksonen DE, Niskanen L, Uusitupa M, Hanninen O, Sen CK. Altered antioxidant enzyme defences in insulin-dependent diabetic men with increased resting and exercise-induced oxidative stress. Acta Physiologica Scandinavica.1997;161:195-201.

[112] Davison GW, George L, Jackson SK, Young IS, Davies B, Bailey DM, et al. Exercise, free radicals, and lipid peroxidation in type 1 diabetes mellitus. Free Radical Biology and Medicine. 2002;33(11):1543-51.

[113] Leehey DJ, Moinuddin I, Bast JP, Qureshi S, Jelinek CS, Cooper C, et al. Aerobic exercise in obese diabetic patients with chronic kidney disease: a randomized and controlled pilot study. Cardiovascular Diabetology. 2009;8:62.

[114] US Renal Data System: USRDS 2008 Annual Data Report Bethesda: National Institutes of Health, National Institute of Diabetes and Digestive and Kidney Diseases; 2008.

[115] Stratton IM, Cull CA, Adler AI, Matthews DR, Neil HAW, Holman RR. Additive effects of glycaemia and blood pressure exposure on risk of complications in type 2 diabetes: a prospective observational study (UKPDS 75). Diabetologia. 2006;49(8):1761-9.

[116] Rossing K, Christensen PK, Hovind P, Parving HH. Remission of nephrotic-range albuminuria reduces risk of end-stage renal disease and improves survival in type 2 diabetic patients. Diabetologia. 2005;48(11):2241-7.

[117] Giugliano D, Ceriello A, Paolisso G. Oxidative stress and diabetic vascular complications. Diabetes Care. 1996;19(3):257-67.

[118] Lee,EY, Lee MY, Hong SW, Choon Hee Chung CH, Hong SY. Blockade of oxidative stress by vitamin C ameliorates albuminuria and renal sclerosis in experimental diabetic rats. Yonsei Medical Journal. 2007; 31; 48(5): 847–855.

[119] Trachtman H, Futterweit S, Maesaka J, Ma C, Valderrama E, Fuchs A, et al. Taurine ameliorates chronic streptozocin-induced diabetic nephropathy in rats. American Journal of Physiology-Renal Fluid and Electrolyte Physiology. 1995;269(3):F429-F38.

[120] Craven PA, derubertis FR, Kagan VE, Melhem M, Studer RK. Effects of supplementation with vitamin C or E on albuminuria, glomerular TGF-beta, and glomerular size in diabetes. Journal of the American Society of Nephrology. 1997;8(9):1405-14.

[121] Bursell SE, Clermont AC, Aiello LP, Aiello LM, Schlossman DK, Feener EP, et al. High-dose vitamin E supplementation normalizes retinal blood flow and creatinine clearance in patients with type 1 Diabetes. Diabetes Care. 1999;22(8):1245-51.

[122] Poortmans JR, Labilloy D. The influence of work intensity on postexercise proteinuria. European Journal of Applied Physiology and Occupational Physiology. 1988;57(2):260-3.

[123] Heathcote KL, Wilson MP, Quest DW, Wilson TW. Prevalence and duration of exercise induced albuminuria in healthy people. Clinical and Investigative Medicine. 2009;32(4):E261-E5.

[124] Poortmans JR, Geudvert C, Schorokoff K, deplaen P. Postexercise proteinuria in childhood and adolescence. International Journal of Sports Medicine. 1996;17(6):448-51.

[125] Poortmans JR, Vanderstraeten J. Kidney-function during exercise in healthy and diseased humans - an update. Sports Medicine. 1994;18(6):419-37.

[126] Ala-Houhala I. Effects of exercise on glomerular passage of macromolecules in patients with diabetic nephropathy and in healthy subjects. Scandinavian Journal of Clinical and Laboratory Investigation.1990;50:27–33.

[127] Huttunen NP, Kaar ML, Puukka R, Akerblom HK. Exercise-induced proteinuria in children and adolescents with type-1 (insulin dependent) diabetes. Diabetologia. 1981;21(5):495-7.

[128] Feldtrasmussen B, Baker L, Deckert T. Exercise as a provocative test in early renal-disease in type-1 (insulin-dependent) diabetes - albuminuric, systemic and renal hemodynamic-responses. Diabetologia. 1985;28(7):389-96.

[129] Kruger M, Gordjani N, Burghard R. Postexercise albuminuria in children with different duration of type-1 diabetes mellitus. Pediatric Nephrology. 1996;10(5):594-7.

[130] Kurdak H, Sandikci S, Ergen N, Dogan A, Kurdak SS. The effects of regular aerobic exercise on renal functions in streptozotocin induced diabetic rats. Journal of Sports Science and Medicine. 2010;9(2):294-9.

[131] Kutlu M, Naziroglu M, Simsek H, Yilmaz T, Kukner AA. Moderate exercise combined with dietary vitamins C and E counteracts oxidative stress in the kidney and lens of streptozotocin-induced diabetic rat. International Journal for Vitamin and Nutrition Research. 2005;75(1):71-80.

[132] Turner RC, Holman RR, Cull CA, Stratton IM, Matthews DR, Frighi V, et al. Intensive blood-glucose control with sulphonylureas or insulin compared with conventional treatment and risk of complications in patients with type 2 diabetes (UKPDS 33). Lancet. 1998;352(9131):837-53.

[133] Keenan HA, Costacou T, Sun JK, Doria A, Cavellerano J, Coney J, et al. Clinical factors associated with resistance to microvascular complications in diabetic patients of extreme disease duration - The 50-year medalist study. Diabetes Care. 2007;30(8):1995-7.

[134] Fong DS, Aiello LP, Ferris FL, Klein R. Diabetic retinopathy. Diabetes Care. 2004;27(10):2540-53.

[135] Ziegler D, Sohr CGH, Nourooz-Zadeh J. Oxidative stress and antioxidant defense in relation to the severity of diabetic polyneuropathy and cardiovascular autonomic neuropathy. Diabetes Care. 2004;27(9):2178-83.

[136] Low PA, Nickander KK. Oxygen free-radical effects in sciatic-nerve in experimental diabetes. Diabetes. 1991;40(7):873-7.

[137] Stevens MJ, Lattimer SA, Kamijo M, Vanhuysen C, Sima AAF, Greene DA. Osmotically-induced nerve taurine depletion and the compatible osmolyte hypothesis in experimental diabetic neuropathy in the rat. Diabetologia. 1993;36(7):608-14.

[138] Hermenegildo C, Raya A, Roma J, Romero FJ. Decreased glutathione-peroxidase activity in sciatic-nerve of alloxan-induced diabetic mice and its correlation with blood-glucose levels. Neurochemical Research. 1993;18(8):893-6.

[139] Nickander KK, mcphee BR, Low PA, Tritschler H. Alpha-lipoic acid: Antioxidant potency against lipid peroxidation of neural tissues in vitro and implications for diabetic neuropathy. Free Radical Biology and Medicine. 1996;21(5):631-9.

[140] Low PA, Nickander KK, Tritschler HJ. The roles of oxidative stress and antioxidant treatment in experimental diabetic neuropathy. Diabetes. 1997;46:S38-S42.

[141] Ozkaya YG, Agar A, Hacioglu G, Yargicoglu P. Exercise improves visual deficits tested by visual evoked potentials in streptozotocin-induced diabetic rats. Tohoku Journal of Experimental Medicine. 2007;213(4):313-21.

[142] Ozkaya YG, Agar A, Hacioglu G, Yargicoglu P, Abidin I, Senturk UK. Training induced alterations of visual evoked potentials are not related to body temperature. International Journal of Sports Medicine. 2003;24(5):359-62.

[143] Kumar JSS, Menon VP. Effect of diabetes on levels of lipid peroxides and glycolipids in rat-brain. Metabolism-Clinical and Experimental. 1993;42(11):1435-9.

[144] Aguiar, A.S. & Pinho, R.A. (2007). Effects of physical exercise over the redox brain state. The Revista Brasileira de Medicina do Esporte. 2007;13(5): 355-360.

[145] Floyd RA, Carney JM. Age influence on oxidative events during brain ischemia reperfusion. Archives of Gerontology and Geriatrics. 1991;12(2-3):155-77.

[146] Maiese K, Chong ZZ, Shang YC. Mechanistic insights into diabetes mellitus and oxidative stress. Current Medicinal Chemistry. 2007;14(16):1729-38.

[147] Alipour M, Salehi I & Ghadiri Soufi F. (2012). Effect of Exercise on Diabetes-Induced Oxidative Stress in the Rat Hippocampus. Iran Red Crescent Medical Journal. 2012;14(4): 222-228.

[148] Kucukatay V, Hacioglu G, Ozkaya G, Agar A, Yargicoglu P. The effect of diabetes mellitus on active avoidance learning in rats: The role of nitric oxide. Medical Science Monitor. 2009;15(3):BR88-BR93.

[149] Tahirovic I, Sofic E, Sapcanin A, Gavrankapetanovic I, Bach-Rojecky L, Salkovic-Petrisic M, et al. Brain antioxidant capacity in rat models of betacytotoxic-induced experimental sporadic Alzheimer's disease and diabetes mellitus. Journal of Neural Transmission-Supplement. 2007(72):235-40.

[150] Ozkaya GY, Aydin H, Toraman NF, Kizilay F, Ozdemir O, Cetinkaya V. Effect of strength and endurance training on cognition in older people. Journal of Sports Science and Medicine. 2005;4(3):300-13.

[151] Cetin E, Top EC, Sahin G, Ozkaya YG, Aydin H, Toraman F. Effect of vitamin e supplementation with exercise on cognitive functions and total antioxidant capacity in older people. Journal of Nutrition Health & Aging. 2010;14(9):763-9.

[152] Gomez-Pinilla F, Ying Z, Roy RR, Molteni R, Edgerton VR. Voluntary exercise induces a BDNF-mediated mechanism that promotes neuroplasticity. Journal of Neurophysiology. 2002;88(5):2187-95.

[153] Rosa EF, Takahashi S, Aboulafia J, Nouailhetas VLA, Oliveira MGM. Oxidative stress induced by intense and exhaustive exercise impairs murine cognitive function. Journal of Neurophysiology. 2007;98(3):1820-6.

Chapter 16

Reactive Oxygen Species Act as Signaling Molecules in Liver Carcinogenesis

María Cristina Carrillo, María de Luján Alvarez, Juan Pablo Parody, Ariel Darío Quiroga and María Paula Ceballos

Additional information is available at the end of the chapter

http://dx.doi.org/10.5772/45966

1. Introduction

Reactive Oxygen Species (ROS) were viewed as the ''bad'' molecules of cells for a long time, but in the recent years, several lines of evidence indicate the contrary: ROS are essential participants in cell signaling and regulation depending on their concentration.

At present it is well established that ROS signaling is an important factor of many gene- and enzyme-catalyzed processes. ROS signaling is responsible for activation or inhibition of numerous processes catalyzed by protein kinases, phosphatases, and many other enzymes although these reactions proceed by heterolytic (non-free radical) mechanisms [1]. Therefore, ROS signaling can initiate both inhibition and activation of tumor formation. This fact might be of utmost importance for the development of anticancer treatment by the drugs possessing both prooxidant and antioxidant properties.

In this chapter, we summarize a series of experiments that have allowed us to establish the role of oxidative stress in the early development of liver cancer process and the effects of cytokines on the modulation of this process.

Through a series of *in vivo* and *in vitro* experiments we are able to describe:

- The oxidative stress status of a preneoplastic liver
- The modulating effect of Interferon α-2b (IFN α-2b) on this oxidative status that triggers the apoptotic mechanism in hepatic cells
- The role of TGFβ1 in the whole process
- The participation of FOXO transcription family proteins in the programmed cell death activated by IFN α-2b and TGFβ1

2. Experimental models of liver cancer development

Hepatocellular carcinoma (HCC) is a malignant solid tumor that arises from the major cell type in the liver: the hepatocyte. HCC is the most common type of primary hepatic tumor; it represents approximately 6% of all malignancies and is the fifth most common tumor worldwide [2].

Nearly all types of primary liver tumors known to occur in humans can be reproduced by chemicals in laboratory animals, especially in rats [3]. In experimental carcinogenesis, preneoplastic foci of altered hepatocytes (AHF) emerge weeks or months before the appearance of hepatocellular adenomas and HCCs [4,5] and this has also been discovered in human with hepatocellular neoplasms and/or cirrhosis [6]. This fact has led to the development of a number of *in vivo* systems for the study of early neoplasia in rat liver [7,8]. The initiation-promotion or two-stage model of cancer development mimics the early events of the latent period of human carcinogenesis. Several two stages models have been developed, including the protocols of Solt and Farber [9], Ito *et al.* [10] and Rao *et al.* [11], that comprise necrogenic doses of carcinogens or other models such as the protocols of Peraino et al. [12] and Pitot et al. [13] that use low, non toxic doses of carcinogens.

In this context, the initiation stage of cancer development can be produced in rat liver by the administration of diethylnitrosamine (DEN) [9–11], a complete carcinogen that produces DNA ethylation and mutagenesis [13]. Necrogenic doses of DEN cause massive hepatic necrosis followed by regeneration [14] and would be expected to cause not only increased gene expression related to regeneration, but also increased expression related to oncogene mutation. Administration of promoting agents causes selective enhancement of the proliferation of initiated cell populations over non-initiated cells in the target tissue [5].

Accordingly, we have developed a two-phase model of liver preneoplasia in rat: basically, the animals are initiated with two necrogenic doses of DEN and subsequently 2-acetylaminofluorene (2-AAF) is administered as promoting agent. The experimental protocol takes six weeks, and at the end of the treatment animals show 5% of liver tissue occupied by microscopic preneoplastic foci. A diagram of the experimental model is shown in Figure 1.

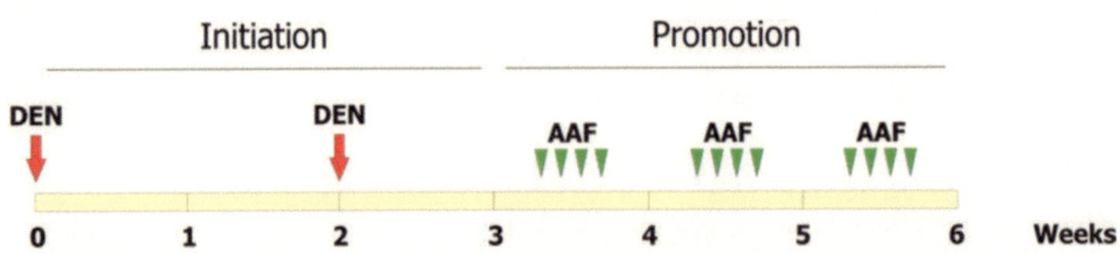

Figure 1. Two-phase or initiation-promotion (IP) model of rat chemical hepatocarcinogenesis. Initiation stage is performed by the administration of 2 necrogenic doses of diethylnitrosamine (DEN, 150 mg / kg body weight, intraperitoneal), separated by 2 weeks. A week after the last injection of DEN, the promotion phase begins by the administration of 2-acetylaminofluorene (2-AAF, 20 mg / kg body weight) by gavage, 4 days per week during 3 weeks. At the end of the sixth week, rat livers show microscopic preneoplastic foci.

The presence of preneoplastic foci in this two-phase (initiation-promotion, IP) hepatocarcinogenic model was determined using rat Pi class isoenzyme of glutathione S-transferase (GST) as a foci marker [15]. This isoenzyme has been described as the most effective single marker of hepatic preneoplasia in the rat [16], and immunohistochemical detection of Pi class GST is the most widely used method for identification, quantitation and assessment of rat AHF [17].

3. GSTs and liver preneoplasia: Our first studies

GSTs are a family of multifunctional dimeric enzymes with an important role in detoxification processes of several xenobiotics, including anticancer drugs, carcinogens and mutagens [18–20]. These enzymes catalyze the nucleophilic attack of reduced glutathione (GSH) on electrophilic compounds [19,21].

Thus, GSTs are part of a cellular defense system which also includes GSH levels (and enzymes related to its biosynthesis) and proteins involved in the uptake of drugs and in the excretion of glutathione conjugates [22]. In the liver, among the several cytosolic classes of GSTs, Pi class GST (GST P), is particularly interesting because its expression in the adult tissue is associated with preneoplastic and neoplastic development [23]. In addition, increased expression of GST P was found to be associated with resistance of tumor tissues to several cytostatic drugs [24,25].

There is a significative increase of GST P in preneoplastic livers. This enzyme has shown to be the more efficient isoenzyme in the catalysis of conjugation of ethacrynic acid (EA) with GSH. How does this enzyme act in the preneoplastic condition?

EA, an electrophilic loop diuretic drug, causes hepatotoxicity through lipid peroxidation mediated by its oxidative metabolism [26,27]. This drug has a preferential conjugation with GSH either spontaneous or GST catalyzed, reducing its intracellular levels and consequently favoring oxidative stress in isolated hepatocytes [27]. The glutathione conjugate of EA (EA-SG) is a substrate of human multidrug-resistance protein 2 (MRP2) and probably of rat Mrp2 [28]. Thus, it has been suggested that EA-SG is excreted through this active canalicular transport protein into bile [29]. In addition, EA and EA-SG (as well as many others α,α-unsaturated carbonyl derivatives and their glutathione conjugates) are important *in vivo* and *in vitro* inhibitors of several human and rat GSTs activities [19,30,31].

As was stated above, at the inactivation step, GSTs are playing a major role by catalyzing the conjugation reaction of GSH with the drug and leading to the inactivation of the therapeutic agent. EA and EA-SG have been proved to be good inhibitors of GSTs activities [19,30,31]. For this reason, we evaluated the enzymatic and cellular *in vitro* response to EA in isolated hepatocytes from preneoplastic rat livers, which present high levels of GST P, and analyzed the role of the GSTs/GSH system and Mrp2 (as a measure of the multidrug resistance) in these cells [15].

Results showed that hepatocytes from IP animals presented higher levels of cell viability than control hepatocytes in the presence of EA. In accordance with this data, IP hepatocytes showed lower levels of release of lactate dehydrogenase and alanine aminotransferase than control cells throughout the incubation time, indicating higher levels of cellular integrity. This suggests that hepatocytes from IP animals are more resistant to the cytotoxic effects of EA than control cells.

Control hepatocytes suspensions showed increased levels of lipid peroxidation measured through the quantification of TBARS (Thiobarbituric Acid Reactive Substances, [32]) production in a time- and dose-dependent manner in the presence of EA. This data was expected since oxidative metabolism of the drug and a subsequent lipid peroxidation was already described as part of the EA-induced toxicity [27]. However, IP hepatocytes suspensions did not show increased levels of lipid peroxidation during incubation at all times and EA-doses evaluated.

The higher basal levels of TBARS observed in preneoplastic hepatocytes could be attributed to the hepatocarcinogenic treatment, since it was described that lipid hydroperoxides are produced by some promotion regimens [33]. The unchanged levels during EA incubation are probably a consequence of both GST P activities: lipid peroxidase [34] and catalysis of EA conjugation with GSH, avoiding its oxidative metabolism.

Although intracellular total GSH (tGSH) levels decreased in both hepatocytes suspensions without EA, initial tGSH levels showed a mild although significantly higher value in hepatocytes from IP animals. This fact suggests that the small extra pool of tGSH is certainly an advantageous factor to prevent susceptibility to oxidative stress.

GST P has shown to be the more efficient isoenzyme in the catalysis of conjugation of EA with GSH [35,36] and may have a role in the detoxification of lipid hydroperoxides through its selenium-independent peroxidase activity [34]. We observed decreased levels of both Alpha and Mu class GSTs in preneoplastic hepatocytes. This fact, and the presence of GST P in hepatocytes from IP rats, gives to this induced isoenzyme a compensatory role in these cells. Based on the measurement of total GST activity and in data obtained from other publications [37,38], we have proposed that GST P could be playing a major role in the defense system against the cytotoxic effects of EA in our preneoplastic model. At high EA concentrations this resistance is overwhelmed over time, probably as a consequence of inhibition of GST P activity by EA-SG acummulation and depleted levels of intracellular tGSH. In the absence of GST P and GSH, EA may alkylate cell proteins thiols, which might be the major determinants of the cytotoxic effects observed with higher EA concentrations.

It has been demonstrated that MRP2 expression, the canalicular membrane protein reported to be the transporter of GSH and GSH conjugates, is higher in human HCCs than in normal cells [39]. MRP2 increased expression could suggest accelerated GSH depletion and hence, enhanced toxicity of cytotoxic compounds. On the other hand, diminution of MRP2 expression could indicate a preservation of GSH intracellular pool. In spite of the data demonstrated in human HCC, in our preneoplastic model, we observed a 75–85% decrease in the expression of Mrp2 in freshly isolated hepatocytes compared to control cells. Thus, for this reason, reduced levels of Mrp2 in preneoplastic liver cells could contribute to the preservation of intracellular GSH and would result, in addition, in an accumulation of EA-SG and a consequent inhibition of GSTs activities suppressing more GSH consumption.

We also saw increased levels of Mrp2 in intracellular microsomal membrane fraction in a EA-dose dependent manner in both control and IP hepatocytes. This internalization phenomenon was already described [40] and could imply a process which takes place as a feedback mechanism under GSH-consumption conditions at the highest dose of EA. Our results showed that the rate of internalization of Mrp2 with increasing doses of EA was markedly higher in IP hepatocytes, although the initial basal values were significantly lower. To our knowledge, this was the first study evaluating this accelerated internalization process in isolated hepatocytes from preneoplastic rat livers.

In conclusion, hepatocytes of IP rats showed an intrinsic resistance to the cytotoxic effects of low doses of EA and it seems likely that the presence of GST P, the higher levels of GSH, and the lower expression of Mrp2 in the cellular membrane are closely related to this phenotype.

4. Interferon α-2b gets into scene

Human lymphoblastoid IFN α has been shown to have a powerful antiproliferative effect on human hepatoma cell line PLC/PRF/5 in a dose-dependent manner, both *in vitro* and *in vivo*, after implantation in nude mice. Moreover, IFN α inhibits liver regeneration by decreasing DNA and total protein synthesis [41,42].

Considerable expectations in reducing the incidence of HCC were connected with the use of IFN α in antiviral treatment of hepatitis B or C. By now, clinical trials have indeed confirmed a reduced incidence of HCC in IFN α–treated patients with chronic hepatitis B or C [43,44]. In contrast, the benefit derived from IFN α treatment of established HCC remains controversial [45,46]. It is important to deepen the understanding of the action of IFN α on HCC cells, because some patients with hepatitis B– or hepatitis C–related liver diseases may already have small, clinically undetectable preneoplastic foci during IFN α therapy. Experimental studies have shown that IFN α exerts its antiproliferative effects against HCC cell lines *in vitro* by inducing apoptosis and inhibiting cell-cycle progression [47–49]. However, the sensitivity of early-stage HCC to IFN α could not be estimated from the sensitivity of the cell lines that have a larger number of gene abnormalities and higher proliferation capability, whereas the activity of IFN is expected to be minimal [50]. However, it was unknown whether IFN α prevents *in vivo* oncogenesis by expressing these effects in the very-early-stage, clinically undetectable cancer cells.

In this context, we have demonstrated that administration of IFN α-2b during the development of rat liver preneoplasia significantly decreased both number and volume percentage of GST P–positive foci [14]. Particularly, these reductions where observed when IFN α-2b was administered during the initiation phase or during the entire experimental protocol. However, when IFN α-2b was administered during the promotion phase no effect on these parameters could be observed. Thus, the use of IFN α-2b as an antitumor agent was lost when it was administered only at the 2-AAF phase. Nevertheless, we cannot discount that the lack of IFN α-2b effect during the 2-AAF phase reflects some interaction between 2-AAF and the cytokine. Administration of IFN α-2b during the initiation stage seems to be essential to exert inhibitory effects against DEN-initiated hepatic carcinogenesis in the rat.

Contrary to our expectations, the proliferation index (measured by immunohistochemical detection of proliferating cell nuclear antigen or PCNA) in preneoplastic foci was not reduced by treatment with IFN α-2b. On the other hand, the apoptotic index (measured by TUNEL technique) in AHF was significantly increased in the groups that received IFN α-2b. The number of apoptotic cells and bodies in AHF after treatment with IFN α-2b was higher than for control rats. Then, the reduction of both number and volume percentage of AHF in IFN α-2b–treated animals is explained by a greater programmed cell death within the foci.

In regard to the effects of IFN α on the cell cycle progression of various normal and tumor cell lines, most studies have observed inhibitory effects on G1 to S phase transition [51,52]; other studies have shown S phase accumulation in response to treatment with IFN α [48,49]. In our studies, the animals with liver preneoplasia that were treated with IFN α-2b showed a diminution in the percentage of preneoplastic hepatocytes in S phase and an accumulation in the G1 phase. Although apoptosis may be initiated in any phase of the cell cycle, most cells undergo apoptosis primarily in the G1 phase of cycling cells, and there is a positive relationship between apoptosis and cell proliferation [53]. This relationship is explained by the presence of many cell cycle regulators/apoptosis inducers such as p53, which operates at the G1/S checkpoint [54].

In this connection, we examined whether p53 and 3 members of the Bcl-2 family (Bax, Bcl-2, and Bcl-xL), which are important regulators of apoptosis [53] were involved in IFN α-2b–mediated programmed cell death. It is known that p53 down-regulates Bcl-2 [55] and up-regulates Bax genes [56]. The role of the Bcl-2 family in IFN α–induced apoptosis still remains controversial. For example, IFN α–induced apoptosis in cells of hematopoietic and hepatic origins can occur without involvement of the Bcl-2 family [48,57] whereas transfection of IFN α–sensitive cell lines with a Bcl-2 expression vector conferred partial resistance to cell death mediated by IFN α [58]. Our results showed that members of the Bcl-2 family were involved in the apoptotic elimination of preneoplastic hepatocytes after treatment with IFN α-2b. Specifically, treatment with IFN α-2b increased levels of the proapoptotic protein Bax, in parallel with increases of p53 protein levels. In addition, there were decreases in the levels of Bcl-2 and Bcl-xL proteins, which are known to promote cell survival through homodimerization. Bax protein promotes cell death via homodimerization, whereas heterodimerization with either Bcl-2 or Bcl-xL results in cell survival [59,60]. The relative prevalence of Bax and Bcl-xL protein are critical factors influencing cell fate, promoting either survival or death, whose ultimate outcome largely depends on the Bax/Bcl-xL ratio. Thus, apoptosis pathways can be activated under conditions in which Bax protein expression is elevated and/or Bcl-xL protein expression is decreased.

We also observed increased Bax protein translocation into the mitochondria in the animals that received IFN α-2b. It has been established that subcellular localization of Bax protein is an important regulator of apoptosis. Bax is localized in the cytoplasm and translocates to the mitochondria at the early stage of apoptosis. Bax mediates its proapoptotic effects through a channel-forming activity of the mitochondrial membrane, resulting in disruption of mitochondrial function, release of cytochrome *c*, and apoptosis [61].

In brief, our experimental observations led us conclude that preneoplastic hepatocytes in the IFN α-2b–treated rats are "primed" for apoptosis and undergo programmed cell death as a primary result of a substantial increase in the level of mitochondrial Bax protein, producing a further increase in the Bax/Bcl-xL protein ratio.

5. Has TGFβ1 any role in this scenario?

Given its antiproliferative, proapoptotic role in the liver, TGFβ1 could be expected to act as a tumor suppressor. However, various types of neoplastic liver cells respond quite differently to TGFβ1. Whereas some human and rat hepatoma cell lines are sensitive to TGFβ1 [62–64], resistance has been reported for other hepatoma cells [64,65]. In addition, TGFβ1 overexpression seems to be a hallmark of human liver cancer [66]. Thus, the relationship between TGFβ1 and cancer is complex: TGFβ1 may stimulate malignant progression itself; conversely, it can have tumor suppressor activity [67]. The escape of certain hepatoma cells from TGFβ1–induced apoptosis seems to be an important and essential step in malignant progression [68,69]. Moreover, it has been suggested that TGFβ1 overexpression is a late event in human hepatocarcinogenesis [66]. These data indicate that loss of TGFβ1 responsiveness is not an initiating or strongly predisposing event, but rather a late event in carcinogenesis [67,70].

Therefore, it was of interest to study if liver preneoplasia as an early stage of cancer development is still sensitive toward TGFβ1 actions.

Given that the changes of pro- and anti-apoptotic proteins induced by IFN α-2b in rats with liver preneoplasia were similar to those attributed to TGFβ1 in other experimental models [62,63,71], we studied the possibility that TGFβ1 could be involved in the programmed cell death induced by IFN α-2b [72]. Primary, we observed that serum TGFβ1 levels in the animals treated with IFN α-2b were significantly increased. In accordance with this, immunohistochemical studies showed that IFN α-2b treatment significantly augmented the quantity of TGFβ1–positive hepatocytes in preneoplastic livers. At first sight, these findings seemed to indicate that administration of IFN α-2b increased serum TGFβ1 production and the number of TGFβ1–positive hepatocytes. Although the mechanisms by which IFN α-2b treatment induced TGFβ1 in the preneoplastic livers were not completely explored, we observed, using Western blot analysis, that preneoplastic livers expressed higher levels of IFN α receptors than control livers. In addition, IFN α-2b administration in animals subjected to the preneoplastic protocol induced elevated levels of phosphorylated Stat1, indicating activation of the IFN α pathway.

Recent investigations have reported that the induction of apoptosis by endogenous TGFβ1 does not require an overall increase in its hepatic concentration [70]. In view of the fact that TGFβ1 hepatic content may not reflect the induction of apoptosis by this cytokine, we determined the nuclear content of p-Smads-2/3 (critical intracellular transducers of TGFβ1 signaling). We observed high levels of p-Smads-2/3 proteins in the nuclear extracts of IFN α-2b–treated animals. These results corresponded with the increased number of TGFβ1–positive hepatocytes, indicating increased TGFβ1 activation in rats with liver preneoplasia that received IFN α-2b.

Nonparenchymal cells, including Kupffer cells and peritoneal macrophages, are the main source of hepatic TGFβ1 [73,74]. Hepatocytes, however, may synthesize TGFβ1 *in vitro* [75] as well as during hepatocarcinogenesis [66]. During liver preneoplasia, neither peritoneal macrophages nor Kupffer cells secreted detectable levels of TGFβ1 when they were stimulated with IFN α-2b. Conversely, hepatocytes from normal, untreated livers did not secrete TGFβ1 in the absence or presence of IFN α-2b. Nevertheless, hepatocytes from preneoplastic livers produced and secreted detectable levels of TGFβ1 when they were cultured without IFN α-2b stimulus, and IFN α-2b presence in the culture media induced several-fold increases of TGFβ1 production.

In vitro studies with isolated hepatocytes have allowed us to demonstrate that IFN α-2b induces apoptosis in hepatocytes from preneoplastic livers, measured by fluorescence microscopy and caspase-3 activity. These cells also had higher nuclear accumulation of p-Smads-2/3, indicating increased TGFβ1 activation. When anti–TGFβ1 was added to the culture media, TGFβ1 activation and apoptosis induced by IFN α-2b were completely blocked. Therefore, the apoptotic effect of IFN α-2b is mediated by the production of TGFβ1 from hepatocytes.

Thus, our work determined for the first time that endogenous TGFβ1 is implicated in the increased apoptosis into the AHF of IFN α-2b-treated rats. Taken together, these data clearly showed that TGFβ1, which is produced and secreted by hepatocytes from preneoplastic liver under IFN α-2b treatment, stimulates hepatocytes apoptotic cell death in an autocrine/paracrine fashion. This postulated mode of action is in agreement with data published previously [70,76,77]. The reduction of preneoplastic foci by endogenous TGFβ1 early in the carcinogenesis process would likewise protect against tumor formation.

6. Participation of ROS

In a new series of *in vitro* experiments, we proved that IFN α-2b induces the production of TGFβ1 in hepatocytes from preneoplastic livers by activation of NADPH oxidase complex (superoxide-producing enzyme consisting of membrane (gp91phox and p22phox) and cytosolic (p47phox, p67phox, and p40phox) components [78]), and TGFβ1 induces apoptosis through a mechanism linked to the production of ROS by the same oxidase [79]. In order to confirm that the induction of NADPH oxidase activity was the main pathway producing ROS, additional experiments were made using IFN α-2b plus an inhibitor of NADPH oxidase activity, diphenyleneiodonium (DPI). Presence of DPI in the culture media totally blocked the activity of NADPH oxidase, the production of ROS and the subsequent apoptosis induced by IFN α-2b.

ROS production induced by IFN α-2b showed a singular pattern of two peaks: one peak in ROS generation at 1 hour of culture, and another peak at 9 hours. The addition of anti-TGFβ1 to the culture media did not block the production of the first peak of ROS whereas totally blocked the appearance of the second one. On the other hand, when ASC was added to the culture media the production of both peaks was abolished. Based on these findings, the postulated mechanism by which ROS act as signaling molecules in liver preneoplasia is as follow: IFN α-2b induces, via NADPH oxidase activation, an early ROS production that serves as a messenger, promoting TGFβ1 production and secretion. This growth factor triggers the production of more reactive oxygen intermediates, as a late event, by inducing the same enzyme complex. It was demonstrated that synthesis of new protein is required for NADPH activation and subsequent apoptosis [80]. This event shows an additive response in ROS production and imposes the final onset of the apoptotic effect. The presence of ASC in the culture media totally blocked the increase in the activity of the NADPH oxidase complex, ROS production and the final apoptotic effect induced by IFN α-2b.

Once the source of ROS was assessed, we analyzed the cellular antioxidant defenses and their behavior during the studied times. We observed a reduction in tGSH levels from 7 hours of culture onwards. For that reason we studied if any form of glutathione was being exported out of the cell, and whether the biosynthetic GSH capacity was altered. We found an increase in oxidize glutathione (GSSG) levels probably due to the oxidation of the reduced form within the cytosol, and its exportation to the culture media, possibly in order to protect cells from a shift in the redox equilibrium. IFN α-2b treatment resulted in the loss of GSH biosynthetic capacity since glutamate cysteine ligase (GCL) activity was decreased at 7 hours of culture and a rapid decrease of the mRNA expression of the catalytic subunit of GLC (GCLC) through a mechanism mediated by TGFβ1 was also observed. Moreover, it was found that IFN α-2b-induced apoptosis in hepatocytes from rat preneoplastic livers is accompanied by the cleavage and loss of GCLC protein, through a mechanism mediated by TGFβ1.

A decrease in the antioxidant enzymes catalase (CAT) and superoxide dismutase (SOD) activities was observed when hepatocytes were treated with IFN α-2b. On the other hand, treatment with anti-TGFβ1 or ASC totally blocked the decrease in CAT and SOD enzymatic activities. These findings indicate that IFN α-2b induced the decrease in enzymatic CAT and SOD activities by a mechanism mediated by ROS and TGFβ1. These enzymes probably protect hepatocytes from the initial IFN α-2b-induced burst of ROS and this may be the reason for the rapid decrease of the first peak of ROS.

These results confirmed that the perturbation of the redox status produced by the IFN α-2b induction of NADPH oxidase complex triggered TGFβ1 synthesis and secretion and assessed the downregulation of antioxidative systems. Similar data have been reported by Herrera et al. [80] when they treated fetal rat hepatocytes with TGFβ1.

Since ASC abolished all the apoptotic effects induced *in vitro* by IFN α-2b, we determined the relevance of ROS on the onset of the apoptotic process *in vivo*, in the whole preneoplastic liver. IFN α-2b plus ASC treatment of rats with liver preneoplasia abrogated the apoptotic effect induced by IFN α-2b, leading to no reduction on size/number of foci. Interestingly, foci volume was almost twice higher in the animals that received IFN α-2b plus ASC than in IFN α-2b-treated rats. This result highlights the importance of ROS signaling during the beneficial effects of IFN α-2b treatment of hepatic preneoplasia. In this regard, it was found that ASC at low concentrations stimulates growth of malignant cells [81], while inhibits their growth at high doses [82]. At the present time, many cancer patients combine some forms of complementary and alternative medicine therapies with their conventional therapies. The most common choice of these therapies is the use of antioxidants such as vitamin C. It must be assumed that any antioxidant, used to reduce toxicity of tumor therapy on healthy tissue, has the potential to decrease effectiveness of cancer therapy on malignant cells [83]. Some data suggest that antioxidants can ameliorate toxic side effects of therapy without affecting treatment efficacy, whereas other data suggest that antioxidants interfere with radiotherapy or chemotherapy [83].

In summary, we demonstrated that increase in ROS levels turns on the process of programmed hepatocytes death, leading to the elimination of these malignant cells. The inhibition of ROS production with an antioxidant such as ASC in the co-treatment with IFN α-2b may be not a beneficial therapy for the prevention of preneoplastic foci.

7. Is p38 MAPK implied in the process?

p38 MAPK pathway has been implicated in a wide range of cellular functions. However, it is now well established that p38 MAPK activation and its role depends on the cellular context, on the specific stimuli, and on the specific p38 MAPK activated isoform [84]. There are controversies about the role of p38 MAPK in apoptosis. It has been shown that p38 MAPK signaling promotes cell death [85,86], whereas it has also been shown that p38 MAPK cascades enhance survival [87,88], cell growth [89], and differentiation [90]. Furthermore, it has been reported that p38 MAPK participates on the estradiol-mediated inhibition of apoptosis in endothelial cells [91], while participates on the apoptosis induced by thrombospondin-1 [92], or by high leves of D-glucose in the same cells [93]. It is believed that p38 MAPK mediates its apoptotic effects through the phosphorylation of proteins of the apoptotic pathways [94].

Previous reports in hematopoietic cells have shown that IFN α and TGFβ1 play their growth inhibitory effects through activation of the p38 MAPK pathway via phosphorylation (activated p38 MAPK or p-p38 MAPK) [95]. However, these effects are primarily ascribed to G1 cell cycle arrest and not to induction of apoptosis. Others have suggested that during the TGFβ1-induced apoptosis in fetal rat hepatocytes, ROS activates p38 MAPK not by induction of apoptosis, but mediating ROS regulation of TGFβ1-gene expression [96]. On the other hand, it was demonstrated that inactivation of p38 MAPK pathway in cultured mice fibroblasts promotes tumor development [97]. Moreover, it was demonstrated that treatment with an inhibitor of p38 MAPK activation, induced carcinogenesis in mice resistant to tumor development, indicating the leading role of p38 MAPK in the regulation of tumor growth [98].

Using *in vivo* studies we could demonstrate that rats subjected to a 2-phase model of chemical hepatocarcinogenesis have less hepatic p38 MAPK activation than control rats, determined as p-p38 MAPK levels [79]. This is in agreement with Honmo et al. [99] that showed that 2-AAF administration induces a decrement of p38 MAPK activation promoting tumor development.

Another important finding of the *in vivo* studies was the effect of IFN α-2b on the activation of p38 MAPK in rat preneoplastic livers. Preneoplastic animals treated with IFN α-2b showed similar p-p38 MAPK levels to those in controls. In this connection, cultured hepatocytes from preneoplastic livers treated with IFN α-2b plus SB-203580 (inhibitor of α and β isoforms of p38 MAPK), totally blocked the IFN α-2b-induced apoptosis. It is clear that activation of p38 MAPK pathway plays a key role in promoting apoptosis after IFN α-2b treatment in our model of experimental preneoplasia. It was previously reported that IFN α suppresses the growth of leukemia cell progenitors through activation of p38 MAPK, which leads to cell cycle arrest in different phases [100].

We demonstrated that IFN α-2b induces an early production of ROS (first peak), in hepatocytes from preneoplastic livers. Then, ROS stimulate the production and secretion of TGFβ1 from hepatocytes, which in turn, generates a new burst of ROS (second peak). These oxygen radicals act as signaling mediators of the onset of the IFN α-2b-induced apoptosis.

Activation of p38 MAPK after IFN α-2b stimulus occurred preceding each increment in ROS generation and so, the particular pattern of two peaks was also functioning for p38 MAPK activation. Interestingly, treatment with ASC was able to block only the second peak, indicating that early activation of the pathway was independent of ROS, while late activation depended on ROS produced by endogenous TGF-β1. Treatment with anti-TGFβ1 completely blocked the second p38 MAPK, demonstrating that TGF β1 induces activation of p38 MAPK through ROS, as previously reported in fetal rat hepatocytes [96].

Another relevant issue is the activation of transcription factors by p38 MAPK. Cell signaling pathway activation could be transmitted to the nucleus in different ways, depending on the stimulus. To assess whether activation of p38 MAPK transmitted the IFN α-2b stimulus to the nucleus, we analysed phosphorylation status of specific p38 MAPK transcription factors CREB/ATF-1 and ATF-2. Our findings documented that early p38 MAPK activation under IFN α-2b stimulus mainly activates the transcription of ATF-2-regulated genes, whereas the late signal of p38 MAPK activation is transmitted to the nucleus mainly by the phosphorylation of CREB/ATF-1. Moreover, it can be also inferred that early phosphorylation of ATF-2 may be dependent on activation of p38 MAPK by IFN α-2b, while late phosphorylation of CREB/ATF-1 may be dependent on activation of p38 MAPK by TGFβ1.

8. Relationship between p38 and NADPH oxidase

We inferred that p38 MAPK activation is essential for NADPH oxidase to function in preneoplastic hepatocytes treated with IFN α-2b, because the presence of p38 MAPK inhibitor SB-203580 totally blocked the activation of the enzyme [101]. Cytosolic component of NADPH oxidase complex, p47phox got phosphorylated following the same pattern as p38 MAPK induction and ROS generation: an early, first increment and a late, second increase. The first increase of p47phox phosphorylation by IFN α-2b was independent of ROS, since ASC did not block such phosphorylation. However, it was dependent of p38 MAPK activation, since it was blocked by SB203580. This is a very interesting finding since it suggests that p-p38 MAPK phosphorylates p47phox, initiating the activation of NADPH oxidase in cells from preneoplastic livers. Analysis at higher times demonstrated that late phosphorylation of p47phox was completely blocked by anti-TGFβ1 or ASC, evidencing the participation of TGF β1 and ROS in this process. Studies of p47phox translocation from cytosol to plasma membrane were consistent with the phosphorylation findings.

It is clear that in liver preneoplasia there is a positive cross-talk between IFN α-2b, TGFβ1 and p38 MAPK pathways. Taken altogether, evidence indicates that p38 MAPK pathway plays a critical role in the generation of the suppressive effects of IFN α-2b, as well as TGFβ1 in the very early stages of hepatic neoplasia. There is strong indication that this signaling cascade acts as a converging signaling point for signaling pathways activated by different cytokines to mediate apoptotic or suppressive signals. These findings may have important clinical implications, as improving the pharmacological development of better drugs for the prevention and treatment of hepatic illness such as cancer.

9. How are IFN α and TGFβ1 signaling pathways connected?

Interactions between TGFβ and other cytokines signaling pathways have been extensively studied, particularly the cross-talk between TGFβ/Smad and IFN γ/Stat signaling in their antagonistic role on collagen deposition and fibrosis [102–107]. However, despite the fact that TGFβ plays a crucial role in cancer, little is known about TGFβ signaling interactions during this process. An investigation in hepatoma cells have described a cross-talk between Il-6 and TGFβ signaling [108] and another study in a melanoma cell line normally resistant to IFN α, have demonstrated that co-stimulation with IFN α and TGFβ induces antiproliferative activity [109].

As was stated above, the relationship between TGFβ and cancer is complex: it functions as a tumor suppressor in early epithelial carcinogenesis, but often becomes prooncogenic in late stages of tumor progression [110]. Autocrine TGFβ1 is known to suppress tumorigenesis and tumor progression in normal and early transformed cells, but it can also promote the survival of various cancer cells [111]. Besides, dysregulation of the downstream effectors of TGFβ has been described in late steps of promotion stage, indicating that may contribute to the progression of preneoplastic lesions [112].

We demonstrated that during liver preneoplasia TGFβ1 has a beneficial role, promoting apoptotic death of AHF. Therefore, we attempted to get more insight into the relationship between IFN α-2b and autocrine TGFβ1 in preneoplastic rat livers. Many *in vitro* cell systems are good tools to explain related actions of distinct types of cytokines in various biological signaling pathways, but they are not physiological. However, the study of IFN α-2b and TGFβ1 signals interactions in hepatocytes derived from the whole preneoplastic liver may be relevant for understanding the mechanisms operating in patients with chronic hepatitis B or C treated with IFN α-2b, who already have small, clinically undetectable preneoplastic liver foci during therapy.

The obtained results provided evidence for the integration of TGFβ1 and IFN α-2b signaling pathways during the development of liver carcinogenesis. IFN α-2b treatment of hepatocytes from preneoplastic livers produced a rapid activation of IFN α signaling, with increased p-Stat1 levels. Subsequently, autocrine TGFβ1 produced under IFN α-2b stimulus was able to induce the activation of TGFβ1/Smad signaling pathway, determined by nuclear content of p-Smad2/3 and confirmed by the use of specific TGF β1 signaling inhibitors (anti-TGFβ1 and SB-431542) [113].

A critical mechanism for regulating the cellular response to cytokines resides at the level of receptor expression. TGFβRII plays a key role in receptor activation and subsequent TGFβ1 signal propagation, functioning both to bind ligand and to activate TGFβRI. Disorders of TGFβRII expression lead to various diseases. For example, reduction of TGFβRII levels contributes to the resistance of tumor cells to TGFβ [114].

We observed that TGFβRII was up-regulated at mRNA and protein levels. This induction was mediated by autocrine TGFβ1, since it was blocked by inhibitors of TGFβ1 signaling. This is an outstanding finding, since TGFβ1-dependent regulation of TGFβRII has not been previously reported.

Inhibitory Smad7 is a key component of TGFβ1 signals. Its expression is not only induced by TGFβ, but also controlled by, for example, IFN γ [102,107]. Therefore, Smad7 is considered as a protein involved in the fine-tuning of the cellular responses to the TGFβ family by integrating various signaling pathways. However, in our model, Smad7 did not show changes in its protein levels, at least during the studied times. Furthermore, Smad7 protein levels in hepatocytes from preneoplastic livers were significantly reduced with respect to their levels in hepatocytes from normal livers. So, additional experiments of Smad7 induction by phorbol 12-myristate 13-acetate (PMA) were performed in order to evaluate if the decreased Smad7 levels showed in preneoplastic livers may contribute in TGFβ1 signaling activation. Results showed that this possibility seems unlikely; provided that Smad7 protein reached similar levels to those in normal hepatocytes, and TGFβ1 signaling continued activated. These experiments indicated that Smad7 protein is not directly related with TGFβ1 and IFN α signals interaction in hepatocytes from preneoplastic livers.

Another decisive aspect in signaling pathways relationships is the availability of certain co-activators for interacting with specific transcription factors. The cofactor p300 is an important component of the transcriptional machinery that integrates TGFβ/ IFN γ-induced signals [115].

In normal fibroblasts exposed to IFN γ and TGFβ simultaneously, activated Stat1 and activated Smad2/3 compete each other for limiting p300. IFN γ-activated Stat1 appears to sequester p300, thereby disrupting TGFβ-induced interaction of p300 with Smad2/3. Ectopic p300 rescues stimulation in the presence of IFN γ, suggesting that p300 acts as an integrator of IFN γ/Stat1 and TGFβ/Smad2/3 signals [103]. In addition, Inagaki et al. [116] have demonstrated that IFN α antagonizes TGFβ/Smad-induced hepatic fibrosis by competition between Stat1 and Smad3 for binding to p300 protein.

In our study, we found that IFN α-2b induced a direct interaction between activated Stat1 and p300 in hepatocytes from preneoplastic livers. Furthermore, activated Smad2/3 induced by autocrine TGFβ1 were able to physically associate with p300. In addition, levels of p300 in hepatocytes from preneoplastic livers were significantly higher than in normal hepatocytes. Together, these findings suggested that in hepatocytes from preneoplastic livers, the intracellular signals triggered by TGFβ1 and IFN α-2b are integrated at the nuclear level, where p-Stat1 and p-Smad2/3 are capable of interact with p300, present in no restrictive cellular amounts.

It was recently found that TGFβ signals potentiate Il-6 signaling in hepatoma cells. This cross-talk occurs by physical interactions between Stat3 and Smad3, bridged by p300 [108]. In our model of liver preneoplasia we did not observe physical interaction between Stat1 and Smad3, but it seems to be enough p300 protein available to interact with p-Stat1 on one hand, and with p-Smad2/3 on the other, leading to the activation of TGFβ1 and IFN α signaling simultaneously.

In fact, we have described for the first time a positive cross-talk between IFN α and TGFβ1 signaling.

10. Summary # 1

In these series of experiments, it was demonstrated that NADPH oxidase complex is activated when IFN α-2b binds to type I receptor. This binding produces early amounts of ROS. ROS, in turn, trigger TGFβ1 production and secretion. TGFβ1, when binding to its receptor, also induces NADPH oxidase complex activation, and, besides, decreases the antioxidant defenses of the cell. Moreover, we demonstrated that p38 MAPK activation is essential for NADPH oxidase to function.

Furthermore, ROS initiate mitochondrial apoptosis directly and/or acting by the Bcl-2 family proteins inducing a mitochondrial permeability transition pore (MPTP), releasing cytochrome c and activating caspase 3. TGFβ1 could induce, as a late event, the activation of caspase 8, which, in turn, induces a higher MPTP through activation of Bid, another Bcl-2 family member [117]. A graphic outline of these concerns is shown in Figure 2.

Altogether, our results demonstrate that the oxidative stress induced in preneoplastic liver by IFN α-2b is able to trigger the apoptotic mechanism and brings into the play another key cytokine in the cancer process: TGFβ1.

11. Targeting the Wnt/β-catenin signaling pathway

Among the growth factor signaling cascades dysregulated in HCC, evidences suggest that the Wnt/Frizzled-mediated signaling pathway plays a key role in hepatic carcinogenesis. Aberrant activation of the signaling in HCC is mostly due to dysregulated expression of the Wnt/β-catenin signaling components. This leads to the activation of the β-catenin/TCF dependent target genes, which control cell proliferation, cell cycle, apoptosis or motility. It has been shown that disruption of the Wnt/β-catenin signaling cascade displayed anti-cancer properties in HCC [118].

For this reason, we determined the status of the Wnt/β-catenin/TCF pathway in the preneoplastic stage and evaluated the possible effects of IFN α-2b on this pathway.

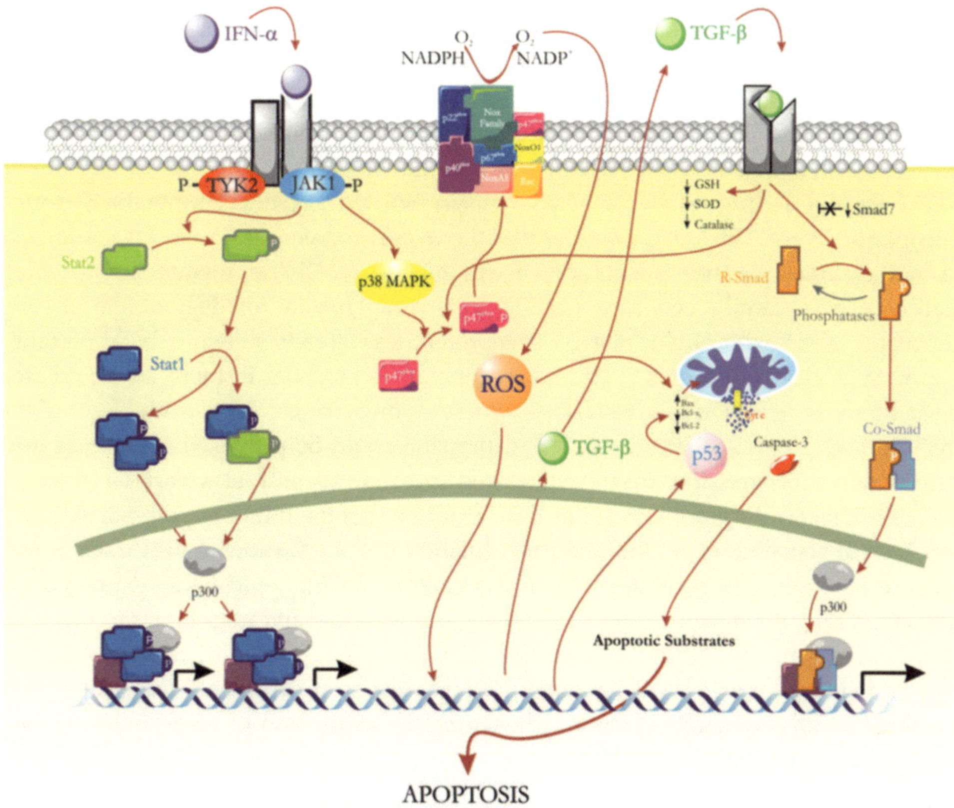

Figure 2. Graphic outline of IFN α-2b, TGFβ1, p38 MAPK, NADPH oxidase and ROS interactions in liver preneoplasia.

The major findings of our studies were related to the impairment of the canonical Wnt/β-catenin/TCF pathway in a very early stage of hepatic carcinogenesis. In addition, we demonstrated that *in vivo* IFN α-2b treatment produces an attenuation of TCF transcriptional activity and enhances FOXO transcriptional activity in preneoplastic livers.

The common denominator of an abnormal Wnt signaling is the stabilization and accumulation of unphosphorylated β-catenin in the cytoplasm of a cell. Eventually, this allows entry of unphosphorylated β-catenin into the nucleus where it promotes the transcription of a subset of genes implicated in cellular proliferation. This β-catenin stabilization was demonstrated in our two-phase carcinogenic model, where plasma membrane delocalization and cytoplasmic accumulation of β-catenin were observed [119]. Moreover, significant reductions of phosphorylated β-catenin levels were found in IP animals. Since total β-catenin (phosphorylated and unphosphorylated) protein levels were preserved in all studied groups, these results indicate a lower phosphorylation rate of cytoplasmic β-catenin in IP rats.

We have also found up-regulation of TCF target genes Cyclin D1, MMP-7, Axin 2, and SP5 in preneoplastic livers. Up-regulation of Cyclin D1 was predicted since this protein is an important regulator of cell cycle progression, and its activity is required for G1 to S-phase transition. Overexpression of this gene has been associated with the development and progression of several cancers [120]. In addition, it has been reported that overexpression of Cyclin D1 in tumor cells contributes with their resistance to cytotoxic drugs [121]. In fact, inhibition of Cyclin D1 enhances the effects of several chemotherapeutic agents [121]. In agreement with these results, we have previously described (see *GSTs and liver preneoplasia: our first studies*) a drug-resistance phenotype in isolated hepatocytes obtained from rat preneoplastic livers. Thus, it is possible that the overexpression of Cyclin D1 could play a role in the drug-resistance phenotype of this model. MMP-7, a member of the matrix metalloproteinase family, acts as a specific proteolytic enzyme for degradation of certain components of the extracellular matrix. This protein was already shown to be important for the growth of early adenomas [122] and its function is essential in more advanced stages such as tumor progression and metastasis, where an invasive growth is a highlight of these steps [123,124]. Hence, enhanced MMP-7 expression could be proposed as an indicator of potential tumor progression, invasiveness, and metastatic ability at a very early stage of hepatocarcinogenic development. It has been reported that the tumor suppressor Axin 2 is a target of Wnt signaling [125,126]. The up-regulation of Axin 2 showed in IP rats, which is known to be a negative regulator of free β-catenin [127,128], could be an expression of a feedback preservation mechanism of the preneoplastic tissue, and might not be sufficient to prevent cytoplasmic β-catenin accumulation. SP5, a member of the SP1 transcription factor family and known target of Wnt signaling [129] was also over-expressed. This protein seems to work as a transcriptional repressor, preventing the expression of genes involved in cell cycle G1 phase arrest such as p21 [129].

In order to determine the involvement of a mutated β-catenin protein in the activation of this pathway as was described for HCC [130–133], we performed a direct sequencing of amplicons encoding a region of exon 2 of rat liver β-catenin gene. Our results demonstrated that this sequence had no deletion or point mutations in any of the studied groups.

Even with a wild-type β-catenin, the pathway can also be triggered because of alterations in other components of the cascade signaling. The Frizzled protein family acts as a seven-span transmembrane receptor for Wnt proteins. It was recently reported an up-regulation of the Frizzled-7 receptor in the presence of wild-type β-catenin in four murine transgenic models of hepatocarcinogenesis [134] and in human HCC [135] with activation of the Wnt/β-catenin/TCF pathway. Therefore, it was suggested that overexpression of Frizzled-7 could lead or contribute to activation of Wnt signaling. The obtained data showed a marked increase of this receptor in preneoplastic livers at mRNA and protein levels. Since it was reported that Frizzled-7 is also a target gene of the Wnt/β-catenin/TCF pathway [136], we presume that overexpression is rather a consequence than a cause of abnormal activation of the Wnt/β-catenin/TCF pathway.

Once we demonstrated that the Wnt/β-catenin/TCF pathway is activated in preneoplastic rat livers, we analyzed the effects of IFN α-2b treatment. Results showed that *in vivo* IFN α-2b administration did not prevent β-catenin delocalization and cytoplasmic accumulation; however, it certainly attenuates activation of the canonical Wnt/β-catenin/TCF pathway as measured by four TCF target genes. The transcription levels of these genes were similar to controls in IP animals that received IFN α-2b.

In addition, IFN α-2b-treated IP rats showed that Frizzled-7 levels remained unchanged compared to control animals. These results reinforced our hypothesis that Frizzled-7 up-regulation occurs as a result of the abnormal activation of the studied pathway.

In an attempt to get more insight into the regulation of Wnt/β-catenin/TCF pathway, FOXO transcription family has come into scene. Recent studies reported that FOXO interacts with β-catenin in a competitive manner with TCF, particularly under cellular oxidative stress conditions [137,138]. Taking this into consideration and the fact that *in vivo* IFN α-2b treatment induces endogenous ROS formation in preneoplastic livers, we analyzed interactions between β-catenin with TCF4 and FoxO3a and association of these transcription factors with their corresponding target gene promoters. Co-immunoprecipitation assays showed that β-catenin/TCF4 interaction effectively occurs in preneoplastic livers and administration of IFN α-2b not only attenuates this interaction but also promotes β-catenin/FoxO3a association. Using ChIP assay, we verified that interaction of FoxO3a with the promoter region of its target gene is enhanced in preneoplastic livers treated with IFN α-2b. On the other hand, TCF4 remains associated with SP5 gene promoter region in all studied groups. It is known that TCF4 contains a conserved domain that binds DNA irrespective of its interaction with β-catenin; however, the transcriptional activity is blocked by the presence of a family of transcriptional repressors [139,140]. TCF4 must bind β-catenin for its transactivation and this interaction was verified by co-immunoprecipitation assays. In addition, it has been demonstrated that interaction of β-catenin with FOXO enhances its transcriptional activity [137,138], so we measured the expression of p130, a FOXO target gene whose main function is related to the maintenance of cell cycle arrest. Furthermore, it was suggested that p130 may exert a proapoptotic effect on certain tumor samples [141]. We found up-regulation of p130 transcript in preneoplastic livers treated with IFN α-2b. These findings suggest that IFN α-2b treatment in preneoplastic livers decreases β-catenin/TCF interaction and consequently reduces TCF transcriptional activity probably via ROS induction. Furthermore, IFN α-2b-induced ROS production could stimulate β-catenin/FOXO interaction, thereby favoring cell cycle arrest and apoptosis. In agreement with this proposal, recent unpublished results from our group demonstrate the participation of ROS in these events.

Collectively, our data demonstrate that the canonical Wnt/β-catenin/TCF signaling pathway is activated at a very early stage of the development of the hepatocarcinogenic process, even with a wild-type β-catenin. More importantly, *in vivo* IFN α-2b treatment could be an efficient therapy to attenuate Wnt/β-catenin/TCF signaling promoting diminution of preneoplastic foci by an apoptotic process. A graphic outline of these concerns is shown in Figure 3.

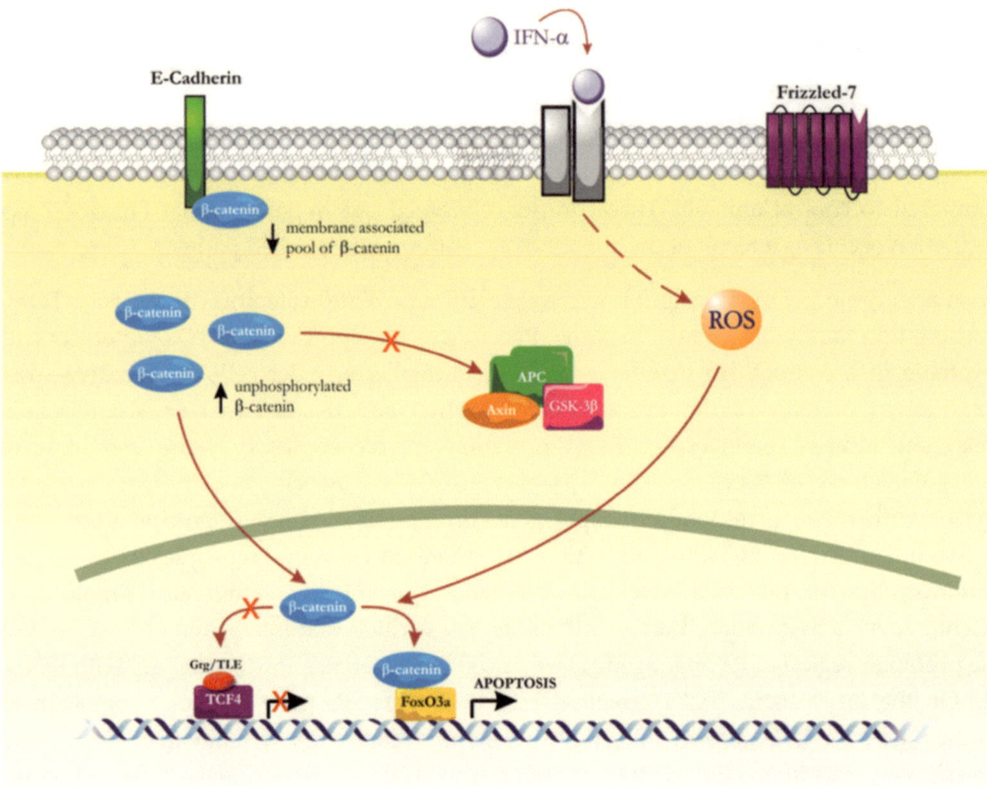

Figure 3. Graphic outline of IFN α-2b, Wnt/β-catenin pathway and ROS/FOXO interactions in liver preneoplasia.

12. Studies in HCC cell lines

The elucidation of the signals induced by IFN α and TGFβ in human liver tumor cells, and their possible cross-talks with other intracellular signals, would have relevance in the future design of therapeutic tools to balance the cellular responses in favor of liver tumor suppression. To gain mechanistic insights into these cooperative signals, we analyzed the effects of IFN α-2b and TGFβ1 on Wnt/β-catenin pathway and Smads intermediates in HepG2/C3A and Huh7 HCC cell lines. We could demonstrate that IFN α-2b or TGFβ1 stimulations not only decreased cellular proliferation but also increased apoptotic cell death [142]. The apoptotic and anti-proliferative effects of both cytokines separately have already been reported in HepG2 and Huh7 [143–145]. More interestingly, we demonstrated that the combined treatment increased these effects. Until now, combined treatment with both cytokines has only been used to analyze their impact on proliferation in human melanoma cell lines [109]. Treatments impact on Wnt/β-catenin pathway was analyzed, together with the analysis of the effects of IFN α-2b and TGFβ1 on Smads proteins. Insufficient information is available concerning TCF4/Smads association and their impact on carcinogenesis in HCC cell lines. Labbé et al. [146] and Letamendia et al. [147] reported the interaction between Smads 2, 3 and 4 and TCF/LEF in HepG2. Additionally, treatment with TGFβ1 in HepG2 reduced the amount of Smad4 protein bound to TCF/LEF and this was associated with the capacity of TGFβ1 of inhibiting cell proliferation [148]. To date, no study on IFN α and Smads has been carried out. More insight could be gained by analyzing the amount of each Smad protein inside the β-catenin/TCF4 nuclear complex, since all Smads coexist in this complex and the balance between them could contribute to the overall cell response by differently regulating gene expression as suggested by Edlund et al. [149].

Our findings clearly showed a negative modulation of IFN α-2b and TGFβ1 on Wnt/β-catenin pathway. This attenuation was evidenced by a decrease in β-catenin and Frizzled-7 receptor proteins levels in C3A and Huh7 and by a diminution in the amount of β-catenin bound to TCF4. Stimulation with both cytokines also caused a decrease in Smads protein contents and their association with TCF4. This effect on Smads proteins seems to be linked to the decrease of β-catenin. Finally, the inhibition of β-catenin/TCF4/Smads complexes formation may have a critical role in slowing down oncogenesis, since the overall action of IFN α-2b and/or TGFβ1 treatments on both HCC cell lines was the diminution in cellular proliferation and the increase in apoptotic cell death. In conclusion, our results support the efficacy of inhibiting Wnt/β-catenin pathway in HCC cell lines through an IFN α-2b and TGFβ1 combined treatment, proving that is effective against either wild-type or truncated β-catenin. These findings open a wide therapeutic option for patients with HCC.

13. Summary # 2

The presented data suggest a model in which IFN α-2b provides a link between TGFβ1 and Wnt signaling pathways and the oxidative stress/FOXO pathway. The stress caused by IFN α-2b treatment might strengthen the interaction between FOXO and β-catenin and potentially inhibit the interaction with TCF and Smads. The inhibition of β-catenin/TCF4/Smads complexes formation may have a critical role in slowing down oncogenesis. These findings may have important clinical implications, since β-catenin, Smads, TCF, and FOXO arise as molecular targets for novel therapies that can modify their interactions favoring cellular apoptosis over proliferation in patients that underwent a potential carcinogenic hepatic injury.

14. Concluding remarks: Oxidative stress as a critical factor in cancer therapy

Preneoplastic hepatocytes are more resistant to oxidative stress than normal ones. Nevertheless, we demonstrated that increase in ROS levels triggered by IFN α-2b enhances the process of programmed hepatocytes death, leading to the elimination of malignant cells. The study of the mechanism of IFN α-2b-induced apoptosis led to demonstrate a link between TGFβ1 and Wnt signaling pathways and the oxidative stress/FOXO pathway.

In conclusion, reactive oxygen species emerge as key mediators in the context of using cytokines as therapeutic agents in the treatment of human liver diseases, so the use of antioxidants could have the potential to decrease effectiveness of the therapy.

Author details

María Cristina Carrillo*, María de Luján Alvarez, Juan Pablo Parody,
Ariel Darío Quiroga and María Paula Ceballos
Institute of Experimental Physiology (IFISE-CONICET), Faculty of Biochemistry and Pharmacological Sciences, National University of Rosario, Rosario, Argentina

Acknowledgement

The authors kindly thank BioSidus Laboratory and Cassará Laboratory for the generous gift of recombinant IFN α-2b.

The work presented in this chapter was supported by research grants from Consejo Nacional de Investigaciones Científicas y Técnicas (CONICET) and from Agencia Nacional de Promoción Científica y Tecnológica (ANPCyT).

15. References

[1] Afanas'ev IB (2009) Signaling Mechanisms of Oxygen and Nitrogen Free Radicals. Boca Raton, FL, USA: CRC Press, Taylor & Francis Group. 212 p.
[2] Motola-Kuba D, Zamora-Valdés D, Uribe M, Méndez-Sánchez N (2006) Hepatocellular carcinoma. An overview. Ann Hepatol. 5(1):16–24.
[3] Stewart H. (1975) Comparative aspects of certain cancers. In: Becker F, editor. Cancer. A Comprehensive Treatise. New York: Plenum Press. pp. 303–74.
[4] Williams GM (1989) The significance of chemically-induced hepatocellular altered foci in rat liver and application to carcinogen detection. Toxicol Pathol.17(4 Pt 1):663–72; discussion 673–4.
[5] Pitot HC, Campbell HA, Maronpot R, Bawa N, Rizvi TA, Xu YH, Sargent L, Dragan Y, Pyron M (1989) Critical parameters in the quantitation of the stages of initiation, promotion, and progression in one model of hepatocarcinogenesis in the rat. Toxicol Pathol. 17(4 Pt 1):594–611; discussion 611–2.
[6] Altmann HW. (1994) Hepatic neoformations. Pathol Res Pract. 190(6):513–77.
[7] Farber E, Sarma DS (1986) Chemical carcinogenesis: the liver as a model. Pathol Immunopathol Res. 5(1):1–28.
[8] Goldsworthy TL, Hanigan MH, Pitot HC (1986) Models of hepatocarcinogenesis in the rat--contrasts and comparisons. Crit Rev Toxicol.17(1):61–89.
[9] Solt D, Farber E. (1976) New principle for the analysis of chemical carcinogenesis. Nature. 263(5579):701–3.

* Corresponding Author

[10] Ito N, Tsuda H, Tatematsu M, Inoue T, Tagawa Y, Aoki T, Uwagawa S, Kagawa M, Ogiso T, Masui T, et al. (1988) Enhancing effect of various hepatocarcinogens on induction of preneoplastic glutathione S-transferase placental form positive foci in rats--an approach for a new medium-term bioassay system. Carcinogenesis. 9(3):387–94.
[11] Rao PM, Nagamine Y, Roomi MW, Rajalakshmi S, Sarma DS (1984) Orotic acid, a new promoter for experimental liver carcinogenesis. Toxicol Pathol. 12(2):173–8.
[12] Peraino C, Staffeldt EF, Carnes BA, Ludeman VA, Blomquist JA, Vesselinovitch SD (1984) Characterization of histochemically detectable altered hepatocyte foci and their relationship to hepatic tumorigenesis in rats treated once with diethylnitrosamine or benzo(a)pyrene within one day after birth. Cancer Res. 44(8):3340–7.
[13] Pitot HC, Barsness L, Goldsworthy T, Kitagawa T (1978) Biochemical characterisation of stages of hepatocarcinogenesis after a single dose of diethylnitrosamine. Nature. 271(5644):456–8.
[14] de Luján Alvarez M, Cerliani JP, Monti J, Carnovale C, Ronco MT, Pisani G, Lugano MC, Carrillo MC. (2002) The in vivo apoptotic effect of interferon alfa-2b on rat preneoplastic liver involves Bax protein. Hepatology. 35(4):824–33.
[15] Parody JP, Alvarez M de L, Quiroga A, Ronco MT, Francés D, Carnovale C, Carrillo MC (2007) Hepatocytes isolated from preneoplastic rat livers are resistant to ethacrynic acid cytotoxicity. Arch Toxicol. 81(8):565–73.
[16] Imai T, Masui T, Ichinose M, Nakanishi H, Yanai T, Masegi T, Muramatsu M, Tatematsu M (1997) Reduction of glutathione S-transferase P-form mRNA expression in remodeling nodules in rat liver revealed by in situ hybridization. Carcinogenesis. 18(3):545–51.
[17] Pitot HC (1990) Altered hepatic foci: their role in murine hepatocarcinogenesis. Annu Rev Pharmacol Toxicol. 30:465–500.
[18] Coles B, Ketterer B (1990) The role of glutathione and glutathione transferases in chemical carcinogenesis. Crit Rev Biochem Mol Biol. 25(1):47–70.
[19] Mannervik B, Danielson UH (1988) Glutathione transferases--structure and catalytic activity. CRC Crit Rev Biochem. 23(3):283–337.
[20] Morrow CS, Cowan KH (1990) Glutathione S-transferases and drug resistance. Cancer Cells. 2(1):15–22.
[21] Keen JH, Jakoby WB (1978) Glutathione transferases. Catalysis of nucleophilic reactions of glutathione. J Biol Chem. 253(16):5654–7.
[22] Hayes JD, McLellan LI (1999) Glutathione and glutathione-dependent enzymes represent a co-ordinately regulated defence against oxidative stress. Free Radic Res. 31(4):273–300.
[23] Satoh K, Kitahara A, Soma Y, Inaba Y, Hatayama I, Sato K (1985) Purification, induction, and distribution of placental glutathione transferase: a new marker enzyme for preneoplastic cells in the rat chemical hepatocarcinogenesis. Proc Natl Acad Sci U S A. 82(12):3964–8.
[24] Armstrong DK, Gordon GB, Hilton J, Streeper RT, Colvin OM, Davidson NE (1992) Hepsulfam sensitivity in human breast cancer cell lines: the role of glutathione and glutathione S-transferase in resistance. Cancer Res. 52(6):1416–21.

[25] Batist G, Tulpule A, Sinha BK, Katki AG, Myers CE, Cowan KH (1986) Overexpression of a novel anionic glutathione transferase in multidrug-resistant human breast cancer cells. J Biol Chem. 261(33):15544–9.

[26] Tolman KG, Gray PD, el Masry S, Luther RR, Janicki RS (1989) Toxicity of uricosuric diuretics in rat hepatocyte culture. Biochem Pharmacol. 38(7):1181–4.

[27] Yamamoto K, Masubuchi Y, Narimatsu S, Kobayashi S, Horie T (2002) Toxicity of ethacrynic acid in isolated rat hepatocytes. Toxicol In Vitro. 16(2):151–8.

[28] Evers R, Kool M, van Deemter L, Janssen H, Calafat J, Oomen LC, Paulusma CC, Oude Elferink RP, Baas F, Schinkel AH, Borst P (1998) Drug export activity of the human canalicular multispecific organic anion transporter in polarized kidney MDCK cells expressing cMOAT (MRP2) cDNA. J Clin Invest.101(7):1310–9.

[29] Tirona RG, Tan E, Meier G, Pang KS (1999) Uptake and glutathione conjugation of ethacrynic acid and efflux of the glutathione adduct by periportal and perivenous rat hepatocytes. J Pharmacol Exp Ther. 291(3):1210–9.

[30] Iersel ML, Ploemen JP, Struik I, van Amersfoort C, Keyzer AE, Schefferlie JG, van Bladeren PJ (1996) Inhibition of glutathione S-transferase activity in human melanoma cells by alpha,beta-unsaturated carbonyl derivatives. Effects of acrolein, cinnamaldehyde, citral, crotonaldehyde, curcumin, ethacrynic acid, and trans-2-hexenal. Chem Biol Interact. 102(2):117–32.

[31] Ploemen JH, van Ommen B, van Bladeren PJ (1990) Inhibition of rat and human glutathione S-transferase isoenzymes by ethacrynic acid and its glutathione conjugate. Biochem Pharmacol. 40(7):1631–5.

[32] Ohkawa H, Ohishi N, Yagi K (1979) Assay for lipid peroxides in animal tissues by thiobarbituric acid reaction. Anal Biochem. 95(2):351–8.

[33] Sato K (1989) Glutathione transferases as markers of preneoplasia and neoplasia. Adv Cancer Res. 52:205–55.

[34] Meyer DJ, Beale D, Tan KH, Coles B, Ketterer B (1985) Glutathione transferases in primary rat hepatomas: the isolation of a form with GSH peroxidase activity. FEBS Lett. 184(1):139–43.

[35] Mannervik B, Alin P, Guthenberg C, Jensson H, Tahir MK, Warholm M, Jörnvall H (1985) Identification of three classes of cytosolic glutathione transferase common to several mammalian species: correlation between structural data and enzymatic properties. Proc Natl Acad Sci U S A. 82(21):7202–6.

[36] Tahir MK, Guthenberg C, Mannervik B (1989) Glutathione transferases in rat hepatoma cells. Effects of ascites cells on the isoenzyme pattern in liver and induction of glutathione transferases in the tumour cells. Biochem J. 257(1):215–20.

[37] Kuzmich S, Vanderveer LA, Walsh ES, LaCreta FP, Tew KD (1992) Increased levels of glutathione S-transferase pi transcript as a mechanism of resistance to ethacrynic acid. Biochem J. 281(Pt 1):219–24.

[38] Morrow CS, Smitherman PK, Townsend AJ (1998) Combined expression of multidrug resistance protein (MRP) and glutathione S-transferase P1-1 (GSTP1-1) in MCF7 cells and high level resistance to the cytotoxicities of ethacrynic acid but not oxazaphosphorines or cisplatin. Biochem Pharmacol. 56(8):1013–21.

[39] Nies AT, König J, Pfannschmidt M, Klar E, Hofmann WJ, Keppler D (2001) Expression of the multidrug resistance proteins MRP2 and MRP3 in human hepatocellular carcinoma. Int J Cancer. 94(4):492–9.

[40] Ji B, Ito K, Sekine S, Tajima A, Horie T (2004) Ethacrynic-acid-induced glutathione depletion and oxidative stress in normal and Mrp2-deficient rat liver. Free Radic Biol Med. 37(11):1718–29.

[41] Favre C, Carnovale CE, Monti JA, Carrillo MC (2001) Inhibition by interferon alpha-2b of rat liver regeneration: effect on ornithine decarboxylase and total protein synthesis. Biochem Pharmacol. 61(12):1587–93.

[42] Rosemberg S (1997) Principles of cancer management: biologic therapy. In: De Vita V, Hellman S RS, editors. Cancer: Principles and Practice of Oncology. Philadelphia: Lippincott-Raven. pp. 349–73.

[43] Ikeda K, Saitoh S, Suzuki Y, Kobayashi M, Tsubota A, Fukuda M, Koida I, Arase Y, Chayama K, Murashima N, Kumada H (1998) Interferon decreases hepatocellular carcinogenesis in patients with cirrhosis caused by the hepatitis B virus: a pilot study. Cancer. 82(5):827–35.

[44] Yoshida H, Shiratori Y, Moriyama M, Arakawa Y, Ide T, Sata M, Inoue O, Yano M, Tanaka M, Fujiyama S, Nishiguchi S, Kuroki T, Imazeki F, Yokosuka O, Kinoyama S, Yamada G, Omata M (1999) Interferon therapy reduces the risk for hepatocellular carcinoma: national surveillance program of cirrhotic and noncirrhotic patients with chronic hepatitis C in Japan. IHIT Study Group. Inhibition of Hepatocarcinogenesis by Interferon Therapy. Ann Intern Med. 131(3):174–81.

[45] Lai CL, Lau JY, Wu PC, Ngan H, Chung HT, Mitchell SJ, Corbett TJ, Chow AW, Lin HJ (1993) Recombinant interferon-alpha in inoperable hepatocellular carcinoma: a randomized controlled trial. Hepatology. 17(3):389–94.

[46] Llovet JM, Sala M, Castells L, Suarez Y, Vilana R, Bianchi L, Ayuso C, Vargas V, Rodés J, Bruix J (2000) Randomized controlled trial of interferon treatment for advanced hepatocellular carcinoma. Hepatology. 31(1):54–8.

[47] Dunk AA, Ikeda T, Pignatelli M, Thomas HC (1986) Human lymphoblastoid interferon. In vitro and in vivo studies in hepatocellular carcinoma. J Hepatol. 2(3):419–29.

[48] Yano H, Iemura A, Haramaki M, Ogasawara S, Takayama A, Akiba J, Kojiro M (1999) Interferon alfa receptor expression and growth inhibition by interferon alfa in human liver cancer cell lines. Hepatology. 29(6):1708–17.

[49] Murphy D, Detjen KM, Welzel M, Wiedenmann B, Rosewicz S (2001) Interferon-alpha delays S-phase progression in human hepatocellular carcinoma cells via inhibition of specific cyclin-dependent kinases. Hepatology. 33(2):346–56.

[50] Gutterman JU (1994) Cytokine therapeutics: lessons from interferon alpha. Proc Natl Acad Sci U S A. 91(4):1198–205.

[51] Creasey AA, Bartholomew JC, Merigan TC (1980) Role of G0-G1 arrest in the inhibition of tumor cell growth by interferon. Proc Natl Acad Sci U S A. 77(3):1471–5.

[52] Roos G, Leanderson T, Lundgren E (1984) Interferon-induced cell cycle changes in human hematopoietic cell lines and fresh leukemic cells. Cancer Res. 44(6):2358–62.

[53] Soini Y, Pääkkö P, Lehto VP (1998) Histopathological evaluation of apoptosis in cancer. Am J Pathol. 153(4):1041–53.

[54] Schulte-Hermann R, Grasl-Kraupp B, Bursch W (1995) Apoptosis and hepatocarcinogenesis. In: Jirtle J, editor. Liver Regeneration and Carcinogenesis. San Diego: Academic. pp. 141–78.

[55] Miyashita T, Harigai M, Hanada M, Reed JC (1994) Identification of a p53-dependent negative response element in the bcl-2 gene. Cancer Res. 54(12):3131–5.

[56] Miyashita T, Reed JC (1995) Tumor suppressor p53 is a direct transcriptional activator of the human bax gene. Cell. 80(2):293–9.

[57] Sangfelt O, Erickson S, Castro J, Heiden T, Einhorn S, Grandér D (1997) Induction of apoptosis and inhibition of cell growth are independent responses to interferon-alpha in hematopoietic cell lines. Cell Growth Differ. 8(3):343–52.

[58] Rodríguez-Villanueva J, McDonnell TJ (1995) Induction of apoptotic cell death in non-melanoma skin cancer by interferon-alpha. Int J Cancer. 61(1):110–4.

[59] Sedlak TW, Oltvai ZN, Yang E, Wang K, Boise LH, Thompson CB, Korsmeyer SJ (1995) Multiple Bcl-2 family members demonstrate selective dimerizations with Bax. Proc Natl Acad Sci U S A.92(17):7834–8.

[60] Yang J, Liu X, Bhalla K, Kim CN, Ibrado AM, Cai J, Peng TI, Jones DP, Wang X (1997) Prevention of apoptosis by Bcl-2: release of cytochrome c from mitochondria blocked. Science. 275(5303):1129–32.

[61] Nechushtan A, Smith CL, Hsu YT, Youle RJ (1999) Conformation of the Bax C-terminus regulates subcellular location and cell death. EMBO J. 18(9):2330–41.

[62] Shima Y, Nakao K, Nakashima T, Kawakami A, Nakata K, Hamasaki K, Kato Y, Eguchi K, Ishii N (1999) Activation of caspase-8 in transforming growth factor-beta-induced apoptosis of human hepatoma cells. Hepatology. 30(5):1215–22.

[63] Yamamoto M, Fukuda K, Miura N, Suzuki R, Kido T, Komatsu Y (1998) Inhibition by dexamethasone of transforming growth factor beta1-induced apoptosis in rat hepatoma cells: a possible association with Bcl-xL induction. Hepatology. 27(4):959–66.

[64] Buenemann CL, Willy C, Buchmann A, Schmiechen A, Schwarz M (2001) Transforming growth factor-beta1-induced Smad signaling, cell-cycle arrest and apoptosis in hepatoma cells. Carcinogenesis. 22(3):447–52.

[65] Gressner AM, Lahme B, Mannherz HG, Polzar B (1997) TGF-beta-mediated hepatocellular apoptosis by rat and human hepatoma cells and primary rat hepatocytes. J Hepatol. 26(5):1079–92.

[66] Rossmanith W, Schulte-Hermann R (2001) Biology of transforming growth factor beta in hepatocarcinogenesis. Microsc Res Tech. 52(4):430–6.

[67] Kanzler S, Meyer E, Lohse AW, Schirmacher P, Henninger J, Galle PR, Blessing M (2001) Hepatocellular expression of a dominant-negative mutant TGF-beta type II receptor accelerates chemically induced hepatocarcinogenesis. Oncogene. 20(36):5015–24.

[68] Matsuzaki K, Date M, Furukawa F, Tahashi Y, Matsushita M, Sakitani K, Yamashiki N, Seki T, Saito H, Nishizawa M, Fujisawa J, Inoue K (2000) Autocrine stimulatory mechanism by transforming growth factor beta in human hepatocellular carcinoma. Cancer Res. 60(5):1394–402.

[69] Santoni-Rugiu E, Jensen MR, Factor VM, Thorgeirsson SS (1999) Acceleration of c-myc-induced hepatocarcinogenesis by Co-expression of transforming growth factor (TGF)-

alpha in transgenic mice is associated with TGF-beta1 signaling disruption. Am J Pathol. 154(6):1693–700.

[70] Grasl-Kraupp B, Rossmanith W, Ruttkay-Nedecky B, Müllauer L, Kammerer B, Bursch W, Schulte-Hermann R (1998) Levels of transforming growth factor beta and transforming growth factor beta receptors in rat liver during growth, regression by apoptosis and neoplasia. Hepatology. 28(3):717–26.

[71] Huang YL, Chou CK (1998) Bcl-2 blocks apoptotic signal of transforming growth factor-beta in human hepatoma cells. J Biomed Sci. 5(3):185–91.

[72] de Luján Alvarez M, Ronco MT, Ochoa JE, Monti J a, Carnovale CE, Pisani GB, Lugano MC, Carrillo MC (2004) Interferon alpha-induced apoptosis on rat preneoplastic liver is mediated by hepatocytic transforming growth factor beta(1). Hepatology. 40(2):394–402.

[73] Hori Y, Takeyama Y, Ueda T, Shinkai M, Takase K, Kuroda Y (2000) Macrophage-derived transforming growth factor-beta1 induces hepatocellular injury via apoptosis in rat severe acute pancreatitis. Surgery. 127(6):641–9.

[74] Kamimura S, Tsukamoto H (1995) Cytokine gene expression by Kupffer cells in experimental alcoholic liver disease. Hepatology. 22(4 Pt 1):1304–9.

[75] Gao C, Gressner G, Zoremba M, Gressner AM (1996) Transforming growth factor beta (TGF-beta) expression in isolated and cultured rat hepatocytes. J Cell Physiol. 167(3):394–405.

[76] Matsuzaki K, Date M, Furukawa F, Tahashi Y, Matsushita M, Sugano Y, Yamashiki N, Nakagawa T, Seki T, Nishizawa M, Fujisawa J, Inoue K (2000) Regulatory mechanisms for transforming growth factor beta as an autocrine inhibitor in human hepatocellular carcinoma: implications for roles of smads in its growth. Hepatology. 32(2):218–27.

[77] Bissell DM, Wang SS, Jarnagin WR, Roll FJ (1995) Cell-specific expression of transforming growth factor-beta in rat liver. Evidence for autocrine regulation of hepatocyte proliferation. J Clin Invest. 96(1):447–55.

[78] Babior BM (2000) The NADPH oxidase of endothelial cells. IUBMB Life. 50(4-5):267–9.

[79] Quiroga AD, Alvarez MDL, Parody JP, Ronco MT, Francés DE, Pisani GB, Carnovale CE, Carrillo MC (2007) Involvement of reactive oxygen species on the apoptotic mechanism induced by IFN-alpha2b in rat preneoplastic liver. Biochem Pharmacol.73(11):1776–85.

[80] Herrera B, Murillo MM, Alvarez-Barrientos A, Beltrán J, Fernández M, Fabregat I (2004) Source of early reactive oxygen species in the apoptosis induced by transforming growth factor-beta in fetal rat hepatocytes. Free Radical Biol Med. 36(1):16–26.

[81] Park CH (1988) Vitamin C in leukemia and preleukemia cell growth. Prog Clin Biol Res. 259:321–30.

[82] Prasad KN, Kumar R (1996) Effect of individual and multiple antioxidant vitamins on growth and morphology of human nontumorigenic and tumorigenic parotid acinar cells in culture. Nutr Cancer. 26(1):11–9.

[83] Seifried HE, McDonald SS, Anderson DE, Greenwald P, Milner JA (2003) The antioxidant conundrum in cancer. Cancer Res. 63(15):4295–8.

[84] Uddin S, Majchrzak B, Woodson J, Arunkumar P, Alsayed Y, Pine R, Young PR, Fish EN, Platanias LC (1999) Activation of the p38 mitogen-activated protein kinase by type I interferons. J Biol Chem. 274(42):30127–31.

[85] Sarkar D, Su Z-Z, Lebedeva IV, Sauane M, Gopalkrishnan RV, Valerie K, Dent P, Fisher PB (2002) mda-7 (IL-24) Mediates selective apoptosis in human melanoma cells by inducing the coordinated overexpression of the GADD family of genes by means of p38 MAPK. Proc Natl Acad Sci U S A. 99(15):10054–9.

[86] Porras A, Zuluaga S, Black E, Valladares A, Alvarez AM, Ambrosino C, Benito M, Nebreda AR (2005) P38 alpha mitogen-activated protein kinase sensitizes cells to apoptosis induced by different stimuli. Mol Biol Cell. 15(2):922–33.

[87] Liu B, Fang M, Lu Y, Mills GB, Fan Z (2001) Involvement of JNK-mediated pathway in EGF-mediated protection against paclitaxel-induced apoptosis in SiHa human cervical cancer cells. Br J Cancer. 85(2):303–11.

[88] Park JM, Greten FR, Li Z-W, Karin M (2002) Macrophage apoptosis by anthrax lethal factor through p38 MAP kinase inhibition. Science. 297(5589):2048–51.

[89] Juretic N, Santibáñez JF, Hurtado C, Martínez J (2001) ERK 1,2 and p38 pathways are involved in the proliferative stimuli mediated by urokinase in osteoblastic SaOS-2 cell line. J Cell Biochem. 83(1):92–8.

[90] Yosimichi G, Nakanishi T, Nishida T, Hattori T, Takano-Yamamoto T, Takigawa M (2001) CTGF/Hcs24 induces chondrocyte differentiation through a p38 mitogen-activated protein kinase (p38MAPK), and proliferation through a p44/42 MAPK/extracellular-signal regulated kinase (ERK). Eur J Biochem. 268(23):6058–65.

[91] Razandi M, Pedram A, Levin ER (2000) Estrogen signals to the preservation of endothelial cell form and function. J Biol Chem. 275(49):38540–6.

[92] Jiménez B, Volpert OV, Crawford SE, Febbraio M, Silverstein RL, Bouck N (2000) Signals leading to apoptosis-dependent inhibition of neovascularization by thrombospondin-1. Nat Med. 6(1):41–8.

[93] Nakagami H, Morishita R, Yamamoto K, Yoshimura SI, Taniyama Y, Aoki M, Matsubara H, Kim S, Kaneda Y, Ogihara T (2001) Phosphorylation of p38 mitogen-activated protein kinase downstream of bax-caspase-3 pathway leads to cell death induced by high D-glucose in human endothelial cells. Diabetes. 50(6):1472–81.

[94] Grethe S, Ares MPS, Andersson T, Pörn-Ares MI (2004) p38 MAPK mediates TNF-induced apoptosis in endothelial cells via phosphorylation and downregulation of Bcl-x(L). Exp Cell Res. 298(2):632–42.

[95] Verma A, Deb DK, Sassano A, Uddin S, Varga J, Wickrema A, Platanias LC (2002) Activation of the p38 mitogen-activated protein kinase mediates the suppressive effects of type I interferons and transforming growth factor-beta on normal hematopoiesis. J Biol Chem. 277(10):7726–35.

[96] Herrera B, Fernández M, Roncero C, Ventura JJ, Porras A, Valladares A, Benito M, Fabregat I (2001) Activation of p38MAPK by TGF-beta in fetal rat hepatocytes requires radical oxygen production, but is dispensable for cell death. FEBS Lett. 499(3):225–9.

[97] Bulavin DV, Demidov ON, Saito S, Kauraniemi P, Phillips C, Amundson SA, Ambrosino C, Sauter G, Nebreda AR, Anderson CW, Kallioniemi A, Fornace AJ Jr, Appella E (2002) Amplification of PPM1D in human tumors abrogates p53 tumor-suppressor activity. Nat Genet. 31(2):210–5.

[98] Bulavin DV, Phillips C, Nannenga B, Timofeev O, Donehower LA, Anderson CW, Appella E, Fornace AJ Jr (2004) Inactivation of the Wip1 phosphatase inhibits mammary

tumorigenesis through p38 MAPK-mediated activation of the p16(Ink4a)-p19(Arf) pathway. Nat Genet. 36(4):343–50.

[99] Honmo S, Ozaki A, Yamamoto M, Hashimoto N, Miyakoshi M, Tanaka H, Yoshie M, Tamakawa S, Tokusashi Y, Yaginuma Y, Kasai S, Ogawa K (2007) Low p38 MAPK and JNK activation in cultured hepatocytes of DRH rats; a strain highly resistant to hepatocarcinogenesis. Mol Carcinog. 46(9):758–65.

[100] Mayer IA, Verma A, Grumbach IM, Uddin S, Lekmine F, Ravandi F, Majchrzak B, Fujita S, Fish EN, Platanias LC (2001) The p38 MAPK pathway mediates the growth inhibitory effects of interferon-alpha in BCR-ABL-expressing cells. J Biol Chem. 276(30):28570–7.

[101] Quiroga AD, de Lujan Alvarez M, Parody JP, Ronco MT, Carnovale CE, Carrillo MC (2009) Interferon-alpha2b (IFN-alpha2b)-induced apoptosis is mediated by p38 MAPK in hepatocytes from rat preneoplastic liver via activation of NADPH oxidase. Growth Factors. 27(4):214–27.

[102] Ulloa L, Doody J, Massagué J (1999) Inhibition of transforming growth factor-beta/SMAD signalling by the interferon-gamma/STAT pathway. Nature. 397(6721):710–3.

[103] Ghosh AK, Yuan W, Mori Y, Chen Sj, Varga J (2001) Antagonistic regulation of type I collagen gene expression by interferon-gamma and transforming growth factor-beta. Integration at the level of p300/CBP transcriptional coactivators. J Biol Chem. 276(14):11041–8.

[104] Higashi K, Inagaki Y, Fujimori K, Nakao A, Kaneko H, Nakatsuka I (2003) Interferon-gamma interferes with transforming growth factor-beta signaling through direct interaction of YB-1 with Smad3. J Biol Chem. 278(44):43470–9.

[105] Ishida Y, Kondo T, Takayasu T, Iwakura Y, Mukaida N (2004) The essential involvement of cross-talk between IFN-gamma and TGF-beta in the skin wound-healing process. J Immunol. 172(3):1848–55.

[106] Jeong W-I, Park O, Radaeva S, Gao B (2006) STAT1 inhibits liver fibrosis in mice by inhibiting stellate cell proliferation and stimulating NK cell cytotoxicity. Hepatology. 44(6):1441–51.

[107] Weng H, Mertens PR, Gressner AM, Dooley S (2007) IFN-gamma abrogates profibrogenic TGF-beta signaling in liver by targeting expression of inhibitory and receptor Smads. J Hepatol. 46(2):295–303.

[108] Yamamoto T, Matsuda T, Muraguchi A, Miyazono K, Kawabata M (2001) Cross-talk between IL-6 and TGF-beta signaling in hepatoma cells. FEBS Lett. 492(3):247–53.

[109] Foser S, Redwanz I, Ebeling M, Heizmann CW, Certa U (2006) Interferon-alpha and transforming growth factor-beta co-induce growth inhibition of human tumor cells. Cell Mol Life Sci. 63(19-20):2387–96.

[110] Roberts AB, Wakefield LM (2003) The two faces of transforming growth factor beta in carcinogenesis. Proc Natl Acad Sci U S A. 100(15):8621–3.

[111] Lei X, Yang J, Nichols RW, Sun L-Z (2007) Abrogation of TGFbeta signaling induces apoptosis through the modulation of MAP kinase pathways in breast cancer cells. Exp Cell Res. 313(8):1687–95.

[112] Park DY, Lee CH, Sol MY, Suh KS, Yoon SY, Kim JW (2003) Expression and localization of the transforming growth factor-beta type I receptor and Smads in preneoplastic lesions during chemical hepatocarcinogenesis in rats. J Korean Med Sci. 18(4):510–9.

[113] Alvarez MDL, Quiroga AD, Parody JP, Ronco MT, Francés DE, Carnovale CE, Carrillo MC (2009) Cross-talk between IFN-alpha and TGF-beta1 signaling pathways in preneoplastic rat liver. Growth Factors. 27(1):1–11.

[114] Kim SJ, Im YH, Markowitz SD, Bang YJ (2000) Molecular mechanisms of inactivation of TGF-beta receptors during carcinogenesis. Cytokine Growth Factor Rev. 11(1-2):159–68.

[115] Ghosh AK, Varga J (2007) The transcriptional coactivator and acetyltransferase p300 in fibroblast biology and fibrosis. J Cell Physiol. 213(3):663–71.

[116] Inagaki Y, Nemoto T, Kushida M, Sheng Y, Higashi K, Ikeda K, Kawada N, Shirasaki F, Takehara K, Sugiyama K, Fujii M, Yamauchi H, Nakao A, de Combrugghe B, Watanabe T, Okazaki I (2003) Interferon alfa down-regulates collagen gene transcription and suppresses experimental hepatic fibrosis in mice. Hepatology. 38(4):890–9.

[117] Alvarez MDL, Quiroga AD, Ronco MT, Parody JP, Ochoa JE, Monti JA, Carnovale CE, Carrillo MC (2006) Time-dependent onset of Interferon-alpha2b-induced apoptosis in isolated hepatocytes from preneoplastic rat livers. Cytokine. 36(5-6):245–53.

[118] Nambotin SB, Wands JR, Kim M (2011) Points of therapeutic intervention along the Wnt signaling pathway in hepatocellular carcinoma. Anticancer Agents Med Chem. 11(6):549–59.

[119] Parody JP, Alvarez ML, Quiroga AD, Ceballos MP, Frances DE, Pisani GB, Pellegrino JM, Carnovale CE, Carrillo MC (2010) Attenuation of the Wnt/beta-catenin/TCF pathway by in vivo interferon-alpha2b (IFN-alpha2b) treatment in preneoplastic rat livers. Growth Factors. 28(3):166–77.

[120] Gautschi O, Ratschiller D, Gugger M, Betticher DC, Heighway J (2007) Cyclin D1 in non-small cell lung cancer: a key driver of malignant transformation. Lung Cancer. 55(1):1–14.

[121] Kornmann M, Danenberg KD, Arber N, Beger HG, Danenberg PV, Korc M (1999) Inhibition of cyclin D1 expression in human pancreatic cancer cells is associated with increased chemosensitivity and decreased expression of multiple chemoresistance genes. Cancer Res. 59(14):3505–11.

[122] Wilson CL, Heppner KJ, Labosky P a, Hogan BL, Matrisian LM (1997) Intestinal tumorigenesis is suppressed in mice lacking the metalloproteinase matrilysin. Proc Natl Acad Sci U S A.94(4):1402–7.

[123] Powell WC, Knox JD, Navre M, Grogan TM, Kittelson J, Nagle RB, Bowden GT (1993) Expression of the metalloproteinase matrilysin in DU-145 cells increases their invasive potential in severe combined immunodeficient mice. Cancer Res. 53(2):417–22.

[124] Yamamoto H, Itoh F, Hinoda Y, Imai K (1995) Suppression of matrilysin inhibits colon cancer cell invasion in vitro. Int J Cancer. 61(2):218–22.

[125] Yan D, Wiesmann M, Rohan M, Chan V, Jefferson AB, Guo L, Sakamoto D, Caothien RH, Fuller JH, Reinhard C, Garcia PD, Randazzo FM, Escobedo J, Fantl WJ, Williams

LT (2001) Elevated expression of axin2 and hnkd mRNA provides evidence that Wnt/beta -catenin signaling is activated in human colon tumors. Proc Natl Acad Sci U S A. 98(26):14973–8.

[126] Jho E-hoon, Zhang T, Domon C, Joo C-K, Freund J-N, Costantini F (2002) Wnt/beta-catenin/Tcf signaling induces the transcription of Axin2, a negative regulator of the signaling pathway. Mol Cell Biol. 22(4):1172–83.

[127] Behrens J (1998) Functional Interaction of an Axin Homolog, Conductin, with -Catenin, APC, and GSK3. Science. 280(5363):596–9.

[128] Hart MJ, de los Santos R, Albert IN, Rubinfeld B, Polakis P (1998) Downregulation of beta-catenin by human Axin and its association with the APC tumor suppressor, beta-catenin and GSK3 beta. Curr Biol. 8(10):573–81.

[129] Fujimura N, Vacik T, Machon O, Vlcek C, Scalabrin S, Speth M, Diep D, Krauss S, Kozmik Z (2007) Wnt-mediated down-regulation of Sp1 target genes by a transcriptional repressor Sp5. J Biol Chem. 282(2):1225–37.

[130] de La Coste A, Romagnolo B, Billuart P, Renard CA, Buendia MA, Soubrane O, Fabre M, Chelly J, Beldjord C, Kahn A, Perret C (1998) Somatic mutations of the beta-catenin gene are frequent in mouse and human hepatocellular carcinomas. Proc Natl Acad Sci U S A. 95(15):8847–51.

[131] Miyoshi Y, Iwao K, Nagasawa Y, Aihara T, Sasaki Y, Imaoka S, Murata M, Shimano T, Nakamura Y (1998) Activation of the beta-catenin gene in primary hepatocellular carcinomas by somatic alterations involving exon 3. Cancer Res. 58(12):2524–7.

[132] Devereux TR, Anna CH, Foley JF, White CM, Sills RC, Barrett JC (1999) Mutation of beta-catenin is an early event in chemically induced mouse hepatocellular carcinogenesis. Oncogene. 18(33):4726–33.

[133] Calvisi DF, Factor VM, Loi R, Thorgeirsson SS (2001) Activation of beta-catenin during hepatocarcinogenesis in transgenic mouse models: relationship to phenotype and tumor grade. Cancer Res. 61(5):2085–91.

[134] Merle P, Kim M, Herrmann M, Gupte A, Lefrançois L, Califano S, Trépo C, Tanaka S, Vitvitski L, de la Monte S, Wands JR (2005) Oncogenic role of the frizzled-7/beta-catenin pathway in hepatocellular carcinoma. J Hepatol. 43(5):854–62.

[135] Merle P, de la Monte S, Kim M, Herrmann M, Tanaka S, Von Dem Bussche A, Kew MC, Trepo C, Wands JR (2004) Functional consequences of frizzled-7 receptor overexpression in human hepatocellular carcinoma. Gastroenterology. 127(4):1110–22.

[136] Willert J, Epping M, Pollack JR, Brown PO, Nusse R (2002) A transcriptional response to Wnt protein in human embryonic carcinoma cells. BMC Dev Biol. 2:8.

[137] Essers MAG, de Vries-Smits LMM, Barker N, Polderman PE, Burgering BMT, Korswagen HC (2005) Functional interaction between beta-catenin and FOXO in oxidative stress signaling. Science. 308(5725):1181–4.

[138] Hoogeboom D, Essers MAG, Polderman PE, Voets E, Smits LMM, Burgering BMT (2008) Interaction of FOXO with beta-catenin inhibits beta-catenin/T cell factor activity. J Biol Chem. 283(14):9224–30.

[139] Cavallo RA, Cox RT, Moline MM, Roose J, Polevoy GA, Clevers H, Peifer M, Bejsovec A (1998) Drosophila Tcf and Groucho interact to repress Wingless signalling activity. Nature. 395(6702):604–8.

[140] Levanon D, Goldstein RE, Bernstein Y, Tang H, Goldenberg D, Stifani S, Paroush Z, Groner Y (1998) Transcriptional repression by AML1 and LEF-1 is mediated by the TLE/Groucho corepressors. Proc Natl Acad Sci U S A. 95(20):11590–5.

[141] Bellan C, De Falco G, Tosi GM, Lazzi S, Ferrari F, Morbini G, Bartolomei S, Toti P, Mangiavacchi P, Cevenini G, Trimarchi C, Cinti C, Giordano A, Leoncini L, Tosi P, Cottier H (2002) Missing expression of pRb2/p130 in human retinoblastomas is associated with reduced apoptosis and lesser differentiation. Invest Ophthalmol Vis Sci. 43(12):3602–8.

[142] Ceballos MP, Parody JP, Alvarez MDL, Ingaramo PI, Carnovale CE, Carrillo MC (2011) Interferon-α2b and transforming growth factor-β1 treatments on HCC cell lines: Are Wnt/β-catenin pathway and Smads signaling connected in hepatocellular carcinoma? Biochem Pharmacol. 82(11):1682–91.

[143] Damdinsuren B, Nagano H, Kondo M, Natsag J, Hanada H, Nakamura M, Wada H, Kato H, Marubashi S, Miyamoto A, Takeda Y, Umeshita K, Dono K, Monden M (2006) TGF-beta1-induced cell growth arrest and partial differentiation is related to the suppression of Id1 in human hepatoma cells. Oncol Rep. 15(2):401–8.

[144] Ho J, Cocolakis E, Dumas VM, Posner BI, Laporte SA, Lebrun J-J (2005) The G protein-coupled receptor kinase-2 is a TGFbeta-inducible antagonist of TGFbeta signal transduction. EMBO J. 24(18):3247–58.

[145] Murata M, Nabeshima S, Kikuchi K, Yamaji K, Furusyo N, Hayashi J (2006) A comparison of the antitumor effects of interferon-alpha and beta on human hepatocellular carcinoma cell lines. Cytokine. 33(3):121–8.

[146] Labbé E, Letamendia A, Attisano L (2000) Association of Smads with lymphoid enhancer binding factor 1/T cell-specific factor mediates cooperative signaling by the transforming growth factor-beta and wnt pathways. Proc Natl Acad Sci U S A. 97(15):8358–63.

[147] Letamendia A, Labbé E, Attisano L (2001) Transcriptional regulation by Smads: crosstalk between the TGF-beta and Wnt pathways. J Bone Joint Surg Am. 83-A Suppl(Pt 1):S31–9.

[148] Lim SK, Hoffmann FM (2006) Smad4 cooperates with lymphoid enhancer-binding factor 1/T cell-specific factor to increase c-myc expression in the absence of TGF-beta signaling. Proc Natl Acad Sci U S A. 103(49):18580–5.

[149] Edlund S, Lee SY, Grimsby S, Zhang S, Aspenström P, Heldin C-H, Landström M (2005) Interaction between Smad7 and beta-catenin: importance for transforming growth factor beta-induced apoptosis. Mol Cell Biol. 25(4):1475–88.

Chapter 17

Lipid Peroxidation and Antioxidants in Arterial Hypertension

Teresa Sousa, Joana Afonso, Félix Carvalho and António Albino-Teixeira

Additional information is available at the end of the chapter

http://dx.doi.org/10.5772/50346

1. Introduction

This chapter aims at giving a critical overview of the major oxidant and antioxidant changes in arterial hypertension, summarizing the experimental and clinical evidence about the involvement of oxidative stress in the pathophysiology of hypertension, either as a cause or a consequence of this disease. This review also provides a description of the biomarkers commonly used to evaluate lipid peroxidation and antioxidant defenses in experimental and human hypertension. Finally, we review the strategies (antioxidants, antihypertensive drugs) known to prevent or ameliorate oxidative damage, both in animal models of hypertension and hypertensive patients.

2. Pathophysiological role of oxidative stress in arterial hypertension

2.1. ROS sources and oxidative pathways involved in the pathogenesis of hypertension

In aerobic organisms, the beneficial effects of oxygen come with the price of reactive oxygen species (ROS) formation. These highly bioactive and short-lived molecules can interact with lipids, proteins and nucleic acids, causing severe molecular damage. However, living organisms have evolved specific mechanisms to adapt to the coexistence of ROS. In physiological conditions, there is a delicate balance between oxidants and antioxidants that not only protects our cells from the detrimental effects of reactive oxygen species (ROS), but also allows the existence of redox signaling processes that regulate cellular and organ functions. However, the disruption of redox homeostasis, leading to persistent high levels of ROS, is potentially pathological [1, 2]. Besides ROS, another group of molecules collectively designated as reactive nitrogen species (RNS) also exerts important functions in diverse physiological and pathological redox signaling processes. The excess of RNS is often termed nitrosative stress [3, 4].

ROS can be classified into two main categories: free radicals [e.g. superoxide ($O_2^{\cdot-}$), hydroxyl ($HO^{\cdot}$), peroxyl ($ROO^{\cdot}$)], which are highly reactive species due to the presence of one or more unpaired electrons, and non-radical oxidants [e.g. singlet oxygen ($^1O_2$) hydrogen peroxide (H_2O_2), hypochlorous acid (HOCl)] that have generally more specific reactivity and higher stability [3, 5, 6]. RNS include nitric oxide ($^{\cdot}NO$) and nitrogen dioxide radicals ($^{\cdot}NO_2$ and also non radicals such as nitrous acid (HNO_2), peroxynitrite ($ONOO^-$), peroxynitrous acid (ONOOH) and alkyl peroxynitrites (ROONO) [3]. Among biological ROS and RNS, $O_2^{\cdot-}$, H_2O_2, $^{\cdot}NO$ and $ONOO^-$ appear to be especially relevant in neuronal, renal and vascular control of blood pressure [3, 7 ,8] (Table 1). Major sources of ROS (and also RNS) within these systems include, but are not limited to, NADPH oxidases, xanthine oxidase, mitochondrial respiratory chain enzymes, $^{\cdot}NO$ synthases and myeloperoxidase [3, 8, 9].

	Free radicals	Non radical oxidants
ROS	$O_2^{\cdot-}$ $HO^{\cdot}$ $ROO^{\cdot}$	H_2O_2 HClO
RNS	$^{\cdot}NO$	$ONOO^-$

Table 1. Reactive oxidant species involved in cardiovascular and renal physiology or pathophysiology

NADPH oxidases (Nox) are enzyme complexes that catalyze the reduction of molecular oxygen using NADPH as an electron donor. Generally, the product of the electron transfer reaction is $O_2^{\cdot-}$ but H_2O_2 is also rapidly formed from dismutation of Nox-derived $O_2^{\cdot-}$ due to the presence of superoxide dismutase (SOD) in the cells or by spontaneous reaction. Nox-derived ROS have been shown to play a role in host defense and also in diverse signaling processes [10]. The Nox family comprises seven members (Nox1-5 and Duox1-2) with distinct tissue distribution and functions [10, 11]. So far, only Nox1, Nox2 and Nox4 have been shown to play relevant roles in hypertension pathophysiology [5, 8, 10]. These isoforms are localized in major sites of blood pressure control. For example, Nox1, Nox2 and Nox4 are expressed in the central nervous system where they appear to regulate sympathetic nerve activity [8]. Nox2 and Nox4 participate in the regulation of renal functions and contribute to end-organ damage associated with hypertension [8, 12]. In the vasculature, Nox1 controls smooth muscle cell growth and migration, Nox2 contributes to endothelial dysfunction and Nox4 controls vascular smooth muscle cell differentiation and improves endothelial-dependent vasodilatation [8, 13, 14].

Xanthine oxidoreductase has two inter-convertible forms, xanthine dehydrogenase (XDH) and xanthine oxidase (XO), that participate in purines metabolism catalyzing the conversion of hypoxanthine to xanthine and xanthine to uric acid [15, 16]. XDH preferentially uses NAD^+ as an electron acceptor while the oxidase reduces molecular oxygen in a reaction that generates $O_2^{\cdot-}$ and H_2O_2 [15, 16]. The XO form predominates in oxidative stress conditions and may contribute to endothelium dysfunction due to its localization in the luminal surface of vascular endothelium [16, 17]. Besides the production of ROS by XO, both XDH and XO generate uric acid which possesses antioxidant properties, such as scavenging of $ONOO^-$ and $HO^{\cdot}$, prevention of oxidative inactivation of endothelium enzymes and stabilization of Vitamin C

[18-22]. On the other hand, uric acid may also have prooxidant and proinflammatory effects [23, 24]. Indeed, high systemic levels of uric acid are associated with increased cardiovascular disease and poor outcome but it is not clear whether these effects reflect deleterious actions of uric acid or the oxidative damage caused by XO-derived ROS [23, 25].

Mitochondrial respiratory chain enzymes are primary intracellular sources of ROS. More than 90% of the total oxygen consumed by aerobic organisms is utilized by mitochondrial oxidases which produce ATP in a process coupled to the reduction of cellular oxygen to water [26]. About 1-4% of the oxygen used in these reactions is converted to $O_2^{\cdot-}$ and H_2O_2 which can be largely detrimental to mitochondrial functions if not adequately detoxified [26-28]. ROS levels in the mitochondria are regulated by the respiratory rate and manganese SOD [29]. Hypertensive animals have increased mitochondrial ROS production in the vessels, kidney and CNS [30-32].

·NO synthases (NOS) constitute a family of enzyme isoforms (neuronal NOS, nNOS; inducible NOS, iNOS; endothelial NOS, eNOS) that produce ·NO in a reaction that converts L-arginine to L-citrulline [28]. However, in conditions of limited bioavailability of the cofactor tetrahydrobiopterin, or the substrate L-arginine, these enzymes become unstable and reduce molecular oxygen to $O_2^{\cdot-}$ instead of ·NO production (uncoupled NOS) [28, 29]. NOS uncoupling is more often described for eNOS and is triggered by oxidative/nitrosative stress [28, 33]. Numerous experimental studies have shown that arterial hypertension is associated with eNOS dysregulation and endothelial dysfunction [34, 35].

Myeloperoxidase (MPO) is a heme protein secreted by activated neutrophiles and monocytes in inflammatory conditions and produces several oxidizing molecules that can affect lipids and proteins [28, 36]. MPO uses H_2O_2 to produce ROS such as HOCl, chloramines, tyrosyl radicals and nitrogen dioxides [36, 37]. Although MPO-derived ROS have a primary role in microbial killing, they also cause tissue damage in the heart, vessels, kidney and brain and appear to contribute to endothelial dysfunction [37, 40]. Figure 1 illustrates the major sources of ROS and/or RNS generation.

Of all the putative oxidative pathways involved in the pathogenesis of hypertension, the impairment of endothelial-dependent vasorelaxation by $O_2^{\cdot-}$ is by far the most studied [41-44]. In conditions of increased $O_2^{\cdot-}$ bioavailability, this ROS rapidly inactivates endothelial-derived ·NO leading to endothelial dysfunction [41]. In addition, $O_2^{\cdot-}$ may also modulate vascular tone by increasing intracellular Ca^{2+} concentration in vascular smooth muscle cells and endothelial cells [45]. The imbalance between $O_2^{\cdot-}$ and ·NO also affects the renal function, leading to enhanced sodium reabsorption and increased $ONOO^-$ formation, which contributes to tissue damage [12, 46]. In the CNS, elevated $O_2^{\cdot-}$ generation also appears to contribute to hypertension by reducing the cardiovascular depressor actions of ·NO in the rostral ventrolateral medulla [47]. In recent years H_2O_2 has also emerged as a pivotal molecule in the pathophysiology of arterial hypertension [48-50]. Of note, H_2O_2 seems to be even more harmful than $O_2^{\cdot-}$ due to its higher life span and diffusibility within and between cells [7, 51]. Furthermore, the conversion of $O_2^{\cdot-}$ to H_2O_2 appears to be favored in cardiovascular diseases since the expression and activity of SOD is enhanced by inflammatory cytokines in hypertension or in response to the pressor peptide, angiotensin II [7]. Several prohypertensive effects have been described for H_2O_2, such as increased vasoconstriction, vascular hypertrophy and hyperplasia, decreased diuresis and natriuresis and also increased spinal sympathetic outflow [7, 50, 52-58]. Increasing evidence has also shown that H_2O_2 amplifies oxidative stress by stimulating ROS generation by NADPH oxidases, XO and eNOS [7, 51]. In addition, H_2O_2 also appears to enhance the activation of the intrarenal renin-angiotensin system, a major regulator of blood pressure and renal function [49]. Altogether, these effects propagate H_2O_2 generation and prolong the redox pathologic signaling involved in blood pressure dysregulation. The oxidative mechanisms contributing to hypertension are summarized in Table 2.

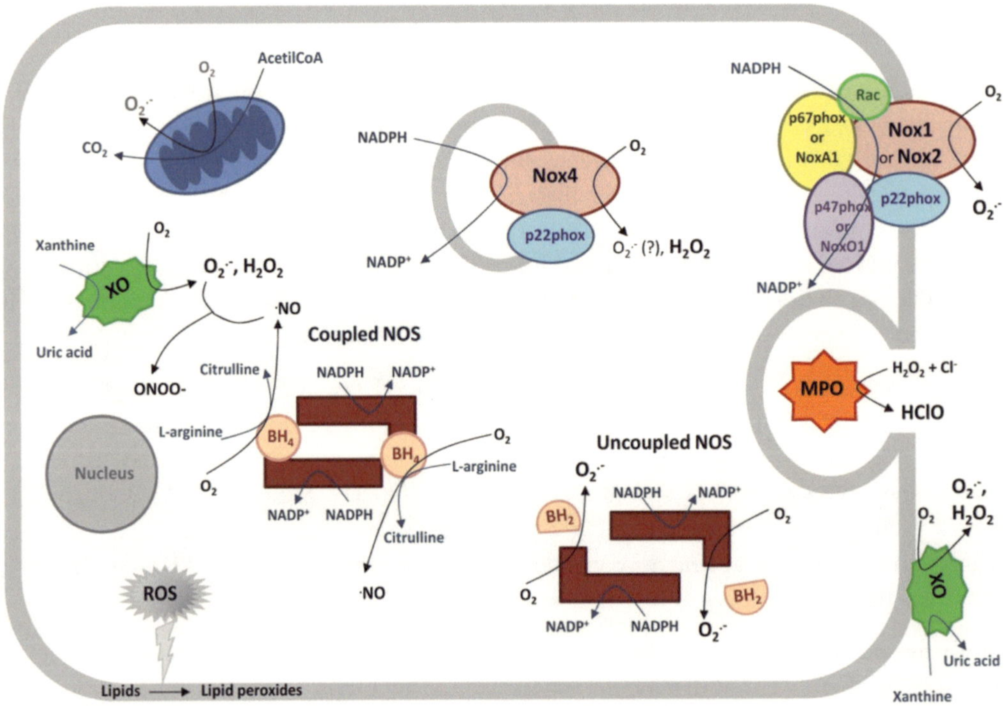

Figure 1. Sources of ROS and/or RNS generation - In normal cells, 1–2% of electrons carried by the mitochondrial electron transport chain leak from this pathway and pass directly to oxygen generating superoxide radical ($O_2^{\cdot-}$) which can be a source of other ROS. $O_2^{\cdot-}$ can also be formed by xanthine oxidase (XO) which catalyzes the oxidation of hypoxanthine and xanthine. All NOX enzymes utilize NADPH as an electron donor and catalyze transfer of electrons to molecular oxygen to generate $O_2^{\cdot-}$ and/or H_2O_2. Nitric Oxide synthases (NOS) generate ·NO and L-citrulline from arginine and O_2. Under pathologic conditions of oxidative stress, or when tetrahydrobiopterin (BH_4) or L-arginine are deficient, NOS enzymes become structurally unstable (uncoupled NOS) resulting in production of $O_2^{\cdot-}$ rather than ·NO. Activated monocytes also secrete a heme enzyme, myeloperoxidase(MPO), that uses H_2O_2 as a substrate to generate products that can oxidize lipids and proteins. One of these oxidants is hypochlorous acid (HOCI) which plays a critical role in host defenses against invading bacteria, viruses, and tumor cells but may also injure normal tissue. Within cell membranes, ROS can trigger lipid peroxidation, a self-propagating chain-reaction that can result in significant tissue damage.

Affected organ	Oxidative stress consequences	Major ROS and RNS involved
Vasculature	Impaired endothelium-dependent vasodilation	$O_2^{\cdot-}$, $\cdot NO$, $ONOO^-$
	Increased vasoconstriction	$O_2^{\cdot-}$, H_2O_2
	Increased hypertrophy and hyperplasia	$O_2^{\cdot-}$, H_2O_2
Kidney	Decreased blood flow	$O_2^{\cdot-}$, H_2O_2
	Increased salt reabsorption	$O_2^{\cdot-}$, H_2O_2
	Tissue damage	$HO\cdot$, HClO, $ONOO^-$
Brain/Spinal cord	Increased sympathetic efferent activity	$O_2^{\cdot-}$, H_2O_2

Table 2. Putative oxidative pathways leading to arterial hypertension

2.2. Evidence for redox changes in experimental and human hypertension

In the last two decades several studies have consistently observed increased oxidative stress in experimental and human arterial hypertension. Studies in diverse experimental models of hypertension have demonstrated raised prooxidant activity and ROS levels, altered antioxidant defenses and increased ROS-mediated damage, both at peripheral and central sites of cardiovascular regulation [8, 33, 59]. In human hypertensive patients there is also evidence of redox dysfunction. $O_2^{\cdot-}$ release from peripheral polymorphonuclear leucocytes is higher in hypertensive patients than in normotensive subjects [60]. Plasma H_2O_2 production is also raised in hypertensive patients. Furthermore, among still normotensive subjects, those with a family history of hypertension have a higher H_2O_2 production [61, 62]. An elevation of several oxidative stress byproducts, such as malondialdehyde, 8-isoprostanes, 8-oxo-2'-deoxyguanosine, oxidized low density lipoproteins, carbonyl groups and nitrotyrosine has also been observed in plasma or serum, urine or blood cells of hypertensive patients [63-66]. Furthermore, both enzymatic and non-enzymatic antioxidant defenses appear to be significantly reduced in human hypertension [65, 67]. Alterations of redox biomarkers in human hypertension are summarized in Table 3.

	Biomarker	**Evaluated in:**	**Alteration in hypertensive patients**	**References**
ROS/RNS	$O_2^{\cdot-}$	Peripheral PMN	↑	[60]
	H_2O_2	Plasma	↑	[61, 62]
		Lymphocytes	↑	[68]
	NOx	Plasma	↓	[69]
		Urine	↓	[70]

	Biomarker	Evaluated in:	Alteration in hypertensive patients	References
Prooxidant enzymes	NADPH oxidase activity p22phox (Nox subunit) mRNA and protein expression	Mononuclear cells Mononuclear cells	↑ ↑	[71] [71]
Oxidative or nitrosative stress byproducts	Malondialdehyde (MDA)/Thiobarbituric acid reactive substances (TBARS)	Plasma Erythrocytes Mononuclear cells and whole blood	↑ ↑ ↑	[67, 72] [64, 73, 74] [65]
	F2-Isoprostane (or 8-isoprostane or 8-epi-PGF2α)	Plasma Urine	↑ ↑	[63, 66, 74] [63, 74, 75]
	8-Oxo-2′-deoxyguanosine		↑ ↑	[65] [76, 77]
	Carbonyl groups	Serum	↑	[64]
	Oxidized low density lipoproteins	Plasma	↑	[63, 78]
	3-Nitrotyrosine	Plasma	↑	[66, 79]
Redox status	GSSG/GSH GSH/GSSG	Mononuclear cells and whole blood Erythrocytes	↑ ↓	[65] [74]
Antioxidants	GSH	Mononuclear cells and whole blood Erytrocytes	↓ ↓	[65] [64]
	Uric acid	Plasma Serum	↑ ↑	[79] [80]
	Vitamin C (ascorbic acid)	Plasma Serum	↓ ↓	[67] [81]
	Vitamin E (α-Tocopherol,	Erytrhocytes	↓	[67]
	Total antioxidant status (TAS) Ferric reducing activity of plasma (FRAP)	Plasma Plasma	↓ ↓	[63, 82] [74, 83]

	Biomarker	Evaluated in:	Alteration in hypertensive patients	References
	SOD activity	Erythrocytes	↓	[64, 74]
		Whole blood and mononuclear cells	↓	[65]
	Catalase activity	Erythrocytes	↑ or ↓	[64, 74]
		Whole blood and mononuclear cells	↓	[65]
	Glutathione peroxidase activity	Erythrocytes	↓	[64, 74]
		Whole blood and mononuclear cells	↓	[65]
	Glutathione-S-transferase activity	Erythrocytes	↑	[64]

NOx- nitrites and nitrates; PMN – Polymorphonuclear leucocytes; GSH – reduced glutathione; GSSG- oxidized glutathione;

Table 3. Altered oxidative/nitrosative stress biomarkers in human arterial hypertension

2.3. Oxidative stress as a cause for arterial hypertension

Whether oxidant imbalance is a cause or a consequence of high blood pressure remains a debatable question. The hypothesis that oxidative stress contributes to arterial hypertension is supported by several lines of evidence: (1) the induction of oxidative stress by the administration of lead or the glutathione synthesis inhibitor, buthionine sulfoximine, or the SOD inhibitor, sodium diethyldithiocarbamate, increases blood pressure in rats [48, 84]; (2) the infusion of H_2O_2 into the renal medulla leads to hypertension [48]; the treatment of hypertensive animals with antioxidants or inhibitors of ROS production prevents or attenuates hypertension [50, 85-87]; (3) the manipulation of genes related to ROS generation or elimination can alter blood pressure [88, 89]; (4) the *in vitro* exposure of cells and tissues to exogenous oxidants reproduces events involved in the pathophysiology of hypertension [43]; (5) systemic and tissue redox dysfunction appears to precede the blood pressure elevation [90].

2.4. Oxidative stress as a consequence of arterial hypertension

Arterial hypertension is associated with oscillatory shear stress and vascular stretch caused by increased vascular pressure. These mechanical forces are known to induce oxidative stress and vascular damage [91]. Furthermore, there is evidence that lowering blood pressure *per se* causes reduction of oxidative stress and improvement in endothelial function [92]. Several antihypertensive drugs with distinct mechanisms of action have been shown to decrease oxidant biomarkers in experimental and human hypertension [93-95]. However, there is limited evidence supporting the use of antioxidants to lower blood pressure in human hypertensive patients [5, 92]. Nevertheless, the failure of these studies does not exclude a role for oxidative stress in human essential hypertension but instead suggests that the antioxidant supplementation approach was not the appropriate therapeutic strategy [96].

3. Biomarkers of redox status in arterial hypertension

The evaluation of redox status may provide valuable information about the pathogenesis and progression of arterial hypertension and related cardiovascular and renal diseases. However, the short lifetime of ROS turns their assessment in animal models and humans a significant challenge, leading to a growing interest in the development and validation of oxidative stress biomarkers. Traditional approaches to evaluate oxidant status have frequently relied on indirect measurements of ROS bioavailability (e.g. evaluation of prooxidant and antioxidant activity, oxidized products from ROS and the GSH/GSSG ratio) as indicators of normal biological processes, pathogenic processes, or pharmacologic responses to therapeutic intervention [9, 96-99].

A biomarker of oxidative stress is classically defined as a biological molecule whose chemical structure has been modified by ROS and that can be used to reliably assess oxidative stress status in animal models and humans [100]. The ideal biomarker of oxidative stress depends on its ability to contribute to an early indication of disease severity and/or its progression, as well as to evaluate therapy efficacy. The measurement of redox status biomarkers may also help to clarify the pathophysiologic mechanisms mediating oxidative injury and may allow the prediction of disease. Ideally, biomarkers of oxidative damage for human studies would be evaluated in specimens that can be collected relatively easily, such as blood or urine. However, to serve these purposes, an ideal biomarker of oxidative damage should fulfill several conditions, such as: a) being a stable product, not susceptible to artifactual induction, oxidation, or loss during sample handling, processing, analysis, and storage; b) having a well-established relationship with the generation of ROS and/or progression of disease; c) allowing direct assessment in a target tissue or being able to generate a valid substitute that quantitatively reflects the oxidative modification of the target tissue; d) being present at concentrations high enough to be a significant detectable product; e) showing high specificity for the reactive species in question and free of erroneous factors from dietary intake; f) being noninvasive; g) being measurable by a specific, sensitive, reproducible and inexpensive assay; h) being measurable across populations; i) being present in concentrations that do not vary widely in the same persons under the same conditions at different times [97].

3.1. Systemic and tissue antioxidant defenses

ROS are involved in many biological processes including cell growth, differentiation, apoptosis, immunity and defense against micro-organisms [1, 101, 102]. Low or moderate concentrations of ROS are beneficial for living organisms. However, high concentrations of ROS can cause direct damage of macromolecules such as DNA, proteins, carbohydrates, and lipids, or disrupt redox signaling and control pathways, leading to a myriad of human diseases [103]. ROS bioavailability is determined by the balance between their production by prooxidant enzymes and their clearance by various antioxidant compounds and enzymes [1]. As defined by Halliwell and Gutteridge, an antioxidant is any substance that, at low concentration, is able to significantly delay or inhibit the oxidation of an oxidizable substrate [104]. Biological antioxidant defenses have evolved to match the diversity of prooxidants and several enzymatic and non-enzymatic molecules exist in cells and body fluids to control ROS levels within the physiological range [105]. The coordinated action of antioxidants results in the interception and deactivation of the damaging species. For example, the radical chain events initiated by free radicals can be terminated by the interaction of radicals with different non-enzymatic antioxidants [e.g. GSH, ascorbic acid, uric acid, α-tocopherol, etc] or prevented by specialized enzymatic defenses such as SOD, catalase and glutathione peroxidase (GPx) [105, 106]. The reduction of antioxidants bioavailability disrupts redox homeostasis leaving organisms more vulnerable to oxidative damage. Therefore, antioxidants may be useful biomarkers for risk stratification and disease prognostication.

3.2. Enzymatic antioxidants defenses

All eukaryotic cells possess powerful antioxidant enzymes which are responsible for neutralizing ROS.The first line of defense against ROS is achieved by SOD which is active in catalyzing the detoxification of $O_2^{\cdot-}$. This radical can be readily converted into H_2O_2 by SOD enzymes present in the cytosol and organelles (Cu,Zn-SOD or SOD-1), mitochondria (Mn-SOD or SOD-2) and extracellular fluids (EC-SOD or SOD-3) [36, 107, 108]. H_2O_2 generated in this reaction can be further decomposed to water and oxygen. This is achieved primarily by catalase in the peroxisomes and also by GPx enzymes in the cytosol and mitochondria [107, 108]. GPx are selenium-containing enzymes whose activity is dependent on GSH availability [108]. Besides neutralizing H_2O_2, GPx also degrades lipid hydroperoxides to lipid alcohols [36]. These reactions lead to the oxidation of GSH to GSSG. Catalase and GPx are differentially required for the clearance of high-levels or low-levels of H_2O_2, respectively [36]. Figure 2 illustrates major antioxidant enzymatic pathways.

In addition to these key antioxidant enzymatic defenses, there are other specialized enzymes with direct and/or indirect antioxidant functions. Glutathione reductase (GR) is responsible for the replenishment of GSH from GSSG disulphide. Glutathione-S-transferase catalyzes the conjugation of GSH with reactive electrophiles and is also involved in the detoxification of some carbonyl-, peroxide- and epoxide-containing metabolites produced within the cell in oxidative stress conditions [109]. Peroxiredoxins are selenium-independent enzymes that decompose H_2O_2, organic hydroperoxides and peroxynitrite [110]. Thioredoxin (Trx) and glutaredoxin (Grx) systems include several enzymes that regulate the thiol-disulphide state of proteins and influence their structure and function [110]. Trx isoforms reduce disulphide bonds in proteins, especially in peroxiredoxins and Trx reductase regenerates the oxidized Trx. Grx protects proteins SH-groups from irreversible oxidation by catalyzing S-glutathionylation and restores functionally active thiols through catalysis of deglutathionylation [110]. Grx enzymes are functionally coupled to GR which reduces the GSSG produced in the deglutathionylation reaction [110].

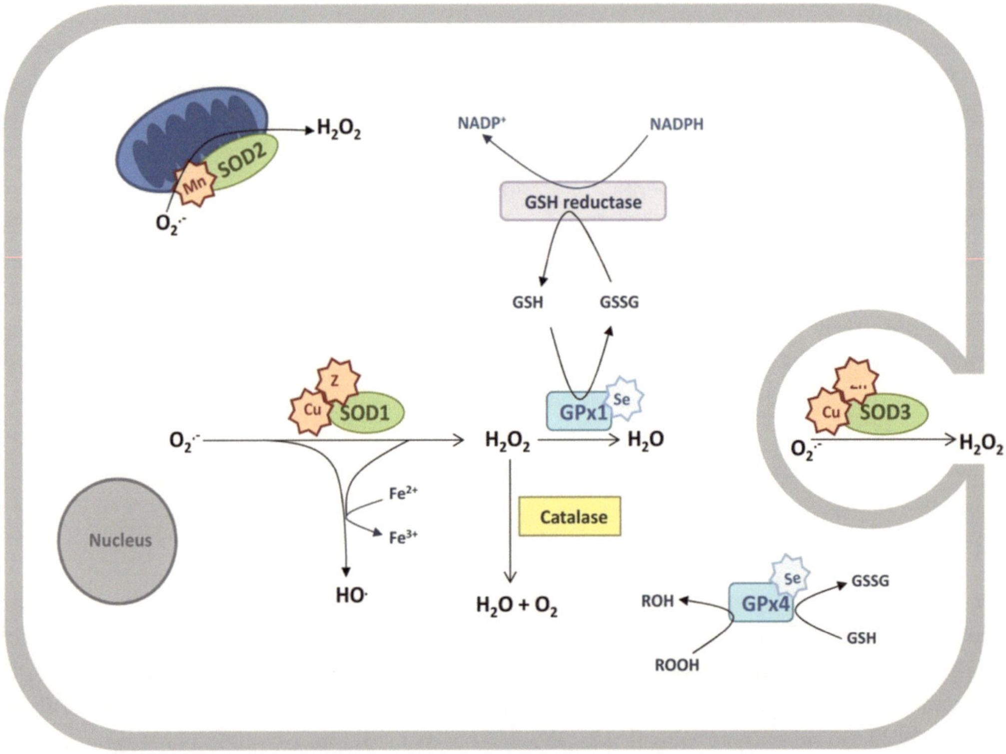

Figure 2. Major antioxidant enzyme defenses

Hypertensive patients have reduced activity and decreased content of antioxidant enzymes, including SOD, GPx, and catalase [43]. However, several studies have also described an adaptive increase in antioxidant enzyme activities in some experimental models of hypertension [50, 111, 112]. The uncoordinated activity of antioxidant enzymes may aggravate oxidative stress. For example, the increased dismutation of $O_2^{\cdot-}$ by SOD significantly increases H_2O_2 concentration, and may lead to deleterious consequences for the tissue in the absence of compensation of catalase and GPx activities [113]. Examples of altered antioxidant defenses in human and experimental hypertension are shown in Table 3 and Table 4, respectively.

Biomarker	Evaluated in:	Alteration	Hypertension Model	Reference
SOD	Brain	↓ expression and activity of Mn-SOD	Spontaneously hypertensive rats (SHR)	[114]
		↓ Cu, Zn-SOD activity	Stroke prone spontaneously hypertensive rats (SHRSP)	[115]
		↓ expression and activity of SOD1 and SOD2	SHR	[116]

Biomarker	Evaluated in:	Alteration	Hypertension Model	Reference
	Kidney	↓ expression of EC-SOD	SHR	[117]
		↑ SOD activity	Angiotensin II (Ang II) induced hypertension	[49]
		↓ expression of SOD1 and SOD3	SHR	[118]
	Arteries	↑ SOD activity	Hypertension induced by renin-angiotensin system (RAS) activation	[50]
		↓ SOD activity	SHR	[119, 120]
		↑ expression and activity of Cu, Zn-SOD and Mn-SOD	SHR	[121]
Catalase	Brain	↓ Catalase expression and activity	SHR	[116]
		↓ Catalase activity	Renovascular hypertensive rat; SHR	[122, 123]
	Kidney	↑ Catalase activity	SHR; Ang II-induced hypertension;	[49, 122]
		↓ Catalase expression	SHR	[124]
		↑ Catalase expression	SHR	[125]
	Arteries	↑ Catalase activity	Hypertension induced by RAS activation	[50]
		↓ Catalase activity	SHR	[120]
GPx	Brain	↓ GPx activity	SHR	[122]
	Kidney	↑ GPx activity	Ang II-induced hypertension	[49]
		↓ GPx expression	SHR	[124, 125]
		↓ GPx activity	SHR	[122]
	Arteries	↑ GPx activity	Hypertension induced by RAS activation	[50]
		↓ GPx activity	SHR	[120]
		↓ GPx expression	Salt- sensitive hypertension (Ovariectomized female rats)	[126]

Table 4. Alterations in major antioxidant enzyme defenses in the brain, kidney and arteries in experimental models of hypertension

3.3. Non-enzymatic antioxidants defenses

Non-enzymatic antioxidants such as GSH, ascorbic acid (vitamin C) and α-tocopherol (vitamin E) play an excellent role in protecting the cells from oxidative damage [4]. GSH has a potent electron-donating capacity that renders GSH both a potent antioxidant *per se* and a conventional cofactor for enzymatic reactions that require readily available electron pairs. In physiological conditions, GSH is present inside the cells mainly in its reduced form and less than 10 percent of total GSH exists in the oxidized form, GSSG [127]. Therefore, intracellular GSH status can be used as a sensitive marker of the cell health and resistance to toxicity. Furthermore, it has been demonstrated that GSH depletion can lead to cell apoptosis [128]. The measurements of GSH and GSSG have been considered useful indicators of the status of oxidative stress [4, 129]. Vitamins E and C are among the major dietary antioxidants. The vitamins have received considerable attention in clinical trials of primary and secondary prevention of cardiovascular diseases (CVD) and cancer. Vitamin E is found in lipoproteins, cell membranes and extracellular fluids. It terminates lipid peroxidation processes and converts $O_2^{\cdot-}$ and $HO^{\cdot}$ to less reactive forms [130]. Vitamin C, a water soluble antioxidant, is found in high concentrations in the adrenal and pituitary glands, liver, brain, spleen and pancreas. It is hydrophilic and can directly scavenge ROS and lipid hydroperoxides. Vitamin C can also restore oxidized vitamin E and can spare selenium [131]. Carotenoids, such as β-carotene are lipid soluble antioxidants that function as efficient scavengers of $^1O_2$ but may also quench $ROO^{\cdot}$ radicals [108]. Uric acid is a highly abundant aqueous antioxidant, considered to be the main contributor for the antioxidant capacity in the plasma [96, 132]. It has the ability to quenche $HO^{\cdot}$ and $ONOO^-$ and may prevent lipid peroxidation [21, 132]. The scavenging of $ONOO^-$ by uric acid is significantly increased in the presence of Vitamin C and cysteine which regenerate the urate radical formed in these reactions. Uric acid also acts as a chelator of iron in extracellular fluids [16]. However, once inside the cells, uric acid appears to exert prooxidant effects. It is not clear whether the correlation between the raised plasma levels of uric acid and cardiovascular risk are due to increased ROS generation by XO or to the prooxidative effects of uric acid itself. Some authors speculate that the increased concentrations of urate might be an adaptive mechanism that confers protection from oxidative damage [132]. It is likely that uric acid effects have different consequences depending on the surrounding microenvironment [21]. Bilirrubin, the end-product of heme catabolism, also appears to function as a chain-breaking antioxidant [133]. Low circulating bilirrubin levels are considered a risk factor for cardiovascular diseases [134]. Plasma albumin, the predominant plasma protein, is also an antioxidant due to its sulfhydryl groups and is able to scavenge MPO-derived chlorinated reactive species and $ROO^{\cdot}$ radicals [108, 135].

The combined antioxidant activities of aqueous- and lipid-antioxidants, including GSH, vitamins, uric acid, bilirrubin, albumin, etc, can be evaluated in the plasma and serum by several assays that measure the ability of the antioxidants present in the sample to inhibit the oxidation of the cation radical $ABTS^+$ [2,2'-azino-bis(3-ethylbenzothiazoline-6-sulphonic acid] (Total Antioxidant Status assay), to reduce a ferric-tripyridyltriazine complex (Ferric Reducing Ability of Plasma, FRAP assay) or to trap free radicals (Oxygen Radical Absorbance Capacity, ORAC assay; Total radical Trapping Parameter, TRAP) [50, 136-138].

The measurement of the overall antioxidant capacity may be more representative of the *in vivo* balance between oxidizing and antioxidant compounds than the evaluation of individual antioxidants [139]. Nevertheless, these assays have also some limitations. First, they correlate poorly with each other as the various antioxidants react differently in each assay. Second, in biological fluids, uric acid appears to account for more than 50% of the total antioxidant activity measured in most assays [108]. However, the putative protective effect of uric acid is debatable [140, 141].

Under conditions of high ROS levels it is expected a decrease of non-enzymatic antioxidants defenses in plasma, since the need for neutralization ROS species implies a higher consumption of endogenous antioxidants. For example, decreased levels of antioxidant vitamins C and E have been demonstrated in newly diagnosed untreated hypertensive patients compared with normotensive control subjects [142-144].

3.4. Systemic, urinary and tissue markers of lipid peroxidation

Measuring oxidative stress in biological systems is complex and requires accurate quantification of ROS or damaged biomolecules. One method to quantify oxidative stress is to measure lipid peroxidation. Lipids that contain unsaturated fatty acids with more than one double bond are particularly susceptible to the action of free radicals. The peroxidation of lipids disrupts biological membranes and is thereby highly deleterious to its structure and function [145]. A large number of by-products are formed during this process and can be measured by different assays. Common biomarkers of lipid peroxidation damage include hydroperoxides, which are primary products generated in the initial stages of lipid peroxidation, and secondary products formed at later lipid peroxidation stages, such as malondialdehyde (MDA) or F2-isoprostanes (Table 5) [146, 147] .The lag time required for the exponential generation of lipid peroxidation products can also be used to evaluate the susceptibility of lipid molecules to free radical damage. Therefore, lipids with higher resistance to oxidative stress exhibit longer lag times than those which are easily attacked by free radicals [147].

3.5. F2-isoprostanes

F2-isoprostanes are prostaglandin F2α isomers primarily produced by free radical-catalyzed peroxidation of the polyunsaturated fatty acid (PUFA), arachidonic acid [97]. Although there is also evidence of F2-isoprostane formation by the action of cyclooxygenase, it is currently assumed that systemic and urinary F2-isoprostanes are mostly derived from free radical-induced lipid peroxidation, independently of cyclooxygenase enzymatic activity. Therefore, F2-isoprostanes have been regarded as reliable biomarkers of oxidative stress. Furthermore, F2-isoprostanes have been shown to exert potent vasoconstrictor effects on animal and human vessels, suggesting a pathogenic role in cardiovascular diseases and have been extensively used as markers of lipid peroxidation in human diseases [74, 75,148]. Their high stability and presence in measurable concentrations in many biological tissues and fluids, under physiological and pathological conditions, has also allowed the establishment of reference intervals and the comparison or monitoring of disease states

[97,149,150]. Urine specimens are particularly suited for F2-isoprostanes measurements. First, the ex vivo formation of F2 isoprostanes is minimized in these samples due to the low urinary lipid content, avoiding the need for time-sensitive sample processing [97, 149, 151, 152]. Second, they provide a noninvasive route for systemic oxidative stress evaluation. Although they can also be locally produced in the kidney, many studies have demonstrated that urinary F2-isoprostanes are mainly derived from free F2-isoprostanes filtered from the circulation [97, 149, 151, 152]. Only hydrolyzed isoprostanes are excreted into the urine whereas blood plasma samples contain both free and esterified isoprostanes. Since plasma samples have considerable amounts of arachidonic acid, the addition of preservatives, such as butylated hydroxytoluene (BHT) and indomethacin, and the storage at -80°C, are recommended to avoid degradation and/or ex vivo formation of F2-isoprostanes [97.

3.6. TBARS

The free radical attack to PUFAs in cellular membranes leads to the disruption of cell structure and function. MDA, one of the end products of these oxidative reactions, can be detected in several biological fluids and tissues and is therefore used as a biomarker of lipid peroxidation and oxidative stress [153]. High MDA levels indicate a high rate of lipid peroxidation [154]. The reaction of MDA with 2-thiobarbituric acid (TBA) is frequently used to estimate oxidative stress [155]. MDA reacts with TBA under conditions of high temperature and acidity generating 2-thiobarbituric acid reactive substances (TBARS) that can be measured either spectrophotometrically or spectrofluorometrically. However, these products can also be formed by sample autooxidation under assay conditions or by cross-reactivity with non-MDA substrates such as bile pigments, proteins, carbohydrates and other aldehydes. Therefore, TBARS measurements often originate doubts due to their limited specificity as markers of lipid peroxidation [156]. Nevertheless, undesirable autooxidation and non-MDA substrates reactivity can be minimized by adding BHT during sample preparation. Plasma TBARS measurements have been reported to correlate with some clinical features of cardiovascular disease, preeclampsia, ischemia/reperfusion, chronic kidney disease and cerebrovascular disorders [157-160]. Since the TBARS assay may overestimate MDA, other methods can be used to evaluate lipid peroxidation products, such as the lipid hydroperoxide (LPO) test. The principle of the LPO test is that in the presence of hemoglobin, lipid hydroperoxides are reduced to hydroxyl derivates with the equimolar production of a methylene blue product, which can be quantified spectrophotometrically [161].

3.7. HNE

The aldehyde 4-hydroxy-2-nonenal (4-HNE) is one of the most cytotoxic products of free radical attack on ω6-PUFA, namely arachidonic and linoleic acids, being able to react with diverse biological molecules such as proteins, peptides, phospholipids and nucleic acids. It also acts as an important mediator of oxidant-induced signaling, cellular proliferation and apoptosis [97, 162]. 4-HNE can be detected in plasma and several biological tissues under physiological conditions but its generation is significantly raised in pathological states associated with oxidative stress [97, 162-164]. Renovascular hypertensive rats showed increased 4-HNE deposition in the intima of injured mesenteric arteries, suggesting the presence of free radical injury and cytotoxicity induced by 4-HNE [163]. A wide diversity of effects have been demonstrated for 4-HNE depending on its concentration. Concentrations below 0.1 microM are within the physiological range and appear to induce chemotaxis and stimulation of guanylate cyclase and phospholipase C [165]. 4-HNE concentrations between 1-20 microM inhibit DNA and protein synthesis and stimulate phospholipase A2. Higher concentrations (100 microM and above) are cytotoxic and genotoxic leading to cell death [165]. Thus, 4-HNE represents a broad indicator of lipid peroxidation.

3.8. Early stage of lipid peroxidation products

Lipid hydroperoxides are the primary products of lipid peroxidation and can further react to form secondary products including aldehydes such as MDA and 4-HNE [166,167]. Therefore, lipid hydroperoxides may be used to evaluate initial stage or acute lipid peroxidation while MDA and 4-HNE appear to be more representative of chronic oxidative stress. Recent reports described that 13-hydroperoxyoctadecadienoic acid (13-HPODE), a precursor to 3-hydroxyoctadecadienoic acid (13-HODE) is able to react with proteins forming adducts by covalently binding to specific amino acid residues. The Hexanoyl-Lysine (HEL) adduct results from the oxidative modification of ω6-PUFAs such as linoleic acid, the predominant PUFA in the human diet, and arachidonic acid [168]. HEL may be another useful biomarker for detecting and quantifying the earlier stages of lipid peroxidation. Monoclonal antibodies and ELISA kits have been developped, and HEL can be detected in oxidatively modified LDL, in human atherosclerotic lesions, human urine and serum. It has been also reported that HEL is formed in rat muscle during exercise and that its formation is inhibited by antioxidants such as flavonoids [169].

The lipid peroxidation biomarkers most commonly evaluated in hypertensive patients or experimental hypertension are shown in Table 3 and Table 5, respectively.

Lipid peroxidation biomarker	Measured in:	Alteration	Experimental model of hypertension	References
MDA	Plasma	↑	SHR	[170]
	Aorta	↑	Salt-loaded SHR	[94]
TBARS	Plasma	↑	Hypertension induced by RAS activation	[50]
	Plasma	↑	Ang II-induced hypertension	[171]
	Plasma, Heart	↑	Mineralocorticoid-induced hypertension	[172]
	Urine	↑	Salt-sensitive hypertension	[173]
	Aorta, Left Ventricle	↑	Ang II-induced hypertension	[174]

Lipid peroxidation biomarker	Measured in:	Alteration	Experimental model of hypertension	References
F2-**Isoprostanes**	Plasma	↑	Salt-sensitive hypertension	[175]
	Plasma	↑	Glucocorticoid-induced hypertension	[87]
	Urine	↑	SHR	[170, 176]
4-HNE	Mesenteric arteries	↑	Renovascular hypertension	[163]
	Aorta	↑	SHRSP	[177]
4-HNE adducts	Blood	↑	SHR	[178]

Table 5. Lipid peroxidation biomarkers in experimental hypertension

3.9. Other prooxidant biomarkers

Besides antioxidants and lipid peroxidation parameters, there are other important indexes of oxidant status. These include the expression and activity of prooxidant enzymes, ROS concentration, byproducts formed by ROS/RNS interaction with DNA (8-hydroxy-2-deoxyguanosine) or proteins (3-nitrotyrosine, carbonyl groups) and redox-sensitive transcription factors such as nuclear factor kappa B (NF-KB). Major sources of cellular ROS include Nox enzymes, mitochondrial electron transport enzymes, uncoupled NOS, XO and MPO. Table 6 summarizes several prooxidant biomarkers evaluated in experimental models of hypertension.

Biomarkers of prooxidant status	Evaluated in:	Alteration	Hypertension model	Reference
Mitochondrial electron transport chain enzymes/ mitochondrial ROS production	Brain	Oxidative impairment of mitochondrial enzymes	SHR	[30]
	Kidney	↑ mitochondrial ROS production	SHR, Mineralocorticoid hypertension	[179, 180]
NADPH oxidase family enzymes (or NOXs)	Brain	↑ NADPH oxidase activity	Salt-loaded SHRSP	[181]
	Kidney	↑ Nox4 expression, ↑ NADPH oxidase activity	Ang II–induced hypertension	[49]
	Arteries	**↑ expression of NAD(P)H oxidase subunits (p67(phox) and gp91(phox)**	Ang II-induced hypertension	[182]
		↑ Nox1 and Nox4 expression	SHRSP	[183]

Biomarkers of prooxidant status	Evaluated in:	Alteration	Hypertension model	Reference
		↑ **NADPH oxidase activity**	Ang II-induced hypertension; Hypertension induced by RAS activation	[50, 184-186]l
eNOS	Arteries	Uncoupled eNOS/ ↑ eNOS-derived ROS	SHR	[34]
	Arteries	↑ eNOS expression/↓ eNOS activity	SHR	[35]
XO	Arteries	↑XO activity	SHR	[187]
	Arteries	↑XO expression	SHR	[188]
H_2O_2	Kidney, Blood /urine	↑ renal production/↑ production in plasma /↑ urinary excretion	Ang II-induced hypertension	[49]
MPO	Arteries	↑ MPO activity	SHRSP	[189]
	Kidney, Heart, Brain	↑ MPO activity	Renovascular hypertension	[190]
GSH/GSSG	Kidney	↓ ratio	SHR	[191]
	Plasma, Heart, kidney	↓ ratio	Salt-sensitive hypertension	[192]
3-nitrotyrosine	Kidney	↑ expression	SHR	[193]
Protein carbonyl groups	Arteries, Heart, Kidney	↑ expression	SHR	[194]
	Kidney	↑ expression	SHR	[195]
8-Hydroxy-2-deoxyguanosine (8-OH-dG)	Arteries, Heart, Kidney,	↑ expression	SHR	[194].
NF-κB	Kidney	↑ activation	SHR; Ang II-induced hypertension	[49, 196]
	Arteries	↑ activation	Ang II-induced hypertension	[197]

Table 6. Other Prooxidant status biomarkers in experimental hypertension

4. Prophylactic and therapeutic strategies to reduce oxidative damage in arterial hypertension

A plethora of studies has demonstrated that hypertension is associated with an imbalance between oxidants and antioxidants that leads to altered cell signaling and oxidative damage.

Therefore, extensive research has been conducted in order to identify the ROS involved in blood pressure dysregulation, as well as the major prooxidant enzymes and antioxidant defenses that contribute to the loss of redox homeostasis in cardiovascular and renal systems. Furthermore, studies on experimental models of hypertension recognized several important neurohumoral stimuli responsible for ROS overproduction and also the main targets for ROS-induced dysfunction [8, 43]. Therapeutic interventions to reduce oxidative stress in hypertension have mostly relied on the administration of drugs that increase antioxidant capacity or inhibit ROS generation. In addition, other strategies aimed at reducing the activation of neurohumoral pathways that stimulate ROS production (upstream mediators) or at blocking/repairing the downstream targets affected by ROS have also been tested [196, 198, 199].

4.1. Targeting oxidative stress in experimental hypertension

The pharmacological modulation of ROS bioavailability in animal models of hypertension has been useful to demonstrate a causative role for oxidative stress in the pathophysiology of hypertension [43, 50]. However, the blood pressure lowering efficacy of these strategies appears to differ when comparing distinct experimental models [48, 50, 85, 200, 201]. This is probably because the development of each animal model was based on a particular etiological factor presumably responsible for human hypertension, such as high salt intake, overactivation of the renin-angiotensin system, genetic factors or renal disease. Since these factors may stimulate different redox pathways, the effectiveness of an antioxidant in one model does not necessarily translate to other models or to human essential hypertension which is known to have a multifactorial nature. Another important observation is that treatments with antioxidants or ROS inhibitors are generally more effective in preventing rather than reversing the hypertension [49, 50, 87, 202]. Indeed, there are several studies demonstrating that ROS activate feed-forward mechanisms that amplify the cardiovascular and renal dysfunction [8, 43, 49, 51]. Once triggered, these pathways may be sufficient to sustain the deleterious effects of oxidative stress even after ROS blockade or elimination [49]. *In vivo* drug treatments targeting oxidative stress in experimental models of hypertension are reviewed below and their effects on blood pressure are summarized in Table 7.

4.1.1. Drugs inhibiting ROS production

Apocynin is a methoxy-substituted catechol (4-hydroxy-3-methoxy-acetophenone), originally extracted from the roots of the tradicional medicinal herb *Picrorhiza kurroa* which has anti-inflammatory properties [203]. Several experimental studies have used apocynin for its ability to inhibit Nox enzymes. The mechanism of inhibition involves the blockade of translocation of cytosolic protein subunits to the membrane which is crucial for the activation of Nox1 and Nox2 [204]. Thus, the effect of apocynin is restricted to inducible Nox enzymes that require cytosolic activators and it does not seem to affect constitutively active Nox isoforms and their putative physiological actions [204]. However, to be an effective Nox inhibitor, apocynin has to undergo a peroxidase-mediated oxidation to be converted into the metabolically active diapocynin [205-207]. The activation of apocynin occurs in the presence of MPO and H_2O_2 [205, 207]. This fact suggests that apocynin may function only in conditions of high inflammatory and prooxidant activity. Apocynin has also been shown to have direct antioxidant properties, being able to scavenge H_2O_2 derived products [205, 207]. However, it can also function as a prooxidant in resting cells [203]. Nevertheless, it is possible that when administered in conditions of enhanced oxidative stress, the protective effect prevails.

Gp91ds-tat is a chimeric peptide that specifically inhibits NADPH oxidase by preventing the assembly of its subunits. It is constituted by a segment of gp91phox (*gp91ds*) important for the interaction of this membrane subunit with the cytosolic subunit, p47phox, and by a *tat* peptide from the HIV virus, which allows the uptake of the peptide into the cell [208, 209]. However, since it is a peptide it may have poor oral bioavailability and may induce sensitization reactions. Furthermore, the *tat* segment may have side effects on cellular signaling and activity [204, 208]. Thus, it is not suitable for long treatments or to clinical use in the treatment of human cardiovascular diseases. Although it was designed to block Nox2, it may also inhibit Nox1 given the substantial degree of homology between the two isoforms [204]. As for apocynin, Nox4 is not likely to be affected by gp91ds-tat since it is constitutively active and does not require the activation of cytosolic subunits [204].

Allopurinol and its metabolite **oxypurinol** are hypoxanthine and xanthine analogs, respectively, that inhibit XO activity [16]. At low concentrations, allopurinol is a competitive inhibitor of XO, while at higher concentrations it behaves as a non-competitive inhibitor [16]. XO rapidly metabolizes allopurinol into oxypurinol, a noncompetitive inhibitor of the enzyme which has a much higher half-life and is therefore responsible for most of the pharmacological effects of allopurinol [16]. In addition, both allopurinol and oxypurinol have intrinsic antioxidant properties, being able to scavenge ROS such as $O_2^{\cdot-}$, $HO^{\cdot}$ and HClO [210-212]. However, these effects appear to require higher doses than those required for XO inhibition [210]. Allopurinol is approved for the treatment of human patients with gout or hyperuricemia, but it has also potential therapeutic application in cardiovascular diseases. Most common adverse effects are nauseas, diarrheas, hypersensitivity reactions and skin rash [16].

4.1.2. Antioxidants

Tempol (4-hydroxy-2,2,6,6-tetramethylpiperidine-1-oxyl) is a membrane-permeable nitroxide that catalyzes the conversion of $O_2^{\cdot-}$ to H_2O_2 thus functioning as a SOD mimetic [213, 214]. Tempol protects the lipids or proteins from oxidative damage and interacts with other antioxidants to promote the reduction of oxidized lipids [214]. The main antihypertensive effect of this drug is related to the reduction of the $O_2^{\cdot-}$ interaction with $^{\cdot}NO$ which improves vasodilation [213, 214]. It also promotes natriuresis by enhancing the vasodilation of renal medullary vessels in a $^{\cdot}NO$ independent manner [214]. Indeed, tempol has been shown to have sympatholytic actions, being able to inhibit afferent, peripheral and central activation of the sympathetic nervous system [214]. These actions are responsible for the rapid fall of blood pressure and heart rate after acute intravenous administration of tempol [214]. Nevertheless, some studies reported that the formation of H_2O_2 by tempol can counteract its vasodilator, natriuretic and antihypertensive effects in models of hypertension where H_2O_2 plays a more prominent role than $O_2^{\cdot-}$ [48, 50, 53]. The co-administration of catalase in these conditions restores the protective action of tempol [48, 50].

N-acetylcysteine (NAC) is a thiol containing compound. It is the acetylated derivative of the aminoacid L-cysteine and a precursor for reduced glutathione (GSH) [215, 216]. It appears to have direct antioxidant action since its free thiol can interact with the electrophilic groups of ROS [215]. However, this effect does not seem likely to occur in vivo because NAC has poor oral bioavailability being rapidly metabolized into GSH, among other metabolites [216]. Thus, the main protective action of NAC is probably related to its role as a GSH precursor, which then detoxifies reactive species either by enzymatic or non-enzymatic reactions [216]. In humans, NAC is approved as a mucolytic agent because it destroys the disulphide bridges of mucoproteins [215]. It is also used as an antidote for acetaminophen poisoning which dramatically depletes hepatic GSH content causing severe damage [217]. NAC may also have potential therapeutical applications in the treatment of heart diseases [218].

Polyethylene glycol-catalase is the conjugated form of the enzyme catalase with polyethylene glycol (PEG) which enhances the stability in aqueous solution, reduces immunogenicity and decreases sensitivity to proteolysis, thus increasing the circulatory half-life of catalase [219]. PEG also enhances the catalase association with cells [219]. The antioxidant effect of PEG-catalase results from the enzymatic degradation of H_2O_2 to water.

Ebselen (2-phenyl-1,2-benzisoselenazol-3[2H]-one) is a lipid-soluble seleno-organic compound that mimics glutathione peroxidase activity, being able to react with H_2O_2 and organic hydroperoxides including membrane-bound phospholipid and cholesterylester hydroperoxides [220]. It appears to reduce lipid peroxidation initiated by hydroperoxides but not free radicals initiators [221]. In addition, ebselen reacts rapidly with $ONOO^-$. The ebselen selenoxid product yielded in this reaction is regenerated to ebselen by GSH, which allows its reutilization as a defense against $ONOO^-$ [222, 223]. Ebselen also directly inhibits inflammation-related enzymes such as 5-lipoxygenase, ·NO synthases, protein kinase C, NADPH oxidase and H^+/K^+-ATPase by reacting with the SH group, leading to the formation of a selenosulphide complex [221]. Some authors have also proposed that the antioxidant and anti-inflammatory actions of ebselen are mediated through interactions with the thioredoxin (Trx) system [220]. Reduced Trx is important for growth and redox regulation by thiol redox control [220]. Ebselen was found to be an excellent substrate for mammalian TrxR and a highly efficient oxidant of reduced Trx. It also seems to function as a Trx peroxidase or peroxiredoxin mimic, thus contributing to the elimination of H_2O_2 and lipid hydroperoxides [220]. Ebselen has been used in clinical trials for the treatment of patients with acute ischemic stroke or delayed neurological deficit after aneurismal subarachnoid hemorrhage [224, 225].

Vitamin C (ascorbic acid) is a water soluble antioxidant found in the body as an ascorbate anion. It acts as a free radical scavenger [226]. Although this effect requires higher concentrations than those achieved in the plasma by oral administration, ascorbate appears to concentrate in tissues in much higher levels than those found in the plasma and can act effectively as a ROS scavenger [227]. In addition, it reduces membrane lipid peroxidation and regenerates Vitamin E [226]. Recent reports also suggest that Vitamin C can suppress NADPH oxidase activity [227].

Vitamin E is a generic term for a group of compounds including tocopherols and tocotrienols. The isoform α-tocopherol appears to be the most abundant in vivo [227]. Vitamin E terminates the propagation of the free radical chain reaction in lipid membranes and inhibits LDL oxidation [226, 227]. Vitamin E can also have non antioxidant actions primarily through the regulation of enzymes involved in signal transduction. Enzymes inhibited by vitamin E include protein kinases C and B, protein tyrosine kinase, lipoxygenases, mitogen activated protein kinases, phospholipase A2 and cyclooxygenase-2. In contrast, vitamin E has stimulatory effects on protein tyrosine phosphatase and diacylglycerol kinase [228]. Both vitamins C and E have been shown to stimulate the activation of NOS activity and increase ·NO synthesis in endothelial cells and thus may contribute to improved endothelial-dependent vasodilation in hypertension [229]. However, although Vitamins C and E are generally considered to be non-toxic, they can undergo oxidation and generate pro-oxidant molecules [226]. Nevertheless, it appears that this is more likely to occur with Vitamin E, especially in the absence of sufficient Vitamin C to regenerate the α-tocopherol radical [227, 230, 231].

Alpha-lipoic acid (1,2-dithiolane-3-pentanoic acid or thioctic acid) has a wide range of effects on cell functions, acting as an antioxidant, a metal chelator and a signaling mediator [232]. Both lipoic acid (LA) and its reduced form dihydrolipoic acid (DHLA), may scavenge HO· and HClO, although neither species is able to neutralize H_2O_2 [232]. DHLA also regenerates Vitamins C and E and does not become a free radical after reacting with these species. Furthermore, LA and DHLA chelate transition metals, thus reducing the metal-catalyzed free radical damage [232]. LA also contributes to improve antioxidant defenses by increasing the intracellular levels of Vitamin C and GSH. Many of LA protective actions have been attributed to its interference in cell signaling processes [232]. For example, LA effect on GSH appears to be mediated by nuclear factor erythroid 2- related factor 2 (Nrf2), an important transcription factor regulating gene transcription through the Antioxidant Response Element. LA was also shown to interact with several kinases and protein phosphatases [232]. Its interaction with components of the insulin signaling cascade also appears to improve glucose disposal in animal models of diabetes and human diabetic patients [232]. In addition, LA improves endothelial ·NO synthesis and endothelial-dependent vasodilation and prevents deleterious modifications of thiol groups in Ca^{2+} channels [232]. It has also important anti-inflammatory effects by inhibiting the activation of NF-KB, a transcription factor that regulates the expression of proinflammatory genes [232].

Pyrrolidine dithiocarbamate (PDTC), a low-molecular weight thiol compound, has the ability to scavenge oxygen radicals and to chelate metals [233, 234]. It may also act as a prooxidant and a thiol group modulator [233]. PDTC has been shown to interfere with the activation of several transcription factors, being a potent inhibitor of NF-KB [233, 234]. PDTC can also activate other signaling pathways, such as the extracellular signal-regulated kinase (ERK), c-Jun N-terminal kinase (JNK) and the transcription factor Heat Shock Factor (HSF) [233, 235].

5, 6, 7, 8-Tetrahydrobiopterin (BH_4) is a key cofactor of NOS [236, 237]. It is involved in the formation and stabilization of eNOS and iNOS [236, 238]. In the absence of BH_4, NOS can become uncoupled and starts producing $O_2^{\cdot-}$ instead of $^{\cdot}NO$ [33, 237]. Furthermore, BH_4 also possesses direct antioxidant activity, being able to scavenge $O_2^{\cdot-}$ and $HO^{\cdot}$ [239]. The protective effects of BH_4 on the development of hypertension appear to be due an increase in eNOS activity, a reduction in $O_2^{\cdot-}$ production and a decrease in iNOS expression [199].

Drug	Antihypertensive effect	Lack of antihypertensive effect
Apocynin	Prevented/attenuated mineralocorticoid-induced hypertension [86, 240] Prevented/reversed glucocorticoid-induced hypertension [241] Prevented/reversed adrenocorticotropic hormone-induced hypertension [242] Prevented the development of Ang II-induced hypertension in mice [186] Prevented the development of renovascular hypertension [243] Prevented the development of hypertension induced by RAS activation [50] Reduced blood pressure in borderline and spontaneous hypertension [244] Attenuated salt-sensitive hypertension [245] Normalized blood pressure in a model of hypertension induced by disruption of dopamine D_2 receptor [246]	Failed to prevent the hypertension induced by chronic infusion of endothelin-1 [200] Failed to prevent hypertension in transgenic mice overexpressing renin or angiotensinogen [247, 248] Failed to prevent Ang II-induced hypertension in rats [249, 250]
Gp91ds-tat	Attenuated the blood pressure rise induced by Ang II in mice [209]	Failed to attenuate salt-sensitive hypertension [251]
Allopurinol	Attenuated salt-sensitive hypertension [252] Prevented glucocorticoid-induced hypertension [253]	Failed to prevent or attenuate mineralocorticoid-induced hypertension [254] Failed to prevent glucocorticoid-induced hypertension [255] Failed to prevent or attenuate adrenocorticotropic-induced hypertension [242] Failed to prevent the development of hypertension induced by the blockade of nitric oxide synthesis [256] Failed to prevent the progression of hypertension in young SHR [257]
Oxypurinol	Reduced blood pressure in SHR [258]	
Tempol	Attenuated hypertension in SHR [201] Prevented the progression of hypertension in salt-loaded SHRSP [259] Attenuated mineralocorticoid-induced hypertension [260]	Failed to prevent Ang II-induced hypertension [264] Failed to attenuate hypertension induced by inhibition of superoxide dismutase [48]

Drug	**Antihypertensive effect**	**Lack of antihypertensive effect**
	Prevented/attenuated glucocorticoid-induced hypertension [87] Attenuated salt-sensitive hypertension [261] Prevented the development of renovascular hypertension [243] Attenuated high-volume hypertension [262] Attenuated hypertension induced by NO inhibition [112] Partially prevented/reversed adrenocorticotropic hormone-induced hypertension [263]	Failed to prevent hypertension induced by RAS activation [50]
NAC	Attenuated hypertension in young SHR [265] Prevented the development of glucocorticoid-induced hypertension [202] Prevented the development of adrenocorticotropic hormone-induced hypertension [266] Markedly reduced salt-sensitive hypertension [267] Prevented/ attenuated hypertension induced by nitric oxide synthesis inhibition [268]	Failed to reduce blood pressure in adult SHR [265] Failed to attenuate glucorticoid-induced hypertension [202] Failed to reverse adrenocorticotropic-induced hypertension [266] Failed to prevent the development of hypertension induced by the blockade of nitric oxide synthesis [256]
PEG-catalase	Prevented the development of hypertension induced by RAS activation [50] Transiently decreased blood pressure in Ang II-hypertensive rats [49] Reduced blood pressure in high-volume hypertension in mice [269]	Lacked a sustained antihypertensive effect in Ang II-induced hypertension [49]
Ebselen	Attenuated the blood pressure rise induced by Ang II in mice overexpressing p22phox in vascular smooth muscle and in littermate control mice [270]	Failed to prevent the development of hypertension induced by the blockade of nitric oxide synthesis [256]
Vitamin C	Prevented the progression of hypertension induced by salt administration in SHRSP and in SHR [229, 271] Attenuated salt-induced hypertension [272, 273]	Failed to prevent adrenocorticotropic hormone-induced hypertension [274]
Vitamin E	Prevented the progression of hypertension induced by salt administration in SHRSP [229] Attenuated hypertension in young SHRSP [275] Attenuated salt-induced hypertension [273]	Failed to prevent adrenocorticotropic hormone-induced hypertension [274]
Lipoic acid	Reduced blood pressure in SHR [276] Prevented fructose-induced hypertension [277] Prevented/attenuated salt-induced hypertension [278] Prevented mineralocorticoid-induced	

Drug	Antihypertensive effect	Lack of antihypertensive effect
	hypertension [279]	
PDTC	Perinatal administration ameliorated hypertension in SHR offsprings [280] Prevented /Reduced hypertension in SHR [32, 196] Attenuated mineralocorticoid-induced hypertension [260]	
BH_4	Suppressed the development of hypertension in SHR [199] Reduced hypertension in SHR [281]	Failed to attenuate hypertension in castrated SHR [281] Failed to prevent the development of adrenocorticotropic hormone in rat [282]

Table 7. Effect of chronic treatment with antioxidants or inhibitors of ROS production on blood pressure

4.2. Antioxidant approaches in human hypertension

Although there is considerable evidence of oxidative stress involvement in the pathophysiology of hypertension, the attempts to demonstrate benefits from antioxidant therapy in human cardiovascular diseases have been very disappointing [5, 96, 283]. Most of the large trials regarding the effects of diet supplementation with Vitamin C, Vitamin E and β-carotene failed to show significant improvements in blood pressure and other cardiovascular endpoints [5, 283]. Furthermore, some of them also led to the conclusion that antioxidant treatment with Vitamin E or β-carotene may even be harmful [283-285]. In contrast, smaller clinical trials have provided some evidence of antioxidant treatment advantages. For example, some studies showed that systemic Vitamin C levels inversely correlates with blood pressure and that Vitamin C supplementation effectively attenuates hypertension [142, 286]. Vitamin E and lipoic acid have also been shown to improve vascular function, though there is not consistent evidence of a blood pressure lowering effect of these agents in human patients [5, 287, 288]. Nevertheless, it has been demonstrated that a high consumption of dietary fruits and vegetables increases plasma antioxidant capacity and reduces blood pressure [289, 290]. Thus, it appears that a diet rich in fruits and vegetables is a better strategy than antioxidant supplementation to improve antioxidant status and cardiovascular health [5]. Overall, the clinical trials with antioxidant supplements have been very unsatisfactory and are in disagreement with the findings obtained in experimental hypertension studies. There are some possible justifications for the disappointing outcomes of these trials. First, the type of the drug used as well as the dose and duration of the therapy might not be adequate [5, 291, 292]. Most trials followed an antioxidant strategy based in the administration of ROS scavengers such as Vitamins C and E. However, these drugs do not neutralize H_2O_2 which has been shown to play a relevant role in the pathophysiology of hypertension and other cardiovascular diseases [5, 7, 48-50]. Furthermore, it is known that human blood and tissues have plenty antioxidants and that several stimuli induce an adaptive increase of enzymatic antioxidant defenses which can mask the benefits of exogenously administered antioxidants [293]. In addition, the antioxidant doses used in most of the experimental studies have been much higher than those tested in human patients [291]. So, there is the possibility that in humans the antioxidants did not achieve effective concentrations to neutralize ROS. Furthermore, it is not known if orally administered antioxidants can reach the precise sites of increased ROS production as oxidative stress is heterogeneously distributed throughout the organs, tissues and cellular compartments [5, 96, 291]. Indeed, the unspecific scavenging of ROS may even interfere with many important physiological functions in a deleterious manner [29, 96]. Another important limitation of most antioxidants tested is that they can exert themselves prooxidant effects in the absence of a coordinated antioxidant response [5, 96]. For example, Vitamin E needs to be regenerated by Vitamin C otherwise it may cause oxidative damage [231].

There are also drawbacks in clinical trials design. In large trials of antioxidant supplementation, patients have not been recruited accordingly to their redox status [5, 294]. It is unlikely that a beneficial effect of antioxidant therapy would be observed in patients without previous evidence of increased oxidative stress. Another important consideration is that these clinical trials often have heterogeneous populations in terms of the etiology of cardiovascular disease [295]. Indeed, most studies have indiscriminately enrolled any patient at cardiovascular risk [294]. This is in obvious contrast to the homogeneous populations analyzed in experimental studies. Furthermore, some of the patients may be at an advanced stage of disease exhibiting irreversible damage insusceptible to antioxidant interventions [5, 29]. It should also be highlighted that many patients enrolled in these studies were already being treated with drugs such as aspirin, lipid-lowering agents and some antihypertensive drugs which can themselves interfere with oxidant status and mask the effects of additional therapy with antioxidants [5, 92, 296, 297].

So far, most interventions aimed at reducing oxidative stress in human hypertension have relied on antioxidant supplementation. However, it is possible that a strategy based on the inhibition of ROS production is more effective than the antioxidant interventions [5, 96]. The disruption of cardiovascular redox status is most likely triggered by an increase in prooxidant activity rather than a reduction in antioxidant defenses. Indeed, many neurohumoral or ambiental prohypertensive stimuli (angiotensin II, aldosterone, high-salt intake) are known to upregulate the expression and activity of prooxidant enzymes [5, 8, 43]. Nevertheless, there are already some studies that investigated the cardiovascular effects of prooxidant enzyme inhibition. Patients treated with allopurinol showed improvements in vascular function [298, 299]. However, a blood pressure lowering effect of this XO inhibitor has been shown only in newly diagnosed hypertensive adolescents and in hyperuricemic patients with normal renal function [300, 301]. Furthermore, the combination of allopurinol with antihypertensive drugs did not provide additional benefits on blood pressure [299]. This is probably because XO is not a major contributor to the development of hypertension, even though its activity may be increased in pathophysiological conditions [5]. Indeed, compelling evidence indicates that NADPH oxidases are the main contributors to ROS overproduction in cardiovascular and renal diseases [5, 8, 302]. Moreover, Nox-derived ROS are known to amplify redox dysfunction by inducing the activation of other prooxidant enzymes, such as XO, mitochondrial enzymes and NOS synthases [7, 51]. Since many antihypertensive drugs block upstream activators of Nox enzymes, it is not surprising that the inhibition of XO by allopurinol does not improve blood pressure control in patients already treated with antihypertensive drugs. To date, no Nox inhibitors have been tested in clinical trials although some specific Nox inhibitors have already been developed and patented [96, 296]. Future strategies to demonstrate the benefits of oxidative stress reduction in cardiovascular diseases should include the testing of specific Nox inhibitors in human patients. Moreover, the development of reliable oxidative stress biomarkers for risk stratification and monitoring of therapy is also highly desirable [96, 296]. Table 8 summarizes the possible reasons for the failure of antioxidants in clinical trials.

Limitations related to the drug treatment	Limitations related to the clinical trials design
Inadequate dose or duration of therapy Lack of effect on non-radical oxidants such as H_2O_2 Lack of effect on prooxidant activity Inaccessibility of ROS scavengers to intracellular sites of increased ROS production Some antioxidants may themselves become prooxidants in the absence of a coordinated antioxidant response Unspecific scavenging of ROS may disrupt physiological functions	Lack of previous evidence of increased redox dysfunction in patients analyzed Heterogeneous populations in terms of the etiology of cardiovascular disease Some patients may be at an advanced stage of disease exhibiting irreversible damage Patients treated simultaneously with drugs that interfere with oxidant status (aspirin, lipid lowering agents, antihypertensive drugs) Lack of validated oxidative stress biomarkers for risk stratification and monitoring of therapy

Table 8. Possible reasons for the failure of clinical trials with antioxidants in cardiovascular diseases

4.3. Antihypertensive treatments with direct and indirect antioxidant effects

It is known that first-line antihypertensive drugs such as angiotensin II receptor blockers (ARB) and angiotensin converting enzyme inhibitors (ACEi) can reduce oxidative stress due to their inhibitory effect on angiotensin II, which is a major stimulus for the activation or upregulation of Nox enzymes [5, 296]. ROS such as $O_2^{\cdot-}$ and H_2O_2 are widely recognized as important downstream mediators of Ang II physiological and pathological effects [303]. Nevertheless, some of these antihypertensive drugs also possess antioxidant effects independently of RAS inhibition. For example, captoptil, a thiol-containing ACEi, is a ROS scavenger and a metal chelator [304]. The ARBs candesartan and olmesartan also exhibit antioxidant effects independent of AT_1 receptor blockade or blood pressure control [305-307]. In addition, other agents belonging to the beta-blocker or calcium channel blocker drug classes have also been shown to exert antioxidant effects unrelated to their blood pressure lowering action. The beta-blockers carvedilol and nebivolol appear to possess ROS scavenging properties as well as inhibitory effects on ROS production, such as the inhibition of Nox activation [308, 309]. In addition, nebivolol also increases $\cdot$NO release from the endothelium, thus attenuating oxidative stress effects on endothelium-dependent vasodilation [309, 310]. The calcium channel blocker lacidipine has also been demonstrated to have a potent antioxidant activity and to reduce the intracellular production of ROS induced by oxidized LDL [311, 312]. Therefore, even though convincing evidence is lacking regarding a clinical therapeutic effect of antioxidants, there is extensive data showing that currently approved antihypertensive treatments have the ability to modify oxidative stress status.

5. Conclusions

Extensive experimental evidence has shown that unbalanced ROS and/or RNS production can disturb several physiological functions, leading to the genesis and progression of arterial hypertension. Many studies have observed marked alterations in direct and indirect oxidative stress biomarkers, such as lipid peroxidation products, prooxidant enzymes and antioxidant defenses. However, most clinical trials with antioxidants have failed to demonstrate a protective effect on blood pressure and cardiovascular function. This does not necessarily exclude a role for oxidative stress in human cardiovascular diseases but instead suggests that other approaches should be addopted to recover redox homeostasis. The inhibition of Nox enzymes appears to be a promising strategy as these enzymes are major sources of ROS overproduction at cardiovascular and renal sites of blood pressure control. Indeed, several drugs already in use for the treatment of hypertension (e.g. ARBs, ACEi, the β-blocker nebivolol) or dyslipidemia (statins) are known to reduce the activation of Nox enzymes. In addition, there is an urgent need to implement universally validated approaches to evaluate oxidative status in human patients. These should cover a broader range of redox biomarkers and would add valuable information for risk stratification and therapeutic monitoring in human patients.

Acknowledgments

"This work was supported by FEDER funds via COMPETE and by national funds through FCT – Portuguese Foundation for Science and Technology [project grant: PTDC/SAU-TOX/114166/2009]."

Author details

Teresa Sousa, Joana Afonso and António Albino-Teixeira
Department of Pharmacology and Therapeutics, Faculty of Medicine, University of Porto, Portugal

Félix Carvalho
REQUIMTE, Laboratory of Toxicology, Department of Biological Sciences, Faculty of Pharmacy, University of Porto, Portugal

6. References

[1] Droge, W. (2002) Free radicals in the physiological control of cell function. Physiol Rev 82: 47-95.

[2] Jones, D.P. (2008) Radical-free biology of oxidative stress. Am J Physiol Cell Physiol 295: C849-868.

[3] Turko, I.V., Murad, F. (2002) Protein nitration in cardiovascular diseases. Pharmacol Rev 54: 619-634.

[4] Valko, M., Leibfritz, D., Moncol, J., Cronin, M.T., Mazur, M., Telser, J. (2007) Free radicals and antioxidants in normal physiological functions and human disease. Int J Biochem Cell Biol 39: 44-84.

[5] Paravicini, T.M., Touyz, R.M. (2008) NADPH oxidases, reactive oxygen species, and hypertension: clinical implications and therapeutic possibilities. Diabetes Care 31 Suppl 2: S170-180.

[6] Pham-Huy, L., He, H., Pham-Huy, C. (2008) Free Radicals, Antioxidants in Disease and Health. Int J Biomed Sci 4: 89-96.

[7] Ardanaz, N., Pagano, P.J. (2006) Hydrogen peroxide as a paracrine vascular mediator: regulation and signaling leading to dysfunction. Exp Biol Med (Maywood) 231: 237-251.

[8] Datla, S.R., Griendling, K.K. (2010) Reactive oxygen species, NADPH oxidases, and hypertension. Hypertension 56: 325-330.

[9] Griendling, K.K., FitzGerald, G.A. (2003) Oxidative stress and cardiovascular injury: Part I: basic mechanisms and in vivo monitoring of ROS. Circulation 108: 1912-1916.

[10] Bedard, K., Krause, K.-H. (2007) The NOX family of ROS-generating NADPH oxidases: Physiology and pathophysiology. Physiol Rev 87: 245-313.

[11] Geiszt, M., Leto, T.L. (2004) The Nox family of NAD(P)H oxidases: Host defense and beyond. J Biol Chem 279: 51715-51718.

[12] Nistala, R., Whaley-Connell, A., Sowers, J.R. (2008) Redox control of renal function and hypertension. Antiox Redox Signal 10: 2047-2089.

[13] Brandes, R.P., Takac, I., Schroeder, K. (2011) No Superoxide-No Stress? Nox4, the Good NADPH Oxidase! Arterioscler Thromb Vasc Biol 31: 1255-1257.

[14] Brown, D.I., Griendling, K.K. (2009) Nox proteins in signal transduction. Free Rad Biol Med 47: 1239-1253.

[15] Berry, C.E., Hare, J.M. (2004) Xanthine oxicloreductase and cardiovascular disease: molecular mechanisms and pathophysiological implications. J Physiol (London) 555: 589-606.

[16] Pacher, P., Nivorozhkin, A., Szabo, C. (2006) Therapeutic effects of xanthine oxidase inhibitors: renaissance half a century after the discovery of allopurinol. Pharmacol Rev 58: 87-114.

[17] Houston, M., Estevez, A., Chumley, P., Aslan, M., Marklund, S., Parks, D.A., Freeman,B.A. (1999) Binding of xanthine oxidase to vascular endothelium - Kinetic characterization and oxidative impairment of nitric oxide-dependent signaling. J Biol

Chem 274: 4985-4994.

[18] Becker, B.F. (1993) Towards the physiological function of uric acid. Free Rad Biol Med 14: 615-631.

[19] Hink, H.U., Santanam, N., Dikalov, S., McCann, L., Nguyen, A.D., Parthasarathy, S., Harrison, D.G., Fukai, T. (2002) Peroxidase properties of extracellular superoxide dismutase - Role of uric acid in modulating in vivo activity. Arterioscler Thromb Vasc Biol 22: 1402-1408.

[20] Hooper, D.C., Spitsin, S., Kean, R.B., Champion, J.M., Dickson, G.M., Chaudhry, I., Koprowski, H. (1998) Uric acid, a natural scavenger of peroxynitrite, in experimental allergic encephalomyelitis and multiple sclerosis. Proc Natl Acad Sci USA 95: 675-680.

[21] Kuzkaya, N., Weissmann, N., Harrison, D.G., Dikalov, S. (2005) Interactions of peroxynitrite with uric acid in the presence of ascorbate and thiols: implications for uncoupling endothelial nitric oxide synthase. Biochem Pharmacol 70: 343-354.

[22] Sevanian, A., Davies, K.J., Hochstein, P. (1985) Conservation of vitamin C by uric acid in blood. J Free Rad Biol Med 1: 117-124.

[23] Johnson, R.J., Kang, D.H., Feig, D., Kivlighn, S., Kanellis, J., Watanabe, S., Tuttle, K.R., Rodriguez-Iturbe, B., Herrera-Acosta, J., Mazzali, M. (2003) Is there a pathogenetic role for uric acid in hypertension and cardiovascular and renal disease? Hypertension 41: 1183-1190.

[24] Waring, W.S., Webb, D.J., Maxwell, S.R.J. (2000) Uric acid as a risk factor for cardiovascular disease. QJM 93: 707-713.

[25] Kirschbaum, B. (2001) Renal regulation of plasma total antioxidant capacity. Med Hypotheses 56: 625-629.

[26] Storey, K.B. (1996) Oxidative stress: Animal adaptations in nature. Braz J Med Biol Res 29: 1715-1733.

[27] Addabbo, F., Montagnani, M., Goligorsky, M.S. (2009) Mitochondria and Reactive Oxygen Species. Hypertension 53: 885-892.

[28] Sugamura, K., Keaney, J.F., Jr. (2011) Reactive oxygen species in cardiovascular disease. Free Radic Biol Med 51: 978-992.

[29] Munzel, T., Gori, T., Bruno, R.M., Taddei, S. (2010) Is oxidative stress a therapeutic target in cardiovascular disease? Eur Heart J 31: 2741-2748.

[30] Chan, S.H., Wu, K.L., Chang, A.Y., Tai, M.H., Chan, J.Y. (2009) Oxidative impairment of mitochondrial electron transport chain complexes in rostral ventrolateral medulla contributes to neurogenic hypertension. Hypertension 53: 217-227.

[31] Dikalova, A.E., Bikineyeva, A.T., Budzyn, K., Nazarewicz, R.R., McCann, L., Lewis, W., Harrison, D.G., Dikalov, S.I. (2010) Therapeutic Targeting of Mitochondrial Superoxide in Hypertension. Circ Res 107: 106-U221.

[32] Elks, C.M., Mariappan, N., Haque, M., Guggilam, A., Majid, D.S., Francis, J. (2009) Chronic NF-{kappa}B blockade reduces cytosolic and mitochondrial oxidative stress and attenuates renal injury and hypertension in SHR. Am J Physiol Renal Physiol 296: F298-305.

[33] Briones, A.M., Touyz, R.M. (2010) Oxidative stress and hypertension: current concepts.

Curr Hypertens Rep 12: 135-142.

[34] Li, H., Witte, K., August, M., Brausch, I., Godtel-Armbrust, U., Habermeier, A., Closs, E.I., Oelze, M., Munzel, T., Forstermann, U. (2006) Reversal of endothelial nitric oxide synthase uncoupling and up-regulation of endothelial nitric oxide synthase expression lowers blood pressure in hypertensive rats. J Am Coll Cardiol 47: 2536-2544.

[35] Vera, R., Sanchez, M., Galisteo, M., Concepcion Villar, I., Jimenez, R., Zarzuelo, A., Perez-Vizcaino, F., Duarte, J. (2007) Chronic administration of genistein improves endothelial dysfunction in spontaneously hypertensive rats: involvement of eNOS, caveolin and calmodulin expression and NADPH oxidase activity. Clin Sci 112: 183-191.

[36] Wassmann, S., Wassmann, K., Nickenig, G. (2004) Modulation of oxidant and antioxidant enzyme expression and function in vascular cells. Hypertension 44: 381-386.

[37] Malle, E., Buch, T., Grone, H.J. (2003) Myeloperoxidase in kidney disease. Kidney Int 64: 1956-1967.

[38] Green, P.S., Mendez, A.J., Jacob, J.S., Crowley, J.R., Growdon, W., Hyman, B.T., Heinecke, J.W. (2004) Neuronal expression of myeloperoxidase is increased in Alzheimer's disease. J Neurochem 90: 724-733.

[39] La Rocca, G., Di Stefano, A., Eleuteri, E., Anzalone, R., Magno, F., Corrao, S., Loria, T., Martorana, A., Di Gangi, C., Colombo, M., Sansone, F., Patane, F., Farina, F., Rinaldi, M., Cappello, F., Giannuzzi, P., Zummo, G. (2009) Oxidative stress induces myeloperoxidase expression in endocardial endothelial cells from patients with chronic heart failure. Basic Res Cardiol 104: 307-320.

[40] Xu, J., Xie, Z., Reece, R., Pimental, D., Zou, M.H. (2006) Uncoupling of endothelial nitric oxidase synthase by hypochlorous acid: role of NAD(P)H oxidase-derived superoxide and peroxynitrite. Arterioscler Thromb Vasc Biol 26: 2688-2695.

[41] Cai, H., Harrison, D.G. (2000) Endothelial dysfunction in cardiovascular diseases - The role of oxidant stress. Circ Res 87: 840-844.

[42] Didion, S.P., Ryan, M.J., Baumbach, G.L., Sigmund, C.D., Faraci, F.M. (2002) Superoxide contributes to vascular dysfunction in mice that express human renin and angiotensinogen. Am J Physiol Heart Circ Physiol 283: H1569-H1576.

[43] Lassegue, B., Griendling, K.K. (2004) Reactive oxygen species in hypertension; An update. Am J Hypertens 17: 852-860.

[44] Rathaus, M., Bernheim, J. (2002) Oxygen species in the microvascular environment: regulation of vascular tone and the development of hypertension. Nephrol Dial Transplant 17: 216-221.

[45] Touyz, R.M. (2003) Reactive oxygen species in vascular biology: role in arterial hypertension. Exp Rev Cardiov Ther 1: 91-106.

[46] Cowley, A.W., Jr. (2008) Renal Medullary Oxidative Stress, Pressure-Natriuresis, and Hypertension. Hypertension 52: 777-786.

[47] Tai, M.H., Wang, L.L., Wu, K.L.H., Chan, J.Y. (2005) Increased superoxide anion in rostral ventrolateral medulla contributes to hypertension in spontaneously hypertensive rats via interactions with nitric oxide. Free Rad Biol Med 38: 450-462.

[48] Makino, A., Skelton, M.M., Zou, A.P., Cowley, A.W., Jr. (2003) Increased renal medullary

H2O2 leads to hypertension. Hypertension 42: 25-30.

[49] Sousa, T., Oliveira, S., Afonso, J., Morato, M., Patinha, D., Fraga, S., Carvalho, F., Albino-Teixeira, A. (2012) Role of H(2) o(2) in Hypertension, Renin-Angiotensin System Activation and Renal Medullary Disfunction Caused by Angiotensin II. Br J Pharmacol: in press.

[50] Sousa, T., Pinho, D., Morato, M., Marques-Lopes, J., Fernandes, E., Afonso, J., Oliveira, S., Carvalho, F., Albino-Teixeira, A. (2008) Role of superoxide and hydrogen peroxide in hypertension induced by an antagonist of adenosine receptors. Eur J Pharmacol 588: 267-276.

[51] Cai, H. (2005) NAD(P)H oxidase-dependent self-propagation of hydrogen peroxide and vascular disease. Circ Res 96: 818-822.

[52] Asghar, M., Banday, A.A., Fardoun, R.Z., Lokhandwala, M.F. (2006) Hydrogen peroxide causes uncoupling of dopamine D1-like receptors from G proteins via a mechanism involving protein kinase C and G-protein-coupled receptor kinase 2. Free Rad Biol Med 40: 13-20.

[53] Chen, Y.F., Cowley, A.W., Jr., Zou, A.P. (2003) Increased H(2)O(2) counteracts the vasodilator and natriuretic effects of superoxide dismutation by tempol in renal medulla. Am J Physiol Regul Integr Comp Physiol 285: R827-833.

[54] Gao, Y.J., Hirota, S., Zhang, D.W., Janssen, L.J., Lee, R. (2003) Mechanisms of hydrogen-peroxide-induced biphasic response in rat mesenteric artery. Br J Pharmacol 138: 1085-1092.

[55] Gao, Y.J., Lee, R. (2001) Hydrogen peroxide induces a greater contraction in mesenteric arteries of spontaneously hypertensive rats through thromboxane A(2) production. Br J Pharmacol 134: 1639-1646.

[56] Lin, H.H., Chen, C.H., Hsieh, W.K., Chiu, T.H., Lai, C.C. (2003) Hydrogen peroxide increases the activity of rat sympathetic preganglionic neurons in vivo and in vitro. Neuroscience 121: 641-647.

[57] Rodriguez-Martinez, M.A., Garcia-Cohen, E.C., Baena, A.B., Gonzalez, R., Salaices, M., Marin, J. (1998) Contractile responses elicited by hydrogen peroxide in aorta from normotensive and hypertensive rats. Endothelial modulation and mechanism involved. Br J Pharmacol 125: 1329-1335.

[58] Thakali, K., Davenport, L., Fink, G.D., Watts, S.W. (2006) Pleiotropic effects of hydrogen peroxide in arteries and veins from normotensive and hypertensive rats. Hypertension 47: 482-487.

[59] Didion, S., Chrissobolis, S., Faraci, F.M. (2008) Oxidative Stress in Hypertension. In: Miwa, Beckman, Muller, editors. Aging Medicine: Oxidative Stress in Aging 3: 229-251.

[60] Kristal, B., Shurtz-Swirski, R., Chezar, J., Manaster, J., Levy, R., Shapiro, G., Weissman, I., Shasha, S.M., Sela, S. (1998) Participation of peripheral polymorphonuclear leukocytes in the oxidative stress and inflammation in patients with essential hypertension. Am J Hypertens 11: 921-928.

[61] Lacy, F., Kailasam, M.T., O'Connor, D.T., Schmid-Schonbein, G.W., Parmer, R.J. (2000) Plasma hydrogen peroxide production in human essential hypertension - Role of heredity, gender, and ethnicity. Hypertension 36: 878-884.

[62] Lacy, F., O'Connor, D.T., Schmid-Schonbein, G.W. (1998) Plasma hydrogen peroxide

production in hypertensives and normotensive subjects at genetic risk of hypertension. J Hypertens 16: 291-303.

[63] Dhawan, V., Jain, S. (2004) Effect of garlic supplementation on oxidized low density lipoproteins and lipid peroxidation in patients of essential hypertension. Mol Cell Biochem 266: 109-115.

[64] Kedziora-Kornatowska, K., Czuczejko, J., Pawluk, H., Kornatowski, T., Motyl, J., Szadujkis-Szadurski, L., Szewczyk-Golec, K., Kedziora, J. (2004) The markers of oxidative stress and activity of the antioxidant system in the blood of elderly patients with essential arterial hypertension. Cell Mol Biol Lett 9: 635-641.

[65] Redon, J., Oliva, M.R., Tormos, C., Giner, V., Chaves, J., Iradi, A., Saez, G.T. (2003) Antioxidant activities and oxidative stress byproducts in human hypertension. Hypertension 41: 1096-1101.

[66] Zhou, L., Xiang, W., Potts, J., Floyd, M., Sharan, C., Yang, H., Ross, J., Nyanda, A.M., Guo, Z. (2006) Reduction in extracellular superoxide dismutase activity in African- American patients with hypertension. Free Rad Biol Med 41: 1384-1391.

[67] Wen, Y., Killalea, S., McGettigan, P., Feely, J. (1996) Lipid peroxidation and antioxidant vitamins C and E in hypertensive patients. Ir J Med Sci 165: 210-212.

[68] Labios, M., Martinez, M., Gabriel, F., Guiral, V., Ruiz-Aja, S., Beltran, B., Munoz, A. (2008) Effects of eprosartan on mitochondrial membrane potential and H(2)O(2) levels in leucocytes in hypertension. J Hum Hypertens 22: 493-500.

[69] Lyamina, N.P., Dolotovskaya, P.V., Lyamina, S.V., Malyshev, I.Y., Manukhina, E.B. (2003) Nitric oxide production and intensity of free radical processes in young men with high normal and hypertensive blood pressure. Med Sci Monit 9: CR304-310.

[70] Sierra, M., Gonzalez, A., Gomez-Alamillo, C., Monreal, I., Huarte, E., Gil, A., Sanchez-Casajus, A., Diez, J. (1998) Decreased excretion of nitrate and nitrite in essential hypertensives with renal vasoconstriction. Kidney Int 54: S10-S13.

[71] San Jose, G., Moreno, M.U., Olivan, S., Beloqui, O., Fortuno, A., Diez, J., Zalba, G. (2004) Functional effect of the p22(phox) -930(A/G) polymorphism on p22(phox) expression and NADPH oxidase activity in hypertension. Hypertension 44: 163-169.

[72] Russo, C., Olivieri, O., Girelli, D., Faccini, G., Zenari, M.L., Lombardi, S., Corrocher, R. (1998) Anti-oxidant status and lipid peroxidation in patients with essential hypertension. J Hypertens 16: 1267-1271.

[73] Kedziora-Kornatowska, K., Kornatowski, T., Bartosz, G., Pawluk, H., Czuczejko, J., Kedziora, J., Szadujkis-Szadurski, L. (2006) Production of nitric oxide, lipid peroxidation and oxidase activity of ceruloplasmin in blood of elderly patients with primary hypertension. Effects of perindopril treatment. Aging Clin Exp Res 18: 1-6.

[74] Rodrigo, R., Prat, H., Passalacqua, W., Araya, J., Guichard, C., Bachler, J.P. (2007) Relationship between oxidative stress and essential hypertension. Hypertension Res 30: 1159-1167.

[75] Minuz, P., Patrignani, P., Gaino, S., Degan, M., Menapace, L., Tommasoli, R., Seta, F., Capone, M.L., Tacconelli, S., Palatresi, S., Bencini, C., Del Vecchio, C., Mansueto, G., Arosio, E., Santonastaso, C.L., Lechi, A., Morganti, A., Patrono, C. (2002) Increased oxidative stress and platelet activation in patients with hypertension and renovascular disease. Circulation 106: 2800-2805.

[76] Espinosa, O., Jimenez-Almazan, J., Chaves, F.J., Tormos, M.C., Clapes, S., Iradi, A., Salvador, A., Fandos, M., Redon, J., Saez, G.T. (2007) Urinary 8-oxo-7,8-dihydro-2 ′- deoxyguanosine (8-oxo-dG), a reliable oxidative stress marker in hypertension. Free Rad Res 41: 546-554.

[77] Rosello-Lleti, E., Garcia de Burgos, F., Morillas, P., Cortes, R., Martinez-Dolz, L., Almenar, L., Grigorian, L., Orosa, P., Portoles, M., Bertomeu, V., Rivera, M. (2011) Impact of Cardiovascular Risk Factors and Inflammatory Status on Urinary 8-OHdG in Essential Hypertension. Am J Hypertens 25: 236-242.

[78] Maggi, E., Marchesi, E., Ravetta, V., Falaschi, F., Finardi, G., Bellomo, G. (1993) Low- density lipoprotein oxidation in essential hypertension. J Hypertens 11: 1103-1111.

[79] Santangelo, L., Cigliano, L., Montefusco, A., Spagnuolo, M.S., Nigro, G., Golino, P., Abrescia, P. (2003) Evaluation of the antioxidant response in the plasma of healthy or hypertensive subjects after short-term exercise. J Hum Hypertens 17: 791-798.

[80] Alderman, M.H., Cohen, H., Madhavan, S., Kivlighn, S. (1999) Serum uric acid and cardiovascular events in successfully treated hypertensive patients. Hypertension 34: 144-150.

[81] Tse, W.Y., Maxwell, S.R.J., Thomason, H., Blann, A., Thorpe, G.H.G., Waite, M., Holder, R. (1994) Antioxidant status in controlled and uncontrolled hypertension and its relationship to endothelial damage. J Hum Hypertens 8: 843-849.

[82] Subash, P., Premagurumurthy, K., Sarasabharathi, A., Cherian, K.M. (2010) Total antioxidant status and oxidative DNA damage in a South Indian population of essential hypertensives. J Hum Hypertens 24: 475-482.

[83] Kashyap, M.K., Yadav, V., Sherawat, B.S., Jain, S., Kumari, S., Khullar, M., Sharma, P.C., Nath, R. (2005) Different antioxidants status, total antioxidant power and free radicals in essential hypertension. Mol Cell Biochem 277: 89-99.

[84] Vaziri, N.D., Wang, X.Q., Oveisi, F., Rad, B. (2000) Induction of oxidative stress by glutathione depletion causes severe hypertension in normal rats. Hypertension 36: 142- 146.

[85] Baumer, A.T., Kruger, C.A., Falkenberg, J., Freyhaus, H.T., Rosen, R., Fink, K., Rosenkranz, S. (2007) The NAD(P)H oxidase inhibitor apocynin improves endothelial NO/superoxide balance and lowers effectively blood pressure in spontaneously hypertensive rats: comparison to calcium channel blockade. Clin Exp Hypertens 29: 287-299.

[86] Beswick, R.A., Dorrance, A.M., Leite, R., Webb, R.C. (2001) NADH/NADPH oxidase and enhanced superoxide production in the mineralocorticoid hypertensive rat. Hypertension 38: 1107-1111.

[87] Zhang, Y., Croft, K.D., Mori, T.A., Schyvens, C.G., McKenzie, K.U., Whitworth, J.A. (2004) The antioxidant tempol prevents and partially reverses dexamethasone-induced hypertension in the rat. Am J Hypertens 17: 260-265.

[88] Dikalova, A., Clempus, R., Lassegue, B., Cheng, G.J., McCoy, J., Dikalov, S., Martin, A.S., Lyle, A., Weber, D.S., Weiss, D., Taylor, R., Schmidt, H., Owens, G.K., Lambeth, J.D., Griendling, K.K. (2005) Nox1 overexpression potentiates angiotensin II-induced hypertension and vascular smooth muscle hypertrophy in transgenic mice. Circulation 112: 2668-2676.

[89] Godin, N., Liu, F., Lau, G.J., Brezniceanu, M.-L., Chenier, I., Filep, J.G., Ingelfinger, J.R., Zhang, S.-L., Chan, J.S.D. (2010) Catalase overexpression prevents hypertension and tubular apoptosis in angiotensinogen transgenic mice. Kidney Int 77: 1086-1097.

[90] Wilcox, C.S. (2005) Oxidative stress and nitric oxide deficiency in the kidney: a critical link to hypertension? Am J Physiol-Regul Integr Comp Physiol 289: R913-R935.

[91] Paravicini, T.M., Touyz, R.M. (2006) Redox signaling in hypertension. Cardiov Res 71: 247-258.

[92] Grossman, E. (2008) Does increased oxidative stress cause hypertension? Diabetes Care 31 Suppl 2: S185-189.

[93] Baykal, Y., Yilmaz, M.I., Celik, T., Gok, F., Rehber, H., Akay, C., Kocar, I.H. (2003) Effects of antihypertensive agents, alpha receptor blockers, beta blockers, angiotensin- converting enzyme inhibitors, angiotensin receptor blockers and calcium channel blockers, on oxidative stress. J Hypertens 21: 1207-1211.

[94] de Cavanagh, E.M., Ferder, L.F., Ferder, M.D., Stella, I.Y., Toblli, J.E., Inserra, F. (2010) Vascular structure and oxidative stress in salt-loaded spontaneously hypertensive rats: effects of losartan and atenolol. Am J Hypertens 23: 1318-1325.

[95] Sugiura, T., Kondo, T., Kureishi-Bando, Y., Numaguchi, Y., Yoshida, O., Dohi, Y., Kimura, G., Ueda, R., Rabelink, T.J., Murohara, T. (2008) Nifedipine improves endothelial function role of endothelial progenitor cells. Hypertension 52: 491-498.

[96] Wingler, K., Hermans, J.J., Schiffers, P., Moens, A., Paul, M., Schmidt, H.H. (2011) NOX1, 2, 4, 5: counting out oxidative stress. Br J Pharmacol 164: 866-883.

[97] Dalle-Donne, I., Rossi, R., Colombo, R., Giustarini, D., Milzani, A. (2006) Biomarkers of oxidative damage in human disease. Clin Chem 52: 601-623.

[98] de Zwart, L.L., Meerman, J.H., Commandeur, J.N., Vermeulen, N.P. (1999) Biomarkers of free radical damage applications in experimental animals and in humans. Free Radic Biol Med 26: 202-226.

[99] Atkinson, A. (2001) Biomarkers and surrogate endpoints: preferred definitions and conceptual framework. Clin Pharmacol Ther 69: 89-95.

[100] Offord, E., van Poppel, G., Tyrrell, R. (2000) Markers of oxidative damage and antioxidant protection: current status and relevance to disease. Free Radic Res 33 Suppl: S5-19.

[101] Hancock, J.T., Desikan, R., Neill, S.J. (2001) Role of reactive oxygen species in cell signalling pathways. Biochem Soc Trans 29: 345-350.

[102] Sohal, R.S., Dubey, A. (1994) Mitochondrial oxidative damage, hydrogen peroxide release, and aging. Free Radic Biol Med 16: 621-626.

[103] Taniyama, Y., Griendling, K.K. (2003) Reactive oxygen species in the vasculature: molecular and cellular mechanisms. Hypertension 42: 1075-1081.

[104] Halliwell, B., Gutteridge, J.M. (1989) Free Rad Biol Med, 2nd edition, Clarendon Press, Oxford, UK.

[105] Sies, H. (1993) Strategies of antioxidant defense. Eur J Biochem 215: 213-219.

[106] Frei, B., Stocker, R., Ames, B.N. (1988) Antioxidant defenses and lipid peroxidation in human blood plasma. Proc Natl Acad Sci U S A 85: 9748-9752.

[107] Shull, S., Heintz, N.H., Periasamy, M., Manohar, M., Janssen, Y.M., Marsh, J.P., Mossman, B.T. (1991) Differential regulation of antioxidant enzymes in response to oxidants. J Biol Chem 266: 24398-24403.

[108] Young, I.S., Woodside, J.V. (2001) Antioxidants in health and disease. J Clin Pathol 54: 176-186.

[109] Hayes, J.D., Pulford, D.J. (1995) The glutathione S-transferase supergene family: regulation of GST and the contribution of the isoenzymes to cancer chemoprotection and drug resistance. Crit Rev Biochem Mol Biol 30: 445-600.

[110] Kalinina, E.V., Chernov, N.N., Saprin, A.N. (2008) Involvement of thio-, peroxi-, and glutaredoxins in cellular redox-dependent processes. Biochemistry (Mosc) 73: 1493- 1510.

[111] Farmand, F., Ehdaie, A., Roberts, C.K., Sindhu, R.K. (2005) Lead-induced dysregulation of superoxide dismutases, catalase, glutathione peroxidase, and guanylate cyclase. Environ Res 98: 33-39.

[112] Sainz, J., Wangensteen, R., Rodriguez Gomez, I., Moreno, J.M., Chamorro, V., Osuna, A., Bueno, P., Vargas, F. (2005) Antioxidant enzymes and effects of tempol on the development of hypertension induced by nitric oxide inhibition. Am J Hypertens 18: 871-877.

[113] Gille, G., Sigler, K. (1995) Oxidative stress and living cells. Folia Microbiol (Praha) 40: 131-152.

[114] Chan, P., Liao, S.S., Hsu, C.T., Lee, Y.S., Tomlinson, B., Kuo, J.S., Cheng, J.T. (1999) Superoxide dismutase gene expression and activity in the brain of spontaneously hypertensive rats and normotensive rats. Chin Med J (Engl) 112: 1119-1124.

[115] Nozoe, M., Hirooka, Y., Koga, Y., Sagara, Y., Kishi, T., Engelhardt, J.F., Sunagawa, K. (2007) Inhibition of Rac1-derived reactive oxygen species in nucleus tractus solitarius decreases blood pressure and heart rate in stroke-prone spontaneously hypertensive rats. Hypertension 50: 62-68.

[116] Chan, S.H., Tai, M.H., Li, C.Y., Chan, J.Y. (2006) Reduction in molecular synthesis or enzyme activity of superoxide dismutases and catalase contributes to oxidative stress and neurogenic hypertension in spontaneously hypertensive rats. Free Radic Biol Med 40: 2028-2039.

[117] Adler, S., Huang, H., Loke, K.E., Xu, X., Tada, H., Laumas, A., Hintze, T.H. (2001) Endothelial nitric oxide synthase plays an essential role in regulation of renal oxygen consumption by NO. Am J Physiol Renal Physiol 280: F838-843.

[118] Simao, S., Gomes, P., Pinto, V., Silva, E., Amaral, J.S., Igreja, B., Afonso, J., Serrao, M.P., Pinho, M.J., Soares-da-Silva, P. (2011) Age-related changes in renal expression of oxidant and antioxidant enzymes and oxidative stress markers in male SHR and WKY rats. Exp Gerontol 46: 468-474.

[119] Potenza, M.A., Addabbo, F., Montagnani, M. (2009) Vascular actions of insulin with implications for endothelial dysfunction. Am J Physiol Endocrinol Metab 297: E568-577. Zheng, H., Yu, Y.S. (2012) Chronic hydrogen-rich saline treatment attenuates vascular

[120] dysfunction in spontaneous hypertensive rats. Biochem Pharmacol 83: 1269-1277.

[121] Ulker, S., McMaster, D., McKeown, P.P., Bayraktutan, U. (2003) Impaired activities of antioxidant enzymes elicit endothelial dysfunction in spontaneous hypertensive rats despite enhanced vascular nitric oxide generation. Cardiovasc Res 59: 488-500.

[122] Polizio, A.H., Pena, C. (2005) Effects of angiotensin II type 1 receptor blockade on the oxidative stress in spontaneously hypertensive rat tissues. Regul Pept 128: 1-5.

[123] Xu, Y., Gao, Q., Gan, X.B., Chen, L., Zhang, L., Zhu, G.Q., Gao, X.Y. (2011) Endogenous hydrogen peroxide in paraventricular nucleus mediates sympathetic activation and enhanced cardiac sympathetic afferent reflex in renovascular hypertensive rats. Exp Physiol 96: 1282-1292.

[124] Fortepiani, L.A., Reckelhoff, J.F. (2005) Increasing oxidative stress with molsidomine increases blood pressure in genetically hypertensive rats but not normotensive controls. Am J Physiol Regul Integr Comp Physiol 289: R763-770.

[125] Kumar, U., Chen, J., Sapoznikhov, V., Canteros, G., White, B.H., Sidhu, A. (2005) Overexpression of inducible nitric oxide synthase in the kidney of the spontaneously hypertensive rat. Clin Exp Hypertens 27: 17-31.

[126] Zhang, L., Fujii, S., Kosaka, H. (2007) Effect of oestrogen on reactive oxygen species production in the aortas of ovariectomized Dahl salt-sensitive rats. J Hypertens 25: 407- 414.

[127] Kosower, N.S., Kosower, E.M. (1978) The glutathione status of cells. Int Rev Cytol 54: 109-160.

[128] Franco, R., DeHaven, W.I., Sifre, M.I., Bortner, C.D., Cidlowski, J.A. (2008) Glutathione depletion and disruption of intracellular ionic homeostasis regulate lymphoid cell apoptosis. J Biol Chem 283: 36071-36087.

[129] Schulz, J.B., Lindenau, J., Seyfried, J., Dichgans, J. (2000) Glutathione, oxidative stress and neurodegeneration. Eur J Biochem 267: 4904-4911.

[130] Singh, U., Devaraj, S., Jialal, I. (2005) Vitamin E, oxidative stress, and inflammation. Annu Rev Nutr 25: 151-174.

[131] Smith, A.R., Visioli, F., Hagen, T.M. (2002) Vitamin C matters: increased oxidative stress in cultured human aortic endothelial cells without supplemental ascorbic acid. FASEB J 16: 1102-1104.

[132] Waring, W.S. (2002) Uric acid: an important antioxidant in acute ischaemic stroke. QJM 95: 691-693.

[133] Stocker, R., Yamamoto, Y., McDonagh, A.F., Glazer, A.N., Ames, B.N. (1987) Bilirubin is an antioxidant of possible physiological importance. Science 235: 1043-1046.

[134] Lin, J.P., Vitek, L., Schwertner, H.A. (2010) Serum bilirubin and genes controlling bilirubin concentrations as biomarkers for cardiovascular disease. Clin Chem 56: 1535- 1543.

[135] Hu, M.L., Louie, S., Cross, C.E., Motchnik, P., Halliwell, B. (1993) Antioxidant protection against hypochlorous acid in human plasma. J Lab Clin Med 121: 257-262.

[136] Benzie, I.F., Strain, J.J. (1996) The ferric reducing ability of plasma (FRAP) as a measure of "antioxidant power": the FRAP assay. Anal Biochem 239: 70-76.

[137] Wayner, D.D., Burton, G.W., Ingold, K.U., Locke, S. (1985) Quantitative measurement of the total, peroxyl radical-trapping antioxidant capability of human blood plasma by controlled peroxidation. The important contribution made by plasma proteins. FEBS Lett 187: 33-37.

[138] Yanes, L., Romero, D., Iliescu, R., Cucchiarelli, V.E., Fortepiani, L.A., Santacruz, F., Bell, W., Zhang, H., Reckelhoff, J.F. (2005) Systemic arterial pressure response to two weeks of Tempol therapy in SHR: involvement of NO, the RAS, and oxidative stress. Am J Physiol Regul Integr Comp Physiol 288: R903-908.

[139] Ghiselli, A., Serafini, M., Natella, F., Scaccini, C. (2000) Total antioxidant capacity as a tool to assess redox status: critical view and experimental data. Free Radic Biol Med 29: 1106-1114.

[140] Kanbay, M., Solak, Y., Dogan, E., Lanaspa, M.A., Covic, A. (2010) Uric acid in hypertension and renal disease: the chicken or the egg? Blood Purif 30: 288-295.

[141] Young, I.S. (2001) Measurement of total antioxidant capacity. J Clin Pathol 54: 339.

[142] Duffy, S.J., Gokce, N., Holbrook, M., Huang, A., Frei, B., Keaney, J.F., Jr., Vita, J.A. (1999) Treatment of hypertension with ascorbic acid. Lancet 354: 2048-2049.

[143] Newaz, M.A., Nawal, N.N., Rohaizan, C.H., Muslim, N., Gapor, A. (1999) alpha- Tocopherol increased nitric oxide synthase activity in blood vessels of spontaneously hypertensive rats. Am J Hypertens 12: 839-844.

[144] Ward, N.C., Hodgson, J.M., Puddey, I.B., Mori, T.A., Beilin, L.J., Croft, K.D. (2004) Oxidative stress in human hypertension: association with antihypertensive treatment, gender, nutrition, and lifestyle. Free Radic Biol Med 36: 226-232.

[145] Gutteridge, J.M., Halliwell, B. (1990) The measurement and mechanism of lipid peroxidation in biological systems. Trends Biochem Sci 15: 129-135.

[146] Halliwell, B., Whiteman, M. (2004) Measuring reactive species and oxidative damage in vivo and in cell culture: how should you do it and what do the results mean? Br J Pharmacol, 142: 231-55.

[147] Vincent, H.K., Taylor, A.G. (2006) Biomarkers and potential mechanisms of obesity- induced oxidant stress in humans. Int J Obes (Lond) 30: 400-418.

[148] Janicka, M., Kot-Wasik, A., Kot, J., Namiesnik, J. (2010) Isoprostanes-biomarkers of lipid peroxidation: their utility in evaluating oxidative stress and analysis. Int J Mol Sci 11: 4631-4659.

[149] Soffler, C., Campbell, V.L., Hassel, D.M.(2010) Measurement of urinary F2- isoprostanes as markers of in vivo lipid peroxidation: a comparison of enzyme immunoassays with gas chromatography-mass spectrometry in domestic animal species. J Vet Diagn Invest. 22: 200-9.

[150] Greco, A., Minghetti, L., Levi, G. (2000) Isoprostanes, novel markers of oxidative injury, help understanding the pathogenesis of neurodegenerative diseases. Neurochem Res 25: 1357-1364.

[151] Kadiiska, M.B., Gladen, B.C., Baird, D.D., Germolec, D., Graham, L.B., Parker, C.E.,Nyska, A., Wachsman, J.T., Ames, B.N., Basu, S., Brot, N., Fitzgerald, G.A., Floyd, R.A.,George, M., Heinecke, J.W., Hatch, G.E., Hensley, K., Lawson, J.A., Marnett, L.J., Morrow, J.D., Murray, D.M., Plastaras, J., Roberts, L.J. 2nd, Rokach, J., Shigenaga, M.K., Sohal, R.S., Sun, J., Tice, R.R., Van Thiel, D.H., Wellner, D., Walter, P.B., Tomer, K.B., Mason, R.P., Barrett, J.C.(2005) Biomarkers of oxidative stress study II: are oxidation products of lipids, proteins, and DNA markers of CCl4 poisoning? Free Radic Biol Med 38:698–710.

[152] Milne GL, Yin H, Brooks JD, et al.: 2007, Quantification of F2-isoprostanes in biological fluids and tissues as a measure of oxidant stress. Methods Enzymol 433:113–126.

[153] Todorova, I., Simeonova, G. Kyuchukova, D., Dinev, D., Gadjeva, V. (2005) Reference values of oxidative stress parameters (MDA, SOD, CAT) in dogs and cats. Comp Clin Path 13: 190–194.

[154] Huszar, G, Vigue, L. (1994) Correlation between the rate of lipid peroxidation and cellular maturity as measured by creatine kinase activity in human spermatozoa. J Androl 15: 71-7.

[155] Janero, D.R. (1990) Malondialdehyde and thiobarbituric acid-reactivity as diagnostic indices of lipid peroxidation and peroxidative tissue injury. Free Radic Biol Med 9: 515- 540.

[156] Yeo, H.C., Helbock, H.J., Chyu, D.W., Ames, B.N. (1994) Assay of malondialdehyde in biological fluids by gas chromatography-mass spectrometry. Anal Biochem 220: 391- 396.

[157] Khaira, A., Mahajan, S., Kumar, A., Saraya, A., Tiwari, S.C., Prakash, S., Gupta, A., Bhowmik, D., Agarwal, S.K. (2011) Endothelial function and oxidative stress in chronic kidney disease of varying severity and the effect of acute hemodialysis. Ren Fail 33: 411- 417.

[158] Rumley, A.G., Woodward, M., Rumley, A., Rumley, J., Lowe, G.D. (2004) Plasma lipid peroxides: relationships to cardiovascular risk factors and prevalent cardiovascular disease. QJM 97: 809-816.

[159] Satoh, K. (1978) Serum lipid peroxide in cerebrovascular disorders determined by a new colorimetric method. Clin Chim Acta 90: 37-43.

[160] Yoneyama, Y., Sawa, R., Suzuki, S., Doi, D., Yoneyama, K., Otsubo, Y., Araki, T. (2002) Relationship between plasma malondialdehyde levels and adenosine deaminase activities in preeclampsia. Clin Chim Acta 322: 169-173.

[161] Tateishi, T., Yoshimine, N, Kuzuya, F. (1987) Serum lipid peroxide assayed by a new colorimetric method. Exp Gerontol 22 (2): 103-111.

[162] Uchida, K. (2003) 4-Hydroxy-2-nonenal: a product and mediator of oxidative stress.Prog Lipid Res 42: 318-343.

[163] Suzuki, H., Nakazato, K., Asayama, K., Masawa, N., Takamata, M., Sakata, N. (2002) The role of oxidative stress on pathogenesis of hypertensive arterial lesions in rat mesenteric arteries. Acta Histochem Cytochem 35: 287-293.

[164] Blasig, I.E., Grune, T., Schonheit, K., Rohde, E., Jakstadt, M., Haseloff, R.F., Siems,W.G. (1995) 4-Hydroxynonenal, a novel indicator of lipid peroxidation for reperfusion injury of the myocardium. Am J Physiol 269: H14-22.

[165] Esterbauer, H., Schaur, R., Zollner, H. (1991) Chemistry and biochemistry of 4- hydroxynonenal, malonaldehyde and related aldehydes. Free Rad Biol Med 11: 81-128.

[166] Michel, F., Bonnefont-Rousselot, D., Mas, E., Drai, J., Therond, P. (2008) [Biomarkers of lipid peroxidation: analytical aspects]. Ann Biol Clin (Paris) 66: 605-620.

[167] Abuja, P.M., Albertini, R. (2001) Methods for monitoring oxidative stress, lipid peroxidation and oxidation resistance of lipoproteins. Clin Chim Acta 306: 1-17.

[168] Kato, Y., Yoshida, A., Naito, M., Kawai, Y., Tsuji, K., Kitamura, M., Kitamoto, N., Osawa, T. (2004) Identification and quantification of N(epsilon)-(Hexanoyl)lysine in human urine by liquid chromatography/tandem mass spectrometry. Free Radic Biol Med 37: 1864-1874.

[169] Kato, Y., Miyake, Y., Yamamoto, K., Shimomura, Y., Ochi, H., Mori, Y., Osawa, T. (2000) Preparation of a monoclonal antibody to N(epsilon)-(Hexanonyl)lysine: application to the evaluation of protective effects of flavonoid supplementation against exercise-induced oxidative stress in rat skeletal muscle. Biochem Biophys Res Commun 274: 389-393.

[170] Duarte, J., Perez-Palencia, R., Vargas, F., Ocete, M.A., Perez-Vizcaino, F., Zarzuelo, A., Tamargo, J. (2001) Antihypertensive effects of the flavonoid quercetin in spontaneously hypertensive rats. Br J Pharmacol 133: 117-124.

[171] Elmarakby, A.A., Imig, J.D. (2010) Obesity is the major contributor to vascular dysfunction and inflammation in high-fat diet hypertensive rats. Clin Sci (Lond) 118: 291-301.

[172] Galisteo, M., Garcia-Saura, M.F., Jimenez, R., Villar, I.C., Zarzuelo, A., Vargas, F., Duarte, J. (2004) Effects of chronic quercetin treatment on antioxidant defence system and oxidative status of deoxycorticosterone acetate-salt-hypertensive rats. Mol Cell Biochem 259: 91-99.

[173] De Miguel, C., Guo, C., Lund, H., Feng, D., Mattson, D.L. (2010) Infiltrating T lymphocytes in the kidney increase oxidative stress and participate in the development of hypertension and renal disease. Am J Physiol Renal Physiol 300: F734-742.

[174] Zhang, G.X., Kimura, S., Nishiyama, A., Shokoji, T., Rahman, M., Abe, Y. (2004) ROS during the acute phase of Ang II hypertension participates in cardiovascular MAPK activation but not vasoconstriction. Hypertension 43: 117-124.

[175] Meng, S., Roberts, L.J., 2nd, Cason, G.W., Curry, T.S., Manning, R.D., Jr. (2002) Superoxide dismutase and oxidative stress in Dahl salt-sensitive and -resistant rats. Am J Physiol Regul Integr Comp Physiol 283: R732-738.

[176] Paliege, A., Pasumarthy, A., Mizel, D., Yang, T., Schnermann, J., Bachmann, S. (2006) Effect of apocynin treatment on renal expression of COX-2, NOS1, and renin in Wistar-Kyoto and spontaneously hypertensive rats. Am J Physiol Regul Integr Comp Physiol 290: R694-700.

[177] Takai, S., Jin, D., Ikeda, H., Sakonjo, H., Miyazaki, M. (2009) Significance of angiotensin II receptor blockers with high affinity to angiotensin II type 1 receptors for vascular protection in rats. Hypertens Res 32: 853-860.

[178] Asselin, C., Bouchard, B., Tardif, J.C., Des Rosiers, C. (2006) Circulating 4- hydroxynonenal-protein thioether adducts assessed by gas chromatography-mass spectrometry are increased with disease progression and aging in spontaneously hypertensive rats. Free Radic Biol Med 41: 97-105.

[179] de Cavanagh, E.M., Toblli, J.E., Ferder, L., Piotrkowski, B., Stella, I., Inserra, F. (2006) Renal mitochondrial dysfunction in spontaneously hypertensive rats is attenuated by losartan but not by amlodipine. Am J Physiol Regul Integr Comp Physiol 290: R1616- 1625.

[180] Zhang, A., Jia, Z., Wang, N., Tidwell, T.J., Yang, T. (2011) Relative contributions of mitochondria and NADPH oxidase to deoxycorticosterone acetate-salt hypertension in mice. Kidney Int 80: 51-60.

[181] Yamamoto, E., Tamamaki, N., Nakamura, T., Kataoka, K., Tokutomi, Y., Dong, Y.F., Fukuda, M., Matsuba, S., Ogawa, H., Kim-Mitsuyama, S. (2008) Excess salt causes cerebral neuronal apoptosis and inflammation in stroke-prone hypertensive rats through angiotensin II-induced NADPH oxidase activation. Stroke 39: 3049-3056.

[182] Cifuentes, M.E., Rey, F.E., Carretero, O.A., Pagano, P.J. (2000) Upregulation of p67(phox) and gp91(phox) in aortas from angiotensin II-infused mice. Am J Physiol Heart Circ Physiol 279: H2234-2240.

[183] Akasaki, T., Ohya, Y., Kuroda, J., Eto, K., Abe, I., Sumimoto, H., Iida, M. (2006) Increased expression of gp91phox homologues of NAD(P)H oxidase in the aortic media during

chronic hypertension: involvement of the renin-angiotensin system. Hypertens Res 29: 813-820.

[184] Fukui, T., Ishizaka, N., Rajagopalan, S., Laursen, J.B., Capers, Q.t., Taylor, W.R., Harrison, D.G., de Leon, H., Wilcox, J.N., Griendling, K.K. (1997) p22phox mRNA expression and NADPH oxidase activity are increased in aortas from hypertensive rats. Circ Res 80: 45-51.

[185] Rajagopalan, S., Kurz, S., Munzel, T., Tarpey, M., Freeman, B.A., Griendling, K.K., Harrison, D.G. (1996) Angiotensin II-mediated hypertension in the rat increases vascular superoxide production via membrane NADH/NADPH oxidase activation. Contribution to alterations of vasomotor tone. J Clin Invest 97: 1916-1923.

[186] Virdis, A., Neves, M.F., Amiri, F., Touyz, R.M., Schiffrin, E.L. (2004) Role of NAD(P)H oxidase on vascular alterations in angiotensin II-infused mice. J Hypertens 22: 535-542.

[187] Suzuki, H., DeLano, F.A., Parks, D.A., Jamshidi, N., Granger, D.N., Ishii, H., Suematsu, M., Zweifach, B.W., Schmid-Schonbein, G.W. (1998) Xanthine oxidase activity associated with arterial blood pressure in spontaneously hypertensive rats. Proc Natl Acad Sci U S A 95: 4754-4759.

[188] DeLano, F.A., Parks, D.A., Ruedi, J.M., Babior, B.M., Schmid-Schonbein, G.W. (2006) Microvascular display of xanthine oxidase and NADPH oxidase in the spontaneously hypertensive rat. Microcirculation 13: 551-566.

[189] Breckwoldt, M.O., Chen, J.W., Stangenberg, L., Aikawa, E., Rodriguez, E., Qiu, S., Moskowitz, M.A., Weissleder, R. (2008) Tracking the inflammatory response in stroke in vivo by sensing the enzyme myeloperoxidase. Proc Natl Acad Sci U S A 105: 18584- 18589.

[190] Ersahin, M., Sehirli, O., Toklu, H.Z., Suleymanoglu, S., Emekli-Alturfan, E., Yarat, A., Tatlidede, E., Yegen, B.C., Sener, G. (2009) Melatonin improves cardiovascular function and ameliorates renal, cardiac and cerebral damage in rats with renovascular hypertension. J Pineal Res 47: 97-106.

[191] Wang, X., Desai, K., Clausen, J.T., Wu, L. (2004) Increased methylglyoxal and advanced glycation end products in kidney from spontaneously hypertensive rats. Kidney Int 66: 2315-2321.

[192] Bayorh, M.A., Ganafa, A.A., Emmett, N., Socci, R.R., Eatman, D., Fridie, I.L. (2005) Alterations in aldosterone and angiotensin II levels in salt-induced hypertension. Clin Exp Hypertens 27: 355-367.

[193] Zhan, C.D., Sindhu, R.K., Vaziri, N.D. (2004) Up-regulation of kidney NAD(P)H oxidase and calcineurin in SHR: reversal by lifelong antioxidant supplementation. Kidney Int 65: 219-227.

[194] Tanito, M., Nakamura, H., Kwon, Y.W., Teratani, A., Masutani, H., Shioji, K., Kishimoto, C., Ohira, A., Horie, R., Yodoi, J. (2004) Enhanced oxidative stress and impaired thioredoxin expression in spontaneously hypertensive rats. Antioxid Redox Signal 6: 89-97.

[195] Tyther, R., Ahmeda, A., Johns, E., Sheehan, D. (2009) Protein carbonylation in the kidney medulla of the spontaneously hypertensive rat. Proteomics Clin Appl 3: 338-346.

[196] Rodriguez-Iturbe, B., Ferrebuz, A., Vanegas, V., Quiroz, Y., Mezzano, S., Vaziri, N.D. (2005) Early and sustained inhibition of nuclear factor-kappaB prevents hypertension in spontaneously hypertensive rats. J Pharmacol Exp Ther 315: 51-57.

[197] Diep, Q.N., El Mabrouk, M., Cohn, J.S., Endemann, D., Amiri, F., Virdis, A., Neves, M.F., Schiffrin, E.L. (2002) Structure, endothelial function, cell growth, and inflammation in blood vessels of angiotensin II-infused rats: role of peroxisome proliferator-activated receptor-gamma. Circulation 105: 2296-2302.

[198] Bayorh, M.A., Ganafa, A.A., Eatman, D., Walton, M., Feuerstein, G.Z. (2005) Simvastatin and losartan enhance nitric oxide and reduce oxidative stress in salt- induced hypertension. Am J Hypertens 18: 1496-1502.

[199] Hong, H.J., Hsiao, G., Cheng, T.H., Yen, M.H. (2001) Supplemention with tetrahydrobiopterin suppresses the development of hypertension in spontaneously hypertensive rats. Hypertension 38: 1044-1048.

[200] Elmarakby, A.A., Loomis, E.D., Pollock, J.S., Pollock, D.M. (2005) NADPH oxidase inhibition attenuates oxidative stress but not hypertension produced by chronic ET-1. Hypertension 45: 283-287.

[201] Schnackenberg, C.G., Wilcox, C.S. (1999) Two-week administration of tempol attenuates both hypertension and renal excretion of 8-Iso prostaglandin f2alpha. Hypertension 33: 424-428.

[202] Krug, S., Zhang, Y., Mori, T.A., Croft, K.D., Vickers, J.J., Langton, L.K., Whitworth, J.A. (2008) N-Acetylcysteine prevents but does not reverse dexamethasone-induced hypertension. Clin Exp Pharmacol Physiol 35: 979-981.

[203] Stefanska, J., Pawliczak, R. (2008) Apocynin: molecular aptitudes. Mediators Inflamm 2008: 106507.

[204] Selemidis, S., Sobey, C.G., Wingler, K., Schmidt, H.H., Drummond, G.R. (2008) NADPH oxidases in the vasculature: molecular features, roles in disease and pharmacological inhibition. Pharmacol Ther 120: 254-291.

[205] Heumuller, S., Wind, S., Barbosa-Sicard, E., Schmidt, H.H., Busse, R., Schroder, K., Brandes, R.P. (2008) Apocynin is not an inhibitor of vascular NADPH oxidases but an antioxidant. Hypertension 51: 211-217.

[206] Touyz, R.M. (2008) Apocynin, NADPH oxidase, and vascular cells: a complex matter. Hypertension 51: 172-174.

[207] Ximenes, V.F., Kanegae, M.P., Rissato, S.R., Galhiane, M.S. (2007) The oxidation of apocynin catalyzed by myeloperoxidase: proposal for NADPH oxidase inhibition. Arch Biochem Biophys 457: 134-141.

[208] Brandes, R.P. (2003) A radical adventure: the quest for specific functions and inhibitors of vascular NAPDH oxidases. Circ Res 92: 583-585.

[209] Rey, F.E., Cifuentes, M.E., Kiarash, A., Quinn, M.T., Pagano, P.J. (2001) Novel competitive inhibitor of NAD(P)H oxidase assembly attenuates vascular O(2)(-) and systolic blood pressure in mice. Circ Res 89: 408-414.

[210] George, J., Struthers, A.D. (2009) Role of urate, xanthine oxidase and the effects of allopurinol in vascular oxidative stress. Vasc Health Risk Manag 5: 265-272.

[211] Kelkar, A., Kuo, A., Frishman, W.H. (2011) Allopurinol as a cardiovascular drug. Cardiol Rev 19: 265-271.

[212] Moorhouse, P.C., Grootveld, M., Halliwell, B., Quinlan, J.G., Gutteridge, J.M. (1987) Al-

lopurinol and oxypurinol are hydroxyl radical scavengers. FEBS Lett 213: 23-28.

[213] Schnackenberg, C.G., Welch, W.J., Wilcox, C.S. (1998) Normalization of blood pressure and renal vascular resistance in SHR with a membrane-permeable superoxide dismutase mimetic: role of nitric oxide. Hypertension 32: 59-64.

[214] Wilcox, C.S., Pearlman, A. (2008) Chemistry and antihypertensive effects of tempol and other nitroxides. Pharmacol Rev 60: 418-469.

[215] Dekhuijzen, P.N. (2004) Antioxidant properties of N-acetylcysteine: their relevance in relation to chronic obstructive pulmonary disease. Eur Respir J 23: 629-636.

[216] Moldeus, P., Cotgreave, I.A., Berggren, M. (1986) Lung protection by a thiol- containing antioxidant: N-acetylcysteine. Respiration 50 Suppl 1: 31-42.

[217] Heard, K.J. (2008) Acetylcysteine for acetaminophen poisoning. N Engl J Med 359: 285-292.

[218] Yilmaz, H., Sahin, S., Sayar, N., Tangurek, B., Yilmaz, M., Nurkalem, Z., Onturk, E., Cakmak, N., Bolca, O. (2007) Effects of folic acid and N-acetylcysteine on plasma homocysteine levels and endothelial function in patients with coronary artery disease. Acta Cardiol 62: 579-585.

[219] Beckman, J.S., Minor, R.L., Jr., White, C.W., Repine, J.E., Rosen, G.M., Freeman, B.A. (1988) Superoxide dismutase and catalase conjugated to polyethylene glycol increases endothelial enzyme activity and oxidant resistance. J Biol Chem 263: 6884-6892.

[220] Zhao, R., Masayasu, H., Holmgren, A. (2002) Ebselen: a substrate for human thioredoxin reductase strongly stimulating its hydroperoxide reductase activity and a superfast thioredoxin oxidant. Proc Natl Acad Sci U S A 99: 8579-8584.

[221] Schewe, T. (1995) Molecular actions of ebselen--an antiinflammatory antioxidant. Gen Pharmacol 26: 1153-1169.

[222] Arakawa, M., Ito, Y. (2007) N-acetylcysteine and neurodegenerative diseases: Basic and clinical pharmacology. Cerebellum: 1-7.

[223] Masumoto, H., Sies, H. (1996) The reaction of ebselen with peroxynitrite. Chem Res Toxicol 9: 262-267.

[224] Saito, I., Asano, T., Sano, K., Takakura, K., Abe, H., Yoshimoto, T., Kikuchi, H., Ohta, T., Ishibashi, S. (1998) Neuroprotective effect of an antioxidant, ebselen, in patients with delayed neurological deficits after aneurysmal subarachnoid hemorrhage. Neurosurgery 42: 269-277; discussion 277-268.

[225] Yamaguchi, T., Sano, K., Takakura, K., Saito, I., Shinohara, Y., Asano, T., Yasuhara, H. (1998) Ebselen in acute ischemic stroke: a placebo-controlled, double-blind clinical trial. Ebselen Study Group. Stroke 29: 12-17.

[226] Gilgun-Sherki, Y., Rosenbaum, Z., Melamed, E., Offen, D. (2002) Antioxidant therapy in acute central nervous system injury: current state. Pharmacol Rev 54: 271-284.

[227] Kizhakekuttu, T.J., Widlansky, M.E. (2010) Natural antioxidants and hypertension: promise and challenges. Cardiovasc Ther 28: e20-32.

[228] Zingg, J.M. (2007) Modulation of signal transduction by vitamin E. Mol Aspects Med 28: 481-506.

[229] Chen, X., Touyz, R.M., Park, J.B., Schiffrin, E.L. (2001) Antioxidant effects of vitamins C and E are associated with altered activation of vascular NADPH oxidase and superoxide dismutase in stroke-prone SHR. Hypertension 38: 606-611.

[230] Carr, A., Frei, B. (1999) Does vitamin C act as a pro-oxidant under physiological conditions? FASEB J 13: 1007-1024.

[231] Carr, A.C., Zhu, B.Z., Frei, B. (2000) Potential antiatherogenic mechanisms of ascorbate (vitamin C) and alpha-tocopherol (vitamin E). Circ Res 87: 349-354.

[232] Shay, K.P., Moreau, R.F., Smith, E.J., Smith, A.R., Hagen, T.M. (2009) Alpha-lipoic acid as a dietary supplement: molecular mechanisms and therapeutic potential. Biochim Biophys Acta 1790: 1149-1160.

[233] Kim, S.H., Han, S.I., Oh, S.Y., Chung, H.Y., Kim, H.D., Kang, H.S. (2001) Activation of heat shock factor 1 by pyrrolidine dithiocarbamate is mediated by its activities as pro-oxidant and thiol modulator. Biochem Biophys Res Commun 281: 367-372.

[234] Schreck, R., Meier, B., Mannel, D.N., Droge, W., Baeuerle, P.A. (1992) Dithiocarbamates as potent inhibitors of nuclear factor kappa B activation in intact cells. J Exp Med 175: 1181-1194.

[235] Min, Y.K., Park, J.H., Chong, S.A., Kim, Y.S., Ahn, Y.S., Seo, J.T., Bae, Y.S., Chung, K.C. (2003) Pyrrolidine dithiocarbamate-induced neuronal cell death is mediated by Akt, casein kinase 2, c-Jun N-terminal kinase, and IkappaB kinase in embryonic hippocampal progenitor cells. J Neurosci Res 71: 689-700.

[236] Knowles, R.G., Moncada, S. (1994) Nitric oxide synthases in mammals. Biochem J 298 (Pt 2): 249-258.

[237] Vasquez-Vivar, J., Kalyanaraman, B., Martasek, P., Hogg, N., Masters, B.S., Karoui, H., Tordo, P., Pritchard, K.A., Jr. (1998) Superoxide generation by endothelial nitric oxide synthase: the influence of cofactors. Proc Natl Acad Sci U S A 95: 9220-9225.

[238] Linscheid, P., Schaffner, A., Schoedon, G. (1998) Modulation of inducible nitric oxide synthase mRNA stability by tetrahydrobiopterin in vascular smooth muscle cells. Biochem Biophys Res Commun 243: 137-141.

[239] Kojima, S., Ona, S., Iizuka, I., Arai, T., Mori, H., Kubota, K. (1995) Antioxidative activity of 5,6,7,8-tetrahydrobiopterin and its inhibitory effect on paraquat-induced cell toxicity in cultured rat hepatocytes. Free Radic Res 23: 419-430.

[240] Park, Y.M., Park, M.Y., Suh, Y.L., Park, J.B. (2004) NAD(P)H oxidase inhibitor prevents blood pressure elevation and cardiovascular hypertrophy in aldosterone-infused rats. Biochem Biophys Res Commun 313: 812-817.

[241] Hu, L., Zhang, Y., Lim, P.S., Miao, Y., Tan, C., McKenzie, K.U., Schyvens, C.G., Whitworth, J.A. (2006) Apocynin but not L-arginine prevents and reverses dexamethasone-induced hypertension in the rat. Am J Hypertens 19: 413-418.

[242] Zhang, Y., Chan, M.M., Andrews, M.C., Mori, T.A., Croft, K.D., McKenzie, K.U., Schyvens, C.G., Whitworth, J.A. (2005) Apocynin but not allopurinol prevents and reverses adrenocorticotropic hormone-induced hypertension in the rat. Am J Hypertens 18: 910-916.

[243] Costa, C.A., Amaral, T.A., Carvalho, L.C., Ognibene, D.T., da Silva, A.F., Moss, M.B.,

Valenca, S.S., de Moura, R.S., Resende, A.C. (2009) Antioxidant treatment with tempol and apocynin prevents endothelial dysfunction and development of renovascular hypertension. Am J Hypertens 22: 1242-1249.

[244] Pechanova, O., Jendekova, L., Vrankova, S. (2009) Effect of chronic apocynin treatment on nitric oxide and reactive oxygen species production in borderline and spontaneous hypertension. Pharmacol Rep 61: 116-122.

[245] Tian, N., Moore, R.S., Phillips, W.E., Lin, L., Braddy, S., Pryor, J.S., Stockstill, R.L., Hughson, M.D., Manning, R.D., Jr. (2008) NADPH oxidase contributes to renal damage and dysfunction in Dahl salt-sensitive hypertension. Am J Physiol Regul Integr Comp Physiol 295: R1858-1865.

[246] Armando, I., Wang, X., Villar, V.A., Jones, J.E., Asico, L.D., Escano, C., Jose, P.A. (2007) Reactive oxygen species-dependent hypertension in dopamine D2 receptor-deficient mice. Hypertension 49: 672-678.

[247] Kopkan, L., Huskova, Z., Vanourkova, Z., Thumova, M., Skaroupkova, P., Maly, J., Kramer, H.J., Dvorak, P., Cervenka, L. (2009) Reduction of oxidative stress does not attenuate the development of angiotensin II-dependent hypertension in Ren-2 transgenic rats. Vascul Pharmacol 51: 175-181.

[248] Liu, F., Wei, C.C., Wu, S.J., Chenier, I., Zhang, S.L., Filep, J.G., Ingelfinger, J.R., Chan, J.S. (2009) Apocynin attenuates tubular apoptosis and tubulointerstitial fibrosis in transgenic mice independent of hypertension. Kidney Int 75: 156-166.

[249] Ceravolo, G.S., Fernandes, L., Munhoz, C.D., Fernandes, D.C., Tostes, R.C., Laurindo, F.R., Scavone, C., Fortes, Z.B., Carvalho, M.H. (2007) Angiotensin II chronic infusion induces B1 receptor expression in aorta of rats. Hypertension 50: 756-761.

[250] Pech, V., Sikka, S.C., Sindhu, R.K., Vaziri, N.D., Majid, D.S. (2006) Oxidant stress and blood pressure responses to angiotensin II administration in rats fed varying salt diets. Am J Hypertens 19: 534-540.

[251] Zhou, M.S., Hernandez Schulman, I., Pagano, P.J., Jaimes, E.A., Raij, L. (2006) Reduced NAD(P)H oxidase in low renin hypertension: link among angiotensin II, atherogenesis, and blood pressure. Hypertension 47: 81-86.

[252] Viel, E.C., Benkirane, K., Javeshghani, D., Touyz, R.M., Schiffrin, E.L. (2008) Xanthine oxidase and mitochondria contribute to vascular superoxide anion generation in DOCA-salt hypertensive rats. Am J Physiol Heart Circ Physiol 295: H281-288.

[253] Wallwork, C.J., Parks, D.A., Schmid-Schonbein, G.W. (2003) Xanthine oxidase activity in the dexamethasone-induced hypertensive rat. Microvasc Res 66: 30-37.

[254] Szasz, T., Linder, A.E., Davis, R.P., Burnett, R., Fink, G.D., Watts, S.W. (2010) Allopurinol does not decrease blood pressure or prevent the development of hypertension in the deoxycorticosterone acetate-salt rat model. J Cardiovasc Pharmacol 56: 627-634.

[255] Ong, S.L., Vickers, J.J., Zhang, Y., McKenzie, K.U., Walsh, C.E., Whitworth, J.A. (2007) Role of xanthine oxidase in dexamethasone-induced hypertension in rats. Clin Exp Pharmacol Physiol 34: 517-519.

[256] Usui, M., Egashira, K., Kitamoto, S., Koyanagi, M., Katoh, M., Kataoka, C., Shimokawa, H., Takeshita, A. (1999) Pathogenic role of oxidative stress in vascular angiotensin-converting enzyme activation in long-term blockade of nitric oxide synthesis in rats. Hyperten-

sion 34: 546-551.

[257] Laakso, J.T., Teravainen, T.L., Martelin, E., Vaskonen, T., Lapatto, R. (2004) Renal xanthine oxidoreductase activity during development of hypertension in spontaneously hypertensive rats. J Hypertens 22: 1333-1340.

[258] Nakazono, K., Watanabe, N., Matsuno, K., Sasaki, J., Sato, T., Inoue, M. (1991) Does superoxide underlie the pathogenesis of hypertension? Proc Natl Acad Sci U S A 88: 10045-10048.

[259] Park, J.B., Touyz, R.M., Chen, X., Schiffrin, E.L. (2002) Chronic treatment with a superoxide dismutase mimetic prevents vascular remodeling and progression of hypertension in salt-loaded stroke-prone spontaneously hypertensive rats. Am J Hypertens 15: 78-84.

[260] Beswick, R.A., Zhang, H., Marable, D., Catravas, J.D., Hill, W.D., Webb, R.C. (2001) Long-term antioxidant administration attenuates mineralocorticoid hypertension and renal inflammatory response. Hypertension 37: 781-786.

[261] Bayorh, M.A., Mann, G., Walton, M., Eatman, D. (2006) Effects of enalapril, tempol, and eplerenone on salt-induced hypertension in dahl salt-sensitive rats. Clin Exp Hypertens 28: 121-132.

[262] Dobrian, A.D., Schriver, S.D., Prewitt, R.L. (2001) Role of angiotensin II and free radicals in blood pressure regulation in a rat model of renal hypertension. Hypertension 38: 361-366.

[263] Zhang, Y., Jang, R., Mori, T.A., Croft, K.D., Schyvens, C.G., McKenzie, K.U., Whitworth, J.A. (2003) The anti-oxidant Tempol reverses and partially prevents adrenocorticotrophic hormone-induced hypertension in the rat. J Hypertens 21: 1513- 1518.

[264] Elmarakby, A.A., Williams, J.M., Imig, J.D., Pollock, J.S., Pollock, D.M. (2007) Synergistic actions of enalapril and tempol during chronic angiotensin II-induced hypertension. Vascul Pharmacol 46: 144-151.

[265] Pechanova, O., Zicha, J., Kojsova, S., Dobesova, Z., Jendekova, L., Kunes, J. (2006) Effect of chronic N-acetylcysteine treatment on the development of spontaneous hypertension. Clin Sci (Lond) 110: 235-242.

[266] Mondo, C.K., Zhang, Y., de Macedo Possamai, V., Miao, Y., Schyvens, C.G., McKenzie, K.U., Hu, L., Guo, Z., Whitworth, J.A. (2006) N-acetylcysteine antagonizes the development but does not reverse ACTH-induced hypertension in the rat. Clin Exp Hypertens 28: 73-84.

[267] Tian, N., Rose, R.A., Jordan, S., Dwyer, T.M., Hughson, M.D., Manning, R.D., Jr. (2006) N-Acetylcysteine improves renal dysfunction, ameliorates kidney damage and decreases blood pressure in salt-sensitive hypertension. J Hypertens 24: 2263-2270.

[268] Rauchova, H., Pechanova, O., Kunes, J., Vokurkova, M., Dobesova, Z., Zicha, J. (2005) Chronic N-acetylcysteine administration prevents development of hypertension in N(omega)-nitro-L-arginine methyl ester-treated rats: the role of reactive oxygen species. Hypertens Res 28: 475-482.

[269] Jung, O., Marklund, S.L., Xia, N., Busse, R., Brandes, R.P. (2007) Inactivation of extracellular superoxide dismutase contributes to the development of high-volume hypertension. Arterioscler Thromb Vasc Biol 27: 470-477.

[270] Weber, D.S., Rocic, P., Mellis, A.M., Laude, K., Lyle, A.N., Harrison, D.G., Griendling,

K.K. (2005) Angiotensin II-induced hypertrophy is potentiated in mice overexpressing p22phox in vascular smooth muscle. Am J Physiol Heart Circ Physiol 288: H37-42.

[271] Nishikawa, Y., Tatsumi, K., Matsuura, T., Yamamoto, A., Nadamoto, T., Urabe, K. (2003) Effects of vitamin C on high blood pressure induced by salt in spontaneously hypertensive rats. J Nutr Sci Vitaminol (Tokyo) 49: 301-309.

[272] Ettarh, R.R., Odigie, I.P., Adigun, S.A. (2002) Vitamin C lowers blood pressure and alters vascular responsiveness in salt-induced hypertension. Can J Physiol Pharmacol 80: 1199-1202.

[273] Tian, N., Moore, R.S., Braddy, S., Rose, R.A., Gu, J.W., Hughson, M.D., Manning, R.D., Jr. (2007) Interactions between oxidative stress and inflammation in salt-sensitive hypertension. Am J Physiol Heart Circ Physiol 293: H3388-3395.

[274] Schyvens, C.G., Andrews, M.C., Tam, R., Mori, T.A., Croft, K.D., McKenzie, K.U., Whitworth, J.A., Zhang, Y. (2007) Antioxidant vitamins and adrenocorticotrophic hormone-induced hypertension in rats. Clin Exp Hypertens 29: 465-478.

[275] Noguchi, T., Ikeda, K., Sasaki, Y., Yamamoto, J., Yamori, Y. (2004) Effects of vitamin E and sesamin on hypertension and cerebral thrombogenesis in stroke-prone spontaneously hypertensive rats. Clin Exp Pharmacol Physiol 31 Suppl 2: S24-26.

[276] Vasdev, S., Ford, C.A., Parai, S., Longerich, L., Gadag, V. (2000) Dietary alpha-lipoic acid supplementation lowers blood pressure in spontaneously hypertensive rats. J Hypertens 18: 567-573.

[277] Vasdev, S., Ford, C.A., Parai, S., Longerich, L., Gadag, V. (2000) Dietary lipoic acid supplementation prevents fructose-induced hypertension in rats. Nutr Metab Cardiovasc Dis 10: 339-346.

[278] Vasdev, S., Gill, V., Longerich, L., Parai, S., Gadag, V. (2003) Salt-induced hypertension in WKY rats: prevention by alpha-lipoic acid supplementation. Mol Cell Biochem 254: 319-326.

[279] Takaoka, M., Kobayashi, Y., Yuba, M., Ohkita, M., Matsumura, Y. (2001) Effects of alpha-lipoic acid on deoxycorticosterone acetate-salt-induced hypertension in rats. Eur J Pharmacol 424: 121-129.

[280] Koeners, M.P., Braam, B., Joles, J.A. (2011) Perinatal inhibition of NF-kappaB has long-term antihypertensive effects in spontaneously hypertensive rats. J Hypertens 29: 1160-1166.

[281] Fortepiani, L.A., Reckelhoff, J.F. (2005) Treatment with tetrahydrobiopterin reduces blood pressure in male SHR by reducing testosterone synthesis. Am J Physiol Regul Integr Comp Physiol 288: R733-736.

[282] Zhang, Y., Pang, T., Earl, J., Schyvens, C.G., McKenzie, K.U., Whitworth, J.A. (2004) Role of tetrahydrobiopterin in adrenocorticotropic hormone-induced hypertension in the rat. Clin Exp Hypertens 26: 231-241.

[283] Kris-Etherton, P.M., Lichtenstein, A.H., Howard, B.V., Steinberg, D., Witztum, J.L. (2004) Antioxidant vitamin supplements and cardiovascular disease. Circulation 110: 637-641.

[284] Bjelakovic, G., Nikolova, D., Gluud, L.L., Simonetti, R.G., Gluud, C. (2007) Mortality in randomized trials of antioxidant supplements for primary and secondary prevention: systematic review and meta-analysis. JAMA 297: 842-857.

[285] Dotan, Y., Pinchuk, I., Lichtenberg, D., Leshno, M. (2009) Decision analysis supports the paradigm that indiscriminate supplementation of vitamin E does more harm than good. Arterioscler Thromb Vasc Biol 29: 1304-1309.

[286] Bates, C.J., Walmsley, C.M., Prentice, A., Finch, S. (1998) Does vitamin C reduce blood pressure? Results of a large study of people aged 65 or older. J Hypertens 16: 925-932.

[287] McMackin, C.J., Widlansky, M.E., Hamburg, N.M., Huang, A.L., Weller, S., Holbrook, M., Gokce, N., Hagen, T.M., Keaney, J.F., Jr., Vita, J.A. (2007) Effect of combined treatment with alpha-Lipoic acid and acetyl-L-carnitine on vascular function and blood pressure in patients with coronary artery disease. J Clin Hypertens (Greenwich) 9: 249- 255.

[288] Rahman, S.T., Merchant, N., Haque, T., Wahi, J., Bhaheetharan, S., Ferdinand, K.C., Khan, B.V. (2012) The Impact of Lipoic Acid on Endothelial Function and Proteinuria in Quinapril-Treated Diabetic Patients With Stage I Hypertension: Results From the QUALITY Study, J Cardiovasc Pharmacol Ther 17: 139-145.

[289] John, J.H., Ziebland, S., Yudkin, P., Roe, L.S., Neil, H.A. (2002) Effects of fruit and vegetable consumption on plasma antioxidant concentrations and blood pressure: a randomised controlled trial. Lancet 359: 1969-1974.

[290] Lopes, H.F., Martin, K.L., Nashar, K., Morrow, J.D., Goodfriend, T.L., Egan, B.M. (2003) DASH diet lowers blood pressure and lipid-induced oxidative stress in obesity. Hypertension 41: 422-430.

[291] Griendling, K.K., FitzGerald, G.A. (2003) Oxidative stress and cardiovascular injury: Part II: animal and human studies. Circulation 108: 2034-2040.

[292] Steinhubl, S.R. (2008) Why have antioxidants failed in clinical trials? Am J Cardiol 101: 14D-19D.

[293] Suzuki, K. (2009) Anti-oxidants for therapeutic use: why are only a few drugs in clinical use? Adv Drug Deliv Rev 61: 287-289.

[294] Violi, F., Loffredo, L., Musella, L., Marcoccia, A. (2004) Should antioxidant status be considered in interventional trials with antioxidants? Heart 90: 598-602.

[295] Yusuf, S., Dagenais, G., Pogue, J., Bosch, J., Sleight, P. (2000) Vitamin E supplementation and cardiovascular events in high-risk patients. The Heart Outcomes Prevention Evaluation Study Investigators. N Engl J Med 342: 154-160.

[296] Montezano, A.C., Touyz, R.M. (2012) Molecular Mechanisms of Hypertension- Reactive Oxygen Species and Antioxidants: A Basic Science Update for the Clinician. Can J Cardiol.

[297] Violi, F., Cangemi, R. (2008) Statin treatment as a confounding factor in human trials with vitamin E. J Nutr 138: 1179-1181.

[298] George, J., Carr, E., Davies, J., Belch, J.J., Struthers, A. (2006) High-dose allopurinol improves endothelial function by profoundly reducing vascular oxidative stress and not by lowering uric acid. Circulation 114: 2508-2516.

[299] Kostka-Jeziorny, K., Uruski, P., Tykarski, A. (2011) Effect of allopurinol on blood pressure and aortic compliance in hypertensive patients. Blood Press 20: 104-110.

[300] Feig, D.I., Soletsky, B., Johnson, R.J. (2008) Effect of allopurinol on blood pressure of adolescents with newly diagnosed essential hypertension: a randomized trial. JAMA 300: 924-932.

[301] Kanbay, M., Ozkara, A., Selcoki, Y., Isik, B., Turgut, F., Bavbek, N., Uz, E., Akcay, A., Yigitoglu, R., Covic, A. (2007) Effect of treatment of hyperuricemia with allopurinol on blood pressure, creatinine clearence, and proteinuria in patients with normal renal functions. Int Urol Nephrol 39: 1227-1233.

[302] Garrido, A.M., Griendling, K.K. (2009) NADPH oxidases and angiotensin II receptor signaling. Mol Cell Endocrinol 302: 148-158.

[303] Griendling, K.K., Ushio-Fukai, M. (2000) Reactive oxygen species as mediators of angiotensin II signaling. Regul Pept 91: 21-27.

[304] Tamba, M., Torreggiani, A. (2000) Free radical scavenging and copper chelation: a potentially beneficial action of captopril. Free Radic Res 32: 199-211.

[305] Chen, S., Ge, Y., Si, J., Rifai, A., Dworkin, L.D., Gong, R. (2008) Candesartan suppresses chronic renal inflammation by a novel antioxidant action independent of AT1R blockade. Kidney Int 74: 1128-1138.

[306] Kadowaki, D., Anraku, M., Tasaki, Y., Taguchi, K., Shimoishi, K., Seo, H., Hirata, S., Maruyama, T., Otagiri, M. (2009) Evaluation for antioxidant and renoprotective activity of olmesartan using nephrectomy rats. Biol Pharm Bull 32: 2041-2045.

[307] Miyata, T., van Ypersele de Strihou, C., Ueda, Y., Ichimori, K., Inagi, R., Onogi, H., Ishikawa, N., Nangaku, M., Kurokawa, K. (2002) Angiotensin II receptor antagonists and angiotensin-converting enzyme inhibitors lower in vitro the formation of advanced glycation end products: biochemical mechanisms. J Am Soc Nephrol 13: 2478-2487.

[308] Dandona, P., Ghanim, H., Brooks, D.P. (2007) Antioxidant activity of carvedilol in cardiovascular disease. J Hypertens 25: 731-741.

[309] Evangelista, S., Garbin, U., Pasini, A.F., Stranieri, C., Boccioletti, V., Cominacini, L. (2007) Effect of DL-nebivolol, its enantiomers and metabolites on the intracellular production of superoxide and nitric oxide in human endothelial cells. Pharmacol Res 55: 303-309.

[310] Gupta, S., Wright, H.M. (2008) Nebivolol: a highly selective beta1-adrenergic receptor blocker that causes vasodilation by increasing nitric oxide. Cardiovasc Ther 26: 189-202. Cominacini, L., Fratta Pasini, A., Garbin, U., Pastorino, A.M., Davoli, A., Nava, C.,

[311] Campagnola, M., Rossato, P., Lo Cascio, V. (2003) Antioxidant activity of different dihydropyridines. Biochem Biophys Res Commun 302: 679-684.

[312] Godfraind, T. (2005) Antioxidant effects and the therapeutic mode of action of calcium channel blockers in hypertension and atherosclerosis. Philos Trans R Soc Lond B Biol Sci 360: 2259-2272.

Chapter 18

Lipid Peroxidation and Reperfusion Injury in Hypertrophied Hearts

Juliana C. Fantinelli, Ignacio A. Pérez Núñez,
Luisa F. González Arbeláez and Susana M. Mosca

Additional information is available at the end of the chapter

http://dx.doi.org/10.5772/45913

1. Introduction

Oxidative stress is characterized by an imbalance between increased exposure to reactive oxygen species (ROS), and antioxidant defenses, comprised of both small molecular weight antioxidants like glutathione, and antioxidant enzymes like superoxide dismutase. ROS cause direct damage to critical biomolecules including DNA, lipids, and proteins. Oxidative stress has been involved in the genesis of hypertension [1, 2] and implicated in the mechanisms of reversible postischemic contractile dysfunction (myocardial stunning), microvascular dysfunction, arrhythmias and cell death [3-6]. In spontaneously hypertensive rats (SHR) there are few reports showing the protective action of antioxidants against ischemia-reperfusion injury [7-9] and specifically in regard to the effects of the scavenger N-(2-mercaptopropionyl)-glycine (MPG) these have not been yet examined.

Ischemic preconditioning (IP) is acknowledged to be an endogenous mechanism of cardioprotection against ischemia and reperfusion injury [10-11]. This intervention is based in that one or more brief periods of ischemia applied previous to a prolonged ischemic period exert beneficial effects on myocardium attenuating the deleterious effects observed in the reperfused myocardium. Although there are some studies showing the beneficial effects of IP in hypertensive animals [12-15], under certain circumstances the effectiveness of that intervention is questioned [16-18]. A recent investigation performed in our laboratory shows that a single cycle of IP attenuated the myocardial stunning produced by 20-min global ischemia in SHR [19] and decreased the lipid peroxidation. Whether this protective action of IP is operating at more extended ischemic period and involves changes in oxidative stress in this rats strain is a point that needs to be clarified.

Therefore, the aim of the present study was to determine if alterations of lipid peroxidation and endogenous antioxidants are linked to myocardial and vascular postischemic damage in ischemic control, preconditioned and MPG treated hearts from SHR.

2. Material and methods

2.1. Isolated heart preparation

Experiments were performed in SHR of 5-month-old following the Guide for the Care and Use of Laboratory Animals published by the US National Institutes of Health (NIH Publication No. 85-23, revised in 1996). Beginning at 12 weeks of age, systolic blood pressure (SBP) was measured weekly in all animals by the standard tail-cuff method [20] following the modifications detailed in a recent paper by Fritz and Rinaldi [21]. Rats were anesthetized with an intraperitoneal injection of sodium pentobarbital (60 mg/kg body wt). The heart was rapidly excised and perfused by the non-recirculating Langendorff technique with Ringer's solution containing (in mmol/L): 118 NaCl, 4.7 KCl, 1.2 $MgSO_4$, 1.35 $CaCl_2$, 20 $NaCO_3H$ and 11.1dextrose. The buffer was saturated with a mixture of 95% O_2-5% CO_2, had a pH 7.4, and was maintained at 37°C. The conductive tissue in the atrial septum was damaged with a fine needle to achieve atrioventricular block, and the right ventricle was paced at 280 ± 10 beats/min. A latex balloon tied to the end of a polyethylene tube was passed into the left ventricle through the mitral valve; the opposite end of the tube was then connected to a Statham P23XL pressure transducer. The balloon was filled with water to provide an end-diastolic pressure (LVEDP) of 8-12 mmHg and this volume remained unchanged for the rest of the experiment. Coronary perfusion pressure (CPP) was monitored at the point of cannulation of the aorta and adjusted to approximately 70 mmHg. Coronary flow (CF), controlled with a peristaltic pump, was 11 ± 2 mL/min. Left ventricular pressure (LVP) and CPP data were acquired by using an analog-to-digital converter and acquisition software (Chart V4.2.3 ADInstruments).

2.2. Experimental protocols

After 10 min of stabilization, hearts from SHR were assigned to the following experimental protocols (Fig. 1):

Non-ischemic control hearts (NIC): Hearts were perfused for 3 hs without any treatment.

Ischemic control hearts (IC): Hearts were subjected to 35 min or 50 min of normothermic global ischemia followed by 2 hours of reperfusion. Global ischemia was induced by stopping the perfusate inflow line and the heart was placed in a saline bath held at 37°C.

Ischemic preconditioning (IP1): A single cycle of 5-min ischemia and 10-min reperfusion was applied previous to the 35-min and 50-min ischemic periods followed by 2-hour reperfusion.

Ischemic preconditioning (IP3): Three cycles of 2-min f ischemia and 5-min reperfusion was applied prior to the 50-min ischemic period followed by 2-hour reperfusion. Previous experiments performed by us showed that three cycles are the fewest for achieving myocardial protection of SHR when global ischemia was extended to 50 min.

MPG: Hearts were treated 10 min before ischemia and during the first 10 min of reperfusion with N-(2-mercaptopropionyl)-glycine (MPG) 2 mM. The administration time for MPG was chosen to attenuate the ROS production during ischemia and reperfusion. The dose was selected according previous experiments performed in our laboratory [22].

Additional experiments were performed (n = 6 for each protocol) to assess the biochemical parameters.

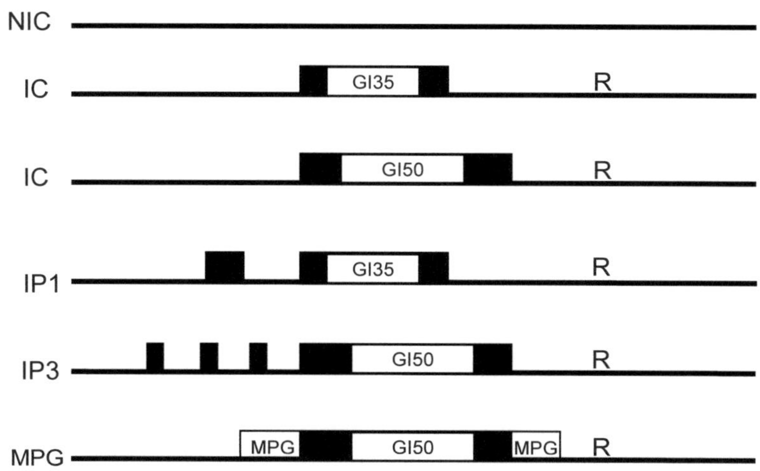

Figure 1. Scheme of the experimental protocols.

2.3. Infarct size determination

Infarct size was assessed by the widely validated triphenyltetrazolium chloride (TTC) staining technique [23]. At the end of reperfusion, atrial and right ventricular tissues were excised and left ventricle (VI) was frozen. The freeze VI was cut into six transverse slices, which were incubated for 5 minutes at 37°C in a 1% solution of triphenyltetrazolium chloride (TTC). To measure myocardial infarction, the slices were weighed and scanned. The infarcted (pale) and viable ischemic/reperfused (red) areas were measured by computed planimetry (Scion Image 1.62; Scion Corp., Frederick, Maryland, USA). Infarct weights were calculated as (A1 × W1) + (A2 × W2) + (A3 × W3) + (A4 × W4) + (A5 × W5) + (A6 × W6), where A is the infarct area for the slice and W is the weight of the respective section. Infarct size was expressed as a percentage of the total area (area at risk, AAR) [24].

2.4. Systolic and diastolic function

Myocardial contractility was assessed by the left ventricular developed pressure (LVDP), obtained by subtracting LVEDP to LVP peak, and maximal velocity of contraction ($+dP/dt_{max}$). The diastolic function was evaluated through LVEDP.

2.5. Assessment of coronary resistance (CR)

CR was calculated as a quotient between CPP and CF and expressed as difference between the values obtained at the end of reperfusion period and that observed in the preischemic period.

2.6. Preparation of tissue homogenate

At the end of reperfusion a portion of VI was homogenized in 5 volume of 25 mM PO_4KH_2 - 140 mM ClK at pH = 7.4 with a Polytron homogenizer. Aliquots of homogenate were used to assess reduced glutathione content (GSH) and thiobarbituric acid reactive substances (TBARS) as an index of lipid peroxidation. The remaining homogenate was centrifuged at 12,000 g for 5 min at 4º C and the supernatant stored at -70 ºC until superoxide dismutase (SOD) activity was assayed.

2.6.1. *Assessment of reduced glutathione (GSH)*

GSH was determined by Ellman's method [25]. This method was based on the reaction of GSH with 5, 5′ dithiobis (2-nitrobenzoic acid) to give a compound that absorbs at 412 nm. GSH levels were expressed as μg/mg of protein.

2.6.2. *Assessment of lipid peroxidation*

TBARS concentration was determined in the supernatant following the Buege and Aust method´s [26]. Absorbance at 535nm was measured and TBARS expressed in nmol/g of tissue using an extinction coefficient of 1.56×10^5 M^{-1} cm^{-1}.

2.6.3. *Measurement of SOD cytosolic activity*

SOD activity was measured by means of the nitroblue tetrazolium (NBT) method [27]. Briefly, the supernatant was added to the reaction mixture of NBT with xanthine-xanthine oxidase, and the SOD activity measured colorimetrically in the form of inhibitory activity toward blue formazan formation by SOD in the reaction mixture.

2.6.4. *Protein determination*

The protein concentration was evaluated by the Bradford method [28] using bovine serum albumin as a standard.

2.6.5. *Correlations*

The relationships between TBARS, GSH and infarct size and CR were determined by linear regression (equation $y = a + b . x$).

2.7. Statistical analysis

Data are presented as mean ± SE and repeated measures of two-way analysis of variance (ANOVA) with Newman-Keuls test were used for multiple comparisons among groups. Relationships were tested for significance using the Pearson correlation coefficient (r). A P value < 0.05 was considered significant.

3. Results

Fig. 2 shows the infarct size in ischemic control and preconditioned hearts from SHR. In non-ischemic control hearts at the end of the 3-hour perfusion the infarct size was approximately 1 % of risk area. After 35-min global ischemia and 2-hour reperfusion, the infarct size was 35 ± 5 %, which was significantly decreased by one cycle of IP (IP1). When ischemia was extended to 50 min, the infarct size (58 ± 5 %) was not reduced by IP1 indicating that this preconditioning protocol is not adequate for protecting that rat strain against reperfusion injury. However, when a larger number of cycles (three in our case) were applied the hearts were protected and the infarct size diminished. A significant reduction of infarct size was also obtained when MPG was added to the perfusate during 10 min before 50-min ischemia and during the first 10 min of reperfusion.

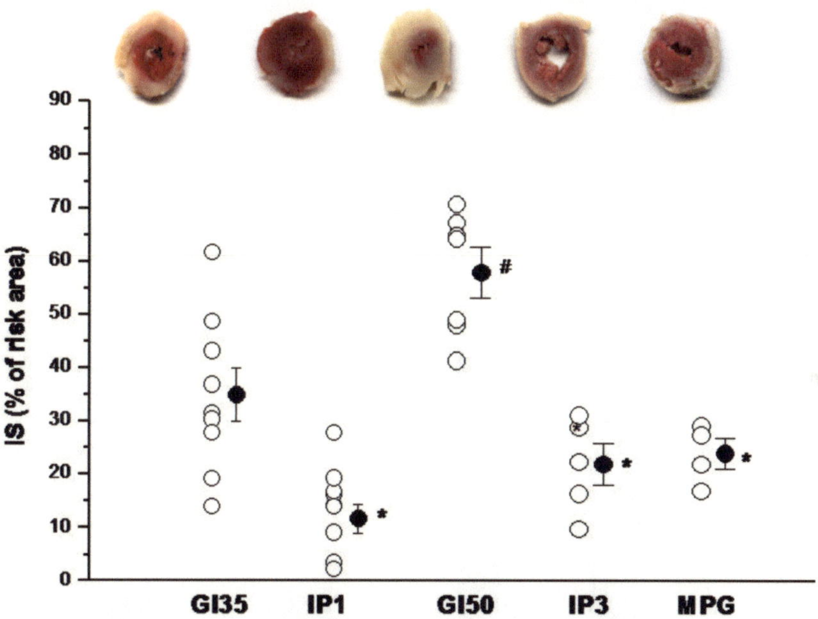

Figure 2. Infarct size (IS), expressed as percentage of risk area, in ischemic control (IC = GI35: 35-min global ischemia; GI50: 50-min global ischemia), preconditioned hearts (IP1= one cycle; IP3= three cycles) and MPG treatment. Note that hearts from SHR showed a higher IS at 50-min compared to 35-min GI. IP1 diminished the IS at 35-min ischemia but it was necessary to apply three cycles (IP3) to protect the hearts when the prolonged ischemia was extended to 50 min and that MPG decreased the IS at a similar value to IP3. * $P < 0.05$ with respect to GI; # $P < 0.05$ with respect to 35-min GI.

At the end of 3-hour non-ischemic hearts exhibited a decrease in contractility of approximately 10 %. After 35-min ischemia and 2-hour reperfusion contractility decreased approximately 90 % with respect to preischemic values. As it is depicted in Fig. 3 the recovery of systolic function was improved by both IP protocols. At the end of the reperfusion period, LVDP and $+dP/dt_{max}$ reached higher values than those obtained in ischemic control hearts. When ischemia was more prolonged (50 min) the postischemic recovery of contractility was scarce (LVDP and $+dP/dt_{max}$ reached values of approximately 2 %) and it was significantly improved by IP3 and MPG treatment.

The diastolic stiffness characterized by LVEDP increased during 35-min and 50-min global ischemia and acquired greater values during reperfusion. These increases were attenuated by both IP protocols and MPG treatment. Fig. 4 shows the changes of LVEDP occurring at 50-min global ischemia in ischemic control and intervened hearts.

The increase in perfusion pressure at constant coronary flow resulted in an increase of coronary resistance. The increases (4.2 ± 0.4 and 7.0 ± 0.9 mmHg/ml x min^{-1} after 35-min and 50-min ischemia, respectively) were significantly attenuated by both IP protocols and MPG treatment (Fig. 5).

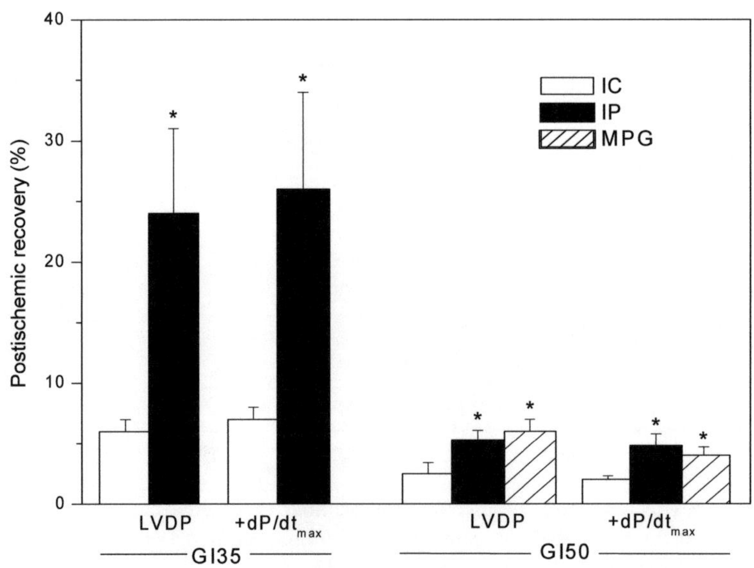

Figure 3. Values of left ventricular developed pressure (LVDP) and maximal velocity of contraction ($+dP/dt_{max}$) at the end of reperfusion period expressed as percentage of preischemic values, in ischemic control (IC = GI35: 35-min global ischemia; GI50: 50-min global ischemia), preconditioned hearts (IP) and MPG treatment. Observe that IP and MPG significantly improved the postischemic recovery of myocardial systolic function at 35-min and 50-min GI. * $P < 0.05$ with respect to IC

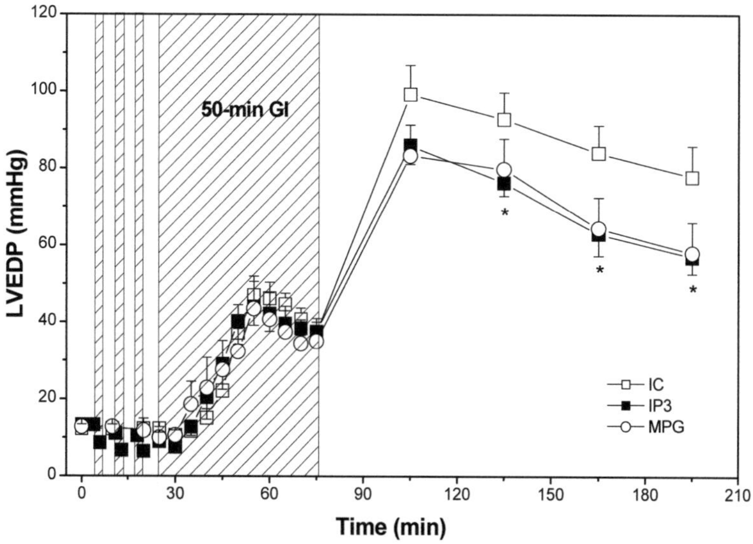

Figure 4. Time course of left ventricular end diastolic pressure (LVEDP) in ischemic control (IC = GI50: 50-min global ischemia), preconditioned hearts (IP) and MPG treatment. The three cycles of IP (IP3) and MPG attenuated in a similar manner the increase of LVEDP detected in IC hearts. * P < 0.05 with respect to IC.

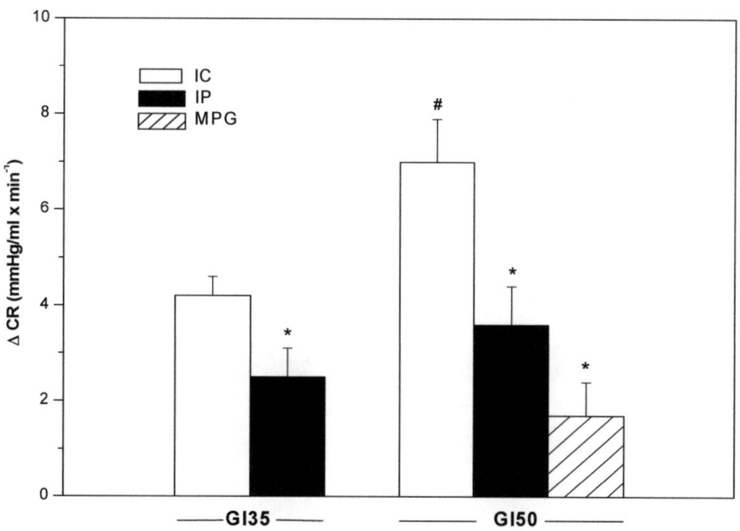

Figure 5. Changes of coronary resistance (CR) at the end of reperfusion in ischemic control (IC = GI35: 35-min global ischemia; GI50: 50-min global ischemia), preconditioned hearts (IP) and MPG treatment. The interventions attenuated the increase of CR detected in IC hearts being MPG the most effective. * P < 0.05 with respect to IC; # P < 0.05 with respect to GI35.

Given that an increase of ROS generation accompanied by a diminution of antioxidants may be responsible for myocardial reperfusion injury [29, 30], we next determined the impact of IP and MPG on myocardial GSH content, a marker of oxidative stress. Fig. 6 shows that

GSH content in non-ischemic hearts (2 ± 0.3 μg/mg prot) was significantly reduced by ischemia and reperfusion. A single or three cycles of IP and MPG treatment were able to preserve part of the GSH content.

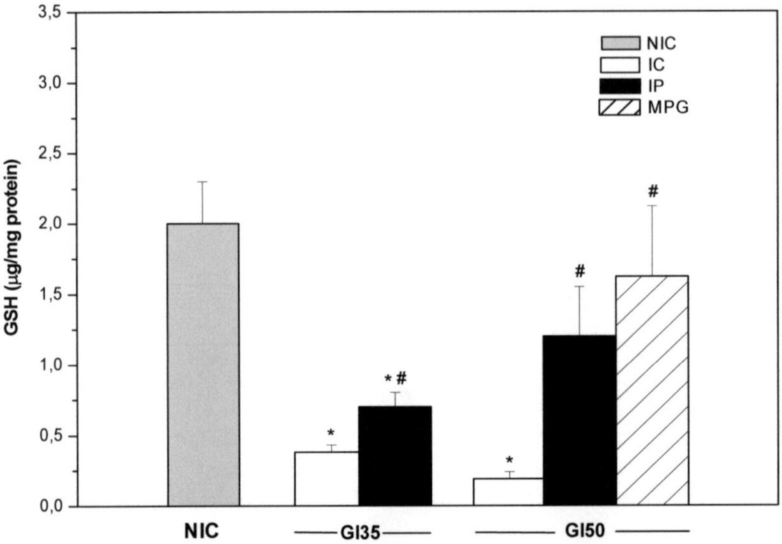

Figure 6. Myocardial reduced glutathione content (GSH, μg/mg protein) in non-ischemic control (NIC), ischemic control (IC = GI35: 35-min global ischemia; GI50: 50-min global ischemia) and preconditioned (IP) and MPG treated hearts. Observe that GSH levels decreased after ischemia and reperfusion in both ischemic periods and were partially preserved by IP and MPG. * $P < 0.05$ with respect to NIC; # $P < 0.05$ with respect to IC.

Moreover, the SOD cytosolic activity increased in ischemic controls hearts and significantly decreased in all intervened hearts (Fig. 7). Both parameters (GSH and SOD) are indicating the presence of oxidative stress caused by ischemia-reperfusion which may be attenuated by IP and MPG treatment.

Since ROS induce membrane lipid peroxidation [29], we determined TBARS content of untreated and treated ischemic-reperfused hearts. Although TBARS determination suffers from potential artifacts associated with sampling, storage and problems caused by the complexity of the biological systems, being easy and reproducible, it is one of the most widely used indexes for assessing oxidative stress. There was an increase in myocardial TBARS content in hearts submitted to ischemia and reperfusion detecting a higher value at 50-min compared to 35-min global ischemia. Preconditioned and MPG treated hearts exhibited lower TBARS levels (Fig. 8).

The analysis of data of the different interventions showed the presence of significant positive correlations TBARS vs IS (Fig. 9, A panel; r = 0,47) and TBARS vs CR (Fig. 9, B panel; n = 0,45) and negative correlations GSH vs IS (Fig. 10, A panel; n = 0,41) and GSH vs CR (Fig. 10, B panel; n = 0,40) in isolated hearts from SHR.

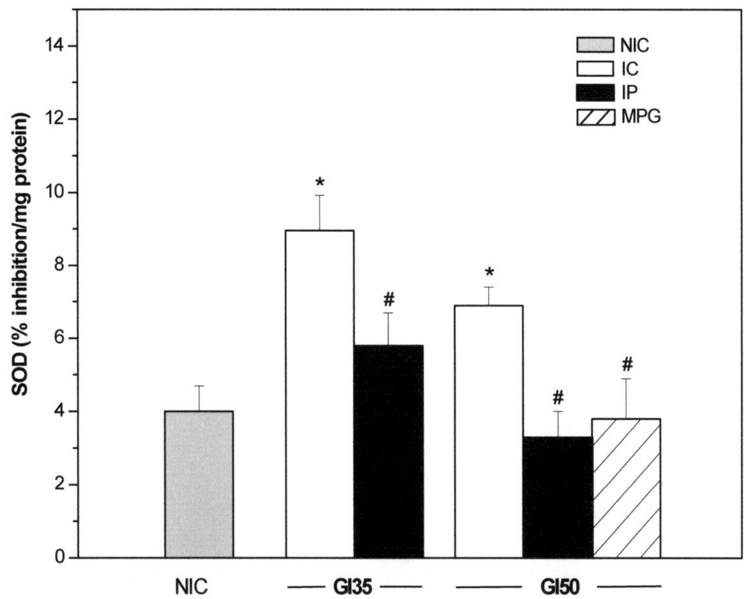

Figure 7. Myocardial SOD cytosolic activity (SOD, % inhibition/mg protein) in non-ischemic control (NIC), ischemic control (IC = GI35: 35-min global ischemia; GI50: 50-min global ischemia), preconditioned (IP) and MPG treated hearts. Note that SOD cytosolic activity increased after 35-min or 50-min GI in comparison to NIC. These increases were attenuated by both interventions (IP and MPG). * P < 0.05 with respect to NIC; # P < 0.05 with respect to IC.

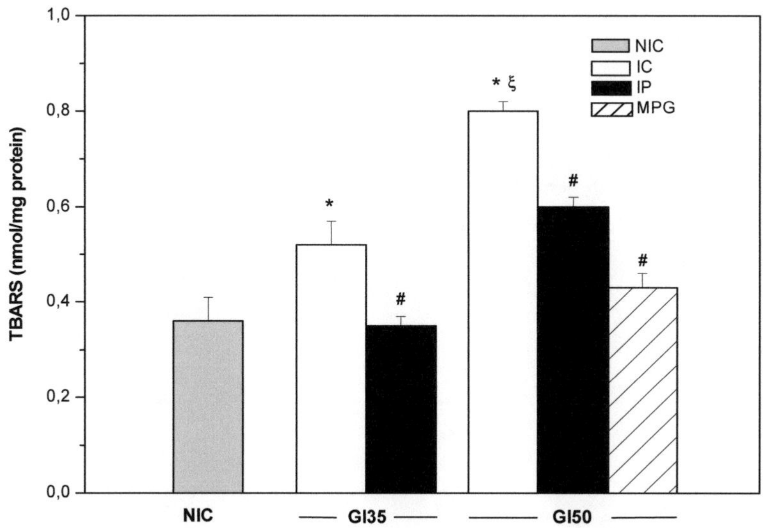

Figure 8. Myocardial thiobarbituric acid reactive substances (TBARS) concentration, expressed in nmol/mg protein in non-ischemic control (NIC), ischemic control (IC = GI35: 35-min global ischemia; GI50: 50-min global ischemia), preconditioned (IP) and MPG treated hearts. An increase of TBARS occurred at the end of reperfusion after the two ischemic periods which were attenuated by IP and MPG. * P < 0.05 with respect to NIC ; # P < 0.05 with respect to IC; ς P < 0.05 with respect to GI35.

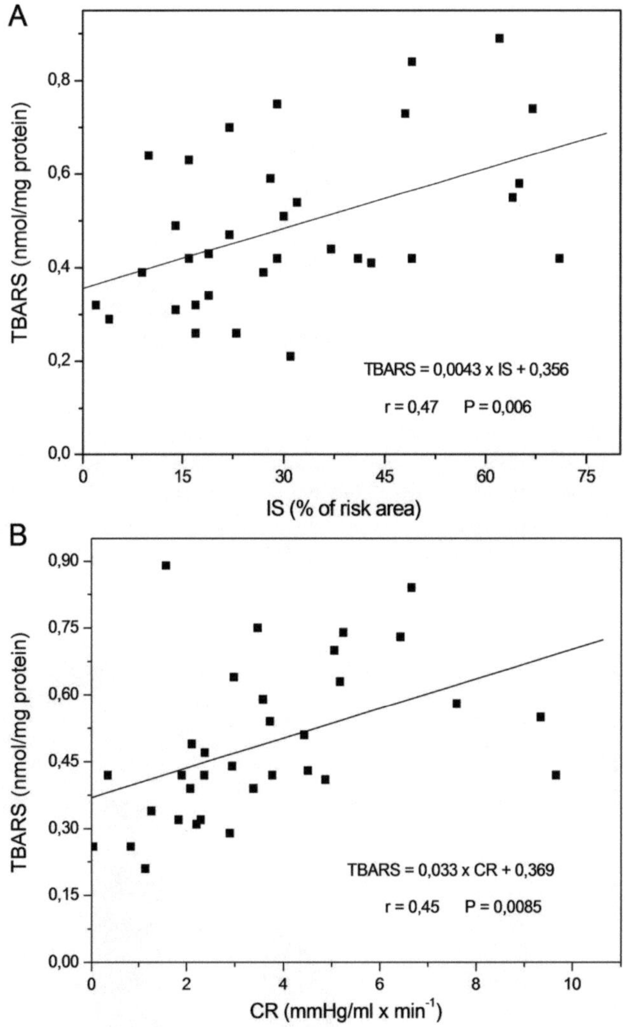

Figure 9. Relationship between TBARS and infarct size (IS, A panel) and TBARS and coronary resistance (CR, B panel) in all experimental situations. The resulting data were fitted to straight line by linear regression. Significant positive correlations between TBARS and IS and CR were found.

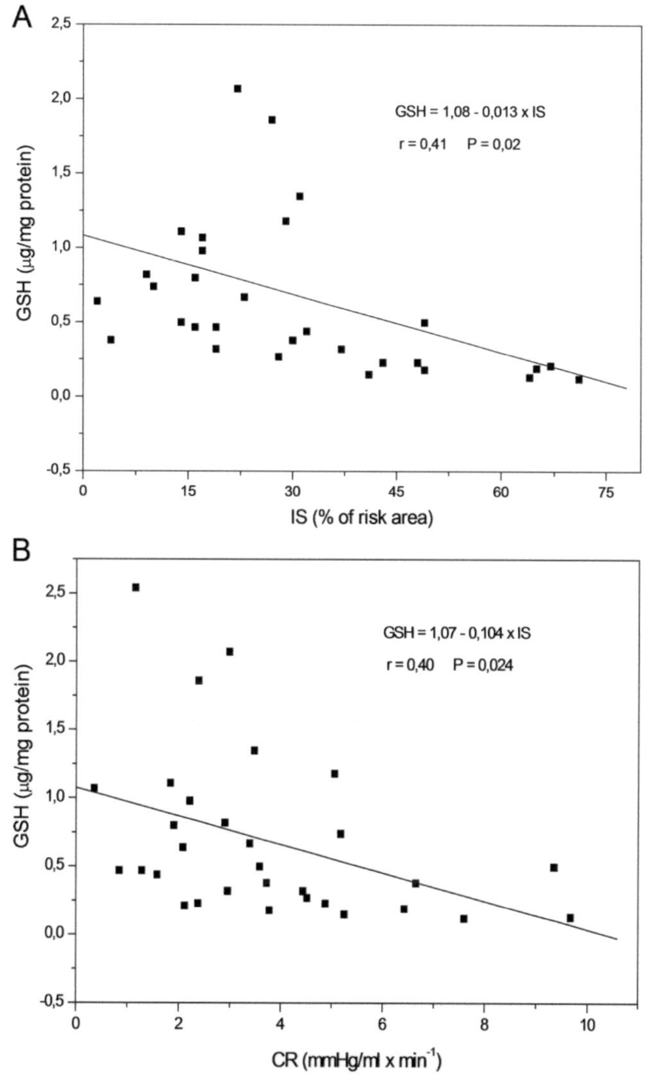

Figure 10. Relationship between GSH and infarct size (IS, A panel) and TBARS and coronary resistance (CR, B panel) in all experimental situations. The resulting data were fitted to straight line by linear regression. Significant negative correlations between GSH and IS and CR were found.

4. Discussion

To our knowledge, this is the first demonstration that the beneficial action of ischemic preconditioning and MPG against ischemia-reperfusion injury is similar in hearts from SHR and is associated with a mitigation of oxidative stress. Thus, our data show the existence of a positive correlation between TBARS concentration-used as an index of lipid peroxidation- and infarct size indicating that it will be found more infarct size when TBARS acquire higher values.

Simultaneously an inverse correlation was detected between GSH content and infarct size indicating that higher levels of GSH are associated to minor infarct size. Both variables (TBARS and GSH) suffered opposite changes due to a possible cause-effect relationship.

These results were also accompanied by changes of SOD cytosolic activity which showed lesser values in preconditioned and MPG treated hearts. Taken together, these data provide evidence to suggest that formation of lipoperoxides is a significant cause of ischemia and reperfusion injury and that the mechanism whereby IP and MPG confer cardioprotection involves, at least in part, an attenuation of those nocive products through a diminution of ROS release and/or production and an improvement of the endogenous antioxidants.

This study clearly shows that hearts from SHR suffer higher irreversible damage at 50-min compared to 35-min global ischemia accompanied with greater impairment of postichemic myocardial function. Thus, at the end of reperfusion the recovery of systolic function was scarce and diastolic stiffness significantly increased in ischemic control hearts. These alterations were attenuated by IP being one cycle of IP (IP1) effective when the ischemic period was 35 min and three cycles (IP3) when the ischemia was extended to 50 min. Thus, although the cardioprotective action of IP in hypertrophied hearts was previously reported [12-15] our study demonstrates that the optimum protocol of IP to protect SHR hearts must be selected according to the duration time of prolonged ischemia. Then, it seems to be possible that the number of IP cycles appears as other key factor for determining the efficacy of IP. Moreover MPG treated hearts in the same way that the preconditioned showed lesser infarct size and improved postischemic recovery of myocardial function in comparison to ischemic control hearts.

Hypertension is associated with an elevation of ROS and frequently with an impairment of endogenous antioxidant mechanisms [30]. These alterations have also been described during ischemia and reperfusion [3, 4, 31-33]. In this study, at the end of reperfusion after ischemic period cardiac tissue showed lesser GSH content, higher TBARS concentration and SOD cytosolic activity in comparison to non-ischemic control hearts. Major changes of GSH and TBARS were detected at 50-min compared to 35-min global ischemia. However, SOD cytosolic activity showed higher increase at 35 min of ischemia. This result may explain the lesser lipid peroxidation found in this experimental group. All these changes were partially reversed by both IP protocols and MPG treatment. Thus, GSH content was higher and SOD cytosolic activity was lower than the values observed in untreated hearts. The favorable changes in GSH and SOD cytosolic activity were reflected in the lower lipid peroxidation (decreased TBARS concentration) observed in preconditioned hearts and in those treated with MPG in comparison to ischemic control hearts. In other words the improvement of the antioxidant systems (SOD and GSH) by IP and MPG treatment were enough to attenuate the oxidative damage detected in untreated hearts. These results suggest that changes of lipid peroxidation and antioxidant systems would be sufficient to promote differences in the cell death and the attenuation of oxidative stress would be considered as a factor contributing to the cardioprotection by IP and MPG treatment in hearts from SHR.

On the other hand, a balance between the production of nitric oxide (NO) and ROS controls the endothelial function [34, 35]. When the NO production is normal its bioavailability may be reduced because of the oxidative inactivation by an excessive production of superoxide ($O_2^{-\cdot}$) in the vascular wall. The available data on the NO system in SHR are limited and apparently contradictory. Increased ROS in SHR have been demonstrated to enhance NO inactivation and reduce NO bioavailability [36], which contributes to the maintenance of hypertension. According to a previous study the peroxynitrite- product of NO and $O_2^{-\cdot}$ combination- may also be involved in maintenance of the high levels of blood pressure in SHR [37. Furthermore in this rats strain was reported that the activity and/or expression of the different nitric oxide synthase (NOS) isoforms would be altered [38-40] which might act as a compensatory mechanism to maintain the production of bioactive NO in the face of increased oxidant stress [41]. In our study, ischemic control hearts showed an increase of coronary resistance at the end of reperfusion compared to pre-ischemic period which was greater after 50-min than 35-min global ischemia. These increases were attenuated by IP and MPG treatment being this last intervention the most effective. Thus, the beneficial effect of IP and MPG on coronary resistance would be attributed to a greater NO availability mediated by an attenuation of oxidative stress. This mechanism could explain the significant correlations between TBARS, GSH and coronary resistance found in this study and reinforces the idea that changes of oxidative stress constitute the basis of myocardial and vascular postischemic alterations.

5. Conclusions

We can conclude that the level of lipid peroxidation and antioxidant defenses are linked to reperfusion injury in hypertrophied hearts from SHR. The finding that IP and MPG reduce the postischemic myocardial and vascular injury as well as levels of TBARS and improve the endogenous antioxidants suggest that the decrease in ROS levels would be the common mechanism of cardioprotection of both interventions.

Author details

Juliana C. Fantinelli, Ignacio A. Pérez Núñez,
Luisa F. González Arbeláez and Susana M. Mosca
Centro de Investigaciones Cardiovasculares, Facultad de Ciencias Médicas,
Universidad Nacional de La Plata, La Plata, Buenos Aires, Argentina

Acknowledgement

This work was supported in part by the grant PICT 1046 from Agencia Nacional de Promoción Científica y Técnica of Argentina to Dr Susana M Mosca.

6. References

[1] Newaz MA, Nawal NN (1999) Effect of gamma-tocotrienol on blood pressure, lipid peroxidation and total antioxidant status in spontaneously hypertensive rats (SHR). Clin. Exp. Hypertens. 21: 1297-1313.

[2] Vaziri ND, Sica DA (2004) Lead-induced hypertension: role of oxidative stress. Curr. Hypertens. Rep. 6: 314–320.

[3] Griendling KK, FitzGerald GA (2003) Oxidative stress and cardiovascular injury: Part II: animal and human studies. Circulation 108: 2034-2040.

[4] Papaharalambus CA, Griendling KK (2007) Basic mechanisms of oxidative stress and reactive oxygen species in cardiovascular injury. Trends Cardiovasc. Med. 17: 48-54.

[5] Ravingerová T, Slezák J, Tribulová J, Džurba A, Uhrík B, Ziegelhöffer A (1999) Reactive oxygen species contribute to high incidence of reperfusion-induced arrhythmias in isolated rat heart. Life Sci. 65: 1927-1930.

[6] Kalaycioglu S, Sinci V, Imren Y, Oz E (1999) Metoprolol prevents ischemia–reperfusion injury by reducing lipid peroxidation. Jpn. Circ. J. 63:718–721.

[7] Abebe W, Liu JY, Wimborne H, Mozaffari MS (2010) Effects of chromium picolinate on vascular reactivity and cardiac ischemia-reperfusion injury in spontaneously hypertensive rats. Pharmacol. Rep. 62: 674-682.

[8] Potenza MA, Gagliardi S, De Benedictis L, Zigrino A, Tiravanti E, Colantuono G, Federici A, Lorusso L, Benagiano V, Quon MJ, Montagnani M (2009) Treatment of spontaneously hypertensive rats with rosiglitazone ameliorates cardiovascular pathophysiology via antioxidant mechanisms in the vasculature. Am. J. Physiol. Endocrinol. Metab. 297(3): 685-694.

[9] Sahna E, Deniz E, Bay-Karabulut A, Burma O (2008) Melatonin protects myocardium from ischemia-reperfusion injury in hypertensive rats: role of myeloperoxidase activity. Clin. Exp. Hyperten. 30: 673-681.

[10] Mosca SM, Gelpi RJ, Milei J, Fernandez Alonso G, Cingolani HE (1998) Is stunning prevented by ischemic preconditioning? Mol . Cell. Biochem. 186: 123-129.

[11] Yellon DM, Downey JM (2003) Preconditioning the myocardium: from cellular physiology to clinical cardiology. Physiol. Rev. 83: 1113–1151.

[12] Speechly-Dick ME, Baxter GF, Yellon DM (1994) Ischaemic preconditioning protects hypertrophied myocardium. Cardiovasc. Res. 28(7):1025-1029.

[13] Boutros A, Wang J (1995) Ischaemic preconditioning, adenosine and bethanechol protect spontaneously hypertensive isolated rat hearts. J. Pharmacol. Exp. Ther. 275: 1148-1156.

[14] Pantos CI, Davos CH, Carageorgiou HC, Varonos DV, Cokkinos DV (1996) Ischaemic preconditioning protects against myocardial dysfunction caused by ischaemia in isolated hypertrophied rat hearts. Basic Res. Cardiol. 91: 444-449.

[15] Nakagawa C, Asayama J, Katamura M, Matoba S, Keira N, Kawahara A, Tsuruyama K, Tanaka T, Kobara M, Akashi K, Ohta B, Tatsumi T, Nakagawa M (1997) Myocardial strech induced by increased left ventricular diastolic pressure preconditions isolated perfused hearts of normotensive and spontaneously hypertensive rats. Basic Res. Cardiol. 92: 410-416.

[16] Ebrahim Z, Yellon DM, Baxter GF (2007) Ischemic preconditioning is lost in aging hypertensive rat heart: independent effects of aging and longstanding hypertension. Exp. Gerontol. 42(8): 807-814.

[17] Ebrahim Z, Yellon DM, Baxter GF (2007) Attenuated cardioprotective response to bradykinin, but not classical ischaemic preconditioning, in DOCA-salt hypertensive left ventricular hypertrophy. Pharmacol. Res. 55(1): 42-48.

[18] Dai W, Simkhovich BZ, Kloner RA (2009) Ischemic preconditioning maintains cardioprotection in aging normotensive and spontaneously hypertensive rats. Exp. Gerontol. 44(5): 344-349.

[19] Fantinelli JC, Mosca SM (2007) Comparative effects of ischemic pre and postconditioning on ischemia-reperfusion injury in spontaneously hypertensive rats (SHR). Mol. Cell . Biochem. 296: 45-51.

[20] Buñag RA (1973) Validation in awake rats of tail cuff method for measuring systolic pressure. J.Appl. Physiol. 34: 279-282.

[21] Fritz M, Rinaldi G (2008) Blood pressure measurement with the tail-cuff method in Wistar and spontaneously hypertensive rats: Influence of adrenergic- and nitric oxide-mediated vasomotion. J. Pharmacol. Toxicol. Methods 58(3): 215-221.

[22] Fantinelli JC, Cingolani HE, Mosca SM (2006) Na^+/H^+ exchanger inhibition at the onset of reperfusion decreases the myocardial infarct size: role of ROS. Cardiovasc. Pathol. 15/4: 179-184.

[23] Fishbein MC, Meerbaum S, Rit J, Lando U, Kanmatsuse K, Mercier JC, Corday E, Ganz W (1981) Early phase acute myocardial infarct size quantification: validation of the triphenyl tetrazolium chloride tissue enzyme staining technique. Am. Heart. J. 101(5): 593– 600.

[24] Suzuki M, Sasaki N, Miki T, Sakamoto N, Ohmoto-Sekine Y, Tamagawa M, Seino S, Marbán E, Nakaya H (2001) Role of sarcolemmal KATP channels in cardioprotection against ischemia/reperfusion injury in mice. J. Clin.Invest. 109: 509–516.

[25] Sedlak J, Lindsay RH (1968) Estimation of total, protein-bound, and nonprotein sulfhydryl groups in tissue with Ellman's reagent. Anal. Biochem. 25:192-205.

[26] Buege JA, Aust SD (1978) Microsomal lipid peroxidation. Methods Enzymol. 52: 302-310.

[27] Beauchamp Ch, Fridovich I (1971) Superoxide dismutase: improved assays and an assay applicable to acrylamide gels. Anal. Biochem. 44: 276-287.

[28] Bradford MM (1976) A rapid and sensitive method for the quantitation of microgram quantitites of protein utilizing the principle of protein-dye binding. Anal. Biochem. 72: 248-254.

[29] Ambrosio G, Flaherty JT, Duilio C, Tritto I, Santoro G, Elia PP, Condorelli M, Chiariello M (1991) Oxygen radicals generated at reflow induce peroxidation of membrane lipids in reperfused hearts. J. Clin. Invest. 87: 2056-2066.

[30] Lassègue B, Griendling KK (2010) NADPH oxidases: functions and pathologies in the vasculature. Arterioscler. Thromb. Vasc. Biol. 30: 653-661.

[31] Bolli R, Jeroudi MO, Patel BS, DuBose CM, Lai EK, Roberts R, McCay PB (1989) Direct evidence that oxygen-derived free radicals contribute to postischemic myocardial dysfunction in the intact dog. Proc. Natl. Acad. Sci. USA 86: 4695-4699.

[32] Leichtweis S, Ji LL (2001) Glutathione deficiency intensifies ischaemia-reperfusion induced cardiac dysfunction and oxidative stress. Acta Physiol. Scand. 172(1): 1-10.

[33] Haramaki N, Stewart DB, Aggarwal S, Ikeda H, Reznick AZ, Packer L (1998) Networking antioxidants in the isolated rat heart are selectively depleted by ischemia-reperfusion. Free Rad. Biol. Med. 25(3): 329-339.

[34] Kodja G, Harrison D (1999) Interactions between NO and reactive oxygen species: pathophysiological importance in atherosclerosis, hypertension, diabetes and heart failure. Cardiovasc. Res. 43: 562-571.

[35] Cai , Harrison DG (2000) Endothelial dysfunction in cardiovascular diseases: the role of oxidant stress. Circ. Res. 87: 840-844.

[36] Shah AM, MacCarthy PA (2000) Paracrine and autocrine effects of nitric oxide in myocardial function. Pharmacol. Ther. 86: 49-86. [37] Kagota S, Tada Y, Kubota Y, Nejime N, Yamaguchi Y, Nakamura K, Kunitomo M, Shinozuka K (2007) Peroxynitrite is involved in the dysfunction of vasorelaxation in SHR/NDmcr-cp rats, spontaneously hypertensive rats. J. Cardiovasc. Pharmacol. 50: 677-685.

[38] Piech A, Dessy C, Havaux X, Feron O, Balligand JL (2003) Differential regulation of nitric oxide synthases and their allosteric regulators in heart and vessels of hypertensive rats. Cardiovasc. Res. 57: 456-467.

[39] Vaziri ND, Ni Z, Oveisi F (1998) Upregulation of renal and vascular nitric oxide synthase in young spontaneously hypertensive rats. Hypertension 31: 1248-1254.

[40] Nava E, Noll G, Luscher TF (1995) Increased activity of constitutive nitric oxide synthase in cardiac endothelium in spontaneous hypertension. Circulation 91: 2310-2313.

[41] Vapaatalo H, Mervaala E, Nurminen ML (2000) Role of endothelium and nitric oxide in experimental hypertension. Physiol. Res. 49: 1–10.

Chapter 19

Lipid Peroxidation by-Products and the Metabolic Syndrome

Nicolas J. Pillon and Christophe O. Soulage

Additional information is available at the end of the chapter

http://dx.doi.org/10.5772/46019

1. Introduction

About twenty-one percent of the air we breathe is composed of oxygen, and our life will not be possible without it. Oxygen is however a toxic, highly reactive molecule which was originally released in atmosphere as a waste product of the first photosynthetic organisms. Its accumulation on Earth indeed led to a massive extinction of living species. Few organisms survived and some developed the ability to use this toxic oxygen to improve the production of energy from carbohydrates. This "oxidative metabolism" was however a double-edge sword as the use of intracellular oxygen generates deleterious oxidative damages. To protect themselves toward this toxicity, those organisms consequently developed several "antioxidants" protection mechanisms which helped them maintain a balance between oxidative damage and efficient use of oxygen to produce energy.

When antioxidant defences are reduced and/or oxidative mechanisms increased, uncontrolled oxidation of cell targets leads to the accumulation of reactive oxygen species (ROS) and a state of "oxidative stress", often deleterious for the cells. This stress is involved in the pathophysiology of several human diseases, and especially in the development of metabolic diseases, even if its causative role remains questionable. A definite increase in oxidative stress biomarkers can be found in obese and diabetic humans as well as in animal model of these diseases. Accumulation of ROS can be deleterious by itself or can induce the oxidation of proteins, nucleic acids and lipids, generating secondary by-products. The specific reactivity of ROS towards polyunsaturated fatty acids (PUFAs) present in cell membranes induces lipid peroxidation, a noxious mechanism producing toxic aldehydes. Among them, malondialdehyde (MDA) and 4-hydroxy-2-nonenal (HNE) have been extensively studied. Originally simple markers of lipid peroxidation, these aldehydes have demonstrated causative roles in the impairment of cellular functions: activation of signalling pathways, apoptosis, and modification of enzyme function. In addition to being hallmarks of oxidative damage, lipid aldehydes could be mediators of oxidative insults, propagating tissue injury and activating cellular stress signalling pathways. Several studies demonstrated the association of obesity and diabetes with lipid peroxidation by-products, and the role of aldehydes in impairment of insulin function and signalling was recently pointed out.

This chapter aims to review the diverse implications of lipid peroxidation by-products in the pathophysiology of metabolic diseases, from evidence of their production during obesity and diabetes to the cellular mechanisms of their toxicity and protection against their deleterious effects.

2. Lipid peroxidation by-products

Under conditions of oxidative stress, excessive production of reactive oxygen species promotes the peroxidation of polyunsaturated fatty acids (PUFA). The resulting accumulation of hydroperoxides, unstable molecules, leads to their non-enzymatic degradation in many compounds, including aldehydes. Quantification of lipid peroxidation in biological samples has been extensively performed with the thiobarbituric acid (TBA) test. TBA detects malondialdehyde (MDA), an end-product of nonenzymatic PUFA oxidative degradation, which has therefore been used for decades as a marker of lipid peroxidation (Gutteridge, 1982). Another aldehyde: acrolein, first attracted attention because of its formation during tobacco combustion and its ubiquitous presence in the environment (Dong & Moldoveanu, 2004). Because of its carcinogenic potential, its role in smoking-related diseases has received extensive attention; however, acrolein is also produced endogenously though lipid peroxidation and its link with oxidative-associated pathologies is now well established. 4-hydroxy-2-alkenals are specific by-products of the oxidation of omega-3 and omega-6 fatty acids. 4-hydroxy-2-nonenal (HNE) is derived from the oxidation of polyunsaturated fatty acids of the n-6 series, mainly linoleic and arachidonic acids, while 4-hydroxy-2-hexenal (HHE) results from the peroxidation of polyunsaturated fatty acids of the n-3 series (mainly docosahexaenoic, eicosapentaenoic and linolenic acid). The peroxidation of arachidonic acid via 12-lipoxygenase leads to the formation of 4-hydroxy-2-dodecadienal (HDDE) (Guichardant *et al.*, 2006).

Name	Molecular Weight (*Da*)	Molecular Formula	Skeletal Formula
Acrolein (prop-2-enal)	56.1	C_3H_4O	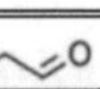
Malondialdehyde (propanedial, MDA)	72.1	$C_3H_4O_2$	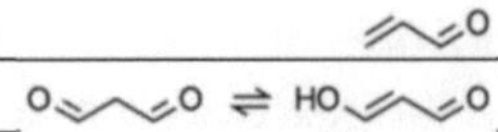
4-hydroxy-2-hexenal (HHE)	114.1	$C_6H_{10}O_2$	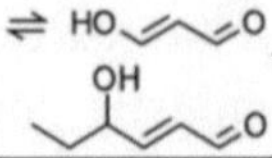
4-hydroxy-2-nonenal (HNE)	156.2	$C_9H_{16}O_2$	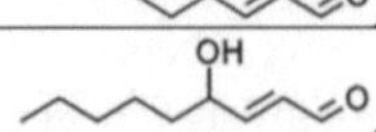
4-hydroxy-2-dodecadienal (HDDE)	196.0	$C_{12}H_{20}O_2$	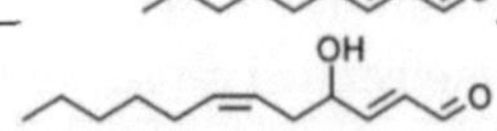

Table 1. α,β-Unsaturated aldehydes produced during polyunsaturated fatty acids oxidation

2.1. Chemistry and reactivity

Acrolein, MDA and 4-hydroxy-alkenals are α,β-unsaturated aldehydes, a class of compounds sharing the general structure C=C–C=O. They are characterised by an aldehyde group (C=O) on carbon 1 and a conjugated double bond (C=C) between carbons 2 and 3 (Table 1). In this structure, the oxygen atom of the carbonyl group increases the polarity of the double bond, which makes α,β-unsaturated aldehydes potent electrophiles. Acrolein has the simplest structure composed of 3 carbons, MDA is a dicarbonyl compound and 4-hydroxy-2-alkenals are characterized by the presence of a hydroxyl group on carbon 4. HHE, HNE and HDDE only differ by the length of their carbon chain and the presence of an additional double bond for the HDDE. In the case of MDA and 4-hydroxy-alkenals, the presence of a second oxygen atom makes the double bond even more reactive.

These aldehydes are part of the "*reactive electrophile species*" able to form covalent adducts with the nucleophilic groups present in DNA, proteins and phospholipids. In physiological conditions, they spontaneously react with the thiol group of glutathione to form Michael adduct by attack of the nucleophilic group of glutathione to the double bond of aldehydes. They can also react with thiol groups present on cysteine residues of certain proteins, leading to impairment of their biological activity. Under certain conditions, especially alkaline pH, aldehydes react with the amine groups present in proteins, nucleic acids and aminophospholipids, leading to Michael adducts. On the other hand, the reaction between a primary amine group and the carbonyl group of the aldehyde leads to the formation of Schiff bases (Schaur, 2003).

Aldehydes produced during lipid peroxidation are precursors of Advanced Lipoxidation End products (ALEs). Together with Advanced Glycation End products (AGEs) generated during glycoxidation, they accumulate in cells and tissues. The "carbonyl stress" is a result of this adduct accumulation, which induces protein dysfunctions and consequent pathological events such as inflammation and apoptosis (Negre-Salvayre *et al.*, 2008).

2.2. Cellular effects

2.2.1. *Cytotoxicity*

Since α,β-unsaturated compounds are strong electrophiles, they exhibit a high cytotoxic and mutagenic potential and have consequently been extensively studied for their effects on cell viability. HNE, HHE and acrolein indeed induce cell death, but the lethal concentration 50 (LC50), concentration that induces the death of 50% of the cells, is subject to variation, depending on the aldehyde, exposure duration and cell type (Table 2). The LC50 for a long exposure (>16 hours) to HNE or HHE is however consistently found 20-60 μM in several cell types, including human lymphoma Jurkat cells, lens epithelial cells, hamster V79-4 cells and muscle cells (Table 2). For acrolein, the same treatment gives a range of LC50 of 5-100 μM in human fibroblasts, human neuroblastoma cells, PC12 chromaffin cells and lymphocytes. MDA-induced cell death is less documented, even if the LC50 range for MDA is found around 1 mM in cortical, endothelial cells and fibroblasts. MDA was also reported to induce cell cycle arrest, which is to relate to cell damage and death (Ji *et al.*, 1998). HDDE appears to be the most toxic lipid aldehyde with a LC50 in endothelial cells in the submicromolar range.

Name	LC50, μM	References
Acrolein (prop-2-enal)	5-100	Poirier *et al.*, 2002; Luo *et al.*, 2005; Jia *et al.*, 2009*b*, 2009*a*
Malondialdehyde (propanedial, MDA)	600-3000	Michiels & Remacle, 1991; Hipkiss *et al.*, 1997; Cheng *et al.*, 2011
4-hydroxy-2-hexenal (HHE)	20-60	Liu *et al.*, 2000; Choudhary *et al.*, 2002; Pillon *et al.*, 2010; Li *et al.*, 2011
4-hydroxy-2-nonenal (HNE)	20-60	Liu *et al.*, 2000; Choudhary *et al.*, 2002; Pillon *et al.*, 2010; Li *et al.*, 2011
4-hydroxy-2-dodecadienal (HDDE)	0.22	Riahi *et al.*, 2010

Table 2. Range of lethal concentration 50 for long-term treatment (>16 hours) with α,β-unsaturated aldehydes. LC50 values were calculated from the indicated references.

On the opposite, very little cell death is detectable for short term treatments (<4 hours), likely because aldehyde-induced cell death involves apoptosis mechanisms not yet occurring during this short period of time. Several studies indeed reported that cell death is induced by aldehydes through apoptosis for low concentration and both apoptosis and necrosis for high doses (Luo *et al.*, 2005; Liu *et al.*, 2010). Acrolein-induced necrosis was described in few studies (Luo *et al.*, 2005), but mitochondrial-driven cell death seems to be the canonical road and was widely studied. Acrolein-induced apoptosis was indeed confirmed in several cell types through DNA fragmentation, phosphatidylserine externalization, poly(ADP-ribose) polymerase cleavage and activation of caspases (Pan *et al.*, 2009; Roy *et al.*, 2010). Last but not least, hydroxyalkenals are potent activators of apoptosis. They both induce DNA fragmentation and activation of caspases in very different cell types (Choudhary *et al.*, 2002; Vaillancourt *et al.*, 2008). In addition, HHE has been shown to decrease the expression/phosphorylation of Bcl-2, while increasing that of Bax, leading to apoptosis of human renal epithelial cells (Bae *et al.*, 2011; Bodur *et al.*, 2012).

Interestingly, the toxicity of aldehydes is highly correlated to their ability to form covalent adducts on proteins. In muscle cells, the lethal concentration 50 (LC50) for 12 different aldehydes was calculated, including HHE and HNE (Pillon *et al.*, 2010). This LC50 was strongly correlated with their respective potency to form covalent adducts on albumin *in vitro* (Figure 1). This demonstrates that the cytotoxicity and likely other biological effects of aldehydes mainly occur through chemical adduction of other biomolecules.

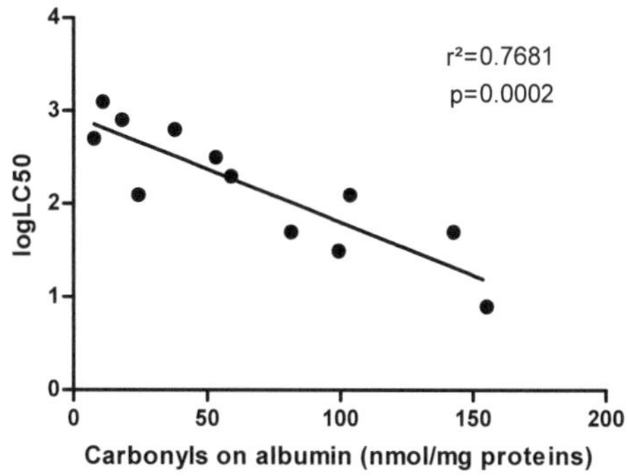

Figure 1. Toxicity is correlated to the adduction ability. Viability of muscle cells was measured in response to 12 different aldehydes[1]. The calculated LC50 was then correlated to their ability to form covalent adducts on bovine serumalbumin (Pillon et al., 2010).

2.2.2. Oxidative stress and ROS production

The classical sequence of events is that oxidative stress triggers lipid peroxidation which in turns produces aldehydes by-products. However, an interesting paradigm was pointed out by several groups: lipid aldehydes are able to induce the production of ROS, and this is thought to be of importance in their deleterious effects. For instance, acrolein treatment produces nitric oxide (Misonou *et al.*, 2006) and induces generation of intracellular oxidants (Luo *et al.*, 2005; Wang *et al.*, 2011). Similarly, accumulation of intracellular ROS was described in cells treated with MDA (Cheng *et al.*, 2011); and unsurprisingly, the 4-hydroxyalkenals HHE and HNE share the same ability. HHE induces ROS in neurons and tubular epithelial cells (Long *et al.*, 2008; Bae *et al.*, 2011) and HNE induces mitochondrial oxidative stress in neurons, vascular muscle, liver and skeletal muscle cells (Uchida *et al.*, 1999; Lee *et al.*, 2006; Pillon *et al.*, 2012). The source of ROS was suggested to be mitochondria, as several aldehydes have been shown to induce a significant decrease in mitochondrial membrane potential (Uchida *et al.*, 1999; Luo *et al.*, 2005).

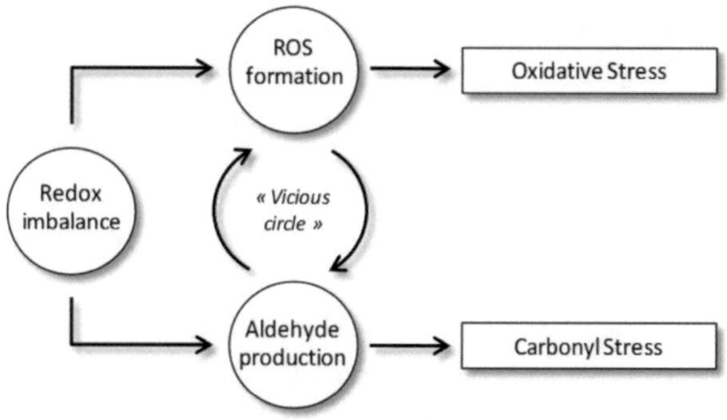

Figure 2. The vicious circle linking ROS and lipid aldehyde production

[1] 4-hydroxy-2-hexenal, 4-hydroxy-2-nonenal, 4-hydroxy-2-dodecenal, 4-hydroxy-2-hexenal dimethylacetal, 4-hydroxy-2-nonenal dimethylacetal, 4-hydroxy-2-dodecenal dimethylacetal, trans-2-hexenal, trans-2-nonenal, trans-2-dodecenal, hexanal, nonanal, dodecanal

This concept is reinforced by the fact that an increase in glutathione pool can prevent the deleterious effects of HNE on both adduct formation and ROS production (Pillon *et al.*, 2012) and that dysfunction of glutathione S-transferase, a major enzyme for aldehyde detoxification, leads to excess 4-hydroxy-2-nonenal and oxidative stress (Kostyuk *et al.*, 2010; Curtis *et al.*, 2010). Based on this body of evidence, lipid by-products can be seen as parts of a vicious circle in which increased ROS production generates aldehydes which further amplify the generation of oxidative species and so on (Figure 2).

2.2.3. Signalling pathways and transcription factors

Consistent with the extensive work done on cytotoxicity, the stress signalling pathways have been primarily pointed out as the main intracellular route activated by aldehydes. The mitogen-activated protein kinases (MAPKs) are indeed activated by several aldehydes including MDA, acrolein and HNE. MDA activates c-Jun N-terminal kinases (JNK) and extracellular signal-regulated kinases (ERK) (Cheng *et al.*, 2011) and acrolein-induced apoptosis occurs through activation of p38 and ERK (Tanel & Averill-Bates, 2007). The activation of these three MAPK has also been described following treatment with HNE (Uchida *et al.*, 1999; Zarrouki *et al.*, 2007; Pillon *et al.*, 2012) and HHE (Je *et al.*, 2004; Bae *et al.*, 2011). Overall, most studies investigated the cytotoxic effects of aldehydes and thus focused on cellular stress pathways; therefore, very little data is available regarding their potential effects on other pathways. Only very recent work shows that HNE interfere with insulin signalling pathway through oxidative stress and adduction of IRS1 and Akt (Demozay *et al.*, 2008; Shearn *et al.*, 2011; Pillon *et al.*, 2012).

Aldehydes regulate gene expression by activating the signalling pathways described above, or by direct modification of transcription factors. Unsurprisingly, aldehyde production regulates the expression of several antioxidant enzymes such as NAD(P)H quinone oxidoreductase-1, Heme oxygenase-1 and glutathione S-transferase (GST). This occurs through the activation of the Nuclear factor (erythroid-derived 2)-like 2 (Nrf2), which drives the expression of these antioxidant enzymes. Aldehyde-induced Heme Oxygenase-1 expression is indeed mediated by the Nrf2 pathway in HUVECs (Lee *et al.*, 2011) and Nrf2 silencing significantly attenuates the induction of this same gene by acrolein (Zhang & Forman, 2008).

Beyond the expression of antioxidant enzymes, cellular growth, apoptosis and inflammatory responses can be induced by aldehydes, involving the activation of the nuclear factor kappa B (NF-κB) family. The effects of acrolein on NF-κB activation are controversial, but its activation by HNE and HHE is well documented. HHE and HNE induce NF-kappaB activation through IKK/NIK pathway, leading to IκB phosphorylation and subsequent proteolysis (Page *et al.*, 1999; Je *et al.*, 2004; Lee *et al.*, 2004). HNE has moreover been shown to induce DNA-binding of NF-κB in vascular smooth muscle cells (Ruef *et al.*, 2001).

The peroxisome proliferator-activated receptor (PPAR) family regulate the expression of genes that encode proteins involved in energy balance. They act as ligand-activated transcription factors and are responsive to the lipid status of the cell, therefore important during high fat diets and obesity. HNE is an intracellular agonist of PPARβ/δ while HHE do not activate this receptor (Coleman *et al.*, 2007). Through this pathway, HNE significantly elevates adiponectin gene expression, concomitant with increased PPAR-γ gene expression and transactivity. Meanwhile, HDDE acts through PPARδ signalling pathways to regulate glucose transport in vascular endothelial cells subjected to hyperglycemia (Riahi *et al.*, 2010) and HNE stimulates insulin secretion from Beta cells through interaction with PPARδ (Cohen *et al.*, 2011).

2.3. Concentration in plasma and tissues

Little data is available concerning aldehydes levels in biological fluids, except for MDA and HNE which have been widely used as lipid peroxidation markers. The concentration of MDA in the plasma of healthy subjects is around 2-5 μM and increases up to 2-fold in type 2 diabetic patients (Figure 4). In the specific context of metabolic diseases, MDA is positively correlated with BMI and waist circumference in obese patients (Furukawa *et al.*, 2004). Depending on the study, HNE concentration has been found to range from 50 nM to 10 μM under normal conditions. This significant variability in concentration according to the authors could be explained partly by the method used (LC/MS, GC/MS ...) and also by the difficulty to measure such reactive derivatives. If HNE has been widely studied, there is however scarce data in the literature regarding the pathophysiological concentrations of HHE and HDDE. Plasma HHE concentration was however found to be around 9 nM in human, and dramatically increases to reach 90 nM after several weeks of a diet rich in omega-3 fatty acids (Calzada *et al.*, 2010). Our group recently showed that HHE concentration was 20 nM in humans and 7 nM in rats and that it increases in both type-2 diabetes patients and type-1 diabetic rats, reinforcing existing evidence for a role of lipid aldehydes in metabolic diseases.

Tissue	Aldehyde	Concentration (μM)	References
Plasma	MDA	1 – 6	See figure 4
	HHE	0.006 – 0.090	(Calzada et al., 2010; Pillon et al, unpublished)
	HNE	0.007 – 11	(McGrath *et al.*, 2001; Selley, 2004; Syslova *et al.*, 2009)
Exhaled breath condensate	HNE	0.25 - 5	(Syslova *et al.*, 2009; Manini *et al.*, 2010)
	MDA	0.63 - 14	(Syslova *et al.*, 2009; Manini *et al.*, 2010)
Ventricular Fluid	HNE	0.2 - 120	(Lovell *et al.*, 1997)
Pancreatic Islets	HNE	23 – 35	(Miwa *et al.*, 2000)

Table 3. Concentration range of aldehydes in healthy plasma and tissues

One should keep in mind that all current quantification methods (HPLC, GC) only assay the free fraction (*i.e.* unreacted) of lipid aldehydes present in samples. Being very reactive, this is certainly not representative of the amount indeed produced from lipid peroxidation, which rapidly react with neighbour targets to form covalent adducts and thus, other non-quantitative methods estimating the amount of HNE have been used. For example, detection of protein adducts by immunohistochemistry has shown a significant increase in HHE and HNE proteins adducts in Parkinson's disease and in chronic liver disease (Yoritaka *et al.*, 1996; Paradis *et al.*, 1997). Of particular interest for this chapter, HNE adducts on plasma albumin are increased in type 2 diabetes (Toyokuni *et al.*, 2000).

3. Insulin secretion and type-1 diabetes

3.1. Lipid aldehydes and insulin secretion

Oxidative stress associated with hyperglycemia is suspected to participate in beta cell dysfunction in terms of insulin synthesis and/or secretion. Diabetic animals display increased levels of lipoperoxidation in pancreas, and HNE concentration was reported to reach up to 35 μM in pancreatic islets of diabetic rats (Miwa *et al.*, 2000). In addition, HNE-modified proteins are increased in the pancreatic beta-cells of Goto-Kakizaki rats, a genetic model of non obese type 2 diabetes (Ihara *et al.*, 1999), as well as in type 2 diabetic patients (Sakuraba *et al.*, 2002). HNE and other lipid peroxidation by-products such as 2-hexenal and 2-butenal inhibit glucose-induced insulin secretion in isolated rat islets. Both glucose utilization and glucose oxidation are blunted in islets after treatment with aldehydes suggesting that they impair glucose-induced insulin secretion through an interference with glycolytic pathway and citric acid cycle (Miwa *et al.*, 2000). Another piece of evidence comes from the exposure of beta cells to NO donors and to interleukin-1 beta, which leads to generation of oxidative stress and lipoperoxidation by-products. MDA and HNE produced under this condition are involved in the activation of an apoptotic program, contributing to the reduction in the beta cell mass (Cahuana *et al.*, 2003). Peroxynitrite indeed triggers lipoperoxidation in the beta-cell line RIN-5-F, and the resulting protein carbonylation is a key factor linking NO-dependent lipoperoxidation and apoptosis (Cahuana *et al.*, 2003). Alloxan, a toxic glucose analogue, has been widely used to generate rodent models of type-1 diabetes, as it selectively destroys insulin-producing cells in the pancreas. Alloxan-induced diabetic rats exhibit increased lipid peroxidation associated with defects in insulin secretion, which can be prevented by the antioxidant S-allyl cysteine therefore restoring insulin secretion and ameliorating the glycaemic control (Augusti & Sheela, 1996).

3.2. Lipid aldehydes and the beta cells: Doctor Jeckyll or Mr Hyde?

HNE was shown to elicit various physiological or physiopathological responses: high concentrations elicit beta cell death and defect in insulin secretion, while lower concentrations act as signalling mediators. In INS-1E beta-cells, elevated glucose levels increase the release of arachidonic acid and linoleic acid from membrane phospholipids and promote their peroxidation to HNE. At non cytotoxic concentrations, HNE behaves as an endogenous ligand for nuclear receptor PPAR-δ, stimulating insulin secretion in beta-cells (Cohen *et al.*, 2011). In rat islet beta-cell-derived RINm5F cells, a recent report demonstrates the involvement of the transient receptor potential (TRP) cation channels in the HNE-induced insulin secretion. Short-term (1h) exposure to HNE induces a transient increase in intracellular calcium concentration and triggers insulin secretion. HNE induces calcium influx through activation of TRP channels (amongst which TRPA1) which appears to be coupled with the L-type voltage-dependent calcium channel, and ultimately insulin secretion (Numazawa *et al.*, 2012). Lipid aldehydes should therefore be considered either as detrimental (>10 μM) or as beneficial (sub micromolar range) depending on their actual tissue concentration.

3.3. Direct adduction of the insulin polypeptide

Under conditions of oxidative stress, insulin, a polypeptidic hormone composed of 51 amino acid residues, is exposed to direct oxidative insult or to modification by lipoperoxidation by-products. Several amino acids are putative sites of adduction, and thus, covalent binding of lipid aldehydes affect the biological actions of this hormone. This applies to acrolein and methylglyoxal, whose fixation on insulin has been shown to reduce both hypoglycemic effects in rats and glucose uptake in 3T3-L1 adipocytes (Jia *et al.*, 2006; Medina-Navarro *et al.*, 2007). HHE and HNE, toxic aldehydes generated during lipid peroxidation, also modify the B-chain of human insulin *in vitro*, predominantly at the His B5 and His B10 residues via Michael adduction (Figure 3). Adduct formation affects the biological activity of insulin *in vivo*, decreasing its hypoglycemic effect in mice and stimulation of glucose uptake in adipose and muscle cells (Pillon *et al.*, 2011).

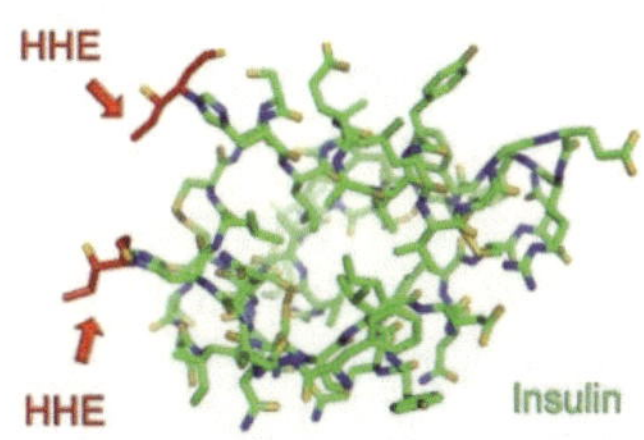

Figure 3. Structure of insulin monomer displaying two HHE adducts on histidine residues. From Pillon et al, 2011.

4. Obesity, insulin resistance and type-2 diabetes

4.1. Lipid peroxidation by-products association with obesity

Obesity is a major factor in the development of metabolic syndrome. After consumption of an energy-dense (i.e. high-fat) diet, plasma HNE levels increase rapidly and significantly within minutes (Devaraj *et al.*, 2008). When consumed regularly, this diet promotes obesity, which suggests a role for HNE very early in the development of obesity. On the other hand, levels of circulating HNE tend to decrease when obese people are maintained on calorie restriction (Johnson *et al.*, 2007), demonstrating that lipid peroxidation is tightly linked to high fat diet and obesity. Furukawa et al. reported that increased oxidative stress in accumulated fat is an important pathogenic mechanism of obesity-associated metabolic syndrome. Production of ROS is indeed selectively increased in white adipose tissue of obese mice and associated with a blunted expression of antioxidant enzymes. In good agreement, fat accumulation correlates with systemic lipid peroxidation in humans (Furukawa *et al.*, 2004), and the plasma concentration of MDA is 1.8 fold higher in subjects with a BMI above 40 kg/m^2 compared to lean individuals (Olusi, 2002). Diet-induced obesity increases tissue and plasma accumulation of ALEs (protein–acrolein and protein–HNE adducts for example), suggesting that obesity is associated with an increase in the formation of lipid peroxidation-derived aldehydes (Baba *et al.*, 2011). A significant accumulation of HNE was noticed in the white adipose tissue of obese mice, where the adipocyte fatty acid binding protein (AFABP also known as aP2) is the soluble protein most highly modified by HNE in this tissue (Grimsrud *et al.*, 2007). In obese mice roughly 7% of the AFABP in adipose tissue is covalently modified by HNE resulting in a decreased binding affinity for fatty acids. Lipid peroxidation is however not restricted to adipose tissue since HNE is also elevated in skeletal muscles of Otsuka Long-Evans Tokushima fatty (OLETF) rat, a model for hyperphagic obesity (Morris *et al.*, 2008). Intracellular triglyacylglycerols accumulate in the muscle of obese humans where it is considered as a pathogenic factor in the development of insulin resistance. In obese compared to endurance-trained subjects, the lipid peroxidation to intracellular triacylglycerols ratio was 4-fold higher suggesting that obesity is associated with increased muscle lipid peroxidation (Russell *et al.*, 2003; Vincent *et al.*, 2006).

Chronic, low grade inflammation of white adipose tissue is a hallmark of obesity and a major contributor to oxidative stress and lipid peroxidation (Wellen & Hotamisligil, 2003). In the expanding adipose tissue, hypertrophied adipocytes contribute to the inflammation by up-regulating the expression and release of pro-inflammatory cytokines. In 3T3-L1 adipocytes, HNE can dose-dependently increase the expression of the inducible cyclooxygenase (COX-2) (Zarrouki *et al.*, 2007) and that of the plasminogen activator inhibitor-1 (PAI-1). In the meantime, HNE decreases the expression of the anti-inflammatory, insulin-sensitizing hormone adiponectin (Soares *et al.*, 2005; Wang *et al.*, 2012), therefore linking lipid peroxidation by-products and chronic inflammation.

4.2. Lipid peroxidation by-products association with type-2 diabetes

Epidemiological studies demonstrates that fasting glycemia is positively correlated with oxidative stress markers such as 8-epi-PGF2α and TBARs and negatively correlated with plasma glutathione (Trevisan *et al.*, 2001; Menon *et al.*, 2004). In type-2 diabetic individuals, 8-epi-PGF2α is positively correlated to the HOMA index for insulin resistance (Gopaul *et al.*, 2001) and urinary acrolein correlates with glycated haemoglobin HbA1c (Daimon *et al.*, 2003). This was confirmed in animal models of insulin resistance which exhibit increased markers of oxidative stress, such as plasma F2-isoprostanes (Laight *et al.*, 1999*a*). In parallel, antioxidant defences are reduced during an oral glucose tolerance test in normal and non-insulin-dependent diabetic subjects (Ceriello *et al.*, 1998), and diabetes is associated with decreased vitamin C and glutathione (Maxwell *et al.*, 1997; Dierckx *et al.*, 2003). Reciprocally, an intensive treatment of diabetes improves circulating levels of H_2O_2 and MDA (Wierusz-Wysocka *et al.*, 1995); and improved insulin sensitivity resulting from exercise and/or dietary restriction is associated with reduced levels of lipid peroxidation products (Reviewed by Vincent *et al.*, 2007). On the other hand, insulin sensitivity can be improved through antioxidant or carbonyl scavenging treatment (Kamenova, 2006; Vincent *et al.*, 2009), demonstrating the tight link existing between oxidative stress, oxidation by-products and insulin resistance.

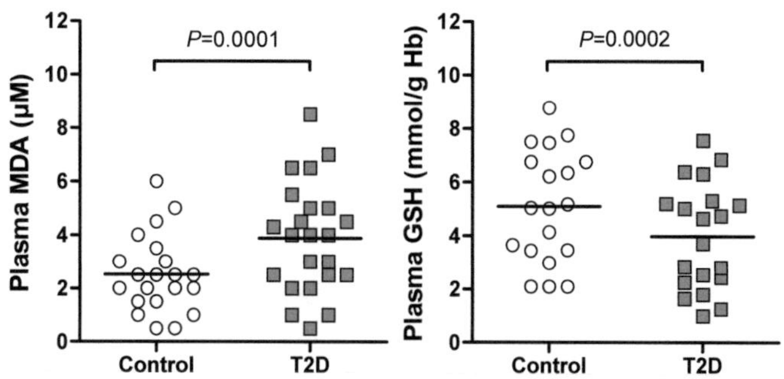

Figure 4. Plasma MDA concentration (µM) and blood GSH content (mmol/g haemoglobin) in healthy and type-2 diabetic (T2D) individuals. Results are a meta-analysis from 23 different publications; one dot represents the mean value obtained in one study[2]. In average, MDA is significantly increased by 60% while GSH is decreased by 25% in T2D compared to healthy subjects (paired student t-test, n=21).

Among α,β-unsaturated aldehydes, only MDA and to some extend HNE have been studied as oxidative stress biomarkers in diabetes, and they are indeed both increased up to 2-fold in both human (Figure 4) and animal models of type-2 diabetes (Wierusz-Wysocka *et al.*, 1995; Dierckx *et al.*, 2003). It has moreover been demonstrated that Type-2 diabetes duration is independently associated with increased levels of lipid peroxidation (Nakhjavani *et al.*, 2010), and our group recently showed an increase in HHE concentration in type-2 diabetes patients, reinforcing existing evidence for the specific role of lipid aldehydes in insulin resistance.

---

[2] Results from Wierusz-Wysocka et al., 1995; Vijayalingam et al., 1996; Feillet-Coudray et al., 1999; Rábago-Velasco et al., 2000; Rizvi & Zaid, 2001; Seghrouchni et al., 2002; Dinçer et al., 2002; Dierckx et al., 2003; Duman et al., 2003; Pasaoglu et al., 2004; Memişoğullari & Bakan, 2004; Skrha et al., 2005; Kurtul et al., 2005; Ozdemir et al., 2005; Mahboob et al., 2005; Saxena et al., 2005; Sampathkumar et al., 2005; Kuppusamy et al., 2005; Lapolla et al., 2007; Sathiyapriya et al., 2007; Singhania et al., 2008; Jain et al., 2009; Nakhjavani et al., 2010; Narasimhan et al., 2010; Shinde et al., 2011; Huang et al., 2011; Bahadoran et al., 2011; Zhang et al., 2011; Pácal et al., 2011; Calabrese et al., 2011 and Rasic-Milutinovic et al., 2012

Aldehydes concentration is increased in several tissues during diabetes, as revealed by increased levels of HNE in pancreas, liver, brain and heart. Pancreatic islets from type-2 diabetic patients are positively stained with HNE, suggesting that oxidative stress through lipid peroxidation could contribute to the reduced beta-cell mass and islet-cell injury (Sakuraba *et al.*, 2002). An increased level of HNE-modified proteins is reported in the pancreatic beta cells of Goto Kakizaki rats as a result of hyperglycaemia (Ihara *et al.*, 1999); and accumulation of HNE is observed in liver of diabetic rats due to the impairment of HNE-metabolizing enzymes (Traverso *et al.*, 1998, 2002). Diabetic mice under a high fat diet exhibit increased HNE adducts levels in temporal lobes relative to control (Lyn-Cook *et al.*, 2009), and HNE conjugation of GLUT3, the glucose transporter present in neurons, is increased in the hippocampus of diabetic rats subjected to stress (Reagan *et al.*, 2000). When db/db obese mice are fed a Western diet containing 21% fat and 0.15% cholesterol, they develop obesity, hyperglycemia, and insulin resistance. In this situation, HNE is significantly elevated in the left ventricular myocardium of diabetic mice compared to their lean littermates (Yamashita *et al.*, 2010). Finally, hyperglycemic Zucker Diabetic Fatty (ZDF) rats exhibited a 8-fold increase in plasma HDDE concentration compared to their lean non diabetic counterparts (Riahi *et al.*, 2010).

4.3. Causative role for lipid peroxidation by-products in the metabolic syndrome

This body of evidence linking oxidative stress with the metabolic syndrome is however only based on correlations and do not decipher the mechanisms and/or the causative role of oxidative stress in diabetes. There are consequently two main hypotheses:

1. First is chronic hyperglycemia which then leads to oxidative stress. In this case, lipid peroxidation products would be by-products of this oxidative stress and a consequence of diabetes, even if contributing to the progression of the disease and its complications.
2. First is oxidative stress which precedes the development of diabetes and plays a causative role in its development. In this hypothesis, deregulation of antioxidant defences and increased oxidative stress would lead to accumulation of secondary by-products consequently inducing insulin resistance.

We will focus in this chapter on the second hypothesis for which evidences have been recently accumulating from human, animal and cell culture studies.

Oxidative stress, through reactive oxygen species can positively and negatively regulate insulin signalling, depending on time, dose, model and free radical used (for review, see Bashan *et al.*, 2009). It is however admitted that prolonged oxidative stress impairs insulin signalling, insulin-induced GLUT4 translocation and glucose uptake. This occurs through several mechanisms including, but not limited to IRS inhibitory phosphorylation, MAPK activation and endoplasmic reticulum stress. This body of evidence for the role of oxidative stress in insulin resistance has been demonstrated in adipocytes, muscle, liver and cardiac cells (Rudich *et al.*, 1998; Bloch-Damti *et al.*, 2006; JeBailey *et al.*, 2007; Singh *et al.*, 2008*a*; Shibata *et al.*, 2010; Tan *et al.*, 2011). Even *in vivo*, a pro-oxidant challenge provokes the onset of type-2 diabetes in insulin resistant rats (Laight *et al.*, 1999*b*) and chronic methylglyoxal infusion by minipump causes pancreatic beta-cell dysfunction and induces type-2 diabetes in Sprague-Dawley rats (Dhar *et al.*, 2011).

The role of lipid aldehydes *in vivo* during metabolic diseases is often indirectly assessed, and the most compelling evidence for a causative role of aldehydes comes from polymorphisms in the glutathione-S-transferase (GST) gene. This family of enzymes is responsible for the detoxification of aldehydes through conjugation to glutathione; and several deletion polymorphisms leading to blunted enzyme activity have been described. Patients carrying certain null GST polymorphisms had up to 3-fold increased incidence of type-2 diabetes mellitus compared to those with normal genotypes (Amer *et al.*, 2011). Accordingly, the expression of the GST4A is blunted in the adipose tissue of obese insulin resistant subjects (Curtis *et al.*, 2010).

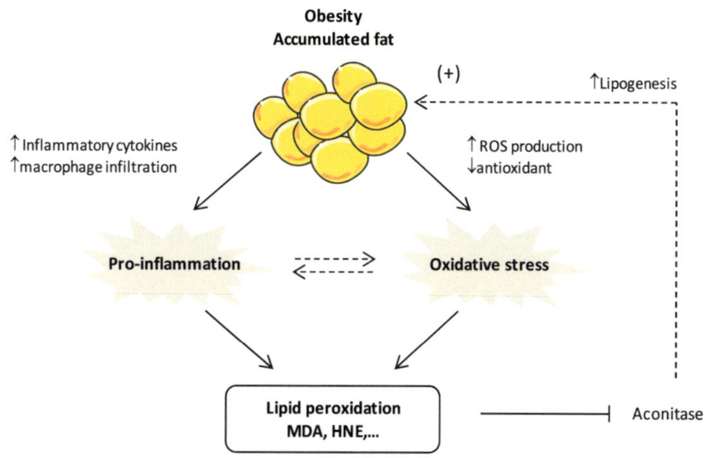

Figure 5. The obesity – lipid peroxidation vicious circle

This same observation was made in animals model, were it has been suggested that excessive production of HNE might be sufficient to cause obesity and the metabolic syndrome. Mice lacking the gene encoding the HNE-conjugating enzyme mGSTA4-4 develop obesity and insulin resistance, unsurprisingly associated with HNE accumulation in multiple tissues (Singh *et al.*, 2008*b*). Cell culture work further shows that dysfunction of glutathione S-transferase or its ablation by siRNA leads to excess HNE accumulation, increased protein carbonylation, oxidative stress, and mitochondrial dysfunction (Kostyuk *et al.*, 2010; Curtis *et al.*, 2010). On the other hand, overexpression of glutathione-S-transferase A4-4 protects against oxidative stress and HNE-induced apoptosis (Vaillancourt *et al.*, 2008). Additional findings suggest that HNE causes fat accumulation by promoting fatty acid synthesis and suppressing fatty acid beta-oxidation. Interestingly, the phenotype of mGSTA4-4 null mice is strain dependent: mGSTA4-4 null mice with 129/sv genetic background exhibit both increased accumulation of HNE and obesity while those with C57Bl6 genetic background are lean and HNE is unchanged (Singh *et al.*, 2008*b*). In good agreement, silencing of the mGSTA4-4 gene in the nematode Caenorhabditis elegans also results in an accumulation of lipid peroxidation by-products and a fatty phenotype (Singh *et* *al.*, 2009). When HNE is experimentally increased in the nematode, either by genetic deletion or through direct exposure, it promotes fat accumulation. The mechanism involves HNE inhibition of mitochondrial aconitase leading to an accumulation of malonyl CoA, precursor of fatty acid synthesis (Zimniak, 2010). Taken together, these data suggest that accumulation of lipid aldehydes and fat deposition could be mutually inductive leading to a vicious circle promoting fat accretion (Figure 5).

The direct causative role of lipid peroxidation by-products in insulin resistance has also been recently evidenced in cell cultures studies. A first study identified that methylglyoxal, an aldehyde by-product of glucose oxidation, can impair the insulin signalling pathways independently of the formation of intracellular reactive oxygen species (Riboulet-Chavey *et al.*, 2006). Then, focusing on lipid peroxidation products, two studies demonstrated that HNE can induce insulin resistance in adipocytes and muscle cell through inhibition of IRS and Akt signalling, as well as insulin-induced glucose uptake (Demozay *et al.*, 2008; Pillon *et al.*, 2012). These two studies identified carbonyl stress (notably IRS1 adduction) and ROS production as the possible mechanisms responsible for HNE effects; a third one being possibly the adduction of Akt2, which inhibits insulin-dependent Akt signalling in HepG2 cells (Shearn *et al.*, 2011). Altogether, these studies strongly suggest that excessive production of aldehydes might be sufficient to cause obesity, diabetes and the metabolic syndrome.

5. Preventing the deleterious effects of aldehydes

It has been known for decades that supplementation in α-lipoic acid in type-2 diabetic subject improves glucose tolerance and insulin sensitivity (Jacob et al., 1996). Similarly, increased intake of Vitamin E in obese and insulin resistant patients reduces fasting glycemia and the HOMA index for insulin resistance, and this is correlated with a decrease in the concentration of peroxides in plasma (Manning et al., 2004). This adds further evidence for a causative role of oxidative stress in the metabolic syndrome, and opens new therapeutic perspectives using antioxidant and/or scavenging of toxic aldehydes.

5.1. Glutathione and enzymatic detoxification

Glutathione (GSH) is a tripeptide (L-γ-glutamyl-L-cysteinyl-glycine) which is present in high concentration in the cytoplasm of living cells: 5-50 nmol/mg proteins (Jungas *et al.*, 2002; Dominy *et al.*, 2007). GSH is an important coenzyme of several enzymatic reactions, and exerts its antioxidant activity mainly through regeneration of vitamin E, and also through direct interaction with free radical and aldehydes. GSH is obviously necessary for the activity of glutathione-S-transferases (GSTs), a family of enzymes responsible for detoxification of electrophile by-products, such as the ones derived from lipid peroxidation, and GSH exhibits indeed a high reactivity for HNE. Though HNE is electrophile enough to spontaneously react with GSH, this reaction is dramatically accelerated via the conjugation process catalyzed by GSTs (Alin *et al.*, 1985). As described above, mice lacking GSTs develop obesity and insulin resistance (Singh *et al.*, 2008*b*), and humans carrying null polymorphisms for GST have a 3-fold increased risk of having type-2 diabetes (Amer *et al.*, 2011), pointing out the important role of GSH in the metabolic syndrome. As addition reaction to GSH contributes to the detoxification of aldehydes, pharmacological strategies to increase glutathione pools or GST activity should be protective against aldehydes. Several studies were indeed successful in protecting cultured cells from the deleterious effects of oxidative stress through an increase in intracellular pools of reduced glutathione. This strategy was particularly efficient to protect the cells against the deleterious effects of HNE (Yadav *et al.*, 2008; Jia *et al.*, 2009*b*; Pillon *et al.*, 2012).

In addition to glutathione-S-transferases, several enzymes are also responsible for the detoxification of aldehydes: aldehyde dehydrogenase, alcohol dehydrogenase (Hartley *et al.*, 1995), aldose reductase (Srivastava *et al.*, 2000) and fatty aldehyde dehydrogenase (Demozay *et al.*, 2008). These enzymes however participate to a lower extent in the metabolism of aldehydes, compared to glutathione and GST. Recent data indicate that if HHE and HNE are both metabolized via glutathione, the effectiveness of detoxification differs for these two molecules. By extension, aldehydes may be metabolized with different affinities and efficiencies by detoxification enzymes, which could explain some differences in their respective toxicities (Long *et al.*, 2010).

5.2. Scavenging

In chemistry, a scavenger is a chemical substance able to remove or inactivate impurities or unwanted reaction products. In living cells, a scavenger is, by extension, a molecule able to inactivate toxic compounds such as ROS and aldehydes, therefore preventing their deleterious effects. In the case of aldehydes, a scavenger would be a strong nucleophile molecule on which HHE, HNE or any other aldehyde would form a covalent adduct. Consequently, most of the aldehyde scavengers are amino- or sulphur-containing drugs such as N-acetyl-cysteine (NAC), hydralazine, S-adenosyl-methionine (SAM) aminoguanidine (AGD) and α-lipoic acid, the latter being tested in several human studies for treatment of type-2 diabetes.

5.2.1. α-lipoic acid

α-lipoic acid is a natural compound found in many foodstuffs (such as potatoes, broccoli and meat), but in rather low amount. The effect of lipoic acid was demonstrated in animal models, where it enhances insulin-stimulated glucose metabolism in skeletal muscle from insulin-resistant rat (Jacob et al., 1996). It was rapidly tested in several clinical studies which demonstrated its beneficial effects in type-2 diabetes through a decrease in fasted blood glucose, enhancement of glucose disposal, improved insulin sensitivity and decreased insulin resistance (Jacob *et al.*, 1995). These results were confirmed *in vitro*, where α-lipoic acid prevents the development of glucose-induced insulin resistance in adipocytes (Greene et al., 2001) and were reproduced by many independent studies. Treatment with lipoic acid decreases oxidative stress in both adipocytes and muscle cells (Rudich et al., 1999; Maddux et al., 2001) and also decreases lipid peroxidation markers in insulin resistant rats (Thirunavukkarasu & Anuradha, 2004), suggesting that the improvement of insulin sensitivity is due to its antioxidant properties. Hence, the current literature supports the use of alpha lipoic acid for the treatment of diabetes complications and it consequently became the first antioxidant supplement used for the treatment of diabetes complications, being already approved in Germany for the treatment of diabetic neuropathy.

5.2.2. N-acetylcysteine (NAC)

N-acetylcysteine is a cysteine derivative and a potent antioxidant. Its properties are mainly due to its thiol group able to reduce free radicals as well as its role as a precursor in the formation of glutathione (Zafarullah *et al.*, 2003). NAC exhibits highly protective scavenging properties against aldehydes and protects against MDA increase and GSH decrease in animal models of insulin resistance. NAC is able to improve insulin sensitivity in healthy rats (Figure 6) and reverses insulin resistance and aldehyde-induced hypertension in rats (Haber *et al.*, 2003). In cell culture studies, NAC can prevent the insulin resistance induced by HNE in muscle cells (Pillon *et al.*, 2012), as well as the one induced by advanced glycation end products in adipocytes (Unoki *et al.*, 2007), thus confirming the important role NAC can play in improving both oxidative stress parameters and insulin resistance.

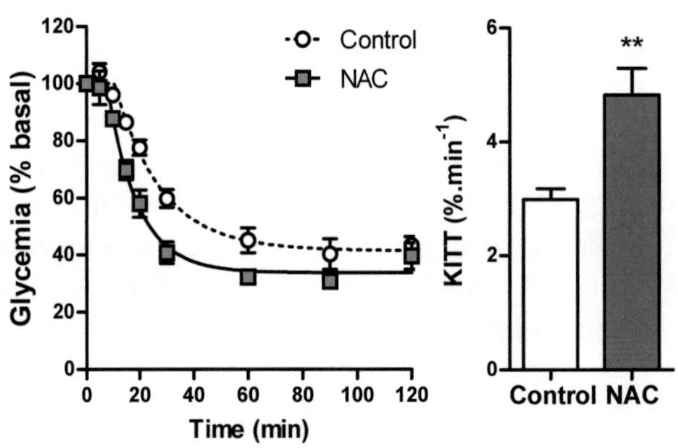

Figure 6. Insulin sensitizing effect of N-acetylcysteine. Wistar rats were given NAC in drinking water for one week (total intake was 225 $mg.kg^{-1}.day^{-1}$). Insulin sensitivity was calculated using a standard insulin tolerance test[3]. Results are average ± SEM from 5 different animals per group, expressed as percent of basal glycemia. From Pillon et al. unpublished results.

---

[3] Animals fasted overnight were then injected intraperitoneally with 0.5 UI/kg body weight of insulin. Plasma glucose was measured from tail vein blood using a glucometer at 0, 20, 40, 60, and 120 min following the injection. Glucose disappearance rate for ITT (K_{ITT}; %/min) was calculated as: $K_{ITT}=(0.693x100)/t_{1/2}$, where $t_{1/2}$ was calculated from the slope of the plasma glucose concentration, considering an exponential decrement of glucose concentration during the 20 min after insulin administration. Higher insulin sensitivity index (K_{ITT}) scores mean higher response of tissues to insulin.

Results from clinical trials in type-2 diabetic subjects show that NAC is able to decrease oxidative stress parameters, increase GSH and decrease plasma VCAM-1 (De Mattia et al., 1998); moreover, long-term N-acetylcysteine administration reduces endothelial activation and is proposed as a potential antiatherogenic therapy (Martina et al., 2008). Despite the fact that NAC can improve insulin sensitivity in women with polycystic ovary syndrome (Fulghesu *et al.*, 2002) and this body of evidence suggesting that NAC may slow down the progression of diabetic complications, to date no clinical trial demonstrated any significant benefit of its supplementation in diabetes.

5.2.3. Aminoguanidine (AGD)

Aminoguanidine is a highly nucleophilic agent which reacts *in vitro* and *in vivo* with aldehydes, therefore protecting against the deleterious effects of ALE precursors (Peyroux & Sternberg, 2006). It is also an antioxidant able to quench hydroxyl radicals and *in vivo*. AGD in drinking water decreases lipid peroxidation in type-1 diabetic rats and rabbits (Ihm *et al.*, 1999). In experimental animal models of diabetes, AGD demonstrates significant effects in protecting against pathological complications, such as diabetic nephropathies, atherosclerosis and neurovascular complications (El Shazly *et al.*, 2009). Consequently, several clinical trials in humans have been designed to evaluate AGD efficiency but they demonstrate only mild effects and were not conclusive, partly because of side-effects, and of weak carbonyl scavenger effects in human vascular tissues (Bolton *et al.*, 2004).

5.2.4. Hydralazine

Primarily used as an antihypertensive drug, hydralazine exhibits a pronounced nucleophilicity and is consequently very efficient in scavenging several aldehydes (acrolein, HNE) and ketones, as well as aldehyde-adducted proteins (Burcham et al., 2002). It is also a powerful antioxidant, able to inhibit the generation of ROS (Münzel *et al.*, 1996). Its scavenging activity *in vivo* was demonstrated by its ability to reverse the formation of HNE and acrolein adducts on tissue proteins in atherosclerotic aortas of hypercholesterolemic animals (Vindis et al., 2006), but its effects on insulin resistance and diabetes are to date uncharacterized.

5.2.5. S-adenosyl-methionine (SAM)

In living organisms, SAM is endogenously synthesized from methionine in every cell, but the liver is the major site of its synthesis and degradation. SAM is an important precursor for cysteine and glutathione production (Lu, 2000) and its involvement in several metabolic pathways makes it essential for a wide spectrum of cellular processes. SAM inhibits both HNE production and adducts formation and efficiently prevents high-fat diet-induced non-alcoholic steatohepatitis in rats (Lieber et al., 2007), even if its direct effects on insulin resistance are still unknown.

6. Conclusion

Lipid peroxidation by-products are associated with metabolic diseases, but their primary role during obesity and diabetes is subject to debate. *In vitro* and *in vivo* animal studies highlighted that an aldehyde challenge, through a state of carbonyl stress, affects several steps involved in the development of obesity and type-2 diabetes (Figure 7). On insulin-sensitive tissues (muscle, adipose tissue), aldehydes lead to insulin resistance, while in pancreas, aldehydes impair insulin secretion. Together with the carbonylation of the insulin peptide itself, aldehydes could contribute to the defects in insulin action, leading to the metabolic syndrome.

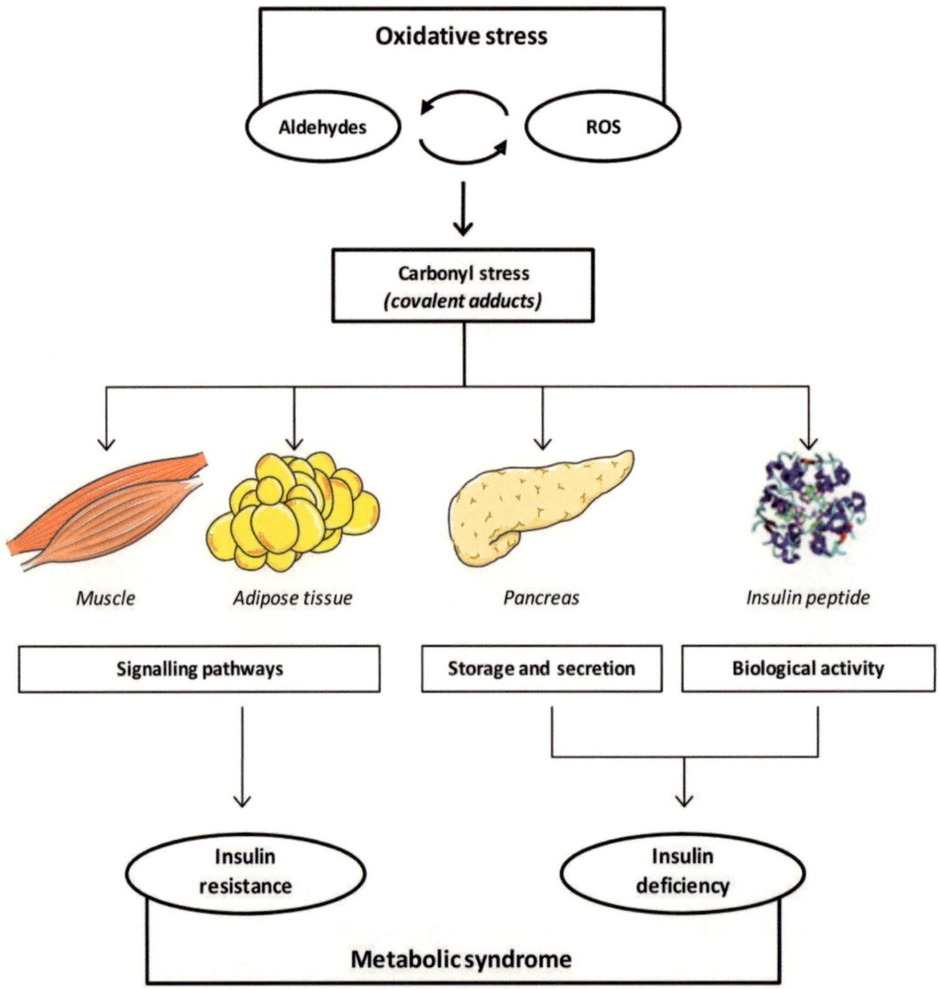

Figure 7. Lipid aldehydes effects leading to the metabolic syndrome.

Despite this evidence, the role of aldehydes is mitigated by the mild effect obtained with antioxidants and/or aldehyde scavengers in the treatment of diabetic complications in human. "Oxidative stress" and its derivatives are nevertheless important in the metabolic syndrome, and prevention or treatment of some of its associated complications could be accessed through reduction of both ROS and toxic aldehydes by-products, as it is the case with lipoic acid.

Author details

Nicolas J. Pillon and Christophe O. Soulage
Université de Lyon, INSA de Lyon, CarMeN, INSERM U1060, Univ Lyon-1, F-69621, Villeurbanne, France

7. References

Alin P, Danielson UH & Mannervik B (1985). 4-Hydroxyalk-2-enals are substrates for glutathione transferase. *FEBS Lett* 179, 267–270.

Amer MA, Ghattas MH, Abo-Elmatty DM & Abou-El-Ela SH (2011). Influence of glutathione S-transferase polymorphisms on type-2 diabetes mellitus risk. *Genet Mol Res*; DOI: 10.4238/2011.October.31.14.

Augusti KT & Sheela CG (1996). Antiperoxide effect of S-allyl cysteine sulfoxide, an insulin secretagogue, in diabetic rats. *Experientia* 52, 115–120.

Baba SP, Hellmann J, Srivastava S & Bhatnagar A (2011). Aldose reductase (AKR1B3) regulates the accumulation of advanced glycosylation end products (AGEs) and the expression of AGE receptor (RAGE). *Chem Biol Interact* 191, 357–363.

Bae EH, Cho S, Joo SY, Ma SK, Kim SH, Lee J & Kim SW (2011). 4-Hydroxy-2-hexenal-induced apoptosis in human renal proximal tubular epithelial cells. *Nephrol Dial Transplant* 26, 3866–3873.

Bashan N, Kovsan J, Kachko I, Ovadia H & Rudich A (2009). Positive and negative regulation of insulin signaling by reactive oxygen and nitrogen species. *Physiol Rev* 89, 27–71.

Bloch-Damti A, Potashnik R, Gual P, Le Marchand-Brustel Y, Tanti JF, Rudich A & Bashan N (2006). Differential effects of IRS1 phosphorylated on Ser307 or Ser632 in the induction of insulin resistance by oxidative stress. *Diabetologia* 49, 2463–2473.

Bodur C, Kutuk O, Tezil T & Basaga H (2012). Inactivation of Bcl-2 through IκB kinase (IKK)-dependent phosphorylation mediates apoptosis upon exposure to 4-hydroxynonenal (HNE). *J Cell Physiol*; DOI: 10.1002/jcp.24057.

Bolton WK, Cattran DC, Williams ME, Adler SG, Appel GB, Cartwright K, Foiles PG, Freedman BI, Raskin P, Ratner RE, Spinowitz BS, Whittier FC & Wuerth J-P (2004). Randomized trial of an inhibitor of formation of advanced glycation end products in diabetic nephropathy. *Am J Nephrol* 24, 32–40.

Burcham PC, Kaminskas LM, Fontaine FR, Petersen DR & Pyke SM (2002). Aldehyde-sequestering drugs: tools for studying protein damage by lipid peroxidation products. *Toxicology* 181-182, 229–236.

Cahuana GM, Tejedo JR, Jimenez J, Ramirez R, Sobrino F & Bedoya FJ (2003). Involvement of advanced lipooxidation end products (ALEs) and protein oxidation in the apoptotic actions of nitric oxide in insulin secreting RINm5F cells. *Biochem Pharmacol* 66, 1963–1971.

Calzada C, Colas R, Guillot N, Guichardant M, Laville M, Véricel E & Lagarde M (2010). Subgram daily supplementation with docosahexaenoic acid protects low-density lipoproteins from oxidation in healthy men. *Atherosclerosis* 208, 467–472.

Ceriello A, Bortolotti N, Crescentini A, Motz E, Lizzio S, Russo A, Ezsol Z, Tonutti L & Taboga C (1998). Antioxidant defences are reduced during the oral glucose tolerance test in normal and non-insulin-dependent diabetic subjects. *Eur J Clin Invest* 28, 329–333.

Cheng J, Wang F, Yu D-F, Wu P-F & Chen J-G (2011). The cytotoxic mechanism of malondialdehyde and protective effect of carnosine via protein cross-linking/mitochondrial dysfunction/reactive oxygen species/MAPK pathway in neurons. *Eur J Pharmacol* 650, 184–194.

Choudhary S, Zhang W, Zhou F, Campbell GA, Chan LL, Thompson EB & Ansari NH (2002). Cellular lipid peroxidation end-products induce apoptosis in human lens epithelial cells. *Free Radic Biol Med* 32, 360–369.

Cohen G, Riahi Y, Shamni O, Guichardant M, Chatgilialoglu C, Ferreri C, Kaiser N & Sasson S (2011). Role of lipid peroxidation and PPAR-δ in amplifying glucose-stimulated insulin secretion. *Diabetes* 60, 2830–2842.

Coleman JD, Prabhu KS, Thompson JT, Reddy PS, Peters JM, Peterson BR, Reddy CC & Vanden Heuvel JP (2007). The oxidative stress mediator 4-hydroxynonenal is an intracellular agonist of the nuclear receptor peroxisome proliferator-activated receptor-beta/delta (PPARbeta/delta). *Free Radic Biol Med* 42, 1155–1164.

Curtis JM, Grimsrud PA, Wright WS, Xu X, Foncea RE, Graham DW, Brestoff JR, Wiczer BM, Ilkayeva O, Cianflone K, Muoio DE, Arriaga EA & Bernlohr DA (2010). Downregulation of adipose glutathione S-transferase A4 leads to increased protein carbonylation, oxidative stress, and mitochondrial dysfunction. *Diabetes* 59, 1132–1142.

Daimon M, Sugiyama K, Kameda W, Saitoh T, Oizumi T, Hirata A, Yamaguchi H, Ohnuma H, Igarashi M & Kato T (2003). Increased urinary levels of pentosidine, pyrraline and acrolein adduct in type 2 diabetes. *Endocr J* 50, 61–67.

Demozay D, Mas J-C, Rocchi S & Van Obberghen E (2008). FALDH reverses the deleterious action of oxidative stress induced by lipid peroxidation product 4-hydroxynonenal on insulin signaling in 3T3-L1 adipocytes. *Diabetes* 57, 1216–1226.

Devaraj S, Wang-Polagruto J, Polagruto J, Keen CL & Jialal I (2008). High-fat, energy-dense, fast-food-style breakfast results in an increase in oxidative stress in metabolic syndrome. *Metab Clin Exp* 57, 867–870.

Dhar A, Dhar I, Jiang B, Desai KM & Wu L (2011). Chronic methylglyoxal infusion by minipump causes pancreatic beta-cell dysfunction and induces type 2 diabetes in Sprague-Dawley rats. *Diabetes* 60, 899–908.

Dierckx N, Horvath G, van Gils C, Vertommen J, van de Vliet J, De Leeuw I & Manuel-y-Keenoy B (2003). Oxidative stress status in patients with diabetes mellitus: relationship to diet. *Eur J Clin Nutr* 57, 999–1008.

Dominy JE, Hwang J & Stipanuk MH (2007). Overexpression of cysteine dioxygenase reduces intracellular cysteine and glutathione pools in HepG2/C3A cells. *Am J Physiol Endocrinol Metab* 293, E62–69.

Dong J-Z & Moldoveanu SC (2004). Gas chromatography-mass spectrometry of carbonyl compounds in cigarette mainstream smoke after derivatization with 2,4-dinitrophenylhydrazine. *J Chromatogr A* 1027, 25–35.

El Shazly AHM, Mahmoud AM & Darwish NS (2009). Potential prophylactic role of aminoguanidine in diabetic retinopathy and nephropathy in experimental animals. *Acta Pharm* 59, 67–73.

Fulghesu AM, Ciampelli M, Muzj G, Belosi C, Selvaggi L, Ayala GF & Lanzone A (2002). N-acetyl-cysteine treatment improves insulin sensitivity in women with polycystic ovary syndrome. *Fertil Steril* 77, 1128–1135.

Furukawa S, Fujita T, Shimabukuro M, Iwaki M, Yamada Y, Nakajima Y, Nakayama O, Makishima M, Matsuda M & Shimomura I (2004). Increased oxidative stress in obesity and its impact on metabolic syndrome. *J Clin Invest* 114, 1752–1761.

Gopaul NK, Manraj MD, Hebe A, Lee Kwai Yan S, Johnston A, Carrier MJ & Anggard EE (2001). Oxidative stress could precede endothelial dysfunction and insulin resistance in Indian Mauritians with impaired glucose metabolism. *Diabetologia* 44, 706–712.

Greene EL, Nelson BA, Robinson KA & Buse MG (2001). alpha-Lipoic acid prevents the development of glucose-induced insulin resistance in 3T3-L1 adipocytes and accelerates the decline in immunoreactive insulin during cell incubation. *Metab Clin Exp* 50, 1063–1069.

Grimsrud PA, Picklo MJ, Griffin TJ & Bernlohr DA (2007). Carbonylation of adipose proteins in obesity and insulin resistance: identification of adipocyte fatty acid-binding protein as a cellular target of 4-hydroxynonenal. *Mol Cell Proteomics* 6, 624–637.

Guichardant M, Bacot S, Moliere P & Lagarde M (2006). Hydroxy-alkenals from the peroxidation of n-3 and n-6 fatty acids and urinary metabolites. *Prostaglandins Leukot Essent Fatty Acids* 75, 179–182.

Gutteridge JM (1982). Free-radical damage to lipids, amino acids, carbohydrates and nucleic acids determined by thiobarbituric acid reactivity. *Int J Biochem* 14, 649–653.

Haber CA, Lam TKT, Yu Z, Gupta N, Goh T, Bogdanovic E, Giacca A & Fantus IG (2003). N-acetylcysteine and taurine prevent hyperglycemia-induced insulin resistance in vivo: possible role of oxidative stress. *Am J Physiol Endocrinol Metab* 285, E744–753.

Hartley DP, Ruth JA & Petersen DR (1995). The hepatocellular metabolism of 4-hydroxynonenal by alcohol dehydrogenase, aldehyde dehydrogenase, and glutathione S-transferase. *Arch Biochem Biophys* 316, 197–205.

Hipkiss AR, Preston JE, Himswoth DT, Worthington VC & Abbot NJ (1997). Protective effects of carnosine against malondialdehyde-induced toxicity towards cultured rat brain endothelial cells. *Neurosci Lett* 238, 135–138.

Ihara Y, Toyokuni S, Uchida K, Odaka H, Tanaka T, Ikeda H, Hiai H, Seino Y & Yamada Y (1999). Hyperglycemia causes oxidative stress in pancreatic beta-cells of GK rats, a model of type 2 diabetes. *Diabetes* 48, 927–932.

Ihm SH, Yoo HJ, Park SW & Ihm J (1999). Effect of aminoguanidine on lipid peroxidation in streptozotocin-induced diabetic rats. *Metab Clin Exp* 48, 1141–1145.

Jacob S, Henriksen EJ, Schiemann AL, Simon I, Clancy DE, Tritschler HJ, Jung WI, Augustin HJ & Dietze GJ (1995). Enhancement of glucose disposal in patients with type 2 diabetes by alpha-lipoic acid. *Arzneimittelforschung* 45, 872–874.

Jacob S, Streeper RS, Fogt DL, Hokama JY, Tritschler HJ, Dietze GJ & Henriksen EJ (1996). The antioxidant alpha-lipoic acid enhances insulin-stimulated glucose metabolism in insulin-resistant rat skeletal muscle. *Diabetes* 45, 1024–1029.

Je JH, Lee JY, Jung KJ, Sung B, Go EK, Yu BP & Chung HY (2004). NF-kappaB activation mechanism of 4-hydroxyhexenal via NIK/IKK and p38 MAPK pathway. *FEBS Lett* 566, 183–189.

JeBailey L, Wanono O, Niu W, Roessler J, Rudich A & Klip A (2007). Ceramide- and oxidant-induced insulin resistance involve loss of insulin-dependent Rac-activation and actin remodeling in muscle cells. *Diabetes* 56, 394–403.

Ji C, Rouzer CA, Marnett LJ & Pietenpol JA (1998). Induction of cell cycle arrest by the endogenous product of lipid peroxidation, malondialdehyde. *Carcinogenesis* 19, 1275–1283.

Jia L, Zhang Z, Zhai L & Bai Y (2009*a*). Protective effect of lipoic acid against acrolein-induced cytotoxicity in IMR-90 human fibroblasts. *J Nutr Sci Vitaminol* 55, 126–130.

Jia X, Olson DJH, Ross ARS & Wu L (2006). Structural and functional changes in human insulin induced by methylglyoxal. *FASEB J* 20, 1555–1557.

Jia Z, Misra BR, Zhu H, Li Y & Misra HP (2009*b*). Upregulation of cellular glutathione by 3H-1,2-dithiole-3-thione as a possible treatment strategy for protecting against acrolein-induced neurocytotoxicity. *Neurotoxicology* 30, 1–9.

Johnson JB, Summer W, Cutler RG, Martin B, Hyun D-H, Dixit VD, Pearson M, Nassar M, Telljohann R, Tellejohan R, Maudsley S, Carlson O, John S, Laub DR & Mattson MP (2007). Alternate day calorie restriction improves clinical findings and reduces markers of oxidative stress and inflammation in overweight adults with moderate asthma. *Free Radic Biol Med* 42, 665–674.

Jungas T, Motta I, Duffieux F, Fanen P, Stoven V & Ojcius DM (2002). Glutathione levels and BAX activation during apoptosis due to oxidative stress in cells expressing wild-type and mutant cystic fibrosis transmembrane conductance regulator. *J Biol Chem* 277, 27912–27918.

Kamenova P (2006). Improvement of insulin sensitivity in patients with type 2 diabetes mellitus after oral administration of alpha-lipoic acid. *Hormones (Athens)* 5, 251–258.

Kostyuk V, Potapovich A, Cesareo E, Brescia S, Guerra L, Valacchi G, Pecorelli A, Raskovic D, De Luca C, Pastore S & Korkina L (2010). Dysfunction of Glutathione S-Transferase Leads to Excess 4-Hydroxy-2-Nonenal and H2O2 and Impaired Cytokine Pattern in Cultured Keratinocytes and Blood of Vitiligo Patients. *Antioxid Redox Signal*; DOI: 10.1089/ars.2009.2976.

Laight DW, Desai KM, Gopaul NK, Anggard EE & Carrier MJ (1999*a*). F2-isoprostane evidence of oxidant stress in the insulin resistant, obese Zucker rat: effects of vitamin E. *Eur J Pharmacol* 377, 89–92.

Laight DW, Desai KM, Gopaul NK, Anggard EE & Carrier MJ (1999*b*). Pro-oxidant challenge in vivo provokes the onset of NIDDM in the insulin resistant obese Zucker rat. *Br J Pharmacol* 128, 269–271.

Lee JY, Je JH, Jung KJ, Yu BP & Chung HY (2004). Induction of endothelial iNOS by 4-hydroxyhexenal through NF-kappaB activation. *Free Radic Biol Med* 37, 539–548.

Lee JY, Jung GY, Heo HJ, Yun MR, Park JY, Bae SS, Hong KW, Lee WS & Kim CD (2006). 4-Hydroxynonenal induces vascular smooth muscle cell apoptosis through mitochondrial generation of reactive oxygen species. *Toxicol Lett* 166, 212–221.

Lee SE, Jeong SI, Kim G-D, Yang H, Park C-S, Jin Y-H & Park YS (2011). Upregulation of heme oxygenase-1 as an adaptive mechanism for protection against crotonaldehyde in human umbilical vein endothelial cells. *Toxicol Lett* 201, 240–248.

Li D, Ferrari M & Ellis EM (2011). Human aldo-keto reductase AKR7A2 protects against the cytotoxicity and mutagenicity of reactive aldehydes and lowers intracellular reactive oxygen species in hamster V79-4 cells. *Chemico-Biological Interactions*; DOI: 10.1016/j.cbi.2011.09.007.

Lieber CS, Leo MA, Cao Q, Mak KM, Ren C, Ponomarenko A, Wang X & Decarli LM (2007). The Combination of S-adenosylmethionine and Dilinoleoylphosphatidylcholine Attenuates Non-alcoholic Steatohepatitis Produced in Rats by a High-Fat Diet. *Nutr Res* 27, 565–573.

Liu W, Kato M, Akhand AA, Hayakawa A, Suzuki H, Miyata T, Kurokawa K, Hotta Y, Ishikawa N & Nakashima I (2000). 4-hydroxynonenal induces a cellular redox status-related activation of the caspase cascade for apoptotic cell death. *J Cell Sci* 113 (Pt 4), 635–641.

Liu X, Yang Z, Pan X, Zhu M & Xie J (2010). Crotonaldehyde induces oxidative stress and caspase-dependent apoptosis in human bronchial epithelial cells. *Toxicol Lett* 195, 90–98.

Long EK, Murphy TC, Leiphon LJ, Watt J, Morrow JD, Milne GL, Howard JRH & Picklo MJ (2008). Trans-4-hydroxy-2-hexenal is a neurotoxic product of docosahexaenoic (22:6; n-3) acid oxidation. *J Neurochem* 105, 714–724.

Long EK, Rosenberger TA & Picklo MJ (2010). Ethanol withdrawal increases glutathione adducts of 4-hydroxy-2-hexenal but not 4-hydroxyl-2-nonenal in the rat cerebral cortex. *Free Radic Biol Med* 48, 384–390.

Lovell MA, Ehmann WD, Mattson MP & Markesbery WR (1997). Elevated 4-hydroxynonenal in ventricular fluid in Alzheimer's disease. *Neurobiol Aging* 18, 457–461.

Lu SC (2000). S-Adenosylmethionine. *Int J Biochem Cell Biol* 32, 391–395.

Luo J, Robinson JP & Shi R (2005). Acrolein-induced cell death in PC12 cells: role of mitochondria-mediated oxidative stress. *Neurochem Int* 47, 449–457.

Lyn-Cook LE Jr, Lawton M, Tong M, Silbermann E, Longato L, Jiao P, Mark P, Wands JR, Xu H & de la Monte SM (2009). Hepatic ceramide may mediate brain insulin resistance and neurodegeneration in type 2 diabetes and non-alcoholic steatohepatitis. *J Alzheimers Dis* 16, 715–729.

Maddux BA, See W, Lawrence JC Jr, Goldfine AL, Goldfine ID & Evans JL (2001). Protection against oxidative stress-induced insulin resistance in rat L6 muscle cells by mircomolar concentrations of alpha-lipoic acid. *Diabetes* 50, 404–410.

Manini P, Andreoli R, Sforza S, Dall'Asta C, Galaverna G, Mutti A & Niessen WMA (2010). Evaluation of Alternate Isotope-Coded Derivatization Assay (AIDA) in the LC-MS/MS analysis of aldehydes in exhaled breath condensate. *J Chromatogr B Analyt Technol Biomed Life Sci* 878, 2616–2622.

Manning PJ, Sutherland WHF, Walker RJ, Williams SM, De Jong SA, Ryalls AR & Berry EA (2004). Effect of high-dose vitamin E on insulin resistance and associated parameters in overweight subjects. *Diabetes Care* 27, 2166–2171.

Martina V, Masha A, Gigliardi VR, Brocato L, Manzato E, Berchio A, Massarenti P, Settanni F, Della Casa L, Bergamini S & Iannone A (2008). Long-term N-acetylcysteine and L-arginine administration reduces endothelial activation and systolic blood pressure in hypertensive patients with type 2 diabetes. *Diabetes Care* 31, 940–944.

De Mattia G, Bravi MC, Laurenti O, Cassone-Faldetta M, Proietti A, De Luca O, Armiento A & Ferri C (1998). Reduction of oxidative stress by oral N-acetyl-L-cysteine treatment decreases plasma soluble vascular cell adhesion molecule-1 concentrations in non-obese, non-dyslipidaemic, normotensive, patients with non-insulin-dependent diabetes. *Diabetologia* 41, 1392–1396.

Maxwell SR, Thomason H, Sandler D, Leguen C, Baxter MA, Thorpe GH, Jones AF & Barnett AH (1997). Antioxidant status in patients with uncomplicated insulin-dependent and non-insulin-dependent diabetes mellitus. *Eur J Clin Invest* 27, 484–490.

McGrath LT, McGleenon BM, Brennan S, McColl D, McILroy S & Passmore AP (2001). Increased oxidative stress in Alzheimer's disease as assessed with 4-hydroxynonenal but not malondialdehyde. *QJM* 94, 485–490.

Medina-Navarro R, Guzmán-Grenfell AM, Díaz-Flores M, Duran-Reyes G, Ortega-Camarillo C, Olivares-Corichi IM & Hicks JJ (2007). Formation of an adduct between insulin and the toxic lipoperoxidation product acrolein decreases both the hypoglycemic effect of the hormone in rat and glucose uptake in 3T3 adipocytes. *Chem Res Toxicol* 20, 1477–1481.

Menon V, Ram M, Dorn J, Armstrong D, Muti P, Freudenheim JL, Browne R, Schunemann H & Trevisan M (2004). Oxidative stress and glucose levels in a population-based sample. *Diabet Med* 21, 1346–1352.

Michiels C & Remacle J (1991). Cytotoxicity of linoleic acid peroxide, malondialdehyde and 4-hydroxynonenal towards human fibroblasts. *Toxicology* 66, 225–234.

Misonou Y, Asahi M, Yokoe S, Miyoshi E & Taniguchi N (2006). Acrolein produces nitric oxide through the elevation of intracellular calcium levels to induce apoptosis in human umbilical vein endothelial cells: implications for smoke angiopathy. *Nitric Oxide* 14, 180–187.

Miwa I, Ichimura N, Sugiura M, Hamada Y & Taniguchi S (2000). Inhibition of glucose-induced insulin secretion by 4-hydroxy-2-nonenal and other lipid peroxidation products. *Endocrinology* 141, 2767–2772.

Morris RT, Laye MJ, Lees SJ, Rector RS, Thyfault JP & Booth FW (2008). Exercise-induced attenuation of obesity, hyperinsulinemia, and skeletal muscle lipid peroxidation in the OLETF rat. *J Appl Physiol* 104, 708–715.

Münzel T, Kurz S, Rajagopalan S, Thoenes M, Berrington WR, Thompson JA, Freeman BA & Harrison DG (1996). Hydralazine prevents nitroglycerin tolerance by inhibiting activation of a membrane-bound NADH oxidase. A new action for an old drug. *J Clin Invest* 98, 1465–1470.

Nakhjavani M, Esteghamati A, Nowroozi S, Asgarani F, Rashidi A & Khalilzadeh O (2010). Type 2 diabetes mellitus duration: an independent predictor of serum malondialdehyde levels. *Singapore Med J* 51, 582–585.

Negre-Salvayre A, Coatrieux C, Ingueneau C & Salvayre R (2008). Advanced lipid peroxidation end products in oxidative damage to proteins. Potential role in diseases and therapeutic prospects for the inhibitors. *Br J Pharmacol* 153, 6–20.

Numazawa S, Takase M, Ahiko T, Ishii M, Shimizu S-I & Yoshida T (2012). Possible Involvement of Transient Receptor Potential Channels in Electrophile-Induced Insulin Secretion from RINm5F Cells. *Biol Pharm Bull* 35, 346–354.

Olusi SO (2002). Obesity is an independent risk factor for plasma lipid peroxidation and depletion of erythrocyte cytoprotectic enzymes in humans. *Int J Obes Relat Metab Disord* 26, 1159–1164.

Page S, Fischer C, Baumgartner B, Haas M, Kreusel U, Loidl G, Hayn M, Ziegler-Heitbrock HW, Neumeier D & Brand K (1999). 4-Hydroxynonenal prevents NF-kappaB activation and tumor necrosis factor expression by inhibiting IkappaB phosphorylation and subsequent proteolysis. *J Biol Chem* 274, 11611–11618.

Pan J, Keffer J, Emami A, Ma X, Lan R, Goldman R & Chung F-L (2009). Acrolein-derived DNA adduct formation in human colon cancer cells: its role in apoptosis induction by docosahexaenoic acid. *Chem Res Toxicol* 22, 798–806.

Paradis V, Kollinger M, Fabre M, Holstege A, Poynard T & Bedossa P (1997). In situ detection of lipid peroxidation by-products in chronic liver diseases. *Hepatology* 26, 135–142.

Peyroux J & Sternberg M (2006). Advanced glycation endproducts (AGEs): Pharmacological inhibition in diabetes. *Pathol Biol* 54, 405–419.

Pillon NJ, Croze ML, Vella RE, Soulère L, Lagarde M & Soulage CO (2012). The Lipid Peroxidation By-Product 4-Hydroxy-2-Nonenal (4-HNE) Induces Insulin Resistance in Skeletal Muscle through Both Carbonyl and Oxidative Stress. Endocrinology 153, 2099-2111.

Pillon NJ, Soulère L, Vella RE, Croze M, Caré BR, Soula HA, Doutheau A, Lagarde M & Soulage CO (2010). Quantitative structure-activity relationship for 4-hydroxy-2-alkenal induced cytotoxicity in L6 muscle cells. *Chem Biol Interact* 188, 171–180.

Pillon NJ, Vella RE, Soulère L, Becchi M, Lagarde M & Soulage CO (2011). Structural and functional changes in human insulin induced by the lipid peroxidation byproducts 4-hydroxy-2-nonenal and 4-hydroxy-2-hexenal. *Chem Res Toxicol* 24, 752–762.

Poirier M, Fournier M, Brousseau P & Morin A (2002). Effects of volatile aromatics, aldehydes, and phenols in tobacco smoke on viability and proliferation of mouse lymphocytes. *J Toxicol Environ Health Part A* 65, 1437–1451.

Reagan LP, Magariños AM, Yee DK, Swzeda LI, Van Bueren A, McCall AL & McEwen BS (2000). Oxidative stress and HNE conjugation of GLUT3 are increased in the hippocampus of diabetic rats subjected to stress. *Brain Res* 862, 292–300.

Riahi Y, Sin-Malia Y, Cohen G, Alpert E, Gruzman A, Eckel J, Staels B, Guichardant M & Sasson S (2010). The natural protective mechanism against hyperglycemia in vascular endothelial cells: roles of the lipid peroxidation product 4-hydroxydodecadienal and peroxisome proliferator-activated receptor delta. *Diabetes* 59, 808–818.

Riboulet-Chavey A, Pierron A, Durand I, Murdaca J, Giudicelli J & Van Obberghen E (2006). Methylglyoxal impairs the insulin signaling pathways independently of the formation of intracellular reactive oxygen species. *Diabetes* 55, 1289–1299.

Roy J, Pallepati P, Bettaieb A & Averill-Bates DA (2010). Acrolein induces apoptosis through the death receptor pathway in A549 lung cells: role of p53. *Can J Physiol Pharmacol* 88, 353–368.

Rudich A, Tirosh A, Potashnik R, Hemi R, Kanety H & Bashan N (1998). Prolonged oxidative stress impairs insulin-induced GLUT4 translocation in 3T3-L1 adipocytes. *Diabetes* 47, 1562–1569.

Rudich A, Tirosh A, Potashnik R, Khamaisi M & Bashan N (1999). Lipoic acid protects against oxidative stress induced impairment in insulin stimulation of protein kinase B and glucose transport in 3T3-L1 adipocytes. *Diabetologia* 42, 949–957.

Ruef J, Moser M, Bode C, Kubler W & Runge MS (2001). 4-hydroxynonenal induces apoptosis, NF-kappaB-activation and formation of 8-isoprostane in vascular smooth muscle cells. *Basic Res Cardiol* 96, 143–150.

Russell AP, Gastaldi G, Bobbioni-Harsch E, Arboit P, Gobelet C, Dériaz O, Golay A, Witztum JL & Giacobino J-P (2003). Lipid peroxidation in skeletal muscle of obese as compared to endurance-trained humans: a case of good vs. bad lipids? *FEBS Lett* 551, 104–106.

Sakuraba H, Mizukami H, Yagihashi N, Wada R, Hanyu C & Yagihashi S (2002). Reduced beta-cell mass and expression of oxidative stress-related DNA damage in the islet of Japanese Type II diabetic patients. *Diabetologia* 45, 85–96.

Schaur RJ (2003). Basic aspects of the biochemical reactivity of 4-hydroxynonenal. *Mol Aspects Med* 24, 149–159.

Selley ML (2004). Increased (E)-4-hydroxy-2-nonenal and asymmetric dimethylarginine concentrations and decreased nitric oxide concentrations in the plasma of patients with major depression. *J Affect Disord* 80, 249–256.

Shearn CT, Fritz KS, Reigan P & Petersen DR (2011). Modification of Akt2 by 4-hydroxynonenal inhibits insulin-dependent Akt signaling in HepG2 cells. *Biochemistry* 50, 3984–3996.

Shibata M, Hakuno F, Yamanaka D, Okajima H, Fukushima T, Hasegawa T, Ogata T, Toyoshima Y, Chida K, Kimura K, Sakoda H, Takenaka A, Asano T & Takahashi S-I (2010). Paraquat-induced oxidative stress represses phosphatidylinositol 3-kinase activities leading to impaired glucose uptake in 3T3-L1 adipocytes. *J Biol Chem* 285, 20915–20925.

Singh I, Carey AL, Watson N, Febbraio MA & Hawley JA (2008*a*). Oxidative stress-induced insulin resistance in skeletal muscle cells is ameliorated by gamma-tocopherol treatment. *Eur J Nutr* 47, 387–392.

Singh SP, Niemczyk M, Saini D, Awasthi YC, Zimniak L & Zimniak P (2008*b*). Role of the electrophilic lipid peroxidation product 4-hydroxynonenal in the development and maintenance of obesity in mice. *Biochemistry* 47, 3900–3911.

Singh SP, Niemczyk M, Zimniak L & Zimniak P (2009). Fat accumulation in Caenorhabditis elegans triggered by the electrophilic lipid peroxidation product 4-hydroxynonenal (4-HNE). *Aging (Albany NY)* 1, 68–80.

Soares AF, Guichardant M, Cozzone D, Bernoud-Hubac N, Bouzaidi-Tiali N, Lagarde M & Geloen A (2005). Effects of oxidative stress on adiponectin secretion and lactate production in 3T3-L1 adipocytes. *Free Radic Biol Med* 38, 882–889.

Srivastava S, Dixit BL, Cai J, Sharma S, Hurst HE, Bhatnagar A & Srivastava SK (2000). Metabolism of lipid peroxidation product, 4-hydroxynonenal (HNE) in rat erythrocytes: role of aldose reductase. *Free Radic Biol Med* 29, 642–651.

Syslova K, Kacer P, Kuzma M, Najmanova V, Fenclova Z, Vlckova S, Lebedova J & Pelclova D (2009). Rapid and easy method for monitoring oxidative stress markers in body fluids of patients with asbestos or silica-induced lung diseases. *J Chromatogr B Analyt Technol Biomed Life Sci*; DOI: 10.1016/j.jchromb.2009.06.008.

Tan Y, Ichikawa T, Li J, Si Q, Yang H, Chen X, Goldblatt CS, Meyer CJ, Li X, Cai L & Cui T (2011). Diabetic downregulation of Nrf2 activity via ERK contributes to oxidative stress-induced insulin resistance in cardiac cells in vitro and in vivo. *Diabetes* 60, 625–633.

Tanel A & Averill-Bates DA (2007). P38 and ERK mitogen-activated protein kinases mediate acrolein-induced apoptosis in Chinese hamster ovary cells. *Cell Signal* 19, 968–977.

Thirunavukkarasu V & Anuradha CV (2004). Influence of alpha-lipoic acid on lipid peroxidation and antioxidant defence system in blood of insulin-resistant rats. *Diabetes Obes Metab* 6, 200–207.

Toyokuni S, Yamada S, Kashima M, Ihara Y, Yamada Y, Tanaka T, Hiai H, Seino Y & Uchida K (2000). Serum 4-hydroxy-2-nonenal-modified albumin is elevated in patients with type 2 diabetes mellitus. *Antioxid Redox Signal* 2, 681–685.

Traverso N, Menini S, Cosso L, Odetti P, Albano E, Pronzato MA & Marinari UM (1998). Immunological evidence for increased oxidative stress in diabetic rats. *Diabetologia* 41, 265–270.

Traverso N, Menini S, Odetti P, Pronzato MA, Cottalasso D & Marinari UM (2002). Diabetes impairs the enzymatic disposal of 4-hydroxynonenal in rat liver. *Free Radic Biol Med* 32, 350–359.

Trevisan M, Browne R, Ram M, Muti P, Freudenheim J, Carosella AM & Armstrong D (2001). Correlates of markers of oxidative status in the general population. *Am J Epidemiol* 154, 348–356.

Uchida K, Shiraishi M, Naito Y, Torii Y, Nakamura Y & Osawa T (1999). Activation of stress signaling pathways by the end product of lipid peroxidation. 4-hydroxy-2-nonenal is a potential inducer of intracellular peroxide production. *J Biol Chem* 274, 2234–2242.

Unoki H, Bujo H, Yamagishi S, Takeuchi M, Imaizumi T & Saito Y (2007). Advanced glycation end products attenuate cellular insulin sensitivity by increasing the generation of intracellular reactive oxygen species in adipocytes. *Diabetes Res Clin Pract* 76, 236–244.

Vaillancourt F, Fahmi H, Shi Q, Lavigne P, Ranger P, Fernandes JC & Benderdour M (2008). 4-Hydroxynonenal induces apoptosis in human osteoarthritic chondrocytes: the protective role of glutathione-S-transferase. *Arthritis Res Ther* 10, R107.

Vincent HK, Bourguignon C & Vincent KR (2006). Resistance training lowers exercise-induced oxidative stress and homocysteine levels in overweight and obese older adults. *Obesity (Silver Spring)* 14, 1921–1930.

Vincent HK, Bourguignon CM, Weltman AL, Vincent KR, Barrett E, Innes KE & Taylor AG (2009). Effects of antioxidant supplementation on insulin sensitivity, endothelial adhesion molecules, and oxidative stress in normal-weight and overweight young adults. *Metab Clin Exp* 58, 254–262.

Vincent HK, Innes KE & Vincent KR (2007). Oxidative stress and potential interventions to reduce oxidative stress in overweight and obesity. *Diabetes Obes Metab* 9, 813–839.

Vindis C, Escargueil-Blanc I, Elbaz M, Marcheix B, Grazide M-H, Uchida K, Salvayre R & Nègre-Salvayre A (2006). Desensitization of platelet-derived growth factor receptor-beta by oxidized lipids in vascular cells and atherosclerotic lesions: prevention by aldehyde scavengers. *Circ Res* 98, 785–792.

Wang L, Sun Y, Asahi M & Otsu K (2011). Acrolein, an environmental toxin, induces cardiomyocyte apoptosis via elevated intracellular calcium and free radicals. *Cell Biochem Biophys* 61, 131–136.

Wang Z, Dou X, Gu D, Shen C, Yao T, Nguyen V, Braunschweig C & Song Z (2012). 4-Hydroxynonenal differentially regulates adiponectin gene expression and secretion via activating PPARγ and accelerating ubiquitin-proteasome degradation. *Mol Cell Endocrinol* 349, 222–231.

Wellen KE & Hotamisligil GS (2003). Obesity-induced inflammatory changes in adipose tissue. *J Clin Invest* 112, 1785–1788.

Wierusz-Wysocka B, Wysocki H, Byks H, Zozulinska D, Wykretowicz A & Kazmierczak M (1995). Metabolic control quality and free radical activity in diabetic patients. *Diabetes Res Clin Pract* 27, 193–197.

Yadav UCS, Ramana KV, Awasthi YC & Srivastava SK (2008). Glutathione level regulates HNE-induced genotoxicity in human erythroleukemia cells. *Toxicol Appl Pharmacol* 227, 257–264.

Yamashita C, Hayashi T, Mori T, Matsumoto C, Kitada K, Miyamura M, Sohmiya K, Ukimura A, Okada Y, Yoshioka T, Kitaura Y & Matsumura Y (2010). Efficacy of olmesartan and nifedipine on recurrent hypoxia-induced left ventricular remodeling in diabetic mice. *Life Sci* 86, 322–330.

Yoritaka A, Hattori N, Uchida K, Tanaka M, Stadtman ER & Mizuno Y (1996). Immunohistochemical detection of 4-hydroxynonenal protein adducts in Parkinson disease. *Proc Natl Acad Sci U S A* 93, 2696–2701.

Zafarullah M, Li WQ, Sylvester J & Ahmad M (2003). Molecular mechanisms of N-acetylcysteine actions. *Cell Mol Life Sci* 60, 6–20.

Zarrouki B, Soares AF, Guichardant M, Lagarde M & Geloen A (2007). The lipid peroxidation end-product 4-HNE induces COX-2 expression through p38MAPK activation in 3T3-L1 adipose cell. *FEBS Lett* 581, 2394–2400.

Zhang H & Forman HJ (2008). Acrolein induces heme oxygenase-1 through PKC-delta and PI3K in human bronchial epithelial cells. *Am J Respir Cell Mol Biol* 38, 483–490.

Zimniak P (2010). 4-Hydroxynonenal and fat storage: A paradoxical pro-obesity mechanism? *Cell Cycle* 9, 3393–3394.

Chapter 20

Region Specific Vulnerability to Lipid Peroxidation in the Human Central Nervous System

Alba Naudí, Mariona Jové, Victòria Ayala, Omar Ramírez, Rosanna Cabré, Joan Prat, Manuel Portero-Otin, Isidre Ferrer and Reinald Pamplona

Additional information is available at the end of the chapter

http://dx.doi.org/10.5772/46022

1. Introduction

Around 100 billion neurons in the human nervous system orchestrate an exceptionally wide range of motor, sensory, regulatory, behavioural, and executive functions. Such diverse functional output is the product of different molecular events occurring in nervous cells and particularly, neurons. Morphologically, central nervous system (CNS) neurons differ in size, number and complexity of dendrites, number of synaptic connections, length of axons and distance across which synaptic connections are established, extent of axonal myelination, and other cellular characteristics. Neuronal diversity is also amplified by the inclusion of chemical specificity on the basis of the neurotransmitters, which they use for chemical transmission or neuromodulation. This great diversity among neuronal populations is a strong indication that although all neurons contain the same genetic code in their genome, each neuronal population has their own gene expression profile. While the diversity of neuronal structures and functions are well documented, what is less appreciated is the diverse response of neurons to stresses and adverse factors during aging or as a result of neurodegenerative diseases. Furthermore, to add complexity to an already heterogeneous landscape, non-neuronal populations, often described as a 'supporting matrix' are recently being recognized as active, information-rich, cellular counterpart in CNS function.

In this scenario of cellular diversity emerges the concept of selective neuronal vulnerability (SNV). SNV is described as the differential sensitivity of neuronal populations in the nervous system to stresses that cause cell damage or death and can lead to neurodegeneration [1,2]. The fact that specific regions of the nervous system exhibit differential vulnerabilities to aging and various neurodegenerative diseases is a reflection of both the specificity in the aetiology of each disease and of the heterogeneity in neuronal (and non-neuronal) responses to cell-damaging processes associated with each of the diseases [3]. The appearance of SNV is not limited to cross-regional differences in the nervous system, as within a single e.g. brain region, such as the hippocampus or the entorhinal cortex, where SNV is manifested as internal, sub-regional differences in relative sensitivities to stress and disease [2,3]. Among these cell-damaging processes one could count inflammatory, proteotoxicity, vascular and many other pathophysiological processes, including an excess of lipid peroxidation (see later).

Oxidative stress (OS) is involved in the basic mechanisms of nervous system aging; whilst an excessive oxidation has been invoked as an etiopathogenic or physiopathologic mechanism for neurodegeneration. Oxidative stress, the result of an imbalance between production of free radicals and the enzymatic or non-enzymatic detoxification of these highly reactive species, is detrimental to cells because free radicals chemically modify lipids, proteins, and nucleic acids. So, it is very important to define the vulnerability of the different neuronal populations in terms of susceptibility to oxidative stress in physiological conditions in order to extent this knowledge to improve our understanding of how this particular form of cell vulnerability causes selective neuronal losses in nervous system, as well as reveal potential molecular and cellular mechanisms that bring about relative resistance or sensitivity of neurons to stresses.

2. Mitochondrial free radical generation and membrane fatty acid composition in the CNS

2.1. Mitochondrial free radical production

Chemical reactions in cells of the nervous system are under strict enzyme control and conform to a tightly regulated metabolic program in order to minimize unnecessary side reactions. Nevertheless, apparently uncontrolled and potentially deleterious reactions occur, even under physiological conditions. Reactive oxygen species (ROS) express a variety of molecules and free radicals (chemical species with one unpaired electron) physiologically generated from the metabolism of molecular oxygen [4]. They are extremely reactive and have damaging effects. Superoxide anion, the product of a one-electron reduction of oxygen, is the precursor of most ROS and a mediator in oxidative chain reactions [4] (**Figure 1**). The character of radical is not circumscribed to oxygen containing species, as nitrogen, chloride and sulphide containing molecules could also play a significant role. Globally, in cells of the CNS the major sites of physiological ROS generation are the complex I and III of the mitochondrial electron transport chain, which contains several redox centers (flavins, iron–sulphur clusters, and ubisemiquinone) capable of transferring one electron to oxygen to form superoxide anion [5,6]. Oxidative damage is a broad term used to cover the attack upon biological molecules by free radicals. ROS attack/damage all cellular constituents [6], but especially biological membranes.

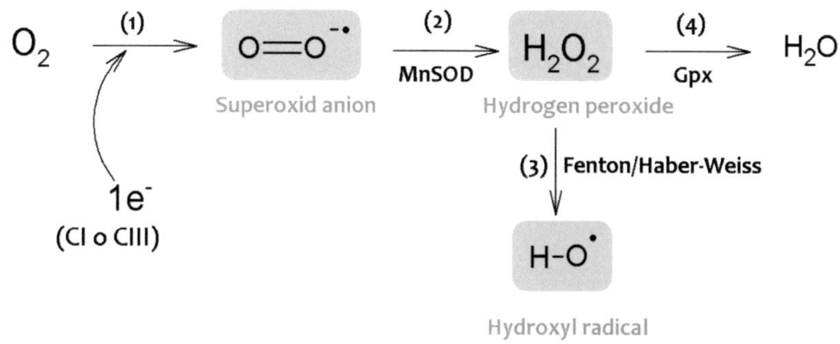

Figure 1. The three main physiological ROS. CI, mitochondrial complex I; CIII, mitochondrial complex III; MnSOD, Manganese superoxide dismutase; Gpx, glutathione peroxidase. MnSOD and Gpx are antioxidants enzymes.

2.2. Membrane fatty acid composition in neural cells

All living organisms have lipid membranes. Biological membranes are dynamic structures that generally consist of bilayers of amphipathic molecules hold together by non-covalent bonds [7]. Phospholipids, consisting of a hydrophilic head group with attached hydrophobic acyl chains, are the predominant membrane lipids and are, from a topographic point of view, asymmetrically distributed across the bilayer. The variation in head groups and aliphatic chains allows the existence of a huge range of different phospholipid species [8,9]. The acyl chains are either saturated, monounsaturated or polyunsaturated hydrocarbon chains that normally vary from 14 to 22 carbons in length (**Figure 2**), with an average chain length strictly maintained around 18 carbon atoms, and a relative distribution between saturated and unsaturated fatty acids of 40:60 (SFA and UFA, respectively) [10]. Polyunsaturated fatty acids (PUFAs) are essential components of cellular membranes that strongly affect their fluidity, flexibility and selective permeability, as well as many cellular and physiological processes [11].

Long-chain polyunsaturated fatty acids are highly enriched in the nervous system. Docosahexaenoic acid (DHA; 22:6n-3; 4,7,10,13,16,19-C22:6), in particular, is the most abundant PUFA in the brain and is concentrated in aminophospholipids of cell membranes. Numerous studies have indicated that this concentration of DHA in the nervous system is essential for optimal neuronal functions. Although the underlying mechanisms of its essential function are still not clearly understood, emerging evidence suggests that unique metabolism of DHA in relation to its incorporation into neuronal membrane phospholipids plays an important role.

Accretion of DHA in the CNS actively occurs during the developmental period, primarily relying on circulating plasma DHA derived from diet or from biosynthesis in the liver [12]. However, local biosynthesis of DHA also occurs in the brain, providing an alternative source of DHA for its accumulation in the brain [13]. It is well established that DHA can be biosynthesized from α-linolenic acid (18:3n-3; 9,12,15-C18:3), a shorter chain n-3 fatty acid precursor, through chain elongation and desaturation processes [14] (**Figure 3**). Linolenic

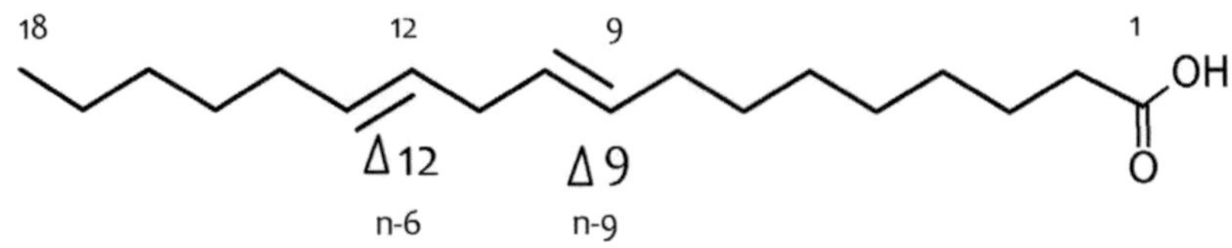

n= chain length= 18 carbons

x= 2 double bounds

Cn:x(n-y) ⟶ C18:2(n-6)

x-y=n-6

position of the first double bound from the methyl end

Figure 2. Fatty acid nomenclature. As an example, the structure of linoleic acid (C18:2n-6) is given.

acid is desaturated to 18:4n-3 (6,9,12,15-C18:4) by Δ6-desaturase, chain-elongated to 20:4n-3 (8,11,14,17-C20:4), and subsequently converted to eicosapentaenoic acid (20:5n-3; 5,8,11,14,17-C20:5) by Δ5-desaturase in the endoplasmic reticulum (ER). Mammalian Δ5- and Δ6-desaturases have been identified and cloned [15]. However, Δ4-desaturase, responsible for making 22:6n-3 directly from 22:5n-3, an elongation product of 20:5n-3, has been identified only in microalgae [16]. In mammals, 22:5n-3 is further elongated to 24:5n-3 (9,12,15,18,21-C24:5) followed by desaturation by Δ6-desaturase to 24:6n-3 (6,9,12,15,18,21-C24:6). Subsequently, 24:6n-3 is transferred to peroxisomes and converted to 22:6n-3 by removing two carbon chains by β-oxidation. DHA thus formed is transferred back to the ER and quickly incorporated into membrane phospholipids by esterification during de novo synthesis or by a deacylation-reacylation reaction. Because biosynthesis of both fatty acids and phospholipids occurs in ER, a particular fatty acid intermediate can be either incorporated into phospholipids or further chain-elongated/desaturated, although the regulation of these processes is still poorly understood. Long-chain n-6 fatty acids are biosynthesized from linoleic acid (18:2n-6; 9,12-C18:2) using the analogous pathway and the same enzyme system (**Figure 3**). In most tissues, the commonly observed long-chain n-6 fatty acid is arachidonic acid (AA; 20:4n-6; 5,8,11,14-C20:4). Docosapentaenoic acid (DPAn-6; 22:5n-6; 4,7,10,13,16-C22:5) produced by further elongation and desaturation of AA and subsequent peroxisomal β-oxidation, is rather a minor component, and yet it accumulates in the brain in place of DHA when the DHA supply is inadequate, especially during developmental periods. The distinctive fatty acid profile in the brain enriched with DHA or DPAn-6 may reflect the brain-specific uptake and/or regulation of fatty acid synthesis and esterification into membrane phospholipids. The liver is considered to be the primary site for biosynthesis of DHA, which becomes available to brain uptake through subsequent secretion into the circulating blood stream. Among neural cells, consisting of neurons, astrocytes, microglia, and oligodendrocytes, the capacity to synthesize DHA has been demonstrated only in astrocytes [13]. Despite the fact that neurons are major targets for DHA accumulation, they cannot produce DHA because of lack of desaturase activity. Cerebromicrovascular endothelia can also elongate and desaturate shorter carbon chain fatty acids. However, they cannot perform the final desaturation step to produce either 22:5n-6 or 22:6n-3 [17].

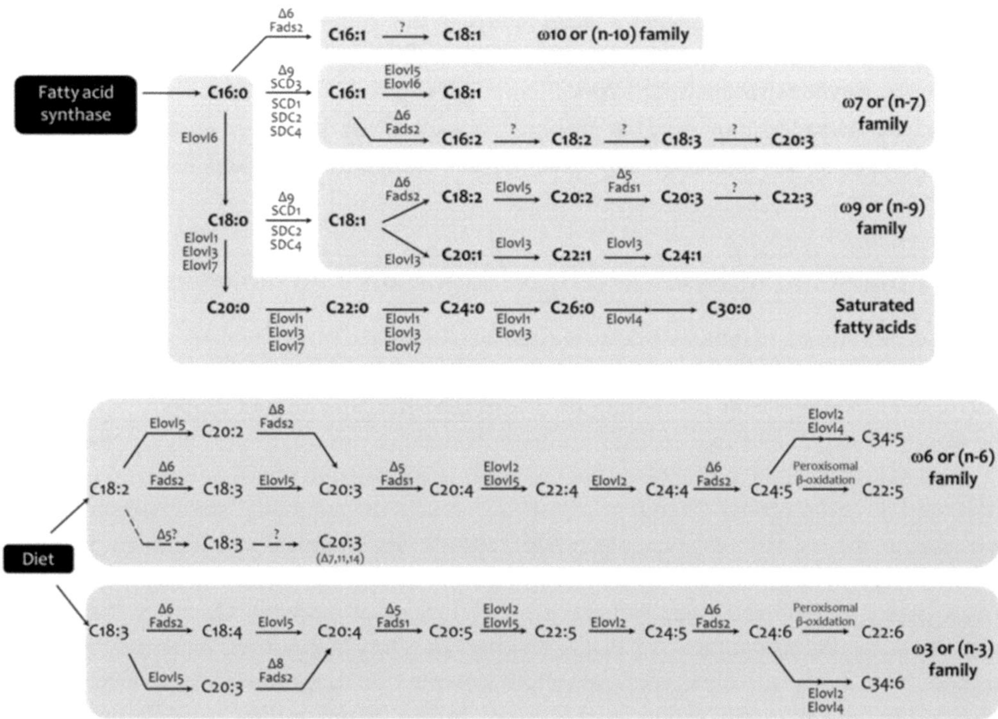

Figure 3. Long chain and very long-chain fatty acid biosynthesis in mammals. The long chain saturated fatty acids and unsaturated fatty acids of the n-10, n-7 and n -9 series can be synthesized from palmitic acid (C16:0) produced by the fatty acid synthase (FAS). Long-chain fatty acids of the n-6 and n-3 series can only be synthesized from precursors obtained from dietary precursors (DIET). Elovl, elongation of very long chain fatty acids (fatty acid elongase); Fads, fatty acid desaturases.

DHA synthesis in astrocytes is negatively influenced by the availability of preformed DHA [18] and thus may represent a quantitatively minor source for the neural DHA accretion when the circulating DHA supply is adequate. Incorporation of circulating DHA across the blood brain barrier appears to be an important route for maintaining adequate levels of DHA in the brain. In agreement with this notion, it has been shown that constant basal turnover of esterified DHA in the brain with unesterified DHA in plasma occurs at an estimated rate of 2–8% per day in adult rats [19]. Generally, it is difficult to deplete DHA from the neural membranes of adult mammals even with a DHA low diet, presumably because of preferential uptake of DHA into the brain to support the basal turnover. In the case of insufficient supply of n-3 fatty acids during development, the loss of DHA does occur but is compensated with DPAn-6 through reciprocal replacement, suggesting a requirement of very long-chain, highly unsaturated fatty acids in neural membranes.

Whether brain DHA is derived from the circulating plasma pool or biosynthesized locally, in the astrocytes, which are situated in close contact with neurons, appear to play an important role in supplying DHA to neurons. DHA can be released readily from astroglial membranes under basal and stimulated conditions, and supplied to neurons. Despite its high abundance in neuronal membranes, DHA is not easily released but is tenaciously retained in the neuronal membranes under the conditions in which AA can be released. Considering the fact that astroglia support neurons by providing neurotrophic factors, DHA supplied by astroglia may also be trophic. Indeed, DHA has been shown to promote neuronal survival [20] and differentiation [21] in both transformed and primary neuronal cells in culture.

3. Membrane unsaturation and lipoxidation-derived molecular damage

The susceptibility of membrane phospholipids to oxidative demise is related to two inherent traits, the physico-chemical properties of the membrane bilayer and the intrinsic chemical reactivity of the fatty acids composing the membrane [10,22]. The first property is related to the fact that oxygen and free radicals are more soluble in the fluid lipid bilayer than in the aqueous solution. Thus, membranes contain an interior organic phase in which the oxygen may tend to concentrate. Therefore, these differences in solubility are important when considering the availability of oxygen/free radicals for chemical reactions inside living systems: organic regions may contain more free radicals than aqueous regions [8] and, consequently, membrane lipids become primary targets of oxidative damage. The second property is related to the fact that PUFA residues of phospholipids are extremely sensitive to oxidation. Every membrane phospholipid contains an unsaturated fatty acid residue esterified to the 2-hydroxyl group of its glycerol moiety. Many of these are polyunsaturated and the presence of a methylene group between two double bonds renders the fatty acid more sensitive to ROS-induced damage. Therefore, the sensitivity of these molecules to oxidation increase exponentially as a function of the number of double bonds per fatty acid molecule [22,23]. Consequently, the high concentration of PUFAs in phospholipids not only makes them prime targets for reaction with oxidizing agents but also enables them to participate in long free radical chain reactions. Reactive free radicals can pull off hydrogen atoms from PUFA side chains. A hydrogen atom (H•) has only one electron. This hydrogen is bonded to a carbon in the fatty acid backbone by a covalent bond. Hence, the carbon from which H• is abstracted now has an unpaired electron (i.e it is a free radical). PUFA side chains (two or more double bonds) are more sensitive to attack by radicals than are SFAs (no double bonds) or monounsaturated fatty acids (MUFA,one double bond) side chains. When C• radicals are generated in the hydrophobic interior of membranes, their most likely fate is combination with oxygen dissolved in the membrane. The resulting peroxyl radical is highly reactive: it can attack membrane proteins and oxidize adjacent PUFA side chains. So, the reaction is repeated and the whole process continues in a free radical chain reaction, generating lipid hydroperoxides [4]. Lipid hydroperoxides are more hydrophilic than unperoxidized fatty acid side chains. They try to migrate to the membrane surface to interact with water, thus disrupting the membrane structure, altering fluidity and other functional properties and making the membrane leaky.

Lipid peroxidation generates hydroperoxides as well as endoperoxides, which undergo fragmentation to produce a broad range of reactive intermediates called reactive carbonyl species (RCS) (**Figure 4**) with three to nine carbons in length, the most reactive being α,β-

unsaturated aldehydes [4-hydroxy-trans-2-nonenal (HNE) and acrolein], di-aldehydes [malondialdehyde (MDA) and glyoxal], and keto-aldehydes [4-oxo-trans-2-nonenal (ONE) and isoketals] [24]. 2-Hydroxyheptanal (2-HH) and 4-hydroxyhexenal (4-HHE) are aldehydic product of lipid peroxidation of PUFAn-6. Additionally, a number of other short chain aldehydes are produced during lipid peroxidation through poorly understood mechanisms. These carbonyl compounds, ubiquitously generated in biological systems, have unique properties contrasted with free radicals. Thus, compared with reactive oxygen and nitrogen species, reactive aldehydes have a much longer half-life (i.e., minutes to hours instead of microseconds to nanoseconds for most free radicals). Further, the non-charged structure of aldehydes allows them to migrate relatively ease through hydrophobic membranes and hydrophilic cytosolic media, thereby extending the migration distance far from the production site. Based on these features alone, these carbonyl compounds can be more destructive than ROS and may have far-reaching damaging effects on target sites within or outside membranes. For the same reason, their long half-life allows them to subserve as second-messenger, signalling for important cellular responses in a mostly unknown fashion.

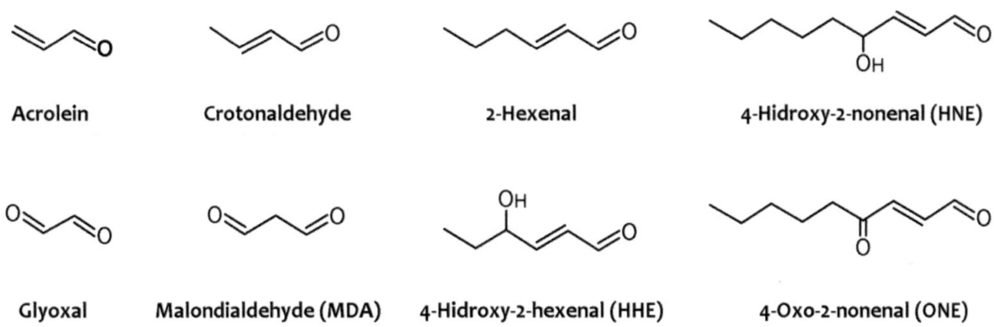

Figure 4. General structures of principal lipoxidative reactive carbonyl species detected in biological systems.

Carbonyl compounds react with nucleophilic groups in macromolecules like proteins, DNA, and aminophospholipids, among other, resulting in their chemical, nonenzymatic, and irreversible modification and formation of a variety of adducts and crosslinks collectively named Advanced Lipoxidation Endproducts (ALEs) [10,25,26] (**Figure 5**) . Thus, by reacting with nucleophilic sites in proteins (belonging basically to Cys, Lys, Arg, and His residues), carbonyl compounds generate ALE adducts such as MDA-Lys, HNE-Lys, FDP-Lys, carboxymethyl-lysine (CML) and S-carboxymethyl-cysteine; and the crosslinks glyoxal-lysine dimer (GOLD), and methylglyoxal-lysine dimer (MOLD), among several others. The accumulation of MDA adducts on proteins is also involved in the formation of lipofuscin (a nondegradable intralysosomal fluorescent pigment formed through lipoxidative reactions). Lipid peroxidation-derived endproducts can also react at the exocyclic amino groups of deoxyguanosine, deoxyadenosine, and deoxycytosine to form various alkylated products. Guanine is, however, the most commonly modified DNA base because of its high nucleophilicity. Some common enals that cause DNA damage, analogously to proteins, are

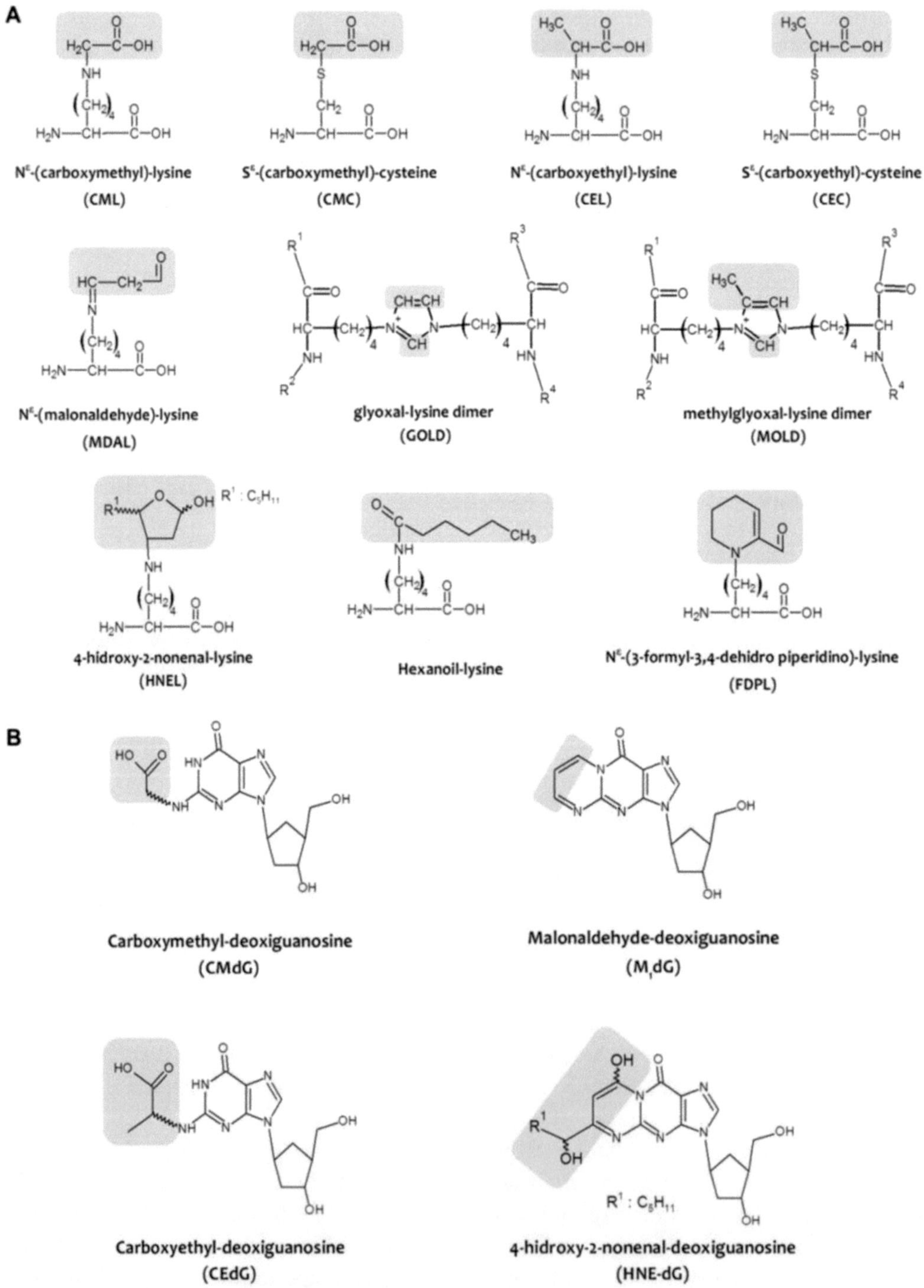

Figure 5. Reactive carbonyl species react with nucleophilic groups in macromolecules (A, proteins; B, DNA) resulting in their chemical, nonenzymatic, and irreversible modification and formation of a variety of adducts and crosslinks collectively named Advanced Lipoxidation Endproducts (ALEs).\

MDA, HNE, and acrolein, among others. Thus, the most common adducts arising from enals are exocyclic adducts such as etheno adducts, and MDA-deoxyguanosine (M1dG). Finally, the amino group of aminophospholipids can also react with carbonyl compounds and initiate some of the reactions ocurring in proteins and DNA, leading to the formation of adducts like MDA-phosphatidylethanolamine, and carboxymethyl-phosphatidylethanolamine [10].

4. ALEs: Molecular and cellular effects

Reactive carbonyl species (RCS) generated during lipid peroxidation reactions exhibit a wide range of molecular and biological effects, ranging from protein, DNA, and phospholipid damage to signaling pathway activation and/or alteration. The detailed mechanisms of 'toxicity' are, however, mostly unknown.

4.1. Molecular damage

Lipoxidation reactions lead to structural and functional changes in proteins [25-27] such as i) alterations in physico-chemical properties (conformation, charge, hydrophobicity, elasticity, solubility, and electrophoretic mobility, among others); ii) decrease/inhibition in enzyme activity and growth factors; iii) alteration of protein degradation; iv) alteration in traffic and processing of proteins; and v) formation of intra- and inter-molecular protein cross-links and aggregates.

DNA lipoxidative damage is present in the genome of healthy humans and other animal species at biologically significant levels similar or even higher that oxidation markers *sensu stricto*. DNA damage is mutagenic, carcinogenic, and have powerful effects on signal transduction pathways [28].

Finally, the amino group of aminophospholipids can also react with carbonyl compounds and initiate some of the reactions occurring in proteins [29]. Biological processes involving aminophospholipids could be potentially affected by this process. Among these processes, it may be highlighted i) asymmetrical distribution of aminophospholpids in cellular and different subcellular membranes; ii) translocation between and lateral diffusion in the membrane; iii) membrane physical properties; iv) biosynthesis and turnover of membrane phospholipids; and v) activity of membrane-bound proteins that require aminophospholipids for their function.

4.2. Cellular adaptive responses

The peroxidation of the PUFA chains of phospholipids generates a complex mixture of carbonyl compounds. Initially, these aldehydes were believed to produce only "cytotoxic" effects associated with oxidative stress (by inducing molecular damage and damaging cellular responses based on inflammatory responses, changes in gene expression and apoptosis) [10], but as depicted above, evidence is increasing in the sense that these compounds can also have specific signaling roles inducing adaptive responses driven to decrease oxidative damage and improve antioxidant defences.

Thus, available studies support the notion that superoxide radical produced by the mitochondrial electron transport chain can cause mild uncoupling of mitochondria by activating the membrane proton conductance by uncoupling proteins (UCPs). Insight into the molecular mechanism by which superoxide radical activates UCPs comes from the finding that the lipid peroxidation product 4-HNE and its homologs induce uncoupling of mitochondria through UCPs and also through the adenine nucleotide translocase [30]. This and other observations support a model in which endogenous superoxide production generates carbon-centred radicals that initiate lipid peroxidation, producing alkenals like 4-HNE that activate UCPs and adenine nucleotide translocase. So, UCPs respond to overproduction of matrix superoxide by catalyzing mild uncoupling, which lower proton motive force and would decrease superoxide production by the electron transport chain. This negative feedback loop will protect cells from ROS-induced damage and might represent the ancestral function of all UCPs.

In addition, RCS can also activate the 'antioxidant response' likely to prevent their accumulation to toxic levels [31]. This signaling cascade culminates in the nuclear translocation of and transactivation by the transcription factor Nrf2, the master regulator of the response [32]. Nrf2 activity is repressed by an inhibitory binding protein, Keap1. Keap1 retains Nrf2 in the cytosol, closely associated with the actin cytoskeleton, and promotes proteasomal degradation of Nrf2 through Cullin3-dependent polyubiquitination. Following exposure to RCS, Keap1 can be directly modified on several cysteine residues, and this modification can promote release of Nrf2. Nrf2 contains a C-terminal basic leucine zipper structure that facilitates dimerization and DNA binding, specifically to the antioxidant response element (ARE). The binding of Nrf2 to the ARE stimulates transcription of downstream cytoprotective genes [32].

5. The selective neuronal vulnerability

The idea that oxygen radicals, especially those of mitochondrial origin, are causally related to the basic aging process is increasingly receiving support from several independent sources [reviewed in 6,33]. Accordingly, the mitochondrial oxygen radical theory of aging apparently fulfils the main characteristics of this natural process: reactive oxygen species (ROS) are endogenously produced at mitochondria under normal physiological conditions, they are produced continuously throughout life (and can thus lead to progressive aging changes), and their deleterious effects on macromolecules may inflict irreversible damage during aging in post-mitotic tissues. The detrimental effects of aging are best observed in postmitotic tissues because cells that are irreversibly damaged or lost cannot be replaced by mitosis of intact ones. Nervous system is considered a postmitotic tissue, and therefore highly susceptible to aging. As this process is involved as a risk factor in most neurodegenerative diseases, and oxidative modifications play a key role in aging, it is often accepted that these diseases should have increased oxidative damage.

In this context, and from an inter-organ comparative approach, two main properties emerge as characteristics that render nervous system as especially sensible to oxidative modification: i) the % free radical leak, and ii) the membrane unsaturation. The % free radical leak (%FRL) refers the fraction (%) of electrons out of sequence which reduce oxygen to oxygen radicals (instead of reducing oxygen to water at cytochrome oxidase) in the mitochondrial respiratory chain. Since two electrons are needed to reduce one molecule of oxygen to H_2O_2, whereas four electrons are needed to reduce one molecule of oxygen to water, the free radical leak is easily calculated by dividing the rate of ROS production by 2 times the rate of oxygen consumption, the result being multiplied by 100. Results show that the higher %FRL corresponds to brain, suggesting that mitochondria are more inefficient in this tissue than in other organs (**Tables 1** and **2,** and **Figure 6**).

		Brain	**Heart**	**Kidney**	**Liver**	**Skeletal muscle**
Pyruvate/malate	State 4	11.2 ± 1.7	28.4 ± 4.7	-	7.8 ± 1.0	22 ± 3
	State 3	23.1 ± 3.0	72.2 ± 5.4	-	19.5 ± 2.4	127 ± 18
Glutamate/malate	State 4	11.5 ± 2.1	-	23.2 ± 4.2	9.4 ± 0.8	-
	State 3	23.6 ± 4.1	-	88.1 ± 13.4	72.1 ± 5.3	-
Succinate/rotenone	State 4	15.8 ± 1.6	85.8 ± 9.4	55.6 ± 11.6	26.2 ± 2.0	97 ± 12
	State 3	24.0 ± 2.0	112. ± 5.3	148.1 ± 27.0	100.9 ± 9.1	239 ± 34

Values are mean ± SEM from 8 different animals. State 4, oxygen consumption in the absence of ADP; State 3, oxygen consumption in the presence of ADP. Data from references: 34-36 and unpublished results.

Table 1. Rates of mitochondrial oxygen consumption (nmoles of O_2/ min ·mg protein) of different organs from male adult rats.

Substrate	**Brain**	**Heart**	**Kidney**	**Liver**	**Skeletal muscle**
Pyr/mal	0.2 ± 0.04	0.24 ± 0.04	-	0.06 ± 0.01	0.084 ± 0.020
Pyr/mal + Rot	1.05 ± 0.13	1.49 ± 0.12	-	0.38 ± 0.04	0.73 ± 0.14
Glut/mal	0.17 ± 0.04	-	0.075 ± 0.05	0.16 ± 0.02	-
Glut/mal + Rot	0.72 ± 0.10	-	0.66 ± 0.12	0.46 ± 0.04	-
Succ + Rot	0.26 ± 0.05	0.51 ± 0.05	0.10 ± 0.04	0.32 ± 0.04	0.31 ± 0.06
Succ	1.72 ± 0.17	0.71 ± 0.12	0.52 ± 0.08	1.00 ± 0.18	1.57±0.31
Succ + AA	2.67 ± 0.31	-	5.97 ± 0.45	3.00 ± 0.41	2.57± 0,30

Values are means ± SEM from 8 different animals. Pyr/mal = pyruvate/malate; Glu/mal = glutamate/malate; Succ, succinate; Rot, rotenone; AA, antimycin A. Data from references: 34-36 and unpublished results.

Table 2. Rates of mitochondrial H_2O_2 production (nanomoles H_2O_2/min mg protein) of different organs from male adult rats.

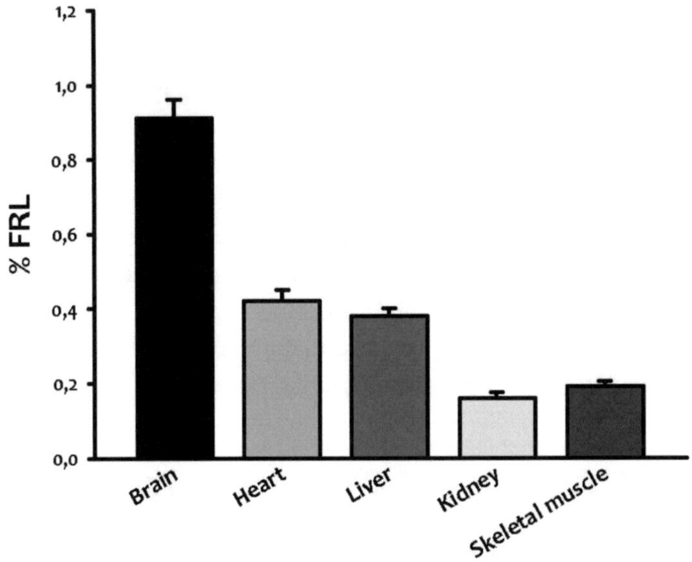

Figure 6. % Free radical leak (% FRL) of different organs from male adult rats. Data from references: 34-36 and unpublished results.

As mentioned above, the susceptibility of biological membranes to oxidative alterations is related to two inherent traits, the physico-chemical properties of the lipid bilayer and the chemical reactivity of the fatty acids which make up the membrane. **Table 3** shows the mitochondrial fatty acid composition (mol%) of different organs. Data clearly indicate that average chain length is maintained around 18 carbon atoms and that the % saturated:unsaturated follows the ratio around 40:60. By contrast, the more relevant

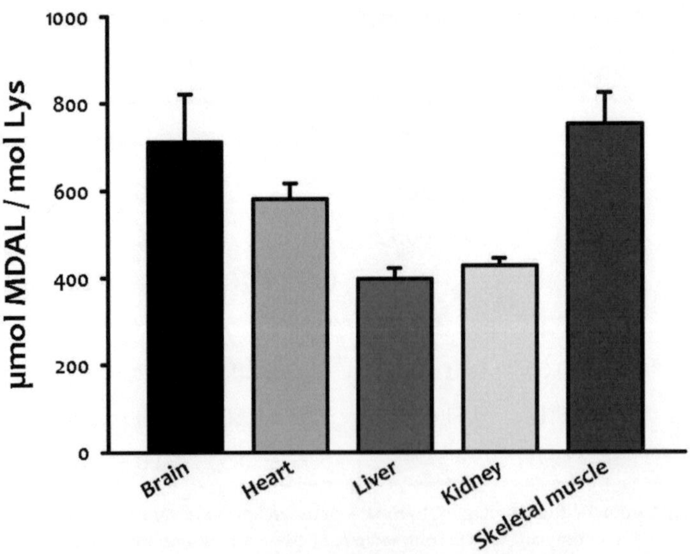

Figure 7. Steady-state level of mitochondrial lipoxidative-derived protein damage in different organs from male adult rats. MDAL, malondialdehyde-lysine. Units: µmol/mol lysine.

differences are droved to the distribution of the different types of PUFAs. Thus, brain is characterized by the presence of the higher content of monounsaturated fatty acids, as well as PUFAn-3, and particularly, the docosahexaenoic acid (22:6n-3) compared to the other postmitotic organs (**Table 3**). This higher PUFAs content which are more susceptible to oxidative damage leads to a higher steady-state level of lipoxidation-derived molecular damage at least at mitochondrial level (**Figure 7**).

	Brain	Heart	Kidney	Liver	Skeletal muscle
14:0	0.14±0.03	1.17±0.03	0.36±0.03	0.18±0.01	0.82±0.14
16:0	10.55±0.46	22.68±0.60	10.00±0.38	17.31±0.44	20.39±0.45
16:1n-7	0.28±0.02	0.91±0.04	0.48±0.09	1.11±0.07	0.67±0.05
18:0	18.12±0.39	26.49±0.34	12.67±0.56	17.19±0.24	18.34±0.73
18:1n-9	24.78±0.31	13.23±0.39	8.97±0.44	9.09±0.17	8.44±0.98
18:2n-6	1.55±0.07	9.27±0.42	14.10±0.42	19.27±0.66	17.73±1.23
18:3n-3	0.12±0.03	0.38±0.01	0.47±0.11	0.18±0.04	0.80±0.12
18:4n-6	2.38±0.08	0.11±0.01	4.67±0.40		
20:0	1.05±0.07	0.20±0.01	0.51±0.07		
20:1n-9	4.12±0.18	0.19±0.02	0.69±0.09		
20:2n-6	0.52±0.05	0.97±0.14	0.92±0.06	0.35±0.01	
20:3n-6	0.10±0.01	0.79±0.06	0.38±0.03	0.27±0.02	
20:4n-6	12.74±0.13	15.69±0.45	36.32±1.21	26.61±0.44	13.96±0.79
20:5n-3	0.14±0.01	0.15±0.01	0.40±0.07	0.35±0.02	1.66±0.44
22:0	1.57±0.26	0.21±0.02	2.24±0.16		
22:4n-6	4.19±0.16	0.73±0.06	0.46±0.02	0.16±0.008	1.54±0.35
22:5n-6	0.68±0.03	1.40±0.22	0.46±0.04	0.28±0.04	1.99±0.23
22:5n-3	2.00±0.16	0.84±0.04	1.56±0.11	0.79±0.04	2.95±0.23
22:6n-3	12.81±0.33	4.28±0.26	1.81±0.19	6.79±0.44	10.65±0.96
24:0	0.14±0.04	0.25±0.04	0.27±0.04		
24:5n-3	0.26±0.07		0.93±0.22		
24:6n-3	1.65±0.16		1.24±0.28		
ACL	19.12±0.05	18.15±0.03	18.97±0.03	18.49±0.01	18.54±0.02
SFA	31.61±0.52	51.02±0.71	26.07±0.62	34.69±0.50	39.56±1.04
UFA	68.38±0.52	48.97±0.71	73.92±0.62	65.30±0.50	60.43±1.04
MUFA	29.19±0.32	14.34±0.44	10.15±0.51	10.20±0.18	9.11±1.01
PUFA	39.19±0.73	34.63±0.94	63.77±1.07	55.09±0.56	51.32±1.41
PUFAn-6	22.18±0.20	28.97±0.92	57.33±1.16	46.97±0.95	36.91±1.18
PUFAn-3	17.00±0.57	5.66±0.25	6.43±0.32	8.12±0.41	14.41±0.88

Values: mean±SEM. N x group: 8. ACL, average chain length; SFA, saturated faaty acids; UFA, unsaturated fatty acids; PUFA n-6/n-3, polyunsaturated fatty acids n-6 or n-3 series; MUFA, monounsaturated fatty acids. Data from references: 34-36 and unpublished results.

Table 3. Mitochondrial fatty acid composition (mol%) of different organs from male adult rats.

Compared to other postmitotic organs, brain seems to be the tissue more susceptible to oxidative damage. However, are there cross-regional differences in the nervous system? Available data seems to indicate that this is the case. Thus, **Table 4** reflects the fatty acid composition in seven different regions from human nervous system samples. The data are conclusive: despite to maintain a stable average chain length (around 18 carbon atoms) and a ratio saurated:unsaturated practically identical (40:60), there is a cross-regional difference with respect to the type of PUFA distribution affecting very specially to both monounsaturated and polyunsaturated fatty acids that seems to be inversely related. In other words, the higher the presence of monounsaturated fatty acids for a given region, the lower the PUFA content. The meaning of this differential distribution remains to be elucidated, but it is evident that determine a differential susceptibility to oxidative damage.

	Frontal Cortex	Occipital Cortex	Hippocampus	Amygdala	Substantia Nigra	Medulla Oblongata	Spinal Cord
14:0	0.51±0.04	0.54±0.05	0.42±0.05	0.56±0.05	0.54±0.04	0.50±0.04	0.48±0.05
16:0	21.13±0.51	20.30±1.70	19.54±0.27	20.69±0.67	12.68±0.39	13.57±0.38	14.75±0.25
16:1n-7	0.98±0.12	0.76±0.19	0.89±0.10	1.48±0.12	1.68±0.18	1.21±0.03	0.94±0.06
18:0	21.35±1.23	19.23±0.41	21.89±0.20	20.19±0.67	24.61±0.27	25.64±0.61	22.07±1.72
18:1n-9	23.78±1.51	28.57±1.16	26.12±0.50	28.40±0.77	33.57±0.19	30.13±0.87	33.51±1.43
18:2n-6	0.77±0.12	0.57±0.07	0.59±0.04	0.53±0.09	0.33±0.06	0.56±0.01	0.96±0.61
18:3n-3	0.14±0.01	0.21±0.04	0.23±0.02	0.08±0.006	0.17±0.003	0.23±0.01	0.34±0.05
20:0	1.25±0.23	2.40±0.52	1.44±0.09	1.71±0.12	4.95±0.12	5.58±0.23	7.62±0.53
20:1	0.19±0.04	0.32±0.05	0.37±0.01	0.28±0.02	0.22±0.02	0.45±0.02	0.75±0.09
20:2n-6	0.23±0.01	0.22±0.01	0.23±0.02	0.20±0.02	0.23±0.03	0.31±0.06	0.28±0.01
20:3n-6	0.63±0.07	0.64±0.04	0.90±0.10	0.60±0.08	0.61±0.02	0.99±0.03	0.87±0.16
20:4n-6	8.28±0.19	6.07±0.35	8.05±0.24	7.42±0.42	3.70±0.29	3.95±0.33	3.13±0.09
22:4n-6	4.77±0.14	5.06±0.69	6.24±0.20	4.79±0.27	4.32±0.15	4.07±0.08	3.16±0.17
22:5n-6	0.67±0.09	0.59±0.09	0.92±0.08	0.94±0.11	0.12±0.007	0.40±0.02	0.24±0.04
22:5n-3	0.26±0.03	0.19±0.02	0.33±0.02	0.13±0.04	0.06±0.006	0.09±0.01	0.38±0.14
22:6n-3	13.68±0.17	13.71±0.71	10.97±0.31	9.99±0.58	9.74±0.28	9.28±0.45	8.84±0.71
24:0	0.38±0.06	0.31±0.05	0.45±0.04	0.47±0.03	0.64±0.06	1.19±0.22	1.00±0.08
24:1n-9	0.93±0.23	0.21±0.03	0.33±0.04	1.46±0.32	1.73±0.18	1.75±0.11	0.59±0.10
ACL	18.60±0.01	18.56±0.04	18.58±0.008	18.48±0.04	18.59±0.005	18.64±0.03	18.52±0.01
SFA	44.64±1.33	42.80±0.74	43.77±0.37	43.65±0.94	43.43±0.34	46.51±0.47	45.95±1.81
UFA	55.35±1.33	57.19±0.74	56.22±0.37	56.34±0.94	56.56±0.34	53.48±0.47	54.04±1.81
MUFA	25.90±1.70	29.88±1.11	27.74±0.52	31.62±1.05	37.22±0.24	33.56±0.92	35.80±1.53
PUFA	29.45±0.46	27.31±0.38	28.48±0.25	24.71±1.02	19.33±0.42	19.92±0.45	18.24±0.43
PUFAn-6	15.36±0.36	13.18±0.25	16.95±0.36	14.49±0.66	9.34±0.26	10.30±0.30	8.67±0.81
PUFAn-3	14.08±0.18	14.12±0.63	11.53±0.34	10.21±0.59	9.99±0.28	9.61±0.46	9.57±0.57

Values: mean±SEM. N x group: 3-9. ACL, average chain length; SFA, saturated faaty acids; UFA, unsaturated fatty acids; PUFA n-6/n-3, polyunsaturated fatty acids n-6 or n-3 series; MUFA, monounsaturated fatty acids.
(#) Unpublished results.

Table 4. Fatty acid composition (mol%) in the human nervous system: a cross-regional comparative approach [#].

This means that saturated and monounsaturated fatty acyl chains (SFA and MUFA) are essentially resistant to peroxidation while polyunsaturates (PUFA) are damaged. Furthermore the greater the degree of polyunsaturation of PUFA the more prone it is to peroxidative damage. Indeed Holman [23] empirically determined (by measurement of oxygen consumption) the relative susceptibilities the different acyl chains (see **Figure 8**): Docosahexaenoic acid (DHA), the highly polyunsaturated omega-3 PUFA with six double bonds is extremely susceptible to peroxidative attack and is eight-times more prone to peroxidation than linoleic acid (LA) which has only two double bonds. DHA is 320-times more susceptible to peroxidation than the monounsaturated oleic acid (OA) [22,23].

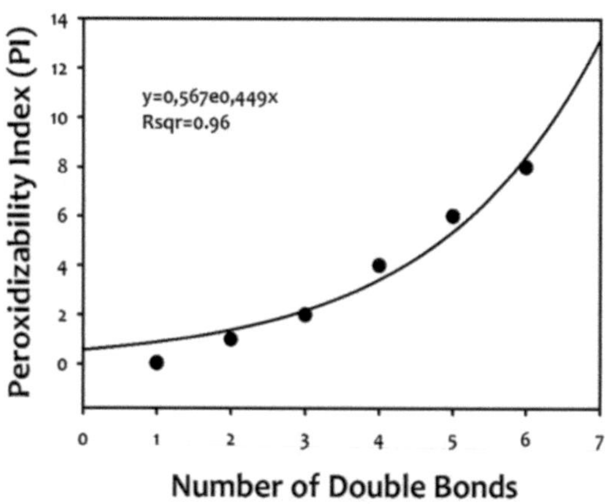

Figure 8. The relative susceptibilities of selected unsaturated fatty acids to peroxidation. Data are from [23], and all were empirically determined as rates of oxygen consumption. They are expressed relative to the rate for linoleic acid (18:2n-6) which is arbitrarily given a value of 1.

Combining the relative susceptibilities of different fatty acids with the fatty acid composition of membrane lipids it is possible to calculate a peroxidizability index[1] (a measure of the susceptibility to peroxidation) for any particular membrane. The peroxidation index of a membrane is not the same as its unsaturation index (sometimes also called its "double bond index") which is a measure of the density of double bonds in the membrane. For example, a membrane bilayer consisting solely of MUFA will have an unsaturation index of 100 and a peroxidation index of 2.5, while a membrane bilayer consisting of 95% SFA and 5% DHA will have an unsaturation index of 30 and a peroxidation index of 40. This means that although the 5% DHA-containing membrane has only 30% the density of double bonds of the monounsaturated bilayer, it is 16-times more susceptible to peroxidative damage. In this context, data clearly show the existence of very important cross-regional differences in peroxidizability index in human central nervous system (**Figure 9**).

[1] *[Peroxidizability Index (PI) = 0.025 x (% monoenoics) + 1 x (% dienoics) + 2 x (% trienoics) + 4 x (% tetraenoics) + 6 x (% pentaenoics) + 8 x (% hexaenoics), while Unsaturation index (UI) = 1 x (% monoenoics) + 2 x (% dienoics) + 3 x (% trienoics) + 4 x (% tetraenoics) + 5 x (% pentaenoics) + 6 x (% hexaenoics)]*

The different cross-regionally PIs observed are due to changes in the type of unsaturated fatty acid that participates in membrane composition. So, there is a systematic redistribution between the types of PUFAs present from highly unsaturated fatty acids to the less unsaturated that is region-specific. Surprisingly, the change shows a gradient that follows the cranio-caudal axis considering the structural organization of the CNS. The mechanism(s) responsible for the cross-regional-related differences in fatty acid profile can be related, in principle, to the fatty acid desaturation pathway, and the deacylation-reacylation cycle. The available delta-5 and delta-6 estimated desaturase activities indicate that they are several folds higher in frontal cortex than in spinal cord (**Figure 10**). The mechanism underlying to the membrane unsaturation regulation could explain the differences in membrane fatty acid composition and, in turn, the peroxidizability index, and suggest a regulatory mechanism region-specific that is expressed differentially in a cranio-caudal axis likely associated to the development process and even the evolution of the central nervous system.

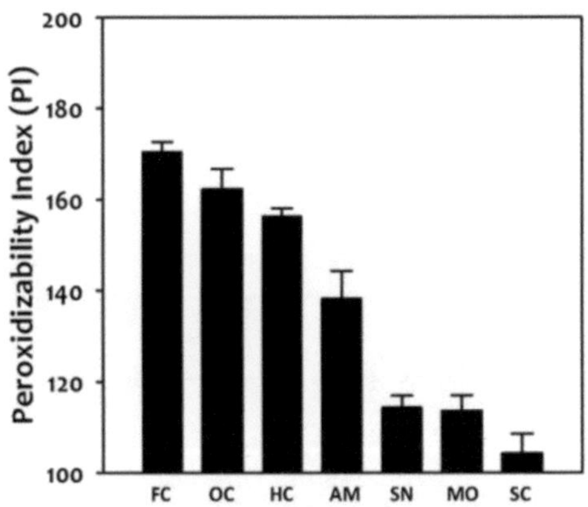

Figure 9. Cross-regional differences in the peroxidizability index in the human central nervous system. FC, frontal cortex; OC, occipital cortex; HC, hippocampus; AM, amygdala; SN, substantia nigra; MO, medulla oblongata; SC, spinal cord.

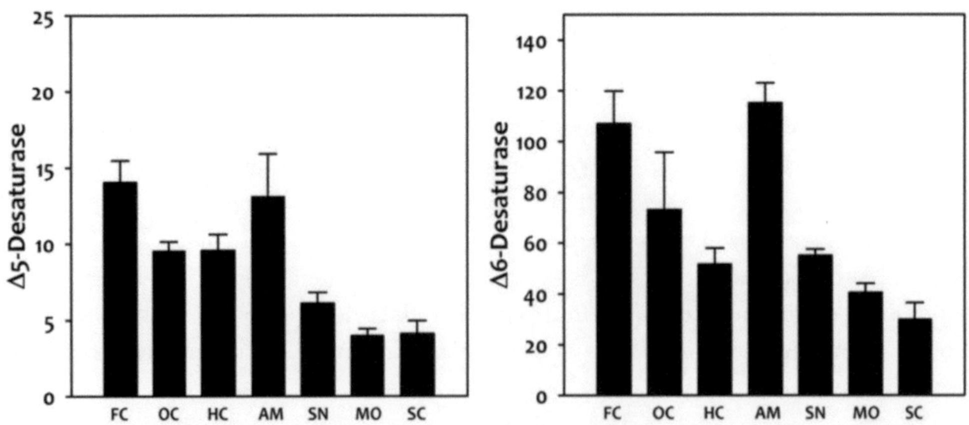

Figure 10. Delta-5 and delta-6 desaturase activities in different regions of the human central nervous system. FC, frontal cortex; OC, occipital cortex; HC, hippocampus; AM, amygdala; SN, substantia nigra; MO, medulla oblongata; SC, spinal cord.

In summary, membrane unsaturation is a key characteristic able to define the selective neuronal vulnerability. In this context, it is plausible to postulate that membrane unsaturation could be a main determinant factor in determining differences in the rate of aging for different regions of the CNS, and in the occurrence of neurodegenerative disorders [e.g., Alzheimer's disease (AD), Parkinson's disease (PD), or amyotrophic lateral sclerosis (ALS), among others] during the sixth, seventh and eighth decades of life. Interestingly, this property is also causally related to the aging process and the lifespan of animal species [6].

6. Conclusions and perspectives

A major goal of research into aging is to extend 'healthspan' by identifying approaches for delaying or preventing age-related diseases. The fact that many individuals maintain a well-functioning nervous system and continue productive lives through their seventies, eighties and even nineties is encouraging. The implication is that if the cellular and molecular mechanisms that determine whether nervous systems adapt positively or develop a disease during aging can be identified, then disease processes can be averted. In this regard, oxidative and metabolic stress and impaired cellular stress adaptation, are mechanisms of aging that render neurons vulnerable to degeneration. On this background of age-related endangerment, genetic and environmental factors likely determine whether a disease process develops. These include causal mutations, more subtle genetic risk factors and environmental factors, including aspects of diet and lifestyle. Because of the cellular and molecular complexity of the nervous system, and the signalling mechanisms that influence neuronal plasticity and survival, the basis of SNV remains elusive. However, available evidence from both an inter-organ comparative approach and cross-regional differences seems to confirm this idea, highlighting membrane unsaturation as a key trait associated with selective neuronal vulnerability. Interestingly, this property is also causally related to the aging process and the lifespan of animal species [6,10], and it is apparently operative in multiple neurodegenerative disorders [3,37]. Currently, most efforts to prevent and treat neurodegenerative disorders are focusing on diet, on lifestyle modification, and on drugs that target disease processes [3,6,33]. Although data on humans is still limited, the emerging evidence that dietary restriction (along with exercise and cognitive stimulation) can bolster neuroprotective mechanisms suggests that diet and lifestyle changes could reduce the risk of neurodegenerative disorders [6,26,33]. Therefore, it seems likely that extension of neural healthspan is possible for most individuals.

Author details

Alba Naudí, Mariona Jové, Victòria Ayala, Omar Ramírez, Rosanna Cabré,
Joan Prat, Manuel Portero-Otin, Isidre Ferrer and Reinald Pamplona
Department of Experimental Medicine, University of Lleida-Biomedical Research Institute of Lleida, Lleida, Spain

Acknowledgement

Work carried out at the Department of Experimental Medicine was supported in part by R+D grants from the Spanish Ministry of Science and Innovation [BFU2009-11879/BFI], and the Autonomous Government of Catalonia [2009SGR735]. The authors declare no conflict of interest.

7. References

[1] Mattson MP, Magnus T (2006) Ageing and neuronal vulnerability. Nat. Rev. Neurosci. 7(4):278-94.

[2] Wang X, Michaelis EK (2010) Selective neuronal vulnerability to oxidative stress in the brain. Front. Aging Neurosci. 2:12.

[3] Martínez A, Portero-Otin M, Pamplona R, Ferrer I (2010) Protein targets of oxidative damage in human neurodegenerative diseases with abnormal protein aggregates. Brain Pathol. 20(2):281-97.

[4] Halliwell B, Gutteridge JMC (2007) Free radicals in biology and medicine. Oxford: Oxford University Press. 851 p.

[5] Barja G. (2007) Mitochondrial oxygen consumption and reactive oxygen species production are independently modulated: implications for aging studies. Rej. Res. 10(2):215-24.

[6] Pamplona R, Barja G (2011) An evolutionary comparative scan for longevity-related oxidative stress resistance mechanisms in homeotherms. Biogerontology, 12(5):409-35.

[7] Vereb G, Szollosi J, Matko J, Nagy P, Farkas T, Vigh L, Matyus L, Waldmann TA, Damjanovich S (2003) Dynamic, yet structured: The cell membrane three decades after the Singer-Nicolson model. Proc. Natl. Acad. Sci. U.S.A. 100(14):8053-8.

[8] Van Meer G, Voelker DR, Feigenson GW (2008) Membrane lipids: where they are and how they behave. Nature Rev. Mol. Cell. Biol. 9(2):112-24.

[9] Dowhan W (1997) Molecular basis for membrane phospholipid diversity: why are there so many lipids? Annu. Rev. Biochem. 66:199-232.

[10] Pamplona R (2008) Membrane phospholipids, lipoxidative damage and molecular integrity: a causal role in aging and longevity. Biochim. Biophys. Acta. 1777(10):1249-62.

[11] Wallis JG, Watts JL, Browse J (2002) Polyunsaturated fatty acid synthesis: what will they think of next? Trends Biochem. Sci. 27(9):467.

[12] Scott BL, Bazan NG (1989) Membrane docosahexaenoate is supplied to the developing brain and retina by the liver. Proc. Natl. Acad. Sci. U. S. A. 86(8):2903-7.

[13] Moore SA, Yoder E, Murphy S, Dutton GR, Spector AA (1991) Astrocytes, not neurons, produce docosahexaenoic acid (22:6 omega-3) and arachidonic acid (20:4 omega-6). J. Neurochem. 56(2):518-24.

[14] Sprecher H (2000) Metabolism of highly unsaturated n-3 and n-6 fatty acids. Biochim. Biophys. Acta. 1486(2-3):219-31.

[15] Nakamura MT, Nara TY (2004) Structure, function, and dietary regulation of delta6, delta5, and delta9 desaturases. Annu. Rev. Nutr. 24:345-76.

[16] Pereira SL, Leonard AE, Huang YS, Chuang LT, Mukerji P (2004) Identification of two novel microalgal enzymes involved in the conversion of the omega3-fatty acid, eicosapentaenoic acid, into docosahexaenoic acid. Biochem. J. 384(Pt 2):357-66.
[17] Moore SA, Yoder E, Spector AA (1990) Role of the blood-brain barrier in the formation of long-chain omega-3 and omega-6 fatty acids from essential fatty acid precursors. J. Neurochem. 55(2):391-402.
[18] Williard DE, Harmon SD, Kaduce TL, Preuss M, Moore SA, Robbins ME, Spector AA (2001) Docosahexaenoic acid synthesis from n-3 polyunsaturated fatty acids in differentiated rat brain astrocytes. J. Lipid Res. 42(9):1368-76.
[19] Rapoport SI, Chang MC, Spector AA (2001) Delivery and turnover of plasma-derived essential PUFAs in mammalian brain. J. Lipid Res. 42(5):678-85.
[20] Akbar M, Calderon F, Wen Z, Kim HY (2005) Docosahexaenoic acid: a positive modulator of Akt signaling in neuronal survival. Proc. Natl. Acad. Sci. U. S. A. 102(31):10858-63.
[21] Calderon F, Kim HY (2004) Docosahexaenoic acid promotes neurite growth in hippocampal neurons. J. Neurochem. 90(4):979-88.
[22] Hulbert AJ, Pamplona R, Buffenstein R, Buttemer WA (2007) Life and death: metabolic rate, membrane composition, and life span of animals. Physiol. Rev. 87(4):1175-213.
[23] Holman RT (1954) Autoxidation of fats and related substances, in: Holman RT, Lundberg WO, Malkin T. Progress in Chemistry of Fats and Other Lipids. London: Pergamon Press. pp. 2-51.
[24] Esterbauer H, Schaur RJ, Zollner H (1991) Chemistry and biochemistry of 4-hydroxynonenal, malonaldehyde and related aldehydes. Free Radic. Biol. Med. 11(1):81-128.
[25] Thorpe SR, Baynes JW (2003) Maillard reaction products in tissue proteins: new products and new perspectives. Amino Acids. 25(3-4):275-81.
[26] Pamplona R (2011) Advanced lipoxidation end-products. Chem. Biol. Interact. 192(1-2):14-20.
[27] Portero-Otin M, Pamplona R (2006) Is endogenous oxidative protein damage involved in the aging process? In: Pietzsch J, editors. Protein Oxidation and Disease. Kerala: Research Signpost. pp 91-142.
[28] Marnett LJ (2002) Oxy radicals, lipid peroxidation and DNA damage. Toxicol. 181-182:219-22.
[29] Portero-Otín M, Bellmunt MJ, Ruiz MC, Pamplona R (2000) Nonenzymatic modification of aminophospholipids by the Maillard reaction in vivo. In: Castell J, Garcia Regueiro JA. Research Advances in Lipids, vol.1. India: Global Research Network. pp. 33-41.
[30] Echtay KS, Esteves TC, Pakay JL, Jekabsons MB, Lambert AJ, Portero-Otín M, Pamplona R, Vidal-Puig AJ, Wang S, Roebuck SJ, Brand MD (2003) A signalling role for 4-hydroxy-2-nonenal in regulation of mitochondrial uncoupling. EMBO J. 22(16):4103-10.
[31] Wakabayashi N, Dinkova-Kostova AT, Holtzclaw WD, Kang MI, Kobayashi A, Yamamoto M, Kensler TW, Talalay P (2004) Protection against electrophile and oxidant stress by induction of the phase 2 response: fate of cysteines of the Keap1 sensor modified by inducers. Proc. Natl. Acad. Sci. U.S.A. 101(7):2040-5.

[32] Copple IM, Goldring CE, Kitteringham NR, Park BK (2008) The Nrf2-Keap1 defence pathway: role in protection against drug-induced toxicity. Toxicol. 246(1):24-33.

[33] Sanz A, Pamplona R, Barja G (2006) Is the mitochondrial free radical theory of aging intact? Antioxid. Redox Signal. 8(3-4):582-99.

[34] Saborido A, Naudí A, Portero-Otín M, Pamplona R, Megías A (2011) Stanozolol treatment decreases the mitochondrial ROS generation and oxidative stress induced by acute exercise in rat skeletal muscle. J. Appl. Physiol. 110(3):661-9.

[35] Caro P, Gomez J, Sanchez I, Naudí A, Ayala V, López-Torres M, Pamplona R, Barja G (2009) Forty percent methionine restriction decreases mitochondrial oxygen radical production and leak at complex I during forward electron flow and lowers oxidative damage to proteins and mitochondrial DNA in rat kidney and brain mitochondria. Rej. Res. 12(6):421-34.

[36] Gomez J, Caro P, Sanchez I, Naudi A, Jove M, Portero-Otin M, Lopez-Torres M, Pamplona R, Barja G (2009) Effect of methionine dietary supplementation on mitochondrial oxygen radical generation and oxidative DNA damage in rat liver and heart. J. Bioenerg. Biomembr. 41(3):309-21.

[37] Ferrer I (2009) Altered mitochondria, energy metabolism, voltage-dependent anion channel, and lipid rafts converge to exhaust neurons in Alzheimer's disease. J. Bioenerg. Biomembr. 41(5):425-31.

Chapter 21

Role of Lipid Peroxidation in the Pathogenesis of Age-Related Cataract

Bojana Kisic, Dijana Miric, Lepsa Zoric and Aleksandra Ilic

Additional information is available at the end of the chapter

http://dx.doi.org/10.5772/45942

1. Introduction

The occurrence and development of cataract affect the decline of visual, working and living comfort. Cataract is the leading cause of blindness, accounting for 50% of blindness worldwide [1]. Cataract is progressive lens opacity in humans of 45 years or more, occurring without any known cause such as trauma, inflammation, hypocalcemia, medications or congenital factors. Risk factors for the occurrence of cataract are numerous: aging, diabetes mellitus, UV radiation, malnutrition, smoking, hypertension, renal disease, and others. Free oxygen radicals and oxidative stress are considered to be an important factor contributing to age-related cataract [1,2]. Oxidative stress has been shown to cause cataract in in vitro models [3]. This hypothesis is supported by studies that examined the anticatarogenic effect of different nutritional and physiological antioxidants [4].

Oxygen does not manifest toxic effects on cells of aerobic organisms in molecular form, but in the form of free oxygen radicals. Free radicals occur in univalent transfer of electrons to molecular oxygen. Due to its biochemical nature, and the low activation energy, they are able to react with biomolecules of all cellular structures, thereby carrying out their chemical and physiological modification. Under physiological conditions, the level of free radicals is controlled by mechanism of antioxidant protection. The balance between the production and catabolism of oxidants by cells and tissue is essential for maintenance of the biologic integrity of the tissue. Ocular tissues contain antioxidants that prevent damage from excessive oxygen metabolites: antioxidant enzymes, proteins, ascorbic acid, glutathione, amino acids cysteine and tyrosine, and other.

Changes in the oxidation of biomolecules can be found in many human diseases of the body, but the cataract is one of the most common diseases, where oxidative modifications of proteins [5] and lipids [1,2] is a dominant metabolic substrate of pathological disorders. Oxidative modification of lens proteins, loss of protein function and the creation of protein aggregates of high molecular mass, which increases the scattering of light, are the main features of age-related cataract [6]. These protein modifications might be caused by oxidative stress resulting in higher levels of reactive oxygen radicals.

2. Sources of reactive oxygen species (ROS) in the lens

2.1. Reactive oxygen species generated in the lens by the UV irradiation

Human lens has several systems of defense from ROS and oxidative stress, which are together responsible for the maintenance of lens transparency and prevention of cataract. But during the life the lens is exposed to multiple sources of oxidative stress, endogenous (altered mitochondrial respiration, respiratory burst of phagocytes, viral infection) and exogenous (UV light, metals,

drugs, cigarette smoke), which can lead to production of reactive oxygen species: superoxide anion ($O_2^{\bullet-}$), hydrogen peroxide (H_2O_2), hydroxyl radical ($HO^{\bullet}$) and others. The lens has a protective role for the other eye structures also, because light and oxygen are synergistically involved in the pathogenesis of cataract. By absorbing the part of the ultraviolet spectrum the lens protects deeper structures of the eye from the harmful effects of the solar radiation, whereby it is only subject to photooxidative damage. Photooxidative stress and the formation of reactive oxygen species by photosensitizing mechanisms are due to absorption of light by the biomolecules of the lens. Specifically, UV irradiation can mediate damage of lens structures, due to: direct absorption of the incident light by the cellular components, resulting in excited state formation and subsequent chemical reaction, and photosensitization mechanisms, where the light is absorbed by endogenous photosensitizers that are excited to their triplet states [6]. The excited photosensitizers can induce cellular damage by electron transfer and hydrogen abstraction processes to yield free radicals or energy transfer with O_2 to yield the reactive excited state, singlet oxygen.

Experiments in organ culture have shown that cataract can be caused by photochemical production of superoxide radicals, hydroxyl radicals and H_2O_2 [7]. Other researchers [8,9] indicate that photochemical generation of reactive species of oxygen in the lens and aqueous and consequent damage to the tissue has been implicated in the genesis of age-related cataract. The fact that the incidence of cataract is higher in the population that is more exposed to sunlight [10] imposes the assumption that photocatalytic conversion of molecular oxygen from ground state to excitatory states, which are highly reactive ($O_2^{\bullet-}$, H_2O_2, $HO^{\bullet}$ and others) occurs. High concentration of ascorbate in the aqueous humor is assumed to represent a kind of filter that prevents the penetration of UV light in the lens and thus protects tissue from oxidative damage, particularly photoinduced damage [11].

Photosensible substance, that absorbs certain wavelengths of light, activates and subtracts hydrogen or electrons from the substrate by converting them into free radicals. In the presence of O_2 , the energy is transferred from excitatory substance and produce $^1O_2$, which can initiate the process of lipid peroxidation. In the ocular tissues numerous substances can initiate photodynamic reactions. These substances are riboflavin, heme derivatives, tryptophan and its oxidation product N-formylkynurenine, lipofuscin, visible pigments (retinol), and photosensible substances of exogenous origin, such as drugs [12]. Key link between photo-oxidation and cataract is that photo-oxidation of thiol groups on lens crystallyne produces disulfide bridges between molecules and, the build-up of these will lead to protein aggregation and hence cataract.

3. Mitochondria as a source of reactive oxygen species in the lens

In the ocular tissues, including the lens, as in other organ systems, ROS are formed in the mitochondria via the electron transport chain where inefficient electron coupling leads to the formation of superoxide anion. Molecular oxygen is tightly bound to the enzyme complex cytochrome C oxidase. However, the bond on the vectors of electrons in the respiratory chain in front of the system cytochrome C oxidase, on the level of NADH-coenzyme-Q reductase and the reduced forms of coenzyme Q, is not that strong and some of transferred electrons can "leak" from the system on molecular oxygen, forming $O_2^{\bullet-}$. Superoxide production is significantly increased during reperfusion of tissues, when the availability of oxygen is increased.

The human lens consists of three metabolically different zones: the epithelium, the cortex and lens nucleus. Epithelial cells and superficial cortical fibers are metabolically most active, and the greatest part of mitochondrion respiration and aerobic glycolysis in the lens occurs in them [13]. One third of total energy produced (ATP) in the lens is produced in epithelial cells under aerobic conditions, while the metabolic activity of nuclear part of the lens is at much lower level. Intense metabolic activity makes epithelial cells susceptible to oxidative damage, especially their membrane pump systems and DNA. Oxidation of unsaturated lipids in epithelial cells could be the initial step that leads to generation of oxidation products. If reactive oxygen species or secondary products of lipid oxidation from the epithelium were to migrate to the fiber cells, it is possible that prolonged accumulation of lipid oxidative products could eventually lead to alterations in fiber cell structure and increased opacity, which leads to the development of cataract.

Thiol (-SH) groups of membrane proteins, the lens epithelial cells, which are significant for regulation of ion transport, are very susceptible to oxidative attack, especially when the concentration if intracellular GSH is reduced. The optimal membrane function of lens epithelial cells depends on reduced state of protein-SH groups. The oxidation of membrane thiol (-SH) groups of the lens cells leads to breakdown of active transport through the membrane, to the increase of membrane permeability and consequently intracellular alternations, which is involved in the development of cataract. The consequence of impaired active transport is also the reduced level of ascorbic acid in the lens. Studies have confirmed that ascorbic acid (AA) levels in human lenses with the development of cataract are reduced [34], and concentration of dehydroascorbic acid (DHA) is increased [14,35]. Timely removal of dehydroascorbic acid from the lens is important because of its potential toxicity as oxidant. Increase of the current concentration of DHA/AA redox balance can be an indicator of oxidative stress in the lens [35].

4. The importance of ascorbic acid in lens

The role of ascorbic acid is important, as a strong reductant and effective scavenger of hydroxyl and superoxide anion radical. Vitamin C has antioxidant, but also prooxidant properties. In which direction will vitamin C work, depends on the concentration of vitamin C, oxygen and the presence of metal ions. Oxidation is the cause of modification of lens proteins which accumulate over a lifetime. Some believe that ascorbate can contribute to protein modifications, react as prooxidant and participate in reactions that generate radicals [14].

These reactions may be caused by light or metal-catalyzed oxidation of endogenous ascorbic acid. It is known that copper and iron are present in micromolar concentrations and that autooxidative processes can occur in the lens. Fenton-type reactions, where H_2O_2 reacts with free metal ions, iron (Fe^{2+}) or copper (Cu^{2+}) to produce the $HO^{\bullet}$ radical, are a major source of oxidative stress initiated by transition metals [15] and are thought to be involved in the formation of cataract [16]. In the presence of metals, especially iron and copper, and oxygen, ascorbic acid is oxidized to dehydroascorbate, which produces hydrogen peroxide and metal is reduced. Hydrogen peroxide can react with reduced metal, generating hydroxyl radical and other reactive oxygen radicals [17]. When copper and protein-bound iron is included in this reaction, the radicals cause oxidative modification of amino-acids that are near the metal. In this way ascorbate can actually become a prooxidant and lead to protein damage via both H_2O_2 and Fenton production of $HO^{\bullet}$. These reactions become important when cells lose their ability to remove metals, making it available for reaction and/or when cells lose their ability to maintain their vitamin C in a reduced form. It is noted that during the aging of lens, as in cataract lenses the concentration of copper and iron increases [18,19]. The data that confirm the level of iron and copper ions is lower in non-cataract lenses and study that compared cortical nuclear and mature cataracts found higher iron levels in the mature cataract [20] suggest that metal ions that mediate the production of $HO^{\bullet}$, may be important in the development of age-related cataract [16].

Experiments on isolated proteins showed that oxidation products of ascorbate (dehydroascorbic acid) can form cross-link with crystalline lens, producing molecules of high molecular weight, which cause light scattering typical for cataract [21]. It is assumed that similar modification of lens proteins occurs in vivo during the development of age-related cataract [22].

5. Lipid peroxidation in the lens

In physiological/controlled conditions the process of lipid peroxidation affects the permeability of cell membranes, the metabolism of membrane lipids and proteins, provides control of cell proliferation, but the adverse effects of this process occurring under conditions of oxidation stress, ie. in conditions of impaired balance of prooxidative and antioxidative factors of the cell. Lipid peroxidation (LPO) is considered a pathogenetic factor of cataractogenesis [1,2,23,24,25]. LPO in the lens may be induced by endogenous or exogenous factors: enzymes, reactive oxygen species, metal ions, UV irradiations, heat, radical-initiating chemicals, drugs. Cell membrane lipids (phospholipids, glycolipids) are the most common substrates of oxidative attacks, and since the cell membranes have lipoprotein structure, the structure of membrane proteins is disturbed at the same time. That causes the disturbance of cell membrane barrier function, leading to a larger entry of calcium and other ions [26]. Structural changes of the cell membrane and its increased permeability change the cell volume and the configuration of the lens, leading to refractory changes that are associated with the early cataract.

Cell membranes are very sensitive to the effects of oxygen radicals, due to the presence of polyunsaturated fatty acids in lipids. Fatty acids in lipids of cell membranes contain a different number of carbon atoms (14 to 24), and the present double bonds are in cis configuration. The presence of double bond in the neighborhood destabilize the bond between the carbon and the hydrogen of methylene group in the chain of polyunsaturated fatty acid, and by subtracting hydrogen from such a methylene group by reactive oxidant, begins the process of oxidative modification of fatty acids - lipid peroxidation [27]. Non-enzymatic peroxidation of polyunsaturated fatty acids is a process that takes place in three stages: initiation, propagation and termination. The intensity of this process, as well as the ability of partial or complete repair of damage, depends on pro/anti-oxidative environment in which this process occurs.

Free radicals formed during lipid peroxidation have a local effect due to short life, but the degradation products of lipid peroxidation can be second messengers of oxidative stress, because of their longer half-life and ability to diffuse from the place of formation. These degradation products, mainly aldehydes, such as malonaldehyde, hexanal, 4-hydroxynonenal or acrolein, have biological roles in cellular signaling, in normal and pathological conditions and in regulation of cell cycle [28]. Due to their chemical reactivity these products can covalently modify macromolecules as nucleic acids, proteins and lipids and consequently exhibit different biological effects. They are also biomarkers of lipid peroxidation and oxidative stress.

During the development of cataract, non-enzymatic lipid peroxidation occurs. It is a nonspecific and uncontrolled process in which the resulting reactive oxygen species readily react with surrounding molecules, leading to intensification of the process, and the damage of the cell membrane. The process LPO in the lens can be initiated by hydroxyl radical, singlet oxygen, peroxyl radical. These radicals seize from the unsaturated fatty acids H^+ from the methyl group ($-CH_2$) in the α- position of the double bond, to form unsaturated fatty acid lipid radical ($L^\bullet$).

By intramolecular rearrangement of double bonds in lipid radicals, conjugated dienes are formed. By adding the molecular oxygen to conjugated dienes, lipoperoxyl radical ($LOO^\bullet$) is formed. Lipoperoxyl radicals have significant oxidative potential, they can further initiate the seizure of hydrogen from the neighbouring unsaturated fatty acid, by which lipid peroxidation enters the phase of propagation and autooxidaton, resulting in formation of lipid hydroperoxyde (LOOH) and new lipid radicals ($L^\bullet$).

Lipid hidroperoxydes (LOOH), as primary molecular products of lipid peroxidation process, are well-soluble, have ability to migrate from the site of development and are important potential sources for the formation of reactive hydroxyl radical (HO•). In the presence of metals with variable valence they may reopen a cascade of lipid peroxidation [27]. In the presence of Fe^{+2} in the classical Fenton-type reactions, by decomposing of LOOH, HO• and alkoxyl radical (LO•) occur. In the presence of Fe^{+3}, LOOH is decomposed to peroxyl radical (LOO•). Alkoxyl and peroxyl radical are responsible for initiation and propagation of LPO processes.

By removing hydroperoxides (LOOH), which occur during the first phase of lipid peroxidation, by activity of glutathione-dependent peroxidase which catalyzes reduction of hydroperoxide into the corresponding alcohol, the reactions of lipid peroxidation propagation can be prevented.

Compounds that react with reactive oxygen species produced during the chain reaction, such as peroxyl and alkoxyl radicals (ROO• and RO•), and lead to the formation of the species that are unable to remove the hydrogen atom from unsaturated fatty acids, are considered to be antioxidant switches of chain reactions, of which the most important is liposoluble α-tocopherol. In series of complex degradation reactions of hydro- and dihydro-peroxides of polyunsaturated fatty acids, many other aldehydes are produced: 4-hydroxy-alkenals, 4-hydroperoxy-alkenals, 4-oxo-alkenals and bis-aldehyde malonaldehyde (MDA) [28]. These aldehydes are highly reactive compounds because they have electrophilic properties, they are less volatile than hydroperoxides, and can diffuse from the place of origin and express their reactivity through biotransformation and adduction to biomolecules (proteins, DNA).

Consequences of peroxidative damage of lens cell membranes are multiple. On the one hand, a number of membrane functions may be altered due to a direct attack by reactive oxygen species on the membrane components responsible for these functions. Indirect consequences of peroxidative damage of membranes are important. Lipid peroxidation modifies the environment of not only membrane proteins and may in this way influence their functional efficiency. The consequences of adduct formation at the protein level is associated with numerous cytotoxic consequences including the disruption of cell signalling, altered gene regulation, inhibition of enzyme activity, mitochondrial dysfunction, impaired energy metabolism. Two different ways of oxidative modifications of cellular constituents have to be considered in cataractogenesis: the direct modification by reactive oxygen species and the indirect modification via reactive products of lipid peroxidation.

Lipid peroxidation products can diffuse across membranes, allowing the reactive aldehyde to covalently modify proteins localized throughout the cell and relatively far away from the initial site of reactive oxygen species (ROS) formation. LPO is implicated in human cataractogenesis because the toxic peroxidation products induce fragmentation of soluble lens proteins and damage vital membrane structures, correlating with an increase in lens opacity and changes in the refractive properties of the lens [23]. The results obtained by the authors, that caused posterior subcapsular cataract in the rabbit, by application of peroxidative products in the vitreous, were published [29].

The resulting lipid peroxidation products, such as, 4-hydroxynonenal (HNE), can form protein-HNE adducts that may result in altered protein functions and can mediate oxidative stress-induced cell death in the lens epithelial cells [30]. Also, the resulting MDA, can react with amino-groups of proteins, forming intra-molecular cross bonds and bind two distinct proteins by forming the inter-molecular bonds [31] and thus affect the structural and functional properties of proteins. MDA is linked to the Lys residues of proteins and enzymes, and forms Schiff bases, for phospholipids, nucleic acids (shown mutagenic properties), and also MDA inhibits a number of thiol dependent enzymes: glucose-6-phosphatase, Na^+-K^+-ATP-ase, Ca^{++}-ATP-ase.

At the level of cell membranes in the lens lipid hydroperoxides induce changes of the lipoprotein structure and permeability [13], and oxidation inhibition of membrane enzymes Na^+/K^+-ATP-ase and Ca^{++}-ATP-ase, which are responsible for osmotic regulation and transport of metabolites. Disturbance of ion transport through the cell membrane is associated with ATP hydrolysis, which increases membrane permeability to protons, as the result of LPO. The Na^+/K^+-ATP- and Ca^{++}-ATP-ases would be directly affected because of their regulation by proton concentration. Other transport proteins would be affected in a secondary sense because they are coupled to ions whose intracellular concentrations are regulated by the ATP-ase. For example, glucose is the main source of energy for lens cells and is typically co-transported with Na^+, and similar transport are known for the import of amino acids across cell membranes. Formation of the lens opacity follows from either osmotic imbalance or a cytoplasmic imbalance of specific cations, in particular Ca^{+2} [32]. Disturbance of function Ca^{++}-ATP-ase and the consequent increase of Ca^{+2} in the lens leads to electrostatic changes in the crystalline, which can disrupt the protein conformation and interaction. Maintenance of calcium homeostasis is critical to lens clarity and cataractous lens has elevated calcium levels. An in vitro binding study indicates that human lens lipids have the capacity to bind nearly all the calcium present in the human lens and that age and cataract diminished the capacity of lens lipids to bind calcium. It is possible that the increased concentration of intracellular Ca^{+2} and reduced ability of lens lipids to bind calcium, to initiate further disturbances that lead to an increase of light scattering from proteins and lipids [33]. Disturbance of the lens cell membrane permeability during the development of cataract, and reduced ability of active transport of substances against the concentration gradient lead to changes in concentration of intracellular compounds and metabolic changes within the cells. This is manifested by lowering the content of GSH, ATP, and other intracellular compounds and electrolytes in lens cells.

Lipid peroxides as potential causes of cataracts, lead to the changes of not only the lens cell membrane, but also of the cytosol, because it serves to reduce concentration of glutathione and cause the change of redox relationship GSH/GSSG [1].

6. Defense against lipid peroxidation in the lens

Low molecular mass compounds which act primarily against peroxyl radicals involved in radical propagation, provide first line of defense against lipid peroxidation in the lens.

These compounds (GSH, ascorbic acid, α- tocopherol) can terminate the propagation of free radical mediated reactions and interrupt the autocatalytic chain reaction of lipid peroxidation. GSH is a major antioxidant in the lens, and helps to reduce proteins, contains a side chain of sulfhydryl (-SH) residue that enables it to protect cells against oxidants. GSH can directly scavenge ROS or enzymatically via two major antioxidant enzyme systems, glutathione peroxidases (GPx) and glutathione S-transferases (GST). Enzymes such as superoxide dismuatase (SOD), catalase (CAT) and GPx can decompose ROS and prevent the damage to cellular constituents and initiation of lipid peroxidation. In the event of ROS induced lipid peroxidation, secondary defense enzymes are involved in the removal of LOOH to terminate the autocatalytic chain of lipid peroxidation and protect membranes. GPx and GST which catalyze GSH-dependent reduction of LOOH through their peroxidase activity are the major secondary defenses in the lens against ROS induced lipid peroxidation. Resynthesis of GSH from the oxidized form is catalyzed by glutathione reductase, where the necessary $NADPH^{+}+H^{+}$ is produced in pentose pathway of carbohydrates [36]. Glutathione reductase (GR) is a control enzyme glutathione-redox cycle and by maintaining of intracellular levels of GSH can affect cation transport systems, lens hydration, sulfhydryl groups of proteins, and membrane integrity. Detoxifying role of GST is reflected in its ability to catalyze reactions of conjugation of reduced glutathione with endogenous electrophiles, mostly by the products of oxidative stress, lipid hydroperoxides and final products of lipid peroxidation [37]. Class μ and π GST isoenzymes are expressed in the human lens, with the π isoenzyme predominating. The highest GST activity occurs in the peripheral and the equatorial cortexes, with the lowest activity in the nucleus [38]. Glutathione peroxidase activity is shown by the enzymes that catalyze the reduction of hydrogen peroxide, organic hydroperoxide and phospholipid hydroperoxide using GSH as a hydrogen donor. Superoxide dismutase (SOD) catalyzes the reaction of dismutation of superoxide anion radicals ($O_2^{-}\bullet$) in the presence of hydrogen donor to hydrogen peroxide and molecular oxygen. By removing the $O_2^{-}\bullet$, SOD prevents the formation of $^{1}O_2$ which can initiate the process of lipid peroxidation.

Aim. Our studies have focused on measuring the products of lipid peroxidation in corticonuclear lens blocks, with different type and different degrees of maturity of age-related cataract. In addition to measuring products of lipid peroxidation in cataract lenses, our study included the determination of activities and ability of lens glutathione peroxidase and glutathione S-transferase to remove hydroperoxides, which are probably involved in the early stages of cataractogenesis and development of mature cataract through oxidative stress.

7. Material and methods – patients

Clinical and biochemical researches were carried out in 101 patients with age-related cataract, 46 women and 55 men. The average age of the group was 72.5 (SD ± 7.9). According to the cataract maturity degree the patients were classified into two groups as follows: age-related cataract incipient (N=41) and matura (N=60). In the group age-related cataract incipient there were 23 patients with posterior subcapsular (PS), 9 patients with nuclear subcapsular (NP) and 9 patients with cortical nuclear (CN) cataract. In the group age-related catataract matura there were 19 patients with cataract which started as a posterior subcapsular, 15 patients which started as a nuclear subcapsular, 16 patients with matura, which started as cortical nuclear and 10 patients diagnosed with matura, which started as cortical cataract.

Samples corticonuclear blocks/parts of lens (without epithelial cells) were obtained from patients undergoing extracapsular extraction of cataracts and used as the test material. Types of cataract were estimated during ophthalmologic examination and confirmed during its extraction. Immediately after sample acquisition, samples were closed in individual capsules and deeply frozen. This research has been conducted following the tenets of the Declaration of Helsinki and approved by the ethics committee of Medical Faculty. Informed consent was provided from all patients after a careful explanation of the aims of the study.

Homogenate of lenses from each group was prepared in 0.2 mol/L potassium phosphate buffer (pH 7.2). For analysis we used supernatant obtained by centrifugation of homogenates at 5000 rpm for 15 min at 4°C.

The concentration of conjugated dienes was measured spectrophotometrically at 233 nm [39].

Lens MDA concentrations were measured as the product of the reaction with thiobarbituric acid (TBA) using a modification of the method Ledwozyw et al [40].

Fluorescent products (lipid- and water-soluble) of lipid peroxidation were determined by spectrofluorimetric analysis at 360/430 (excitation/emission) nm [41].

The concentration of GSH in the sample was determined in the reaction 5,5'-dithiobis-2-nitrobenzoic acid (DTNB) (Ellman's reagent), after removal of proteins by perchloric acid [42].

8. Enzyme assays

The activity of glutathione peroxidase was determined at 412 nm, by the method Chin et al. [43]. The conjugation of GSH with 1-chloro,2-4 dinitrobenzene (CDNB), a hydrophilic substrate, was examined spectrophotometrically at 340 nm to measure glutathione S-transferase activity [44]. One unit of GST was defined as the amount of enzyme required to conjugate 1μmol of CDNB with GSH/min. The activity of glutathione reductase was assayed by the procedure of Glatzle et al. [45]. Superoxide dismutase (SOD) activity was determined by the method Misra and Fridovich [47], based on the inhibition of the adrenochrome during the spontaneous oxidation of adrenaline in basic conditions. The change in absorbance was read at 480 nm on a spectrophotometer. The SOD activity was expressed as kU/g protein (one unit was considered to be the amount of enzyme that inhibited adrenaline auto-oxidation by 50%).

To calculate the specific enzyme activity, protein in each sample was estimated by the method of Lowry et al [46].

9. Results and discussion

Senile cataract is manifested in the later years of life, so it is estimated that the costs of operation would be reduced by 45% if the incidence of age-related cataract could be delayed for ten years [48]. By studying the oxidation changes of lens structures during the development of cataract, we attempted to contribute to clearing up the process of cataractogenesis, of which even today there are many unknowns.

Lipid peroxidation is one of the possible mechanisms of cataractogenesis, caused by excessive production of reactive oxygen species in aqueous environment and reduced antioxidant defense of the lens. By studying the corticonuclear lens block of the patients with age-related cataract were detected increased concentrations in primary molecular products LPO (diene conjugates and lipid hydroperoxides) and end fluorescent LPO products (table 1).

	Age-related cataract incipient (n=41)	Age-related cataract matura (n=60)
Conjugated diens (nmol/g weight of lens)	2.48 ± 0.84*	1.57 ± 0.49
Fluorescent products/ g protein (lipid soluble)	46.59 ± 14.40	70.94 ± 13.21*
Fluorescent products/ g protein (water soluble)	57.53 ± 18.23	103.08 ± 27.81*
MDA (nmol/g weight of lens)	1.81 ± 0.67	3.17± 0.78*

Data is presented as means ± SD *p<0.001

Table 1. Lipid peroxidation products in cataractous lenses

In the group of patients with the incipient cataract, we obtained significantly higher concentration of diene conjugates in the lenses compared to matura cataract (p<0.001) (table 1). This can be explained by the fact that at the early stages of the development of cataract the most intense is the proces of lipid peroxidation, which is either the initiator of cataractogenesis proces or initiated by creation of reactive oxigen types, and continues to affect changes in the lens by its propagation.

In the lenses with the incipient cataract, the concentration of conjugated diens is significantly higher in cortical nuclear cataract (CN) compared with the lenses with posterior subcapsular cataract (PS) (p=0.001) (table 2). Also, in matura cataract the concentration of conjugated diens is the highest in the lenses diagnosed with CN cataract (p<0.05) (table 3).

Cataract incipient (type)	PS (n=23)	NP (n=9)	CN (n=9)
Conjugated diens (nmol/g lens)	2.12 ± 0.55	2.63 ± 0.97	3.24 ± 0.84*
Fluorescent products/g protein (lipid soluble)	40.09 ± 10.18	52.34 ± 13.92‡	57.47 ± 16.33†
Fluorescent products/g protein (water soluble)	50.17 ± 12.88	61.99 ± 14.58	71.90 ± 24.12†
MDA (nmol/g weight of lens)	1.60 ± 0.56	1.74 ± 0.73‡	2.41 ± 0.55*

Data is presented as means ± SD *$p<0.001$, ‡$p<0.05$, †$p<0.01$.
PS - posterior subcapsular, NP - nuclear subcapsular, CN - cortical nuclear

Table 2. Products LPO in lenses with cataract incipient

Cataract matura	Cataract matura began as PS (N=19)	Cataract matura began as NP (N=15)	Cataract matura began as CN (N=16)
Conjugated diens (nmol/g lens)	1.41 ± 0.30	1.46 ± 0.67	1.91 ± 0.39†
Fluorescent products/ g protein (lipid soluble)	67.32 ± 15.45	76.69 ± 14.65‡	74.74 ± 7.49
Fluorescent products/ g protein (water soluble)	94.83 ± 25.43	105.29 ± 24.72	120.68 ± 27.08†
MDA (nmol/g lens)	3.32 ± 1.07	3.23 ± 0.80	3.06 ± 0.48

Data is presented as means ± SD †$p<0.01$, ‡$p<0.05$.

Table 3. Products LPO in lenses with cataract matura

In the presence of free metal ions with variable valence (Fe^{+2} or Cu^{2+}) hydrogen peroxide in Fenton's reaction is translated into highly reactive hydroxyl radical, while the lipid peroxides are translated into peroxyl and alkoxyl radicals. Because of the longer half-life time compared to the alkoxyl radical, peroxyl radical is ideal for propagation of oxidative chain reactions, while the oxidation of alkoxyl radicals produce dihydroperoxide, which is degraded to toxic aldehydes, such as 4-hydroxy-2,3-trans-nonenal, 4-hydroxy-pentanal, short chain malondialdehyde. Such conditions may exist in humane senile lens [49]. MDA and 4-hydroxy-2,3-trans-nonenal by forming of Schiff's bases with amino groups amino acid residues of protein contribute to increase of carbonyl groups content and produce fluorescent products of lipid peroxidation. We have measured significantly higher concentration of these fluorescent products of LPO in the lenses with mature cataract (table

1), and the highest concentration was measured in the lenses with the NP and CN cataract in relation to posterior subcapsular cataract (PS) ($p<0.05$) (table 2,3). Some authors have identified fluorescent Schiff bases in higher concentration in human cataract lenses compared to healthy lenses, resulting from the interaction of reactive carbonyl groups MDA with amino groups of lens membrane phospholipids [50].

By measuring the concentration of MDA in the homogenate of cataract lenses, we obtained significantly higher concentration in the group of patients with matura compared to incipient cataract ($p<0.001$) (table 1). This can be explained by the fact that the malondialdehyde is one of the final products of LPO, which accumulates in the lens during the process of lipid peroxidation and the development of cataract. Author's results with the experimentally induced cataract that have measured significantly higher concentration of MDA in cataract lenses compared to the control group were published [51], as well as author's works that obtained significantly higher concentration of MDA in cataract lenses of diabetics [2], myopic lenses and senile cataract [52]. Reduced activity of glutathione peroxidase enzyme and glutathione S-transferase, that are important for the removal of malondialdehyde, contribute to the increase of its concentration in the lens.

The reasons that cause significantly higher concentration of lipid peroxidation products in lenses with cortical nuclear (CN) and nuclear subcapsular (NP) cataract, in relation to the posterior subcupsular (PS) are numerous. The results of other researchers [49] indicate that during the life time in the lens some kind of "internal" lens barrier is developed between nuclear and corticular parts, which hinders the diffusion of molecules to the nucleus. This barrier prevents the diffusion of antioxidative molecules to the nuclear part, which increases the sensitivity of central part of the lens to oxidative damage. Also, it is possible that unstable prooxidant molecules have longer residency in the central part of the lens. The endogenous lens hromofore, tryptophan metabolites (kynurenine, 3-hydroxykynurenine, 3-hydroxykynurenine glucoside), which are relatively inert photochemically, during oxidative stress and/or aging are formed photochemically active tryptophan metabolite (N-formyl-kynurenine, xanthurenic acid) that have photochemical properties, and also act as an endogenous photosensitizers in the lens [53].

Through photosensible reactions tryptophan products transfer absorbed energy to oxygen, which further leads to a series of cellular changes through the oxidation. With age, the level of free components of UV filters ie. tryptophan derivatives in the lens are reduced, and their binding to lens proteins increases [54]. Tryptophan products are subject non-enzyme deamination, at physiological pH, resulting in α,β-unsaturated ketone, reactive intermediates [55], which can covalently bind to amino acids, usually His, Cys, or Lys residues in proteins of human lenses, or react with the Cys residue of glutathione (GSH) to form GSH-3OHKynG [54]. GSH that is present in the lens in relatively high concentrations may compete with the amino acid residues for the unsaturated ketone derivative of kynurenine, thereby protecting the crystalline from modification. This covalent modification is particularly expressed in the nucleus of the lens containing the older proteins, causing altered transport/diffusion of small molecules in the lens. Specifically, it develops a barrier in to the movement of molecules between metabolically active cortex and inert nucleus. The barrier also restricts the flow of GSH from the cortex, which reduces the concentration of GSH in the nucleus of the lens, so the response to oxidative damages in this part of the lens is also reduced [49]. During the life time, lens fibers are very compactly arranged in nuclear part, with minimal presence of extracellular space. Nuclear plasma of the membrane undergo oxidation damage, whereas the phospholipid molecules modified by oxygen accumulate in the lipid layer, leading to changes in the structure and violate lipid-lipid and protein-lipid interactions in membranes of the lens fibers. This probably contributes to concentration of the lipid peroxidation products to be the highest in the lenses with early nuclear cataract.

Reduced glutathione (GSH) and GSH-dependent enzymes, glutathione peroxidase (GPx) and glutathione S-transferase (GST), are very important in defending the lens structures of the products of lipid peroxidation. One reason for increased production and accumulation of lipid hydroperoxides in the lens with cataract may also be reduced activity of GPx and GST.

The primary biological role of superoxide dismutase (SOD) is to catalyze reaction of dismutations of superoxide anion radicals ($O_2^-\bullet$) in the presence of hydrogen donor to hydrogen peroxide and molecular oxygen. Superoxides can first be degraded into H_2O_2 by SOD, and subsequently, catalyzed into ground-state oxygen and water by catalase and enzymes of the glutathione redox cycle, including glutathione reductase and glutathione peroxidase. At low levels of hydrogen peroxide, the glutathione redox cycle is responsible for protecting against H_2O_2-induced damage and maintaining high levels of GSH in the lens, whereas, at a higher concentration, the principal mechanism for the removal of hydrogen peroxide is catalase. The study of the lens cell culture where the expression of DNK for superoxide dismutase was performed on intact lens cells, showed that the cells with higher activity of SOD resistant to oxidative damage, caused by hydrogen peroxide, superoxide anion radical and UV radiation. Expression of the superoxide dismutase enzyme prevented the beginning of the cataract in lens cells [56].

GSH and other sulfhydryls are particularly important in the protection of thiol (-SH) groups of crystalline and prevent the formation of aggregates which reduce transparency of the lens [57]. The oxidized glutathione (GSSG) is reduced back to GSH by a NADPH-dependent glutathione reductase, which in physiological conditions maintains a high ratio of GSH/GSSG in the lens and other ocular tissues. Oxidative stress, induced by accumulation of LPO products in the lens during the development of cataract, causes consumption of GSH and disruption of redox balance in the lens, so the age-related cataract is associated with progressive reduction of GSH concentration in the lens. Probably with the progression of the cataract than the consumption of GSH against toxic compounds, the synthesis of GSH is reduced, as the result of deficient availability of substrates and reduced activity of enzymes for its synthesis (γ-glutamyl-cysteine syntethase). This reduces the amount of available GSH for optimal function GPx and GST, which causes the peroxide metabolism disorder.

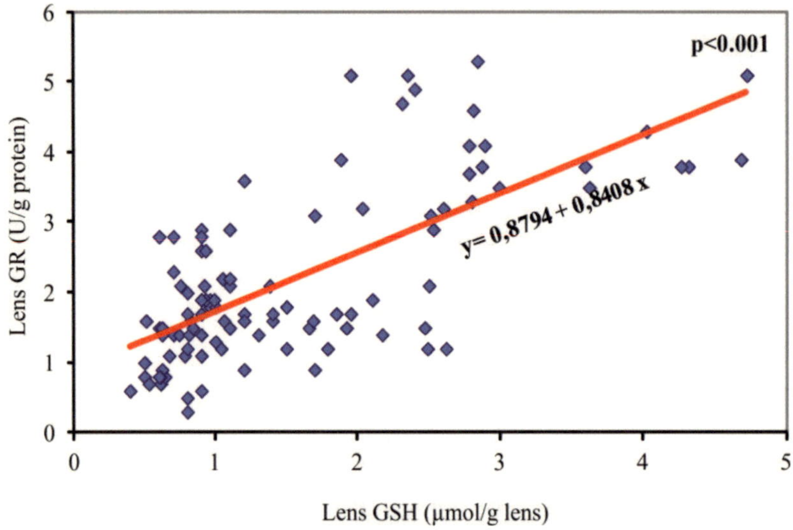

Figure 1. Lens glutathione and glutathione reductase in cataract.

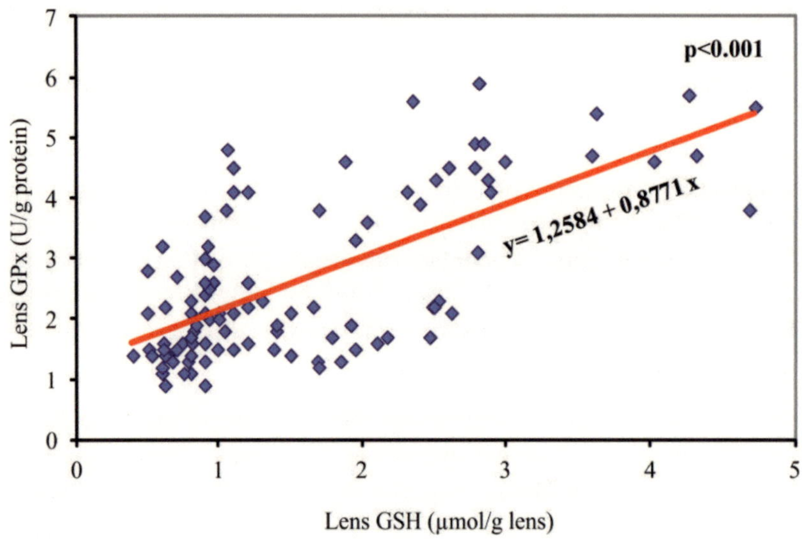

Figure 2. Lens glutathione and glutathione peroxidase in cataract.

The significance of GSH, as an important compound for the function of GPx and GST in the tested lenses, is confirmed also by the positive correlations between the concentration of GSH and glutathione reductase ($r = 0.7011$, $p < 0.001$) (figure 1), between the concentration of GSH and glutathione peroxidase activity ($r = 0.6749$, $p < 0.001$) (figure 2), and between activities of glutathione S-transferase and the concentration of GSH ($r = 0.6379$, $p < 0.001$) (figure 3).

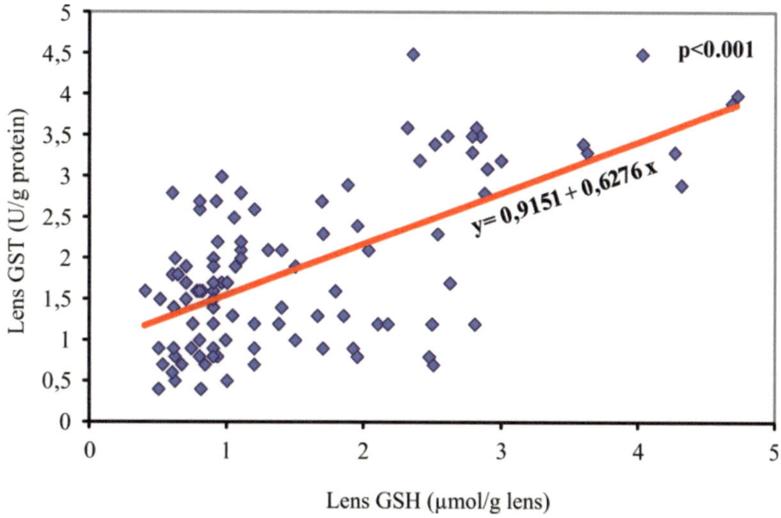

Figure 3. Lens glutathione and glutathione S-transferase in cataract.

Experiments showed that applying the injections of buthionine-sulphoximine as inhibitors of reduced glutathione synthesis, on the newborn rats, causes the development of cataract [58].

	Cataract incipient (n=41)	Cataract matura (n=60)
GSH	2.55 ± 0.9	0.88 ± 0.26*
GPx	3.40 ± 1.47	2.09 ± 0.90*
GR	3.03 ± 1.29	1.61 ± 0.71*
GST	2.46 ± 1.14	1.50 ± 0.67*
SOD	4.13 ± 2.14	2.14 ± 0.91*

Data is presented as means ± SD *p<0.001
GSH (µmol/g weight of lens), GPx, GR, GST (U/g protein), SOD (kU/g protein).

Table 4. Antioxidative defense factors in cataractous lenses

In tested corticonuclear lens blocks with mature cataract, we measured lower activity of GR, GPx and GST enzymes in relation to the incipient (p<0.001) (table 4).

Cataract incipient (type)	PS (n=23)	NP (n=9)	CN (n=9)
GSH	2.99 ± 0.90	2.16 ± 0.55†	1.81 ± 0.42*
GPx	4.58 ± 0.70	2.17 ± 0.53*	1.64 ± 0.17*
GR	4.03 ± 0.66	1.91 ± 0.73*	1.61 ± 0.27*
GST	3.31 ± 0.60	1.61 ± 0.75*	1.14 ± 0.29*
SOD	5.77 ± 1.25	2.34 ± 0.59*	1.71 ± 0.58*

Data is presented as means ± SD *p<0.001, †p<0.01.
GSH (µmol/g weight of lens), GPx, GR, GST (U/g protein), SOD (kU/g protein)

Table 5. Antioxidative defense factors in lenses with cataract incipient

The lowest activity of these enzymes was measured in the lenses with the CN and NP in relation to the PS cataract (table 5). Reduced activity of GR, GPx and GST enzymes is followed by a significant decrease in the concentration of GSH in lens homogenates with matura (0.88 ± 0.26 μmol/g weight), compared to the initial cataract (2.55 ± 0.9 μmol/g weight) ($p<0.001$) (table 4), with lower concentrations of GSH measured in CN and NP compared to the PS cataract (table 5,6).

Cataract matura	Cataract matura began as PS (N=19)	Cataract matura began as NP (N=15)	Cataract matura began as CN (N=16)
GSH	0.98 ± 0.19	0.78 ± 0.20	0.81 ± 0.35
GPx	2.96 ± 0.99	1.89 ± 0.57*	1.44 ± 0.31*
GR	2.26 ± 0.60	1.31 ± 0.49*	1.06 ± 0.47*
GST	2.13 ± 0.50	1.13 ± 0.43*	1.07 ± 0.49*
SOD	2.96 ± 0.65	1.77 ± 0.52*	1.25 ± 0.36*

Data is presented as means ± SD *$p<0.001$.
GSH (μmol/g weight of lens), GPx, GR, GST (U/g protein), SOD (kU/g protein)

Table 6. Antioxidative defense factors in lenses with cataract matura

Also the results of other researchers show that the level of reduced glutathione in the lens decreases with the development of the cataract [24]. Such changes are probably a reflection of oxidative processes, increased by the formation of toxic products of lipid peroxidation, compared to a weakened antioxidative capacity of cataract lenses. In cataract lenses the concentration of GSH is being reduced, since, as a main representative of non-protein thiols, it is included and consumed in oxido-reduction processes in terms of excess oxidized substrates. A possible reason for the consumption of GSH during oxidative stress is its conversion into oxidized form, which can be conjugated with protein thiol groups to form mixed disulphides (PSSG), via a process called protein-S-thiolation [59].

Results of other researchers also show that the human lens with age-related cataract glutathione peroxidase activity was significantly reduced compared with normal lenses. The kinetic study of GPx showed that lipid hydroperoxides achieve saturation of enzymes at a concentration that is approximately 1 mmol ie. that Km GPx is achieved at a concentration of lipid hydroperoxides of 0.434 mmol [60]. Because of these kinetic properties, GPx activity was probably inhibited in age-related cataract by products of lipid peroxidation, using non-competitive inhibition principle. GPx activity, aside from availability of GSH, is affected by other factors such as glutathione reductase activity, the amount of produced $NADPH^{+}+H^{+}$ in the pentose pathway and availability of selenium. The authors who have examined the activity of GPx in the lenses of experimental mice and compared the degree of lens blur, and age of mice with specific activity of GPx, showed significant correlation between decreased activity of GPx and the level of lens blur, as well as the age of mice [61].

Considering the function of GST to catalyze reactions of conjugation of lipid peroxidation products with GSH [62], thereby reducing the toxicity of electrophylic compounds and their reactivity towards nucleophilic groups in biomolecules, it is logical that the activity of the GST measured in incipient cataract is higher (2.46 ± 1.14 U/g protein), because at the beginning of the development of the cataract the most intense is the process of lipid peroxidation. With advancing of the process of cataractogenesis, the amount of GSH is reduced for glutathione-S-transferase, and the enzyme activity is significantly decreased in the lenses with mature cataract (1.5 ± 0.67 U/g protein) ($p<0.001$) (table 4).

In the study of GST activity in epithelial cells of operated cataract lenses, the group of authors showed that lens epithelial cells with cortical nuclear and cortical cataract show complete loss of activity of glutathione-S-transferase [63].

By analyzing the activity of glutathione reductase, significantly higher activity was found in the lenses with the incipient cataract (3.03 ± 1.29 U/g protein) compared to the lenses with mature cataract (1.61 ± 0.71 U/g protein) ($p<0.001$) (table 4). Other researchers have obtained similar results after comparing the activities of GR in cataract lenses and intact lenses of older persons [64]. Glutathione reductase plays a key role in maintaining thiol (-SH) groups in the lens, and this is probably the most important role of this enzyme in maintaining lens transparency. It is possible that lower GR activity in comparison to normal activities, can be one of the causes of lens blur. In addition to its predominantly cortical distribution within lens fiber cells, high susceptibility of GR to post-translational modifications [65] could also be of critical importance for the early dysfunction of GPx and GST under oxidative stress.

Based on the fact that the activity of GPx and GST is focused on degradation of lipid peroxidation products, we tested the relationship between the activity of the antioxidant defense enzyme and products of LPO in cataract lenses.

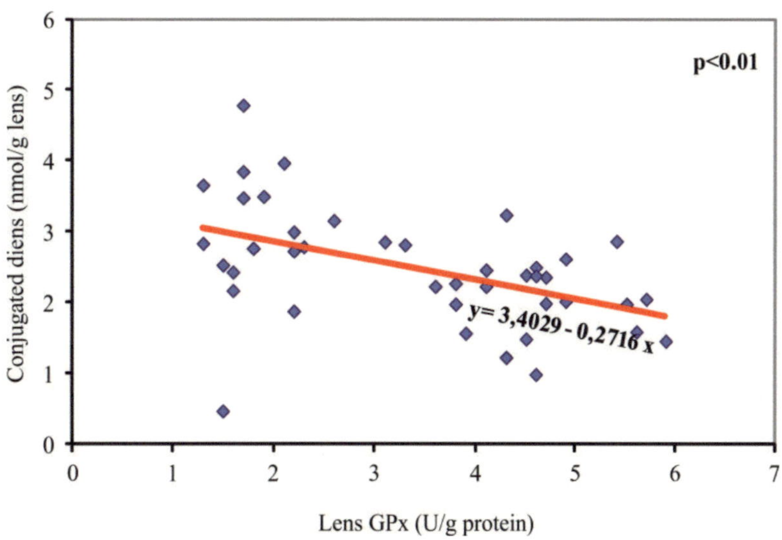

Figure 4. Lens conjugated diens and glutathione peroxidase relationship in cataract.

Reduced specific GPx activity (U/g protein) in the lenses shows significant correlation with the increased concentration of conjugated diens (r= -0.476, $p<0.01$) (figure 4), lipid-soluble fluorescent products (r= -0.429, $p<0.01$) (figure 5), water-soluble fluorescent products (r= -0.367, $p<0.05$) (figure 6) and MDA (r= -0.328, $p<0.05$) (figure 7).

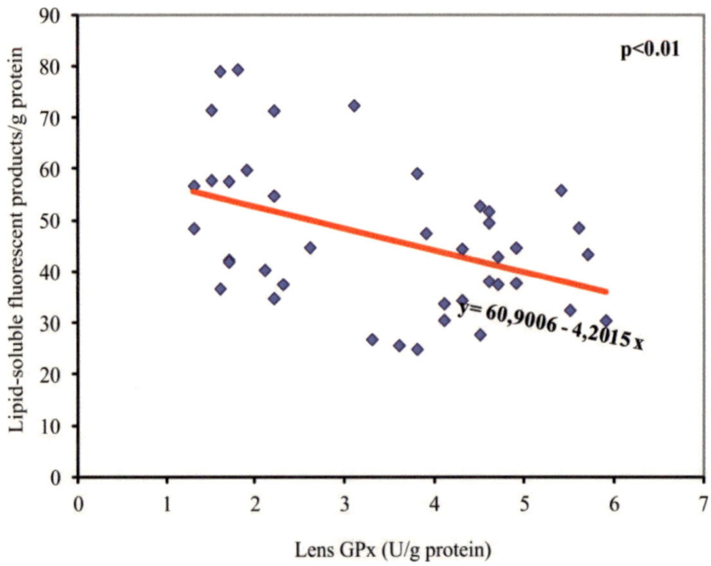

Figure 5. Lens lipid-soluble fluorescent products and GPx relationship in cataract.

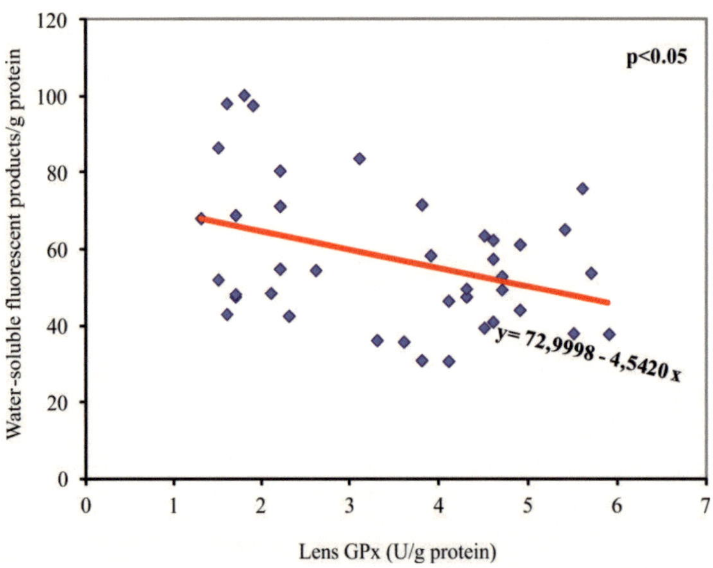

Figure 6. Lens water-soluble fluorescent products and GPx relationship in cataract.

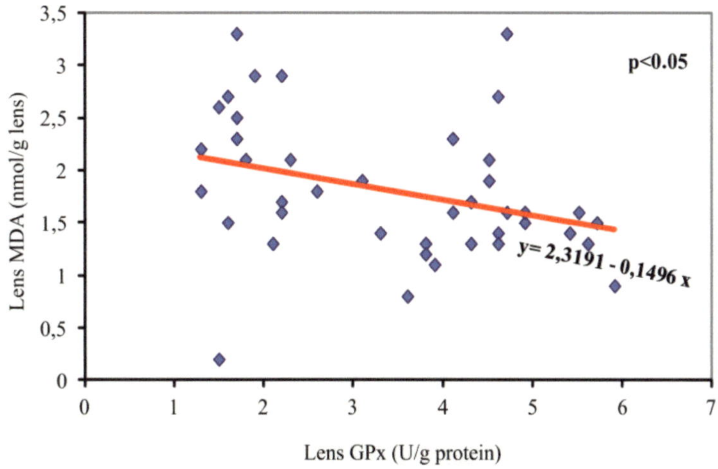

Figure 7. Lens malondialdehide and glutathione peroxidase relationship in cataract.

Also, the reduced GST activity (U/g protein) shows the significant correlation with the increased concentration of total hydroperoxides (r= -0.313, $p<0.05$), conjugated diens (r= -0.465, $p<0.01$) (figure 8), lipid-soluble fluorescent products (r= -0.398, p=0.01) (figure 9), water-soluble fluorescent products (r= -0.347, $p<0.05$) (figure 10) and MDA (r= -0.345, $p<0.05$) (figure 11) in homogenates of the lenses with the incipient cataract.

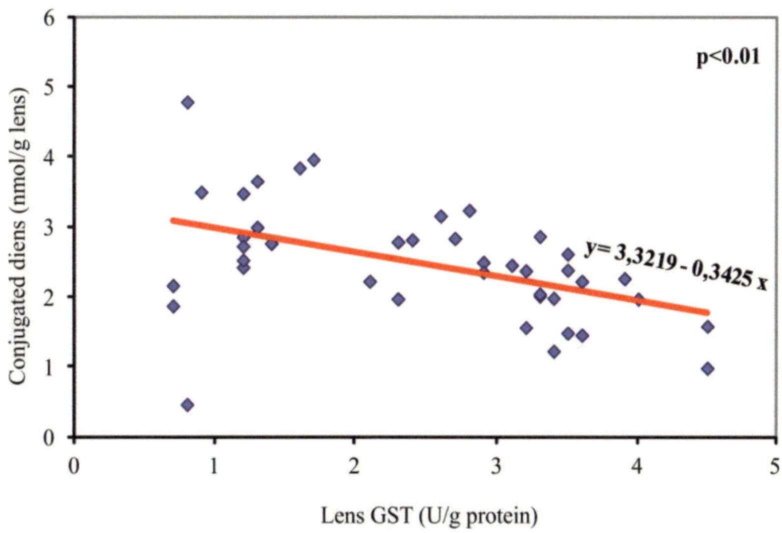

Figure 8. Lens conjugated diens and glutathione S-transferase relationship in cataract.

Probably, the consumption of GSH and other antioxidants in reactions of degradation of lipid peroxidation products, affects the decrease of GPx and GST enzymes activity, and as the process of cataractogenesis progresses towards mature cataract, all lens structures become affected by the changes, which probably leads to the change of enzyme molecules themselves.

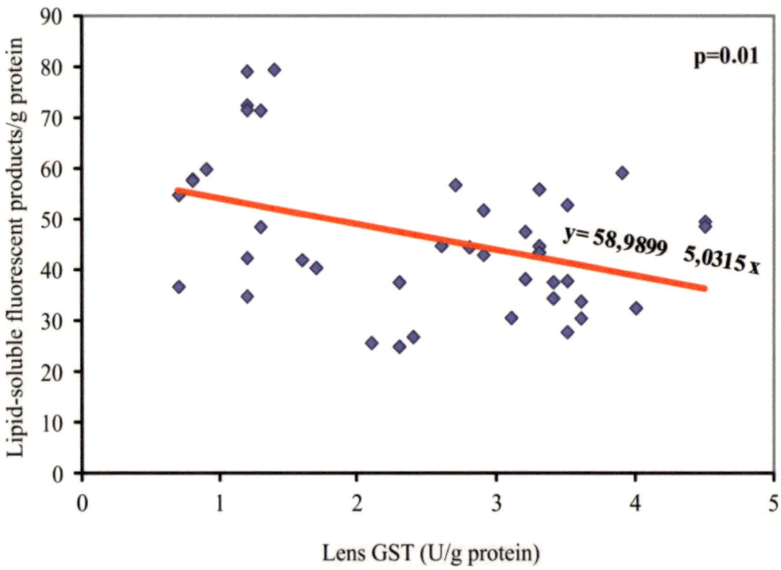

Figure 9. Lens lipid-soluble fluorescent products and GST relationship in cataract.

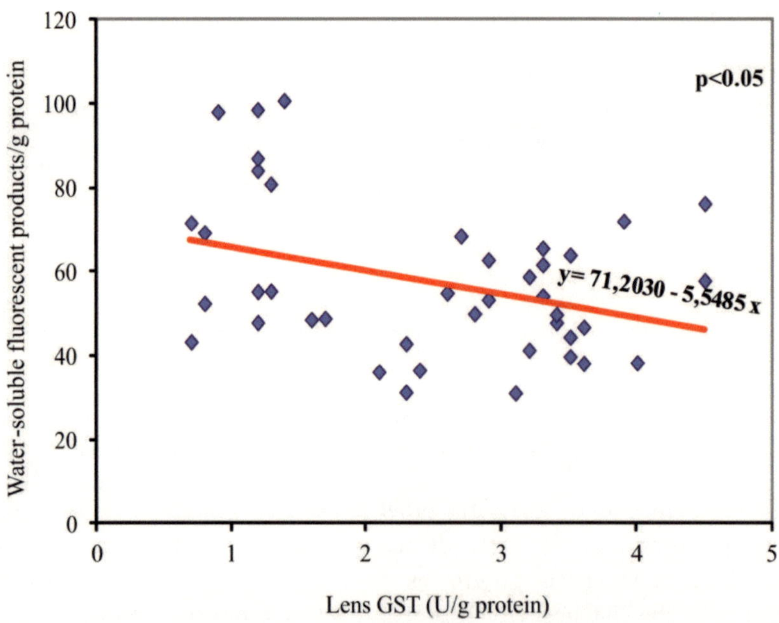

Figure 10. Lens water-soluble fluorescent products and GST relationship in cataract.

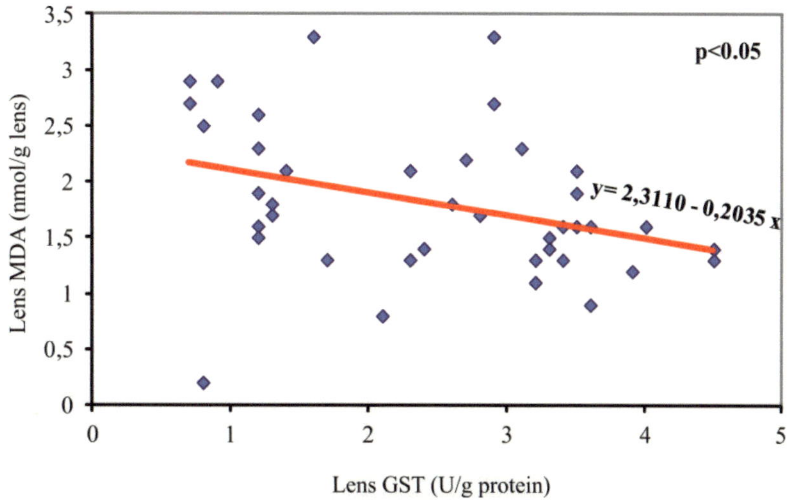

Figure 11. Lens malondialdehide and glutathione S-transferase in cataract.

We measured higher activity of SOD (4.13 ± 2.14 kU/g protein) in lenses with incipient cataract in comparison to mature cataract (2.14 ± 0.91 kU/g protein) ($p<0.001$) (table 2). Such changes are probably a reflection of oxidative processes, increased by formation of toxic products of lipid peroxidation in relation to the weakened antioxidant capacity of cataract lenses. Also, other researchers have measured decreased SOD activity in the lenses of patients with senile and diabetic cataract [66]. The decreased activity of superoxide dismutase in the lenses with age-related cataract may be due to denaturation of enzyme molecules, and/or slow enzyme synthesis. Reduced SOD activity and consequently, the increase of concentration of H_2O_2 lead to the formation of hydroxyl radical from Fenton's type reaction. Subsequently, hydroxyl radical induces the formation of superoxide radical, which may initiate the process of lipid peroxidation in the lens.

10. Conclusion

Based on the results of measuring products of lipid peroxidation and antioxidant enzyme activity in corticonuclear lens blocks with age-related cataract, we can say that the lens structure changes induced by lipid peroxidation may with other risk factors present, affect the beginning or the development of cataract. The changes in redox system are particularly pronounced in cortical nuclear cataract, but are reflected in all parts of the lens, regardless of the initial localization of the lens blur. The lowest level of oxidative stress was detected in posterior subcapsular cataract, so it is possible that it has less importance in the development of PS cataract. The development of cataract can probably be prevented/slowed down by preventing the accumulation of products of lipid peroxidation in the lens and maintaining adequate level of GSH and function of GSH-dependent antioxidant enzymes.

Author details

Bojana Kisic and Dijana Miric
Institute of Biochemistry, Serbia

Lepsa Zoric
Clinic for Eye Diseases, Serbia

Aleksandra Ilic
Institute of Preventive Medicine, Faculty of Medicine, Kosovska Mitrovica, Serbia

11. References

[1] Babizhayev MA (2012) Biomarkers and special features of oxidative stress in the anterior segment of the eye linked to lens cataract and the trabecular meshwork injury in primary open-angle glaucoma: challenges of dual combination therapy with N-acetylcarnosine lubricant eye drops and oral formulation of nonhydrolyzed carnosine. Fundam Clin Pharmacol. 26(1):86-117.

[2] Donma O, Yorulmaz E, Pekel H, Suyugul N (2002) Blood and lens lipid peroxidation and antioxidant status in normal individuals, senile and dibetic cataractous patients. Curr Eye Res. 25(1): 9-16.

[3] Spector A, Wang GM, Wang RR, Garner WH, Moll H (1993) The prevention of cataract caused by oxidative stress in cultured rat lenses. I. H_2O_2 and photochemically induced cataract. Curr Eye Res. 12 (2):163–179.

[4] Yagci R, Aydin B, Erdurmus M, Karadag R, Gurel A, Durmus M, Yigitogly R (2006) Use of melatonin to prevent selenite-induced catarct formation in rat eyes. Curr Eye Res. 31(10): 845-850.

[5] Boscia F, Grattagliano I, Vendemiale G, Micelli-Ferrari T, Altomare E (2001) Protein oxidation and lens opacity in humans. Invest Ophthalmol Vis Sci. 41(9):2461-2465.

[6] Davies MJ, Truscott RJ (2001) Photo-oxidation of proteins and its role in cataractogenesis. J Photochem Photobiol B. 63(1-3):114-125.

[7] Spector A (2000) Oxidative stress and disease. J Ocular Pharmacol. 16:193-201.

[8] McCarty CA, Taylor HR (2002) A review of the epidemilogic evidence linking ultraviolet radiation and cataracts. Dev Ophthalmol. 35: 21-31.

[9] Abraham AG, Cox C, West S (2010) The differential effect of ultraviolet light exposure on cataract rate across regions of the lens. Invest Ophthalmol Vis Sci. 51(8):3919-3923.

[10] Delcourt C, Carrière I, Ponton-Sanchez A, Lacroux A, Covacho MJ, Papoz L and the POLA Study Group (2000) Light Exposure and the Risk of Cortical, Nuclear, and Posterior Subcapsular Cataracts: The Pathologies Oculaires Liées à l'Age (POLA) Study. Arch of Ophthalmol.118(3):385–392.

[11] Kannan R, Stolz A, Ji Q, Prasad PD, Ganapathy V (2001) Vitamin C Transport in Human Lens Epithelial Cells: Evidence for the Presence of SVCT2. Exp Eye Res. 73(2):159-165.

[12] Varma SD, Hegde KR (2010) Kynurenine-induced photo oxidative damage to lens in vitro: protective effect of caffeine. Mol Cell Biochem. 340(1-2):49-54.

[13] Huang L, Tang D, Yappert MC, Borchman D (2006) Oxidation-induced changes in human lens epithelial cells 2. Mitochondria and the generation of reactive oxygen species. Free Radic Biol Med. 41(6):926-936.

[14] Linetsky M, Shipova E, Cheng R, Ortwerth BJ (2008) Glycation by ascorbic acid oxidation products leads to the aggregation of lens proteins. Biochim et Bioph Acta: Molec Basis of Disease. 1782(1):22-34.

[15] Fridovich I (1997) Superoxide anion radical ($O_2^{\cdot-}$), superoxide dismutases, and related matters. J Biol Chem. 272(30):18515-18517.

[16] Garner B, Davies M, Truscott RJ (2000) Formation of hydroxyl radicals in the human lens is related to the severity of nuclear cataract. Exp Eye Res. 70:81-88.

[17] Garland DL (1991) Ascorbic acid and the eye. Am J of Clin Nutrit. 54(6):1198S-1202S

[18] Balaji M, Sasikala K, Ravindran T. (1992) Copper levels in human mixed, nuclear brunescence, and posterior subcapsular cataract. British J of Ophthalmol. 76(11):668-669.

[19] Garner B, Roberg K, Qian M, Eaton JW, Truscott RJ (2000) Distribution of ferritin and redox-active transition metals in normal and cataractous human lenses. Exp Eye Res. 71(6):599-607.

[20] Dawczynski J, Blum M, Winnefeld K, Strobel J (2002) Increased content of zinc and iron in human cataractous lenses. Biol Trace Elem Res. 90(1-3):15-23.

[21] Nagaraj RH, Monnier VM (1995) Protein modification by the degradation products of ascorbate: formation of a novel pyrrole from the Maillard reaction of L-threose with proteins. Biochim et Biophys Acta. 1253(1):75-84

[22] Fan X, Reneker LW, Obrenovich ME, Strauch C, Cheng R, Jarvis SM, Ortwerth BJ, Monnier VM (2006) Vitamin C mediates aging of lens crystallins by the Maillard reaction in a humanized mouse model. Proc of the Nat Acad of Sci of the United States. 103(42):16912-16917.

[23] Ansari NH, Wang L, Srivastava SK (1996) Role of lipid aldehydes in cataractogenesis: 4-hydroxynonenal-induced cataract. Biochem Mol Med. 58(1):25-30;

[24] Zoric L, Elek-Vlajic S, Jovanovic M, Kisic B, Djokic O, Canadanovic V, Cosic V, Jaksic V (2008) Oxidative stress intensity in lens and aqueous depending on age-related cataract type and brunescense. Eur J Ophthalmol. 18(5):669-674.

[25] Kisic B, Miric D, Zoric L, Dragojevic I, Stolic A (2009) Role of lipid peroxidation in pathogenesis of senile cataract. Vojnosanit Pregl. 66(5):371-375.

[26] Stark G (2005) Functional consequences of oxidative membrane damage. J Memb Biol. 205(1):1-16

[27] Halliwell B, Chirico S (1993) Lipid peroxidation: its mechanism, measurement, and significance. Am J Clin Nutr. 57: 715S-725S.

[28] Guéraud F, Atalay M, Bresgen N, Cipak A, Eckl PM, Huc L, Jouanin I, Siems W, Uchida K (2010) Chemistry and biochemistry of lipid peroxidation products. Free Rad Res. 44(10):1098–1124.

[29] Goosey JD, Tuan WM, Garcia CH (1984) A lipid peroxidative mechanism for posterior subcapsular cataract formation in the rabbit. A possible model for cataract formation in tapetoretinal diseases. Invest Ophthalmol Vis Sci. 25:608-612.

[30] Choudhary S, Srivastava S, Xiao T, Andley UP, Srivastava SK, Ansari NH (2003) Metabolism of lipid derived aldehyde, 4-hydroxynonenal in human lens epithelial cells and rat lens. Invest Ophthalmol Vis Sci. 44(6):2675-2682.

[31] Berlett BS, Stadtman ER (1997) Protein oxidation in aging, disease, and oxidative stress. J Biol Chem. 272(33):20313-20316.

[32] Biju PG, Rooban BN, Lija Y, Devi VG, Sahasranamam V, Abraham A (2007) Drevogenin D prevents selenite-induced oxidative stress and calpain activation in cultured rat lens. Mol Vis. 13:1121-1129.

[33] Tang D, Borchman D, Yappert MC, Vrensen GF, Rasi V (2003) Influence of age, diabetes, and cataract on calcium, lipid-calcium, and protein-calcium relationships in human lenses. Invest Ophthalmol Vis Sci. 44(5):2059-2066.

[34] Huang W, Koralewska-Makar A, Bauer B, Akesson B (1997) Extracellular glutathione peroxidase and ascorbic acid in aqueous humor and serum of patients operated on for cataract. Clin Chim Acta. 261(2):117-130.

[35] Kisic B, Miric D, Zoric L, Ilic A, Dragojevic I (2012) Antioxidant Capacity of Lenses with Age-Related Cataract. Oxidative Medicine and Cellular Longevity. Available: http://www.hindawi.com/journals/oximed/2012/467130/.

[36] Linetsky M, Chemoganskiy VG, Hu F, Ortwerth BJ (2003) Effect of UVA Light on the Activity of Several Aged Human Lens Enzymes. Invest Ophthalmol Vis Sci 44(1): 264-74.

[37] Singhal SS, Awasthi S, Srivastava SK, Zimniak P, Ansari NH, Awasthi YC (1995) Novel human ocular glutathione S-transferases with high activity toward 4-hydroxynonenal. Invest Ophthalmol Vis Sci. 36(1):142-150.

[38] Alberti G, Oguni M, Podgor M, Sperduto RD, Tomarev S, Grassi C, Williams S, Kaiser-Kupfer M, Maraini G, Hejtmancik JF (1996) Glutathione S-transferase M1 genotype and age-related cataracts. Lack of association in an Italian population. Invest Ophthalmol Vis Sci. 37(6):1167-1173.

[39] Recknagel RO, Glende EA JR (1984) Spectrophotometric detection of lipid conjugated dienes. Methods Enzymol.105:331-337.

[40] Ledwozyw A, Michalak B, Stepien A, Kadziolka A (1986) The relationship between plasma triglicerides, cholesterol, total lipids and lipid peroxidation products during human atherosclerosis. Clin Chim Acta. 155(3):275-284.

[41] Shimasaki H (1994) Assay of fluorescent lipid peroxidation products. Methods Enzymol. 233:338-346

[42] Beutler E, Duron O, Kelly BM (1963) Improved method for the determination of blood glutathione. J Lab Clin Med. 61:882-888.

[43] Chin PTY, Stults FH, Tapell AL (1976) Purification of rat lung soluble glutathione peroxidase. Biochem Biophys Acta. 445 (3):558-566.

[44] Habig WH, Pabst MJ, Jakoby WB (1974) Glutatione-S-transferases. J Biol Chem. 249:7130-7134.

[45] Glatzle D, Vuillenmir JP, Weber F, Decker K (1974) Glutatione reductase test with whole blood, a convenient procedure for the assesment of the riboflavine status in humans. Experimentia. 30(6):665-667.
[46] Lowry OH, Rosebrough NJ, Farr AL, Randall RJ (1951) Protein measurement with the Folin phenol reagent. J Biol Chem.193(1):265–275.
[47] Misra HP, Fridovich I (1972) The role of superoxide anion in the autoxidation of epinephrine and a simple assay for superoxide dismutase. J Biol Chem. 247(10):3170-3175.
[48] Taylor A, Jacques PF, Epstein EM (1995) Relations among aging, antioxidant status, and cataract. Am J Clin Nutr. 62(6):1439-1447.
[49] Ttuscott RJW (2005) Age-related nuclear cataract – oxidation is the key. Exp Eye Res. 80:251-259.
[50] Bhuyan KD, Master RWP, Bhuyan KC (1996) Crosslinking of aminophospholipids in cellular membranes of lens by oxidative stress in vitro. Biochim Biophys Acta. 1285 (1):21-28.
[51] Gupta SK, Trivedi D, Srivastava S, Joshi S, Halder N, Verma SD (2003) Lycopene attenuates oxidative Stress Induced experimental Cataract Development: An In Vitro and In Vivo Study. Nutrition.19(9):794-799.
[52] Micelli-Ferrari T, Vendemiale G, Grattagliano I, Boscia F, Arnese L, Altomare E, Cardia L (1996) Role of lipid peroxidation in the pathogenesis of myopic and senile cataract. Br J Ophthalmol. 80(9): 840-843.
[53] Roberts JE, Finley EL, Patat SA, Schey KL (2001) Photooxidation of lens proteins with xanthurenic acid: a putative chromophore for cataractogenesis. Photochem Photobiol. 74(5):740-744.
[54] Vazquez S, Aquilina JA, Jamie JF, Sheil MM, Truscott RJ (2002) Novel protein modification by kynurenine in human lenses. J Biol Chem. 277: 4867-4873.
[55] Taylor LM, Aquilina JA, Jamie JF, Truscott RJ (2002) UV filter instability: consequences for the human lens. Exp Eye Res. 75:165-175.
[56] Lin D, Barnett M, Grauer L, Robben J, Jewell A, Takemoto L, Takemoto DJ (2005) Expression of superoxide dismutase in whole lens prevents cataract formation. Molecular Vision 11:853-858.
[57] Ganea E, Harding JJ (2006) Glutathione-related enzymes and the eye. Curr Eye Res. 31(1):1-11.
[58] Li W, Calvin HI, David LL, Wu K, McCormack AL, Zhu GP, Fu SC (2002) Altered patterns of phosphorylation in cultured mouse lenses during development of buthionine sulfoximine cataracts. Exp Eye Res. 75(3):335-346.
[59] Lou MF (2000) Thiol Regulation in the lens. Journ of Ocular Pharmacol and Therapeut. 16 (2):137-148.
[60] Babizhayev MA (1996) Failure to withstand oxidative stress induced by phospholipid hydroperoxides as a possible cause of the lens opacities in systemic diseases and ageing. Biochim Biophys Acta.1315(2):87-99.
[61] Rieger G, Winkler R (1994) Changes of Glutathione Peroxidase Activity in Eye Tissues of Emory Mice in Relation to Cataract Status and Age. Ophthalmologica. 208(1):5-9.

[62] Srivastava SK, Singhal SS, Awasthi S, Pikula S, Ansari NH, Awasthi YC (1996) A glutathione S-transferases isozyme (bGST 5.8) involved in the metabolism of 4-hydroxy-2-trans-nonenal is localized in bovine lens epithelium. Exp Eye Res. 63(3):329-337.

[63] Huang QL, Lou MF, Straatsma BR, Horwitz J (1993) Distribution and activity of glutathione-S-transferse in normal human lenses and in cataractous human epithelia. Curr Eye Res. 12(5): 433-437.

[64] Yan H, Harding JJ, Xing K, Lou MF (2007) Revival of Glutathione in Human Cataractous and Clear Lens Extracts by Thioredoxin and Thioredoxin Reductase, in Conjunction with α-Crystallin or Thioltransferase. Curr Eye Res. 32(5): 455-463.

[65] Linetsky M, Chemoganskiy VG, Hu F, Ortwerth BJ (2003) Effect of UVA light on the activity of several aged human lens enzymes. Invest Ophthalmol Vis Sci. 44(1):264-74.

[66] Ozmen B, Ozmen D, Erkin E, Habif S, Bayindir O (2002) Lens speroxide dismutase and catalase activities in diabetic cataract. Clin Biochem. 35(1):69-72.

Chapter 22

Lipid Peroxidation in Hepatic Fibrosis

Ichiro Shimizu, Noriko Shimamoto,
Katsumi Saiki, Mai Furujo and Keiko Osawa

Additional information is available at the end of the chapter

http://dx.doi.org/10.5772/46180

1. Introduction

Hepatic fibrosis is a complex dynamic process which is mediated by death of hepatocytes and activation of hepatic stellate cells (HSCs). Lipid peroxidation including the generation of reactive oxygen species (ROS), transforming growth factor-β, and tumor necrosis factor-α can be implicated as a cause of hepatic fibrosis.

Damage of any etiology, such as infection with hepatitis C virus (HCV) or hepatitis B virus (HBV), heavy alcohol intake, and iron overload, to hepatocytes can produce oxygen-derived free radicals and other ROS derived from lipid peroxidative processes. Persistent production of ROS constitutes a general feature of a sustained inflammatory response and liver injury, once antioxidant mechanisms have been depleted. The major source of ROS production in hepatocytes is NADH and NADPH oxidases localized in mitochondria (Figure 1). NADH and NADPH oxidases leak ROS as part of its operation. Kupffer cells (hepatic resident macrophages), infiltrating inflammatory cells such as macrophages and neutrophils, and HSCs also produce ROS in the injured liver.

2. Oxidative stress in liver injury

ROS include the free radicals superoxide (O_2^-) and hydroxyl radical (HO^-) and non-radicals such as hydrogen peroxide (H_2O_2). A number of reactive nitrogen species including nitric oxide (NO) and peroxynitrite ($ONOO^-$) are also ROS. Superoxide production is mediated mainly by NADH oxidase. Hydrogen peroxide is more stable and membrane permeable in comparison to other ROS. Thus, hydrogen peroxide plays an important role in the intracellular signaling under physiological conditions. With respect to pathological actions, ROS participate in the development of liver disease. In this regard, hydrogen peroxide is converted into the hydroxyl radical, a harmful and highly reactive ROS, in the presence of transition metals such as iron (Figure 1). The hydroxyl radical is able to induce not only lipid peroxidation in the structure of membrane phospholipids, which results in irreversible modifications of cell membrane structure and function (membrane injury), but DNA cleavage (DNA injury) as well. Such a chain of events due to increased ROS production exceeding cellular antioxidant defense systems are called oxidative stress, inducing cell death.

Malondialdehyde (MDA) and 4-hydroxynonenal (HNE) (Figure 1), end products of lipid peroxidation, are discharged from destroyed hepatocytes into the space of Disse (Figure 2). Cells are well equipped to neutralize the effects of ROS by virtue of a series of the antioxidant protective systems, including superoxide dismutase (SOD), glutathione peroxidase, glutathione (GSH), and thioredoxin. Upon oxidation, GSH forms glutathione disulfide (GSSG).

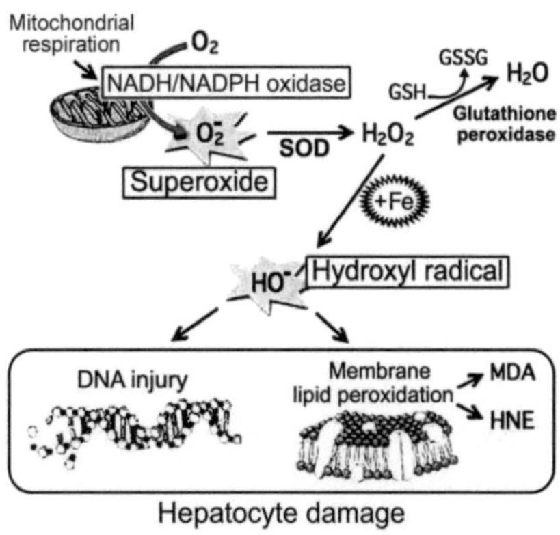

Figure 1. Oxidative stress and hepatocyte damage (Shimizu et al., 2012). A primary source of reactive oxygen species (ROS) production is mitochondrial NADPH/NADH oxidase. Hydrogen peroxide (H_2O_2) is converted to a highly reactive ROS, the hydroxyl radical, in the presence of transition metals such as iron (+Fe) and copper. The hydroxyl radical induces DNA cleavage and lipid peroxidation in the structure of membrane phospholipids, leading to cell death and discharge of products of lipid peroxidation, malondialdehyde (MDA) and 4-hydroxynonenal (HNE) into the space of Disse. Cells have comprehensive antioxidant protective systems, including SOD, glutathione peroxidase and glutathione (GSH). Upon oxidation, GSH forms glutathione disulfide (GSSG).

A single liver injury eventually results in an almost complete resolution, but the persistence of the original insult causes a prolonged activation of tissue repair mechanisms, thereby leading to hepatic fibrosis rather than to effective tissue repair. Hepatic fibrosis, or the excessive collagen deposition in the liver (see next section), is associated with oxidative stimuli and cell death. Cell death is a consequence of severe liver damage that occurs in many patients with chronic liver disease, regardless of the etiology such as HCV/HBV infection, heavy alcohol intake, and iron overload.

3. What is hepatic fibrosis?

At the cellular levels, origin of hepatic fibrosis is initiated by the damage of hepatocytes, followed by the accumulation of neutrophils and macrophages including Kupffer cells on the sites of injury and inflammation in the liver. When hepatocytes are continuously damaged, leading to cell death, production of extracellular matrix proteins such as collagens predominates over hepatocellular regeneration. Overproduced collagens are deposited in injured areas instead of destroyed hepatocytes. In other words, hepatic fibrosis is fibrous scarring of the liver in which excessive collagens build up along with the duration and extent of persistence of liver injury. Hepatic fibrosis itself causes no symptoms but can lead to the end-stage cirrhosis. In cirrhosis the failure to properly replace destroyed hepatocytes and the excessive collagen deposition to distort blood flow through the liver (portal hypertension) result in severe liver dysfunction. Cirrhosis is an important host-related risk factor for the development of hepatocellular carcinoma (HCC) in chronic hepatitis C and B, as well as a major factor predicting a poor response to interferon-based antiviral therapy in chronic hepatitis C. Staging of chronic liver disease by assessment of hepatic fibrosis always is a major function of prognostic interpretation of individual data including liver biopsy. Of the commonly used staging systems, the METAVIR fibrosis score has been widely used (Huwart et al., 2008). The stages are determined by both the quantity and location of the fibrosis. With this score, F0 represents no fibrosis; F1 (mild fibrosis), portal fibrosis without septa; F2 (moderate fibrosis), portal fibrous and few septa; F3 (severe fibrosis), numerous septa without cirrhosis; and F4, cirrhosis (Figure 3), the tissue is eventually composed of nodules surrounded completely by fibrosis.

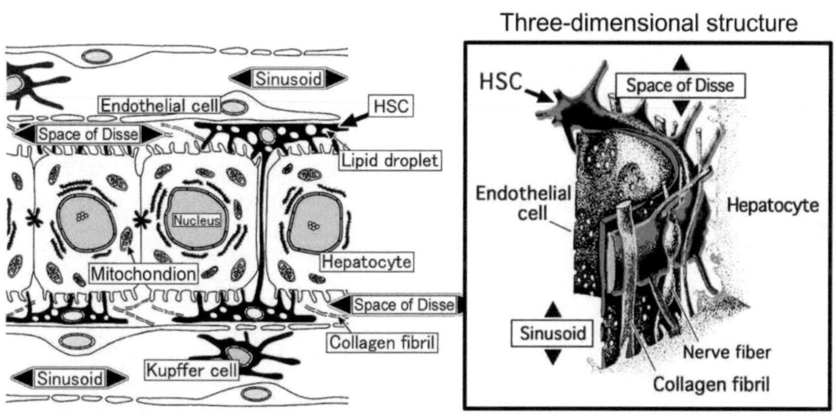

Figure 2. Schema of the sinusoidal wall of the liver (Shimizu et al., 2012). Schematic representation of hepatic stellate cells (HSCs) was based on the studies by Wake (Wake, 1999). Kupffer cells (hepatic resident macrophages) rest on fenestrated endothelial cells. HSCs are located in the space of Disse in close contact with endothelial cells and hepatocytes, functioning as the primary retinoid storage area. Collagen fibrils course through the space of Disse between endothelial cells and the cords of hepatocytes.

4. Activation of HSCs

Normal liver has a connective tissue matrix which includes collagen type IV (non-fibrillary), glycoproteins such as fibronectin and laminin, and proteoglycans such as heparan sulphate. These comprise the low density basement membrane in the space of Disse. Following liver injury there is a 3- to 8-fold increase in the extracellular matrix which is of a high density interstitial type, containing fibril-forming collagens (types 1 and III) as well as fibronectin, hyaluronic acid and proteoglycans. Collagen types 1 and III are major components of the extracellular matrix, which is principally produced by cells known as HSCs. HSCs are located in the space of Disse in close contact with hepatocytes and sinusoidal endothelial cells (Figure 2). Their three-dimensional structure consists of the cell body and several long and branching cytoplasmic processes (Wake, 1999). In the resting liver, HSCs have intracellular droplets containing retinoids. Retinoids refer to a group of chemical compound associated with vitamin A. HSCs contain approximately 50-80% of the whole body stores of retinoids (Blomhoff et al., 1990). In contrast, in the injured liver, HSCs are regarded as the primary target cells for inflammatory and oxidative stimuli, and they are proliferated, enlarged and transformed into myofibroblast-like cells. These HSCs are referred to as activated cells and are responsible for the overproduction of collagens during hepatic fibrosis to cirrhosis. This activation is accompanied by a loss of cellular retinoids, and the synthesis of α-smooth muscle actin (α-SMA), and large quantities of the major components of the extracellular matrix including collagen types I, III, and IV, fibronectin, laminin and proteoglycans. α-SMA is produced by activated HSCs (myofibroblast-like cells) but not by resting (quiescent) HSCs, thereby a marker of HSC activation. Moreover, activated HSCs produce ROS and transforming growth factor-β (TGF-β) (Figure 4). TGF-β is a major fibrogenic cytokine, regulating the production, degradation and accumulation of the extracellular matrix in hepatic fibrosis. TGF-β expression correlates with the extent of hepatic fibrosis (Castilla et al., 1991). This cytokine induces its own expression in activated HSCs, thereby creating a self-perpetuating cycle of events, referred to as an autocrine loop. TGF-β is also released in a paracrine manner from Kupffer cells, endothelial cells, and infiltrating inflammatory cells following liver injury. Similarly, ROS are produced by activated HSCs in response to ROS released from adjacent cells such as destroyed hepatocytes and activated Kupffer cells.

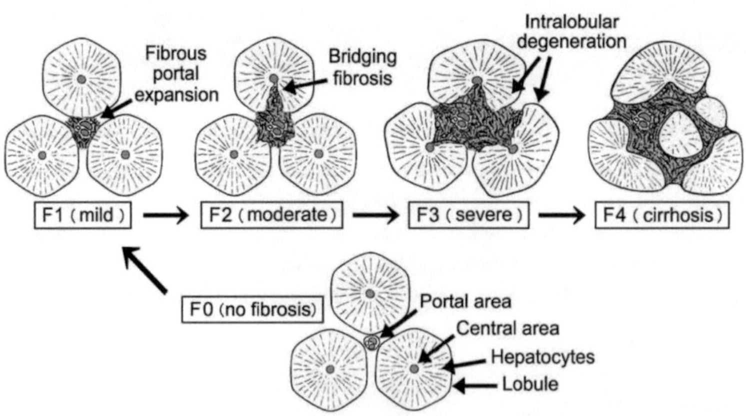

Figure 3. Stages of hepatic fibrosis in chronic hepatitis according to the five stages (0-4) of the METAVIR scoring system (1994). With this score, F0 represents no fibrosis; F1 (mild fibrosis), fibrous expansion of portal areas without septa; F2 (moderate fibrosis), fibrous septa extend to form bridges between adjacent vascular structures, both portal to portal and portal to central, occasional bridges; F3 (severe fibrosis), numerous bridges or septa without cirrhosis; and F4 (cirrhosis), the tissue is eventually composed of nodules surrounded completely by fibrosis.

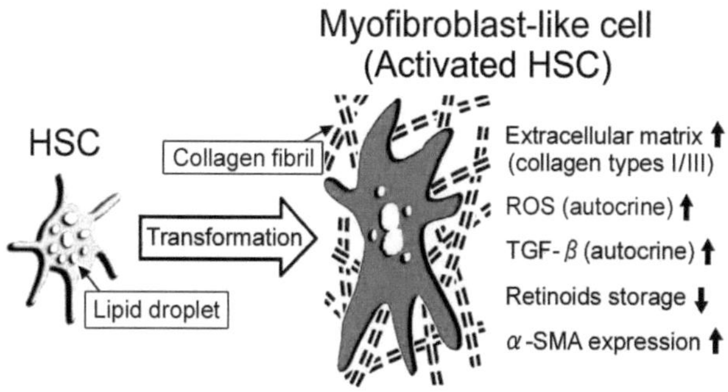

Figure 4. During liver injury, HSCs are proliferated, enlarged and transformed into myofibroblast-like cells (Shimizu, 2001). These activated HSCs produce large quantities of collagens, α-smooth muscle actin (α-SMA), ROS, and transforming growth factor-β (TGF-β), and lose cellular retinoids.

HSCs are activated mainly by ROS, products of lipid peroxidation (MDA and HNE) (Lee et al., 1995; Parola et al., 1993), and TGF-β, which are released from destroyed hepatocytes, activated Kupffer cells and infiltrating macrophages and neutrophils in the injured liver (Figure 5). In addition to ROS, exogenous TGF-β increases the production of ROS, particularly hydrogen peroxide, by HSCs, whereas the addition of hydrogen peroxide induces ROS and TGF-β production and secretion by HSCs (De Bleser et al., 1999). This so-called autocrine loop of ROS by HSCs is regarded as mechanism corresponding to the autocrine loop of TGF-β which HSCs produce in response to this cytokine with an increased collagen expression in the injured liver (Itagaki et al., 2005).

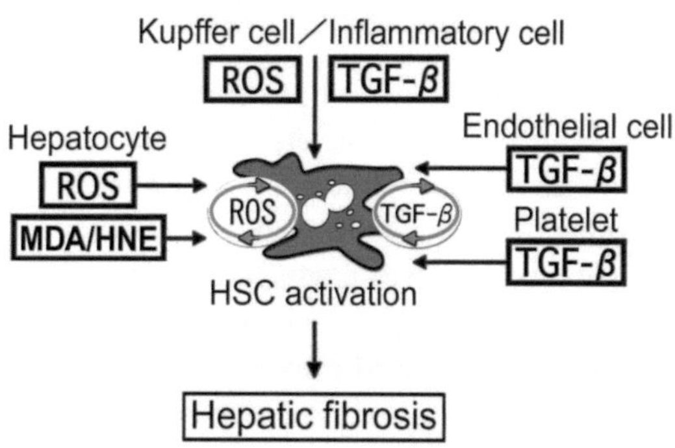

Figure 5. Activation of HSCs. HSCs are activated by such factors as ROS, lipid peroxidation products (MDA and HNE), and TGF-β released when adjacent cells including hepatocyte, Kupffer cells, and endothelial cells are injured. ROS and TGF-β are also produced by HSCs in response to exogenous ROS and TGF-β in an autocrine manner.

Other important factors for HSC activation are platelet-derived growth factor (PDGF) released from platelets, and endothelin-1 from endothelial cells. PDGF is the most potent mitogen. HSCs congregate in the area of injury, through proliferation and migration from elsewhere, in response to the release of PDGF and monocyte chemotactic peptide-1 (MCP-1). MCP-1 is produced by activated Kupffer cells and infiltrating macrophages and neutrophils. The number of activated HSCs also increases after liver injury (Enzan et al., 1994).

At the molecular levels, HSCs express the genes which encode for enzymes such as matrix metalloproteinase (MMP)-1 (interstitial collagenase) (Casini et al., 1994), which digests native fibrillar collagen types I and III, and MMP-2 (Milani et al., 1994), which digests denatured collagen types I and III and native collagen type IV, as well as tissue inhibitors of MMPs (TIMP)-1 and TIMP-2 (Iredale et al., 1992). Imbalance between matrix synthesis and degradation plays a major role in hepatic fibrosis (Shimizu, 2001). Matrix degradation depends upon the balance between MMPs, TIMPs and converting enzymes (MT1-MMP and stromelysin) (Li and Friedman, 1999). Collagen types I and III constitute the main framework of the so-called "fibrillar matrix". The space of Disse is a virtual space constituted by an extracellular matrix network composed of collagen type IV and non-collagenous components such as laminin. The large majority of collagen types III and IV, and laminin are synthesized by HSCs and endothelial cells, whereas all cell types synthesize small amounts of collagen type I. During hepatic fibrosis, however, HSCs become the major extracellular matrix producing cell type, with a predominant production of collagen type I (Maher and McGuire, 1990). In the resting liver, a balance between matrix synthesis and degradation exists, whereas, in the injured liver, the balance is disrupted. The net result of the changes during hepatocyte damage is increased degradation of the normal basement membrane collagen, and reduced degradation of interstitial-type collagens. The latter can be explained by increased TIMP-1 and TIMP-2 expressions relative to MMP-1. The degradative portion of the remodeling process is coordinated by MMPs and TIMPs.

5. Oxidative stress and intracellular pathway

Origin of hepatic fibrosis is initiated by the damage of hepatocytes, resulting in the recruitment of inflammatory cells and platelets, and activation of kupffer cells, with subsequent release of cytokines and growth factors. HSCs are the primary target cells for these inflammatory and oxidative stimuli, because during hepatic fibrosis, HSCs undergo an activation process to a myofibroblast-like cell, which represents the major matrix-producing cell. In the injured liver, hydrogen peroxide seems to act as a second messenger to regulate signaling events including mitogen activated protein kinase (MAPK) activation. The MAPK family includes three major subgroups, extracellular signal regulated kinase (ERK), p38 MAPK (p38), and c-Jun N-terminal kinase/stress activated protein kinase (JNK). MAPK participates in the intracellular signaling to: (1) induce the gene expression of redox sensitive transcription factors, such as activator protein-1 (AP-1) and nuclear factor κB (NF-κB) (Pinkus et al., 1996), (2) stimulate apoptosis (Clement and Pervaiz, 1999), and (3) modulate cell proliferation (Lundberg et al., 2000). ERK and JNK lie upstream of AP-1. JNK and p38 activation are more important in stress responses such as inflammation, which can also activate NF-κB. AP-1 and NF-κB induce the expression of multiple genes involved in inflammation and oxidative stress response, cell death and fibrosis, including proinflammatory cytokines such as tumor necrosis factor-α (TNF-α), interleukin-1 and interleukin-6 and growth factors such as PDGF and TGF-β. TGF-β is a major fibrogenic cytokine, acting as a paracrine and autocrine (from HSCs) mediator as already noted. TGF-β triggers and activates the proliferation, enlargement and transformation of HSCs, but it exerts its inhibitory effect on hepatocyte proliferation (Nakamura et al., 1985).

Since many cytokines exert growth factor like activity, in addition to their specific proinflammatory effects, the distinction between cytokines and growth factors is somewhat artificial. No growth factor or cytokine acts independently. The injured liver, predominantly Kupffer cells and infiltrating macrophages and neutrophils, produces TNF-α, interleukin-1 and interleukin-6. These proinflammatory cytokines may also also inhibit hepatic regeneration. In particular, TNF-α plays a dichotomous role in the liver, where it not only induces hepatocyte proliferation and liver regeneration but also acts as a mediator of cell death (Schwabe and Brenner, 2006). During TNF-α-induced apoptosis in hepatocytes, hydrogen peroxide is an important mediator of cell death (Bohler et al., 2000).

In liver injury of hepatitis virus infection, transgenic mice expressing HBsAg exhibit the generation of oxidative stress and DNA damage, leading to the progression of hepatic fibrosis and carcinogenesis (Hagen et al., 1994; Nakamoto et al., 2004). In addition, HBV X protein changes the mitochondrial transmembrane potential and increases ROS production in the liver (Waris et al., 2001). Moreover, structural and non-structural (NS) proteins of HCV are involved in the production of ROS in an infected liver. HCV core protein is associated with increased ROS, decreased intracellular and/or mitochondrial glutathione content, and increased levels of lipid peroxidation products (Moriya et al., 2001). Glutathione is an antioxidant. NS3 protein of HCV activates NADPH oxidase in Kupffer cells to increase production of ROS, which can exert oxidative stress on nearby cells (Thoren et al., 2004).

6. Conclusion

Hepatic fibrosis is a complex dynamic process which is mediated by death of hepatocytes and activation of HSCs. Lipid peroxidation including the generation of ROS, TGF-β, and TNF-α can be implicated as a cause of hepatic fibrosis. HSCs are regarded as the primary target cells for inflammatory stimuli, and produce extracellular matrix components. HSCs are activated by such factors as ROS, lipid peroxidation products (MDA and HNE), and TGF-β released when adjacent cells including hepatocyte, Kupffer cells, and endothelial cells are injured. ROS and TGF-β are also produced by HSCs in response to exogenous ROS and TGF-β in an autocrine manner. During TNF-α-induced death in hepatocytes, ROS is an important mediator of cell death. The most common cause of hepatic fibrosis is currently chronic HCV/HBV infection.

Understanding the basic mechanisms underlying the ROS-mediated fibrogenesis provides valuable information on the search for effective antifibrogenic therapies.

Author details

Ichiro Shimizu*, Noriko Shimamoto, Katsumi Saiki, Mai Furujo and Keiko Osawa
Showa Clinic, Shin Yokohama, Kohoku-ku, Yokohama, Kanagawa, Japan

* Corresponding Author

7. References

The French METAVIR Cooperative Study Group (1994). Intraobserver and interobserver variations in liver biopsy interpretation in patients with chronic hepatitis C. Hepatology 20:15-20.

Blomhoff R, Green MH, Berg T & Norum KR . (1990). Transport and storage of vitamin A. Science 250:399-404.

Bohler T, Waiser J, Hepburn H, Gaedeke J, Lehmann C, Hambach P, Budde K & Neumayer HH . (2000). TNF-alpha and IL-1alpha induce apoptosis in subconfluent rat mesangial cells. Evidence for the involvement of hydrogen peroxide and lipid peroxidation as second messengers. Cytokine 12:986-991.

Casini A, Ceni E, Salzano R, Milani S, Schuppan D & Surrenti C . (1994). Acetaldehyde regulates the gene expression of matrix-metalloproteinase- 1 and -2 in human fat-storing cells. Life Sci 55:1311-1316.

Castilla A, Prieto J & Fausto N . (1991). Transforming growth factors beta 1 and alpha in chronic liver disease. Effects of interferon alfa therapy. N Engl J Med 324:933-940.

Clement MV & Pervaiz S . (1999). Reactive oxygen intermediates regulate cellular response to apoptotic stimuli: an hypothesis. Free Radic Res 30:247-252.

De Bleser PJ, Xu G, Rombouts K, Rogiers V & Geerts A . (1999). Glutathione levels discriminate between oxidative stress and transforming growth factor-beta signaling in activated rat hepatic stellate cells. J Biol Chem 274:33881-33887.

Enzan H, Himeno H, Iwamura S, Saibara T, Onishi S, Yamamoto Y & Hara H . (1994). Immunohistochemical identification of Ito cells and their myofibroblastic transformation in adult human liver. Virchows Arch 424:249-256.

Hagen TM, Huang S, Curnutte J, Fowler P, Martinez V, Wehr CM, Ames BN & Chisari FV . (1994). Extensive oxidative DNA damage in hepatocytes of transgenic mice with chronic active hepatitis destined to develop hepatocellular carcinoma. Proc Natl Acad Sci U S A 91:12808-12812.

Huwart L, Sempoux C, Vicaut E, Salameh N, Annet L, Danse E, Peeters F, ter Beek LC, Rahier J, Sinkus R, Horsmans Y & Van Beers BE . (2008). Magnetic resonance elastography for the noninvasive staging of liver fibrosis. Gastroenterology 135:32-40.

Iredale JP, Murphy G, Hembry RM, Friedman SL & Arthur MJ . (1992). Human hepatic lipocytes synthesize tissue inhibitor of metalloproteinases-1. Implications for regulation of matrix degradation in liver. J Clin Invest 90:282-287.

Itagaki T, Shimizu I, Cheng X, Yuan Y, Oshio A, Tamaki K, Fukuno H, Honda H, Okamura Y & Ito S . (2005). Opposing effects of oestradiol and progesterone on intracellular pathways and activation processes in the oxidative stress induced activation of cultured rat hepatic stellate cells. Gut 54:1782-1789.

Lee KS, Buck M, Houglum K & Chojkier M . (1995). Activation of hepatic stellate cells by TGF alpha and collagen type I is mediated by oxidative stress through c-myb expression. J Clin Invest 96:2461-2468.

Li D & Friedman SL . (1999). Liver fibrogenesis and the role of hepatic stellate cells: new insights and prospects for therapy. J Gastroenterol Hepatol 14:618-633.

Lundberg AS, Hahn WC, Gupta P & Weinberg RA . (2000). Genes involved in senescence and immortalization. Curr Opin Cell Biol 12:705-709.

Maher JJ & McGuire RF . (1990). Extracellular matrix gene expression increases preferentially in rat lipocytes and sinusoidal endothelial cells during hepatic fibrosis in vivo. J Clin Invest 86:1641-1648.

Milani S, Herbst H, Schuppan D, Grappone C, Pellegrini G, Pinzani M, Casini A, Calabro A, Ciancio G & Stefanini F . (1994). Differential expression of matrix-metalloproteinase-1 and -2 genes in normal and fibrotic human liver. Am J Pathol 144:528-537.

Moriya K, Nakagawa K, Santa T, Shintani Y, Fujie H, Miyoshi H, Tsutsumi T, Miyazawa T, Ishibashi K, Horie T, Imai K, Todoroki T, Kimura S & Koike K . (2001). Oxidative stress in the absence of inflammation in a mouse model for hepatitis C virus-associated hepatocarcinogenesis. Cancer Res 61:4365-4370.

Nakamoto Y, Suda T, Momoi T & Kaneko S . (2004). Different procarcinogenic potentials of lymphocyte subsets in a transgenic mouse model of chronic hepatitis B. Cancer Res 64:3326-3333.

Nakamura T, Tomita Y, Hirai R, Yamaoka K, Kaji K & Ichihara A . (1985). Inhibitory effect of transforming growth factor-beta on DNA synthesis of adult rat hepatocytes in primary culture. Biochem Biophys Res Commun 133:1042-1050.

Parola M, Pinzani M, Casini A, Albano E, Poli G, Gentilini P & Dianzani MU . (1993). Stimulation of lipid peroxidation or 4-hydroxynonenal treatment increases procollagen (I) gene expression in human liver fat-storing cells. Biochem Biophys Res Commun 194:1044-1050.

Pinkus R, Weiner LM & Daniel V . (1996). Role of oxidants and antioxidants in the induction of AP-1, NF-kappaB, and glutathione S-transferase gene expression. J Biol Chem 271:13422-13429.

Schwabe RF & Brenner DA . (2006). Mechanisms of Liver Injury. I. TNF-alpha-induced liver injury: role of IKK, JNK, and ROS pathways. Am J Physiol Gastrointest Liver Physiol 290:G583-G589.

Shimizu I . (2001). Antifibrogenic therapies in chronic HCV infection. Curr Drug Targets Infect Disord 1:227-240.

Shimizu I, Kamochi M, Yoshikawa H & Nakayama Y. (2012). Gender difference in alcoholic liver disease. In: Shimizu I (ed) Trends in alcoholic liver disease research: Clinical and scientific aspects. InTech-Open Access Publisher, Rijeka, Croatia, pp 23-40.

Thoren F, Romero A, Lindh M, Dahlgren C & Hellstrand K . (2004). A hepatitis C virus-encoded, nonstructural protein (NS3) triggers dysfunction and apoptosis in lymphocytes: role of NADPH oxidase-derived oxygen radicals. J Leukoc Biol 76:1180-1186.

Wake K . (1999). Cell-cell organization and functions of 'sinusoids' in liver microcirculation system. J Electron Microsc 48:89-98.

Waris G, Huh KW & Siddiqui A . (2001). Mitochondrially associated hepatitis B virus X protein constitutively activates transcription factors STAT-3 and NF-kappa B via oxidative stress. Mol Cell Biol 21:7721-7730.

Chapter 23

Use of CoA Biosynthesis Modulators and Selenoprotein Model Substance in Correction of Brain Ischemic and Reperfusion Injuries

Nina P. Kanunnikova, Natalya Z. Bashun and Andrey G. Moiseenok

Additional information is available at the end of the chapter

http://dx.doi.org/10.5772/46077

1. Introduction

Acute disruption in brain blood circulation is a widespread cause of death and the most frequent cause of health loss in most countries of the world. About 6 millions of people suffer from stroke every year and this number is constantly increasing. Stroke has a high mortality rate – up to 30% of patients die. Only about 20% of surviving patients manage to return to their previous occupation. Most of patients are unable to take care of themselves and need help of relatives or medical personnel [1, 2]. About 80-85% of all cases of stroke are ischemic strokes. Therefore protecting brain from ischemia-induced damage is in the focus of modern neuropathology and neurosurgery studies, especially due to the increase in the number of neurosurgical operations which might cause additional blood flow impairements.

The severity of injuries of physiological reactions and biochemical processes caused by blood flow impairements depends on the degree of blood flow disruption in brain (fig.1) [1, 3].

Blood brain flow, ml/100 g/min	Parameters
60-80	Standard
35-60	Decrease of protein synthesis, selective gene expression
20-35	Lactate acidosis, cytotoxic edema
10-20	Energic deficiency, glutamate excitoxicity
0-10	Anoxic depolarization, necrosis, apoptosis

Figure 1. Correspondence between brain tissue changes and blood flow disruption

Decline of partial pressure of oxygen, significant decrease of ATP and glucose levels, membrane depolarization, extremely high levels of extracellular glutamate and intracellular calcium ions - all these factors contribute to development of the aforementioned injuries in the nervous tissue [4, 5, 6]. For example, higher level of calcium ions leads to stimulation of phospholipases and proteases, and activation of glutamate NMDA-receptors which in turn increase activity of nNoS and eNOS isoforms. As the result, amplification of lipoperoxidation takes place.

Disruption of electron transfer and oxidative phosphorilation within mitochondria are the first manifestations of ischemia-induced damages in the brain [7, 8, 9, 10], and the basic object of the injuries are presumably mitochondrial membranes [11, 12, 13, 14]. On this basis, ability of the brain to restore its functions following ischemia and reperfusion depends mainly on three processes - depletion of energic resources, excessive accumulation of excitatory amino acids [15, 16, 17], and formation of reactive oxygen species caused by leakage of electrons from intermediate links of respiratory change [18, 19, 20].

2. Features of oxidative stress in brain tissue during ischemia-reperfusion

Brain ischemia causes formation of free radical forms of oxygen which induce damage of neuronal membranes and biomacromolecules, particularly nucleic acids and proteins. Brain tissue has heightened disposition to development of oxidative stress. Brain cell membranes have high concentrations of polyunsaturated fatty acids which are the main substrate of free radical reactions [21, 22, 23]. When a free radical appears in membrane chance for its interaction with fatty acid is increased as a number of unsaturated links is rised. Unsaturated fatty acids provide more fluidity for membranes, therefore their changes caused by more active lipoperoxidation lead to increase of their viscosity and injuries of their barrier functions. It is known that synapse plasmatic membranes contain higher level of polyunsaturated fatty acids than myelin membranes. Many functionally important neuronal proteins are membrane-bound and depend on lipid environment. Simultaneously, system of antioxidant protection in brain has obviously less capacity than in other tissues, and enzymatic components of the system in brain are more sensitive to oxidative action [24, 25].

The second danger of lipoperoxidative activation in brain lies in the fact that disruption of nerve membrane integrity leads to increasing release of "excitotoxic" transmitters, such as glutamate, aspartate, etc [15, 16, 26-28].High rate of biogenic amine metabolism in brain leads to formation of ROS [29]. For example, monoamine oxidase reaction is linked with H_2O_2 formation. This phenomenon may be an additional source for generation of active radical products which are able to initiate lipoperoxidation in the presence of metals with variable valence. Dopamine, its precursor L-DOPA, 5-hydroxytraptamine, and norepinephrine may generate $O^{2\bullet-}$ not only, but quinones/semiquinones, too, which may decrease GSH level and bound with protein SH-groups. Oxidation can be catalized by transitional metal ions. Maximal increase for free radical generation and following activation for lipoperoxidation takes place in postischemic time – during recovery of blood circulation in brain tissue [30-33]. Nevertheless, possibility for formation of free radicals at earlier stage of brain damage exists during ischemia, too [34]. Enhancement of redox state of mitochondrial respiratory chain in these conditions gives an opportunity for oxygen to interact with intermediate components of the respiratory chain, for example ubisemiquinone [7]. This process takes the path of one-electron reduction of molecular oxygen and leads to formation of superoxide-anion. Consequently, heightened formation of free radicals may take place in tissues with insufficient blood circulation and decreased partial oxygen pressure [35]. Studies in mice and rats with genetic deficiency of superoxide dismutase confirmed the important role of free radicals in neuronal death/survival during brain ischemia [36, 37].In the postischemic period (during recirculation) when oxygen actively absorbs by brain tissue, oxygen radicals generation is caused by activation of enzymatic processes, too (arachidonic acid cascade, xantine oxydase system, activation of NADPH-oxydase in polymorphonuclear leucocytes) [38-40].

The fact that ischemia itself is unable to increase level of lipoperoxidation intermediate products is not surprising because during hypoxia amount of molecular oxygen is insufficient for observable activation of lipoperoxidation in brain tissue. Nevertheless, in these conditions the amount of hydroxyperoxides is increased and lag period for lipoperoxidation activation becomes shorter, which serves as evidence for decrease of antioxidant protection and increased formation of superoxide oxygen anion in brain tissue [41]. Herewith level of endogenic antioxidants in brain may be unchanged [42, 43]. For example, α-tocoferol level in rat brain is unchanged following 80 min after occlusion of middle cerebral artery and subsequent reperfusion. Unchanged levels of antioxidants were observed following bilateral occlusion of arteria carotis and reperfusion in gerbils, too [44, 45].

Recently a concept on polyfunctional physiologic role of free radicals in organism and in brain especially, is declared [42, 43, 46-48]. On the one hand, they act on key cell enzymes and receptors inside cells and cause destructive processes in tissues. On the other hand, they play a role of second messengers and may help with cell adaptive reactions to changed environmental conditions. Therefore low efficiency of antioxidant therapy by substances which bound free radical excess during ischemic injury treatment is not surprising [49, 50, 51]. In addition, these drugs have low bioavailability and must be used for at least several weeks before any effect can be observed [52, 53, 54-56].

It is necessary to emphasize that the main defense from excessive amounts of free radicals formed within cells is the action of antioxidant protection enzymes, such as selenoproteins, but not the action of low molecular wieght antioxidants. Under normal conditions these enzymes are sufficient for maintenance of low safe levels of free radicals, but in reperfusional conditions their activity is insufficient for maintenance pro- and antioxidant balance. Earlier attempts to administer substances of superoxide dismutase and catalase enzymes to animals in experimental models were unsuccessful because they poorly penetrate blood-brain barrier and cell membranes [51, 53].

Further research of neuroprotection in this direction is not very promising because activation of lipoperoxidation in phospholipid structures of nerve cell membranes is eliminated by the system of superoxide dismutase – catalase to a small degree [53]. Detoxication in these structures is primarily carried by enzymes of glutathione cycle, selenium-cystein-comprising glutathione peroxidases [49, 54, 55, 57-59].

3. Role of glutathione in mechanisms of antioxidant protection in brain

Glutathione cycle is the most important antioxidant system in brain cells [59-66]. Glutathione protects cells against reactive oxygen, nitrogen and other species. As an antioxidant it is involved in the detoxication of malonic dialdehyde, 4-hydroxy-2-nonenal and other products of lipoperoxidation. The glutathione couple GSH/GSSG takes part in maintaining cellular redox status [67, 68]. Glutathione is presumably a key participant of the defense system in brain cell [69-71].

Increased level of oxidized form of glutathione and changes of glutathione system activity occur at early stages of oxidative stress and may be marks of the severity of oxidative stress [71-75]. Hydroxyl radical and nitric oxide or peroxynitrite may interact directly with GSH leading to GSSG formation. Hydrogen peroxide may be removed by catalase or by glutathione peroxidase [76].

GSH is present in cytoplasma, endoplasmatic reticulum, nucleus, and mitochondria. In most of the compartments GSH is found predominantly in highly reduced state (about 99% of the total level of glutathione). Glutathione peroxidase is localized mainly in cytosol, too, whereas catalase is found mostly in peroxisomes. The affinity of glutathione peroxidase for H_2O_2 is one to two orders of magnitude higher than that of catalase, and catalase is less active in brain than in other tissues.

GSH is synthesized from cysteine, glutamate and glycine. Neurones have lower GSH level and use a more limited list of substrates for GSH synthesis, but they use glutamine for GSH synthesis more effectively than astrocytes because they have glutaminase for formation of glutamate from glutamine. Neurones can not absorb cystine but they actively carry off cysteine [65], so availability of cysteine influences at the GSH level inside neurones. At the same time maximal rate for GSH synthesis within astrocytes is observed in the presence of glutamate, cystine and glycine.

GSSG restores by glutathione reductase into GSH in the presence of NADPH (salvage cycle), which originated mainly from pentose phosphate pathway of metabolism. About 3-5% of oxygen in brain is consumed this way.

In physiological conditions GSSG level both in neurones and astrocytes reach no more than 1% of total content of glutathione in tissue but during oxidative stress it may be about 40% of total value of the glutathione in the astrocyte culture. One astrocyte cell may effectively protect 20 neurons from peroxides but lack of glucose greatly decreases capacity of astrocytes to bind peroxides. It has been shown that pentose phosphate pathway in astrocyte culture is very sensitive to peroxide action.

Decrease of total glutathione content and decline of GSH/GSSG ratio are indicators of the severity of oxidative stress in ischemic brain tissue [67]. It is known that decrease of GSH level leads to aggravation of ischemia-induced injuries, while increase of its level leads to opposite result. Glutamate may facilitate decrease of GSH level because it inhibits use of cysteine which is required for glutathione synthesis by cells [67]. Genetic failure of a cell glutathione peroxidase makes rats more susceptible to neurotoxins and brain ischemia [68]. Excessive glutathione peroxidase expression in transgenic mice leads to prevention of irreversible hypoxia-induced changes. Decline of GSH concentration may weaken the stability of an organism to hypoxia both by inactivation of pentose phosphate way enzymes as by inhibition of thioenzymes of tissue respiration chain [76, 77]. These disruptions cause development of energy deficiency which is the main chain of biochemical mechanism of tissue hypoxia. In addition thiol-disulfide metabolism changes may form the basis for mechanisms of disconnection of the oxidation and phosphorilation processes [78-80]. As a result, use of oxygen in biological oxidation processes may be broken and become a base for pathogenetic component of intiation and generalization of oxidative stress.

4. Role of energy metabolism changes in mechanisms of brain tissue ischemia-induced injuries

Brain is very sensitive to disruptions of energy metabolism processes beacause brain tissue requires constant supply of energy substrates whereas sources for energy formation in brain are rather limited, turnover of metabolism is high, and metabolism is dependent on aerobic oxidation of glucose and constant supply of oxygen in a great extent [81, 82] Maintenance for electric neuron activity and rate of impulse passage depend directly on presence and availability of energy substrates, too [83, 84].

Brain tissue cannot metabolize fatty acids therefore the main source for energy formation in brain is glucose. Nevertheless, during focal brain ischemia, increase of glucose level does not help cells to prevent ischemic injuries and also promotes their structural and functional damage [85]. Mechanisms of these changes include shift of pH to an acid side inside cell, increase of permeability of blood-brain barrier, infiltration of brain parenchyma by neutrophiles, accumulation of extracellular glutamate, and unfavorable corticosterone action. Intensive metabolism of glucose in the penumbra region may promote increasing acidic reaction of the medium, promote attraction of neutrophiles in the region.

Limitation of metabolic consumption of glucose in ischemic brain tissue may have protective effect, especially in such conditions when its metabolism will be faster or other source of fuel will be used. Possibilities for replacement of glucose in brain tissue are rather limited [86-89]. Lactic acid may be an alternative source for energy formation in brain in certain conditions because the glucose is metabolized presumably in glial cells whereas in neurones energy metabolism is based presumably on lactate oxidation [90, 91]. There is some evidence that the process is particularly important for maintenance of vital functions during postischemic time. For example, decrease of lactate transfer through plasmatic mebrane in brain following preliminary whole ischemia causes neurone injuries. From clinic practice it is known that consumption of lactate or pyruvate during brain ischemia show neuroprotective actions of the substances [90, 91]. Presence of adequate concentrations of pyruvate facilitate for maintenance of stable level of membrane potentials and proton gradient on vesicular membranes [87, 92].

Brain ischemia is different from other types of ischemia because oxygen deficiency causes significant changes in the oxidating process of energy substrates which are present in brain in suffucient quantities [20, 42]. Anaerobic glycolisis as alternative way for energy supply is not substantial for supporting ATP stock in nerve tissue during compensated and decompensated stages of hypoxia [11].

Aerobic energy formation is the basic process for nerve tissue, but starts to fail before oxygen concentration falls below critical level, because hypoxia influences kinetic properties of respiratory chain enzymes. During early stages of ischemia energy functions of mitochondria already start changing: conjugacy of oxidative phosphorilation process and regulatory control by ATP becomes weaker, rate of inphosphorilated respiration increases. Shift of ratio NAD/NADH occurs to the side of NADH, as a result final stages of the Krebs' cycle are inhibited, and activation of succinate oxidase stage takes place. This way allows to maintain oxidative phosphorilation and respectively macroergic substance production at sufficient level for some time. "Oxygen hunger" already at early stages of hypoxia leads to beginning of relative "substrate hunger" - energy substrates are not being oxydated while they are still available. This is a characteristic property for ischemia [93].

There are only 2 ways of restoration of brain metabolism after stroke and hypoxia – restoration of NAD-dependent part of the Krebs' cycle and stimulation of alternative path of metabolism, succinate oxidation.

Succinic acid is an intermediate of the cytric cycle which supports formation of macroergic phosphates and reductive equivalents in the conditions with physical loadings and stress [94-97]. Oxidation of succinic acid is the most potent energy process inside mitochondria, and during stress this process becomes even more important due to succinate dehydrogenase activation. Depletion of endogenic succinic acid may be a reason explaining inability of tissue to maintain reaction of activation of energy processes for a long time [98]. If NADH and CO_2 are present in excess, conversion of reactions of second part of the cytrate cycle in which NADH is consumed takes place - from oxaloacetate to succinate, and that fact supports reactions of the first part of the cytrate cycle reactions which require oxidized NAD and promote for additional accumulation of succinate.

Another important result of the bioenergic hypoxia is damage of ion pump action and ion imbalance in the form of excessive accumulation of intracellular calcium, sodium, chlorine [99]. The intracellular calcium excess leads to activation of phospholipase A, damage of cell membranes and release of arachidonic acid take place. As a result, lipoperoxidation activates and causes following cell membrane damages, neuron depolarization and release of excitotoxic amino acids, especially glutamate, in extracellular space [99].

Thereby adaptive effects of succinate derivatives are related to their property to induce compensatory metabolic flows in mitochondrial respiratory chain ("succinate oxidase" way) in extreme conditions, to provide replenishment for cytoplasmatic pool by reduced forms of NAD and NADP, to accelerate ATP formation, change over energy formation from NAD-dependent to FAD-dependent way, eliminate an excess of acetyl-CoA, support activity of the Krebs' cycle in hypoxic conditions, stabilize membrane potential of mitochondria and cell membranes. Advantage of succinate oxidase way versus NAD-dependent substrates in competition for respiratory chain is amplified in the hypoxia conditions because flavines (flavoproteins) continue in oxygenated form longer than pyridine nucleotides.

Disruption of energy metabolism can be mainly observed at the stage of succinate formation. That may be caused by oxidative stress-induced changes of stable state of plasmatic and mitochondrial membranes and changes in activity of membrane-bound enzymes of the Krebs cycle and GABA bypass [100, 101]. Significant activation of the GABA bypass enzymes takes place during ischemia-reperfusion which not only causes raise of succinate formation, but also leads to increasing formation of gamma-hydroxybutyric acid through reductase reaction. GHBA has protective effect against changes of energy formation processes in brain tissue during hypoxia [102]. As activity of glutamate dehydrogenase in brain is rather low compared with other tissues, GABA bypass plays a key role in compensatory maintenance of succinate level sufficient for adequate metabolism in "succinate oxidase" way in different extremal situations, for example in brain ischemia-reperfusion conditions.

Succinic acid derivatives are effective modern antioxidants in the brain because succinate regulates activity of SDH in the Krebs' cycle and restores activity of respiratory mitochondrial chain not only, but increases microcirculation in tissues.

5. Role of CoA in mechanisms of neuroprotection in brain ischemia-reperfusion conditions

Beneficial effects of precursors of CoA biosynthesis, such as pantothenic acid and its derivatives, include protection from lipoperoxidation and supporting membrane structure, and these effects have been observed in radiation injury, miocard ischemia, diabetes mellitus, CCl_4 -intoxication, heavy hypothermia, etc [103-106]. Protective action of pantothenate derivatives have been reported in situations accompanied by oxidative stress, for example, in experimental ischemia-reperfusion of myocardium. It has been shown that antioxidative and membrane-protective effects of the pantothenate derivatives are accompanied by an activation of biosynthesis CoA system and increasing of intracellular level of a free CoA [108, 109]. Presumably, the mechanism of cell protection is CoA-dependent or realized through CoA-(acyl-CoA)-dependent biochemical reactions, including rise of intracellular glutathione level and maintenance of its redox status.

It is believed that the physiological function of CoA system is participation in formation of redox potential of glutathione and proteins, redox signaling and maintenance of biological membrane stability, especially in brain tissue [104, 105, 106].

The CoA biosynthesis system is a group of very stable continuously active self-regulated processes focused on maintaining stability of intramitochondrial CoA-SH (up 70-80% of the total cell value). This function maintains constant flow of oxidative substrates and their effective using for ATP formation in the citric acid cycle [109, 110, 111].

The lesser CoA pool in cytosol where acyl-CoA is used in biosynthetic processes (biosynthesis of phospholipids, fatty acids) is studied to a lesser extent. "Turnover" pool of the coenzyme takes part in reactions of carnitine-dependent transfer of fatty acid residues and acetate through mitochondrial membranes [105]. The main events for interrelations between specifically bound cytosolic CoA-S-S-protein, CoA-S-S-glutathione, free and proteidized glutathione take place within cytosolic compartment (including endoplasmatic reticulum) presumably due to thiol-disulfide-exchange reactions which provide stability during limited variations of redox potential and support a realization of redox sygnaling. Based on this hypothesis, the capability of CoA biosynthesis precursors in low concentrations (0.1-1 мM) or in vivo experiments prevents lipoperoxidation activation, damages of membrane integrity initiated by different physical or chemical factors. The obligatory condition of the above-mentioned defensive effect is biotransformation of pantothenate derivatives into CoA and significant increase of intracellular GSH level. The process is highly specific because homopantothenic acid which is similar to pantothenic one in terms physical and chemical properties can not transform into CoA, does not increase intracellular glutathione level and does not protect plasmatic membrane stability in cell culture [112].

Additional effects of the CoA precursors in defense of lipoperoxidative activation have also been observed. These include rapid initiation of lipid biosynthesis from labeled precursors, positive influence on mitochondrial energy parameters, as well as protection against apoptosis activation caused by free radical oxidation products [113, 114]. Redox sygnaling process controls the initiation and direction of these processes. The redox potential is determined by the glutathione system predominantly [105, 115]. This data may confirm that the CoA biosynthesis system is the most important factor of intracellular stability of GSH level [103, 105].

Maintenance of sufficiently high CoA biosynthesis activity has an important role in brain because acetyl-CoA is used not only as the main way for glucose intake into the Krebs cycle but is also a substrate for acetylcholine synthesis. The relationship between ability for CoA biosynthesis and activity of acetylcholine metabolism within cholinergic neurones may be an important factor in modulation of their sensitivity to damaging influences [103].

Among the necessary conditions for successful biosynthesis of acetyl-CoA, are presence of CoA precursors inside mitochondria, and also the presence of carnitine which transfers acetyl radicals into mitochondria. Under oxidative stress conditions when a lot of lipoperoxide products are released from membranes as a result of phospholipase activation, the potential for CoA sequestration increases, which includes appearance of hard to metabolize acyl-CoA derivatives. Under these conditions the role of carnitine increases. CoA and L-carnitine are among the key factors of intramitochondrial metabolism of fatty and organic acids, and relationship between their levels represents an essential mechanism for cytosol-mitochondrial process of acyl residue activation and transfer [110, 116]. Based on the main localization of a total CoA within mitochondrial matrix, while carnitine is located in cytosol, the molar ratio of CoA/carnitine may have significant functional role for decrease of long-chain acyl-CoA in cytosol and their accelerated utilization in a β-oxidation process.

Generation and use of succinyl-CoA in mitochondria have a special role for mitochondrial oxidation regulation during oxidative stress caused by ischemia-reperfusion [107, 108]. Chances for alternative succinyl-CoA biosynthesis increase significantly when CoA biosynthesis processes are activated in cytosol. In view of this, data on the effects of carnitine on the activity of the key enzyme of CoA biosynthesis, namely patothenate kinase, has high significance. It has been shown that L-carnitine cancels out inhibitory action of physiological concentrations of dephospho-CoA, CoA-SH, and acetyl-CoA on pantothenate kinase. This enables directed regulation of CoA-dependent metabolic processes following simultaneous injection of carnitine substances and pantothenate derivatives - precursors of CoA biosynthesis - namely, panthenol [117, 118].

Study of changes of CoA level during ischemia or ischemia-reperfusion showed markedly stable ratio and levels of free CoA, short-chain acyl-CoA, and on the whole the acid-soluble fraction of CoA, in hemispheres during ischemic damage [117]. Following 2-3 h of brain ischemia, the free CoA level declines. This diminishment with simultaneous decrease of the acid-soluble CoA fraction achieves maximal reduction within 24 h under continued conditions of reperfusion (reoxygenation). These results confirm significance of the CoA system in pathogenic mechanisms of reoxygenation-reperfusion syndrome development.

CoA is one of the fundamental metabolism factors, and its biosynthesis and catabolism are subject to rigid control on the cell level. Therefore, as a rule, changes of particular CoA forms may happen only under extreme conditions and after prolonged period of time, sufficient to cause imbalance in metabolism regulation systems in the cell. Such imbalance starts to influence the CoA system during ischemia no earlier than one hour after occlusion of arteria carotis.

Data on the key role of the CoA biosynthesis system in maintaining redox potential of the glutathione system, neuronal membrane stability and defense of nitroperoxide acyl-CoA gives rationale to the use of CoA precursors in treatment of ischemia and ischemia-reperfusion-induced damages in the brain tissue.

6. Role of selenium in mechanisms of antioxidant protection for brain

Selenium is an essential microelement in different brain functions [119-124]. Neuroprotective potential of selenium is realized through the expression of selenoproteins: glutathione peroxidase, thioredoxine reductases, methionine sulfoxide reductases, selenoproteins P and R, which participate in regulation of the oxidation-reduction state of the neurones and glial cells under both physiologic conditions, and during oxidation stress [125, 126, 128]. Selenium regulates antioxidative processes in the CNS, protects brain tissue from neurodegenerative injuries during Alzheimer and Parkinson diseases, prione diseases, has antiischemic and angiogenic actions, etc. Insufficient level of selenium intensifies damages of neuron functions and structure caused by different endogenic and exogenic affections and leads to some neurodegenerative pathologies [122, 129-132].

The biological role of selenium is explained by the selenium presence in active sites of selenium-related enzymes [121, 133], which protect brain tissue during oxidative stress. Expression extremely diverse Se-containig proteins is observed in the brain. Selenoprotein P is required for transfer of selenium into the brain, and the brain selenium level is strictly dependent from an expression of selenoprotein [129]. Activity of Se-dependent enzymes in the brain is maintained at rather stable levels even during profound selenium deficiency, owing to the presence of unique Se-transport system in the brain (proteins containing selenium-cysteine, Se-transported protein of a Golgi apparatus). This system achieves its maximum value in hypothalamus.

Injections of selenium-containing compounds lead to an increase of activities of glutathione peroxidase and thioredoxin reductase, decrease of lipoperoxidation processes, cell defense from apoptosis [122, 126, 127]. Selenium ions activate oxidative-reductive enzymes of mitochondria and microsomes, take part in ATP synthesis, in electron transfer from hemoglobin to oxygen, maintain cysteine turnover, enhance α-tocoferol action.

7. Metabolic approaches to correction for brain ischemia-reperfusion-caused injuries

Steady advances in the neurosciences have elicidated the pathophysiological mechanisms of brain ischemia and have suggested many therapeutic approaches to achieve neuroprotection in the acutely ischemic brain that are directed at specific injury mechanisms [134-136]. Nevertheless, methods of protection of ischemia and reperfusion-induced damages are still lacking [51, 137, 138]. Search for new ways of neurodefense during brain ischemia-reperfusion is necessary due to the absence of sufficient protective activity in the most substances with specific focus in clinical conditions: controlling excitotoxic effects of neurotransmitter amino acids (modulators of glutamic acid receptor activity and Ca-channels), regulating redox status of cells, as well as presence of high toxicity in the most anti-ischemic medicines [51, 139-141]. In the past two decades, numerous attempts were made to use different substances with the effect on Ca level in a cell and glutamate extracellular level, aiming to apply these as drugs for ischemia-induced injuries treatment, but they have not been successful in men [51, 139-141]. For example, in experimental models, blockators of NMDA- and AMPA-receptors of glutamate exhibit high protective action, but they have strong side-effects and weak protective effects in humans, especially blockators of NMDA-receptors. The role of glutamate in neurotoxic phenomena during ischemia is known to be significant, but usage of glutamate receptor antagonists is rather problematic [51]. There are ongoing studies of Mg substances which block NMDA-receptors, as well as with blockators of AMPA-receptors. There have been recent proposals to combine usage of several drugs with different mechanisms of action. All of the above drugs have a common property – rather high toxicity. Therefore, usually a certain combination of drugs is applied in order to minimize their toxicity and maximize effectiveness [140].

Substances for metabolic therapy may be particularly useful during treatment of brain blood circulation injuries in the case of their simultaneous application with specific medicines because they have no toxic effects and may be used safely for prolonged period [51, 142, 143]. Apart from these drugs, compounds for so-called restoration therapy may be used. Their effects include restoration of metabolism and blood flow in damaged region. Application of the metabolic substances that help to maintain energy metabolism and redox status of glutathione system may be useful for remedying damages to the brain after ischemia-reperfusion [51, 143]. Previously we have shown high efficacy of pantothenic acid derivatives – CoA precursors, as a means of protecting cell membranes from different types of oxidative stress [117, 118, 144]. D-panthenol presents an important substance in this respect because it penetrates into the brain through blood-brain barrier easily and is converted into pantothenic acid, 4-phosphopantothenic acid, CoA, and after that into acyl-CoA (acetyl-, malonyl-, succinyl-CoA), which have high metabolic activity. These effects create the preconditions for stabilization of CoA-dependent processes of membrane phospholipid biosynthesis, neurotransmitter biosynthesis, regulation of energic processes, etc [118].

The efficacy of panthenol as a neuroprotector within a stroke model in rats has been demonstrated [117, 118, 144]. Panthenol not only decreased the volume of infarction, but also diminished neurological deficiency in animals [103]. Fairly high protector activity of D-panthenol was observed in respect to changes of energic metabolism and glutathione system activity during brain ischemia-reperfusion. Protective effects of pantothenic derivatives is not related to their action as free radical scavengers, however. They act primarily as CoA precursors, whereas CoA accelerates various metabolic pathways, such as biosynthesis of glutathione, which constitutes one of the main systems of cell protection against oxidative stress.

Succinic acid is essential for keeping energy formation processes stable in the brain under extreme conditions [94-97]. Consequently, injections of panthenol and succinate following brain ishemia-reperfusion stabilize levels of lipoperoxidation in blood and in brain hemispheres, stabilize levels of protein SH-groups in blood, lead to significant decrease of the GSSG level and normalization of glutathione enzyme activities, as well as glutamate and glutamine metabolism in the brain to control values [118]. D-panthenol and succinate ammonium injection served to partially remedy the injuries and restore these parameters to their normal levels, especially if administered together. These effects are likely linked to activation of succinyl-CoA biosynthesis.

Attempts were made to use selenium-containing compounds for prevention of ischemia-indused injuries, such as ebselen (2-phenyl-1,2-benzisoselenozol-3), which imitated glutathione peroxidase activity [146-148]. However, under clinical conditions the ebselen was not effective. Di-(3-methylpyrazolil-4)-selenide (selecor) imitates effects of selenoproteins, has low toxicity, and satisfactory bioavailability. Additional injections of selecor increase effects of the panthenol and succinate, especially on the lipoperoxidation parameters and activities of glutathione system and selenium-bound enzymes, on ischemia-reperfusion- induced injuries [149].

Effects of D-panthenol and succinate on decrease of lipoperoxidation activities contribute to the overall protective effects of the composition. However, it is evident that metabolic actions of the substances are related to their capasity for regulation of energy metabolism and mitochondrial respiration activity, restoration of the CoA-SH level and cell redox status, membrane-protective activity of the panthenol [118]. Addition of di-(3-methylpyrazolil-4)-selenide (selecor) to D-panthenol and succinate does has limited effect on protective antioxidant properties of the composition. It is likely that this provides additional evidence for significance of specifically metabolic effects of the composition. Increase of selenium level in blood plasma, which may contribute in maintaining of antioxidant activity of glutathione system, takes place in the absence of selenoprotein substrates, after injections of panthenol and succinate. Nevertheless, addition of a selenium source to panthenol and succinate strengthened protective potential of the substances with respect to changes for enzyme activities of glutamate and glutamine which play an important role in maintaining of energy supply and detoxication in ischemic brain tissue and confirms the antiischemic effect of the substances. Effects of di-(3-methylpyrazolil-4)-selenide may be explained less by selenium supply as a selenoprotein component rather than by its modeling of selenoprotein activity, as is known to be the case with ebselen [146-148].

Therefore it is expected that the tested substances, such as panthenol, succinic acid, selecor, and potentially other metabolic therapy drugs may have high efficacy as neuroprotectors in brain ischemia and reperfusion-induced damages.

Author details

Nina P. Kanunnikova and Natalya Z. Bashun
Department of Zoology and Physiology of Men and Animals,
Yanka Kupala's Grodno State University, Grodno, Republic of Belarus

Andrey G. Moiseenok
National Center for Foodstuffs, National Academy of Sciences of Belarus, Minsk, Republic of Belarus

8. References

[1] Gusev EI, Skvortsova VI (2001) Brain ischemia. Moscow. Medicina. 328 p (rus).

[2] Breton RR, Rodriguez JCG (2012) Excitoxicity and oxidative stress in acute ischemic stroke. Acute ischemic stroke. Ed.JCG Rodriguez: 29-58.

[3] Heiss WD (1983) Flow threshold of functional and morphological damage of brain tissue. Stroke. 14: 329-331.

[4] Choi DW (1995) Calcium: still center stage in hypoxic-ischemic neuronal death. Trends Neurosci. 18: 58-60.

[5] Rayevsky KS, Bashkatova VG (1996) Oxidative stress, apoptosis and brain damage. Neirokhimia 13(1): 61-64 (rus).

[6] Rothman SM, Olney JW (1987) Excitotoxicity and the NMDA receptor // Trends Neurosci. 10: 299-302.

[7] Doubinina EE (2003) Role of oxidative stress in pathological states of nervous system. Uspekhi Funkc.Neirokhimii.Ed. Dambinova SA, Arutiunian AV. StPetersburg.: 285–301 (rus).

[8] Lazarewicz J, Strosznajder J, Gromek A (1972) Effects of ischemia and exogenic fatty acids on the energy metabolism in brain mitochondria. Bull.Acad.Pol.Sci.Biol. 20: 599-606.

[9] Strosznajder J (2005) Czynniki biochemiczne inicjuace niedokrwienie. Mozg a niedokrwienie. Pod red. J. Strosznajder, Z. Czernicki. Wydawnictwo PLATAN. Krakow: 7-15.

[10] Zavalishin IA, Zakharova MN (1996) Oxidative stress is a general mechanism of damages in nerve system diseases. Zh. Neurol. Psychiatr. 96(2): 111–114 (rus).

[11] Lin MT, Beal MF (2006) Mitochondrial disfunction and oxidative stress in neurodegenerative diseases. Nature. 443: 787-795.

[12] Andreyev AYu, Kushnareva YuE, Starkov AA (2005) Mitochondrial metabolism of reactive oxygen species. Biochemistry. 70 (2): 200–214.

[13] Beal MF (1996) Mitochondria, free radicals, and Neurodegeneration. Curr. Opin. Neurobiol. 6: 661–666.

[14] Shinder AF, Olson EC, Spitzer NC, Montal M (1996) Mitochondrial disfunction is a primary event in glutamate neurotoxicity. J.Neurosci. 16: 6125-6133.

[15] Sopala M, Danysz W (2005) Rola ukladu glutaminianergicznego w ischemicznym uszkodzeniu mozgu // Mozg a niedokrwienie / Pod red. J. Strosznajder, Z. Czernicki // Wydawnictwo PLATAN. Krakow: 47-65.

[16] Benveniste H et al (1984) Elevation of extracellular concentrations of glutamate in rat hippocampus during transient cerebral ischemia monitored by intracerebral microdialysis. J.Neurochem. 43: 1369-1374.

[17] Choi DW (1988) Glutamate neurotoxicity and diseases of the nervous system. Neuron 1: 623-634.

[18] Pelligrini-Giampietro DE, Cherici G, Alesiani M, Carla V, Moroni F. Excitatory amino acid release and free radical formation may cooperate in the genesis of ischemia-induced neuronal damage. J.Neurosci. 10: 1035-1041.

[19] Boldyrev AA (2003) Role of active species of oxygen in neuron vital functions. Uspekhi physiol.nauk. 34(3): 21–34 (rus).

[20] Boldyrev AA, Kukley ML (1996) Free radicals in normal and ischemic brain. Neirokhimia. 13(4): 271–278 (rus.).

[21] Kinuta Y, Kikuchi H, Ishikawa M, Kimura M, Itokawa Y (1989) Lipid peroxidation in focal cerebral ischemia. J.Neurosurg. 71: 421-429.

[22] Boldyrev AA (1998) Mechanisms of brain defense from oxidative stress. Priroda. 3: 26-34 (rus).

[23] Hayashi M. (2009) Oxidative stress in developmental of brain disorders. Neuropathology. 29(1): 1–8.

[24] Yamamoto M [et al.] (1983) A possible role of lipid peroxidation in cellular damages caused by cerebral ischemia and the protective effect of alpha-tocopherol administration. Stroke. 14: 977-982.

[25] White BC et al (2000) Brain ischemia and reperfusion: molecular mechanisms of neuronal injury. J.Neurol.Sci.179(S1-2): 1-33.

[26] Siesjo BK, Zhao Q, Pahlmark K, Siesjo P, Katsura K, Folbergrova J (1995) Glutamate, calcium and free radicals as mediators of ischemic brain damage. Ann.Thorac.Surg. 59: 1316-1320.

[27] Dugan LL, Choi DW (1994) Exitotoxicity, free radicals and cell membrane changes. Ann. Neurol. 35: 17-21.

[28] Oh SM, Betz L (1991) Interaction between free radicals and excitatory amino acids in the formation of ischemic brain edema in rats. Stroke 22: 915-921.

[29] Zeevalk GD, Bernard LP, Nicklas WJ (1998) Role of Oxidative Stress and the Glutathione Systems in Loss of dopamine Neurons Due to Impairment of Energy Metabolism. J. Neurochem. 70(4): 1421–1429.

[30] Fiodorova TN, Boldyrev AA, Gannushkina IV (1999) Lipoperoxidation in experimental brain ischemia. Biokhimia. 64(1): 94–98 (rus).

[31] Kukley ML, Stvolinsky SL, Shavratsky VCh (1995) Lipoperoxidation in rat brain during ischemia. Neirokhimia. 12(2): 28–35 (rus).

[32] Rafols JA, Krause GS (2000) Brain ischemia and reperfusion: molecular mechanisms of neuronal injury. J.Neurol.Sci.179(S1-2): 1-33.

[33] Sakamoto A, Ohnishi ST, Ohnishi T, Ogawa R (1991) Relationship between free radical production and lipid peroxidation during ischemia-reperfusion injury in the rat brain. Brain Res. 554: 186-192.

[34] Kitagawa RM [et al.] (1990) Free radical generation during brief period of cerebral ischemia may trigger delayed neuronal death. Neuroscience. 35: 551–558.

[35] Fridovich I (1995) Superoxide radical and superoxide dismutases. Annu.Rev. Biochem. 64: 97-112.

[36] Chan PH (2004) Oxidant stress as a molecular switch in neuronal death/ survival after stroke. J.Neurochem. 88(Suppl): S6-4.

[37] Murakami K, Kondo T, Kawase M, Sato S, Chen SF, Chan PH (1998) Mitochondrial susceptibility to oxidative stress exacerbated cerebral infarction that follows permanent focal cerebral ischemia in mutant mice with manganese superoxide dismutase deficiency. J Neurosci. 18: 205-213.

[38] Clemens JA (2000) Cerebral ischemia: gene activation, neuronal injury, and the protective role of antioxidants. Free Rad.Biol.Med. 28: 1526-1531.
[39] Nita D Al et al. (2001) Oxidative damage following cerebral ischemia depends on reperfusion – a biochemical study in rat. Journal Cell. Mol. Med. 5(2): 163–170.
[40] Skvortsova VI [et al.] (2007) Oxidative stress and oxygen status in ischemic stroke. Zh. Nevrol. Psikhiatr. Im. S.S. Korsakova. 107(1): 30–36 (rus).
[41] Turrens JF (1997) Superoxide production by the mitochondrial respiratory chain. Biosci.Rep. 17: 3-8.
[42] Boldyrev AA (1995) Dual role of free radical species of oxygen in ischemic brain. Neirokhimia. 12(3): 3–13 (rus).
[43] Gulyaeva NV, Yerin AN (1995) Role of free radical processes in development of neurodegenerative diseases. Neirokhimia. 13(2): 3–15 (rus).
[44] Hall ED, Braughler JM, Pazak KE (1993) Hydroxyl radical production and lipid peroxidation parallels selective post-ischemic vulnerability in gerbil brain. J. Neurosci. Res. 45(1): 107–112.
[45] Oliver CN [et al.] (1990) Oxidative damage to brain proteins loss of glutamine synthesis activity and production of free radicals during ischemia–reperfusion induced injury to gerbil brain. Proc. Natl. Asad. Sci. USA. 87: 5144–5147.
[46] Halliwell B (1992) Reactive oxygen species and the central nervous system. Free Radical in the brain. Aging. Neurological and Mental Disorders/ Eds Packer L, Christen Y. Springer – Verlag. Berlin, N.–Y., London: 21–41.
[47] Halliwell B (2006) Oxidative stress and neurodegeneration: where are we now? J. Neurochem. 97: 1634–1658.
[48] Halliwell B (2001) Role of free radicals in the neurodegenerative diseases: therapeutic implications for antioxidant treatment. Drugs Aging. 18(3): 685–716.
[49] Gupta R, Singh M, Sharma A (2003) Neuroprotective effect of antioxidants on ischaemia and reperfusion-induced cerebral injury. Pharmacol. Res. 48(Is.2): 209–215.
[50] Halliwell B. (1996) Antioxidants in human health and disease. Ann. Rev. Nutr. 16: 33–50.
[51] Grieb P (2005) Szanse na skuteczne neuroprotekcyjne leczenie udaru niedokrwiennego mozgu. Mozg a niedokrwienie. Pod red. J. Strosznajder, Z. Czernicki. Wyd. PLATAN. Krakow: 237-260.
[52] Gruener N [et al.] (1994) Increase in superoxide dismutase after cerebrovascular accident. Life Sci. 54(11): 711–713.
[53] Kumari NK, Panigrahi M, Prakash Babu P (2007) Changes in endogenous antioxidant enzymes during cerebral ischemia and reperfusion. Neurol. Res. 29(8) 877–883.
[54] Suslina ZA et al (2000) Antioxidant therapy in ischemic insult. Zh. Neurol. Psychiatr. 100(10): 34-38 (rus).
[55] Shapoval GS, Gromovaya VF (2003) Mechanisms of antioxidant protection during action of reactive oxygen species. Ukr.Biokhim.Zh. 75(2): 5–13 (rus).

[56] Valko M, Leibfritz D, Moncol J, Cronin MTD, Mazur M, Telser J (2007) Free radicals and antioxidants in normal physiological functions and human disease. Int. J. Biochem.Cell Biol. 39: 44-84.

[57] Kalinina EV et al. (2010) Modern conceptions on antioxidant role of glutathione and glutathione-dependent enzymes. Vestnik RAMN. 3: 46–54 (rus).

[58] Sies H (1997) Oxidative stress: oxidants and antioxidants. Exp. Physiology. 82: 291-295.

[59] Sies H (1999) Glutathione and its role in cellular functions. Free Radical Biology and Medicine. 27(Iss.9–10): 916–921.

[60] Schulz JB Lindenau J, Seyfried J, Dichgans J (2000) Glutathione, oxidative stress, and neurodegeneration. Eur. J. Biochem. 267(16): 4904–4911.

[61] Lorenc-Koci E. (2003) Neuroprotekcyjne wlasciwosci glutationu I ich znaczenie w schorzeniach neurodegeneracyjnych. Neuroprotekcji. XX Zimowa Szkola Instytutu Farmakologii PAN. Pod red. M.Smialowskiej. – Mogilany: 169-185.

[62] Cooper AJL, Kristal BS (1997) Multiple roles of glutathione in the central nervous system. J.Biol.Chem. 378: 793-802.

[63] Cruz R, Almaguer Melian W, Bergado JA Rosado J (2003) Glutathione in cognitive function and neurodegeneration. Rev. Neurol. 36(9): 877–886.

[64] Dringen R (2000) Metabolism and functions of glutathione in brain. Prog. Neurobiol. 62: 649-671.

[65] Dringen R, Hirrlinger J (2003) Glutathione pathways in the brain Biol. Chem. 384(4): 505–516.

[66] Lushchak VI (2012) Glutathione homeostasis and functions: potential targets for medical interventions. J Amino Acids. 2012: http://www.hindawi.com/journals/jaa/2012/736837/

[67] Jones DP (2002) Redox potential of GSH/GSSG couple: assay and biological significance. Methods in enzymology. 348: 93-112.

[68] Makarov P, Kropf S, Wiswedel I, Augustin W, Schild L (2006) Consumption of redox energy by glutathione metabolism contributes to hypoxia/reoxygenation-induced injury in astrocytes. Mol. Cell Biochem. 286 (1-2): 95-101.

[69] Bains JS, Shaw CA (1997) Neurodegenerative disorders in humans: The role of glutathione in oxidative stress-mediated neuronal death. Brain Res. Rev. 25: 335–358.

[70] Townsend DM, Tew KD, Tapiero H (2003) The importance of glutathione in human disease. Biomed.Pharmacotherapy. 57(3): 145-155.

[71] Mari M, Morales A, Colell A, Garcia-Ruiz C, Fernandez-Checa JC (2009) Mitochondrial glutathione, a key survival antioxidant. Antioxidants and Redox Signalling. 11(11):2685-2700.

[72] Beer SM, Taylor ER, Brown SE et al. (2004) Glutaredoxin 2 catalyzes the reversible oxidation and glutathionylation of mitochondrial membrane thiol proteins: implications

for mitochondrial redox regulation and antioxidant defense. J Biol Chem. 279(46): 47939-47951.

[73] Kumar C, Igbaria A, D'Autreaux B et al. (2011) Glutathione revisited: a vital function in iron metabolism and ancillary role in thiol-redox control. EMBO J. 30(10): 2044-2056.

[74] Kramer K et al. (1992) Glutathione mobilization during cerebral ischemia and reperfusion in the rat. Gen. Pharmacol. 23(1): 105–108.

[75] Bessonova LO, Verlan NV, Kolesnichenko LS ДС (2008) Role of glutathione system in antioxidant protection during complex pathology of hypoxic genesis. Sibirskij Med.Zh. 6: 19–21 (rus).

[76] Mizui T, Kinouchi H, Chan PH (1992) Depletion of brain glutathione by buthionine sulfoximine enhances cerebral ischemic injury in rats. Am. J. Physiol. 262(2): 313–317.

[77] Islecel S [et al.] (1999)Alterations in superoxide dismutase, glutathione peroxidase and catalase activities in experimental cerebral ischemia-reperfusion. Res. Exp. Med. 199: 167–176.

[78] Anderson MF, Sims NR (2002) The effects of focal ischemia and reperfusion on the glutathione content of mitochondria from rat brain subregions. J. Neurochem. 81: 541–549.

[79] Holmgren A (2000) Antioxidant function of thioredoxin and glutaredoxin systems. Antioxidant Redox Signal.2: 811-820.

[80] Shivakumar BR, Kolluri SV, Ravindranath V (1995) Glutathione and protein thiol homeostasis in brain during reperfusion after cerebral ischemia. J. Pharmacol. Exp. Ther. 274(3): 1167–1173.

[81] Shivakumar BR, Kolluri SR, Ravindranath V (1992) Glutathione homeostasis in brain during reperfusion following bilateral carotid artery occlusion in the rat. Moll. Cell. Biochem. 111(1–2): 125–129.

[82] Suh SW, Shin BS, Ma H, Van Hoecke M, Brennan AM, Yenari MA, Swanson RA (2008) Glucose and NADPH oxidase drive neuronal superoxide formation in stroke. Ann.Neurol. 64: 654-663.

[83] Zauner A, Daugherty WP, Bullock MR, Warner DS (2002) Brain oxygenation and energy metabolism. Neurosurgery 51: 289-301.

[84] Erecinska M, Nelson D, Chance B. (1991) Depolarization induced changes in cellular energy production // Proc. Natl Acad.Sci.USA. 88: 7600–7604.

[85] Kozuka M (1995) Changes in brain energy metabolism, neurotransmitters, and choline during and after incomplete cerebral ischemia in spontaneously hypertensive rats. Neurochem. Res. 20(1): P. 23–30.

[86] Martin RL, Lloyd HG, Cowan AI (1994) The early events of oxygen and glucose deprivation: setting the scene for neuronal death? Trends Neurosci 17: 251-257.

[87] Elman I. et al. (1999) The effects of pharmacological doses of 2-deoxyglucose on cerebral blood flow in healthy volunteers. Brain Res. 815: 243-249.

[88] Gonzalez-Falcon A. et al. (2003) Effects of pyruvate administration on infarct volume and neurological deficits following permanent focal cerebral ischemia in rats // Brain Res. 990: 1-7.

[89] Suzuki M et al.(2001) Effect of of beta-hydroxybutyrate, a cerebral function improving agent, on cerebral hypoxia, anoxia and ischemia in mice and rats. Japan J.Pharmacol. 87: 143-150.

[90] Wey J, Cohen DM, Quast MJ (2003) Effects of 2-deoxy-d-glucose on focal cerebral ischemia in hyperglicemic rats // J.Cereb.Blood Flow Metab. 23: 556-564.

[91] Bouzier-Sore AK et al (2003) Lactate is a preferential oxidative energy substrate over glucose for neurons in culture. J.Cereb.Blood Flow Metab. 23: 1298-1306.

[92] Chih CP, Lipton P, Roberts EL (2001) Do active cerebral neurons really use lactate rather than glucose? Trends Neurosci. 24: 573-578.

[93] Hsueh-Meei Huang [et al.] (2003) Inhibition of the α-ketoglutarate dehydrogenase complex alters mitochondrial function and cellular calcium regulation. Biochim. Biophys.Acta.1637: 119–126.

[94] Niizuma K, Endo H, Chan PH (2009) Oxidative stress and mitochondrial disfunction as determinants of ischemic neuronal death and survival. J.Neurosci. 109 (Suppl.1): 133-138.

[95] Kondrashova MN (1991) Cooperation for processes of transaminations and oxidation of carbonic acids in different cell functional states. Biokhimia. 56(3): 388-405 (rus).

[96] Kondrashova MN, Grigorenko EV, Babski AM, Khazanov VA (1987) Homeostasis of physiologic functions at the level of mitochondria. In: Mol.mechanisms of cell homeostasis. Novosibirsk: 40–66 (rus).

[97] Khazanov VA (2004) Past, present, and future of bioenergic pharmacology. In: Regulators of energic metabolism. Clinical and pharmacologic aspects. Tomsk: 3-7 (rus).

[98] Khazanov VA (1997) Oxidation of succinic acid in brain mitochondria. In: Succinic acid in medicine, food industry, agriculture. Ed. Kondrashova MN. Pushchino: 74–78 (rus).

[99] Mayevsky EI, Rosenfeld AS, Vazatashvili MV, Tchilaya SM (1997) Possibility for oxidation of exogenic succinic acid in vivo. In: Succinic acid in medicine, food industry, agriculture. Pushchino: 52–57 (rus).

[100] Lazarewicz J, Salinska E (2005) Udzial ionow wapnia w pathologii niedokrwiennej mozgu. Mozg a niedokrwienie. Pod red. J. Strosznajder, Z. Czernicki // Wydawnictwo PLATAN. Krakow: P.16-46.

[101] Villa RF, Gorini A, Hoyer S (2009) Effect of aging and ischemia on enzymatic activities linked to Krebs' cycle, electron transfer chain, glutamate and amino acids metabolism of free and intrasynaptic mitochondria of cerebral cortex. Neurochem.Res. 28: P.347–351.

[102] Kanunnikova NP, Omeyanchik SN, Bashun NZ, Shalavina EG, Doroshenko EM, Zolotukhin MM, Karayedova LM, Artiomova OV, Moiseenok AG (2003) Metabolism of

GABA abd glutamate in the rat brain in experimatnal ischemia and reperfusion. Neirokhimia. 20(3): 196-200 (rus).

[103] Mutuskina EA, Zarzhetski YuV, Trubina IE, Avrushchenko MA, Volkov LV, Onufriev MN, Lazareva NA, Stepanichev My, Gulyaeva NV, Gurvich AM (1997) Influence of ammonium succinate on functional, biochemical, and morphologic parameters of the central nervous system recovery in rats after 10 min break of blood flow. In: Succinic acid in medicine, food industry, agriculture. Pushchino: 145-150 (rus).

[104] Moiseenok AG, Komar VI, Khomich TI, Kanunnikova NP, Slyshenkov VS (2000) Pantothenic acid in maintaining thiol and immune homeostasis. Biofactors. 1: 53-55.

[105] 104] Brass EP (1994) Overview of Coenzyme A metabolism and its role in cellular toxicity. Chem. Biol. Interact. 90(3): 203-214.

[106] Moiseenok AG et al. (2004) CoA biosynthesis is an universal mechanism for conjugation of exogeneity and multiplicity of pantothenic acid functions. Ukr. Biokhim.Zh. 76(4): 68–81 (rus).

[107] Moiseenok AG (2003) Pantothenic acid: from universal distribution to universal functions. In: Biochemistry, pharmacology, and clinical use of pantothenic acid derivatives. Grodno: 107-114 (rus).

[108] Omelyanchik SN, Kanunnikova NP, Bashun NZ (2007) Changes of CoA pool structure in experimental brain ischemia and reperfusion and their correction by CoA biosynthesis precursors. Vestnik GrDU. 2(2): 89-93 (rus).

[109] Rabin O [et al.] (1997) Changes in cerebral acyl–CoA concentrations following ischemia–reperfusion in awake gerbils. J.Neurochem. 68: 2111–2118.

[110] Leonardi R, Zhang Yong-Mei, Rock ChO., Jackowski S (2005) Coenzyme A: Back in action. Progress in Lipid Res. 44: 125-153.

[111] Moiseenok AG, Rolevich IV, Kanunnikova NP, Khomich TI, Slyshenkov VS, Omelyanchik SN (1998) Simultaneous use of D-pantothenic acid substances and carnitine: foundation, and perspectives of use in prophylactic and therapeutic treatment. In: Panthenol and other derivatives of pantothenic acid: biochemistry, pharmacology, and medical use. Grodno: 130-137 (rus.).

[112] Grishina EV et al. (2007) Inhibition of mitochondrial energetics by CoA derivatives. Toxic effects of saturated fatty acids. In: Reception and intracellular signalling. Pushchino: 215-218 (rus).

[113] Slyshenkov VS, Rakowska M, Moiseenok AG, Wojtczak L. (1995) Pantothenic acid and its derivatives protect Ehrlich ascites tumor cells against lipid peroxidation. Free Rad. Biol. Med.19(66): 767-772.

[114] Slyshenkov VS, Piwocka K, Sikora E, Wojtczak L (2001) Pantothenic acid protects jurkat cells against ultraviolet light-induced apoptosis. Free Rad. Biol. Med. 30(11): 1303-1310.

[115] Slyshenkov VS, Dymkowska D, Wojtczak L (2004) Pantothenic acid and pantothenol increase biosynthesis of glutathione by boosting cell energetics. FEBS Letters 569: 169-172.

[116] Etensel B. et al. (2007) Dexpanthenol attenuates lipid peroxidation and testicular damage at experimental ischemia and reperfusion injury. Pediatr. Surg. 23(2): 177-181.

[117] Kanunnikova NP, Bashun NZ, Raduta EF, Slyshenkov VS, Omelyanchik SN (2007) Influence of panthenol and carnitine on parameters of lipoperoxidation and glutathione in experimental brain ischemia-reperfusion. In: Ecolog.Antropologia. Minsk: 154-156 (rus).

[118] Omelyanchik SN, Kanunnikova NP, Slyshenkov VS, Bashun NZ (2003) Role of panthenol in correction of some damages caused by experimental brain ischemia-reperfusion. In: Biochemistry, pharmacology, and clinical use of pantothenic acid derivatives. Grodno: 144-150 (rus).

[119] Bashun NZ, Raduta HF, Balash ZhI, Kirvel PCh, Sushko LI, Kanunnikova NP, Moiseenok AG (2007) Correction of postischemic disturbances in the hemispheres of the brain using precursors of succinyl-CoA biosynthesis. Neurochemical Journal. 1(3): 249-252.

[120] Schweizer U, Brauer A, Kohrle J, Nitsch R, Savaskan NE (2004) Selenium and brain function: a poorly recognized liaison. Brain Res. Rev. 45: 164–178.

[121] Brauer AU, Savaskan NE (2004) Molecular actions of selenium in the brain: neuroprotective mechanisms of an essential trace element. Rev.Neurosci. 15: 1-19.

[122] Gromova OA, Rebrov VG, Solupayeva LV (2004) Biological role of selenium. In: Selenium substances and health. Ed. Sanockij IV. Moscow: 12–42 (rus).

[123] Arteel GE, Sies H (2001) The biochemistry of selenium and the glutathione system. Environmental Toxicol.Pharmacol. 10: 153–158.

[124] Whanger PD (2001) Selenium and the brain: a review. Nutr. Neurosci. 4(2): 81–97.

[125] Baraboj VA (2004) Selen: biological role and antioxidant activity. Ukr.Biokhim.Zh. 76(1): 23-31(rus).

[126] Tapiero H, Townsend DM, Tew KD (2003) The antioxidant role of selenium and seleno–compounds. Biomedicine and Pharmacoterapy. 57: 134–144.

[127] Chen J, Berry MJ (2003) Selenium and selenoproteins in the brain and brain diseases. J. Neurochem. 86(1): 1–12.

[128] Steinbrenner H [et al.] (2006) Involvement of selenoprotein P in protection of human astrocytes from oxidative damage. Free Radical Biol. Med. 40(9): 1513–1523.

[129] Ozbal S [et al.] (2008) The effects of selenium against cerebral ischemia-reperfusion injury in rats. Neurosci. Lett. 438(Is.3): 265–269.

[130] Holmgren A, Lu J (2010) Thioredoxin and thioredoxin reductase: current research with special reference to human disease. Biochem. Biophys. Res.Communs. 396(1): 120-124.

[131] Ansari M.A. [et al.] (2004) Selenium protects cerebral ischemia in rat brain mitochondria. Biol. Trace Elem. Res. 101(1): 73–86.

[132] Yousuf S [et al.] (2007) Selenium plays a modulatory role against cerebral ischemia-induced neuronal damage in rat hippocampus. Brain Res. 114(25): 218–225.

[133] Huang K., Lauridsen E, Clausen J (1994) The uptake of Na-selenite in rat brain. Localization of new glutathione peroxidases in the rat brain. Biol. Trace Elem.Res. 46(Is.1–2): 91–102.

[134] Moiseenok AG (2004) Efficiency of sodium selenite and dimethyldiprosolilselenid (selecor) in protection of development of oxidative stress during endogenic intoxication. In: Selenium substances and health. Ed. Sanotsky. Moscow: 79–88 (rus).

[135] Moustafa RR, Baron J-C (2008) Pathophysiology of ischaemic stroke: insights from imaging, and implications for therapy and drug discovery. Br J Pharmacol 153: S44-S54.

[136] Adamek D (2003) Neuroprotekcja w urazach mozgu I rdzenia. Neuroprotekcija. XX Zimova Szkola Farmakol. PAN. Pod red. Smialowskiej M. Mogilany: 107-125.

[137] Zaleska MM, Mercado MLT, Chavez J, Feuerstein GZ, Pangalos MN, Wood A (2010) The development of stroke therapeutics: promising mechanisms and translational challenges. Neuropharmacol. 56: 329-341.

[138] Ginsberg MD (1995) Neuroprotection in brain ischemia: an update (Part II). The Neuroscientist. 1(3): 164-175.

[139] Cheng YD, Al-Khoury L, Zivin JA (2004) Neuroprotection for ischemic stroke: two decades of success and failure. NeuroRes1:36-45.

[140] Zhang ZG, Chopp M (2009) Neurorestorative therapies for stroke: uderlying mechanisms and translation to the clinic. Lancet Neurol 8: 491-500.

[141] Mehta SL, Manhas N, Raghubir R (2007) Molecular targets in cerebral ischemia for developing novel therapeutics. Brain Res. Rev 54: 34-66.

[142] Bacigaluppi M, Hermann DM (2008) New targets of neuroprotection in ischemic stroke. Scientific World Journal. 13(8): 698–712.

[143] 142] Szczudlik A (2003) Neuroprotekcja jako kierunek leczenia niedokrwienia mozgu. Neuroprotekcji. XX Zimowa Szkola Instytutu Farmakologii PAN. Pod red. M.Smialowskiej. - Mogilany: 99-105.

[144] Kanunnikova NP (2009) Role of metabolic therapy in complex treatment of ischemic and reperfusion injuries in brain. Zh. Grodn.Gos.Med.Univ. 4: 16-19 (rus).

[145] Slyshenkov VS, Shevalye AA, Moiseenok AG (2007) Pantothenate Prevents Disturbances in the Synaptosomal Glutathione System and Functional State of Synaptosomal Membrane under Oxidative Stress Conditions. Neurochem. J. 1(3): 235–239.

[146] Onufriev MV, Stepanichev MYu, Lazareva NV, Katkovskaya IN, Tishkina AO, Moiseenok AG, Gulyaeva NV (2010) Panthenol as neuroprotectant: study in a rat model of middle cerebral artery occlusion. Neurochem.J. 4(2): 148-152.

[147] Imai H, Graham DJ, Masayasu H, Macrae IM (2003) Antioxidant ebselen reduces oxidative damage in focal cerebral ischemia. Free Rad.Biol.Med. 34: 56-63.

[148] Parnham M, Sies H (2000) Ebselen: prospective therapy for cerebral ischaemia. Expert Opin.Invest.Drugs. 9: 607-619.

[149] Yamagata K [et al.] (2008) Protective effect of ebselen, a seleno-organic antioxidant on neurodegeneration induced by hypoxia and reperfusion in stroke-prone spontaneously hypertensive rat. Neuroscience. 153(2): 428–435.

[150] Kanunnikova NP, Raduta EF, Yelchaninova VA, Katkovskaya IN, Kovalenchik IL, Lukiyenko EP, Pehovskaya TA, Bashun NZ, Moiseenok AG (2010) Influence of panthenol, succinate ans selecor on changes of activities of lipoperoxidation and glutathione system in rat brain and blood after brain ischemia and reperfusion. Novosti med.-biol.nauk. 2(4): 156-162 (rus).

Chapter 24

Lipid Peroxidation and Polybrominated Diphenyl Ethers – A Toxicological Perspective

Mary C. Vagula and Elisa M. Konieczko

Additional information is available at the end of the chapter

http://dx.doi.org/10.5772/50345

1. Introduction

Understanding the structure and composition of plasma membrane is important as it guards the integrity and function of the cell as a whole. Lipid peroxidation is a process in which the lipids of the cell, particularly the membrane lipids, are degraded by oxidation resulting in the disruption of the entire cell. All cells are enclosed by a plasma membrane which serves as a boundary and it is the primary barrier between cell's internal and external environments. It is semi-permeable and plays a major role in regulating the transport of molecules into and out of the cells. This property of the plasma membrane is known as selective permeability. The fluid mosaic model describes the arrangement of the lipids and the proteins in the plasma membrane [1]. The lipids are smaller in size than the proteins. However, there are many more lipids than there are proteins. The lipids give the membranes their basic shape. Membrane lipids are amphipathic molecules, with hydrophilic and hydrophobic ends [2]. These lipids arrange in two layers (bilayer) such that the hydrophobic ends of one layer touches the hydrophobic ends of the other lipid layer. The hydrophilic ends of the phospholipids are in contact with the aqueous external and internal environments. Membrane lipids include phospholipids, cholesterol, sphingomyelin, triacylglycerides and glycolipids [3]. Several different types of phospholipids are found in cellular membranes: phosphatidylcholine (the most abundant), phosphatidylserine, phosphatidylinositol and phosphatidylethanolamime [4]. Each species of organisms has its own unique combination of these different phospholipids in its cellular membranes.

The proteins of the membrane float in the lipid bilayer. There are fewer proteins than lipids in the plasma membrane. However, they are much bigger than the lipids. The proteins can be associated with the lipids of the bilayer in one of the three ways [5]. Integral proteins are amphipathic molecules that span the entire width of lipid bilayer. Their hydrophilic and hydrophobic portions allow them to interact with both lipids and the aqueous environments on either side of the lipids. Membrane associated proteins are associated with either external or internal surfaces of the membrane. These proteins often contain a lipid anchor which helps to hold them in the membrane. Peripheral proteins are attached to the inner surface of the membrane through covalent bonds to another protein. For example, the proteins that link the plasma membrane to the cytoskeleton. Plasma membrane proteins have several functions [6]. They serve as transporters, receptors, enzymes and as anchors, tying the cell to the external environment or to other cells. Carbohydrates attached to some proteins serve as identification markers and help in the differentiation of self from non-self cells [7].

The composition and structure of the plasma membrane described above is affected by free radicals. So the next section deals with free radicals, on how they are generated in the cells and their effects on cellular macromolecules.

2. Free radicals and Reactive Oxygen Species (ROS)

A free radical may be defined as any atom or molecule that has one or more unpaired electrons in its outermost shell. Free radicals were discovered in biological material about 50 years ago [8] and since then, they have generated great curiosity for their involvement in disease and aging. Oxygen free radicals or reactive oxygen species (ROS) are defined as any free radical molecules containing oxygen. Some examples of the most common ROS include: hydrogen peroxide (H_2O_2), hydroxyl ($^{\bullet}OH$), and superoxide ($O_2^{\bullet\,-}$) anion. Under normal conditions, ROS are produced as necessary intermediates in white blood cells to defend against invading pathogens, and they also play a role in intracellular signaling. Owing to their unpaired electrons in their outermost shell, ROS are highly reactive and unstable and they exert damage on biological structures. Free radicals react quickly with other molecules and "steal" electrons to acquire stability. The molecule which lost its electron to a free radical will become a free radical itself, thus beginning a chain reaction. Free radicals modify the structure of other molecules and cause oxidative damage. Once the process is started, if not quenched in time, it can eventually result in cell damage. Additionally, excessive amounts of free radicals are generated when tissues are damaged, for example, in times of tissue hypoxia, exposure to smoke, ultraviolet radiation and pollutants.

2.1. Generation of reactive oxygen species

Most of the energy required to fuel metabolic functions of aerobic organisms is produced at the cellular level in the mitochondria via the electron transport chain. In addition to energy, reactive oxygen species (ROS) are also generated from electron transport chain as part of normal cellular metabolic reactions. In this transport chain, ROS are formed at the mitochondrial inner membrane. Under normal conditions, the oxygen molecule acts as a terminal electron acceptor in the electron transport chain and is reduced to form water. However, sometimes electrons leak prematurely and reduce oxygen to form ROS. The electron transfer between ubiquinone-cytochrome b is the most probable site of ROS formation [9]. Microsomes and membranes of the nucleus also involve electron transportation through the cytochromes P_{450} and B_5 which can produce free radicals [10]. The following few paragraphs discuss the generation of various ROS.

2.2. Superoxide anion ($O_2^{\bullet -}$)

When ground state oxygen accepts a single electron a superoxide anion ($O_2^{\bullet -}$) is formed.

$$O_2 + e^- \rightarrow O_2^{\bullet -}$$

Under normal physiological conditions cell organelles, namely, mitochondria, microsomes and peroxisomes, generate $O_2^{\bullet -}$. The half-life of superoxide anion is longer than that of another potent ROS, hydroxyl radical ($OH^{\bullet}$), and it can react with biological molecules for longer times [10]. Phagocytic cells produce superoxide during a phenomenon known as the "respiratory burst" which occurs when they encounter pathogens. Phagocytic cells known to produce superoxide anions are macrophages, monocytes, neutrophils and eosinophils [11]. Superoxide anions can trigger the formation of hydroxyl and peroxyl radicals, which in an acidic environment can form hydrogen peroxide (H_2O_2).

2.3. Hydrogen peroxide (H_2O_2)

The addition of a second electron to $O_2^{\bullet -}$ gives rise to the peroxide ion, O_2^{2-}. At physiological pH any peroxide ion formed is protonated to hydrogen peroxide (H_2O_2).

$$2O_2^{\cdot -} + 2H^+ \rightarrow H_2O_2 + O_2$$

Hydrogen peroxide can then diffuse far from the site of its production to other sites where its biological effects may be greater. The diffusion range is extended by H_2O_2 carriers formed spontaneously by hydrogen bonding with compounds such as amino and dicarboxylic acids, peptides, proteins, nucleic acid bases, and nucleosides. Equilibrium exists between an adduct-forming compound and H_2O_2. The hydrogen peroxide adducts (HPAs) retard the decomposition of H_2O_2 up to several hundredfold [12]. The overall charge on an HPA influences the cytotoxic and clastogenic effects of H_2O_2. The adducts, especially L- Histidine, play an important role in the stabilization and reduction in the reactivity of the hydrogen peroxide thereby preventing single strand breaks of the DNA in cell free DNA systems [13].

2.4. Hydroxyl radical ($^{\bullet}OH$)

The hydroxyl radical ($^{\bullet}OH$) can be formed by the homolytic fission of the O-O bond of the H_2O_2 molecule. Simple mixing of iron (II) salts with the H_2O_2 also forms the $^{\bullet}OH$ radical. This reaction was first reported by Fenton in 1894 and it is called Fenton's reaction. The hydroxyl radicals can also be produced by Haber-weiss reactions in the cells, in which H_2O_2 interacts with superoxide radical [10]. Additionally, copper salts can also react with H_2O_2 to generate $^{\bullet}OH$ radicals. All the three ways of $^{\bullet}OH$ radical generation is shown below:

By the decomposition of H_2O_2 by Fenton's reaction:

$$Fe^{+2} + H_2O_2 \rightarrow Fe^{+3} + {}^{\cdot}OH + OH^-$$

and H_2O_2 interaction with superoxide radical by the Haber-Weiss reaction:

$$O_2^{\cdot -} + H_2O_2 \rightarrow O_2 + H_2O + {}^{\cdot}OH$$

The hydroxyl radical can also be formed when copper salts react with H_2O_2 to make hydroxyl radicals as shown below:

$$Cu^{+} + H_2O_2 \rightarrow Cu^{2+} + {}^{\cdot}OH + OH^{-}$$

Although the hydroxyl radical ($OH^{\cdot}$) has the shortest half-life, it is one of the most reactive of all ROS [14, 15]. Due to this property it reacts with almost all macromolecules found in the cells such as proteins, nucleic acids (RNA and DNA), phospholipids, sugars and carbohydrates, and exerts deleterious effects on them.

2.5. Free radicals and the damage to cellular components

Excess generation of free radicals combined with the failure of cells to scavenge them through protective mechanisms (antioxidative agents and enzymes) results in oxidative damage. Oxidative damage is witnessed in the form of free radical mediated attack on macromolecules within their vicinity and causing disruption of these molecules. Severe oxidative stress can even result in apoptosis and cell death [16]. Free radical mediated damage to proteins, carbohydrates, nucleic acids and lipids is described below.

2.6. Free radical damage to proteins

The oxidization reaction of free radicals on amino acids causes changes in the physical properties of proteins that they compose. These physical changes are of three types: fragmentation, aggregation and susceptibility to enzyme digestion [17]. The fragmentation of albumin and collagen due to free radical mediated oxidization is a classic example of this phenomenon [18, 19]. Selective fragmentation at specific amino acids, especially at proline, histidine and arginine, is observed due to their close association with transition metals. The second type of physical change, protein aggregation, is believed to be caused predominantly by hydroxyl radicals due to their ability to form cross-links between the constitutive amino acids. Lastly, gross alteration in the conformation of protein structure may be the reason for the susceptibility of these altered proteins to enzyme digestion [10].

2.7. Free radical damage to carbohydrates

Hydroxyl radicals generated by Fenton reaction are reported to induce damage on simple carbohydrates [20]. It is also reported that oxidation of glucose may be a means of quenching hydroperoxide radicals while also generating a source of oxygen free radicals [21]. Another study indicated that rapid auto-oxidation of glucose results in the formation of dicarbonyl and H_2O_2 under physiological conditions [22]. Thus, oxidized glucose can react with proteins in a process called glucosylation or glucation leading to the disruption in the function of these macromolecules. Glucosylation of certain long-lived proteins by a non-enzymic reaction with free glucose may contribute to ageing.

2.8. Free radicals and the alteration of gene expression

Free radicals are extensively implicated for their role in DNA damage and alteration in the gene expression leading to various types of cancers [23]. It has been reported that about twenty types of changes are induced by ROS in the DNA. The damage in the DNA's deoxyguanosine residues is estimated to be in the range of 8 to 83 residues per million deoxyguanosine residues. This damage increases with age and seems to affect organs such as the liver, kidney and spleen while leaving brain tissue relatively undamaged [24]. Mitochondrial DNA experiences greater damage due to proximity to the ROS generated from the electron transport chain of mitochondrial membranes. Free radical induced damage to DNA includes a range of specifically oxidized purines and pyrimidines, alkali labile sites, single strand breaks and instability formed directly or by repair processes [25,26]. The pyrimidine residues, cytosine and thymine, of DNA are the most susceptible to the attack by hydroxyl radical, followed in decreasing levels of susceptibility by adenine, guanine and deoxyribose sugar [10]. Some of these modified bases have been found to possess mutagenic properties. Therefore, if not repaired they can lead to carcinogenesis.

2.9. Free radicals and lipid peroxidation

Free radicals steal electrons from the cell membrane's polyunsaturated fatty acids (PUFAs) and initiate an attack called lipid peroxidation. The peroxidation of lipids occurs in three steps, namely, initiation, propagation and termination. Free radicals target the carbon-carbon double bond of PUFAs, as this double-bond weakens the carbon-hydrogen bond, permitting easy dissociation of the hydrogen by a free radical. During the first step, a single electron is stripped by a free radical from hydrogen atoms associated with the carbon at the double-bond leaving the carbon with a single electron. Molecular arrangements occur in an attempt to stabilize, leading to the formation of conjugated dienes. In the propagation phase, the conjugated diene readily reacts with molecular oxygen and transitions to a hydroperoxide radical that begins stealing electrons from other lipid molecules, resulting in a lipid hydroperoxide [11]. The hydroperoxide molecules undergo decomposition with the aid of transition metals into alcoxyl and peroxyl radicals that, in turn, can initiate a chain reaction and further propagate lipid peroxidation and cause damage to the cell. The three phases or steps of lipid peroxidation are shown in the following equations:

Initiation reaction

$$\text{Lipid} - \text{H} + \text{OH} \rightarrow \text{Lipid} + \text{H}_2\text{O}$$

Hydroperoxide formation

$$\text{Lipid} + \text{O}_2 \rightarrow \text{Lipid-O}_2$$

Lipid hydroperoxide

$$\text{Lipid-O}_2 + \text{Lipid} - \text{H} \rightarrow \text{Lipid} + \text{Lipid-O}_2\text{H}$$

The important fatty acids that undergo peroxidation are: linoleic acid, arachidonic acid, and docosahexaenoic acids [14]. The hydroxyl radical damages cell membranes and other lipoproteins by lipid peroxidation. It is important to note that the low density lipoprotein based lipid peroxidation plays a significant role in atherosclerosis [27].

The generation of free radicals and their action on macromolecules thus leads to disruption of cellular functions. The following section discusses chemicals that are present in the body to protect against these damaging radicals.

3. Antioxidative protection against reactive oxygen species (Free radicals)

All aerobic cells have an armory of chemicals known as antioxidants which are capable of counteracting the damaging effects of oxidative free radicals and inhibiting the oxidation of other molecules in the body. Based on the mechanism of their action, these antioxidants are divided into two categories: chain breaking antioxidants and preventive antioxidants. Vitamin E, glutathione, vitamin C and beta-carotene are some examples of the chain breaking type as they donate their own electron to the existing free radical and prevent the continuation of oxidation chain reaction. These antioxidants are found in the plasma and interstitial fluid [28, 29]. The preventative category of antioxidants consists of enzymes that scavenge the free radicals before they initiate an oxidative attack.

Another way to categorize these antioxidants is based on their solubility in lipids or water. The antioxidants which are lipid soluble are located in the cellular membrane and lipoproteins, while the water soluble ones are located in the aqueous environments, such as the cytoplasm inside cells and the blood [30]. The following few sections discuss some of the important antioxidants of cells.

3.1. Glutathione

Glutathione (GSH) is one of the important cellular antioxidants. It is made up of three amino acids namely, cysteine, glutamic acid and glycine. All the cells of the body have the ability to synthesize glutathione, but it is found in high concentrations in the liver, lungs, and intestinal tract. The important functions of the glutathione include: anti-oxidation, detoxification, strengthening the immune system and signal transduction under stressful conditions. The diverse functions of GSH are due to the sulfhydryl group in cysteine, enabling it to chelate and detoxify harmful substances [31]. Additionally, glutathione also plays a role in processing medications and cancer-causing compounds (carcinogens), and in building DNA, proteins, and other important cellular components. Its usefulness as an antioxidative agent lies in its ability to donate electrons, and in that process it is oxidized to glutathione disulfide (GSSG) form. By donating an electron to a free radical, GSH can quench the unstable and highly reactive free radical and thus can prevent free radical mediated oxidative damage in the cells. GSSG is reduced to GSH by NADPH, an electron donor in this reaction. This reduction reaction is catalyzed by glutathione reductase. Thus the ratio between reduced and oxidized glutathione serves as an indicator of cellular toxicity mediated by oxidative damage. Large amounts of GSH are produced and stored in the liver, where it is used to detoxify harmful compounds. GSH levels are found to decline with age leaving the body susceptible to the damage by free radicals.

3.2. Vitamin C, vitamin E and beta carotene

Vitamin C is a water soluble antioxidant which plays an important role in protecting the body against oxidative damage. Chemically, it is a carbohydrate-like substance. Human beings, unlike other mammals, do not have the ability to synthesize Vitamin C. Therefore, it must be furnished through the diet. Vitamin C is easily oxidized and many functions in the body depend on this property. It stops the free radical chain reaction by donating an electron and prevents the reaction from progressing.

Carotenoids are a class of cellular antioxidants. Carotenoids are referred as provitamin A as they can potentially yield Vitamin A. Free radicals which can damage DNA and cause diseases like macular degeneration, coronary artery disease and cancer are neutralized by these carotenoids. Beta-carotene is perhaps the best known carotenoid. It was isolated in 1831 and its structure was determined in 1931.

Vitamin E is a fat soluble compound which also can impede oxidation to some degree [32]. Though there are eight forms in nature, only alpha-tocopherol form is functional in human beings. Vitamin E stops the production of ROS formation when fat undergoes oxidation. In addition to the above mentioned vitamins, essential trace elements, selenium, zinc, manganese, and copper, also play a critical role as antioxidants.

3.3. Antioxidant enzymes

Antioxidant enzymes inside the cells are the main defense against the free radicals. The most important antioxidant enzymes present inside the cells are superoxide dismutase (SOD), catalase, glutathione peroxidase (GPx), and glutathione S-transferase (GST). SOD catalyzes the dismutation of superoxide anion to O_2 and H_2O_2. This reaction is considered to be the body's primary defense against free radicals as it prevents further generation of free radicals [29]. SOD is found in high levels in the liver, adrenal glands, kidney and spleen [33]. GPx, catalyzes the reduction of hydroperoxides and serves to protect the cell from hostile free radicals. Catalase, another antioxidant enzyme, along with GPx, is essential for removing hydrogen peroxide formed during oxidation reactions, and is found in high levels in liver, kidney and red blood cells. Both catalase and GPx enzymes exhibit a great degree of cooperativity in their action. Glutathione S-Transferase (GST) is yet another antioxidant enzyme primarily responsible for cellular defense mechanism against ROS. GST metabolizes xenobiotics, whether they are of endogenous or exogenous origin [34]. GST along with glutathione and GPx neutralizes free radicals and lipid hydroperoxides especially at low levels of oxidative stress [35]. The undesirable end products of lipid peroxidation are thus detoxified by GST. All of these enzymes prove to be indispensable in the cellular antioxidant defense mechanism and they are all necessary for the survival of the cell, even in normal conditions. An appropriate equilibrium between antioxidant enzyme activities is vital to ensure the cell survival during increased oxidative stress.

However some exogenous factors, such as environmental contaminants, have the potential to disrupt this equilibrium and cause incredible damage to the cells. The following section describes a specific group of brominated flame retardants called polybrominated diphenyl ethers (PBDEs).

4. Polybrominated diphenyl ethers: A class of brominated flame retardants

Fire accidents in the United States have the potential to cause severe economic problems for the individuals and for society at large. Each year, several thousand Americans die, several thousand more are injured, and millions of dollars of loss are all due to fires. To recover from these losses, individuals, their insurance companies, and society as a whole spend significant resources on temporary housing, groceries and meals, medical costs, psychological support, and relocation costs, to name a few expenditures. These ancillary costs may be 10-times higher than the cost to rebuild structures actually gutted by fire. To give a rough idea, the annual losses from natural calamities such as floods, hurricanes, tornadoes, earthquakes, etc., all put together in the United States equal just a fraction of total loss due to fires [36]. To comply with fire safety regulations in the United States manufacturers must follow strict guidelines in order to reduce the flammability of their products. For years, manufacturers have added chemicals to plastics and fabrics in order to reduce combustibility and increase overall safety of the products. Different types of brominated flame retardants have the promise of slowing down fire and hence are being extensively used. Figure 1 shows the chemical structures of four typical bromine compounds, namely, polybrominated biphenyls (PBBs), polybrominated diphenyl ethers (PBDEs), hexabromocyclododecane (HBCDs) and tetrabromobisphenol A (TBBPAs) [37].

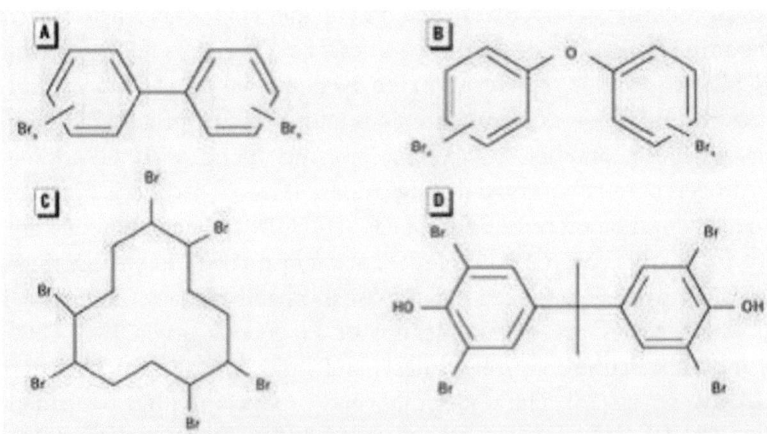

Figure 1. Chemical structures of BFRs. (A) PBBs, (B) PBDEs, (C) HBCD, and (D) TBBPA [37]

In the past few decades, use of PBDEs has become quite commonplace and these chemicals are added in large quantities to many commercial and household products, such as computers, TVs, foam mattresses, and carpets, to inhibit ignition and prevent fire accidents [37]. At high temperatures PBDEs liberate bromine atoms, which are effective at hindering the basic oxidizing reactions that drive combustion. Ever since PBDEs were introduced into the market, deaths due to fire accidents dropped significantly. Functionally, these flame retardants are similar to polychlorinated biphenyls (PCBs) that were banned from production in the U.S in the late 1970s due to potential adverse health effects and a persistent bio-accumulative nature [38]. Like PCBs, the PBDEs class contains 209 possible congeners. During the synthesis of PBDEs, a bromination reaction occurs on the biphenyl rings as one or more bromine atoms are attached to the biphenyl rings giving rise to one of the possible 209 possible congeners of PBDEs. The maximum number of bromine atoms that can attach to biphenyl rings is 10 and this brominated flame retardant is called decaBDE. Commercially these chemicals are available in three mixtures, pentaBDEs, octaBDEs and decaBDEs. Congeners with 1-5 bromine atoms are referred as lower BDEs and those with 6 - 10 are referred as higher BDEs.

4.1. Production of PBDEs

PBDEs were introduced into the market in the U.S. in the 1970s and peaked in the late 1990s. The production of various PBDEs in different regions of the world is shown in Table 1[37].

BDE	Americas	Europe	Asia	Rest of the World
Deca BDE	24,500	7,600	23,000	1,050
Octa BDE	1,500	610	1,500	180
Penta BDE	7,100	150	150	100
Total BDEs	33,100	8,360	24,650	1,330

Data from BSEF (2001)

Table 1. BFRs produced (metric tons) in various regions of the world [37]

Table 2 shows the percent composition of various commercial PBDE mixtures. Currently, deca BDE constitutes 97% of the total production of PBDEs worldwide and it is the only BDE which is in use in the US [39].

Product	Tetra	Penta	Hexa	Hepta	Octa	Nona	Deca
Penta	24-38	50-60	4-8				
Octa			10-12	44	31-35	10-11	<1
Deca						<3	<97

Table 2. Percent composition of commercial PBDE flame retardant mixtures [39].

Many research findings indicate that lower brominated congeners manifest greater affinity for lipids and accumulation. Hence they are more toxic and pose a greater health risk to humans and livestock than the higher brominated congeners [40].

4.2. Exposure to PBDEs

PBDEs are lipophilic and resistant to chemical and physical degradation. Due to their persistent nature, they have become widespread environmental contaminants and about 97% of the American adult population has detectable levels of PBDEs [41]. Over the last 30 years the levels of these compounds in humans have increased and their levels in the US population are 10-100 times higher than levels measured in the Europe and Asia. According to a recent report, their levels in humans may be leveling off or decreasing due to their ban in many states of the US and European countries[39]. They are reported to be present in sediments, soil, outdoor and indoor air, household dust, foods, birds, fish, and terrestrial organisms [42].They have been detected in human serum, adipose tissue, and breast milk, and long-term exposure to these contaminants may pose a human health risk, especially to the developing fetus.

Despite their ban in many countries, vast amounts of these compounds are found to persist in existing consumer products, potentially contributing to environmental and human burdens for years to come. When released into the environment PBDEs adsorb rapidly onto solid particles such as soil or sediments due to their low water solubility and vapor pressure. Therefore, sediments and sewage sludge serve as a sink for these pollutants. These contaminants are still bioavailable even after adsorbing to the sediment particles [43]. PBDEs exposure to human being occurs through inhalation of household and workplace dust and by eating PBDE-contaminated foods. Exposure to PBDEs is nearly impossible to avoid as they are found in air, water and food. The fetus is also exposed to these compounds in utero, and after birth the newborn continues to be exposed from breast milk, where the toxins are transferred from the mother to the baby. A 25-year Swedish study found that the concentration of PBDEs in breast milk doubled every five years during the 25-year period [44]. The PBDE levels in breast milk from North American women are much higher than in breast milk from Swedish women, indicating greater exposure to PBDEs in this continent.

4.3. Dietary intake of PBDEs

The dietary intake of PBDEs from food in humans between ages 0-60 years are presented in Table 3. The intake of PBDEs was the highest in the first year during nursing. Breast milk has the highest levels of PBDEs. The intake of PBDEs per kg body weight reduced gradually with age. Children have higher PBDE intake than adults due to higher food intake per kg body weight [45].

4.4. Adverse effects of PBDEs in laboratory animals

Few toxic studies on PBDEs are reported in laboratory animals [46,47]. Some studies report these compounds to be toxic, carcinogenic and endocrine disruptors [48]. There are some reports indicating that PBDEs can induce disruption in spontaneous behavior, reduced habituation, impaired memory and learning in rats exposed during critical developmental period [46, 47]. BDE-99 is a penta congener and it has been found to induce hyperactivity and impaired spermatogenesis in rats. Additionally, two recent in-vitro studies showed the induction of oxidative stress in rat neurons exposed to BDE-47 (another congener of penta-BDE congener) in primary cultured rat hippocampal neurons [49] and human cells [50] serving as note-worthy examples that call for more careful assessment of these toxic substances.

Age in Years	0-1	2-5	6-11	12-19	20-39	40-59	>60
Diary	0	427	293	372	250	203	175
Meat	0	1,839	1,170	1,721	1530	1,408	1,094
Fish	0	280	232	300	309	357	467
Eggs	0	69	38	55	48	53	58
Margarine	0	6	6	7	6	8	9
Butter	0	30	17	19	22	15	22
Human Milk	306,560	0	0	0	0	0	0
Total PBDE		2652	1755	2774	2164	2,084	1857

Table 3. Daily intake of PBDEs from food (pg/kg body weight) [45]

5. Lipid peroxidation during PBDE exposure

Despite many studies, exact mechanism of toxicity of PBDEs is not yet clear and there is no definite explanation as to how PBDEs trigger the above mentioned deleterious effects. There appeared a school of thought that PBDEs, by reason of their electron affinity favor free radical formation in various tissues [51]. In the face of increased free radical generation the antioxidant enzymes namely, SOD, GPx, Catalase, and GST, are expected to give protection to the cells against the oxidative damage. An appropriate equilibrium between the antioxidant enzyme activities and free radicals is vital to ensure the cell's survival during increased oxidative stress. In the process of investigating the toxic mechanism authors have surmised oxidative stress induction in mice exposed to BDE-209, one of the most abundantly used flame retardant worldwide. This is based on a recent study showing the induction of oxidative stress in rat neurons exposed to BDE-47, a penta congener [49]. To interpret the adverse effects, both an assessment of the magnitude of free radical production and a careful evaluation of the threshold of protection in terms of antioxidant enzyme activity is necessary. Antioxidant enzymes such as SOD, GPx, GST, and catalase could also be the key indicators that could disclose the sequence of biochemical dysfunctions and alterations in signal transduction brought about by BDE-209. In the following few paragraphs the research protocols that one of the authors (MCV) adopted to investigate the biological markers for oxidative stress following BDE-209 exposure are explicated. The study is performed on the liver and brain tissues of adult mice. The liver is a major bio-transforming organ and brain is the most sensitive target to lipid soluble toxicants. Studying these two organs would give a clear picture of how this particular BDE is handled by these two vital organs. Further, since the metabolism and excretion of many PBDEs or for that matter any xenobiotics are tissue, species, sex and age dependent [52] it is intended to understand the effects of BDE-209 on these two vital organs; thus the hepato- and neuro-toxic mechanisms could be simultaneously studied and correlated.

5.1. Materials and methods

5.1.1. Chemicals

BDE- 85- CS (certified standard) was procured from Cambridge Isotope laboratories, Inc., Andover, MA, USA. Assay kits for the determination of lipid hydroperoxides, SOD, GPx, catalase, and GST were purchased from Cayman Chemical Company, Ann Arbor, Michigan, USA. BCA™ protein assay kit was purchased from Pierce Biotechnology Inc., Rockford IL, USA, for the protein assay of the sample. All other chemicals needed for the experiments were obtained from Sigma Aldrich Company.

5.1.2. Animals

Adult male Swiss Webster mice were obtained from Taconic Inc (Hudson, New York) and used for oxidative stress studies. Mice weighing 33±3 g were maintained at $23^0C\pm2^0C$ with 12-hour light/12-hour dark photoperiods. Mice were provided food pellets and water ad libitum and were allowed to acclimatize for 15 days. The guidelines set forth by the Institutional Animal Care and Use Committee, Gannon University, Erie, Pennsylvania, were followed in the maintenance and use of mice. Mice were grouped into two groups randomly (n = 9). Group 1 was injected intraperitoneally with 0.25mg/kg body weight of BDE-209 for four consecutive days and group 2 served as control and received same volume of vehicle solution, corn oil. Oral route of administration was not considered in this study as the aim of this work is to study the biochemical alterations induced at this particular concentration and administering via oral route results in elimination of some fraction of this chemical through feces. The present dosage of BDE-209 was selected based on some reports on other penta-BDE congeners in rodents [42,52]. In addition, the dosage used in the present study also is commensurate with the average levels of PBDEs (except BDE-209) found in household dust [53]. On the fifth day mice were sacrificed and liver and brain tissues were separated, washed in phosphate-buffered saline (PBS), and weighed. For all the biochemical assays, a 5% tissue homogenate of liver or brain was prepared using appropriate homogenizing buffers. The protocols that came with the assay kits were used for all biochemical assays and they are described below. The protein content of all the enzyme samples is assayed using Pierce® BCA Protein Assay Kit. Bovine serum albumin was used as standard.

5.1.3. Lipid hydroperoxides

Lipid peroxidation is quantified by a new method which directly quantifies the lipid hydroperoxides utilizing the redox reactions with ferrous ions.The liver and brain tissues from control and experimental mice were homogenized in nanopure water and the lipid hydroperoxides from the homogenate were extracted into chloroform. According to the protocol accompanying the lipid hydroperoxide assay kit, supplied by the Cayman Chemical Company, this chloroform extract was then used for the determination of lipid hydroperoxides.

5.1.4. *Superoxide Dismutase (SOD)*

The liver and brain tissues from control and experimental mice were homogenized in 4-(2-hydroxyethyl)-1-piperazineethanesulfonic acid (HEPES) buffer containing 1 mmole/L ethylene glycol tetraacetic acid (EGTA), 210 mmole/L mannitol and 70 mmole/L sucrose and centrifuged. The supernatant was used for the assay of SOD activity and was determined from the quantity of tetrazolium salt formed. One unit of SOD is defined as the amount of enzyme needed to exhibit 50% dismutation of superoxide anion.

5.1.5. *Glutathinoe Peroxidase (GPx)*

GPx activity was measured indirectly by a coupled reaction with glutathione reductase. Oxidized glutathione, produced upon reduction of hydroperoxide by GPx is recycled to its reduced state by glutathione reductase and NADPH. The oxidation of NADPH to $NADP^+$ is accompanied by a decrease in absorbance at 340 nm. The rate of decrease in the A_{340} is directly proportional to the GPx activity in the sample. The tissues from control and exposed mice were homogenized in cold Tris- HCl with 5 mmole/L EDTA and 1 mmole/L dithiothreitol (DTT). The homogenate was centrifuged and the supernatant was used for the enzyme assay. The protocol described in GPx assay kit was followed.

5.1.6. *Catalase*

A tissue 5% homogenate of liver or brain was prepared in cold potassium phosphate buffer containing 1mmole/L EDTA. The supernatant that surfaced was used for catalase assay as described in the assay protocol provided by Cayman Chemical Company. The catalase assay kit used the peroxidatic function of CAT for determination of enzyme activity. The enzyme reacts with methanol in the presence of an optimal concentration of H_2O_2. The formaldehyde produced is measured spectrophotometrically (◎ = 540 nm) with 4-amino-3-hydrazino-5-mercapto-1,2,4-triazole (Purpald) as the chromogen. Purpald specifically forms a bicyclic heterocycle with aldehydes, which upon oxidation changes from colorless to a purple color. One unit is defined as the amount of enzyme that will cause the formation of 1.0 nmol of formaldehyde per minute at 250°C.

5.1.7. *Glutathione S- transferase (GST)*

Tissues from control and experimental mice were homogenized in cold potassium phosphate containing 2 mmole/L EDTA. Homogenates were centrifuged and the supernatants used for the assay of Glutathione S-transferase, GST. GST activity was measured from the quantity of 1-Chloro-2, 4-Dinitrobenzene (CDNB) conjugated to glutathione. The one unit of enzyme conjugates 1.0 nmole of CDNB with reduced glutathione/min/mg.

5.1.8. *Statistical treatment of the data*

Data presented in the Tables 4 and 5 are mean ± standard deviation (SD) of 9 animals per group. The data were subjected to analysis of variance (ANOVA) and when means were significantly different they were further subjected to Tukey test. The percentage variation from controls was calculated and presented in the following results section. A value of $P < 0.05$ was considered significant.

Oxidative Stress Marker	Control	BDE-209	% change
LPO	2.06 ± 0.63	2.94 ± 1.33	+42.72*
SOD	13.89 ± 1.77	19.25 ± 2.56	+38.58*
GPx	103.43 ± 14.77	85.92 ± 5.61	-16.93*
Catalase	58.78 ± 14.22	44.54 ± 5.9	-24.22▪
GST	0.22 ± 0.07	0.17 ± 0.05	-22.73▪

Table 4. The lipid hydroperoxides and antioxidant enzyme activities in the brain tissue. Values are mean ± Standard deviation ($n = 9$). *Significant at $p < 0.05$, ▪ not significant

Oxidative Stress Marker	Control	BDE-209	% change
LPO	22.52 ± 2.76	27.31 ± 1.98	+21.67*
SOD	33.37 ± 1.59	30.69 ± 1.77	-8.03▪
GPx	198.77 ± 26.87	143.78 ± 7.8	-28.01*
Catalase	134.75 ± 18.34	92.78 ± 10.54	-31.15*
GST	4.32 ± 0.9	2.97 ± 0.3	-31.3*

Table 5. The lipid hydroperoxides and antioxidant enzyme activities in the liver tissue. Values are mean ± Standard deviation ($n = 9$). *Significant at $p < 0.05$, ▪not significant

6. Results

The overall short term effect of BDE-209 over a period of four days is a remarkable disruption in the oxidant/antioxidant equilibrium. First, the levels of lipid hydroperoxides in the experimental tissues, liver and brain, have increased considerably. The percent increase in the levels of lipid hydroperoxides in liver was 27.67 % ($p < 0.05$) and in brain was 42.72% ($p < 0.05$) over controls. These results are shown in Table 4 and 5. The activities of SOD showed varied response in brain and liver. In liver, though there was a decrease in activity by 8.03 %, when compared to controls, it was not statistically significant. However in brain there was significant (($p < 0.05$) increase in activity (38.58%). Another anti-oxidant enzyme activity measured was the GPx. The activity of this enzyme has been reduced in both brain (16.93%) and liver tissues (28.01%). These percentage changes also were significant at $p < 0.05$ for both the tissues.

The enzyme catalase also suffered a similar reduction in activity over a four day exposure to BDE-209. Both in the brain and the liver tissues the catalase activity was lowered. In the brain there was a 24.22% of reduction and in liver it went down by 31.15%. Just like GPx and catalase, GST also can reveal oxidative damage. In the present work GST activities were decreased in exposed liver by 31.3% ($p < 0.05$) and in brain by about 22.73% (*NS*). All the above results clearly exhibited differential susceptibility of mice liver and brain tissues to BDE-209 – induced oxidative stress.

7. Discussion

The results of this study clearly indicate the susceptibility of mice brain and liver tissues to BDE-209 induced damage when exposed to 0.25 mg/kg body weight for four consecutive days. The oxidative stress markers namely, lipid hydroperoxides, were found to increase in both the exposed liver and brain tissues. The increase in the lipid hydroperoxide levels could be due to an immediate perturbation in metabolizing enzymes in liver [54].The degree to which this can happen is yet open for further research. Liver being an important metabolizing organ hydrolyzes PBDEs in its attempt to eliminate them from the body. In this process the metabolites formed might possess greater toxicity than the parent BDE [55], thus making the liver prone to excessive damage. In the present study, BDE-209 exposed brains also showed significant increase in lipid hydroperoxides. Similar observations of increased peroxidation in the brain regions of adult rats exposed to two different concentrations of BDE-99 [42] lend support to this observation. In another study an increased peroxidation was noted in rat hippocampal neurons during BDE-47 exposure [49]. Additionally, there are reports on BDE-99 exposure in rats causing an increased ROS production in the tissues [47,55]. The mechanism responsible for ROS generation during PBDEs is not clear; however, it is suggested that PBDEs are electron acceptors under standard conditions prevailing in biological systems [51]. Reports on PBDE exposure and their involvement in the release of [3H] arachidonic acid, protein kinase-C translocation, and disruption in calcium homeostasis [56,57] could further offer an explanation for the mechanism of ROS formation [49]. Thus the present observation of increased lipid hydroperoxide levels in the liver and brain tissues of mice substantiates higher lipid peroxidation process and oxidative stress following BDE-209 exposure.

The objective of the second part of this study was to examine how antioxidative enzymes responded to BDE-209 exposure in mice tissues. Although lipid hydroperoxides and lipid peroxidation have increased in the tissues, there was no corresponding increase in all the protective antioxidant enzymes studied, namely, SOD, GPx, catalase and GST in BDE-209 exposed mice tissues with the exception of brain SOD. The elevated lipid hydroperoxides under normal conditions should trigger an increase in antioxidant enzyme activity to suit the increased need to quench them. But in the present study there was a decrease in the activities of GPx, GST and catalse in the liver tissue. A decrease in catalase reflects an inability of the liver to remove hydrogen peroxide. As mentioned already, both catalase and GPx show high cooperativity in their action [58]. During the periods of increased hydroperoxide formation catalase activity depends on GPX activity to keep from being inactivated. On account of decreased GPx activity in BDE-209 exposed tissues of mice, the catalase might be inactivated by higher concentration of hydroperoxides. Glutathione-S-transferase is another antioxidant enzyme whose activity was reduced in liver tissue during BDE-209 exposure. BDE-99 exposure in rats showed a similar response of decreased antioxidant enzymes at 0.6 mg/kg bw [42]. In another report on BDE-99, rat liver and kidney showed an upsurge of lipid peroxidation and unstable antioxidant system [55], rendering support to our observations. The overall lowered activities of SOD, GPx and catalase in liver denote that liver tissue is not able to quench increased ROS and oxidative stress. These findings thus show that BDE- 209 brings about an oxidative stress in liver tissue not only through increased lipid hydroperoxide formation but also by reduced antioxidative enzyme protection.

The responses in the brain tissues were different with respect to SOD, catalase, and GST. Unlike in liver, SOD activity in brain was elevated indicating the presence of some degree of protection in the brain tissue against the oxidative damage. Though catalase and GST activities showed reduction (a trend as seen in the liver tissue) the values however, were not statistically significant. This variation in responses in brain compared to liver suggests that brain tissue was able to cope with increased lipid hydroperoxides to some degree, by way of increased SOD activity. The differences in responses in liver and brain tissues could be due to their intrinsic biochemical and functional differences. Brain with its high rate of O2 utilization, high levels of polyunsaturated lipids tends to respond differently [59]. A report on GST of rat brain regions after BDE-99 treatment showed similar response [47]. Further, the dosage used in this study could have resulted in this response in the brain tissue. A study [42] on the adult rat brains exposed to 2 different concentrations of BDE-99 showed contrasting results on TBARS (indicative of oxidative stress) levels. At lower dosage of exposure, the levels of TBARS were found to be reduced and at higher dosage they were increased. There are reports on rats showing decreased antioxidative enzymes in brain tissue during BDE-99 indicating that not all congeners have the same impact on the tissues. Another observation made by the authors (MCV) is BDE-209 is much less toxic compared to BDE-85 [60].

In summary, exposing mice to BDE-209 for four days has brought about an increased lipid hydroperoxide production as seen in the liver and brain tissues. The consequent peroxidation process has significantly impeded the activity of antioxidant enzymes, GPx, GST and catalase, in liver. Brain tissue on the other hand, showed elevated SOD, lowered GPx without significantly reducing catalase and GST activities. The results of this study indicate that BDE-209 congener of PBDEs has grave potential to disrupt the activity of antioxidant enzymes and induce oxidative stress. Additionally, liver seemed to be more susceptible to this toxicant than the brain. These differences in responses in these two tissues could be due to their inherent biochemical and functional differences.

8. Conclusion

Plasma membrane covers a cell and acts as a barrier between the internal and external environment thereof. It also regulates the movement of molecules across it. Chemically, the plasma membrane is a bi-layer composed of phospholipids and proteins. Membrane lipids form the bulk of the membrane itself and are more abundantly found than the proteins; however, they are smaller in size. Lipids serve as semipermeable barrier allowing only those materials that are lipid soluble and non-polar in nature to freely cross the plasma membrane. The membrane proteins give structural support and also have several functions including serving as transporters, receptors, enzymes and anchors. Carbohydrates which are attached to the membrane proteins serve as identification markers and help in the differentiation of self from non-self. Thus the plasma membrane is a very important structure of the cell and disruption in its structure leads to disruption in cellular function. The membrane lipids are prone to disruption and loss of selective permeability needed for normal metabolism especially through lipid peroxidation by free radicals.

Free radicals are atoms or molecules with one or more unpaired electrons in their outermost shell. Free radicals with oxygen are called reactive oxygen species (ROS). Examples of ROS include hydrogen peroxide, hydroxyl radicals, and superoxide anions. From time to time free radicals are generated as part of aerobic life during normal metabolism and in some instances these radicals are intentionally generated as part of normal signal transduction process. Superoxide anions are generated in cell organelles such as mitochondria and microsomes. These anions can react with biological molecules the most due to their longer half-life. Another most reactive free radical of the cell is hydroxyl radical. This radical is formed by the hemolytic fission of hydrogen peroxide. Free radicals, due to their high reactivity, attack other macromolecules within their vicinity and cause disruption of these molecules. When these free radicals attack proteins they change their biophysics thereby leading to functional disruption. They also earned notoriety in disrupting DNA and gene expression. It is reported that twenty types of alterations are introduced in the DNA by these ROS leading to variations in the gene expression. Last but not the least, ROS also can attack lipids especially the membrane lipids of the cell and cause extensive damage to the cellular function. Free radicals steal electrons from the polyunsaturated fatty acids (PUFAs) and other lipoproteins of the cells and initiate an attack called lipid peroxidation. The peroxidation of lipids occurs in three steps, namely, initiation, propagation and termination. In particular, lipid peroxidation characterized by low density lipoproteins plays an important role in atherosclerosis.

All cells have antioxidants in place to counteract the harmful effects of free radicals. Antioxidants are of two types, chain breaking antioxidants and preventive type. Vitamins such as E and C are examples for chain breaking type as they donate electrons to the existing free radicals and lessen their degree of reactivity to prevent further oxidative damage. The second category of antioxidants (the preventive type) consists of enzymes that scavenge the free radicals before they initiate oxidative attack. Glutathione is an important antioxidant of the cells whose function includes anti-oxidation, strengthening immune system, building DNA, and detoxification of medications and cancer causing compounds. The diverse functions of glutathione are mainly due to the sulfhydryl group in one of its amino acids known as cysteine. Vitamins C, E and provitamin A also play a role in protecting the body against the free radical induced damage. There are many enzymes inside the cells with antioxidant property. The most important antioxidant enzymes present inside the cells are superoxide dismutase (SOD), catalase, glutathione peroxidase, and glutathione S-transferase. SOD catalyzes the dismutation of superoxide anion to O_2 and H_2O_2. GPx catalyzes the reduction of hydroperoxides and serves to protect the cell from the hostile free radicals. Catalase, another antioxidant enzyme, along with GPx, is essential for removing the hydrogen peroxide formed during oxidation reactions. GST along with glutathione and GPx neutralizes free radicals and lipid hydroperoxides especially at low levels of oxidative stress.

Polybrominated diphenyl ethers (PBDEs) are organic brominated flame retardants that were introduced into the market in 1970s and are added in large quantities to many commercial and household products such as computers, TVs, foam mattresses, carpets, etc., to inhibit ignition and prevent fire accidents. Ever since they were introduced into the market, deaths due to fire accidents abated significantly. Commercially these chemicals are available in three mixtures, pentaBDEs, octaBDEs and decaBDEs. Currently, deca BDE constitutes 97% of the total production of PBDEs worldwide and it is the only BDE which is in use in the US. PBDEs are lipophilic and resistant to chemical and physical degradation. Due to their persistent nature, they have become widespread environmental contaminants. Avoiding exposure to PBDEs is nearly impossible as they are ubiquitously found in air, water and food. Few toxic studies on PBDEs are reported in laboratory animals. There are some reports indicating that PBDEs can induce disruption in spontaneous behavior, reduced habituation, impaired memory and learning in rats exposed during critical developmental period. When the toxic effect of BDE-209 was explored in the mice at 0.25 mg/kg body weight, it became clearly evident that this particular BDE triggers lipid hydroperoxide generation and increased lipid peroxidation process. Additionally, the antioxidant enzymes were diminished in liver, but in brain the levels of SOD increased without any change in catalase and GST. These results show that BDE-209 has the ability to elicit serious oxidative stress in mice with increased lipid peroxidation and decreased antioxidant protection. Research is underway to understand the effects of BDE-209 on the expression of genes and to find out if it are responsible for impairment of cell cycle regulation and causing apoptosis.

Author details

Mary C. Vagula* and Elisa M. Konieczko
Biology Department, Gannon University, Erie, PA, USA

Acknowledgement

The lead author (MCV) thanks Gannon University for granting the award of Faculty Research Grant. Authors also thank Prof. Gustafson and Dr. Masters for financial support and encouragement.

9. References

[1] Singer S, Nicholson G (1972) The fluid mosaic model of the structure of cell membranes. Science. 175(4023): 720-731.

* Corresponding Author

[2] van Meer, G (2006) Cellular lipidomics. EMBO Journal. 24:3159-3165.
[3] Simons K, Sampaio J (2011) Membrane ogranization and lipid rafts. Cold Spring Harbor Perspectives in Biology. 3(10):a004697.
[4] Henneberry A, Wright M, McMaster C (2002) The major sites of cellular phospholipid synthesis and molecular determinants of fatty acid and lipid group specificity. Molecular and Cellular Biology. 13:3148-3161.
[5] Eisenberg D (1984) Three-dimensional structure of membrane and surface proteins. Annual Reviews in Biochemistry. 53:595-623.
[6] Driscoll P, Vuidepot A (1999) Peripheral membrane proteins. FYVE sticky fingers. Current Biology. 9:R857-R860.
[7] Gahmberg C, Tolvanen M (1996) Why mammalian cell surface proteins are glycoproteins. Trends in Biochemal Science 21:308-311.
[8] Commoner B, Townsend J, Pake G (1954) Free radicals in biological materials. Nature 174: 689–691.
[9] Tyler D (1975) Polarographic assay and intracellular distribution of superoxide dismutase in rat liver. Biochemistry Journal. 147:493-504.
[10] Ramos de Andrade Jr. D, Becco de Souza R, Alves dos Samtos S, Ramos de Andrade R (2005) Oxygen free radicals and pulmonary disease. Journal of Brazilian Pneumology. 31(1): 1-17.
[11] Gutteridge M, Halliwell B (1984) Oxygen toxicity, oxygen radicals, transition metals and disease. Biochemistry Journal. 219:1-14.
[12] Schibert J, Wilmer J (1991) Does hydrogen peroxide exist "free" in biological systems? Free radical biology. 11(6): 545-55.
[13] Oya-Ohta Y, Ochi T, Komoda Y, Yamamoto K (1995) The biological activity of hydrogen peroxide. VI. Mechanism of the enhancing effects of L-histidine: the role of the formation of a histidine-peroxide adduct and membrane transport. Mutation Research. 326(1):99-107
[14] Yu B (1994) Cellular defenses against damage from reactive oxidative species. Physiology Reviews. 74:139-62.
[15] Del Maestro R (1980) An approach to free radicals in medicine and biology. Acta Physiology Scandanavia. 492:153-68.
[16] Lennon S, Martin S, Cotter T (1991) Dose-dependent induction of apoptosis in human tumour cell lines by widely diverging stimuli. Cell Proliferation. 24 (2): 203–14.
[17] Griffiths H, Unsworth J, Blake D, Lunec J (1988) Free radicals in chemistry, pathology and medicine. London: Richelieu, pgs. 439-54.
[18] Marx G Chevion M (1986) Site-specific modification of albumin by free radicals. Reaction with copper (II) and ascorbate. Biochemistry Journal. 236:397-400.
[19] Wolff S, Dean S (1986) Fragmentation of protein by free radicals and its effect on their susceptibility to enzymic hydrolysis. Biochemistry Journal. 234:399-403.
[20] Morelli R, Russo-Volpe S, Bruno N, Lo Scalzo R (2003). Fenton-dependent damage to carbohydrates: free radical scavenging activity of some simple sugars. Journal of Agricultural and Food Chemistry 51(25): 7418-7425.

[21] Sagone Jr. A, Greenwald E, Kraut J, Bianchine J, Singh D (1983) Glucose: role as free radical scavenger in biological systems. Journal of Laboratory and Clinical Medicine. 101:97-104.

[22] Wolff S, Crabbe M, Thornalley P (1984) The autoxidation of glyceraldehydes and other simple monosaccharides. Exp basel. 40: 244-6.

[23] Ames B (1983) Dietary carcinogens and anticarcinogens. Oxygen radicals and degenerative diseases. Science. 221:1256-64.

[24] Fraga C, Shigenaga M, Park, J Degan, P, Ames B (1990) Oxidative damage to DNA during aging: 8-hidroixy-deoxyguanosine in rat organ DNA and urine. Proceedings of the National Academy of Sciences USA. 87:4533-7.

[25] Dizdaroglu M, Jaruga P, Birincioglu M, Rodriguez H (2002) Free radical-induced damage to DNA: mechanisms and measurement. Free Radicals in Biology and Medicine. 32:1102-15.

[26] Cooke M, Evans M, Dizdaroglu M, Lunec J (2003) Oxidative DNA damage: mechanisms, mutation, and disease. FASEB Journal 17:1195-214.

[27] Kerr M, Bender C, Monti E (1996) An introduction to oxygen free radicals. Heart and Lung: the Journal of Acute and Critical Care. 25(3):200-208.

[28] McDermott J (2000) Antioxidant nutrients: current dietary recommendations and research update. Journal of American Pharmacology Association. 11(1):28-42.

[29] Scheibmeir H, Christensen K, Whitaker S, Jegaethesan J, Clancy R, Pierce J (2005) A review of free radicals and antioxidants for critical care nurses. Intensive and Critical Care Nursing. 21:24-28.

[30] Clark S (2002) The biochemistry of the antioxidants revisited. Nutrition in Clinical Practice. 17:5-17.

[31] Jozefczak M, Remans T, Vangronsveld J, Cuypers A (2012) Glutathione is a key player in metal-induced oxidative stress defenses. International Journal of Molecular Science. 13(3):3145-75.

[32] Traber M, Vitamin E. In: Shils M, Shike M, Ross A, Caballero B, Cousins R (eds.) Modern Nutrition in Health and Disease, 10th ed. Baltimore, MD: Lippincott Williams Wilkins; 2006. P. 396-411.

[33] Halliwell B (1996) Antioxidants in human health and disease. Annual Review of Nutrition. 16:33-50.

[34] Sheweita S (1998) Heavy metal-induced changes in the glutathoine levels and glutathione reductase/glutathione-S-transferase activities in the liver of male mice. International Journal of Toxicology. 17:383-392.

[35] Yan H, Harding J (1997) Glycation induced inactivation and loss of antigenicity of catalase and superoxide dismutase. Biochemical Journal. 338:599-605.

[36] Fire in the United States, 2003-2007. Fifteenth edition, October 2009. FEMA. Executive summary. accessed at ://www.usfa.fema.gov/downloads/pdf/statistics/fa_325.pdf

[37] Birnbaum, L, Staskal, D (2004) Brominated flame retardants: Cause for concern? Environmental Health Perspectives. 112: 9–17.

[38] Hites R (2004) Polybrominated diphenyl ethers in the environment and in people: A meta-analysis of concentrations. Environmental Science Technology. 38: 945–956.

[39] Kimbrough K, Johnson W, Lauenstein G, Christensen J, Apeti D (2009) An Assessment of Polybrominated Diphenyl Ethers (PBDEs) in Sediments and Bivalves of the U.S. Coastal Zone. Silver Spring, MD. *NOAA Technical Memorandum* NOS NCCOS 94. 87.

[40] Gandhi N, Bhavsar S, Gewurtz S, Tomy G (2011) Can biotransformation of BDE-209 in lake trout cause bioaccumulation of more toxic, lower-brominated PBDEs (BDE-47, -99) over the long term? Environment International. 37(1):170-7.

[41] de Wit C (2002) An overview of brominated flame retardants in the environment. Chemosphere. 46 (5): 583–624.

[42] Belles M, Alonso V, Linares V (2010) Behavioral effects and oxidative status in brain regions of adult rats exposed to BDE-99. Toxicology Letters. 194(1-2): 1–7.

[43] Voorspoels S, Covaci A, Maervoet J, Schepens P (2004) PBDEs in marine and freshwater sediments from Belgium: levels, profiles and relations with biota. Journal of Environmental Monitoring. 6:914-918.

[44] Eriksson P, Jakobsson E, Fredriksson A (2004) Brominated Flame Retardants: A Novel Class of Developmental Neurotoxicants in our Environment? Environmental Health Perspectives 109(9): 903-908.

[45] Schecter A, Papke O, Harris T, Tung K, Musumba A, Olson J, Birnbaum L (2006) Polybrominated diphenyl ether (PBDE) levels in an expanded market basket survey of u.s. food and estimated pbde dietary intake by age and sex. Environmental Health Perspectives. 114(10) doi:10.1289/ehp.9121.

[46] Fischer C, Fredriksson A, Erilsson P (2008) Coexposure of neonatal mice to a flame retardant PBDE 99 (2,2'4,4'5-pentabromodiphenyl ether) and methyl mercury enhances developmental neurotoxic defects. Toxicology Science. 101(2):275-285.

[47] Cheng J, Gu J, Ma J, Chen X, Zhang M, Wang W (2009) Neurobehavioral effects, redox responses and tissue distribution in rat offspring developmental to BDE-99. Chemosphere 75: 963–968.

[48] McDonald T (2002) A perspective on the potential health risks of PBDEs. Chemosphere. 46: 745-755.

[49] He P, He W, Wang A, Xia T, Xu B, Zhang M, Chan X (2008) PBDE-47 induced oxidative stress, DNA damage and apoptosis in primary cultured rat hippocampal neurons. Neurotoxicology. 29(1):124-129.

[50] Kawashiro Y, Fukata H, Sato K, Aburatani H, Takigami H,Mori C (2009) Polybrominated diphenyl ethers cause oxidative stress in human umbilical vein endothelial cells. Human and Experimental Toxicology. 28(1):703-713.

[51] Zhao Y, Tao F, Zeng E (2008) Theoretical study on the chemical properties of polybrominated diphenyl ethers. Chemosphere. 70(5): 901–907.

[52] Chen L, Lebetkin E, Sanders J, Burka L (2006) Metabolism and disposition of 2,2',4,4',5 pentabromodiphenyl ether (BDE-99) following single or repeated administration to rats or mice. Xenobiotica. 36(6):515-534.

[53] Toms L, Hearn L, Kennedy K (2009) Concentrations of polybrominated diphenyl ethers in the matched samples of human milk, duct and indoor air. Environment International. 35(6):864-869.

[54] Wolkers H, Boily F, Fink-Gremmels J, Bavel B, Hammill M, Primicerio R (2009) Tissue specific contaminant accumulation and associated effects on hepatic serum analytes and cytochrome p450 enzyme activity in hooded seals (cystophora cristata) from the gulf of St. Lawrence. Archives of Environmental Contamination. 56: 360–370.

[55] Albina M, Alonso V, Linares V (2010) Effects of exposure to (BDE-99) on oxidative status of liver and kidney in adult rats. Toxicology. 271(1-2): 51–56.

[56] Kodavanti P, Derr-Yellin E (2002) Differential effects of polybrominated diphenyl ethers and polychlorinated biphenyl on [3 H] arachidonic acid release in rat cerebellar granule neurons. Toxicological Sciences. 68(2):451-457.

[57] Kodavanti P, Ward T 2005. Differential effects of commercial polybrominated diphenyl and polychlorinated biphenyl mixtures on intracellular signaling in rat brain in vitro. Toxicological Sciences. 85(2):952-962.

[58] Baud O, Greene A, Li J, Wang H, Volpe J, Rosenberg P (2004) Glutathione peroxidase–catalase cooperativity is required for resistance to hydrogen peroxide by mature rat oligodendrocytes. The Jouranl of Neuroscience. 24(7):1531-1540.

[59] Sayre, L, Perry G, Smith M (2008) Oxidative stress and neurotoxicity. Chemical Research in Toxicology. 21, 172–188.

[60] Vagula M, Kubeldis N, Nelatury C (2011) Effects of BDE- 85 on the Oxidative Status and Nerve Conduction in Rodents. International journal of toxicology. 30(4): 428-434.